International Conference on Chemistry and Material Science (IC2MS 2017)

Advancing to the Frontier of Innovation in Chemistry and Material Science

IOP Conference Series: Materials Science and Engineering
Volume 299

Malang, Indonesia
4 – 5 November 2017

ISBN: 978-1-5108-5838-1
ISSN: 1757-8981

Printed from e-media with permission by:

Curran Associates, Inc.
57 Morehouse Lane
Red Hook, NY 12571

Some format issues inherent in the e-media version may also appear in this print version.

Printed by Curran Associates, Inc. (2018)

For permission requests, please contact the Institute of Physics
at the address below.

Institute of Physics
Dirac House, Temple Back
Bristol BS1 6BE UK

Phone: 44 1 17 929 7481
Fax: 44 1 17 920 0979

techtracking@iop.org

Additional copies of this publication are available from:

Curran Associates, Inc.
57 Morehouse Lane
Red Hook, NY 12571 USA
Phone: 845-758-0400
Fax: 845-758-2633
Email: curran@proceedings.com
Web: www.proceedings.com

Table of contents

Volume 299

International Conference on Chemistry and Material Science (IC2MS) 2017

4–5 November 2017, Malang, East Java, Indonesia

Accepted papers received: 10 January 2018
Published online: 7 February 2018

Preface

Papers

ANALYTICAL CHEMISTRY

BIOCHEMISTRY

INORGANIC CHEMISTRY

MATERIALS CHARACTERIZATION

MOLECULAR SCIENCE AND COMPUTATIONAL CHEMISTRY

NANOMATERIALS AND NANO DEVICES

PHYSICAL CHEMISTRY

RENEWABLE ENERGY AND FUEL CELL

SYNTHETIC AND CATALYTIC MATERIALS

IOP Conf. Series: Materials Science and Engineering **299** (2018) 011001 doi:10.1088/1757-899X/299/1/011001

The INTERNATIONAL CONFERENCE ON CHEMISTRY AND MATERIAL SCIENCE (IC2MS) 2017

"Advancing to the frontier of innovation in chemistry and material science"

4-5 November 2017

IJEN SUITES RESORT & CONVENTION

MALANG, INDONESIA

IOP Conf. Series: Materials Science and Engineering **299** (2018) 011001 doi:10.1088/1757-899X/299/1/011001

PREFACE

On behalf of the Organizing Committee, I would like to welcome all of you, keynote speakers, invited speakers, distinguished guests, and participants, to the **International Conference on Chemistry and Material Science (IC2MS) 2017**. This conference is organized by Chemistry Department, Brawijaya University, and is held for the first time. The IC2MS runs for 2 days, 4-5 November 2017, and consists of 5 mini symposia, they are:

- Analytical and Inorganic Chemistry
- Biochemistry and Organic Chemistry
- Synthetic and Catalyst Materials, Renewable Energy and Fuel Cells, Molecular Science & Computational Chemistry
- Materials Characterization
- Nanomaterials & Nanodevices and Physical Chemistry

The accepted papers of the conference will be published in IOP Conference Series: Materials Science and Engineering. Around 200 people are attending this conference. They consist of 150 presenting participants, 50 non-presenting participants, 5 keynote speakers, and 6 invited speakers. In terms of country of origin, the participants of the IC2MS are coming from 8 countries, including Indonesia, Brunei Darussalam, Thailand, Taiwan, Japan, Kazakhstan, Australia, and Singapore.

The IC2MS should bring advantages for all participants. The IC2MS attendees can exchange and share their experiences and research results on all aspects of chemistry and material sciences. The IC2MS also provides a premier interdisciplinary platform for researchers, practitioners and educators to present and discuss the most recent innovations, trends, and concerns as well as practical challenges encountered and solutions adopted in the fields of chemistry and material sciences.

The ICM2S could not become a reality without the help and assistance of many parties. Thus, in this occasion I would like to sincerely thank the Rector of Brawijaya University, Dean of Faculty of Mathematics and Natural Sciences Brawijaya University, Head of Chemistry Department, Brawijaya University, all members of the Organizing Committee, Ijen Suites Hotel, and all sponsors, who have provided meaningful help and assistance for the implementation of this conference.

Malang, 4 November 2017
Chairman of the Organizing Committee
Anna Safitri, Ph.D

IOP Conf. Series: Materials Science and Engineering **299** (2018) 011001 doi:10.1088/1757-899X/299/1/011001

CONFERENCE PHOTOGRAPHS

IOP Conf. Series: Materials Science and Engineering **299** (2018) 011001 doi:10.1088/1757-899X/299/1/011001

IOP Conf. Series: Materials Science and Engineering **299** (2018) 011001 doi:10.1088/1757-899X/299/1/011001

FUNDING ACKNOWLEDGEMENT

1. **BRAWIJAYA UNIVERSITY**

2. **CRC-ASMAT (COLLABORATIVE RESEARCH CENTER-ADVANCED SYSTEM AND MATERIAL TECHNOLOGY)**
   http://www.crg-asmat.ub.ac.id

3. **PT Ditek Jaya**

4. **DJ Labware**

IOP Conf. Series: Materials Science and Engineering **299** (2018) 011001 doi:10.1088/1757-899X/299/1/011001

LIST OF COMMITTEE

Steering Committee

- **Prof Mohammad Bisri**

Rector, Brawijaya University

- **Adi Susilo, PhD**

Dean, Faculty of Science, Brawijaya University

- **Masruri, Ph.D**

Head of Chemistry Department, Brawijaya University

International Scientific Committee

- **Prof Hideki Tanaka**

Okayama University, Japan

- **Prof Shoji Motomizu**

Okayama University, Japan

- **Prof Lim-Lee Wah**

Gifu University, Japan

- **Prof Taufiq-Yap Yun Hin**

Universiti Putra Malaysia, Malaysia

- **Prof Duangjai Nacapricha**

Mahidol University, Thailand

- **Dr. Sc. Akhmad Sabarudin**

Brawijaya University, Indonesia

- **Dr. Sc. Lukman Hakim**

Brawijaya University, Indonesia

Organizing Committee

IC2MS Chair	: Anna Safitri, PhD
Secretary	: Zubaidah Ningsih, Ph.D
Finance	: Dr. Sc. Siti Mariyah Ulfa (Coordinator)
Secretariat	: Ellya Indahyanti, M. Eng. (Coordinator)
Scientific Division	: Dr. Sc. Lukman Hakim
	Dr. Sc. Akhmad Sabarudin

IOP Conf. Series: Materials Science and Engineering **299** (2018) 011001 doi:10.1088/1757-899X/299/1/011001

Abstract Reviewer	: Dr. Edi Priyo Utomo
	Dr. Warsito
	Dr. Uswatun Hasanah
	Dr. Hermin Sulistyarti
	Dr. Ani Mulyasuryani
	Dr. Diah Mardiana
	Dr.rer.nat. Rachmat Triandi Tjahjanto
	Dr. Tutik Setianingsih
	Dr. Sasangka Prasetyawan
	Anna Roosdiana, M.App.Sc.
Program	: Dr. Arie Srihardyastutie (coordinator)
Banquet	: Sri Wardhani, M.Si (coordinator)
Website and Publication	: Yuniar Ponco Prananto, Ph.D (coordinator)
Logistics	: Danar Purwonugroho, M.Si (coordinator)

IOP Conf. Series: Materials Science and Engineering **299** (2018) 011002 doi:10.1088/1757-899X/299/1/011002

Peer review statement

All papers published in this volume of *IOP Conference Series: Materials Science and Engineering* have been peer reviewed through processes administered by the proceedings Editors. Reviews were conducted by expert referees to the professional and scientific standards expected of a proceedings journal published by IOP Publishing.

IOP Conf. Series: Materials Science and Engineering **299** (2018) 012001 doi:10.1088/1757-899X/299/1/012001

Determination of glucose in human urine by cyclic voltammetry method using gold electrode

Riyanto[1*] and Muh. Supwatul Hakim[1]

[1]Department of Chemistry, Faculty of Mathematics and Natural Sciences, Universitas Islam Indonesia, Jalan Kaliurang KM 14,5 Sleman Yogyakarta, Indonesia.

E-mail: riyanto@uii.a.c.id

Abstract. This study has been the determination of glucose in human urine by cyclic voltammetry method using gold electrode. The gold electrode was prepared using gold wire with purity 99.99%, size 1.0 mm by length and wide respectively, connected with silver wire using silver conductive paint. The effect of electrolyte, pH and glucose concentration has been determined to produce the optimum method. The research showed the KNO_3 is a good electrolyte for determination of glucose in human urine using gold electrode. The effect of glucose concentration have the coefficient correlation is $R^2 = 0.994$. The results of the recovery using addition method showed at range95-105%. As a conclusion isa gold electrode is a good electrode for electrochemical sensors to the determination of glucose in human urine.

1. Introduction

Innovation to improve the quality of medical diagnostic equipment is very necessary to produce tools at affordable prices, easy to mass-produced, easy to operate, have good resistance and environmentally friendly. The demand for medical diagnostic tools for early detection of various diseases such as diabetes and kidney caused by increased uric acid and urea levels is increasing. The increase in demand is due to the increasing number of patients both nationally and internationally. Prevention is better than treatment, the detection of glucose, uric acid and urea is needed to prevent the occurrence of the disease in the community.

Diabetes mellitus (DM) is a condition of disruption of metabolism in the body due to the inability of the body to make or supply the hormone insulin resulting in increased blood glucose levels, exceeding the normal value. If left unchecked, diabetes will be the leading cause of end-stage renal disease, heart attack, stroke, disease, and other complications. Based on WHO data, Indonesia currently ranks fourth as a country with a total diabetes mellitus reaching about 8 million people and estimated the number exceeds 21 million in the year 2025.

Quantitative analysis of glucose is essential in clinical chemistry [1]. Glucose is usually only analyzed by methods involving specific enzymes for glucose so it is not easily disturbed by other sugars that may be present. Various enzyme immobilization techniques may be used, including adsorption on the solid buffer. This is due to the simple sol-gel process conditions and the possibility to design as needed. Sol-gel flexibility permits the formation of sensors as monoliths and thin layers that can be paired with optical fibers or deposited on the electrodes, as well as nanoparticles.

The weakness of the enzyme-based sensors is that the tool has a usage limit and requiring storage conditions. Therefore, research to create a current-based sensor for analysis or nonenzymatic glucose sensors is necessary [2]. Many researchers have been development of sensor glucose non enzyme using various electrode is graphene [3], N,N-Dibromo Dimethylhydantoin [4], platinum nanoparticle [5], nickel-cobalt [6], nickel modified indium [7], nickel [8], Ag and NiO-Ag [9], Pt-Au [1], gold [10,11,12].

In this study has been done cyclic voltammetry was used to evaluate the electrocatalytic activity of the sensor against the oxidation of nonenzymatic glucose in various electrolyte solutions. Non-enzyme glucose sensors are highly resistant to poisoning by chloride ions and are able to overcome interference from uric acid and ascorbic acid up to concentrations of 0.05 mM and 0.5 mM, respectively. In this paper, the results of research on the electrochemical response of Au electrode in various electrolytes are KNO_3, NaOH, KOH, buffer phosphate solution pH 4, 7 and 10. This study also conducted a study of the effect of glucose concentration on the peak oxidation and reduction currents. This study has produced data on the sensitivity and selectivity of Au electrodes for glucose analysis. Au metal can be used as a working electrode in the manufacture of nonenzyme glucose sensors.

2. Experimental Section

2.1. Material and Instrumentation

The instrument used in this research is a set of analysis tools cyclic voltammetry PGSTAT 100 N 100 V/250 mA (Metrohm Autolab) equipped with Nova Software and pH meter (Hanna). The materials used in this study were glucose (Merck), KNO_3 (Merck), KOH (Merck), buffer phosphate solution pH 4, 7 and 10, NaOH (Merck), KH_2PO_4 (Merck), distillate water and human urine. Human urine was collected from student.

2.2. Preparation of Gold Electrode

Au metal with purity 99.99% (Sigma Aldrich) cut with length and width 1 cm x 0.5 cm. Au metal is connected to the silver wire with silver conductive paint. Silver wire is inserted in the perforated glass. Silver conductive paint is covered with epoxy so that it comes into contact with only Au metal solution. Figure 1 shown is the design of gold electrode for cyclic voltammetry of glucose analysis.

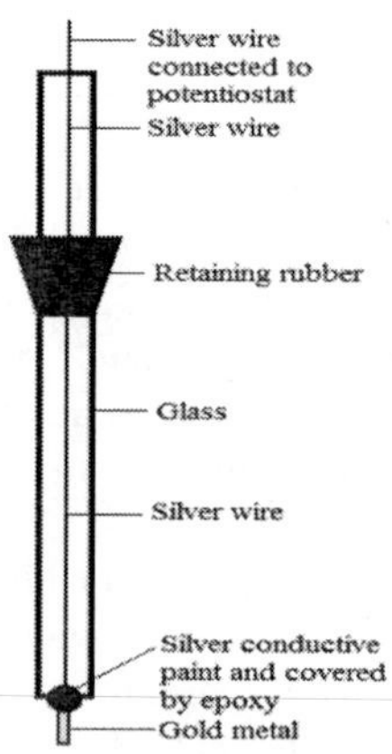

Figure 1. Design of gold electrode for glucose analysis using cyclic voltammetry

2.3. Cyclic voltammetry of Gold in Glucose in KOH, NaOH, and Buffer solution pH

Electrolysis cell and Au electrode are cleaned with distillate water and dried. As much as 5 mL of glucose 1M and 5 mL of 0.1 M KNO_3 solution were included in the electrolysis cell. The platinum electrode is the counter electrode, Au as the working electrode and Ag/AgCl as the reference electrode connected to the potentiostat. The steps were repeated using different electrolyte solutions is KOH, NaOH and buffer phosphate solution pH 4, 7 and 10.

3. Result and Discussion

3.1. Cyclic Voltammetry Glukosa using Au electrode

Figure 2 shows the results of analysis with cyclic voltammetry glucose at Au electrode in KNO_3 solution (without glucose) and Figure 1B cyclic voltammogram glucose solution 0.001 M with electrolyte KNO_3 0.1 M. Figure 2 shows that the reversible voltammogram is that starts from a potential of -1.0 volts moving towards a positive direction then moves towards a more negative potential and back to potential -1.0 volts. In the peak voltammogram of the electrolyte solution showed that at a positive potential there is a peak of the electrolyte at a potential of 0.5 V. The peak for glucose and electrolyte at the same potential of 0.5 V and there is an anodic peak rise in current. The peak is the oxidation of glucose. The reaction of glucose oxidation on the surface of the Au electrode is shown in Figure 3. Figure 3 shows that glucose is oxidized on the surface of the Au electrode to gluconolactone. Glucose also undergoes a reduction shown by the appearance of a cathodic peak current at a potential of 0.35 V.

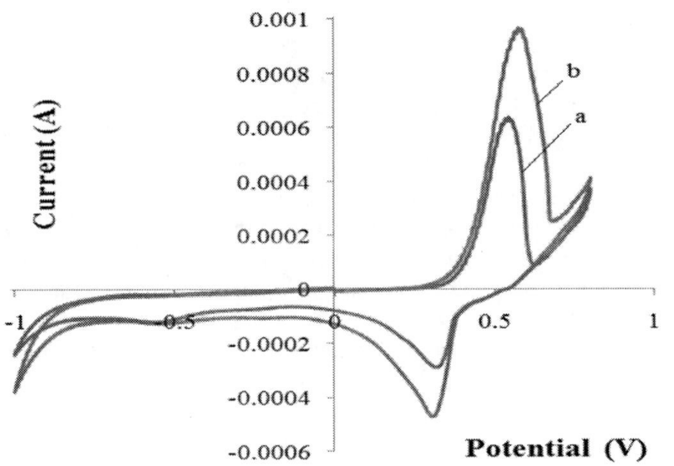

Figure 2. Cyclic voltammogram of (a) 0.1 M KNO_3 (without glucose), (b) 0.1 M KNO_3 + glucose 0.001 M glucose at scan rate 100 mVs^{-1}

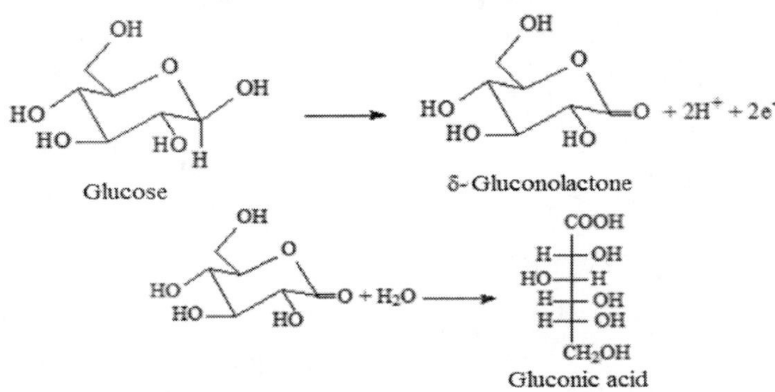

Figure 3. Redox reaction of glucose in gold electrode surface [11]

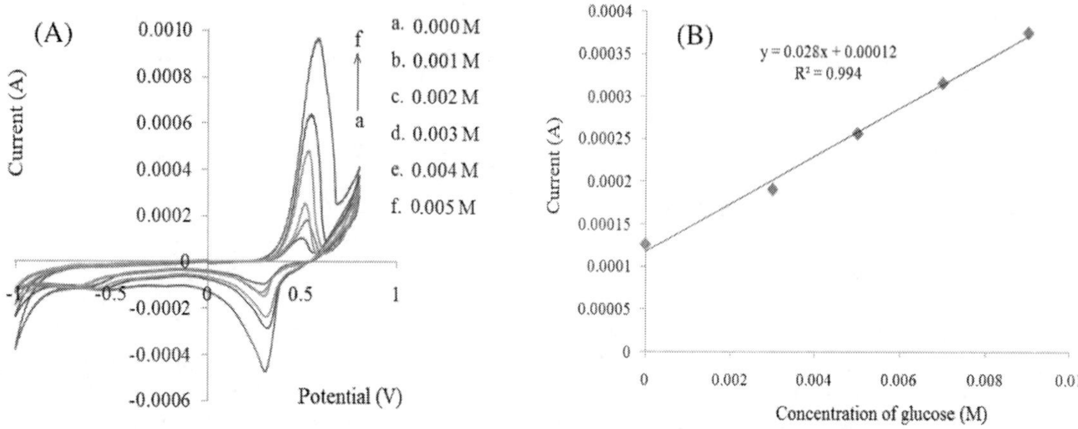

Figure 4. Cyclic voltammogram of various concentration of glucose at Au electrode in 0.1 M KNO_3(without glucose) (a) and (b-f) is 0.1 M KNO_3 + (0.001-0.005 M), at scan rate 100 mVs^{-1}

The method to ensure that the anodic current peak at 0.5 V potential is the oxidation of glucose, then cyclic voltammetry glucose with different concentrations is performed. Figure 2 shows that the greater the glucose concentration, the peak potential of oxidation and the peak potential of glucose reduction, the greater. The results of this study in accordance with the applicable formula for cyclic voltammetry are $i_p = 2.686x10^5n^{3/2}AcD^{1/2}v^{1/2}$.

Where i_p is the peak current A: electrode area, c: concentration, v: scan rate and D: diffusion coefficient. The above formula has a constant value that is, n, A, D and v so that it can be simplified to i_p = kc, the peak current is directly proportional to the concentration. Figure 4B shows the calibration curve of the effect of glucose concentration (M) on the anodic peak current. The anodic peak current rise is directly proportional to glucose concentration. Kumary et al. (2013) have research and that the linearity of the anodic peak current of glucose is directly proportional to the concentration [4]. Figure 3 shows that the correlation coefficient or R^2 = 0.993. KNO_3 is the best electrolyte for glucose analysis with gold electrodes.

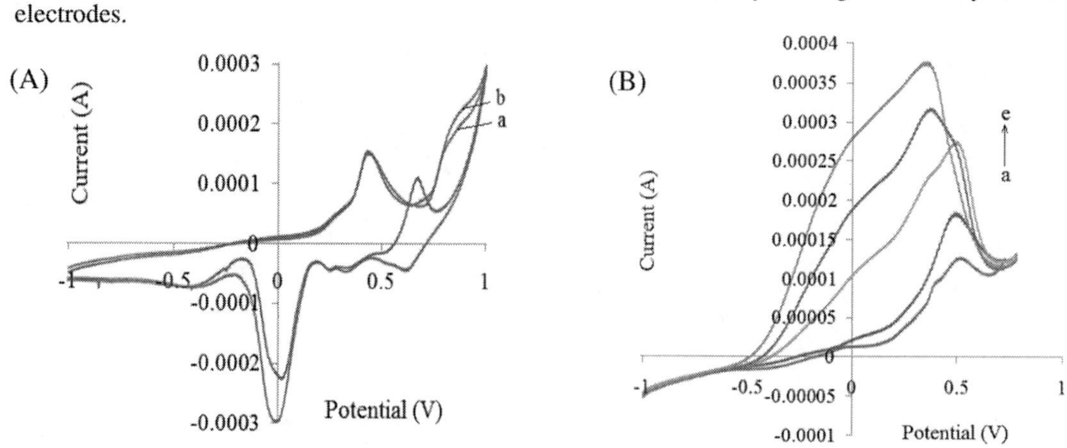

Figure 5. Cyclic voltammogram of (A) (a) 0.1 M NaOH (without glucose), (b) 0.1 M NaOH + glucose 0.001 M glucose. (B) Cyclic voltammogram of various concentration of glucose at Au electrode in 0.1 M NaOH solution in glucose (a-e) 0.003-0.009 M, at scan rate 100 mVs^{-1}

IOP Conf. Series: Materials Science and Engineering **299** (2018) 012001 doi:10.1088/1757-899X/299/1/012001

Figure 5 shows a cyclic voltammogram of glucose with NaOH electrolyte. Figure 5A shows the cyclic voltammogram comparison between the electrolyte NaOH and glucose. The peak of glucose oxidation is difficult to identify because of the nearly same voltammogram form. Figure 5B shows a cyclic voltammogram with various variations of glucose concentration. Figure 5 B shown is the greater the concentration of glucose, the higher the peak oxidation current. Glucose oxidation peaks occur at a potential of 0.4 V. Figure 6 shows cyclic voltammogram glucose with KOH electrolyte. The resulting cyclic voltammogram did not differ significantly between the NaOH and KOH electrolytes. The significant difference between the NaOH and KOH electrolytes is between Figure 5B and Figure 6B. Figure 5B shows a cyclic voltammogram having an oxidation peak and has no reduction peak. Figure 6B shows a cyclic voltammogram having an oxidation peak and a reduction peak.

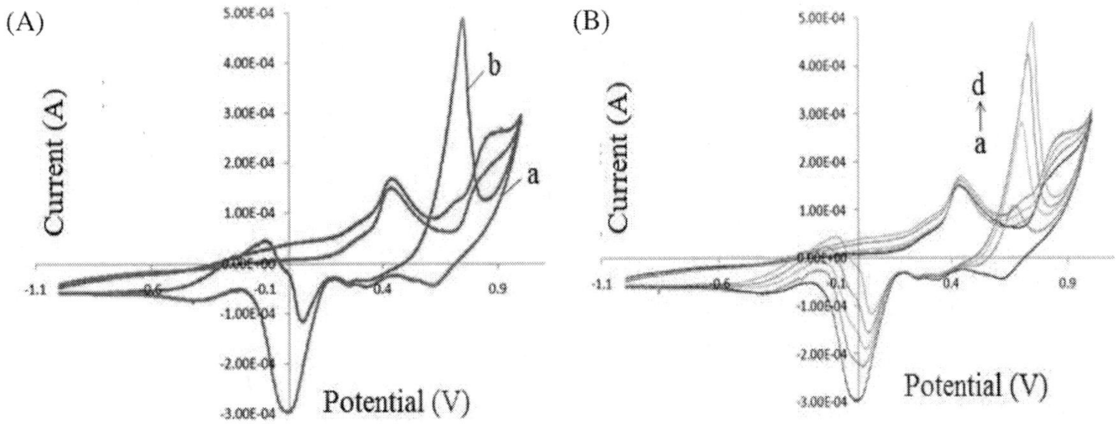

Figure 6. Cyclic voltammogram of (A) (a) 0.1 M KOH (without glucose), (b) 0.1 M NaOH + glucose 0.001 M glucose, (B) Cyclic voltammogram of various concentration of glucose at Au electrode in 0.1 M NaOH solution in glucose (a-e) 0.001-0.004 M, at scan rate 100 mVs^{-1}

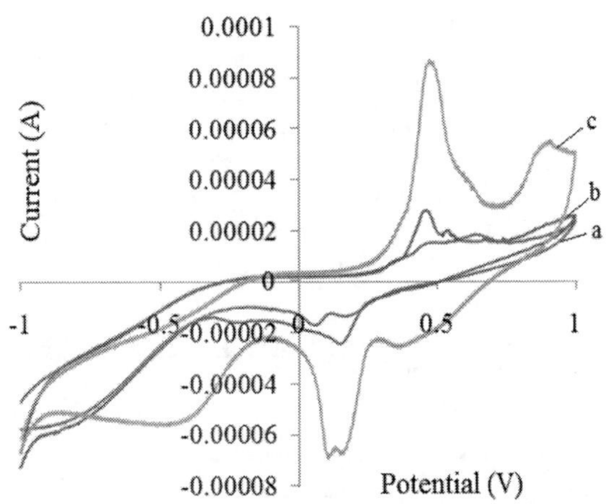

Figure 7. Cyclic voltammogram of 0.001 M glucose in buffer phosphate solution with various pH solutions(a) 4 (b) 7 (c) 10

Determination of effect pH on glucose analysis with Au electrode was performed at pH 4, 7, and 10. The tests were performed to determine the effect of pH on glucose response. Cyclic voltammogram comparison pH 4, 7, and 10 can be seen in Figure 7. Effect of pH on anodic and cathodic response can be seen at pH 7 and pH 10. Figure 7 (a) shown the buffer phosphate pH 4 shows a low anodic peak. The Au metal at the acidic solution has corrosion. Cyclic voltammogram at buffer phosphate pH 7 shows the anodic peak rise and at phosphate buffer pH 10 shows the highest anodic peak. The buffer phosphate pH 10 is an electrolyte suitable for glucose analysis with Au electrode.

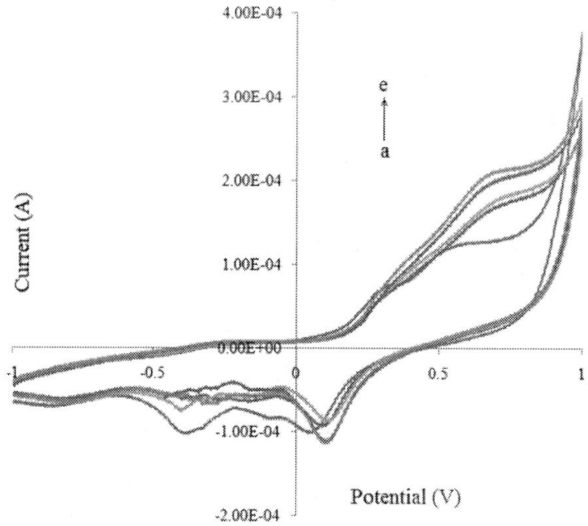

Figure 8. Cyclic voltammogram of (a) 0.1 M NaOH + human urine (without glucose), (b) 0.1 M NaOH + human urine + glucose 0.003 M, (c) 0.1 M NaOH + human urine + glucose 0.005 M (d)0.1 M NaOH + human urine + glucose 0.007 M (e) 0.1 M NaOH + human urine+ glucose 0.009 M, at scan rate 100 mVs^{-1}

Figure 8 shows a cyclic voltammogram by addition method to determine the concentration of glucose in the urine. Glucose concentration in human urine was determined using anodic peak with a glucose concentration of 0.003-0.009 M in 0.1 M NaOH electrolyte and scan rate 100 mVs-1. The cyclic voltammogram of the analysis is presented in Figure 8. Table 1 shows the results of recovery of determination of glucose in human urine using cyclic voltammetry at Au electrode by the KNO_3 electrolyte. Determination of glucose in human urine by cyclic voltammetry at Au electrode by KNO_3 electrolyte has very good recovery value that is in the range 95-105%.

Table 1. Recovery of determination of glucose in human urine using cyclic voltammetry at Au Electrode by KNO_3 electrolyte

Urine Sample	Glucose added (mM)	Glucose founded (mM)	Recovery (%)
Sample 1	3.00	2.99	99.67
Sample 2	5.00	5.12	102.40
Sample 3	7.00	6.79	97.00
Sample 4	9.00	8.78	97.56

4. Conclusion

KNO_3 is a good electrolyte for determination of glucose in human urine using gold electrode. The effect of glucose concentration has the coefficient correlation is $R^2 = 0.994$. The results of the recovery showed the recoveryat range 95-105% with addition method. As a conclusion isa gold electrode is a good electrode for electrochemical sensors to the determination of glucose in human urine.

References

[1] Zhu X Li C Zhu X and Xu M 2012 *Int. J. Electrochem. Sci.*, **7** 8522
[2] Park S Boo H and Chung T D 2006 *Analytica Chimica Acta*, **556** 46
[3] Kang X Wang J Wu H Liu J 2010 Aksay I A and Lin Y 2010 *Talanta* **81** 754
[4] Kumary K G and Letha R 1997 *J. Pharm. Biomed. Anal.* **15** 1725
[5] Han X Zhu Y Yang X and Li C 2010 *Microchimica Acta,* **171** 233
[6] Wang L Lu X Ye Y Sun L and Song Y 2013 *Electrochim. Acta.* **114** 484
[7] Tian H Jia M Zhang M and Hu J 2013 *Electrochim. Acta.* **96** 285
[8] Sun A Zheng J and Sheng Q 2012 *Electrochim. Acta.* **65** 64
[9] Ding Y Wang Y Su L Zhang H and Lei Y 2010 *J. Mater. Chem.* **20** 99
[10] Aoun S B 2013 *J. Mater. Environ. Sci.* **4** 68
[11] Pasta M Mantia F L and Cui Y 2010 *Electrochimica Acta.* **55** 61
[12] Pasta M Ruffo R Falletta E Mari C M and Pina C D 2010 *Gold Bulletin.* **43 1**

Acknowledgements

This research was supported by Ministry of Research, Technology and Higher Education of the Republic of Indonesia, Directorate General of Research and Development Reinforcement, Directorate of Research and Community Services through *Penelitian Terapan Unggulan Perguruan Tinggi* Research Grant 2017 for the financial support. We would like to thank the Integrated Instrumentation Laboratory and the Environmental Quality Laboratory at the Islamic University of Indonesia for the ease of use of the equipment to make this research successful.

IOP Conf. Series: Materials Science and Engineering **299** (2018) 012002 doi:10.1088/1757-899X/299/1/012002

Peat Water Purification by Hydroxyapatite (HAp) Synthesized from Waste Pensi (*Corbicula moltkiana*) Shells

Matlal Fajri Alif[1*], Wandha Aprillia[1], Syukri Arief[1]

Department of Chemistry, Faculty of Mathematics and Natural Sciences, Andalas University, Indonesia

*mfalif@fmipa.unand.ac.id

Abstract. Hydroxyapatite (HAP) were synthesized from Pensi (*Corbicula moltkiana*) sheels by hydrothermal method and used as adsorbent for peat water purification. Batch adsorption experiments were performed to investigate the effects of various factors such as contact time, adsorbent dosage, and pH. The obtained materials were characterized by powder X-ray diffraction (XRD), Fourier transform infrared (FTIR) spectroscopy and scanning electron microscope (SEM). Results showed that HAP calcined at 900°C (HAP900) and 1000°C (HAP1000) have a poorly crystalline shape. HAP900 also contain Tetracalsium Phosphate (TTCP) with a Ca/P molar ratio 2.18, while HAP 1000 contain HAp with a Ca/P molar ratio 1.67. Optimum condition for peat water purification with HAP900 and HAP1000 were both achieved at 1 hours, 1 grams adsorben mass at pH 2. SEM micrographs show that after purification, the surface of HAP were covered by organic compounds from peat water.

1. Introduction

In Indonesia, tropical peatlands cover almost 10% of the land surface mostly at low altitudes in coastal and sub-coastal areas on the islands of Sumatra, Borneo, and New Guinea (West Papua, Papua).The extent of peat land, resulting difficult for people to obtain clean water. Peat water is surface water or ground water is widely available in tidal areas, marshy and lowlands, maroon, sour (high acidity level), and has a high organic content. Peat water which contains organic substances (humic acid, fulvic acid and humin) are not qualified to supply as drinking water and needs of households. The existence of humic acid and fulvic acid in water can cause environmental and health serious problems.

Hydroxyapatite (HAP) have been widely used as implant materials due to its its high biocompatibility, good bioaffinity and their similar properties with natural bones, and gained great importance in the field of orthopedics and in the development of dental materials [1-2]. The calcium to phosphorous molar ratio (stoichiometric Ca/P ratio), must be controlled at about 1.67. HAP with the stoichiometric formula $Ca_{10}(PO_4)_6(OH_2)$ and a Ca/P molar ratio=1.67 is the most similar to the inorganic part of bones and teeth. Deviation of the calcium/phosphorus ratio from 1.67 can arise during preparation, leading to the formation of other compounds such as α-tricalcium phosphate and β-tricalcium phosphate [3–5]. Extraction of HAP from natural resources is promising as an alternative choice for the production of HAP.

Corbicula moltkiana or Pensi in local name, is a freshwater clams found in West Sumatera, Indonesia. It is a popular mollusk-based dishes in the local region, resulting in abundant amounts of waste shells every year. *Corbicula* shells (CS) are a rich source for calcium carbonate ($CaCO_3$), a material with great value to be converted to HAP.

2. Experimental

2.1. Materials preparation
CS biowaste was collected from local market around lake Singkarak, West Sumatera, Indonesia. CS were properly cleaned with detergent solution, washed thoroughly with tap water followed by soaking it with 0.1 M HNO_3 (Merck) to ensure the removal of remove attached materials such as soft tissue. After washing several times with distilled water, the CS was dried at 105°C for 5 h in oven, then crushed into powder with a grinder and finally sieved with a 50-mesh sieve.

2.2. Synthesis of CaHAP
The CS was calcined in an electrical muffle furnace and maintained at various temperatures (900 and 1000°C) for 5 h, the CS transformed into calcium oxide by releasing carbon dioxide. For synthesis, a stoichiometric value of ammonium di-hydrogenphosphate ($NH_4H_2PO_4$, 99.9%; Aldrich) was dissolved in distilled water and added to 10g pyrolised CS. After heating, the samples were cooled naturally inside the furnace. The reaction was carried out in a 500 mL beaker glass (under atmospheric pressure), and continuously mixed at a speed of 300 rpm using a mechanical stirrer. The temperature was maintained at approximately 90°C for 6 h. The slurry is dried in an oven at 120 °C for 5 h and calcined at 600 °C for 2 h.

2.3 Measurement of peat water
Measurement of peat water absorption spectrum was performed with a UV-Vis spectrophotometer (Thermo Scientific GENESYS 10S Series, USA), at 220-400 nm region. Purification is indicated by the changing pattern of absorption spectra in the UV-Vis due to the reduced organic compounds such as humic acid. Absorbance of each sample was measured both before and after purification.

3. Results and Discussion

3.1 X-ray diffraction analysis

Fig. 1 shows the X-ray diffraction patterns of the synthesized natural HAP at different temperature. As shown in this figure, HAP samples synthesized had similar XRD patterns. The crystalline phases identified mostly correspond to HA (ICSD-157481). All the diffraction peaks were assigned to low-crystalline HAP.

The broad peaks of the XRD patterns indicate that the HAP particles had small sizes and low crystallinity [6]. Poorly crystalline calcium phosphate apatite crystals indicate that they are similar to the crystals of bones and did not dissolve in the physiologic solutions [7]. At 900°C a tetracalcium phosphate (TTCP) peak was observed. TTCP is the only calcium phosphate phase with a Ca/P ratio greater than stoichiometric HAP [8].

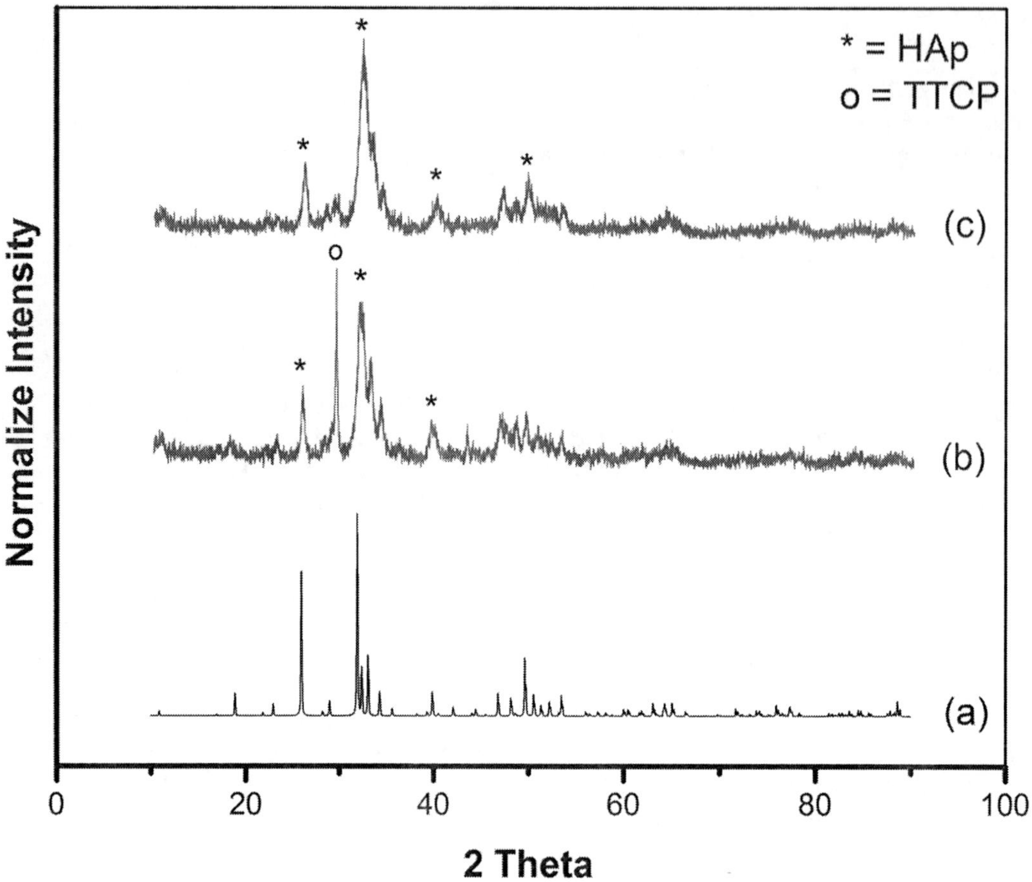

Figure 1. XRD patterns of HAP (a) HA (ICSD-157481), (b) HAP900, (c) HAP1000

3.2 FT-IR analysis

The FT-IR spectra of CS biowaste and HAP are plotted in Fig. 2. Based on Fig. 2. (a) shows the absorption with a strong peak at 1461.89 cm^{-1} indicating the presence of carbonate (CO_3) group and the peak at 1974.47 cm^{-1} indicate the presence of CH bending. C-N group (854.31 cm^{-1}) and C-C (707.49 cm^{-1}) indicates that the CS contains amine and carbonate groups.

The spectra of HAP900 and HAP1000 in Fig. 2 (b) and (c) show phosphate groups at 1024. 07 cm-1 and 1027 cm^{-1} each with sharp peaks. The carbonate group at 1453.01 cm^{-1} at HAP900 and 1447.29 cm^{-1} appears on the CS HAP1000. The peak at 2196.74 cm^{-1} and 2200.25 cm^{-1} shows the uptake of HOH. The FTIR analysis results show that the synthesized HAP is not pure in the presence of carbonate groups into the crystal lattice HAP. This is also due to CO_2 absorption in the air during sample preparation [9].

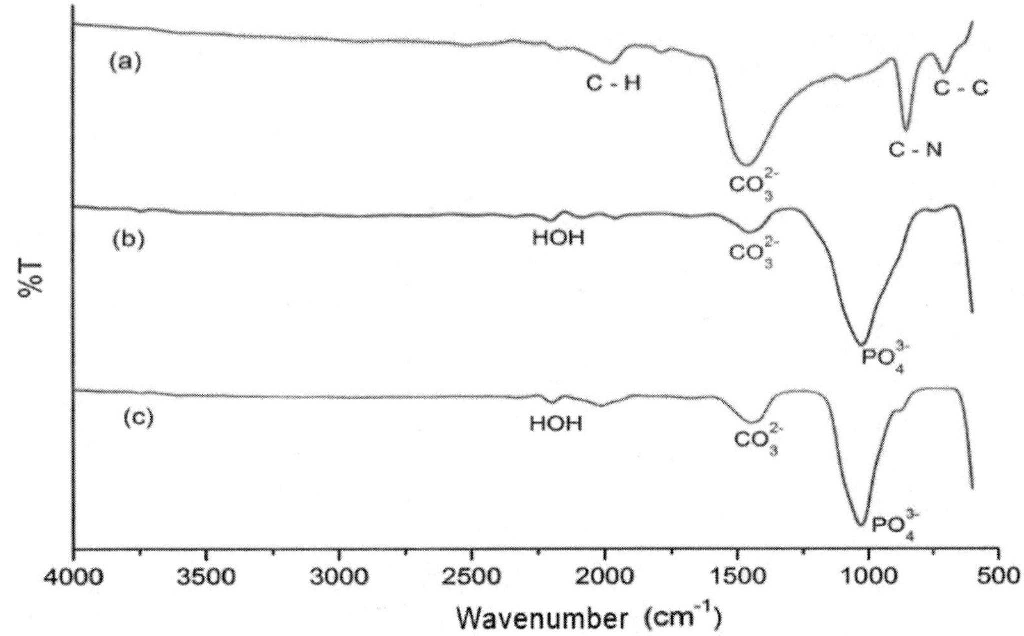

Figure 2. FT-IR spectra of (a) CS biowaste (b) HAP900 and (c) HAP1000.

3.3 UV-Vis analysis

The content of organic compounds in peat water is dominated by humic acid (HA). The properties of humic acid molecules can be described by UV-Vis spectroscopy widely as a simple and informative method [10]. The UV-Vis spectrum of humic acid is wide and monotonically decreases with increasing wavelength. The absorbance of the UV region spectrum can be used to analyze HAP [11].

UV absorption at 280 nm represents the overall aromatisity, since the electron transition π-π * occurs in this UV region (270-280 nm) such as phenolic, benzoic acid, aniline derivative, poly and polycyclic aromatic hydrocarbons with two or more rings and also absorbents at 280 nm can be used to characterize peat humidity [12]. To find out the reduction of humic acid contained in peat water, this is done by comparing the absorbance at 280 nm before and after clearance. This measurement aims to determine the reduced content of humic acid contained in peat water. The lower the absorbance of peat water the more the humic acid content is reduced.

HAP900 and HAP1000 are added 0.5 grams into the erlenmeyer and then added 10 mL of peat water. Stirring at 250rpm was carried out for a various contact time (0.5, 1, 2, 3, 4 h). The mixture was filtered and the filtrate was analyzed by UV-Vis Spectrophotometer at λ 220-400 nm range. The measurements of the UV spectrum of peat water and after purification can be seen in Fig. 3.

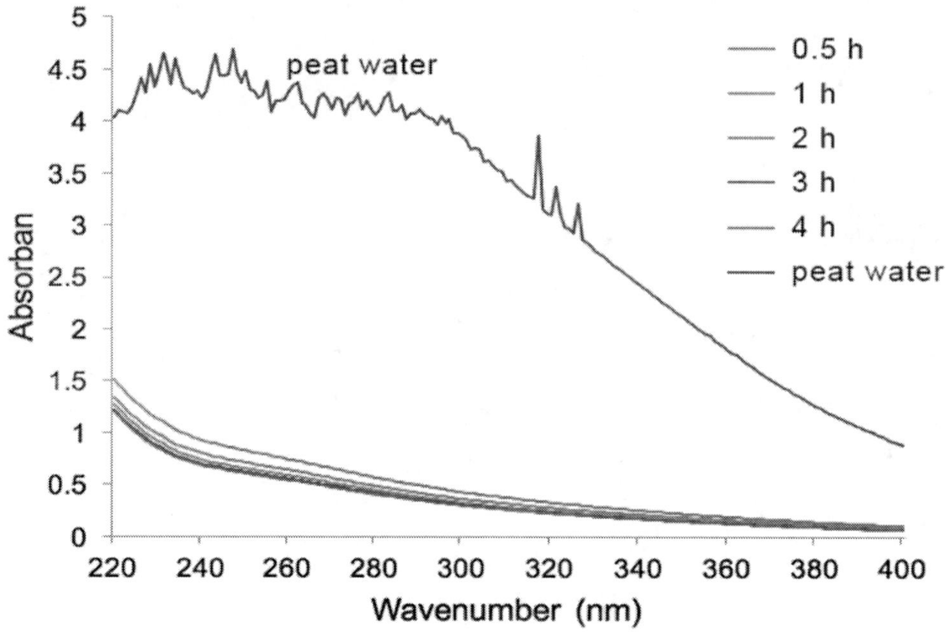

Figure 3. UV spectra of peat water before and after purification by HAP.

3.4. SEM analysis

Surface morphology of the synthesized hydroxyapatite before and after peat water purification were analyzed by SEM and are shown in Fig.4. The morphology of HAP900 and HAP1000 (Fig. 4 (a) and (b)) at 3000 x magnification were seen to have agglomeration with variable size pore distributed evenly. HAP900 and HAP1000 experience changes in the surface morphology after purification as shown in figure 4C and 4D. The surface pores of HAP900 and HAP1000 are covered by organic compounds originated from peat water.

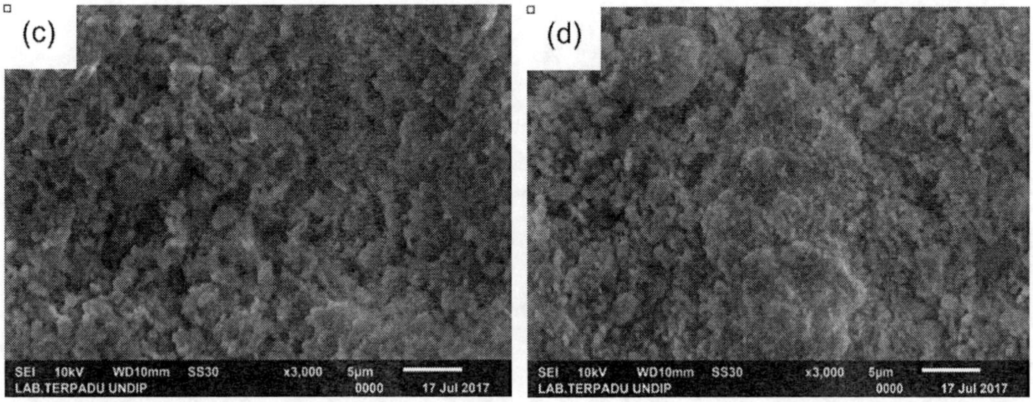

Figure 4. SEM images of (a) HAP900, (b) HAP1000, (c) and (d) HAP900 and HAP 1000 after peat water purification treatment respectively.

4. Conclusion

CS biowaste can be used for hydroxyapatite material by hydrothermal method. Based on XRD analysis HAP900 have TTCP phase with a molar ratio of Ca / P 2.18. HAP1000 was dominated with HAP (molar ratio of Ca / P 1.67). Both HAP900 and HAP1000 shows an amorphous crystal structure. HAP from CS biowaste was able to purify peat water. Optimum condition for peat water purification by HAP900 and HAP1000 were both achieved at 1 hours, 1 grams adsorben mass at pH 2.

References

[1] Vahabzadeh S, Roy M, Bandyopadhyay A and Bose S 2015 Phase stability and biological property evaluation of plasma sprayed hydroxyapatite coatings for orthopedic and dental applications *Acta Biomater.* **17** 47–55

[2] Rau J V, Cacciotti I, Bonis A De, Fosca M, Komlev V S, Latini A, Santagata A and Teghil R 2014 Fe-doped hydroxyapatite coatings for orthopedic and dental implant applications *Appl. Surf. Sci.* **307** 301–5

[3] Li Y, Weng W and Tam K C 2007 Novel highly biodegradable biphasic tricalcium phosphates composed of a-tricalcium phosphate and b-tricalcium phosphate *Acta Biomater.* **3** 251–4

[4] Boutinguiza M, Pou J, Comesaña R, Lusquiños F, De Carlos A and León B 2012 Biological hydroxyapatite obtained from fish bones *Mater. Sci. Eng. C* **32** 478–86

[5] Pon-On W, Suntornsaratoon P, Charoenphandhu N, Thongbunchoo J, Krishnamra N and Tang I M 2016 Hydroxyapatite from fish scale for potential use as bone scaffold or regenerative material *Mater. Sci. Eng. C* **62** 183–9

[6] Kim H, Kim Y, Park S, Rey C, Lee H, Glimcher M J and Ko J S 2000 Thin film of low-crystalline calcium phosphate apatite formed *Biomaterials* **21** 1129–34

[7] Kim H, Rey C and Glimcher M J 1995 Isolation of Calcium-Phosphate Crystals of Bone by Non-Aqueous Methods at Low Temperature *J. Bone Miner. Res.* **10** 1589–601

[8] Moseke C and Gbureck U 2010 Tetracalcium phosphate : Synthesis , properties and biomedical applications *Acta Biomater.* **6** 3815–23

[9] Wei W, Yang L, Zhong W, Cui J and Wei Z 2015 Poorly crystalline hydroxyapatite: A novel adsorbent for enhanced fulvic acid removal from aqueous solution *Appl. Surf. Sci.* **332** 328–39

[10] Shirshova L T, Ghabbour E A and Davies G 2006 Spectroscopic characterization of humic acid fractions isolated from soil using different extraction procedures *Geoderma* **133** 204–16

[11] Uyguner C S and Bekbolet M 2005 Evaluation of humic acid photocatalytic degradation by UV – vis and fluorescence spectroscopy *Catal. Today* **101** 267–74

[12] Chin Y, Alken G and Loughlins E O 1994 Molecular Weight, Polydispersity, and Spectroscopic Properties of Aquatic Humic Substances *Environ. Sci. Technol.* **28** 1853–8

Acknowledgement

This research was supported by the Grant from the Faculty of Mathematics and Natural Sciences of Andalas University with contract no.22/UN.16.03.D/PP/FMIPA/2017. The conference grant provided by Andalas University is gratefully acknowledged.

IOP Conf. Series: Materials Science and Engineering **299** (2018) 012003 doi:10.1088/1757-899X/299/1/012003

Use of *Eichhornia crassipes* modified Nano-chitosan as a biosorbent for lead (II), cadmium (II), and copper (II) ion removal from aqueous solutions

A. H. Alkaff, H. Hendri, I. H. Farozy*, M. Annisa, R.P. Aritonang

Chemistry Department, University of Indonesia, Depok, West Java, Indonesia.

*ikko.haidar@sci.ui.ac.id

Abstract. Industrial waste in a major city poses a considerable threat to water environment from the accumulation of heavy metals. Additionally, uncontrolled growth of *Eichhornia crassipes* will also damage the water environment by lowering the levels of dissolved oxygen. Therefore, we conduct research to not only treat industrial waste in water but also reduce the population of *E. crassipes* in water. We made this biosorbent by mixing *E. crassipes* with nano-chitosan in various compositions. Its absorptivity was tested against single metal solutions of lead (II), cadmium (II), and copper (II) to get the best biosorbent composition. The chosen biosorbent then went through an adsorptivity test against a mixture of three solutions, with each test was carried at various pH. The best biosorbent composition is the mixture of 1 g of *E. crassipes* with 30 mL of nano-chitosan 0.01%, while adsorption tests in single or three metals solution show that the biosorbent performs better in neutral pH.

1. Introduction

Pollution is a major problem that plagued major cities in Indonesia such as DKI Jakarta, because it gave serious problems to people, as well as the environment, with one example being water pollution. One of the biggest contributors toward water pollution is a waste being disposed into the water without processing, especially industrial waste that still contains heavy metals. Such waste is hazardous because the heavy metals it contains can cause a massive degradation of water quality, as well as cause long-term harm to the residents due to the accumulation of heavy metals, especially on fish from such waters.

Heavy metal waste is one of the main problems plaguing DKI Jakarta. With many of industries disposing their waste into waters without proper processing, which cause severe pollution on waterways around Jakarta. It is a problem because many industries that did not properly process their waste before disposal have heavy metal content, which is known for their potential of causing long-term harm, and are already accumulated in sediments and water at worryingly high concentrations. Such as the waters in Western parts of Jakarta, the concentrations of metals Pb (Lead) are 8.49-31.22ppm; Cu (Copper) = 13.81-193.75 ppm, and Cd (Cadmium) = <0.001-0.47 ppm. It is a problem because the safe limit for Pb is 0 mg/L, or in other words, no lead is allowed at all, while for Cu is 1.3 mg/L and Cd is 0.005 mg/L [1]. It is a clear and present danger to residents because such metals in water will cause long-term harm to people.

On the other hand, water hyacinth (known to locals as eceng gondok. Biological name: *E. crassipes*) is a floating wild water flora that grows rapidly on freshwater surfaces, with growth rates as high as 220 kg/ha/day [2]. With its rapid growth and spread, water hyacinth can cause problems because an overpopulation of water hyacinth will bring harm, such as siltation of rivers due to remains of dead water hyacinths sinking into the river bottom, and also causes dissolved oxygen in the water to drop dramatically, which disturbs other organisms in the river.

While water hyacinth has been utilized for applications such as furniture, biogas, fertilizer, and livestock feeds, but there are still more potential applications for the water hyacinth. One of which is a biosorbent for absorbing heavy metals such as lead (Pb), cadmium (Cd), and copper (Cu). It is possible because of their cellulose content capable of absorbing such metals [3]. Cellulose itself is an organic compound consisting of long chains of glucose units which is bind by 1,4-β-glucosidic bonds, which can interact well with adsorbates thanks to the high amount of –OH [4]. The water hyacinth would then be modified with nano-chitosans to make a cost-effective biosorbent but is capable of adsorbing heavy metals effectively.

Chitosan and chitin are commercially valuable biopolymers which are commonly used in industry. Chitin is used as base ingredients in the industry such as pharmacology, agriculture, membrane makers, textile, cosmetics, etc. On the other hand, chitosan is produced from deacetylation process of chitin, which is commonly found as a major constituent of the exoskeleton of crustaceous marine animals [5]. Chitosan itself is also often used in water purification, a chelating agent, color binding agents, and also used as adsorbents of heavy metals.

Adsorption is a process where the dissolved substance in a solution is coagulated by the surface of the adsorbent, in which a chemical-physical bond is occurring between the substance and its adsorbent. Adsorption is divided into two kinds, physical and chemical adsorption (physisorption and chemisorption) [6]. This phenomenon should be distinguished from absorption, in which adsorption is a surface phenomenon while absorption is a volumetric phenomenon [7].

Physisorption is a reversible process involving van der Waals and hydrogen bonding with the adsorbent [8], while chemisorptions involves the formation of chemical bonding which consists of covalent and electrostatic bonding which makes the adsorbed substance harder to release and makes the process less reversible [9]. Factors affecting adsorption are as follows: contact time, the area of the adsorption media, adsorbate solubility, adsorbate molecule size, pH of the solution, and the temperature.

2. Experiment

2.1. Materials

Water hyacinths were picked up from a lake near the Faculty of Mathematics and Natural Science in University of Indonesia, Depok. Chitosan powder, glacial acetic acid, $Pb(NO_3)$, $Cd(NO_3).4H_2O$, and $Cu(NO_3).3H_2O$ were supplied by Merck, Germany. All chemical reagents used in this experiment were of analyitical grade and did not need further purification. Deionized water (Aquadest) was used as solvent for all solutions

2.2. Biosorbent from water hyacinth synthesis

Water hyacinth was washed with aquadest to remove contaminants that stuck to the leaves and stem. It was then roughly blended using an ordinary blender and then dried at room temperature. After drying, it was then blended again until it became a fine powder. The powder was sieved with 60 mesh sieve to get a finer powder.

2.3. Nano-chitosan synthesis

Nano-chitosan was synthesized by dissolving chitosan powder in glacial acetic acid while stirring it with a magnetic stirrer for 3 h at room temperature. The glacial acetic acid is stirred first, with the chitosan was slowly added to the solution. This process is done at room temperature (25^{0}C). The stirrer is heated to facilitate dissolution of chitosan powder.

2.4. Biosorbent modification with nano-chitosan

The biosorbent was added into the synthesized nano-chitosan, and then stirred with magnetic stirrer. The resulting mixture was centrifuged, and the sediment was dried. Biosorbent mass was varied at 1 g, 0.75 g, and 0.5 g. The optimum mass was then determined.

10 mL nano-chitosan was then added for each biosorbent mass (1 g, 0.75 g, and 0.5 g) for each addition. The mixture was then stirred for 2 h, centrifuged, and the resulting residue was dried. The steps were repeated for two and three times. The maximum biosorbent mass was determined by Prestige 21 FTIR by Shimadzu, Japan, while the percent of nano-chitosant that interact with biosorbent was determined by gravimetry.

About 0.05 gr of biosorbent at each nano-chitosan addition was then put into a crucible, and burned in a furnace for two hours, then stored in a desiccator for 15 min. The crucible was then weighed and compared with empty crucible weight. The weight difference was the weight of the chitosan. The process is repeated until a constant weight is achieved.

2.5. Biosorbent application on single metal solution

Biosorbent application tests on the single metal solution were applied for Pb (II), Cd (II), and Cu (II). The single metal solutions were prepared as solution (50 mL) with each concentration of 1500 ppm, 300 ppm, and 600 ppm respectively. The application was done by inserting 0.1 g of the optimum biosorbent into the column. Then, 50 mL of metal solution was poured in. The solution which passed the biosorbent is then held in a container and then analyzed with Shimadzu AA-7000 AAS by Shimadzu, Japan. These procedures were done for Pb, Cd, and Cu metals.

2.6. Biosorbent application on mixture of metals solution

The application test was done by preparing a mixture solution by pipetting each 50 mL of Pb 4500 ppm, Cd 900 ppm, and Cu 1800 ppm solutions into a 150 mL volumetric flask, and then 100 mL of the solution was then used, so the concentration of each metal were the same with the single metal solutions. 50 mL of the mixture was then taken and flowed into a column with biosorbent. The solutions were held in a container and then analyzed by AAS. From this process, it was determined that the optimum composition is a mixture consisting of 1 g of *E. crassipes* in 30 mL of nano-chitosan 0.01%.

3. Results and Discussion

3.1. Results

3.1.1. pH variation Modified biosorbent adsorption on the single metal solution is done at pH 5 (acid) and pH 7 (neutral). Higher pH is not investigated due to formation of precipitate of metal ions, which cannot be adsorbed by the adsorbent.

3.1.2. Adsorption application for single metal solution According to Figure 1, it is known that the best adsorption is done against Pb metal, which is in accordance with HSAB (Hard-soft acid-base) theory that states hard acids are likely to interact with hard bases, so do the soft acids that are likely to interact with soft bases [10]. Because Pb and Cu are borderline acids, they are more easily bound to OH groups of cellulose and NH_2 groups of chitosan, which are both hard bases compared to Cd metal. However, Cd gives higher adsorption results than Cu which is not in accordance to the HSAB theory, because Cd is a soft acid and therefore would be less likely to bind with OH and NH_2. This error can be attributed to systematic errors in the experiment.

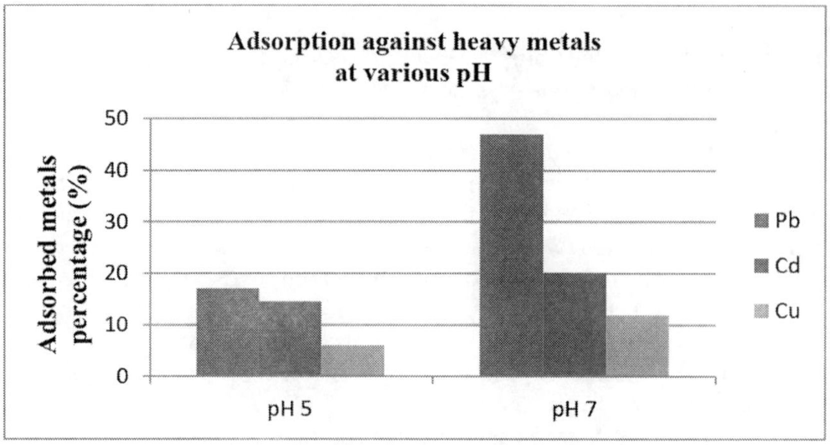

Figure 1. Adsorption percentage of single metal solutions against Pb, Cd, and Cu metals at various pH

Adsorption application for a mixture of three metals solution According to Figure 2, it is known that Pb has the best adsorption compared to Cd and Cu metals. The biosorbent also shows better adsorption against Cu metal compared to Cd metal, which is in accordance with HSAB theory, because Pb and Cu are borderline acids while Cd is a soft acid. In the mixture solution, the adsorption capacity is decreased compared to single solution, because there are more metals in the solution to be adsorbed. Also, it is shown that the adsorption process is better at pH 7 (neutral) because at acidic pH, there are H^+ ions that can degrade the adsorption capabilities of the adsorbent by making the active sites less available for metal ions [11].

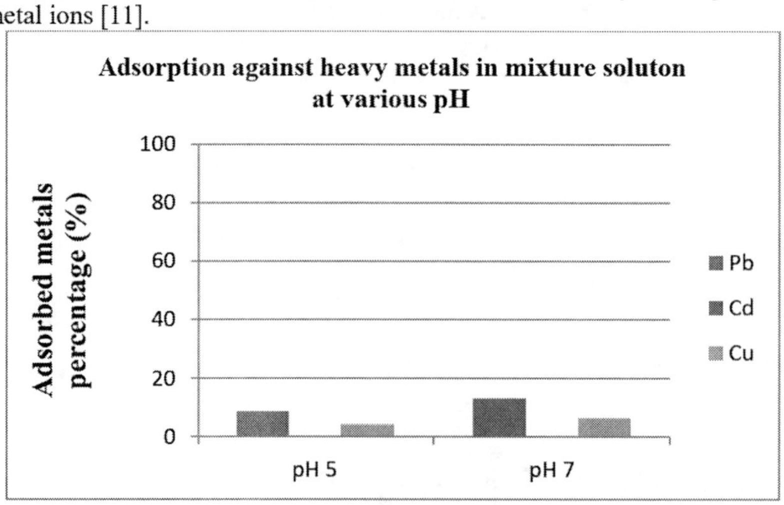

Figure 2. Adsorption percentage of mixed metal solutions against Pb, Cd, and Cu metals at various pH

4. Conclusion

It is concluded from our research that it indeed is possible to synthesize a biosorbent from nano-chitosan modified water hyacinth to adsorb heavy metals. The synthesized biosorbent performs best in neutral pH compared to acidic pH, because at acidic pH, the H^+ ions will degrade the adsorption abilities of the biosorbent. While in basic pH, the biosorbent cannot function properly because basic pH will cause the formation of solids which cannot be adsorbed. During adsorption tests, the adsorbent adsorbs Pb better than Cu and Cd, which can be attributed to HSAB theory, and the adsorbent capacity in mixture solution is less than single metal solutions because there are more kinds of metals to adsorb.

References

[1] Aryo S 2009 *Analysis of Cd, Pb, and Hg Heavy Metal Contents in Water and Sediments on Kamal Muara Waters, North Jakarta* (Bogor: Bogor Agriculture Institute)
[2] Bich T N T, Lu Ki O, Dieu T N T and Yi-Hsu J 2017 *J. of the Taiwan Inst. of Chem. Engineers* **71** 55-61
[3] Jia-Chuan Z, Hou-Qi L, Hui-Min F, Wen-Wei L, Michael Hon-Wah L, Paul Kwan-Sing L and Han-Qing Y 2016 *Environmental Polluton* **219** 837-45
[4] Dedy M 2013 *Biocomposites of Banana Peel Nanochitosan for Adsorption Heavy Metal ion* Pb(*II*), *Cd*(*II*), *and Cu*(*II*) *Isotherm, Kinetics, and Thermodynamics Studies* (Depok: University of Indonesia)
[5] Jiao C, Shaouyu L, Zhe Z, Xuxia Z, Xinming L, Piao N and Mingzhu L 2018 *Sci. of the Total Env.* **613-14** 829-39
[6] Eliya N K 2009 *Oseana Popular Arts Science Magazine* **34** 1-7
[7] H Z Hassan and A A Mohamad 2012 *Renewable and Sust. Energy Reviews* **16** 2516-38
[8] K Fukutani and T Sugimoto 2013 *Progr. in Surface Sci.* **88** 279-348
[9] C Minot and A Markovits 1998 *J. of Molecular Str.* **424** 119-34
[10] Richard F D and Sally J D 2005 Acid-Base Theory *Organic Chemistry* (Walla Walla: Daley & Daley) chapter 5 pp 205-242
[11] Fu-Qiang A, Rui-Yan W, Min L, Tuo-Ping H, Jian-Feng G and Zhi-Guo Y 2017 *React. and Funct.Polymers* **118** 42-50

Characteristics of Alcian-blue Dye Adsorption of Natural Biofilm Matrix

A Kurniawan[1,3,4], T Yamamoto[2], Sukandar[1,3], and Guntur[1,3]
[1] Coastal and Marine Research Centre, Brawijaya University, 65145, Malang, Indonesia
[2] Graduate School of Life Sciences, Ritsumeikan University, 525-8577, Kusatsu, Japan
[3] Faculty of Fisheries and Marine Science, Brawijaya University, 65145, Malang, Indonesia
[4] Corresponding author. E-mail: andi_k@ub.ac.id

Abstract. In this study, natural biofilm matrices formed on stones have been used for the adsorption of Alcian blue dye. Alcian blue is a member of polyvalent basic dyes that largely used from laboratory until industrial dying purposes. The adsorption of the dye onto the biofilm matrix has been carried out at different experimental conditions such as adsorption isotherm and kinetic of adsorption. The electric charge properties of biofilm matrix and its changes related to the adsorption of Alcian blue have been also investigated. Moreover, the results of Alcian blue adsorption to the biofilm were compared to those onto the acidic and neutral resin. The kinetics of adsorption result showed that the adsorption of the Alcian blue dye reached to a maximum adsorption amount within 60 minutes. The adsorption amount of Alcian blue to biofilm increased monotonously, and the maximum adsorption amount was greater compared to the resins. On the contrary, Alcian blue did not attach to the neutral resin having no electric charge. It seems that Alcian blue attached to the acidic resins due to electrostatic attractive force, and the same seems to be the case for adsorption of Alcian blue to biofilm. The adsorption of Alcian blue to the biofilm and acidic resins fitted to Langmuir type indicates that the binding of Alcian blue to the biofilm and acidic resins occurred in a monolayer like form. The maximum adsorption amount of Alcian blue on the biofilm (0.24 mmol/dry-g) was greater than those of acidic resin (0.025 mmol/dry-g). This indicates that the biofilm has many more sites for Alcian blue attachment than acidic resins. According to the result of this study, the biofilm matrix can be a good adsorbent for dye such as Alcian blue or other dyes that causing hazards in nature.

1. Introduction

The dyes are lost in industries effluents such as textile, leather, paper and plastics every year during the dyeing and finishing operations [1]. This problem becomes serious concern due to their effects on human and ecosystems [2,3]. The dangerous of dyes for ecosystems come from the high visibility, recalcitrance and undesirability, and thus, the removal of the dyes becomes a critical important effort to maintain clean environment. One of the dyes that largely use is alcian blue. This dye is a member of polyvalent basic dyes that largely used from laboratory until industrial dying purposes. When the dyes release to the environment such as aquatic ecosystems, the dyes may be accumulated in the interface surface including organism life on the surface.

Organism that found almost on the every interface in aquatic ecosystems is microbes formed biofilm. Biofilm consists of various microorganisms imbedded in a matrix composed of microbially produced extracellular polymer attaching to a surface [4,5]. Biofilm is formed when bacteria and other microorganisms attach to a surface and then replicate with producing extracellular polymer substrates [6,7].

The previous studies have shown that biofilm has high sorption capacities for many substances [8,9, 10,11]. The main driving force of the sorption process is an electrostatic interaction between the biofilm and the substances. Hence, the dyes such as alcian blue should be attracted and accumulated onto natural biofilm matrix. However, the study concerning this process has been rarely conducted. In the present study, the characteristics of alcian blue adsorption to natural biofilm matrix are investigated. According to the result of this study, the biofilm matrix can be a good adsorbent for dye such as alcian blue that causing hazards in aquatic ecosystems.

2. Materials and Method

2.1. Study area and sample preparation

Biofilm that used in this study was collected from Lake Biwa, Japan. This lake is located in the central part of Japanese Archipelago, having a surface area of 670 km^2 and total storage volume of 2.75×10^{10}m^3. The sample used in this study was collected from the shore of the southern basin (Akanoiwan) of Lake Biwa.

Samples for the characterization of the biofilm polymers were collected in November 2012 (winter) as the previous study founded that the biofilm thickness reaches the maximum amount in these seasons [6]. Stones were taken from the depth of 30 – 50 cm and brought back to the laboratory in a plastic container filled with nearby lake water; the container was maintained at 4°C.

The biofilm on the surfaces of the stones were removed using a toothbrush and suspended in sterilized distilled water. The biofilm pellets were prepared by centrifuging (8,000×g at 4°C for 10 min) the biofilm suspensions. The pellets of biofilm used as a sample in adsorption experiment were washed six times with 10 mM phosphate-buffered saline (PBS) (NaCl, 0.526 g; $Na_2HPO_4 \cdot 12H_2O$, 0.358 g; distilled water, 1000 mL) of pH 7 and stored at -40°C until ion adsorptions analyses were conducted.

2.2. Maintaining biofilm wet/dry weight

The pellet of biofilms (approximately 1 wet-g) were taken and then dehydrated for 3 days until the weight was stable to give a dry weight. This data was used to convert the dry weight per the wet weight of biofilm polymer.

2.3. Electrophoretic Mobility

Electrophoretic mobility of biofilm polymer was investigated based on Kurniawan *et al.*, 2012 [6]. The pellet of biofilm polymer was washed three times as follows. The pellet was resuspended in 40 mL of 10 mM NaCl aqueous solution. The suspension was centrifuged (8,000 × g at 4°C for 10 min), and the supernatant was discarded. The obtained biofilm pellet (ca. 0.03 g) was suspended in 1 mL of 10 mM NaCl aqueous solution. The suspension was mixed vigorously with a vortex for 5 min, then sonicated (2510J-MT, Yamato Scientific, Tokyo, Japan; 42 kHz, 125 W) for 10 min, followed by the vortex for 10 s. The obtained suspension was mixed with 10 mM of PBS at a ratio of 1:19 and used to analyze the electric charge of the biofilm polymer. The electrophoretic mobility (EPM) of the biofilm was measured on a ZETASIZER Nano-Z (Malvern Instruments, Ltd., Worcestershire, England) in PBS varying in pH values from 2.0 to 9.0. The pH of the buffer was adjusted with 20 mM of HCl or NaOH. The pHs of PBS before and after the addition of biofilm and after measurement of EPM were recorded.

2.4. Kinetics of Adsorption

One wet-g of biofilm pellet was resuspended in 100 mL of 100 mM PBS of pH 7. The suspension was mixed vigorously with a vortex for 5 min, and then sonicated for 10 min, followed by the vortex for 30 s. Then, 6.0 mL of 10 mM alcian blue was added to the suspension. The temperature of the suspension was maintained at 25°C using a water thermostat, and mixed well using magnetic stirrer. The aliquots of the suspension were taken after 1 min – 60 min, then centrifuge (8,000 × g at 4°C for 3 min) to separate the supernatant and the pellet. The adsorbed amount of alcian blue by biofilm polymer calculated from the difference of absorbance at 614 nm wavelength between controls (only PBS and alcian blue) and supernatants. The experiment was also conducted using one dry-g of acidic resin (SuperQ-650M; Toyopearl) and one dry-g of neutral resin (HW-65; Toyopearl) as an adsorbent.

2.5. Adsorption Isotherm

Several sets of biofilm suspensions were prepared (1 set was 0.1 wet-g of biofilm pellet resuspended in 5 mL of 100 mM PBS of pH 7). The suspension was mixed vigorously with a vortex for 5 min, and then sonicated for 10 min, followed by the vortex for 30 s. Then, a 10 mM of alcian blue solution, prepared with suspended alcian blue 8GX in 40 mL of 100 mM PBS of pH 7, was added to the biofilm suspension with various volumes (100 – 2000 μL). The temperature of the suspension was maintained at 25°C using a water thermostat. After 5 minutes, the biofilm pellet and the supernatant was separated using centrifugation (8,000 × g at 25°C for 3 min). The absorbance of alcian blue in the supernatant (abs-S) was measured calorimetrically by measuring the absorbance at 614 nm (the maximum absorption wavelength of alcian blue).

In the preliminary experiment, it was investigated that the supernatant (yellow supernatant) of the biofilm suspension also can adsorb alcian blue. Therefore, this supernatant was used as a control in the alcian blue adsorption experiment by biofilm polymer. To obtain the yellow supernatant, parts of the sets of biofilm suspension were centrifuged (8,000 × g at 4°C for 3 min). The further steps were same with the experiment that used biofilm pellet as a sample. The result was the absorbance of alcian blue in the supernatant of control (abs-C). The adsorbed amount of alcian blue by biofilm polymer was calculated from the difference between abs-C and abs-S that equal with the adsorbed amount of alcian blue by biofilm polymer. The adsorption experiments (kinetics of adsorption and adsorption isotherm) were also conducted using one dry-g of acidic resin and neutral resin. The resins used were TOYOPEARL CM-650M and HW-65, respectively.

3. Result and Discussion

3.1. Electric charge properties of biofilm polymers

The EPM value of biofilm was measured (Fig. 1). The result was used to analysis the surface charge distribution of biofilm polymer and estimates the functional groups in biofilm polymer. The EPM of biofilm ranged from negative to positive values as a function of pH. It seems that both negative and positive charges exist on biofilm polymer. In the lower pHs the EPM of biofilm showed greater positive value. This indicates the decreasing of negative charge; the ionization of functional group such as carboxylic acids seems to be depressed at lower pH [6].

The EPM changed greatly around pH 4. This indicates there are some functional groups that have maximum proton adsorption ability around pH 4. This fact strengthens the estimation that carboxylic groups exist within BF polymer. To analyze the influent of alcian blue attachment on the electric charge characteristics of biofilm polymers, the EPM dependence on pH of the intact and alcian blue stained biofilm were compared (Fig. 1). EPM of alcian blue stained biofilm showed greater positive value than intact biofilm. It was indicating that biofilm polymers adsorb alcian blue which carries positive charge, and thus, makes biofilm polymers to have more positively charged sites. Alcian blue that designed to be adsorbed by electrostatic forces could attach to some surfaces having oppositely charges sites. This reveals that one of the main factors that promote adsorption of the dye to BF polymer is electrostatic attractive force.

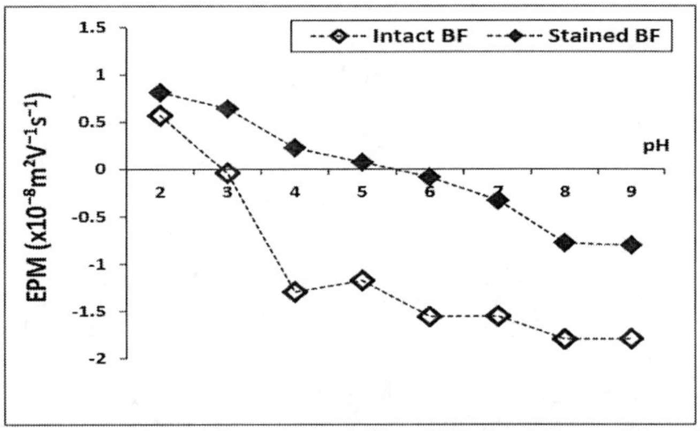

Figure 1. Electrophoretic mobility (EPM) of intact (♦) and alcian blue stained (◊) biofilms. EPM values were measured in from pH 2 until pH 9.

3.2. Time course of adsorption

The time courses of alcian blue adsorption to biofilm, acidic resins and neutral resin were investigated (Fig. 2). The adsorption of alcian blue to various resins occurred quite quickly, i.e., reached to a maximum adsorption amount within a few minutes. This result and the smaller adsorption amount of alcian blue on a neutral resin, indicate that the driving force of alcian blue adsorption is an electrostatic attractive force between the positively charged alcian blue and the negatively charged resin surface.

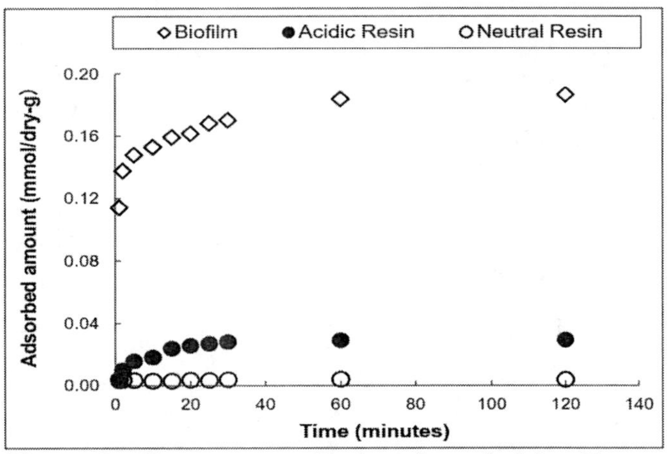

Figure 2. Time courses of adsorption of alcian blue to biofilm (◊), acidic resin (•) and neutral resin (○).

The adsorption of alcian blue to biofilm was also a quick process (Fig. 2), indicating that the adsorption of the dye to biofilm is a process as in the case of ion exchange resins. The adsorption amount of alcian blue to biofilm increased monotonously, and the maximum adsorption amount was greater compared to the resins. Thus, it can be deduced that the sites in biofilm for alcian blue adsorption have similar electrostatic properties and their number per dry gram is far greater than that of ion exchange resins.

3.3. Adsorption isotherm

Alcian blue (positively charged) attached to the biofilm and acidic resins (negatively charged on their surfaces) (Fig. 3). The adsorbed amount of alcian blue leveled off at higher concentration of the dye. On the contrary, alcian blue did not attach to the neutral resin having no electric charge (Fig. 3). These results indicate that alcian blue attached to the acidic resins due to electrostatic attractive force, and the same seems to be the case for adsorption of alcian blue to biofilm [11,12,13].

The maximum adsorption amount and the adsorption equilibrium constant for alcian blue adsorption were calculated using a variant of the Langmuir isotherm equation [13,14,15], as described below.

$$\frac{C}{N} = \frac{1}{(N_{max})b} + \frac{C}{N_{max}}$$

The plot of C/N against C yields a straight line with a slope of 1/Nmax and a y-axis intercept of 1/(Nmax)b, and thus, the values of Nmax (the maximum amount of adsorbed ion; mmol dry-g–1) and b can be calculated [13].

The adsorption of alcian blue to the biofilm and acidic resins fitted to Langmuir type adsorption (Fig. 4). Thus, it seems that the binding of alcian blue to the biofilm and acidic resins occurred in a monolayer like form. The maximum adsorption amount of alcian blue on the biofilm (ca. 0.24 mmol/dry-g) was greater than that of acidic resin (0.025 mmol/dry-g). This indicates that the biofilm has many more sites for alcian blue attachment than acidic resins.

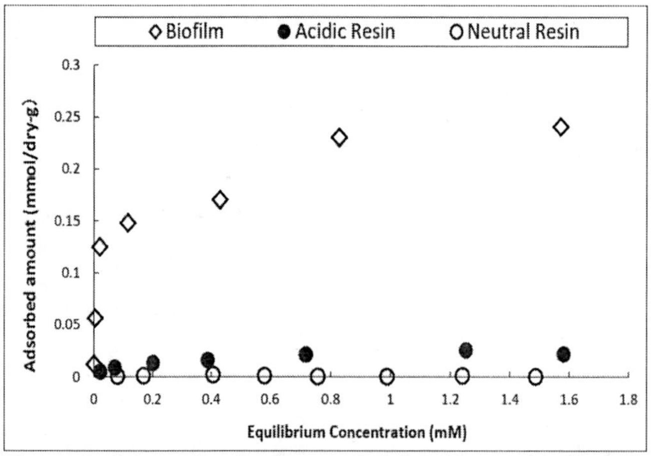

Figure 3. Adsorption isotherm of alcian blue to biofilm (◊), acidic resin (•) and neutral resin (○).

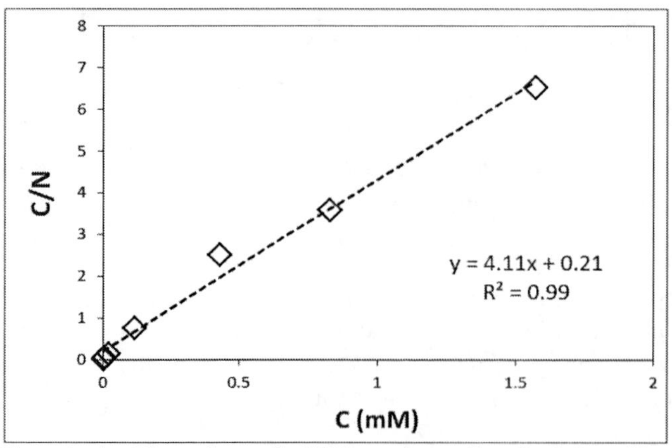

Figure 4. The plot of C/N against C of alcian blue adsorption to biofilm based on the variant of Langmuir isotherm model.

4. Conclusion

The present study revealed the characteristics of alcian blue adsorption to the biofilm polymers. The adsorption amount of alcian blue to biofilm increased monotonously, and the maximum adsorption amount was greater compared to the resins. On the contrary, alcian blue did not attach to the neutral resin having no electric charge. It seems that alcian blue attached to the acidic resins due to electrostatic attractive force, and the same seems to be the case for adsorption of Alcian blue to biofilm. The adsorption of Alcian blue to the biofilm and acidic resins fitted to Langmuir type indicates that the binding of Alcian blue to the biofilm and acidic resins occurred in a monolayer like form. The maximum adsorption amount of Alcian blue on the biofilm (0.24 mmol/dry-g) was greater than those of acidic resin (0.025 mmol/dry-g). This indicates that the biofilm has many more sites for Alcian blue attachment than acidic resins. According to the result of this study, the biofilm matrix can be a good adsorbent for dye such as Alcian blue that causing hazards in nature.

Reference

[1] Roosta M, Ghaedi M, Daneshfar A, Sahraei R and Asghari A 2014 *Ultrason. Sonochem.* **21**: 242–52
[2] Ran Y, Dawei L, Aimin L and Hu Y 2018 *Appl. Clay. Sci.* **151**: 20–8
[3] Mridusmita G and Prodeep P 2017 *J. Environ. Chem. Eng.* **5**: 3508–17
[4] Costerton JW, Lewandowski Z, Caldwell DE, Korber DR and Lappin SHM 1995 *Annu. Rev. Microbiol.* **49**:711–45
[5] Hiraki A, Tsuchiya Y, Fukuda Y, Yamamoto T, Kurniawan A and Morisaki H 2009 *Microbes Environ.* **24**:265–72
[6] Kurniawan A, Yamamoto T, Tsuchiya Y and Morisaki H 2012 *Microbes Environ.* **27**:399–406
[7] Kjelleberg S and Givskov M 2007 In: Kjelleberg S, Givskov M (eds) *The Biofilm Mode of Life: Mechanisms and Adaptations*, Original, Horizon Bioscience, Norfolk UK, p 5–21
[8] Kurniawan A and Yamamoto T 2013 *Procedia Environ. Sci.* **17**:179–87
[9] Kurniawan A and Fukuda Y 2015 *Microbiol. Indones.* **9**:106–12
[10] Li J, Bishop PL 2004 *Water Sci. Technol.* **49**:237–45
[11] Pengfei S, Cai H, Sheng W, Li W, Xin Z, and Yuhua Z 2016 *Colloids Surf B Biointerfaces* **139**:164–70
[12] Kurniawan A, Tsuchiya Y, Eda S and Morisaki H 2015 *Colloid Surf B Biointerfaces* **136**:22–6
[13] Freifelder D 1985 *Principles of physical chemistry with application to the biological sciences*, 2nd ed., Jones and Bartlett Publisher, Boston
[14] Gadd, G.M 2009 *J. Chem. Technol. Biotechnol.* **84**:13–28
[15] Volesky B 200) Water Res. **41**:4017–29

Acknowledgments

This study is a part of research project supervised by Prof. Hisao Morisaki from Graduate School of Life Sciences, Ritsumeikan University. The authors are grateful to Prof. Makoto Ikenaga from Kagoshima University and Assoc. Prof. Yuki Tsuchiya from Nihon University for their persistent help during this research project.

IOP Conf. Series: Materials Science and Engineering **299** (2018) 012005 doi:10.1088/1757-899X/299/1/012005

Determination of Parabens by Injection-Port Derivatization Coupled With Gas-Chromatography-Mass Spectrometry and Matrix Solid Phase Dispersion

Rosalina Djatmika[1,2,3*], Wang-Hsien Ding,[2]Hermin Sulistyarti[3]

[1] Department of Pharmacy, Sekolah Tinggi Ilmu Kesehatan Karya Putra Bangsa, Jl. Raya Tulungagung-Blitar KM 4, Tulungagung, Indonesia
[2] Department of Chemistry, National Central University, 300 Jhong-Da Road, Jhong-Li, Taoyuan, Taiwan
[3] Depertment of Chemistry, Faculty of Mathematics and Natural Sciences, Brawijaya University, Jl.Veteran 65145, Malang, Indonesia

Email: rosalina.djatmikaa@gmail.com

Abstract. A rapid determination of four parabens preservatives (methyl paraben, ethyl paraben, propyl paraben, and butyl paraben) in marketed seafood is presented. Analytes were extracted and purified using matrix solid phase dispersion (MSPD) method, followed by Injection port acylation gas chromatography–mass spectrometry (GC-MS) with acetic anhydride reagent. In this method, acylation of parabens was performed by acetic anhydride at GC injection-port generating reduction of the time-consuming sample-processing steps, and the amount of toxic reagents and solvents. The parameters affecting this method such as injection port temperature, purge-off time and acylation (acetic anhydride) volume were studied. In addition, the MSPD influence factors (including the amount of dispersant and clean-up co-sorbent, as well as the volume of elution solvent) were also investigated. After MSPD method and Injection port acylation applied, good linearity of analytes was achieved. The limits of quantitation (LOQs) were 0.2 to 1.0 ng/g (dry weight). Compared with offline derivatization commonly performed, injection port acylation employs a rapid, simple, low-cost and environmental-friendly derivatization process. The optimized method has been successfully applied for the analysis of parabens in four kind of marketed seafood. Preliminary results showed that the total concentrations of four selected parabens ranged from 16.7 to 44.7 ng/g (dry weight).

1. Introduction

Parabens (esters of 4-hydroxybenzoates) are employed as antimicrobial preservatives in food, industrial, pharmaceutical and personal care products due to their broad spectrum of action against numerous microorganisms, biodegradability, efficiency in wider pH range, non-volatility, and no color [1-3]. Parabens were considered quite low toxicity compounds. Nevertheless, some studies have suggested that parabens acted as endocrine disruptors which affect the functions of the male reproductive system, potentially impacted on female reproductive and enhanced the risk of breast cancer [4,5]. Earlier researchers have also exposed the significant relationship between urinary paraben concentration and sperm DNA damage [6].

Parabens were potentially polute aquatic system through recreational activities pathways such as showering, laundering, washing, swimming and bathing in rivers or lake. They entered in aquatic food chain and bioaccumulate in aquatic biota because of their high lipophilic nature and degradation resistance. Previous studies found parabens in aquatic systems, including environmental water, sediment and sludge [1,7]. The finding parabens in human urine, blood, breast milk and serum proved that human has been exposed parabens [1,8].

Parabens concentration in aquatic biota assumed are low to pose, lead some difficulties regarding the limitation of methods and instrumentation. As the result, the pretreatment process was needed to enrichment analytes before determined by instrumentation. Matrix Solid Phase Dispersion (MSPD) is the most common technique for pretreatment biota samples since it has many advantages, such as extraction and clean-up is integrated into a single step, thus making the procedure simple, low-cost, and convenient [9].

Gas Chromatography-Mass Spectrometry (GC–MS) is well suited for the quantity identification organic volatile compound due to its high chromatographic resolution capacity and reproducible ionization efficiency. Otherwise, derivatization process before GC analysis is needed to establish compound more amenable to standard GC-MS analysis, by improving volatility, thermal stability, and increasing chromatographic performance [10]. Derivatization reactions are frequently performed off-line, otherwise off-line derivatization need multi-step reactions, the procedure was laborious, tedious, time-consuming, and use toxic and harmful reagents [11]. Injection-port derivatization (IPD) was developed method to derivatize analytes and enhance the analytical efficiency of organic compounds in short time reaction and need small amount of organic solvent [12].

This paper demonstrate a injection-port derivatization using GC–MS instrument and matrix solid phase dispersion as a pretreatment method for determination of parabens in seafood. Acetic anhydride was a derivatization reagent used in this study. The derivatization factors affecting the derivatization process efficiency, such as, the injection port temperature, purge-off time, and derivatization reagent volume, were examined.

2. Experimental

2.1 Chemicals and reagent.

All chemicals and solvents are analytical grade and used without further purification. Standard of Methylparaben (MP), Ethylparaben (EP), Propylparaben (PP), and Butylparaben (BP) (purify 99%) are supplied by Alfa Aesar. P-Terphenyl-d_{14} (purify 98%, used as internal standard), was supplied by Sigma-Aldrich. Acetonitrile (purity 99.9%) are supplied by Sigma-Aldrich. Sodium sulfate anhydrous (purity > 99%) was supplied by Fluka. Activated magnesium silicate (Florisil, <200 mesh) was supplied by Sigma- Aldrich. Supelclean ENVI-18 (C_{18}) was supplied by Supelco. Silica gel (70-230 mesh) was supplied by Merck. Acetic Anhydride was supplied by Sigma-Aldrich. Methyl alcohol (purity 99.9 %) supplied by Merck. Deionized water was further purified using Millipore Elix 10 RO system and a Millipore Synergy UV system (Millipore SAS, Molsheim, France).

2.2 Sample Preparation using MSPD

Two freshly killed fishes (perch) were purchased from a local fish market in Chung-Li city, Taiwan. The fish samples were stored in a thermo-insulator box and then brought to the laboratory. The fishes were washed several times with Milli-Q deionized water. The fish skin were removed and the muscle tissue was cut into small pieces, and then homogenized in a blender. The homogenate was then freeze-dried for 3 days, and ground into powder. The lyophilized sample was stored in sealed container at -20°C.

A portion of powdered fish sample (0.5 gram) was mixed with 0.5 gram of anhydrous sodium sulfate and 1.0 gram of Florisil as a dispersant and placed in polypropylene centrifuge tube, then shaked in tube shaker to homogenize this mixture. This blend was transferred to a polypropylene SPE cartridge containing mixture 1.5 gram of silica + C18 (w/w, 9:1), which was packed at the bottom as the clean-up co-sorbent. The analytes were eluted with 10 mL of acetonitrile by gravity flow. The extract was evaporated to dryness by a gentle stream of nitrogen at room temperature, then re-dissolved with 50 μL of methanol. A small fraction of extract (10 μL) was placed in micro-vial, added 1 μL internal standard, 1 μL derivatization reagent and subjected to online acylation GC–MS analysis. This procedure is appropriate with previous MSPD studies [9,13,14].

2.3 GC-MS analysis and derivatization

The GC-MS analysis was performed using a Varian 450 GC directly connected to a Varian 220 ion-trap mass spectrometer (Walnut Creek, CA, USA) operating in the SIS (Selected Ion Storage) mode. A ChromatoProbe (Varian) and a temperature-programmed injector (liner: 3.4 mm i.d.) were used to introduce large-volume samples for injection-port acylation. ChromatoProbe device with a disposable micro-vial was used in these experiments. The micro vial containing 10 μL sample extract, 1 μL internal standard and 1 μL acetic anhydride was placed into a ChromatoProbe vial holder, then positioned in the GC injection port.

The temperature of the injection- port acylation was maintained at 90 °C for 1.5 min then rapidly increased to 300 °C (the temperature ramping rate: 120 °C/min). Separations were carried out in a DB-5MS capillary column (30 m × 0.25 mm i.d., 0.25 μm film thickness; Agilent, Santa Clara, CA, USA). The GC oven was programmed as follows: 2.5 min at 60 °C, elevated 30°C/min to 200°C hold for 2 min and at 40 °C min−1 to 280 °C. The GC-MS interface and the ion trap temperatures were set at 270°C and 220°C, respectively. The carrier gas used was Helium (99.999%) at a constant flow of 1 mL/min. The temperature of the transfer line was set at 280 °C. Full-scan EI spectra were acquired under the following conditions: scan time 1 s, mass range 100–500 m/z, ion trap temperature 200 °C.

3. Result and Discussion

3.1 Parabens Derivatization

Parabens derivatization is needed to improve parabens volatility thus it could be amenable for GC-MS analysis. In this study, injection-port derivatization with acylation reagent was applied. Acetic anhydride is one of popular acylation reagent which commonly used for injection-port derivatization. The derivatization reaction between parabens and acetic anhydride is described in Figure 1.

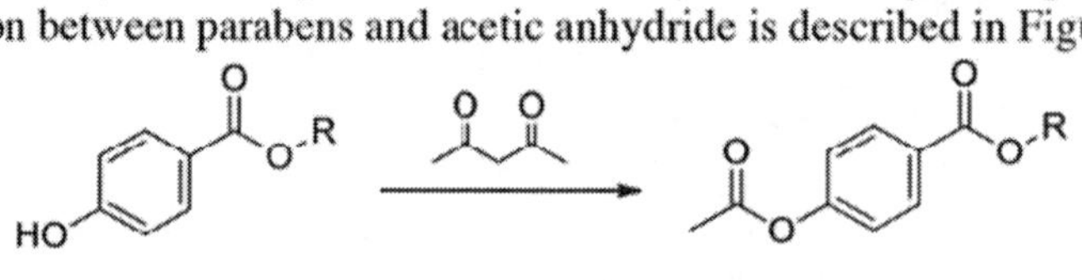

Figure 1 The paraben derivatization reaction by acetic anhydride

The paraben derivatization reaction by acetic anhydride involved the replacement of active hydrogen in hydroxyl group with the acyl group from acetic anhydride and converted into ester group. The ester group attached in derivative of parabens could improve the volality of analytes. The ester group in parabens derivative has no hydrogen atom attached directly to an oxygen atom. Therefore, it is incapable of engaging in intermolecular hydrogen bonding thus has considerably lower boiling points than originally parabens.

An overview of the retention times, and the ions used to quantitate the SIS signal was provided in Table 1. The success of injection-port acylation was confirmed by appearance molecular ion peak at m/z 195,208, 222, and 236 for derivates of MP, EP, PP, and BP, respectively, which represented the molecular weight of the parabens derivatives. The fragment at m/z 121 for fourth parabens derivatives was observed, which is confirmed by the loss of $-COCH_3$ (acetyl group) and $-O(CH_2)n$ ($n = 1$ to 4, for derivatives MP, EP, PP and BP, respectively). Moreover, the fragment at m/z 151 for derivative-methyl paraben was attributed to the loss of $-COCH_3$.; the fragment at m/z 137 for other three parabens was attributed to the loss of $-COCH_3$ (acetyl group) and $-(CH_2)n$ (n = 2 to 4, for EP, PP and BP, respectively).

Table 1 The logarithm of the octanol/water coefficient, retention time, molecular ion and fragment ion of parabens observed in mass spectra

Analytes	Log K_{ow}	t_R (min)	Molecular ion	Fragment ion
Methylparaben	1.91	7.645	195	121+ 151
Ethylparaben	2.34	8.040	208	121 + 137
Propylparaben	2.94	8.713	222	121 + 137
Butylparaben	3.50	9.517	236	151 + 137

3.2 Evaluation of Injection-Port Acylation

Injection-port acylation was evaluated in order to study the influence of various parameters on the acylation derivatization process, three parameter (injection-port temperature, purge-off time, and the volume of acylation reagent) was examined. The RRF (Relative Response Factors) was a parameter used to compare derivatization efficiency to evaluate the best compromise condition of each variable. RRF is the ratio of analyte response factor and internal standard response factor.

3.2.1 Selection of injection port temperature

Injection-port temperature was an important parameter in the derivatization efficiency due to thermally catalyse the derivatization reaction process [11]. The different injection-port temperature ranging from 80 to 110°C (at 10 °C increments) were evaluated. As shown in Figure 2, the RRF value of parabens derivatives increased gradually from 80 to 90°C indicating that parabens was derivatized well as the higher temperature could overcome the energy barrier of the reaction and steric interference, thus made increasing of the derivatization reaction efficiency [12]. Therefore, in this study, 90°C was chosen as the optimised injection-port temperature. The RRF value was significantly decreased gradually from 90 to 110°C. It was corresponding with some previous study which reported that too high temperature of injection-port may cause the derivative decomposition [11,12.13].

3.2.2 Selection of Purge-off Time

The other parameter affect the injection port acylation process is the purge-off time. The purge-off time had a correlation with detection sensitivity. A long purge-off time time was related with long introduction of majority of analytes into the column which guarantee that all the derivatives was derivatized well. Otherwise, solvent tailing was observed when the purge-off time was too long which followed the decreasing detection sensitivity [12].

In this study, purge-off time was evaluated for 0.5; 1.0; 1.5; 2.0 and 3.0 min (shown in Figure 3). It was clearly observed that the RRF value increased with increasing purge-off time from 0.5 min 1.5 min followed the lowering RRF value in 2.0 min. Thus, 1.5 min was chosen as the purge-off time.

3.2.3 Selection of acylation reagent volume

Acylation reagent volume had also effect to the derivatization efficiency. The excessive of acylation reagent could disturb separation of analytes once the excess of derivatization reagents was detected [15]. In the other side, insufficient of acylation reagent cause incomplete derivatization, poor separation and unreliable quantification [16]. The volume of *acylation reagent* was optimized by evaluating various *acylation reagent* volume: 0.5; 1.0; 2.0 to 3.0 μL. As shown as Figure, the poor RRF value was observed when 0.5; 2.0 to 3.0 μL were used. Therefore, 1.0 μL was selected as the best volume of acylation reagent.

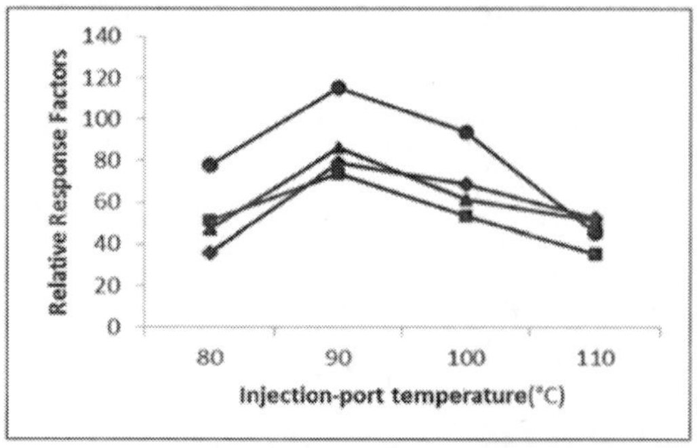

Figure 2. Effect of injection-port temperature. MP (●), EP (■), PP (▲) and BP (♦).

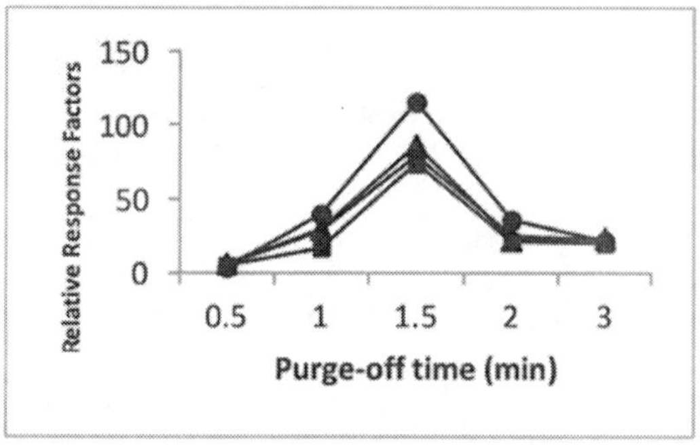

Figure 3. Effect of injection-port temperature. MP (●), EP (■), PP (▲) and BP (♦).

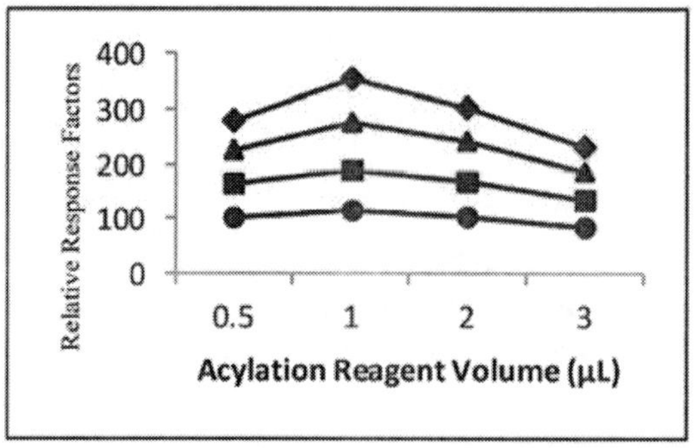

Figure 4. Effect of acylation reagent volume. MP (●), EP (■), PP (The paraben derivatization reaction by acetic anhydride (▲) and BP (♦).

3.3 Optimization of MSPD

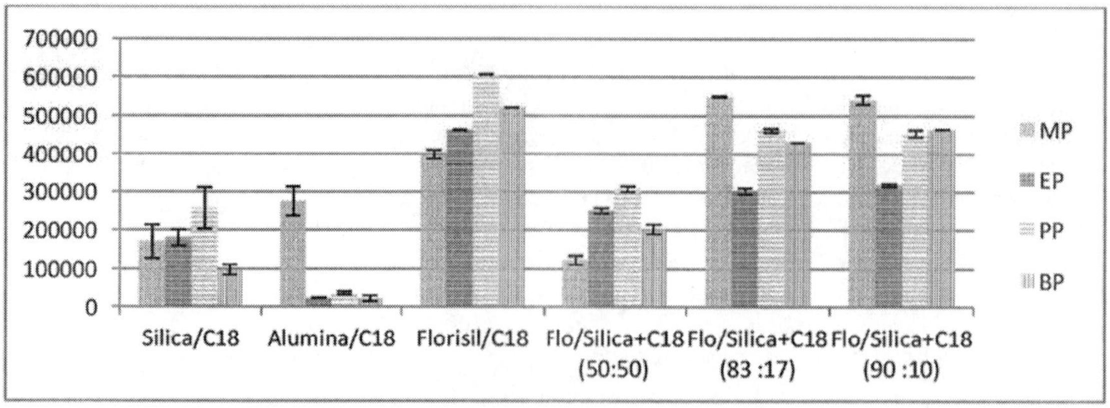

Figure 5. Effects of Florisil (as a dispersant) combined with various clean-up co-sorbents on the recoveries of analytes extracted by MSPD technique.

One of the most attractive advantage of MSPD procedure is analytes can be extracted from the sample and separated from lipid and other interfering species in a single step. In order to achieve this aim, experimental conditions like type of dispersants, co-sorbents and extraction solvent were optimized. All the optimatization were determined based on relative peak area of analytes to the internal standard, from the average of three replicate measurements.

The preliminary experiments were carried out using several combinations of $C_{18,}$ silica, Florisil, alumina, activated carbon (AC) as dispersants and/or co-sorbents. For this purpose, 0.5-gram of spiked fish sample was dispersed was mixed with 0.5 g of anhydrous sodium sulfate and dispersed with 1.0 gram of Florisil using a vortex for 2 min. Acetonitrile was applied as elution solvent in this first study corresponding with previous study which use acetonitrile to elute polar compound in fish sample [9]. Samples were prepared and extracted by MSPD method as described in session 2.2. The experimental results of dispersants and/or co-sorbents types were presented in Figure 5. The used of Florisil/C18 as dispersant/co-sorbent gave good recoveries. Otherwise, a light yellow coloration was present on the drying extract, perhaps as a result of colored polar interfering species co-elution. The intense peaks also appeared in the GC–MS chromatograms, and interfered with peaks quantification of the target analytes that cause lower recoveries were observed. Therefore, the silica gel was added to co-sorbent to improve the chromatograms and eliminate the colored polar interferences. Figure 5 shows that better recoveries and precisions were achieved when Florisil as dispersant and the mixing ratio of silica gel + C18 (9:1,w/w) as co-sorbent were applied. This may be due to the high ratio silica gel capable to receive and retain polar species resulting cleaner extracts and better chromatograms for peaks quantitation. These results are in agreement with previous findings reported by Gar´cia de Llasera and Reyes–Reyes [17].

The second experiment, type of elution solvent was tested. Elution solvent could travel down the analytes through the cartridge. Selecting type of elution solvent was performed using four solvent with enhancing polarities:, dichloromethane, ethyl acetate, methanol, and acetonitrile. 1 gram Florisil, silica gel + C18 (9:1,w/w) and 10 mL of each elution solvent were used.

3.4 Method Performance

In the interest to determine the feasibility and efficiency of the evaluated injection-port derivatization and MSPD method, the analytical characteristic, such as, linearity, repeatability, reproducibility, LODs, and LOQs were investigated (summarized in Table 2). As shown in table 2, good linearity (ranging from 0.4 to 500 ng/g) was confirmed by the r^2 which higher than 0.998. Instrumental quantification detection limit (LOQ) was defined for signal-to-noise (S/N) ratio of 10, ranged from 0.2 ng/g to 1.0 ng/g. The limit of detection (LOD) was confirmed at the S/N of 3, ranged from 0.06 ng/g to 0.30 ng/g.

Precision of the whole method was evaluated with relative standard deviation (RSD) value of intra-day and inter-day analysis. Under repeatability condition (intra-day, 5 extractions performed in a day), RSD values remained below 8% for all analytes. Reproducibility (inter-day, 20 extractions performed 4 days with 5 extraction in a day) RSD value range from 3% to 8%. The method accuracy was determined by average recovery of all the analyte, range from 98% to 118%. These results denoted that the optimized injection-port derivatization give an excellent linearity, reproducibility, repeatability, accuracy and precision for determination of parabens.

Table 2 Performance of the proposed method

Analytes	Linear Range (ng/g)	r^2	LOQ (ng/g)	LOD (ng/g)	Intra-day (n=5)	Inter-day (n=20)	Recovery (%)
MP	0.4-500	0,9999	0.2	0.06	107[a](6)[b]	111[a](5)[b]	98 (6)[a]
EP	0.4-500	0,9996	0.4	0.12	93 (6)	102 (8)	98 (2)
PP	0.4-500	0,9996	0.4	0.12	112 (3)	106 (6)	84 (1)
BP	0.4-500	0,9994	1.0	0.30	101 (7)	102 (3)	118 (1)

[a] Average spiked recovery (final spiked concentration of 10 ng/g).
[b] Relative standard deviation (% RSD) of spiked recovery.

3.5 Application in real samples

The optimized injection port derivatization and MSPD method was applied to assess the seafoods in order to evaluate method feasibility and applicability. Four kinds of seafoods commonly consumed and commercially sold were pretreated as described in section 2.2. Table 3 listed the recovery of unspiked sample under 10 ng/g of final concentrations. Almost four kinds of parabens detected in four seafood samples. However, propylparaben didn't find in tilapia fish sample. The finding parabens in fish samples represent that the optimize MSPD method is sensitive enough to allow the quantification of target analytes.

Table 3 Concentration (ng/g, dry weight) of parabens detected in seafood samples.

Samples	Analytes			
	MP	EP	PP	BP
Shrimp				
Unspike conc. (ng/g)[a]	10.75 (2)[b]	8.01 (1)	5.45 (4)	7.44 (1)
Cod				
Unspike conc. (ng/g)	11.47 (0.7)	5.63 (7)	6.81 (0.1)	5.60 (9)
Tilapia				
Unspike conc. (ng/g)	6.23 (4)	5.47 (2)	n.d.*	5.02 (5)
Perch				
Unspike conc. (ng/g)	18.53 (7)	15.12 (3)	4.89 (9)	6.24 (3)

*n.d., not detected at LOQ
[a] Original concentration (ng/g) of analytes found in fish samples (n = 3)
[b] Relative standard deviation (%RSD) are given in parentheses (n = 3)

4 Conclusion

A rapid, effective and simple MSPD method coupled with on-line derivatization GC-MS was demonstrated and optimized to determine trace level parabens in seafood samples. The optimized MSPD method does not involve high-priced instrumentation and take moderate volume of organic solvent, amount of dispersant and clean-up co-sorbent. Furthermore, MSPD employs a single clean-up step thus make this method simple and rapid in extraction. Compared with off-line derivatization, on-line acylation which coupled with MSPD method provides time-saving and environment-friendly that can be reacted in hot injection port without requiring further treatment process. On the whole, the proposed method appears to be a good technique for determination of hydroxylated semi-volatile trace pollutant in biota sample as it provided acceptable recoveries and good sensitivity of real samples

Acknowledgment
This study was supported by a grant from the Ministry of Science and Technology of Taiwan under contract No. MOST 103-2113-M-008-005-MY2. We thank the Instrumental Center of NationalCentral University for instrumentation support. We also thank Erica Ding, who checked over the English and grammar usage within this manuscript.

Reference
[1] Azzouz, A., Rascón, A. J., & Ballesteros, E 2016 Simultaneous determination of parabens, alkylphenols, phenyl phenol, bisphenol A and triclosan in human urine, blood and breast milk by continuous solid-phase extraction and gas chromatography–mass spectrometry. *Journal of Pharmaceutical and Biomedical Analysis* **119** 16-26.
[2] Kajornkavinkul, S., Punrat, E., Siangproh, W., Rodthongkum, N., Praphairaksit, N., & Chailapakul, O 2016 Graphene/polyvinylpyrrolidone/polyaniline nanocomposite-modified electrode for simultaneous determination of parabens by high-performance liquid chromatography. *Talanta* **148** 655-660.
[3] Molins-Delgado, D., Díaz-Cruz, M. S., & Barceló, D 2016 Ecological risk assessment associated to the removal of endocrine-disrupting parabens and benzophenone-4 in wastewater treatment *Journal of Hazardous Materials* 310 143-151.
[4] Oishi, S 2002 Effects of propyl paraben on the male reproductive system. *Food and Chemical Toxicology* **40** 1807-1813.
[5] Shanmugam, G., Ramaswamy, B. R., Radhakrishnan, V., & Tao, H 2010 GC–MS method for the determination of paraben preservatives in the human breast cancerous tissue. *Microchemical Journal* **96** 391-396.
[6] Kang, H.-S., Kyung, M.-S., Ko, A., Park, J.-H., Hwang, M.-S., Kwon, J.-E., Suh, J.-H., Lee, H.-S., Moon, G. I., Hong, J.-H., & Hwang, I. G 2016 Urinary concentrations of parabens and their association with demographic factors: A population-based cross-sectional study. *Environmental Research* **146** 245-251.
[7] Rocío-Bautista, P., Martínez-Benito, C., Pino, V., Pasán, J., Ayala, J. H., Ruiz-Pérez, C., & Afonso, A. M 2015 The metal–organic framework HKUST-1 as efficient sorbent in a vortex-assisted dispersive micro solid-phase extraction of parabens from environmental waters, cosmetic creams, and human urine *Talanta* **139** 13-20.
[8] Hines, E. P., Mendola, P., von Ehrenstein, O. S., Ye, X., Calafat, A. M., & Fenton, S. E 2015. Concentrations of environmental phenols and parabens in milk, urine and serum of lactating North Carolina women *Reproductive Toxicology* **54** 120-128.
[9] Tavares, R. S., Martins, F. C., Oliveira, P. J., Ramalho-Santos, J., & Peixoto, F. P 2009. Parabens in male infertility—Is there a mitochondrial connection? *Reproductive Toxicology*, **27** 1-7.
[10] Ocaña-González, J.A., Villar-Navarro, M., Ramos-Payán. Maria., Fernández-Torres. R., Bello-López. M.A 2015 New developments in the extraction and determination of parabens in cosmetics and environmental samples *Analytica Chimica Acta* **858** 1-15
[11] Tsai, D.-Y., Chen, C.-L., & Ding, W.-H 2014 Optimization of matrix solid-phase dispersion for the rapid determination of salicylate and benzophenone-type UV absorbing substances in marketed fish. *Food Chemistry* **154** 211-216.
[12] Canosa, P., Rodríguez, I., E. Rubí, M. R., & Cela, R 2008 Simplified sample preparation method for triclosan and methyltriclosan determination in biota and foodstuff samples. *Journal of Chromatography A* **1188** 132–139.
[13] Bowdena, J. A., Colosia, D. M., Mora-Montero, D. C., Garrett, T. J., & Yost, R. A 2009 Enhancement of chemical derivatization of steroids by gas chromatography/mass spectrometry (GC/MS). *Journal of Chromatography B* **877** 3237–3242.
[14] Wang, Q., Ma, L., & Chen-ru Yin, L. X. 2013 Developments in injection port derivatization.

Journal of Chromatography A **1296** 25–35.

[15] Wu, J., Hu, R., Yue, J., Yang, Z., & Zhang, L 2009 Determination of fecal sterols by gas chromatography–mass spectrometry with solid-phase extraction and injection-port derivatization *Journal of Chromatography A* **1216** 1053–1058.

[16] Zhan, J., Li, J., Liu, D., Liu, C., Yang, G., Zhou, Z., & Wang, P 2016 A simple method for the determination of organochlorine pollutants and the enantiomers in oil seeds based on matrix solid-phase dispersion *Food Chemistry* **194** 319-324.

[17] M.P. Gar´cia de Llasera, M.L. Reyes-Reyes, A validated matrix solid-phasedispersion method for the extraction of organophosphorus pesticides frombovine samples *Food Chemistry* 2009 **114** 1510–1516.

IOP Conf. Series: Materials Science and Engineering **299** (2018) 012006 doi:10.1088/1757-899X/299/1/012006

Spectrophotometric determination of cobalt in horse urine using 2-(5-bromo-2-pyridylazo)-5-[N-n-propyl-N-(3-sulfopropyl)amino]aniline as chromogenic reagent

Arfidyaninggar Septia Rinda[*,1,2], **Kanchana Uraisin**[1,2], **Akhmad Sabarudin**[3], **Duangjai Nacapricha** [1,2], **and Prapin Wilairat**[1,4]

[1]Flow Innovation-Research for Science and Technology Laboratories (Firstlabs.), and [2]Department of Chemistry and Center for Innovation in Chemistry, Faculty of Science, Mahidol University, Rama VI Road, Bangkok 10400, Thailand
[3]Research Center for Advanced System and Material Technology, Brawijaya University, Jl Veteran Malang, 65145, Indonesia
[4]National Doping Control Centre, Mahidol University, Rama VI Road, Bangkok 10400, Thailand

[*]E-mail: arfidyarinda@gmail.com

Abstract. Cobalt has been reported for being abused as an illegal doping agent due to its ability as an erythropoiesis-stimulating agent for enhancing performance in racehorses. Since 2015, cobalt is listed as a prohibited substance by the International Federation of Horseracing Authorities (IFHA) with a urinary threshold of 0.1 μg cobalt per mL urine. To prevent the misuse of cobalt in racehorse, a simple method for detection of cobalt is desirable. In this work, the detection of cobalt is based on the spectrometric detection of the complex formation between cobalt(II) and 2-(5-bromo-2-pyridylazo)-5-[N-n-propyl-N-(3-sulfopropyl)amino]aniline at pH 4. The absorbance of the complex is monitored at 602 nm. The metal:ligand ratio of the complex is 1:2. The calibration graph was linear in the range of 0 – 2.5 μM {Absorbance = (0.0825 ± 0.0013)[Co^{2+}] + (0.0406 ± 0.0003), r^2 = 0.999} and the detection limit (3 SD of intercept)/slope) was 0.044 μM. The proposed method has been successfully applied to horse urine samples with the recoveries in the range 91 – 98%.

1. Introduction

Since first observations of cobalt and its salts as an erythropoiesis-stimulating agent in 1929 [1], cobalt became an alternative agent to treat anemia [2, 3]. Cobalt stabilizes the hypoxia-inducible transcription factor (HIF) which activates erythropoietin (Epo) expression resulting in increase of red blood cells (RBCs) [4-6]. However, due to the adverse effects associated with this therapy, cobalt is no longer used for the treatment of anemia [4, 5, 7].

In view of cobalt's ability as an erythropoiesis-stimulating agent, presumption about the illicit administration of cobalt as a performance-enhancer has been raised. The fact that cobalt salts can be orally taken and are affordable can lead to abuse of cobalt salt as an illegal doping agent [4, 5]. Recent reports have confirmed that cobalt is being abused in equine sports [5, 8]. Excessive administration of cobalt can be acutely toxic which can promote several health issues and even lead to death [8, 9]. In horse racing, cobalt is listed as a prohibited substance according to International Federation of

Horseracing Authorities (IFHA) and the urinary threshold is 0.1 μg cobalt per mL urine or 1.7 μM [10]. For this reason, to prevent the misuse of cobalt in racehorse, simple detection of cobalt in urine is needed.

A standardized analytical methodology for the determination of cobalt in horse urine is not available [8]. Only a few papers reported methods for determination of cobalt either in horse urine or horse blood/serum samples [5, 6, 11]. According to Ho *et al.* [5], Knych *et al.*[6], and Popot *et al.*[11], the cobalt threshold in horse urine is based on total cobalt concentration which have been quantified by inductively coupled plasma – mass spectrometry (ICP-MS).

In general, determination of cobalt in various samples had been developed with mass-spectrometry-based detection, including ICP-MS, gas chromatography-mass spectrometry (GC-MS) [12], and electrospray ionization tandem mass spectrometry (ESI-MS-MS) [13]. Even though mass spectrometry-based method provides low limit of detection, they are complicated methods, high-cost, and require skill operator. Another common method which is widely used is spectrophotometry, which has been employed for the determination of cobalt(II) using 2-(5-Bromo-2-pyridylazo)-5-[*N*-*n*-propyl-*N*-(3-sulfopropyl)amino]aniline (5-Br-PSAA) as a chromogenic reagent. Compared to mass spectrometry-based method, spectrophotometry is more convenient due to its lower cost and simplicity.

5-Br-PSAA (Figure 1) is a reagent which forms colored complex with metals including cobalt, nickel, iron, copper, and palladium [14, 15]. Cobalt forms complex with 5-Br-PSAA at the optimum pH range of 3 – 4.5 and with maximum absorption at 602 nm [14, 16]. According to Horiguchi *et al.*, the molar absorptivity of this complex is 8.8 x 10^4 L mol^{-1} cm^{-1} at 602 nm, which is considered a high value [14]. In addition, 5-Br-PSAA is water-soluble reagent which is more convenient to use than other organic reagents. By using water-soluble reagent, the need for solvent extraction can be avoided [16]. Moreover, 5-Br-PSAA is commercially available at reasonable costs.

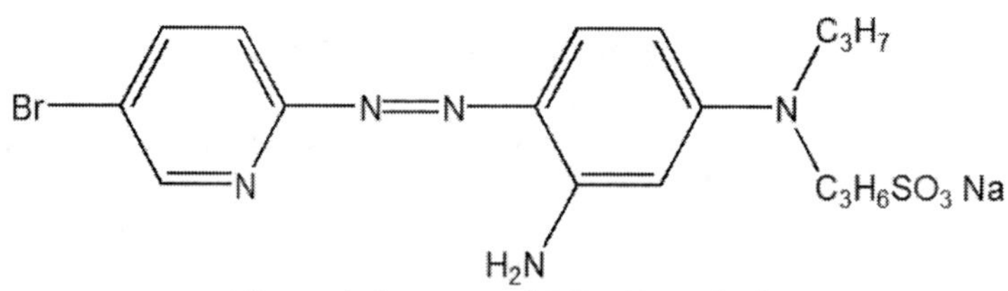

Figure 1. Structure of 5-Br-PSAA [14].

This paper describes the study of the determination of cobalt in horse urine using spectrophotometric method based on complex formation between cobalt(II) and 5-Br-PSAA as water-soluble reagent for simple, sensitive, and rapid analysis.

2. Experimental

2.1. Reagents

A stock standard solution of cobalt(II) was prepared by dissolving 0.1010 g of cobalt(II) chloride (97%) (Sigma-Aldrich, United States) in 25 mL of 2% of hydrochloric acid. The concentration of the stock cobalt solution was calculated from the equation c = A/εb, where c is the concentration of the stock solution of cobalt (mol L^{-1}), A is the absorbance of the stock solution of cobalt, ε is the molar absorptivity (L mol^{-1} cm^{-1}), and b is the path length of the cuvette (1.00 cm). The molar absorptivity was determined by measuring the absorbance at 509 nm of 0.017 M of a standard solution of cobalt(II) (standard for AA and ICP-OES, PerkinElmer, N9303766). The concentration of the stock cobalt solution was found to be 0.0175 M. Working solutions of cobalt were prepared by appropriate dilution of the standard solution of cobalt(II) with 0.05 M of acetate buffer pH 4.

A 0.001 M 5-Br-PSAA stock solution was prepared by dissolving 0.024 g of 2-(5-Bromo-2-pyridylazo)-5-[N-n-propyl-N-(3-sulfopropyl)amino]aniline, sodium salt (Dojindo Laboratories, Japan) in 50 mL of water.

A 0.042 M of acetic acid was prepared by diluting 1.21 mL of acetic acid glacial (Lab Scan, Thailand) to 500 mL of water. A 0.523 g of sodium acetate (Merck, Germany) was dissolved into 500 mL of acetic acid 0.042 M. These solutions were mixed to prepare a 0.05 M of acetate buffer solution pH 4.

2.2. Sample preparation
Horse urine samples were obtained from National Doping Control Centre, Mahidol University. The sample was first passed through a Sep-Pak C18 plus short cartridge (360 mg sorbent, 55-105 µm particle, Waters, Milford, United States) to remove coloured matrix. The cartridge was conditioned with methanol, followed by deionized water.
A UV-visible spectrophotometer (Lambda 25, Perkin Elmer, United States) was used to measure absorbance of the complex at 602 nm.

3. Results and Discussions
The UV-visible absorption spectrum of the cobalt(II)−5-Br-PSAA complex was measured in the range of 300 – 700 nm at pH 4 (see Figure 2). The maximum absorption of free 5-Br-PSAA is at 457 nm. A maximum absorbance of the complex is observed at 602 nm where the free reagent has minimum absorbance.

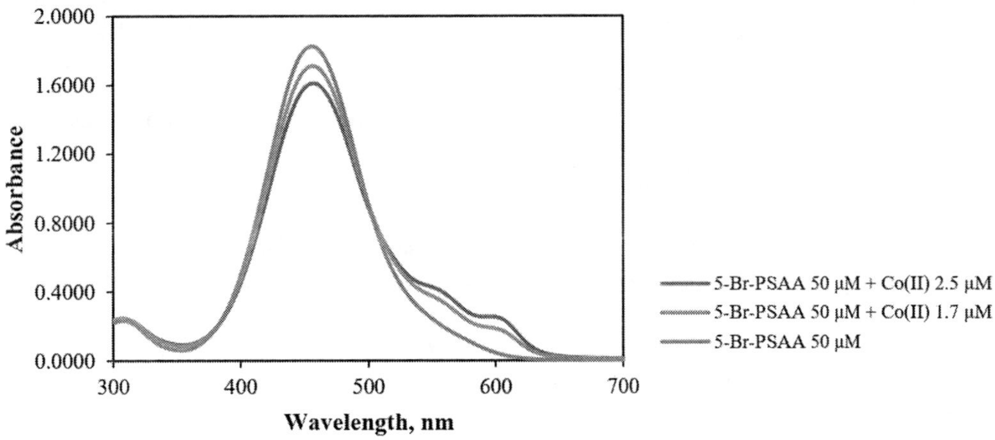

Figure 2. Absorption spectra of cobalt(II)−5-Br-PSAA complex.

3.1. Job's plot of cobalt(II)−5-Br-PSAA complex
The Job's method was applied to determine the ratio between cobalt(II) *(M)* and 5-Br-PSAA *(L)*.

$$nM + pL \rightarrow M_nL_p \quad (1)$$

A series of solution, where the total of mole M and mole L is constant, were prepared. Therefore, the total concentration of cobalt(II) and 5-Br-PSAA was also constant when the ratio was continuously varied. The maximum absorbance of M_nL_p complex corresponds to the maximum concentration of M_nL_p complex [17]. Thus,

$$\frac{n}{p} = \frac{X_M}{X_L} \quad (2)$$

Where, X_M is mole fraction of M and X_L is mole fraction of L.

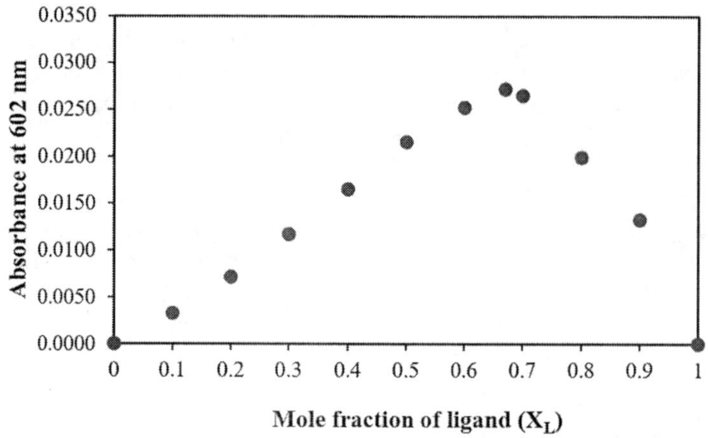

Figure 3. Modified job's plot after baseline subtraction.
Conditions: Co(II) + 5-Br-PSAA = 5 µM; acetate buffer 0.05 pH 4.

The results in Figure 3 show that the maximum absorbance is observed when the mole fraction of ligand X_L is 0.67 and X_M is 0.33 (Figure 3). Therefore, the composition of complexes obtained is Co(II) : 5-Br-PSAA = 1 : 2. The result was in good agreement with the previous work of Iburaim *et al.*[18].

3.2. Calibration and detection limit

Calibration graph for cobalt is linear (Figure 4) for 0 – 2.5 µM (r^2 = 0.999), with the equation:

$$\text{Absorbance (602 nm)} = (0.0825 \pm 0.0013)[Co^{2+}] + (0.0406 \pm 0.0003) \quad (3)$$

Limit of detection (LOD) of cobalt solution was calculated from [19]

$$LOD = intercept + 3\ S_{y/x} \quad (4)$$

LOD was 0.044 µM. In this work the urine sample from the Sep-Pak clean-up was diluted 10 fold with the acetate buffer. This method is suitable for analysis of cobalt in horse sample because urinary cobalt threshold after dilution is 0.17 µM (0.01 µg/mL).

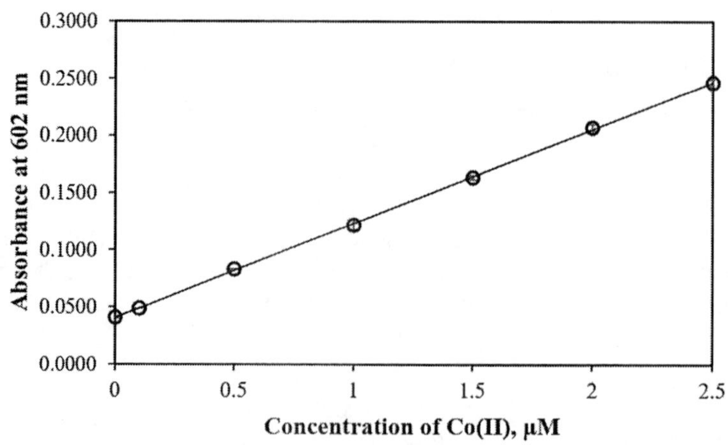

Figure 4. Calibration graph of Co(II) (0, 0.1, 0.5, 1.0, 1.5, 2.0, 2.5 µM) obtained at 602 nm.
Conditions: 5-Br-PSAA 50 µM, acetate buffer 0.05 M pH 4.

3.3. Interference study (nickel and zinc)

Recent reports found a significant amount of undeclared nickel in performance enhancing products [20,21]. Thevis *et al* in their pilot study reported that the mean urinary nickel level was 0.1 (± 0.086) μM from 200 equine doping control urine samples. Therefore, the interference of nickel(II) was studied since nickel(II) also form a complex with 5-Br-PSAA.

Besides, the interference of zinc was also studied. Although zinc(II) does not form coloured complex with 5-Br-PSAA, it may affect the quantitation of the cobalt-complex when the urinary zinc excretion is high. Normally, the level of urinary zinc excretion in horse does not change with excess of zinc dietary intake [22].

The results are summarized in Table 1. Both nickel(II) and zinc(II) does not significantly interfere at levels up to 1.7 μM.

Table 1. Interference of nickel(II) and zinc(II) on the determination of 1.7 μM cobalt(II).

Added (μM)		Found (μM)	Recovery (%)
Cobalt(II) 1.7 μM	Nickel(II) 1.7 μM	1.716 ± 0.003	101
Cobalt(II) 1.7 μM	Zinc(II) 1.7 μM	1.674 ± 0.001	98

Conditions: 5-Br-PSAA 50 μM, acetate buffer 0.05 M pH 4, absorption at 602 nm.

3.4. Application

The method was applied to the determination of Co(II) in horse urine samples. After sample pre-treatment using the C18 cartridge, 1.7 μM Co(II) were spiked into the diluted horse urine samples[a]. Table 2 shows that the recoveries ranged between 91 – 98 %.

Table 2. Recovery study of cobalt determination in horse urine samples.

Horse urine sample	Concentration of diluted sample (μM)[a,b]	Added (μM)	Found (μM)[b]	Recovery (%)
AA	Not detected	1.7	1.663 ± 0.023	98
AI	0.107 ± 0.011	1.7	1.647 ± 0.016	91
AW	0.058 ± 0.002	1.7	1.640 ± 0.056	93
AT	Not detected	1.7	1.596 ± 0.036	94
AO	Not detected	1.7	1.572 ± 0.016	92

Conditions: 5-Br-PSAA 50 μM, acetate buffer 0.05 M pH 4, absorption at 602 nm.
[a] The sample was diluted ten fold
[b] Mean of concentration ± SD, n = 3

4. Conclusions

In the present work, a simple, convenient, and rapid spectrophotometric method have been successfully developed for the determination of cobalt using 5-Br-PSAA as water-soluble reagent. This method is sensitive and suitable to be applied to horse urine. Therefore, we propose that this spectrophotometric method can be an alternative method for screening of cobalt in equine sport.

Acknowledgement

The scholarship from Indonesia Endowment Fund for Education namely "Lembaga Pengelola Dana Pendidikan" (LPDP) given to Miss A. S. Rinda is gratefully acknowledged.

References

[1] Waltner K and Waltner K 1929 Kobalt and Blut *Klin. Wochenschr* **8** 313A
[2] Berk L, Burcehenal, J H and Castle W B 1949 Erythropoietic effect of cobalt in patients with or without anemia *N. Engl. J. Med.* **240** 754-61
[3] Krug O, Kutscher D, Piper T, Geyer H, Schänzer W and Thevis M 2014 Quantifying cobalt in doping control urine samples – a pilot study *Drug Test. Anal.* **6** 1186-90
[4] Ebert B and Jelkmann W 2014 Intolerability of cobalt salt as erythropoietic agent *Drug Test. Anal.* **6** 185-9
[5] Ho E N M, Chan G H M, Wan T S M, Curl P, Riggs C M, Hurley M J and Sykes D 2015 Controlling the misuse of cobalt in horses *Drug Test. Anal.* **7** 21-30
[6] Knych H K, Arthur R M, Mitchell M M, Holser I, Poppenga R, Smith L L, Helm M N, Sams R A, Gaskill C L 2015 Pharmacokinetics and selected pharmacodynamics of cobalt following a single intravenous administration to horses *Drug Test. Anal.* **7** 619-25
[7] Simonsen L O, Harbak H and Bennekou P 2012 Cobalt metabolism and toxicology – A brief update *Sci. Total Environ.* **432** 210-5
[8] Brewer K, Maylin G A, Fenger C K, Tobin T 2016 Cobalt use and regulation in horseracing: a review *Comp. Exerc. Physiol.* **12** 1-10
[9] Balzer M 2015 Horse welfare is the issue with use of cobalt *Aust. Vet. J.* **93** N2
[10] International Federation of Horseracing Authorities 2016 *International Agreement on Breeding, Racing, and Wagering* 1-80 http://www.horseracingintfed.com/resources/2016Agreement.pdf
[11] Popot M-A, Ho E N M, Stojiljkovic N, Bagilet F, Remy P, Maciejewski P, Loup B, Chan G H M, Hargrave S, Arthur R M, Russo C, White J, Hinks P, Pearce C, Ganio G, Zahra P, Batty D, Jarrett M, Brooks L, Prescott L-A, Baily-Chouriberry L, Bonnaire Y and Wan T S M 2017 Interlaboratory trial for the measurement of total cobalt in equine urine and plasma by ICP-MS *Drug Test. Anal.*
[12] Aggarwal S K, Kinter M and Herold D A 1992 Determination of cobalt in urine by gas chromatography-mass spectrometry employing nickel as an internal standard *J. Chromatogr.* **576** 297-304
[13] Minakata K, Suzuki M, Suzuki O 2008 Application of electrospray ionization tandem mass spectrometry for the rapid and sensitive determination of cobalt in urine *Anal. Chim. Acta*, **614** 161-4
[14] Horiguchi D, Saito M, Imamura T and Kina K 1983 Water-soluble pyridylazoaniline reagents for the spectrophotometric determination of metals: Determination of iron(II) with 2-(5-Bromo-2-pyridylazo)-5-(*N*-propyl-*N*-sulfopropylamino)aniline *Anal. Chim. Acta* **151** 457-63
[15] Sakai T and Ohno N 1988 Flow-injection spectrophotometric determination of palladium in catalyst and dental alloys with 2-(5-Bromo-2-pyridylazo)-5-(N-propyl-N-sulfopropylamino)aniline *Anal. Chim. Acta* **214** 271-7
[16] Yamane T and Ishimizu C 1991 Simultaneous differential kinetic determination of cobalt and nickel based on on-line complex formation and ligand substitution reaction using a stopped-flow FIA system *Mikrochim. Acta* **1**, 1 121-9
[17] Dash D C 2011 *Analytical Chemistry* (New Delhi: PHI Learning Private Limited)
[18] Iburaim A, Itagaki M and Watanabe K 2001 Flow injection analysis of cationic method using a teflon capillary tube *Bunseki Kagaku* **50** 739-45
[19] Miller J N and Miller J C 2005 *Statistics and Chemometrics for Analytical Chemistry* 5th Ed. (Essex: Pearson Education Limited)
[20] Thevis M, Krug O, Piper T, Geyer H and Schänzer W 2015 Solutions advertised as erythropoiesis-stimulating products were found to contain undeclared cobalt and nickel species *Sports Med.*, **36** 82-4
[21] Thevis M, Machnik M, Schenk I, Krug O, Piper T, Geyer H, Schänzer W, Düe M, Bondesson U and Hedeland M 2016 Nickel in equine sports drug testing – pilot study results on urinary nickel concentrations *Rapid Commun. Mass Spectrom.* **30** 982-4

[22] Lewis L D 2005 *Feeding and Care of The Horse* 2nd Ed. (Oxford: Blackwell Publishing Ltd)

The use of Whatman-31ET paper for an efficient method for radiochemical purity test of ^{131}I-Hippuran

Amal Rezka Putra*, Maskur, Yono Sugiharto, Chairuman, Adang Hardi Gunawan, and Rohadi Awaludin

Center for Radioisotope and Radiopharmaceutical Technology, BATAN, Tangerang Selatan, 15314, Banten, Indonesia

*E-mail: amalrezka@batan.go.id

Abstract. Current chromatography methods used for radiochemical purity test of ^{131}I-Hippuran is time consuming. Therefore, in this study we explored several static and mobile phases in order to have a chromatography method which is accurate and efficient or less time consuming. In this study, stationary phases (Whatman-1, 31ET, and 3MM papers) and several mobile phases were explored to separate ^{131}I-Hippuran from its impurity ($^{131}I^-$). The results of this study showed that the most efficient chromatography system for measurement of radiochemical purity of ^{131}I-Hippuran was by using Whatman-31ET paper and n-butanol: acetic acid: water (4:1:1) as a static phase and mobile phase respectively. Developing time for this method was of approximately 75.7 ± 2.7 minutes. The result of radiochemical purity (%RCP) of ^{131}I-Hippuran measured with this chromatography system either using Whatman-1 or Whatman-31ET paper strips was 98.7%. The short size of Whatman-31ET paper strip (1 x 8 cm) was found to have shorter developing time compared to that of long size paper. This system showed a good separation of ^{131}I-Hippuran from its impurities and gave %RCP of 98.1% ± 0.04% with developing time approximately 44.3 ± 9.4 minutes. The short size of Whatman-31ET paper strips was found to be more efficient compared to that of Whatman-1 and Whatman-3MM paper strips in term of developing time.

1. Introduction

Labelled compounds have been widely used for diagnostic or therapeutic purposes. One of them is ^{131}I-Hippuran which is used for diagnosis of renal function [1]. In Indonesia, ^{131}I-Hippuran has been routinely used in several hospitals namely Ulin Hospital, Kalimantan Tengah; Hasan Sadikin Hospital, Bandung; Annur Specialist Surgery Hospital and Bethesda Hospital, Yogyakarta; POLRI Hospital, Jakarta; and Kartini Hospital, Jepara [2]. The utilization of the ^{131}I-Hippuran must be supported with an effective and efficient of its quality control. Labelled compounds must meet with the certain requirements for their physicochemical and biological qualities. The physicochemical quality of the labelled compound includes osmolality, pH, clarity, radionuclide purity, and radiochemical purity (RCP) [1,3,4], while the biological quality includes sterility, pyrogenecity, and biodistribution in testing animals [5].

Synthesis of ^{131}I-Hippuran and its optimization also radiochemical purity test method had been performed and reported [2]. Radiochemical purity (RCP) of ^{131}I-Hippuran (OIH) is associated with percentage of ^{131}I bind to Hippuran and radiochemical impurities is associated with the percentage of free ^{131}I in form of iodide ions ($^{131}I^-$) and ^{131}I-o-iodobenzoic acid (OIB). The RCP data of the ^{131}I-Hippuran and its impurities are shown in Table 1.

Table 1. The RCP of ^{131}I-Hippuran (OIH), ^{131}I-o-iodobenzoic acid (OIB), and Iodide ion ($^{131}I^-$) [1,2,6,7,8,9,10,11,12,13]

Stationary phase	Mobile phase	Rf	(%) RCP of ^{131}I-Hippuran	Developing time/ size of paper	Ref.
Chromatography paper	Acetic acid: chloroform (1:9)	OIH = 0.2	>96.00	-	[1]
Whatman-1	n-butanol: acetic acid: water (4:1:1)	I^- = 0.00 – 0.30 OIH = 0.70 – 0.90	>95.00	-	[2]
Whatman-1	n-butanol: acetic acid: water (4:1:1)	I^- = 0.09-0.23 OIH= 0.82-0.92 OIB = 0.88-0.95	-	30cm =16 h	[6]
Whatman-1	Benzene: acetic acid: water (2:2:1)	I^- = 0.00-0.05 OIH= 0.48-0.52 OIB = 0.88-0.92	-	30cm =16 h	[6]
Whatman-3MM	Sodium acetate: glacial acetic acid: water (10:10:100)	I^- = 0.83	-	-	[7]
TLC cellulose	n-butanol: glacial acetic acid: water (6:1.5:2.5)	I^- = 0.35 m-IH = 0.57 m-IB = 0.72	-	-	[8]
Whatman-1	n-butanol: water: glacial acetic acid (120:50:30)	OIH= 0.94 OIB= 0.92	>99.00	15 cm = 18 h	[9]
Silica Gel G-plates	n-butanol : 1N acetic acid (1:1)	OIH= 0.72 OIB= 0.60	-	15cm	[9]
TCL silica gel	Benzene: acetic acid:water: n-butanol (5:5:2:1.5)	-	98.00 – 99.00	-	[10]
TLC; silica gel F 1500LS 254	Methanol 80%	I^- = 0.00 OIH= 0.40 OIB = 0.80 - 0.90	93.74 – 99.62 different batch	9 cm = 50 minutes	[11]
TLC	Methanol: water: acetic acid (30:70:0.5)	-	99.20 ± 0.20	-	[12]
TLC 5716 plates	Benzene: acetic acid: water (2:2:1)	I^- = 0.00 - 0.100 OIH = 0.50 - 0.60 OIB = 0.90 - 1.00	94.00	1-2 h	[13]

RCP test of the ^{131}I-Hippuran generally performed by a paper chromatography method using Whatman-1 paper as stationary phase and n-butanol: acetic acid: water (4:1:1) solvent mixture as mobile phase [2,6]. In the USP 30/ NF 35 is stated that the %RCP of the ^{131}I-Hippuran must be over than 97% when measured by using Whatman-1 as static phase and benzene: glacial acetic acid: water (2:2:1) as a mobile phase [14]. However, this method is considered inefficient because its long developing time (150 minutes). Therefore, there is a need to find out a new or modified method or system with a shorten developing time in the RCP test of ^{131}I-Hippuran. The modified method can be carried out by modifying the mobile or stationary phase used in the paper chromatography method.

Several types of the paper commonly used in the paper chromatography method are Whatman-1, Whatman-3MM, and Whatman-31ET papers. The Whatman-31ET paper is one of the fast flow chromatography paper compared to that of Whatman-1 and Whatman-3MM papers [15]. Figure 1 shows the retention factor (Rf) of ^{131}I-Hippuran (OIH) and iodide ion ($^{131}I^-$). The Rf value of ^{131}I-Hippuran approximately 0.70 – 0.92, and for $^{131}I^-$ approximately of 0.00 to 0.30 [2,6]. The Rf value was calculated by the Formula 1 [9]:

$$Rf = \frac{migrationdistanceofsubstance}{migrationdistanceofsolvent} \quad (1)$$

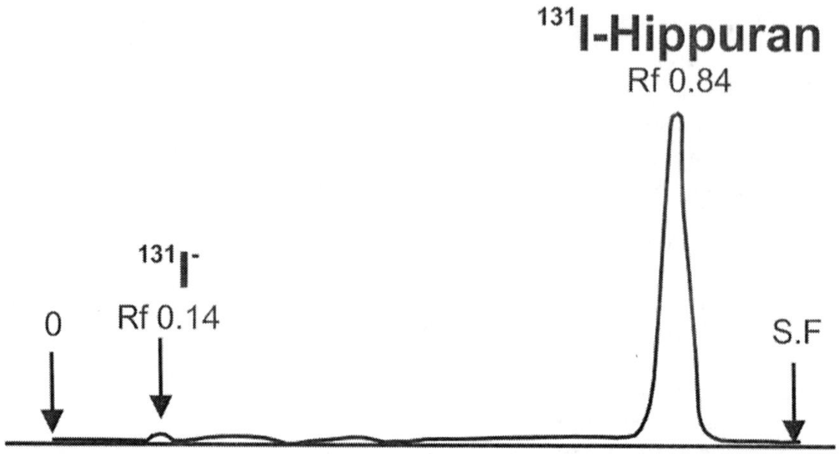

Figure 1. The chromatogram of ^{131}I-Hippuran using Whatman No.1 paper as static phase and n-butanol: acetic acid: water (4:1:4) as mobile phase [6].

The aim of this study is therefore to determine the most efficient paper chromatography method based on various mobile phases or various Whatman paper (Whatman-1, Whatman-31ET, and Whatman-3MM papers), based on the required %RCP of ^{131}I-Hippuran.

2. Methods

2.1. The materials
Labelled compound of ^{131}I-Hippuran was produced by The Center for Radioisotope and Radiopharmaceutical Technology – BATAN, water (IPHA Laboratories), n-butanol for analysis 99.5%, acetic acid for analysis 99.8%, benzene for analysis, acetone for analysis 99.8%, propanol for analysis, ethanol absolute for analysis 99.8%, ammonia 25% (Merck), plastic film, strips of 1 x 11 cm Whatman-1 paper (Schleicher & Schuell), strips of 1 x 11 cm Whatman-3MM (Sigma-Aldrich), strips of 1 x 8 cm; 1 x 10 cm; 1 x 11 cm; 1 x 12 cm; 1 x 16 cm; and 1 x 21 cm Whatman-31ET paper (GE Healthcare Life Sciences).

2.2. The equipment
Equipment used this study were glass cylindrical tank, stopwatch, gamma counter (Caprac-t), micropipette (Eppendorf).

2.3. The procedure
The ^{131}I-Hippuran was spotted 2 cm from the bottom of the chromatographic paper strips and then developed in a closed 10 x 29 cm glass cylindrical tank, which contained the developing solvent. The solvent is allowed to migrate up to 1 cm from the top of the paper strips. They were immediately removed, left to dry at room temperature, wrapped in a plastic film. The paper strips were then cut (1 cm/ portion) and then counted on gamma counter (Caprac-t) [16]. The measurement of repeatability was carried out in the same experimental conditions, tools, place, and short interval of time trial. The same procedure is also done for various mobile phases, stationary phases and size of paper.

2.4. Data analysis

The data comparison performed using the independent t-test, especially in order to know the significantly difference between two samples with homogeneous data. The t-test of independent two-sample with equal variance can be calculated in the Formula 2 and 3 [17,18].

$$t = \frac{\bar{x}_1 - \bar{x}_2}{Sp\sqrt{\frac{1}{n_1}+\frac{1}{n_2}}} \quad (2)$$

where

$$Sp = \sqrt{\frac{(n_1-1)S_1^2+(n_2-1)S_2^2}{n_1+n_2-2}} \quad (3)$$

Where

t = t-value,
x_1 = average data of sample 1,
x_2 = average data of sample 2,
Sp = pooled standard deviation,
S_1 = sample varian 1,
S_2 = sample varian 2,
n_1 = sample size 1, and
n_2 = sample size 2

The percentage of RCP of the ^{131}I-Hippuran data processing which were analysed using various Whatman papers performed by t-test. The difference of % RCP of ^{131}I-Hippuran are significant if the t-value of t-test statistic is equal to or greater than 2.10, the critical value in this case (significance level (P') = 0.05 with number of degrees freedom (Ø) = 18).

3. Results and Discussion

The RCP of the ^{131}I-Hippuran was determined by a paper chromatography method using various mobile and stationary (Whatman-1 paper) phases. The RCP results of the ^{131}I-Hippuran are showed in Table 2.

Table 2. The results of ^{131}I-Hippuran RCP test using Whatman-1 paper with various mobile phases

Mobile phase	Rf			Developing Time (minute)
	$^{131}I^-$	^{131}I-Hippuran	OIB	
benzene: acetic acid: water (2:2:0.5)	0.0 – 0.2	0.6 – 0.7	Not defined	242
n-butanol: acetic acid: water (4:1:1)	0.10	0.8 – 0.9	-	130
n-butanol: acetic acid: water (8:1:1)	0.0 – 0.1	0.8 – 0.9	-	133
n-butanol: acetic acid: water (1:1:1)	0.4 – 0.5	0.8	-	147
acetone : acetic acid: water (4:1:1)	0.9 – 1.0	0.9	-	65
propanol : acetic acid: water (4:1:1)	0.0 – 0.2	0.7 – 0.9	-	120
ethanol: water: ammonia 25% (25:3:4)	0.5 – 0.7	0.5 – 0.7	-	95
ethanol: water: ammonia 25% (10:1:10)	0.8	0.8 – 0.9	-	105

It can be seen in Table 2 that the mobile phases of benzene: acetic acid: water (2:2:0.5), n-butanol: acetic acid: water (4:1:1) and n-butanol: acetic acid: water (8:1:1) gave a good separation between ^{131}I-Hippuran (Rf ~ 0.6 – 0.9) and $^{131}I^-$ was (Rf ~ 0.0 – 0.2), while OIB was not visible in mobile phase benzene: acetic acid: water (2:2:0.5). The mobile phase of acetone: acetic acid: water (4:1:1), ethanol: water: ammonia 25% (25:3:4) and ethanol: water: ammonia 25% (10:1:10) however showed a bad separation between ^{131}I-Hippuran and its impurity $^{131}I^-$. The use of n-butanol: acetic acid: water (1:1:1) and propanol : acetic acid: water (4:1:1) mobile phases gave a tail separation, therefore it could not be used in RCP test of ^{131}I-Hippuran. The standard method where n-butanol: acetic acid: water (4:1:1) used as a mobile phase, gave a good separation between ^{131}I-Hippuran (Rf ~ 0.8 – 0.9) and its impurity ,$^{131}I^-$, (Rf ~ 0.10). This system also has relatively short developing time (130 minutes). Hippuran (Figure 2) is a polar compound which has an amine group (secondary amine) and methylene [2]. Polar molecules have a high attraction for polar solvent such as n-butanol and acetic acid is able in forming a strong hydrogen bonding, which resulted in an acceleration of the migration of polar compound [19].

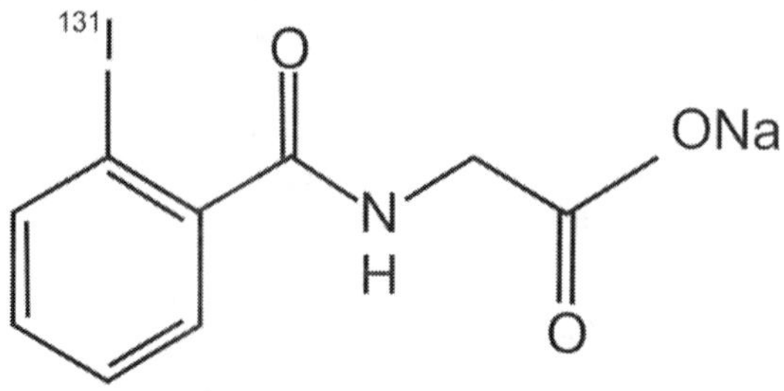

Figure 2. Molecule structure of ^{131}I-Hippuran

The chromatograms of ^{131}I-Hippuran which were developed on various stationary phases (Whatman-1, 31ET and 3MM papers) with mobile phase of n-butanol: acetic acid: water (4:1:1) are presented in Figure 3 and Table 3.

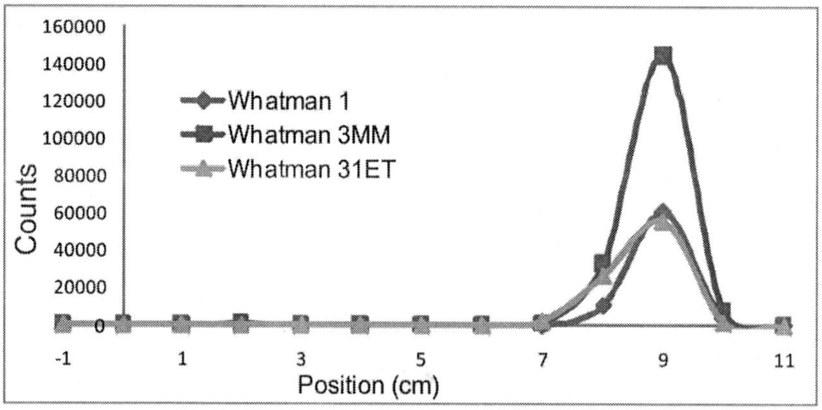

Figure 3. Chromatograms of ^{131}I-Hippuran developed using various Whatman papers as stationary phases and n-butanol: acetic acid: water (4:1:1) as a mobile phase

Table 3. Chromatography systems of ^{131}I-Hippuran based on various Whatman papers as stationary phases and n-butanol: acetic acid: water (4:1:1) as a mobile phase

Stationary phase	%RCP	Developing time (minute)	Description
Whatman No. 1	98.7 ± 0.9%	158.3 ± 6.2%	Standard method
Whatman No.31 ET	98.7 ± 0.7%	75.7 ± 2.7%	Fast flow
Whatman No. 3MM	98.7 ± 0.8%	142.0 ± 3.7%	Medium flow

It can be seen from Figure 3 that all types of Whatman papers gave Rf value of approximately 0.9 for ^{131}I-Hippuran. The use of different Whatman paper (Whatman-31ET, Whatman-1 and Whatman-3MM) gave no significant effect on the Rf value of ^{131}I-Hippuran. Figure 3 also shows that the use Whatman-3MM paper in the above mentioned chromatography system gave a higher peak counts than the other papers because the Whatman-3MM paper is thicker and has higher capacities therefore it is able to hold more sample [20]. The %RCP of the ^{131}I-Hippuran measured using for either with

Whatman-1 or and Whatman-31ET papers were higher than 98%. The Whatman-31ET paper showed faster developing time of 75.7 ± 2.7 minutes compared to that of Whatman-1 (158.3 ± 6.2 minutes) for ^{131}I-Hippuran as Whatman-31ET paper has a looser fiber network [20].

The difference in %RCP of the ^{131}I-Hippuran measured using Whatman-1 and 31ET paper was analyzed using t-test. The results were shown in Table 4. The percentage of relative standard deviation (%RSDs) for Whatman-1 paper and Whatman-31ET paper were of 0.30% and 0.22%, respectively which indicated that the value of %RSD for both papers were in range of the precision standard. It should be noted that the required (%RSDs) should less than 1%. Overall, t-test showed that the %RCP results of ^{131}I-Hippuran obtained using either Whatman-1 or Whatman-31ET paper is similar, as there was no significant statistical difference (t-value = 0.40).

Table 4. The t-test of % radiochemical purity between Whatman-1 paper and Whatman-31ET paper

	%RCP	
	Whatman-1	Whatman-31ET
1	99.10	99.513
2	99.25	99.001
3	99.43	98.940
4	98.73	99.037
5	99.30	98.870
6	98.85	98.673
7	99.28	99.028
8	98.86	99.172
9	98.64	99.183
10	98.66	99.135
Mean	99.01	99.055
SD	0.30	0.22
%RSD	0.30 %	0.22 %
t-value	0.40	
t-table value P'=0.05; Ø=18	2.10	

The experiments using various size of Whatman-31ET paper were also performed in order to get paper chromatography system which has more efficient developing time. The shorter size of chomatography paper strip migh offer a faster developing time. However, a paper chromatography system is not only has short developing time but is also able to properly separate ^{131}I-Hippuran from its impurities.

Percentage of RCP and developing time of ^{131}I-Hippuran measured using Whatman-31ET paper with size of 1 x 8 (short size) up to 1 x 21 cm (long size) as stationary phases and n-butanol:acetic acid: water (4:1:1) as mobile phase are shown in Table 5. The Rf value of this method showed in Figure 4. Based on Table 5, it be seen that there are no different of the %RCP and the Rf value of ^{131}I-Hippuran when measured using 1 x 8 (short size), 1 x 11 (standard method) and 1 x 21 cm (long size) Whatman-31ET paper strips. The %RCPs of ^{131}I-Hippuran measured with those strips were >98%. The results also showed that there are no correlation between size of paper strips and Rf value of the ^{131}I-Hippuran. The long size Whatman paper extends the migration distance of substance so that it can nicely separate ^{131}I-Hippuran from its impurities. The Rf value of ^{131}I-Hippuran from this experiment was of 0.9 – 1.0. This value is almost the same with the reported Rf for this paper chromatography system Rf approximately 0.70 – 0.92 [2,6]. The shortes developing time for this RCP test of ^{131}I-Hippuran using the above mentioned chromatography was of ~ 44.3 ± 9.4 minutes which was showed by 1 x 8 cm size Whatman-31ET paper strips. This chromatography system was found to be more efficient compared to that of the USP 30/ NF 35 standard method with developing time of 150 minutes [14].

Table 5. The %RCP and the developing time of ^{131}I-Hippuran measured using various size of paper Whatman-31ET paper strips

Size of Paper (cm)	%RCP	Developing time (minute)
1 x 8	98.1 ± 0.0%	44.3 ± 9.4%
1 x 10	98.1 ± 0.1%	65.3 ± 0.8%
1 x 11	98.7 ± 0.7%	75.8 ± 2.7%
1 x 12	98.4 ± 0.1%	101.3 ± 11.5%
1 x 16	98.2 ± 0.1%	167.0 ± 2.2%
1 x 21	98.0 ± 0.1%	314.0 ± 2.2%

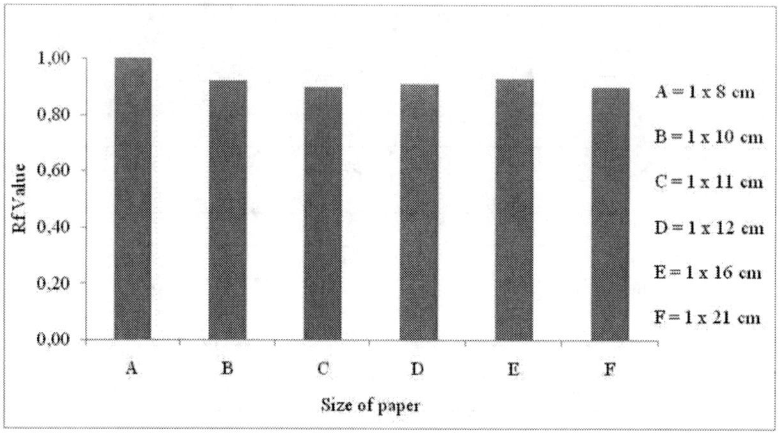

Figure 4. Rf Value of ^{131}I-Hippuran using various size of Whatman-31ET paper strips

4. Conclusions

Paper chromatography method for measuring the radiochemical purity of ^{131}I-Hippuran has been studied. The most suitable mobile phase for this system was found to the mixture of n-butanol: acetic acid: water (4:1:1) which showed a good separation between ^{131}I-Hippuran from its impurities. This study also found that the most efficient stationary phase for the above mentioned chromatogaraphy system Whatman-31ET paper strips (1 x 8 cm) which has a developing time of 44.3 ± 9.4 minutes. This method has been to be efficient in term of developing time and comparable with other chromatography systems which have been reported elsewhere.

References

[1] Huang L S*et al*2014 Microwave Assistance of Labeling Hippuric Acid by I-131 *J. Appl. Radiat. Isot.***89** 53–57

[2] Maskur*et al* 2015 *Proc. of the Nat. Sem. on Chem.*(Surabaya)(Indonesia)pp11–18

[3] Kilian K*et al*2014 Synthesis, Quality Control and Determination of Metallic Impurities in ^{18}F-Fludeoxyglucose. *Reports Pract. Oncol. Radiother***9** 22–31

[4] Dziel T*et al* 2015 Radionuclidic Purity Tests in Process F Radiopharmaceuticals Production *J. Appl. Radiat. Isot.***109** 242–246

[5] Pathuri G*et al* 2016 Synthesis and In Vivo Evaluation of Ortho- [^{124}I] Iodohippurate for PET Renography in Healthy Rats *J. Appl. Radiat. Isot.***115** 251–255.

[6] Varga L Kelemen I and Kovach A 1968 Paper Chromatographic Study of the Radioactive Contaminants of ^{131}I-Hippuran *J. Nucl. Med.***9**(**12**) 604–608

[7] Pluchet E and Lederer M 1960 Adsorption Paper Chromatography of Inorganic Anions in Acetate Buffers *J. Chromatogr.***3** 290–296

[8] Elias H Arnold C and Kloss G 1973 Preparation of ^{131}I-Labelled m-Iodohippuric Acid and its Behaviour in Kidney Function Studies Compared to o-Iodohippuric Acid *J. Appl. Radiat. Isot.***24** 463–469.

[9] Belkas E P Hiotellis E and Dassiou C 1975 Hippuran-^{131}I; Preparation and Purity Control *J. Appl. Radiat. Isot.***26** 629–630

[10] Kaspersen F M and Westera G 1979 Radioiodinated o-Iodobenzoic Acid as Impurity in Hippuran Preparations *Int. J. Appl. Radiat. Isot.***31** 97–99

[11] Hammermaier A Reich E and Bogl W 1986 Radiochemical Purity and In Vitro Stability of Commercial Hippurans *J. Nucl. Med.***27(6)** 850–854

[12] Mertens J*et al* 1987 New fast preparation of ^{123}I labelled radiopharmaceuticals *Eur. J. Nucl. Med.***13(7)** 380–381

[13] Hawkins L*et al* 1992 A Rapid Method for the Preparation of ^{123}I-Iodo-Hippuric Acid *Eur. J. Nucl. Med.***7** 58–61

[14] Anonim 2011 *United States Pharmacopeia 35/ National Formulary 30*(2nd vol. Rockville)(The United States Pharmacopeial Convention) pp 3522-3523

[15] Ponto J A 2011 Evaluation of Alternative Methods for Radiochemical Purity Testing of Indium-111 Capromab Pendetide *J. Am. Pharm. Assoc.***51(3)** 359–362

[16] Wilson ID 2000 Paper Chromatography. In: Reference Module in Chemistry, Molecular Sciences and Chemical Engineering (UK:Academic Press) pp 397-404

[17] Zhang J and Pan M 2016 A High-Dimension Two-Sample Test for The Mean Using Cluster *Comput. Stat. Data Anal.***97** 87–97

[18] Kim C*et al*2013 An Efficient and Effective Wrapper Based on Paired t-Test for Learning Naive Bayes Classifiers from Large-Scale Domains *Procedia Comput. Sci.***23** 102–112

[19] Ahuja S 2003 *Chromatography and Separation Science, Separation Science and Technology* 4th ed(United States of America: Academic Press) pp 102-112

[20] Sherma J and Fried B 1984 *Flat bed techniques. In: New Comprehensive Biochemistry* 8th ed. (USA:Elsevier Science Publisher) pp 363-413

Acknowledgement
This work was supported by the Center for Radioisotope and Radiopharmaceutical Technology-BATAN. The authors would like to thank to Yayan Tahyan and Enny Lestari for their expertise assistances

The Effect of Temperature and Concentration of Foaming Agent to the β-Carotene Content in Product Derived from Carrots

Qonitah Fardiyah*, Barlah Rumhayati, Yuniesti Husnul Khotimah

Department of Chemistry, Faculty of Mathematics and Natural Sciences, Brawijaya University, Malang, Indonesia

Email: fardiyah@ub.ac.id

Abstract. Carrot *(Daucus carota L)* is vegetable that contain body essetial vitamins, especially β-carotene. In this research, the essense of fresh carrots are taken and processed to carrot powder using foam mat drying method. This research aims to study the effect of temperature and concentration of foaming agent to the β- carotene content in product derived from carrots. The temperature variation that used in this research are 40°C, 50°C, 60°C and 70°C, while he variation of foaming agent (tween 80) are 0,01% (v/v); 0,1%(v/v); 0,2%(v/v) and 0,3%(v/v). The results shows that the maximum drying temperature is 50°C with β-carotene content 10.55 mg/kg and the maximum concentration of foaming agent (tween 80) is 0.2%(v/v) with β- carotene content 10.36 mg/kg. *Keywords: carrot, foam mat drying, tween 80, β- carotene.*

1. Introduction

Carrots are one of the plated vegetables that are rich in alpha and betha carotene, Vitamin C, B, and minerals. The carotenes are essential for body as provitamin A. Besides, carrot also containing vitamin C, vitamin B and mineral especially calcium and phosphor [1].

The purpose of making carrot powder is to improve the quality of carrot products so that its utilization is more widespread and the storage time of carrot products becomes longer. In the process of making carrot powder requires special conditions for the content of carotene in carrots is not lost due to the nature of β- carotene sensitivity to oxygen and light. Therefore, in the process of making carrot powder required an effective and efficient drying method is the foam mat drying method. .

In this method, foaming agent is needed to accelerate the drying process, decreasing the water content and crumb powder. Foaming agent increasing the evaporation of water although the temperature is lower than water boiling point, dried product at 50 - 80 °C contain 2-3% water [6].

Refer to Karim [7], foam-mat drying method have several advantages, such as easy to apply and cheaper than spray drying and freeze drying methods. According to Kudra and Ratti [8], foam-mat drying method have advantages, such as, faster drying, lower operation temperature, high quality product, good color and taste, and high solubility. Foam-mat drying method used to produce dry product from liquid that high sugar content.

Drying process in this method can be done by adding foaming agent. Foaming agent is a food additive that used to form and protect disperse homogenity of gas phase in solid or liquid food. Foaming agent that usually used are egg, tween 80, monostearicgliserol, xanthan gum, microchrystalline cellulose, ethyl methyl cellulose. Foaming agent addition aims to increase the surface, decrease the surface tension, increasing pore, increasing water evaporation, and keeping material quality [9].

2. Experiment

2.1. Materials and Equipment
Material that used in this research are chentenay carrot, aquades, maltodextrin, citric acid p. a., tween 80 p. a., petroleum ether p.a., β- carotene powder p. a. (Merck Sigma).

2.2. Procedure
The absorbance of β- carotene was measured by Shimadzu spectrophotometer. The standard solution of β- carotene (1000 ppm) was prepared by dissolving 50 mg β- carotene p. a. in 50 mL petroleum ether, then diluted to 500 ppm and 100 ppm. Standard solutions of β- carotene p.a., 5 ppm, 10 ppm, 15 ppm, 20 ppm, 25 ppm, and 30 ppm were made to make the standard curve of β- carotene [10].

Product derived from carrot was produced by destructing 125 gram carrot with 60 ml aquades and then the essence was taken. Then 0.2%(v/v) tween 80, 2.5 gram maltodextrin and 0.0125 gram citric acid were added and shaken untill foaming. The foam placed in glass plate then dried at temperature variation 40°C, 50°C, 60°C and 70°C. After the optimum drying temperature acquired, the product derived from carrot was prepared again with foaming agent concentration variation 0,05%(v/v); 0,1%(v/v); 0,2%(v/v) and 0,3%(v/v) and dried at optimum temperature

2.3. Uv-Vis Analysis
The acquired carrot powder at every variation of temperature and foaming agent concentration were analysis by spectrometer UV- Vis by diluting 50 mg carrot powder into 25 mL petroleum ether and quantified at maximum wavelength 448.9 nm.

3. Results and Discussion
The Effect of Temperature to β- Carotene Content in Product Derived from Carrot

In this method, tween 80 used as foaming agent in the solution, and maltodextrin used to stabilize the foam so the drying process is fast and β- carotene not destroyed. The variation of drying temperature that used in this research are 40°, 50°, 60° and 70°C with tween 80 concentration 0.2%(v/v). The results shows at **Table 1**.

Table 1. The β- carotene content in product derived from carrot as a function of temperature with tween 80 concentration 0.2%(v/v)

Temperature (°C)	Contained of β- caroten mg/kg			Contained of β- caroten mg/kg average
	1	2	3	
40	3.89	3.9	4.11	3.97
50	10.72	10.5	10.43	10.55
60	7.30	7.99	8.73	8.01
70	3.87	3.15	3.04	3.35

And the curve of β- carotene content in sample as a function of temperature can be seen at **Figure 1**.

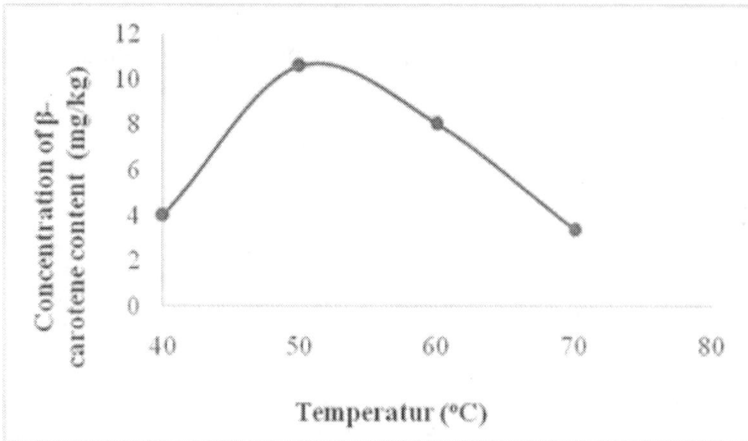

Figure 1. The effect of drying temperature to β- carotene content in carrot powder

Table 1 and Figure 1 shows that at temperature 400°C, the β- carotene content in carrot powder is lower than in 50°C. It can be caused by the higher amount of water content in 40°C so there are more micelle that dissolved in water. The higher amount of micelle dissolved in water means lower extracted micelle in petroleum ether so the amount of β- carotene is lower. At 60°C and 70°C, β-carotene content in carrot powder is decreases due to the isomerization of β- carotene. A higher temperature, β- carotene changes from cis to trans configuration. The changes of configuration makes the maximum wavelength shift of β- carotene so it is not quantified at maximum wavelength used.

The β- carotene content also can be seen from the colour of the carrot powder and carrot powder solutions in Figure 2 and Figure 3.

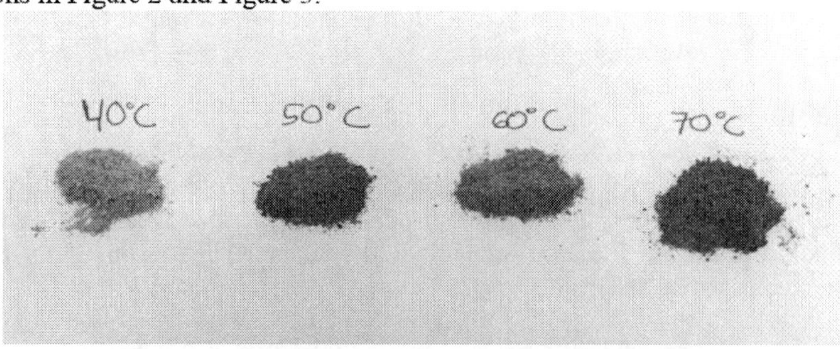

Figure 2. Carrot powder in variations of temperature

Figure 3. Carrot powder solutions in variations of temperature

The Effect of Foaming Agent Concentration to The β-Carotene Content in Product Derived from Carrot

According to the applied temperature in preparation of carrots before, the optimum temperature for drying is 50^{0}C. After the optimum temperature obtained, the carrots preparation to powder was did in various tween 80 concentration. Variation of tween 80 that used are 0.05%(v/v); 0.1%(v/v); 0.2%(v/v) and 0.3%(v/v). The β- carotene content as a function of tween 80 concentration can be seen in Table 2.

Table 2. The β- carotene content in product derived from carrot as a function of tween 80 concentration

Tween 80 Concentration % (v/v)	β- carotene content (mg/kg)			Concentration of β- carotene content(mg/kg)
	1	2	3	
0.05	2.85	2.88	2.66	2.79
0.1	6.92	5.73	7.81	9.12
0.2	10.54	10.42	10.13	10.36
0.3	8.33	7.27	6.6	7.40

And the curve of β- carotene content in sample as a function of tween 80 concentration can be seen at Figure 4.

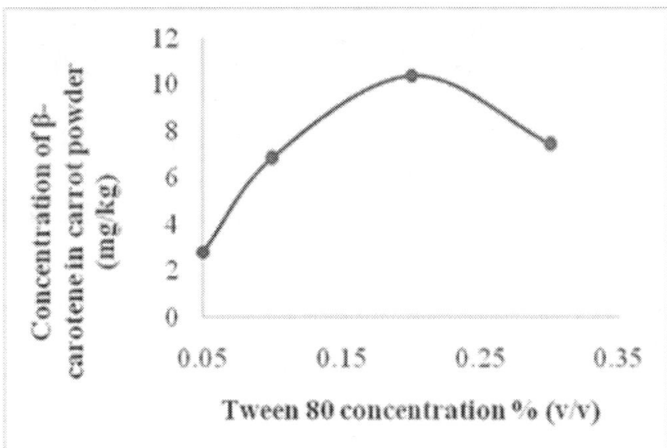

Figure 4. The effect of tween 80 concentration of β- carotene content in carrot powder.

According to Figure 4, the optimum tween 80 concentration is 0.2%(v/v). In this concentration, tween 80 has best performance as surfactant. The addition of surfactant in the solution made the surface tension lower. At certain concentration, the surface tension is constant even the concentration increased. The addition of surfactant correspond to the formation of micelle. The concentration when micelle formed is called as critical micelle concentration (CMC). The surface tension is lowering till CMC reached. Then, the surface tension is constant which is shows that the surface is saturated and micelle in dynamic stability with its monomer [11].

According to Attwood and Florence [12], tween 80 is non ionic surfactant that its alkyl side does not have a charge. In this research, tween 80 can lowering the surface tension of carrot essence solutions to form micelle so the β- carotene molecules will trapped in micelle and diluted to medium. The solution becomes foam medium and accelerate the drying process.

β- caroten content is increases at concentration tween 80 0.2%(v/v). It caused by the formation of micelle at the CMC. Nevertheless, the β- caroten content is decreases when the micell concentration adjusted to 0.3%(v/v). This can be understood because at higher concentration, tween 80 becomes foam destroying agent which decrease the formed foam so β- caroten dilute to the solution. It proved by the colour of carrot powder and the solutions of its powder as can be seen at Figure 5 and Figure 6.

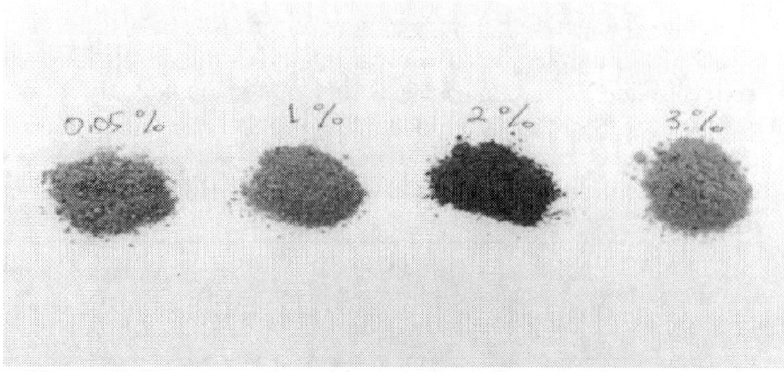

Figure 5. Carrot powder with variation of tween 80 concentration

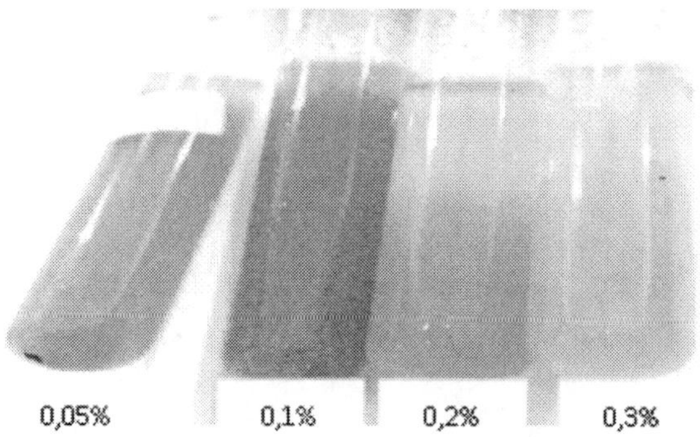

Figure 6. Solution of product derived from carrot with variation of tween 80

4. Conclusion

The results shows that the optimum drying temperature is $50^{0}C$ with β- carotene content 10.55 mg/kg and the optimum concentration of foaming agent (tween 80) is 0.2%(v/v) with β- carotene content 10.36 mg/kg.

References

[1] Rubatzky VE, and Yamaguchi M 1997 SayuranDunia 2 Prinsip, Produksi, dan Gizi Penerbit ITB Bandung
[2] Takahata Y, Noda T, and Nagata T 1993 *J. Breed* **43** 421-427.
[3] Lanier JJ, and Sistrunk WA 1979 *Journal of Food Science* 44 **(2)** 374-380.
[4] Mohamed,. S and R. Hussein 1994 Effect of Low Termperature Blanching, Cyleing-HCl, N-Acetyl-L-Cyteine, Na Meta Bisulphite and Drying Temperatures On The Firmness And Nutrientcontent Of Dried Carrots *Jurnal Processing Preservat.*
[5] Muflihah R., Sri Kumalaningsih., and Imam Santoso, 2012, *Jurnal Teknologi Pertanian* **13** no.2 125-137.
[6] Kumalaningsih, S., Suprayogi, and B. Yuda , 2005, **Techno Pangan** : Membuat Makanan Siap SAji, Trubus Agrisarana, Surabaya
[7] Karim, A.A and C.C Wai, 1999, Foam Mat Drying of Starfruit (Averrhoa Caramba L.) Puree., Stability Air Drying Characteristics, *Food Chemistry*, **64** : 337-343
[8] Kudra, T. and C. Ratti 2006 *J.Canadian Biosystem Engineering* **48** 327–332.

[9] Zubaidah, E. and D. Widyaningtyas 2009 Jurnal Teknologi dan Industri Pangan 14 **(3)** 234239.
[10] Nielsen SS 1995 Introduction to The Chemical Analysis of Food New Yor Chapman and Hall.
[11] Genaro, R.A, 1990, Rhemingstons Pharmaceutical Science, 18th 267 ed, Mack Printing Company Easton Pennyslvania, USA.
[12] Attwood, D., and Flourence,A.T. 1985 Surfactan System 1st Ed. Champman and Hall, London, New York.

Method Development for Pesticide Residue Analysis in Farmland Soil using High Perfomance Liquid Chromatography

Marselina Theresia Djue Tea[1], Akhmad Sabarudin[1], and Hermin Sulistyarti [1,2]
[1]Department of Chemistry, Faculty of Science, University of Brawijaya, Malang, 65145, Indonesia
[2]Central of Low Cost and Automated Method and Instrumentation Analysis (LCAMIA), Chemistry Department, Faculty of Science, University of Brawijaya, Malang 65145, Indonesia

Corresponding email: hermin@ub.ac.id

Abstract. A method for the determination of diazinon and chlorantraniliprole in soil samples has been developed. The analyte was extracted with acetonitrile from farmland soil sample. Determination and quantification of diazinon and chlorantraniliprole were perfomed by high perfomance liquid chromatography (HPLC) with an UV detector. Several parameters of HPLC method were optimized with respect to sensitivity, high resolution of separation, and accurate determination of diazinon and chlorantraniliprole. Optimum conditions for the separation of two pesticides were eluent composition of acetonitrile:water ratio of 60:40, 0.4 mL/min of flow rate, and 220 nm of wavelength. Under the optimum conditions, diazinon linearity was in the range from 1-25 ppm with R^2 of 0.9976, 1.19 mgL^{-1} LOD, and 3.98 mgL^{-1} LOQ; while the linearity of chlorantraniliprole was in the range from 0.2-5 mgL^{-1} with R^2 of 0.9972, 0.39 mgL^{-1} LOD, and 1.29 mgL^{-1} LOQ. When the method was applied to the soil sample, both pesticides showed acceptable recoveries for real sample of more than 85%: thus, the developed method meets the validation requirement. Under this developed method, the concentrations of both pesticides in the soil samples were below the LOD and LOQ (0.577 mgL^{-1} for diazinon and 0.007 mgL^{-1} for chlorantraniliprole). Therefore, it can be concluded that the soil samples used in this study have neither diazinon nor chlorantraniliprole.

Keywords: HPLC, Pesticide, Diazinon, Chloranthraniliprole

1. Introduction

Pesticides Pesticides (herbicides, fungicides, insecticides or acaricides) have an important role in agricultural development and guarantees of increased agricultural production. The use of these chemical compounds affects soil, water, and food contamination, leading to the accumulation of contaminants into the environment that can enter the human food chain [1]. Pesticides are found in soil [2], which causes the accumulation of pesticide residue in the soil that takes a long time to degrade causing soil contamination and causing the decline of biodiversity of organisms in the soil [3]. Contact with the residue of the pesticides will cause adverse effects to health, including cancer, depression, respiratory disorders, neurologic effects and dermatological disease [3].

Diazinon is an organophosphate insecticide with molecular weight of 304.35 g/mol, with boiling point 83 - 94^{0}C, vapor pressure of 4.6 x 10-5 mmHg at a temperature of 10^{0}C, 1.4 x 10^{-4} mmHg at a temperature of 20^{0}C, and 1.1 x 10^{-3} mmHg at 400C, water solubility 0.004% at 20^{0}C and dissolved in organic solvent. The diazinon structure with its chemical name O,O-diethyl-O-(2-isopropyl-6-methyl-4-4-pyrimidyl) phosphorothioate. Diazinon absorbs UV light at a wavelength of 247 nm [4].

Chlorantraniliprole with a new action mode is called "Ryanodine Receptors Activators" [5] [6] and is a stomach poison and a contact poison. Symptoms indicated by insects due to the application of chlorantraniliprole insecticide are paralysis, stop eating and within a few days can lead to death [4]. Chlorantraniliprole with chemical name 3-bromo-N- [4-chloro-2-methyl-6 - [(methyl amino) carboryl] phenyl] -1- (3-chloro-2-pyridynil) -1H-pyrazole-5-carboxamide absorbs UV light at a wavelength of 260 nm [7].

The disadvantage of the liquid-solid extraction technique that used in the determination of residual pesticides in the soil are the use of large amounts of solvent and a long time to analysis, but it can be solved by other extraction methods namely Ultrasonic Solvent Extraction (USE) [8]. USE has several advantages: speed, selectivity, and low use of the solvent. USE increases the efficiency of extraction but because of its low selectivity it is followed by a clean up to obtain a clean extract [8]. For the detection and quantification of pesticide residues, one used high performance liquid chromatography (HPLC).

Chromatography is a method in which the mixed components are separated on the adsorbent column in the flow system. For chromatographic separation, samples are included in the mobile phase flowing through the stationary phase [9]. HPLC was used for the analysis of some thermolabile or highly polar compounds and also on compounds with high molecular weight. The success of an analysis is also strongly influenced by the accuracy of the chromatographers in selecting and using column, the stationary phase, and the mobile phase [10]. The advantages of separation with HPLC compared to conventional methods include fast analysis times, low cost, and able to analyze compounds with low stability [11]. The method of analysis related to pesticide residue content ever done is the determination of chlorantraniliprole pesticide residue using solid phase extraction method (SPE) and HPLC using UV detector at 260 nm wavelength, this method has 200 pg detection limit and more than 97% recovery [7]. Determination of diazinon pesticide residues in water and soil samples is using HPLC using UV detector at 245 nm wavelength; this method has a quantification limit of 0.1 ng/ml water sample whereas in soil samples of 1 ng/ml and % recovery ranges 86 to 102% [12].

The research conducted optimization of eluent composition, flow rate, and wavelength. Furthermore, the validity test, consisting of linearity test, accuracy test, precision test, limit of detection (LOD), and limit of quantification (LOQ).

2. Methods

2.1Chemicals

The materials used in this research were: a soil sample was taken from Bumiaji Sub-district, Batu City, Malang. 99% acetonitrile (Merck), standard diazinon pestanal™ 98.6% (Sigma-Aldrich), pestanal® 99.2% (Sigma-Aldrich) chlorantraniliprole standard. The tools that used in this research were: Genesys 10S UV-VIS spectrophotometer, ultrasonic cleaner (model: 008; unit size: 176 x 108 x 130 mm; tank size: 150 x 85 x 63 mm; volt: AC 220-240 V 50 Hz; ultrasonic power: 50 W; frekuensi: 40 kHz); timer: 0-30 min; full capacity: 800 mL), magnetic stirer, syringe pump and high performance liquid chromatography (Shimtzu Coorporation, Kyoto Japan) consisting of prominence degassing unit (DGU-20A5R), prominence liquid chromatography (LC-20AD), prominence comunication bus module (CBM-20A), prominence UV/VIS detector (SPD-20AV), column Lichrospher® 100 RP-8, 10 μm, 250 mm x 4 mm i.d.

2.2Procedure

2.2.1 Preparation of standard solutions

The diazinon standard was prepared by dilution process using acetonitrile solvent until the standard solution of 25 mgL^{-1} was obtained. The chlorantraniliprole standard was prepared by dilution using acetonitrile solvent until a chlorantraniliprole solution was obtained at a concentration of 5 mgL^{-1}. The standardized was 25 mgL^{-1} standard diazinon and 5 mgL^{-1} standard chlorantraniliprole.

2.2.2. Optimization
This method was optimized under condition of composition of the mobile phase of acetonitrile/water from 50-80% for isocratic elution system, flow rate (0.2; 0.4; 0.6; 0.8; 1.0 mL/min), and wavelenght UV detector (220-260 nm).

2.2.3. Sample preparation
The soil sample was taken from the apple farm area of Bumiaji Sub-district, Batu City. The soil was taken at 0-10 cm deep, sieved and stored in airtight plastic at ± 4^0C until the soil sample would be analyzed.The soil sample weighed 20 g and 60 ml of acetonitrile was added to the soil. The solution was put in a 100 ml beaker glass. It was stirred for 1 hour and sonicated for 2 minutes. The extract was filtered using 0.2 μm filter paper.

2.2.4. Optimization of LC-UV method
Each standard solution of prepared pesticide was taken as ± 1 mL using a 1 cc syringe. It was subsequently injected at HPLC with the condition of separation HPLC instrument as follow:

Temperature	: room temperature
Column	: Lichrospher® 100 RP-8, 10 μm, 250 mm x 4 mm i.d
Flow rate	: 0.4 mL/min
Detector	: UV 240 nm
Mobile phase	: acetonitrile:water (80:20), (70:30), (60:40), (50:50)

The optimum eluent composition of the separation result was determined by the values of R, N, k ', and α of the formed chromatograms.

Determination of optimum flow rate was done with separation condition of HPLC instrument, with selected eluent composition (optimum condition), with variation of flow rate (0.2; 0.4; 0.6; 0.8; 1.0 mL/min). The optimum flow rate of the separation results was determined by the values of Rs, N, k ', and α of the formed chromatograms.

Determination of the optimum wavelength was performed by the separation condition of the HPLC instrument, with eluent composition and the selected flow rate (optimum condition), with the variation of wavelength (220, 230, 240, 250, and 260 nm). The optimum wavelength of the separation result is determined by the formed chromatogram.

For qualitative and quantitative analysis, the dilute extract soil was injected in LC-UV system under optimum mobile phase of acetonitrile:water (60:40), optimum flow rate (0.4 mL/min), and maximum wavelength (220 nm). Qualitative analysis was done by comparing the retention time of standard MET with the retention time of peaks form Myrmeleon sp. Chromatogram and by spiking method, where the diazinon and chlorantraniliprole may be determined by increasing peak area by standard diazinon and chlorantraniliprole injection. Spiking method was conducted by injecting 20 μL of diazinon standard and 40 μL of chlorantraniliprole standard into 2 mL of soil extract followed by homogenization with sonication for 300 seconds, and then filtration through a 0.2 μm filter paper before injection into the LC-UV system. The presence of diazinon and chlorantraniliprole were detected by the peak with higher intensity (area) compared to chromatogram peak from the un-spiked sample. Quantitative analysis of diazinon and chlorantraniliprole in the soil extract were done using spiking. The concentration of diazinon and chlorantraniliprole were determined by comparing the peak area of diazinon and chlorantraniliprole before and after spiking with a known concentration of diazinon and chlorantraniliprole standard solution.

3. Results and Discussion

3.1 Optimization of LC-UV method
The chromatogram of diazinon and chlorantraniliprole separation resulted in four variations of acetonitrile:water composition can be seen in Figure 1. Figure 1 showed that the more acetonitrile solvent used, the faster the retention time of the compound. This was caused by nonpolar columns and

polar solvents so that the more non-polar compounds would be retained longer in the column. At 60:40 of mobile phase compositions, both compounds separated well at the retention time of each compound, chlorantraniliprole 9.8 and diazinon 18.64 and a short retention time of fewer than 20 minutes.

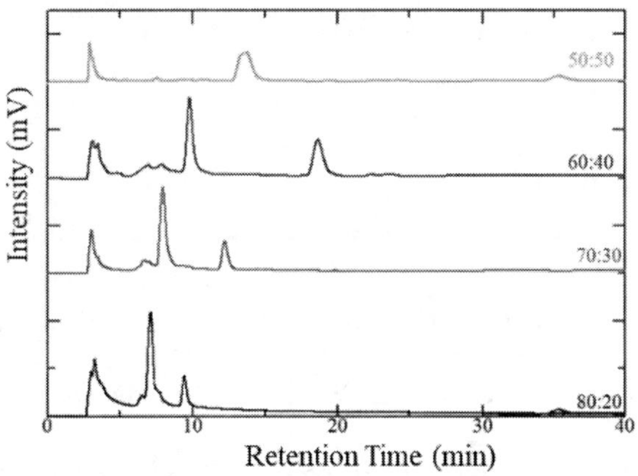

Figure 1. Chromatogram Standard of Chlorantraniliprole (1) dan Diazinon (2) (Separation conditions: Lichrospher® 100 RP-8, 10 μm, 250 mm x 4 mm i.d; V.sampel 2 μL; Flow rate: 0.4 mL/min; λ = 240 nm)

Figure 2 showed the effect of the flow rate on the retention time of diazinon and chlorantraniliprole compounds.

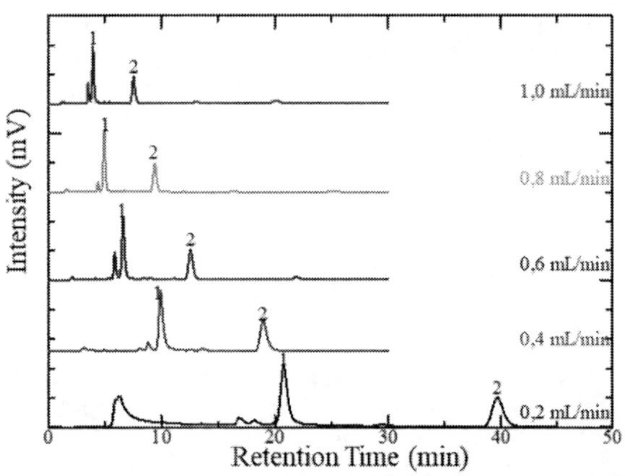

Figure 2. Chromatogram Standard of Chlorantraniliprole (1) dan Diazinon (2) (Separation conditions: Lichrospher® 100 RP-8, 10 μm, 250 mm x 4 mm i.d; V.sample 2 μL; Composition of Acetonitrile:Water (60:40); λ = 240 nm)

Figure 2 showed that faster the flow rate caused smaller standard peak area. So it may be determined the optimum flow rate conditions for the separation of diazinon and chlorantraniliprole was 0.4 mL/min. Where the fast analysis time is below 20 minutes and the large peak compound area was 550134 for chlorantraniliprole and 427586 for diazinone.

Figure 3 showed the influence of wavelength on peak intensity diazinon and chlorantraniliprole.

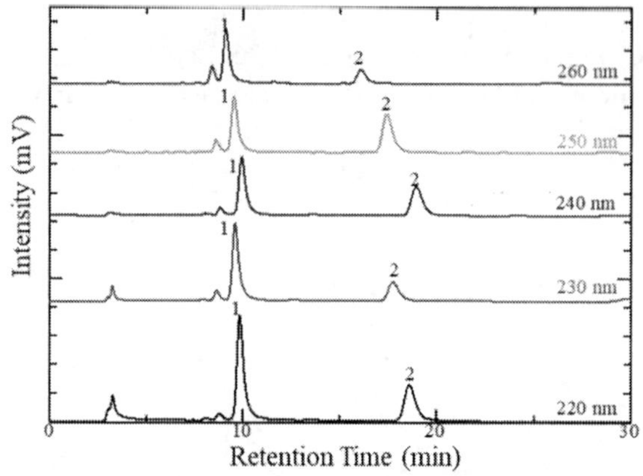

Figure 3. Chromatogram Standard of Chlorantraniliprole (1) dan Diazinon (2) (Separation Condition: Column Lichrospher® 100 RP-8, 10 µm, 250 mm x 4 mm i.d; V.sample 2 µL; Composition of Acetonitrile:Water (60:40); Flow rate: 0.4 mL / min)

From Figure 3 may be seen that the optimum wavelength for diazinon as well as chlorantraniliprole was at 220 nm of wavelength because it exhibited high sensitivity compared to other wavelengths so that both components (diazinon and chlorantraniliprole) may be detected.

3.2 Validation method

Validation method was done in optimum condition. Standard curve diazinon and chlorantraniliprole may be seen in Figure 4. Method validation obtained diazinon linearity was in the range from 1-25 mgL^{-1} with R^2 of 0.9976, 1.19 mgL^{-1} LOD, and 3.98 mgL^{-1} LOQ; whilst the linearity of chlorantraniliprole was in the range from 0.2-5 mgL^{-1} with R^2 of 0.9972, 0.39 mgL^{-1} LOD, and 1.29 mgL^{-1} LOQ. When the method was applied to the soil sample, both pesticides showed acceptable recoveries for real sample of more than 85% (Table 1) thus, the developed method met the validation requirement.

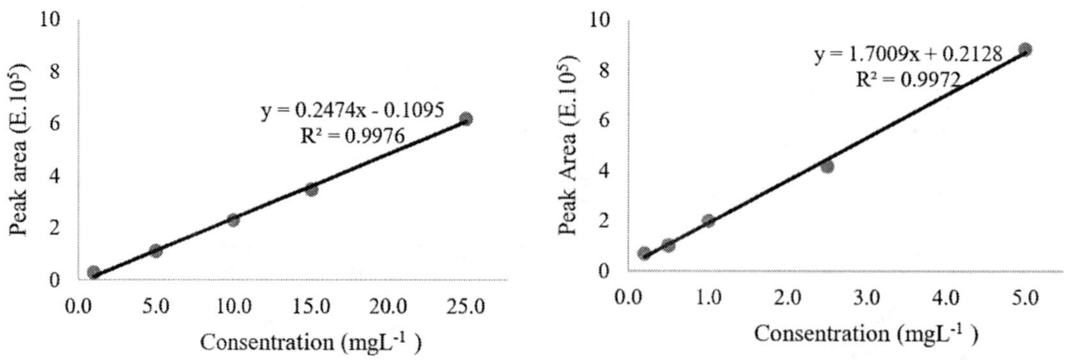

(a) (b)

Figure 4. Standard Curve Diazinon (a) and Chlorantraniliprole (b)

3.3Analysis of pesticide

Qualitative analysis of diazinon and chlorantraniliprole in soil samples using spiking method was by adding standard diazinon and chlorantraniliprole solutions which were known to concentrate into soil samples. Figure 5 showed that there is no peak increase at the same retention time was the standard retention time is dispersed into the soil sample extract. Whereas after a quantitative analysis of the diazinon and chlorantraniliprole in soil samples, the concentrations of both pesticides in the soil samples were below the LOD and LOQ (0.577 mgL^{-1} for diazinon and 0.007 mgL^{-1} for chlorantraniliprole). Therefore, it can be concluded that the soil samples used in this study have neither diazinon nor chlorantraniliprole.

Table 1. % Recovery of Pesticides in Soil (n = 3)

Compound	Soil sample	Concentration added (mgL^{-1})	Concentration measured (mgL^{-1})	% *recovery*
Diazinon	A	0	0,17	-
	B	5	0,19	87,73 %
Chlorantraniliprole	A	0	0,06	-
	B	1	0,07	95,65 %

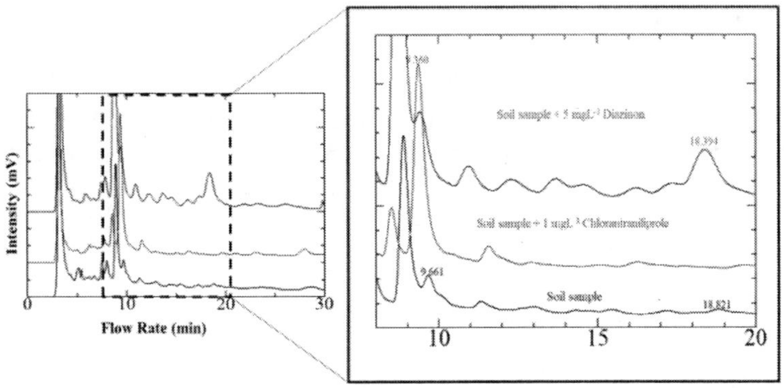

Line	tR (min)	Area
Black	9.661	12788
	18.821	3786
Red	9.360	230705
Blue	18.394	125170

Figure 5. The Chromatogram of The Soil Sample, The Soil Sample Spiked with 5 mgL^{-1} Diazinon, and The Soil Sample Spiked with 1 mgL^{-1} Chlorantraniliprole

4. Conclusion

Optimum conditions for the separation of two pesticides were eluent composition of acetonitrile:water ratio of 60:40, 0.4 mL/min of flow rate, and 220 nm of wavelength. Under the optimum conditions, diazinon linearity was is between 1 and 25 mgL^{-1} with R^2 of 0.9976, 1.19 mgL^{-1} LOD, and 3.98 mgL^{-1} LOQ; whilst the linearity of chlorantraniliprole was between 0.2 and 5 mgL^{-1} with R^2 of 0.9972, 0.39 mgL^{-1} LOD, and 1.29 mgL^{-1} LOQ. When the method is applied to the soil sample, both pesticides show acceptable recoveries for real sample of more than 85%: thus, the developed method meets the validation requirement. The concentrations of both pesticides in the soil samples are below the LOD and LOQ (0.577 mgL^{-1} for diazinon and 0.007 mgL^{-1} for chlorantraniliprole). Therefore, it can be concluded that the soil samples used in this study have neither diazinon nor chlorantraniliprole.

References

[1] P. Tette, L. Guidi, M. Gloria, and C. Fernandes (2016), Pesticides in Honey: A Review on Chromatographic Analytical Methods, *Talanta*, **149**, 124–141.

[2] C. Sanchez-Brunete, B. Albero, and J. Tadeo (2004), Multiresidue Determination of Pesticides in Soil by Gas Chromatography-Mass Spectrometry Detection, *J. Agric. Food Chem.*, **52**, 1445–1451.

[3] Indrianingsih, A., Nisa, K., Wahono, S. (2007), Analisis Residu Pestisida dalam Tanah dan Umbi Bawang Merah di Lahan Pasir Sanden, Bantul, Yogyakarta dengan Kromatografi Gas, di Seminar Nasional Rekayasa Kimia Dan Proses 2007, 181–185.

[4] E. Hartanti, F. Mahatmanti, and E. Susatyo, (2012), Sintesis Kitosan_Bentonit serta Aplikasinya Sebagai Penurun Kadar Insektisida Jenis Diazinon," *Indones. J. Chem. Sci.*, **1**, 111–115.

[5] Cordova, D., Benner, E. A., Sacher, M. D., Rauh, J. J., Sopa, J. S., Lahm, G. P., Tao, Y.

(2006), Anthranilic Diamides: A New Class of Insecticides with A Novel Mode of Action, Ryanodine Receptor Activation, *Pestic. Biochem. Physiol.*, **84**, 196–214.

[6] Lahm, George P., Selby, Thomas P., Freudenberger, John H., Stevenson, Thomas M., Myers, Brian J., Seburyamo, G., Smith, Ben K., Flexner, L., Clark, Christopher E., Cordova, D. (2005), Insecticidal Anthranilic Diamides: A New Class of Potent Ryanodine Receptor Activators," *Bioorg. Med. Chem. Lett.*, **15**, 4898–4906.

[7] P. Xu, Y. Ren, Z. Zhou, A. Liu, and H. Zhang (2010), Determination of Chlorantraniliprole in Vegetables, Fruits, and Grains by SPE, *Chromatographia*, **72**, 763–766.

[8] C. Lesueur, M. Gartner, A. Mentler, and M. Fuerhacker, (2008), Comparison of Four Extraction Methods for The Analysis of 24 Pesticides in Soil Samples with Gas Chromatography-Mass Spectrometry and Liquid Chromatography – Ion Trap – Mass Spectrometry, *Talanta*, **75**, 284–293,.

[9] Moldoveanu, S., David, V., (2013), Essentials In Modern HPLC Separations, USA: Elsevier.

[10] Rouessac, F. and Rouessac, A., (2007), Chemical Analysis Modern Instrumentation Methods and Thecniques, Edisi kedua, Prancis.

[11] Nurhamidah, (2005), Penentuan Kondisi Optimum HPLC untuk Pemisahan Residu Pestisida Imidakloprid. *Ilmu-Ilmu Pertanian Indonesia*, **7(2)**, 87–93.

[12] M. E. Sánchez, R. Méndez, X. Gómez, and J. Martín-Villacorta, (2003), Journal of Liquid Chromatography & Related Determination of Diazinon and Fenitrothion in Environmental Water and Soil Samples by HPLC, *Liq. Chromatogr. Relat. Technol.*, **26**, 483–497.

Molecularly Imprinted Polymers Chitosan-Glutaraldehyde for Monosodium Glutamate

Ani Mulyasuryani,[1] Edi Haryanto,[1] Hermin Sulistyarti,[1] Barlah Rumhayati[1]

Analytical Chemistry Laboratory of Chemistry Department, Brawijaya University, Jl.Veteran, Malang, Indonesia

Email: mulyasuryani@ub.ac.id

Abstract. Chitosan has been used as a functional monomer in the synthesis of molecularly imprinted polymers (MIP) for monosodium glutamate (MSG). MIP is made from a mixture of 5 g chitosan, 50 mg glutaraldehyde and 2 g MSG, MIP is formed as flakes and beads. MIPs are identified by the FTIR spectrum, SEM image and their adsorption capabilities. MIP flakes and beads have no structural differences if they are based on FTIR or SEM spectra, but MIP adsorption capacity of beads higher than flakes. Adsorption capacity of MIP flakes is 548 mg/g and MIP beads 627 mg/g.

1. Introduction

MIP is a technique for designing an artificial receptor that is a hollow polymer or molecular trace due to the disposal of mold molecules. Molecular Imprinting Technology (MIT) MIT is based on the formation of a complex between an analyte (template) and a functional monomer. In the presence of a large excess of a cross-linking agent, a three-dimensional polymer network is formed. After polymerization process, the template is removed from the polymer leaving specific recognition sites complementary in shape, size and chemical functionality to the template molecule. Usually, intermolecular interactions like hydrogen bonds, dipole–dipole and ionic interactions between the template molecule and functional groups present in the polymer matrix drive the molecular recognition phenomena. Thus, the resultant polymer recognizes and binds selectively only the template molecules [1].

MIP has a very high selectivity that can be applied to the extraction (separation) and the development of sensors. Solid Phase Extraction (SPE) is another important area of application of MIPs in analytical chemistry [2-7]. MIP for Solid-phase extraction (MISPE) has been applied both in on-line and off-line procedures. MIPs particles, used as selective sorbent materials, can be packed in an HPLC pre-column for the on-line mode and in a cartridge between two frits for the off-line mode [8].

One of the potentially functional materials is chitosan. Chitosan, a natural muco-polysaccharide with similar structure characteristics to cellulose, has been considered as one of thermo promising materials due to its biodegradability, biocompatible and non-toxicity [8–11]. Being a natural chiral compound, chitosan is also a multifunctional polymer containing large number of amine groups together with hydroxyl groups capable of assembling with template molecules such as amino acids through forming hydrogen bonds [12–14]. However, the poor physical/mechanical properties and high degree of swelling of chitosan in aqueous system limits its practical application in MIPs preparations. One way to overcome this limitation is the chemical cross-linking. Many of bi-functional reagents are frequently used as cross-linking agents such as glutaraldehyde [10], epichlorohydrin [15], sulfuric acid [16] and glyoxal [17].

In this study we described the preparation of new enantio-selective chitosan/glutaraldehyde molecularly imprinted resin in aqueous system with monosodium glutamate as a template molecule. The microscopic structure and the surface area of the resin were characterized. The method of determining MSG is done indirectly as ammonium (NH_4^+) by spectrophotometric .

2. Methods

2.1 Preparation of molecularly imprinted [19]

The molecularly imprinted MSG chitosan/glutaraldehyde was prepared as follows: 10 g of the previously prepared chitosan was dissolved in 100ml of 2% acetic acid aqueous solution at 60 ◦C. After removal of the insoluble impurities by filtration, a 50ml of 0.5M MSG were added into the chitosan solution. The mixture was stirred for 2 h, and then 25ml of glutaraldehyde solution was added to the above solution and mechanically stirred for another 1 h. At the end of this period the resulted gelatinous resin was collected, washed with 0.01M HCl and dried in an oven at 60°C for 24 h. Then, the dry resin was ground in a mortar and sieved to get particles with a size about 120 mesh. To prepare the desired adsorbent for MSG separation from dilute aqueous solutions, the imprint template MSG molecules were leached out by stirring 3 g of MIP particles in 1000 ml of 0.1M HCl solution for 24 h. This step was repeated for three times to assure maximum extraction of imprint MSG molecules which may cross-link with the chitosan based network during the reaction to avoid its leaching in subsequent adsorption and extraction experiments.

2.2 Adsorption and extraction steps

In adsorption step, 0.1 g of imprinted polymer particles were added into a 25 ml aqueous solution containing 7.5 mg of MSG, the pH was adjusted to 6 ; 7 ; 8 value and stirred for 90 minutes. The solution was filtered and remaining particles on the filter were washed with a 50 ml aqueous solution at the same pH to remove the solution trapped within the polymer particles and non-specifically adsorbed solute. Extraction step was performed by replacing the washed polymer particles into a 100ml HCl (0.1–0.001M) solution with an adjusted pH, stirred for 90 minutes, at 150 rpm

3. Result and Discussion

MIPs are made by the same method but differ in their form, MIP A is a flake while MIP B is a form of beads. Structurally MIP A and B have no chemical structure difference, this can be seen from the FT-IR spectrum in Figure 1.

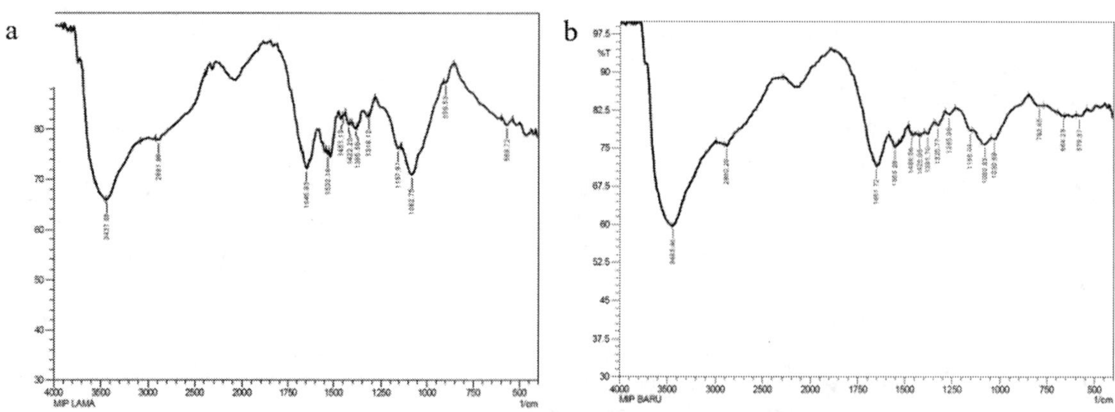

Figure 1. FT-IR MIP flakes (a) and MIP beads (b)

Based on the morphology of microscopic scanning results both MIPs also have no significant differences. This is evident from the SEM image in Figure 2. Therefore, MIP surface morphology

IOP Conf. Series: Materials Science and Engineering **299** (2018) 012010 doi:10.1088/1757-899X/299/1/012010

testing adsorptive MSG is only performed on MIP B alone, as shown in Figure 4.2, (c) and (d). From the SEM results it can be seen that the adsorption of MIP to MSG has certain patterns according to the shape of the MSG molecule.

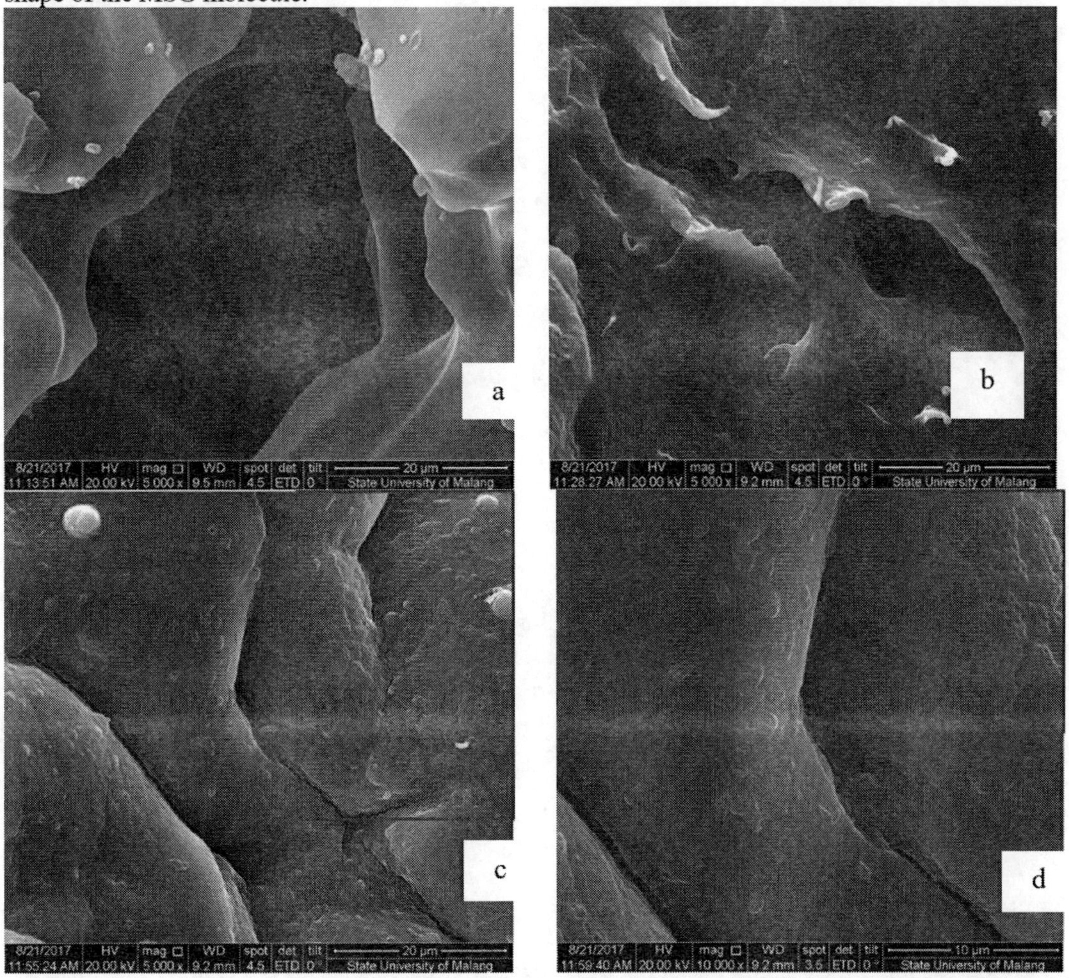

Figure 2. MIP morphology of Scanning Electron Microscopic (SEM) images for MIP flakes (a); MIP beads (b) and MIP beads that adsorb MSG (c) and (d).

The determination of the optimum contact time on MSG adsorption in MIP was done without pH adjusting, the contact time are 30, 60, 90 and 120 minutes. The purpose of this experiment is to determine the time required by MIP and MSG to reach adsorption equilibrium, it is the optimum time of MSG adsorption by MIP.

IOP Conf. Series: Materials Science and Engineering **299** (2018) 012010 doi:10.1088/1757-899X/299/1/012010

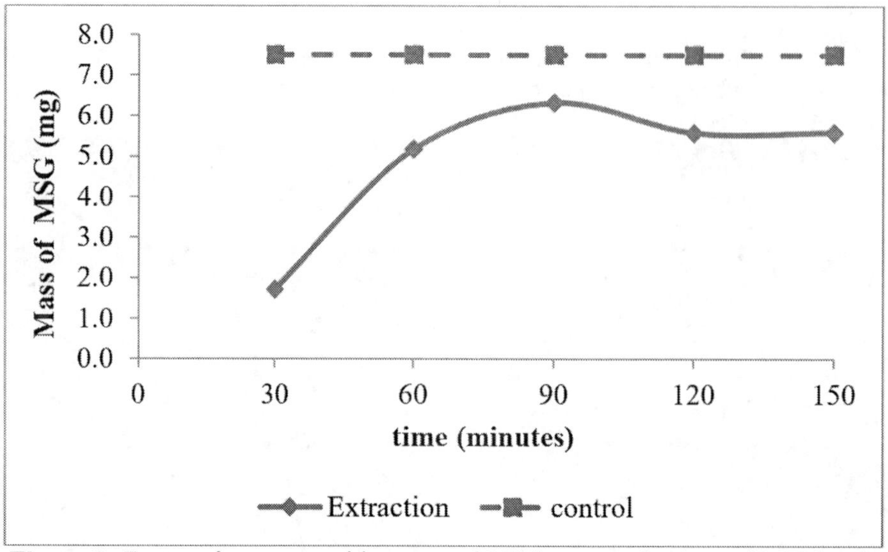

Figure 3. Contact time curve with extracted MSG amount (mg), used MIP beads

Based on the results shown in Figure 3, it is known that an increase in the number of MSG molecules adsorbed by MIP ranges from 60 to 90 minutes. This is due to the increasing diffusion of MSG molecules into the pores of MIP. The adsorption of MSG in the range of 90 to 120 minutes tends to decrease. This is caused by the desorption velocity is greater than the adsorption speed so that there is a decrease in the adsorption of MSG. In this experiment the equilibrium occurred at a contact time of 90 minutes with a 100% adsorption, equilibrium reach when the adsorption rate was equal to desorption.

The pH of the solution is an important factor in determining the optimum conditions of adsorption. The degree of acidity affects the solubility of the compound as well as the active groups present in MIP. So an experiment was conducted to determine the optimum condition of MSG molecule adsorption by MIP of chitosan-glutaraldehyde carried out at pH 6, 7 and 8 using MIP type A, B and C. MIP type C is MIP beads whose contain of glutaraldehyde more.

Based on the results shown curve in Figure 4, it is known that both MIP A and B result in significant adsorption increase from pH 6 to pH 8. This indicates that the pH effect on MSG adsorption by both types of MIPs. Based on the results it is known that the optimum pH of MSG adsorption by both types of MIP is pH 8 with adsorption percentage of 89.91% (flakes) and 100% (beads). At a pH greater than pK_a, of 4.25 to 9.10 monosodium glutamate will be negatively charged while chitosan as a functional monomer tends to be positively charged, so monosodium glutamate will tend to be absorbed in positively charged chitosan pores.

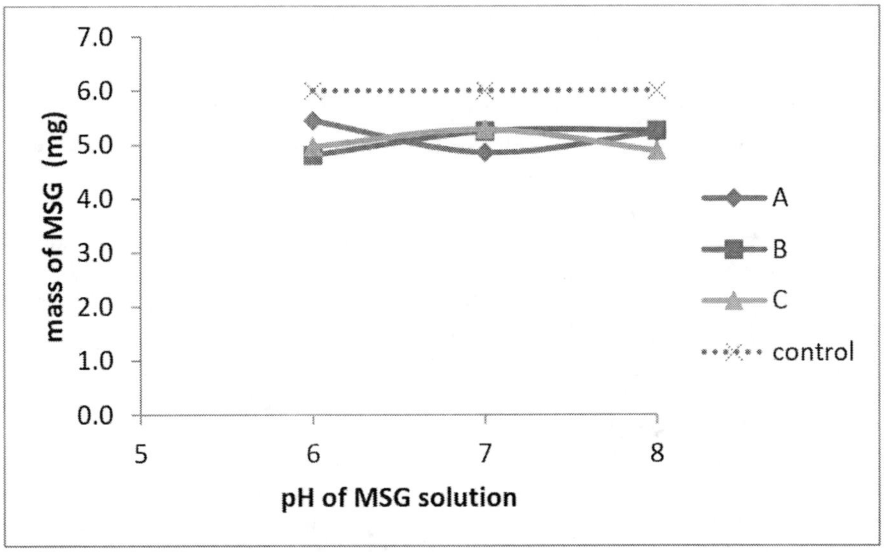

Figure 4. The curve of the relationship between pH and MSG mass is extracted by MIP types A, B and C

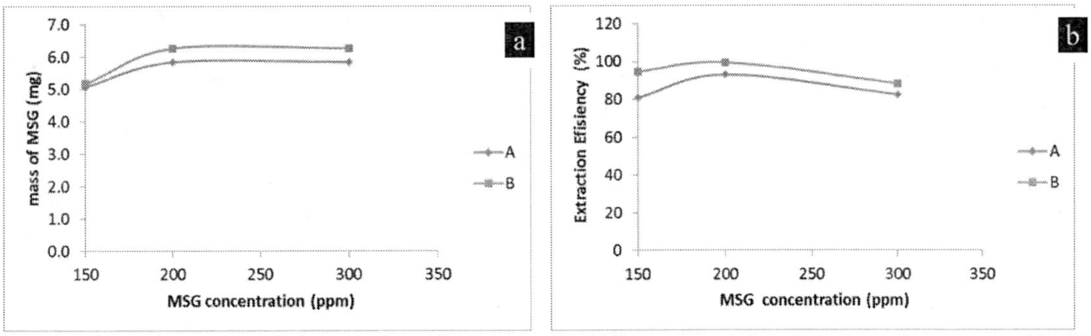

Figure 5. The curve of the relationship between MSG concentration (ppm) to the extracted MSG mass (a) and % E (b).

The determination of MIP adsorption capacity on MSG is done by studying the effect of MSG concentration. This is done on the optimum conditions of adsorption at pH 8 and contact time 90 minutes. The adsorption capacity is expressed in mg of MSG per g MIP. For this, extraction for some MSG concentrations was then calculated by the extracted MSG mass or% E. In Figure 5 there was shown a relationship curve between the MSG concentrations to the MSG mass of the extraction. The extracted MSG mass would have remained if the adsorption capacity had been achieved, although MSG concentrations were increased and vice versa for % E, would decrease with increasing MSG concentrations. Based on Figure 5, the adsorption capacity can be determined that is 585 mg / g for MIP flakes and 627 mg / g for MIP beads. Thus MIP beads have a higher adsorption capacity than MIP flakes.

4. Conclusion

Selective MIP of MSG can be made from chitosan with glutaraldehyde as a crosslinker. Differences in the making of MIP textually ie flakes and beads do not produce the chemical structure difference based on FT-IR and SEM images. MIP flake adsorption capacity of 585 mg / g and MIP beads 627 mg/g at adsorption time 90 minutes at pH 8.

Acknowledgments
The research is funded by DPP/SPP FMIPA Brawijaya University 2017

Refferences

1. Ramstrom, O., Mosbach, K., 1999, *Curr. Opin. Chem. Biol.*, **3**, 759–764.
2. Tamayo, F.G., Turiel, E., Martín-Esteban, A., 2007, *J. Chromatogr. A* , **1152,** 32–40.
3. Lasáková, M., Jandera, P., 2009, *J. Sep. Sci.*, **32**, 788–812.
4. Pichon, V., 2007, *J. Chromatogr. A* , **1152**, 41–53.
5. Caro, E., Marce, R.M., Borrull, F., Cormack, P.A.G., Sherrington, D.C., 2006, *Trends Anal. Chem.*, **25,** 143–154.
6. He, C., Long, Y., Pan, J., Li, K., Liu, F., 2007, *J. Biochem. Biophys. Meth.*, **70,** 133–150.
7. Turiel, E., Martin-Esteban, A., 2010, *Anal. Chim. Acta*, **668**, 87–99.
8. Pichon, V., Chapuis-Hugon, F, 2008, *Anal. Chim. Acta*, **622**, 48–61.
9. Lee, H.S., Hong, J., 2000, *Chromatogr. A*, **868**, 189–196.
10. Kumar, M.N.V.R., Muzzarelli, R.A.A., Muzzarelli, C., Sashiwa, H., Domb, A.J., 2004, *Chem. Rev.*, **104**, 6017–6084.
11. Jameela, S.R., Jayakrishnan, A., 1995, *Biomaterials*, **16** , 769–775.
12. Kim, J.H., Kim, J.H., Jegal, J., Lee, K.H., 2003, *J. Membr. Sci.* **213**, 273–283.
13. Aburto, J., Borgne, S.L., 2004, *Macromolecules*, **37,** 2938–2943.
14. Xia, Y.Q., Guo, T.Y., Song, M.D., Zhang, B.H., Zhang, B.L., *Biomacromolecules* , **6,** 2601–2606
15. Su, H., Wang, Z.X., Tan, T.W., 2005, *J. Chem. Technol. Biotechnol.*, **80**, 439–444.
16. Tan, T.W., He, X.J., Du, W.X., 2001, *J. Chem. Technol. Biotechnol.*, **76**, 191–195.
17. Mukoma, P., Jooste, B.R., Vosloo, H.C.M., 2004, *J. Power Sources* , **136,** 16–23.
18. Guinesi, L.S., Cavalheiro, E.T.G., 2006, *Carbohydr. Polym.*, **65**, 557–561.
19. Monier, M.,. El-Sokkary, A.M.A., 2010, *Inter. J. Biol. Macromolecules*, **47,** 207–213.

The Development of Analytical Method for the Determination of Azelaic Acid Content in Cosmetic Cream Products

E Lusianti[1,2], R Wibowo[1] and S Hudiyono[1]

[1]Department of Chemistry, Faculty of Mathematics and Natural Sciences, Universitas Indonesia, Depok, Indonesia

[2]National Quality Control Laboratory of Drug and Food, National Agency of Drug and Food Control Indonesia, Jakarta, Indonesia

E-mail: sumi.hudiyono@sci.ui.ac.id

Abstract. Azelaic acid is one of the substances that has anti-acne and skin lightening effects which is often added to cosmetics. In the acne treatment, the azelaic acid is generally used with a concentration of 20% in cream formulation and 15% in gel. The use at concentrations below 10% is not recommended because it does not work effectively. While the use of above 10% is categorized as a medical treatment. In Indonesia, the Head of the National Agency of Drug and Food Control (BPOM) has issued Regulation No. 18 of 2015 on the Technical Requirements of Cosmetics Ingredients Annex V stating that the azelaic acid is banned in cosmetics. However, until this research began the BPOM has not had a valid method to identify it in cosmetics. Consequently, surveillance of such ingredient in products is hard to do. In this research, the fatty acid standard analysis method of AOAC International was modified and validated to be used in the laboratory. The method of analysis involves heating the cream preparations dissolved with methanol and then added BF_3-methanol catalyst, followed by extraction and analysis using GCMS. The validation of method shows that the calibration curve is linear with correlative value of 0.9997. The method is fairly sensitive with 0.02% detection limit, and fairly precision with relative standard deviation (RSD) of between 0.626-0.961% and fairly accurate which the recovery percentage is 99.85% at range 98.27-100.72%. In sum the results demonstrate that the method can be used as a routine analysis method for laboratory testing.

Key words: Azelaic acid, Dimethyl azelate, Derivatization, BF_3-Methanol, GCMS, Validation of analytical methods.

1. Introduction

Azelaic acid (1,7-heptane-dicarboxylic acid) with the molecular formula of $C_9H_{16}O_4$ is a saturated dicarboxylic acid. It is naturally present in yeasts Pityrosporum ovale, rice grains, milk and castor oil. Figure 1 shows the structure of azelaic acid. Azelaic acid is used for the local treatment of various types of cystic acne, including inflammation caused by acne, as an alternative to topical treatment with retinoic, benzoyl peroxide and erythromycin.

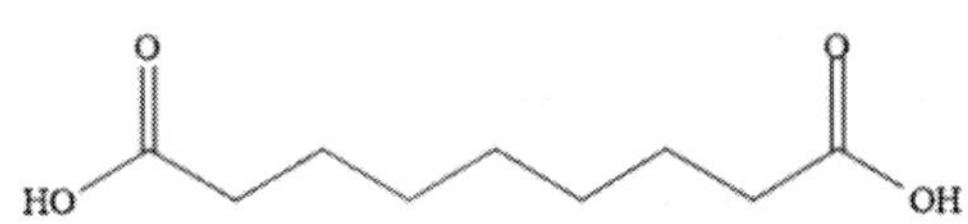

Figure 1. The structure of azelaic acid

Determination of dicarboxylic acid posesses a great challenge because of the absence of chromophores and fluorophore in this compound [1]. This compound also has a low volatility, so it needs derivatization before it is analyzed by gas chromatography [2].

This paper describes the development of a method to analyze azelaic acid in cosmetic products by means of derivatization through boron trifluoride (BF_3) in methanol esterification and the results were analyzed by gas chromatography mass spectrometry (GCMS). Derivatization methods with these reagents are easy and inexpensive to form esters which most often used to form methyl esters by reaction with an acid [3].

2. Experimental

2.1. Chemicals and samples

Azelaic acid, purity > 99% of Sigma Aldrich is used as standard. Methyl docosanoate, purity > 99% of Sigma Aldrich is used as an internal standard to minimize the possibility of deviation during the injection process into gas chromatography. 10% boron trifluoride in methanol (BF_3-methanol) solution from Sigma Aldrich is used as a degrading reagent for analysis with gas chromatography. LC grade methanol from Merck is used as a solvent to prepare standard solutions and test solutions. N-Hexane is used in extraction. Sodium chloride is used to produce saturated solutions which also used in the extraction. Sodium sulfate anhydrate pro-analysis degrees from Merck is used to pull water from the extract solutions. 90 mm diameter filter paper from Whatman is used to filter the solutions. Anti-acne cream sample which does not contain azelaic acid is self-made in the laboratory by following cosmetic cream-making procedures. The formulations contain compositions commonly present in cosmetic samples such as emulsifiers, refiners, moisturizers, pH regulators and solvents.

2.2. Test Method

2.1.1. Preparation of Standard Solution and Test Solution. The derivatization is carried out by following the fatty acid and oil derivatization methods listed in AOAC [4] with some adjustment of pipette 2 ml of the solution and transferred into the threaded cap erlenmeyer, 4 mL of BF_3-methanol 10% is added into the solution, afterward the solution is heated over the water bath at 60 ^{o}C for 10 minutes. The solution is cooled until the room temperature, then is moved into separating funnel. 10 mL of saturated NaCl is added into the solution and shaken for 15 seconds. 10 mL n-Hexane is added into the solution, then followed by extraction for 1 min. Next the n-Hexane layer is taken. Hereinafter the solution is strained through the filter paper that has been added sodium sulfate anhydrate. The filtrate is then taken as much as 1 mL. The filtrate is inserted into the vial auto sampler and 0.5 mL of internal standard solution is added, followed by homogenization and analysis with GCMS.

2.1.2. Instrument. Analysis GCMS is carried out on GCMS-QP 2010 plus system (Shimadzu) with AOAC 2000 autosampler. The GCMS experimental conditions for azelaic acid content as shown in Table 1.

Table 1. GCMS Experimental Conditions

Column	VF-wax MS (30 m, 0.25mm i.d., 0.25 μm film thickness)
Injector temperature	250 ^{o}C
Carrier gas	Helium
Carrier gas flow	1 mL/min
Split ratio	Splitless
Oven programme	150 ^{o}C
	20 ^{o}C/min 205 ^{o}C 2 min
	20 ^{o}C/min 250 ^{o}C 5 min
	Total run time: 14.75 min
MS Parameters:	
Ionization source	EI

Electron energy	70 Ev
Source temperature	180 °C
Interface temperature	250 °C
Fragment scan range	30 m/z – 220 m/z
Solvent cut time	4.01 minutes

2.1.3. Validation Method Parameters. The validation method is done by evaluating specificity, precision, accuracy, linearity, limit of detection and limit of quantitation [5].

3. Results and Discussion

3.1. The Study of BF_3-Methanol Volumes and Reaction Temperatures

The reaction between an azelaic acid (a carboxylic acid) with methanol (alcohol) requires BF_3 as an acid catalyst which produces dimethyl azelate (ester) is called an esterification reaction, as shown in Figure 2. Esterification involves heating the carboxylic acid with an acid catalyst in an alcohol solvent. The catalyst protonates an oxygen atom of the COOH group, making the acid much more reactive to nucleophiles. An alcohol molecule (CH_3OH) then combines with the protonated acid, to yield the ester product (R-COO-CH_3) with the lost of one water molecule. Esterification is a reversible reaction. To obtain a high yield of the ester, the equilibrium should be shifted towards the ester side. One technique for achieving it is by using one of the reagents excessively.

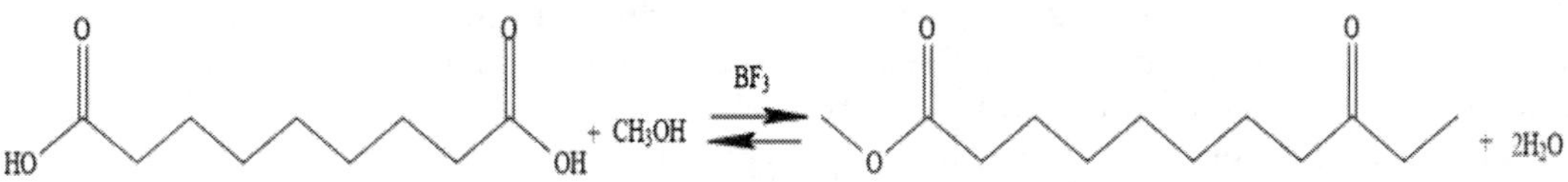

Figure 2. Esterification reaction of azelaic acid

From the equation of the reaction it is apparent that an appropriate amount of BF_3-methanol is required for the right shift reaction and the proper reaction temperature to accelerate the reaction time but not reduce the amount of the present. The result of the analysis is shown in Figure 3.

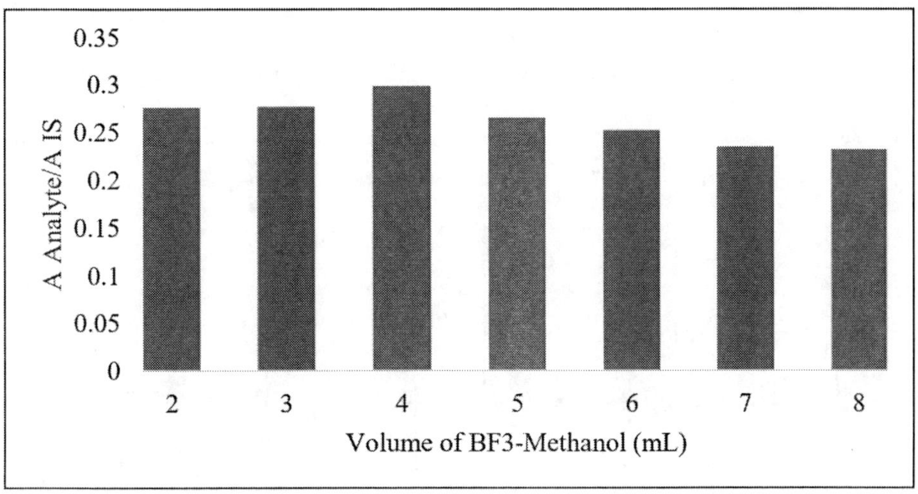

(a)

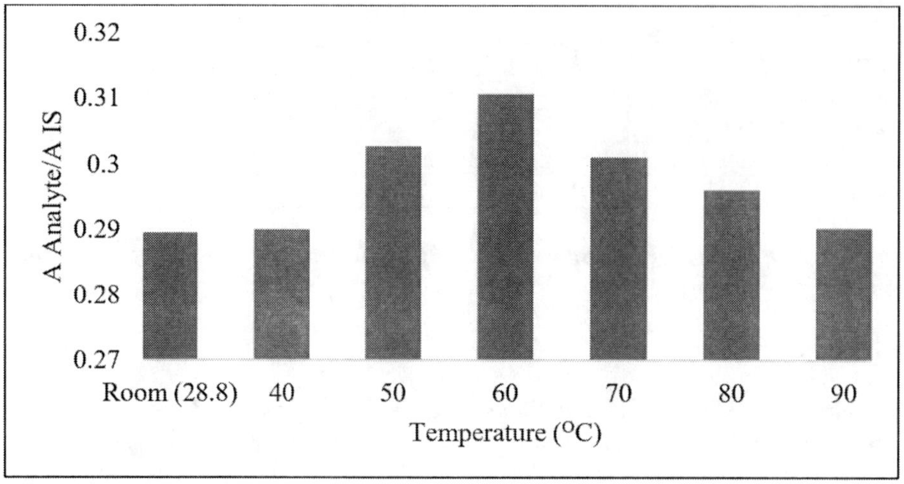

(b)

Figure 3. Effects of volume (a) and temperature (b) on yield of the derivatization results

From Figure 3 it is clearly seen that 4 mL of BF_3-methanol and heating at 60 °C was the optimal condition for this derivatization method.

3.2. Mass Spectral Analysis

According to the analysis performed with GCMS the retention time of dimethyl azelate was in the range 6.0 to 6.3 respectively (Figure 5(b)). The analyte spectrum match to the reference spectrum of NIST. The mass spectrum and reference mass spectrum of the dimethyl azelate are shown in Figure 4. From the mass spectra profile is obtained that the parent peak is $[M+1]^+$ 217. The most abundant ions are in the high mass range, *m/z* 185 ($[M\text{-}CH_3O]^+$), *m/z* 152 ($[M\text{-} 2 \times CH_3O]^+$), *m/z* 143 ($[M\text{-}CH_3OCOCH_2]^+$), *m/z* 124 ($[M\text{-}(CH_3OCO{+}CH_3O{+}2H]^+$), and *m/z* 111 ($[M\text{-}(CH_3OCOCH_2{+}CH_3O{+}H]^+$). Ions equivalent to *m/z* 156 ($[M\text{-}CH_3OCO{+}H]^+$) seem to be characteristic of shorten-chain dibasic esters [6].

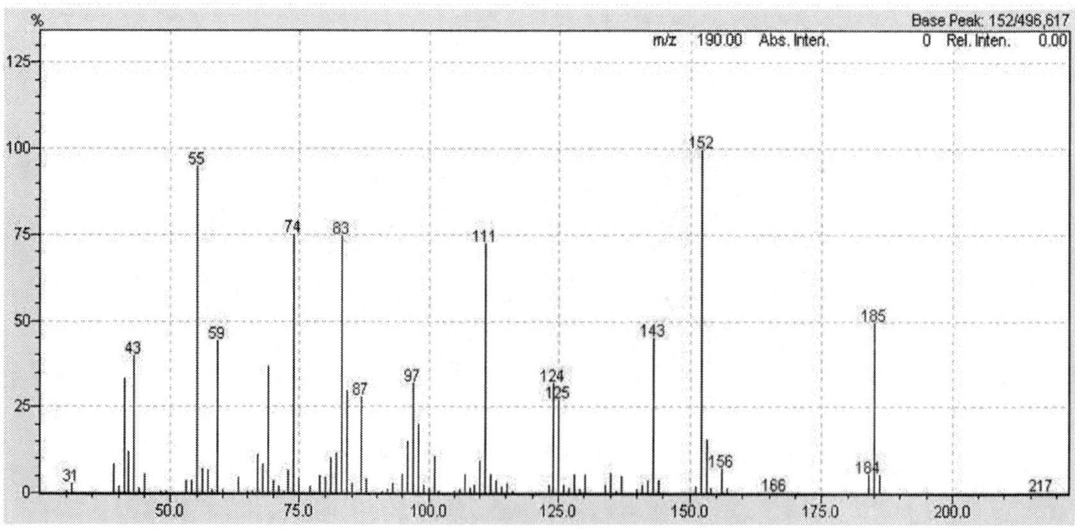

(a)

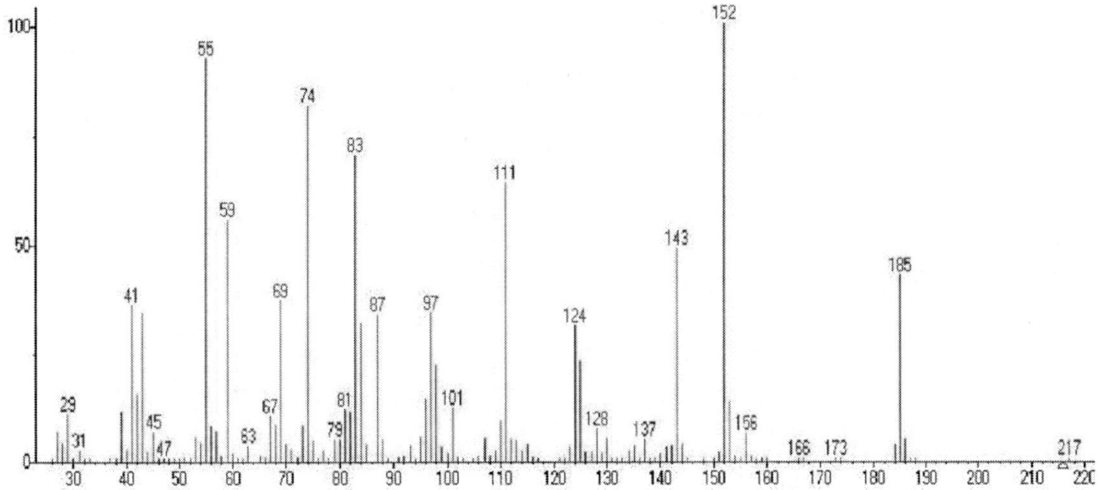

(b)

Figure 4. (a) Mass spectrum of azelaic acid dimethyl ester and (b) Reference mass spectrum of azelaic dimethyl ester in NIST

3.3. Selectivity

Selectivity is the ability of a method to explicitly test the analyte in question by the presence of other or predicted components such as contamination, degradation, and other components that may be present in the sample matrix. The specificity of proposed method is justified by the chromatograms of placebo, azelaic acid with internal standard methyl docosanoate and spiked sample solution under same chromatographic conditions show in Figure 5. The placebo is an anti-acne cream sample without azelaic acid and internal standard methyl docosanoate while the spiked sample is spike of placebo with azelaic acid and internal standard methyl docosanoate. The placebo did not interfere in determination of Azelaic acid in commercial cream. The retention time for azelaic acid was about 6.118 minutes and for internal standard was about 12.718 minutes.

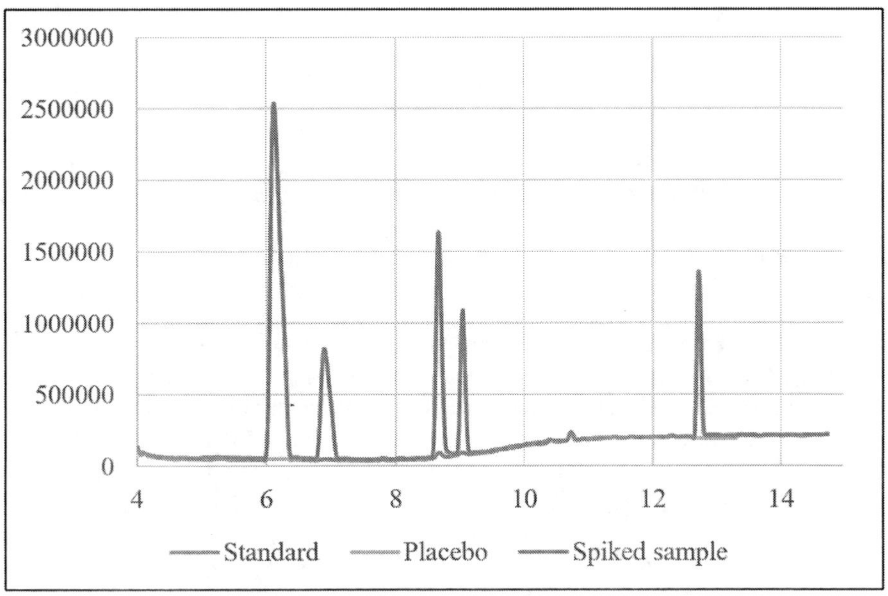

Figure 5. Chromatogram of solutions azelaic acid with internal standard (methyl docosanoate), placebo and spiked sample

3.4. Linearity

To view the level of linearity in the working range of the method according to the range of objective observation under the terms of the contents of analyte in the sample, the determination of linearity is required. This determination is carried out by making a spiked sample solution with at least five concentration levels, each of two replications in the working range of the method. The coefficient value of linear regression variance (Vx0) is calculated by the equation:

$$Vx0 = \frac{Sy/x}{b \,.\, x_{average}}$$

where b is slope and $X_{average}$ is the average content of azelaic acid in spiked sample solution.

The criteria for linearity acceptance is at a condition where the regression value of r ≥ 0.995 and the value of Vx0 < 5%. From the calculation it is obtained that value of r = 0.9997 (linear) for the range of content 63.44 µg/mL – 380.64 µg/mL as in the Figure 6 and Vx0 = 1.6%.

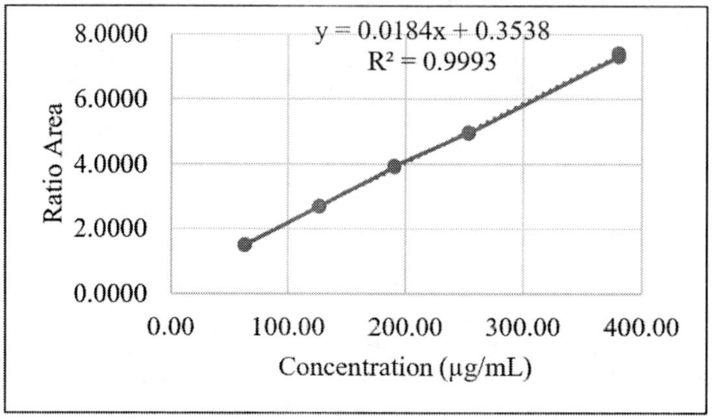

Figure 6. Linearity regression curve

3.5. Accuracy

Determination of accuracy is obtained by measuring the spiked sample solution at three levels of analyte concentration within the working range of the method (calibration curve). Each level consists of three spiked sample solutions.

The accuracy value is calculated based on the recovery value (%R), that is by the formula:

$$\% R = \frac{\text{Measurable content}}{\text{Actual content}} \times 100 \%$$

It is obtained that the recovery value of azelaic acid in spiked sample solutions with a range of 2.5% - 7.5% is 98.27% -100.72%. Based on the requirements listed in [5], it meets the requirements, for which the analyte content in the sample ≥ 1% is 97-103%.

3.6. Precision

Precision determination was done by measuring the spiked sample solution at three levels of analyte concentration within the working range of the method (calibration curve). Repeat 3 times for each spiked sample solution, and calculated standard deviation (SD) and relative standard deviation (% RSD).

The results of this research showed that range of % RSD is 0.122% - 2.220%. Based on the requirements listed in [5], it meets the requirements, for which the analyte contents in the sample ≥ 1% RSD should be < 2.7%.

3.7. Detection Limit (LOD) and Quantitation Limit (LOQ)

The detection limit (LOD) and quantitation limit (LOQ) in this research were determined by using the calibration curve method of spiked sample solution with five concentration levels. The LOD and LOQ values are determined by the formula:

$$\mathrm{LOD} = \frac{3 \cdot S_a}{b} \quad \text{and} \quad \mathrm{LOQ} = \frac{10 \cdot S_a}{b}$$

S_a is standard deviation of intercept and b is slope.

From the research results, the method of analysis produces LOD equal to 0,02% and LOQ equal to 0,07%.

4. Conclusions

The validation of azelaic acid analysis method by derivatization and GCMS analysis developed in this study provides good results. Therefore, this method can be adopted as a routine analysis in the laboratory.

Acknowledgment

The author would like to thank you for the research facility that has been provided by National Quality Control Laboratory of Drug and Food, National Agency of Drug and Food Control Indonesia.

References

[1] M. Garelnabi, D. Litvinov, and S. Parthasarathy 2010 Evaluation of a gas chromatography method for azelaic acid determination in selected biological samples vol. **2**, no. 9

[2] A. Chisvert, J. Sisternes, Á. Balaguer, and A. Salvador 2010 A gas chromatography – mass spectrometric method to determine skin-whitening agents in cosmetic products," *Elsevier*, vol. **81** 530–536

[3] F. Orata, 2007 Derivatization Reactions and Reagents for Gas Chromatography Analysis

[4] G. W. Latimer Jr. 2016 Official Methods of Analysis of AOAC International, 20 th. Maryland: AOAC International

[5] AOAC Official Methods of Analysis 2016 Appendix F: Guidelines for Standard Method Performance Requirements

[6] W. W. Christie 2014 Mass Spectra of Derivatives of Dicarboxylic Fatty Acids," AOCS Lipid Library, Scotland, p. 2

Evaluation of Heavy Metal Exposure to Soil and Paddy Plant around the Closed Municipal Solid Waste Landfill: Case Study at Gunung Tugel Landfill, Banyumas-Central Java

Kasam*), Suphia Rahmawati, Fajri Mulya Iresha, Dhandhun Wacano, Ida Farida Fauziah, Muhammad Afif Amrullah

Environmental Engineering Study Program
Faculty of Civil Engineering and Planning
Universitas Islam Indonesia, Yogyakarta, Indonesia

*) E-mail : kasam@uii.ac.id

Abstract. This work was focused on assessing the exposure of heavy metal from closed municipal solid waste (MSW) landfill on soil and paddy plants. This study aimed to determine heavy metal content whether at the soil in the around Gunung Tugel landfill included and accumulated in the paddy plant tissues. The investigated metals include chromium (Cr), copper (Cu), cadmium (Cd), iron (Fe), and zinc (Zn). The samples were acid-digested before the desired elements were measured using Atomic Absorption Spectrophotometry (AAS). The results are presented as distribution map of the landfill area based on the total heavy metals content distribution in the soil and paddy plants. The samples shown that the concentrations of heavy metals around Gunung Tugel landfill are 6.27-34.71 mg/kg, 0.17-0.42 mg/kg, 28.29-48.69 mg/kg, 18,997.26-32,572.29 mg/kg, 342.74-834.49 mg/kg, 136.10-290.14 mg/kg at the top soil and 0.00-1.70 mg/kg, 0.00-0.26 mg/kg, 0.79-10.46 mg/kg, 13.88-61.46 mg/kg, 18.79-50.56 mg/kg, 87.27-273.22 mg/kg at the paddy for Cr, Cd, Cu, Fe, Mn, and Zn respectively. According to the results, The Gunung Tugel landfill is not a direct source of heavy metal pollution at paddy plant in the landfill area, but through surface water and soil media. Rainfall around landfill is quite high ie more 2000 mm/year of rainfall and soil permeability is 1.0 cm/sec.

Keywords: MSW, Landfill, Heavy metal, Exposure

1. Introduction

In the developing countries of Asia, due to the lack of technical knowledge, economic and infrastructural arrangements, 70-90 % of the solid waste is disposed in open dumps [1]. The MSW in landfill have been generating of leachate which is polluting the environment. In most landfills, leachate is composed of the liquid that has entered the landfill from external sources, such as surface drainage and rainfall and the liquid produced from the decomposition of the wastes, if any [2]. Organic and inorganic pollutants, including heavy metals are the major categories of pollutants [3]. Unlike other pollutants, heavy metals are hazardous, since they cannot be degraded by biological or chemical means leading to harmful effects on ecosystems, including bioaccumulation [4].

Heavy metals are pollutants that are classified as toxic to the environment, even at minute concentrations. Although different sources of heavy metals exist, metals also occur naturally in the environment as ores and few other complexes; hence, several metals play useful roles in the environment, particularly in plant growth [5-7]. However, when heavy metals exist in excessively high concentrations, pollution occurs and can be catastrophic because of its effect on microbes.

The problem of heavy metal pollution is becoming more serious with increasing industrialization and the disturbance of natural biogeochemical cycles. Unlike organic compounds, heavy metals are essentially non-biodegradable and, therefore, accumulate in the environment. The accumulation of heavy metals in soils and waters poses a risk to the environmental and human health [8]. Landfills accumulate great amounts of waste and are an important source of liquid effluents, named leachates, which can have an adverse impact on the environment when released in an uncontrolled manner [9].

Heavy metals cause toxicities and other harmful effects not only in humans and animals but also in plants and soil microorganisms. Heavy metals disrupt many biochemical and physiological activities in bacteria, including growth, development, enzyme and hormone production. Indole acetic acid (IAA) is one of the most important hormones in plants, which is secreted by both bacteria and plants [10]. Plants are under the continual threat of changing climatic conditions that are associated with various types of abiotic stresses. In particular, heavy metal contamination is a major environmental concern that restricts plant growth. Plants absorb heavy metals along with essential elements from the soil and have evolved different strategies to cope with the accumulation of heavy metals [11]. The metals only contaminate the environment for now since their concentrations are below permissible limits. Solid waste that deposited on the surface adds metallic contaminants to the soil on which economic crops are cultivated. It is necessary to undertake regular environmental impact study to assess waste dumpsites. The higher concentration of copper in vegetation than in soils is attributed to direct contact rather than absorption from the soil [12].

The objective of this study was to determine the total concentration of heavy metals in rice from paddy plant growing in contaminated sites and to investigate the exposure distribution of selected heavy metals from rice from paddy plan and soils.

2. Materials and methods

2.1. Study Area

The soil and paddy plant samples were collected from a landfill site in Banyumas Region in Central Java Province, Indonesia named Gunung Tugel Landfill, as shown in Figure 1. The investigated landfill is located in the Kedungrandu Village, Patikraja Sub Region. This landfill was operated in 1983 with capacity around 282 m^3 per day, and is being used to dispose of mixed municipal waste and the disposal method is open dumping method.

The type of soil in Gunung Tugel landfill is ultiosol, thus allowing leachate to seep and pollute the groundwater of the surrounding population [13]. Ultisol is a land with a sourly argill horizon with low base saturation. The saturation of the base at a depth of less than 1.8 m from the soil surface is <35%. This soil texture is clay to sandy clay, bulk density between 1.3-1.5, and slow to moderate permeability [14]. Meanwhile, that Ultisol soil reaction is generally acid to very acid (pH 3.1-5.0) [15].

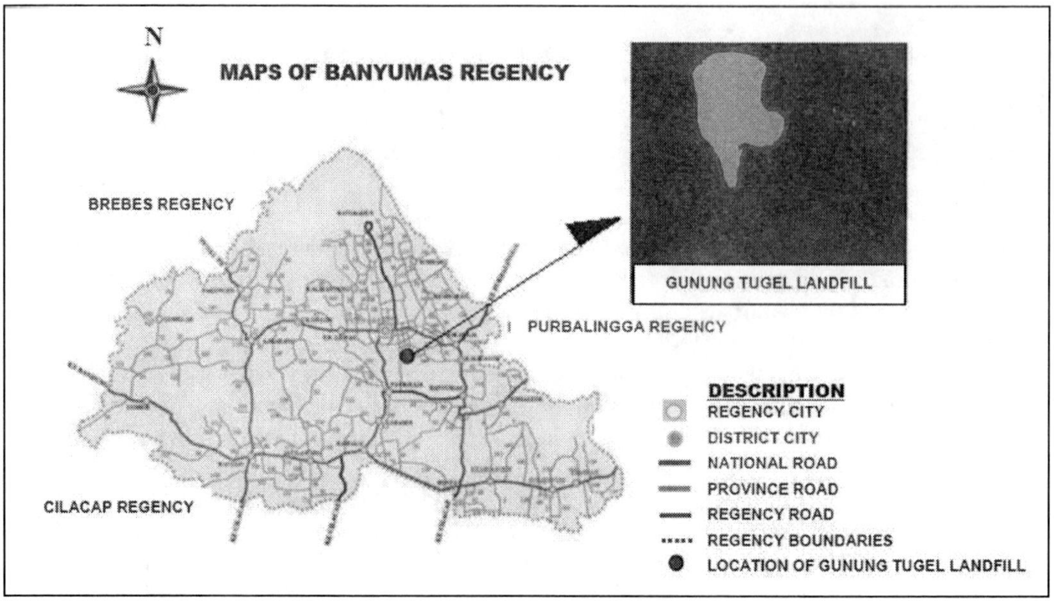

Figure 1. Location of Gunung Tugel Landfill

2.2. Sampling and Sample Preparation

The Sample was divided in 2 parts; soil and rice samples. Soil sampling method used in this research is rigid grid method for landfill area and stratified method for rice field area. Rigid grid method is a method that determining sampling point based on coordinate line which form square (grid) with interval distance 60 x 60 m with depth 50 cm. Soil samples were collected using stratified method. This method based on topography condition divided by zone in paddy field area with depth 50 cm. From the observation by using Quantum Geographic Information System (QGIS) software, 12 sampling points cover the area around landfill that has been closed and 6 sampling points covering rice field area. The location of the sampling point to be used can be seen in Fig. 2.

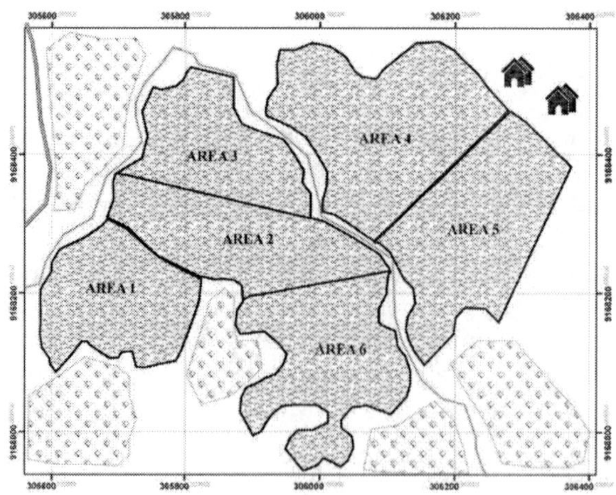

Figure 2. Sampling Point: Soil and Rice Sample in Paddy Field

Soil samples from each sample point were taken about 500g. Meanwhile, for the rice, samples were taken about 290-600g. Soil samples were dried under the sun for one day. Five hundred grams of dried samples were crushed and sieved using 100 mesh. One gram of each fine soil sample was destruct by adding 100 ml aquades and 10 ml of HNO_3 (P) mixture. The destruction was conducting until the samples was clear using using electric stove under the fume hood. The solution was filtered with filter Ø 200 mm, Ø 150 mm and sieve Ø 70 mm, then diluted again with aquades to 100 ml then shaken until homogen and added again with 5 ml of HNO_3 (P), then make destruction again by using electric stove in the cabinet acid until clear, after the solution is clear then the solution filtered with Ø70 mm sieve, added aquades up to 25 ml then the solution shaken until homogeneous and allowed to reach room temperature then stored in sterile bottles before heavy metal analysis. Analysis of heavy metals by AAS were performed at wavelength of 228,80 nm, 228,80 nm, 324,80 nm, 248,30 nm, 279,50 nm, 213,90 nm, 217 nm for Cr, Cd, Cu, Fe, Mn, Zn, Pb respectively.

2.3. *Analytical Methods*
Heavy metal parameters to be analyzed for presence in soil and rice samples are Cu, Fe, Mn, and Zn which are the essential metals, and Cr, Cd, and Pb which are non-essential metals for biological systems of living things. Heavy metals in a sample can be analyzed by AAS flame [16].

2.4. *Calibration Curve*
The first thing to do is preparing 6 (six) 100-ml flasks. Then each flask is filled with 0; 0.5; 1.0; 1.5; 2.0 and 2.5 mL of 1000 ppm Chromium standard solution, each of the flasks was added with aquades until half of the volume of the flask, then the solution was repeatedly repelled, and aquades were added to the boundary marker. The absorbance of each standard solution was measured using an Atomic Absorption Spectrophotometer on a wavelength (λ) of 228.80 nm. This procedure was repeated for Cadmium, Copper, Iron, Manganese, Zinc and Lead calibration curves using the parent solution of Cadmium, Copper, Iron, Manganese, Zinc and Lead 1000 ppm and 228.80 nm, 324.80 nm, 248.30 nm, 279.50 nm, 213.90 nm, 217.00 nm, respectively.

2.5. *Data Analysis*
The Analysis of the data is obtained by mapping of heavy metal dispersion. The mapping is done by using Quantum Geographic Information System (QGIS) software. After knowing the concentration of heavy metals at each sampling point then determine the plotting based on the quality standard of heavy metal, where each sampling point will represent the area inside, so it can assist in the analysis of the distribution of heavy metals in the soil.

There are three international standards that are used for comparison of heavy metals analysis of Cd, Cu, Cr, Zn, and Pb in soils, namely Environmental Protection Authority of Australia (EPAA) in 2012, Environmental Protection Ministry of China (EPMC) 2015. While the quality standards for heavy metals Fe and Mn use the quality standards of the United States Environmental Protection Agency (US EPA) in 2012. In other haand, for rice sample, standard for Cr and Cd are from National Food Safety Standard of China 2012. Cu, Zn, and Pb from Food Standards Australia New Zealand 2002. The last, for Fe and Mn from USDA Food Compotition Databases 2017.

Each area is recorded in coordinate with UTM (Universal Transverse Mercator) format. The coordinates will be displayed on the map, so that it can be known that the location of high or low hazard according to the class determined according to the respective quality standards of the heavy metals tested by the classification using the scoring method. The classification of concentration levels used in the scoring method can be seen in Table 1.

After the concentration level classification in each parameter of heavy metals, then done overlay with scoring method shown in Table 1. Each area determines the degree of heavy metal content by using the number of scores from each parameter. It aims to determine the area containing the highest to the lowest heavy metals. Score classification to determine the concentration with the largest classification to the smallest of each sampling point can be seen in Table 2.

Table 1. Concentration Level Classification

Class	Colour	Score
Very Low	Dark Green	1
Low	Light Green	2
Medium	Yellow	3
High	Orange	4
Very High	Red	5

Table 2. Classification and Score of Exposure Concentration

Class	Colour	Score
Very Low	Dark Green	< 7
Low	Light Green	8 – 14
Medium	Yellow	15 – 21
High	Orange	22 – 28
Very High	Red	> 29

3. Result and Discussion

3.1. *Soil in Landfill and Paddy Field*

Based on testing of samples taken both soil and rice from paddy field around the land of Gunung Tugel landfill indicate heavy metal content with various concentration. Table 3 and 4 shows that the soil and rice around Gunung Tugel landfill has been exposed with heavy metals. It is known that the concentration range of Cd from 0.17 - 0.42 mg/kg with an average value of 0.32 mg/kg. While the concentrations of other heavy metals are 24.03 – 48.69 mg/kg, 971- 34.71 mg/kg, 18997.26 mg/kg, 342.74 – 834.49 mg/kg and 136.10 – 290.00 mg/kg for Cu, Cr, Fe, Mn, and Zn, respectively. There are parameters of heavy metals in the soil exceeding:

Table 3. Concentration of Heavy Metals in Soil around Gunung Tugel Landfill

Location	pH	Concentration of Heavy Metals (mg/kg)					
		Cd	Cu	Cr	Fe	Mn	Zn
Area 1	6.00	0.37	40.55	34.71	18997.26	342.74	163.76
Area 2	6.00	0.42	39.47	18.61	19217.73	636.20	136.10
Area 3	5.00	0.37	48.69	16.41	32572.30	709.78	170.72
Area 4	5.00	0.25	28.29	6.27	31151.74	629.69	190.19
Area 5	6.00	0.35	24.03	14.53	19564.74	552.37	188.91
Area 6	6.00	0.17	37.12	9.71	19639.77	834.49	290.14
Average	5.67	0.32	36.36	16.71	23523.92	617.54	189.97

Table 4. Concentrations of Heavy Metal On Rice from Paddy Plant around Gunung Tugel Landfill

Location	Concentration of Heavy Metals (mg/kg)					
	Cr	Cd	Cu	Fe	Mn	Zn
Area 1	0.00	0.05	1.37	21.06	45.66	273.22
Area 2	0.88	0.13	2.55	18.99	50.56	115.21
Area 3	1.70	0.04	10.43	26.80	49.73	149.82
Area 4	0.76	0.00	2.05	20.12	39.16	88.06
Area 5	1.32	0.04	1.06	61.46	21.10	87.27
Area 6	0.57	0.26	0.79	13.88	18.73	87.43
Average	0.87	0.09	3.04	27.05	37.49	133.50

Table 5 shows the concentration level score of the map results of each parameter of heavy metals according to quality standard:

Table 5. Score and Classification of Heavy Metal Concentration Level on Soil in Paddy Plant Area

Location	Each Parameter Score							Total	Class
	Cd	Cu	Cr	Fe	Mn	Zn	Pb		
Area 1	3	4	3	2	1	4	1	18	Medium
Area 2	3	4	2	2	2	3	1	17	Medium
Area 3	3	4	2	3	2	4	1	19	Medium
Area 4	2	3	1	3	2	4	1	16	Medium
Area 5	3	2	2	2	2	4	1	16	Medium
Area 6	2	4	1	2	2	5	1	17	Medium

Figure 3 shows a map of the overlay and classification of heavy metal concentration levels on soils in the landfill and paddy field according to score level. The concentration of heavy metals in the soil is in the medium grade with a score range of 16 to 19, indicating that the overall heavy metal concentration on the soil in the paddy field area is evenly distributed. The concentration of heavy metals in the soil in the paddy fields is not only from the landfill, but can result from excessive use of fertilizers and pesticides. This can be seen in area 1 which is not adjacent to the landfill but has a score of 18 higher in areas other than in area 3. Area 1 uses more river water adjacent to the area, but there is a possibility that leachate water polluting area 3 flows into area 1 and accumulates in the area, so that the concentration of heavy metals in the soil in area 1 results from the accumulation of fertilizer and pesticide use and irrigation water that has been contaminated with leachate water. While in area 3, most likely pollution occurs due to leachate water from landfills that contaminate surface water used for irrigation. The exposure of heavy metals in the paddy fields is evenly distributed, since concentration scores are not too far between areas and have only moderate classes overall. The concentration level of heavy metals in the soil in the area of medium paddy fields can also be caused by the absorption of heavy metal content to rice crops, so that heavy metal content can be more accumulated and pollute rice plants.

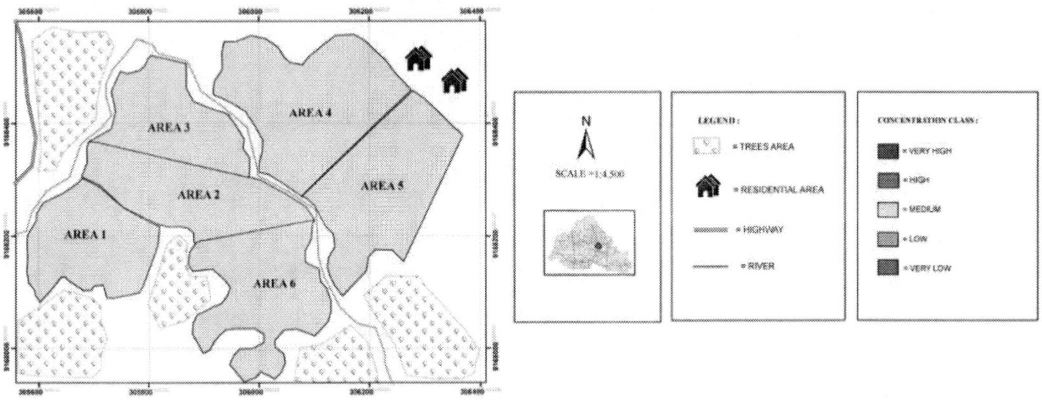

Figure 3. Map of Heavy Metal Exposure of Soil on Paddy Plant

This mapping aims to show which areas have the highest contamination levels to the smallest, it is based on map overlay of each heavy metal parameter that has been classified using the scoring method. The following table shows the concentration level score based on the map of each heavy metal parameter in the rice sample:

Table 6. Scoring Result of Heavy Metal Concentration Level on Rice Samples

Area	Each Parameter Score							Total	High
	Cr	Cd	Cu	Fe	Mn	Zn	Pb		
Area 1	1	1	5	5	5	5	3	25	High
Area 2	4	3	5	5	5	5	3	30	Very High
Area 3	5	1	5	5	5	5	3	29	Very High
Area 4	4	1	5	5	5	5	3	28	High
Area 5	5	1	5	5	5	5	3	29	Very High
Area 6	3	5	3	5	5	5	3	29	Very High

Based on Table 6, area 2 is the area with the highest number of scores of 30, included in the category of very high contamination level. However, other areas of 3, 5 and 6 are also in the same category. This is further illustrated by the map in Figure 4 which shows that all areas have a red color which means the area has a score between 29 – 35. The relatively similar hazard level indicates that the source of heavy metal contamination in rice is not only from landfill, but from other sources such as excessive use of fertilizers and pesticides by farmers. Evident from the level of hazard area 1 located in zone A with irrigation location is before the landfill is equal to area 4 located in zone B with irrigation location is located after the landfill landslide marked with high color orange category.

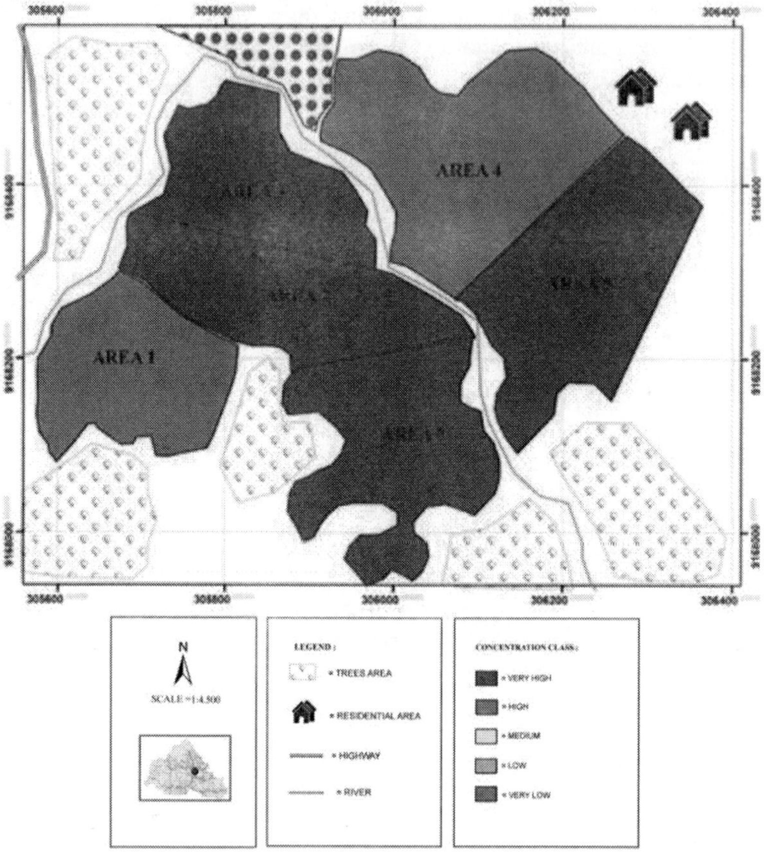

Figure 4. Map of Heavy Metal Exposure of Rice in Paddy Plant

4. Conclusions

Based on the results of analysis and discussion of this study can be concluded the following points:

1, Soil and paddy plants around Gunung Tugel landfill have been exposed to heavy metals with varying degrees. The average concentration in soil are 0.3236.36, 16.71, 23523.92, 617.54, 189.97 mg/kg and paddy plant are 0.87, 0.09, 3.04, 27.05, 37.49, 133.50 for Cd, Cu, Cr, Fe, Mn, Zn mg/kg respectively.

2, Overlay of heavy metal content on the soil in the landfill area has medium class and the paddy plant has high to very high. Exposure of heavy metals in the soil tends to be higher in the center of the landfill area, but in the paddy plants more evenly distributed. It can be caused the absorption of heavy metal to rice plants, so that accumulated in rice plants

5. Acknowledgement

This research has been supported by Bappeda Banyumas Regency, We would like to thank the regency of Banyumas and the Technical Services of the regency of Banyumas, We would also like to express our great appreciation to the management and analist of the Environmental Quality Laboratory Islamic University of Indonesia for the chance to test the samples according to the high standar,

References

[1] Visvanathan C, Tubtimthai O and Kuruparan P 2004, Proc. Int. Conf. on APLAS Kitakyushu, Japan,

[2] Tchobanoglous G, Theisen H, and Vigil S A 2004 *Integrated Solid Waste Management, Engineering Principles and Management Issues*, McGraw-Hill, New York,

[3] Boyd R S and Rajakaruna N 2013 *Heavy Metal Tolerance, in D, Gibson (Ed,), Oxford Bibliographies in Ecology*, New York: Oxford University Press

[4] Montazer-Rahmati M M et al 2011 J. Hazardous Materials 185 pp 401–407,

[5] Sing J et al 2011 J. Toxicol, Environ, Chem, 93 pp 462–473,

[6] Galal T M and Shehata H S 2015 J. Ecol, Indic, 54 pp 108–115,

[7] Jayanthi Bet al 2016 J. Catena 147 pp 25–31,

[8] Adamcov D et al 2016□J. Chemosphere 185 pp 1011-1018,

[9] Koda E et al 2016 J. Sustainability 8 1253,

[10] Seneviratne M et al 2016 J. Botany 105 pp 19–24,

[11] Mustafa G and Komatsu S 2016 J. Biochimica et Biophysica Acta 1864 pp 932–944,

[12] Ideriah T J K et al 2010 J. Appl, Sci, Environ, Manage, 14 pp 101 – 109,

[13] Santoso S P, Lestari S and Windyartini D S 2010 *Biosorpsi Kadmium pada Leacheate TPA Gunung Tugel Menggunakan Biomassa Sargassum cinereum (Biosorption of Cadmium on Gunung Tugel Landfill Leachate*, Biology Faculty of Jenderal Soedirman University

[14] Hardjowigeno and Sarwono 1993 *Klasifikasi tanah dan pedogenesis (Classification of soil and pedogenesis)* Akademika Pressindo Jakarta

[15] Prasetyo B H and Suriadikarta D A 2006 *Karakteristik, Potensi, dan Teknologi Pengelolaan Tanah Ultisol untuk Pengembangan Pertanian Lahan Kering di Indonesia (Characteristics, Potential, and Technology of Ultisol Soil Management for Dryland Agricultural Development in Indonesia)*, Litbang Pertanian

[16] National Indonesia Standard no, 13-6974-2003, Determination of concentration of Pb, Cu, Zn, Fe, Mn, dan Cd by Atomic Absoprtion Sepctropothometry (AAS).

Recombinant Protein Production from TPO Gen Cloning and Expression for Early Detection of Autoimmune Thyroid Diseases

Aulanni'am Aulanni'am[1,2*] Dyah Kinasih Wuragil[2], Djoko Wahono Soeatmadji[3], Zulkarnain[3], and Agung Pramana W. Marhendra[1]

[1)] Faculty of Sciences, Brawijaya University

[2)] Faculty of Veterinary Medicine, Brawijaya University

[3)] Faculty of Medicine, Brawijaya University

Research Group of Biopeptide, Immunodetection and Vaccine (BIODET Vaccine) Brawijaya University

*Corresponding author: aulani@ub.ac.id

Abstract. Autoimmune Thyroid Disease (AITD) is an autoimmune disease that has many clinical symptoms but is difficult to detect at the onset of disease progression. Most thyroid autoimmune disease patients are positive with high titre of thyroid autoantibodies, especially thyroid peroxidase (TPO). The detection AITD are still needed because these tests are extremely high cost and have not regularly been performed in most of clinical laboratories. In the past, we have explored the autoimmune disease marker and it has been developed as source of polyclonal antibodies from patient origin. In the current study, we develop recombinant protein which resulted from cloning and expression of TPO gene from normal person and AITD patients. This work flows involves: DNA isolation and PCR to obtain TPO gene from human blood, insertion of TPO gene to plasmid and transformation to *E. coli* BL21, Bacterial culture to obtain protein product, protein purification and product analysis. This products can use for application to immunochromatography based test. This work could achieved with the goal of producing autoimmune markers with a guaranteed quality, sensitive, specific and economically. So with the collaboration with industries these devices could be used for early dete

Keywords: recombinant protein, TPO gene, Autoimmune thyroid diseases (AITD)ction of the diseases in the community.

1. Introduction

Autoimmune thyroid disease (AITD) is a disorder in the activity of the thyroid gland. Its most commonly found in women, and is still a problem that has not been addressed comprehensively because of delays in diagnosis. The incidence of AITD in the world is reported to be more prevalent in adult women than in men, and affects approximately 2-5% of the world's population [1]. (Based on US and European population screening data, AITD reportedly found in 27% of adult women with increasing frequency at age above 50 years and only 7% occurred in adult males [2]. Ministry of Health the Republic of Indonesia in 2013 indicates that thyroid abnormalities still occupy the top 5 non-communicable diseases in Indonesia, with prevalence of about 0.4-0.7% [3]. Manifestations and complications caused by autoimmune thyroid abnormalities have been widely reported, especially during pregnancy will produce poor outcomes for the mother and foetus [4,5,6].In order to support one of Indonesian's government program of Sustainable Development programs (SDG's), Improving maternal and child health quality is implemented by early detection and treatment of AITD. Early detection and good treatment of pregnant women with AITD will manifest to improved fetal growth. Pregnant women with AITD rarely show typical clinical symptoms, making it difficult to identify early in pregnancy. If not treated early enough will increase the risk of miscarriage and preterm birth [7], low birth weight [6], as well as manifestations of disability and impaired fetal brain development [4]. Therefore, a proper routine screening is needed to identify potential pregnant women with AITD, so that early care can be undertaken by a doctor prior the manifestation of maternal and fetal health occurs.

In the blood serum of patients with AITD found several antibodies that can be used as a serological marker, such as thyroid peroxidase (TPO), thyroid stimulating hormone receptor (TSHR) and thyroglobulin (TG). The formation and enhancement of autoantibodies titre to both TPO and TSHR appear to be more dominant and directly involved in AITD patho-mechanism. Antibodies titre to TPO and TSHR can be detected earlier before alteration of T3 and T4 hormone levels. It is appropriate to monitor both hypothyroid and hyperthyroid conditions, which is generally without any specific signs that appear. This phenomenon supports to create a design for early diagnosis in the phase of pre-disease of AITD to determine the progression of the disease towards prevention as well. A tool to diagnose the disease at early stage is more focused to detect the presence of autoimmune markers of TPO, which has a specificity and sensitivity better than available commercial detection kit. This work proposes to produce recombinant protein of thyroid peroxidase (TPO) as a rapid test material for early detection AITD.

2. Methods

This study has been running for more than 15 years to look for markers of autoimmune diseases and develop it into a Rapid Test, conducted at several laboratories in UB, which has been certified by ISO and KAN. The uses of a sample of human blood were approved by the Research Ethics Committee, Brawijaya University No. 109/EC/KEPK/04/2012).

2.1. Chemical and sample collection

Sample was taken from 20 of patient's blood serum which had been confirmed with gold standard of Mab-human Thyroid peroxidase (TPO), phosphate buffer pH 7, phenimethylsulfonil fluoride (PMSF), sodium etilene diamineacetate (NaEDTA), sodium dihydrogen phosphate monohydrate ($NaH_2PO_4.H_2O$), disodium hydrogen phosphate (Na_2HPO_4), glycerol, sodium hydroxide (NaOH), chloric acid, ammonium sulphate ($(NH_4).2SO_4$), Barium Chloride ($BaCl_2$), Tris-Cl, Tween-20, polyacrilamide, bis-acrylamide, Comassie-blue R-250, Sephadex-G75, tris-glycine buffer, pottasium carbonate, TPO gene primer; Forward TPO: 5' ATG GAATTC ATG GCTTCTTGTACAACGGGTTC 3' and Reverse TPO: 5' ATG GTCGAC TTA GTCGTGGTC GATGTATTGTC 3', PstI restriction enzyme, DNAeasy blood and tissue kit (Qiagen), agarose, Oneshoot® cell, Super Optimum Broth medium with catabolite repression (SOC) Medium, Luria Berthani medium and agar Medium from Oxoid, *E. coli BL21* previously preserved in our laboratory, pQE-T4, distilled water are used throughout this works.

2.2. Instruments

Centrifuge used was Tomy High Speed Sentrifuge MX-305, horizontal electrophoresis used from Biorad, miniprotean electrophoresis used from Biorad electrophoresis, Biorad-Mini transblot, thermocycler used C1000 thermal cycler from Biorad, and Shaking Waterbath from Memmert, Biologic Duolflow from Biorad, and Gel documentation from Biostep.

2.3. DNA isolation and PCR to obtain TPO gene from human blood
DNA isolation was conducted based on protocol of DNAeasy blood and tissue kit (Qiagen). Then it was continued with amplification of DNA target gene with programs of: pre-denaturation at 95 °C for 7 min; 35 cycles of denaturation at 95 °C for 30 sec, annealing on 60 °C for 20 sec, and elongation at 72 °C for 30 sec, then it was maintained at 60 °C. The gene encoding TPO was then separated by PstI enzyme, then it was analyzed with agarose gel electrophoresis. The used of human samples for this research were approved by Ethical Committee of Faculty Medicine Brawijaya University and were completed with inform consent (No. 109/EC/KEPK/04/2012).

2.4. Insertion of TPO gene to plasmid and Transformation to E. coli BL21
2 μL of PCR product was mixed with a 0.50 of salt solution and 0.50 μL of pQE-T7 vector, then it were blended carefully and incubated for 5 min at room temperature (22-23 °C). Before the transformation process, the waterbath was set at 42 °C, S.O.C medium was warmed to room temperature, then the tube then be applied was warmed to 37 °C for 30 min. 1 vial of Oneshot® cell then immersed in ice for each transformation. 2 μL of pQE T-7® cloning reaction from the previous stage then was added to the One Shot® then mixed slowly, then it was incubated in ice for 5 to 30 min.

The cells were given a heat shock for 45 seconds at 42 °C without being shaken, then the tube was transferred to ice for 2 min, added 250 μL S.O.C. medium at room temperature. Tubes were closed tightly and centrifuged at 200 rpm. A total of 10-50 μL of the transformed was cultured on a selectively preheated agar plate and incubated overnight at 37 ° C. To ensure even small breeding, 20 μL added S.O.C. medium.

2.5. Bacterial culture to obtain protein product
400 μL of LB medium was acclimatized at 37 C for 1 h, then it was added with 100 μL cells which passed of colonic selection. Then it was incubated at 37 °C with at 250 rpm shift speed, 24 h.

2.6. Protein purification and product analysis
Competent cell culture media added with PBST containing 4mM PMSF as much as 5 times the volume. The homogenized sample was then centrifuged at 6000 rpm, 4 °C for 15 min, and supernatant was collected. Supernatant then was added with cold ethanol with a ratio of 1:1, kept for 24 h in the refrigerator. Then it was centrifuged at 10,000 rpm for 10 minutes at 4 °C. Ethanol then was removed, the precipitate was dried and added with Tris Cl buffer of 50 μl. The recombinant TPO protein iwas stored in -20 ° C. To obtain TPO recombinant protein with uniform size then the recombinant product was purified using Biologic Duo Flow Biorad

2.7. Application to immunochromatography based test
The application of recombinant TPO to reverse-slow immunochromatography based test is conducted based on this schematic ilustration in Figure 1. This components consist of 2 (two) main parts sections are test area which consits of polyester membrane, nitrocellulose membrane to place the test components are control line (goat anti mouse IgG) as well as test (recombinant protein of TPO) and sample pad. The observation area is consist of absorbent paper and signal reagent area to put signal reagent and buffer to let the immunochromatography system works.

Rapid tests are generated based on the concept in Figure 1. The series of detection devices to be used using nitrocellulose membranes, polyester membranes and adsorbent paper on the plastic base. Nitrocellulose membrane was used as a place of attachment antigen (recombinant human protein TPO) and the control line (goat anti mouse IgG), on the top and bottom membranes attached polyester with a width of 0.5 cm, to cover the nitrocellulose membrane. Antigen attached to a linear manner over the nitrocellulose membrane as a test line. while goat anti-mouse IgG attached to a nitrocellulose membrane as a standard or control lines. A total of 20 microliters of serum sample is dripped onto polyester pads near the control zone. Then dropped as much as 10 microliters of a solution of colloidal gold on polyester over the observation window and dried. A total of 1 drop of buffer solution is dripped onto the same pads and 2 drops of buffer on a polyester cushion coloidal gold signal reagent, then incubated a few moments to penetrate the sample and buffer boundaries. Test card immediately closed and the observed results of the examination of the observation window 20 to 30 minutes later. A positive result achieved if it appears 2 red lines on the control line and a test line. Negative results achieved if the red lines appear first on the control line. The results appear invalid if tidal lines at all or only a line appears on the test [10,11].

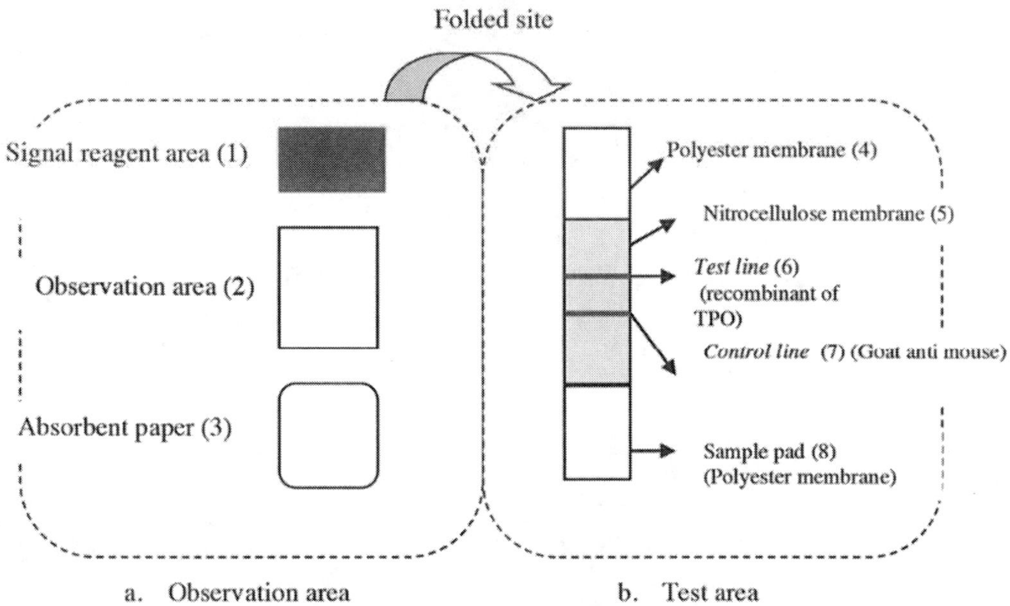

Figure 1 Schematic illustration of reverse flow-immunochromatography based test for detecting TPO antibody from sera

3. Result and discussions

A pair of primers used to detect the high-sensitizing protein coding of TPO gene in DNA of AITD patients by PCR method, and below is PCR product resulted from Forward TPO: 5' ATG GAATTC ATG GCTTCTTGTACAACGGGTTC 3' and Reverse TPO : 5' ATG GTCGAC TTA GTCGTGGTC GATGTATTGTC 3'. TPO gene characterization by PCR method that has a gene length of ± 700 bp (Figure 2).

This amplification process plays a role in the provision of TPO gene encoding as the main antigenic substance in the test in the form of recombinant proteins. The plasmid choosen in this work also have a role in gene transcription while the gene were inserted and transporter to cells. The principal components of plasmid vector having influence to gene transcription include the origin of replication (ORI), promoter, antibiotic selection marker and transcription terminators [8]. Below the schematic illustration of plasmid reconstruction to obtain TPO gene insertion area (Figure 3).

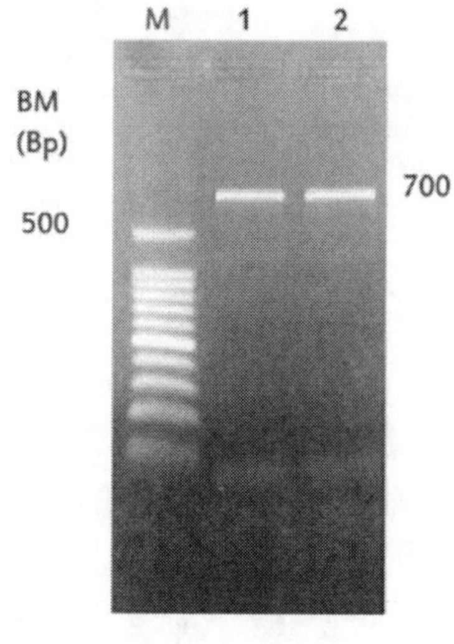

Figure 2 PCR product of TPO gene amplification (Agarose 2%); M: Marker, C: 1-2: sample

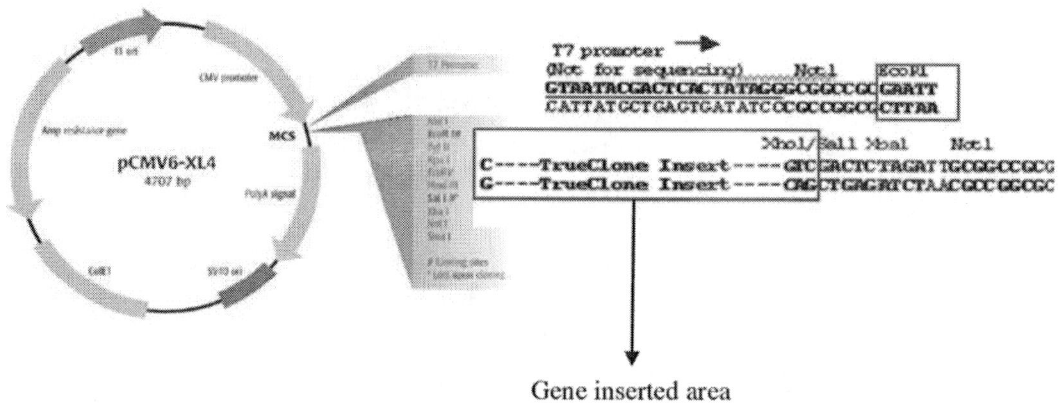

Figure 3 Plasmid reconstruction for TPO-gene insertion

After the plasmid reconstruction, TPO gene is inserted into the Plasmid vector with DNA ligase. After the insertion is successful then the Plasmid is transformed to a competent BL / 21 cell for the cloning process. After a competent cell was transformed, it was continued for selection with IPTG and X-GAL administration to obtain white and blue colonies. Recombinant protein then was isolated from competent cells of *E coli* BL21 then it was characterized with Western Blotting. Then the protein with molecular weight approximately 51-52 kDa were purified to apply to immunochromathography test strip [9].

The main advantage of the immunochromatography method is the simplicity of the test, which mixes the specimen with complex particles with antibodies, inserts the mixture into the membrane and waits for several minutes (10-15 minutes) until the chromatographic process is complete [12].

4. Conclusion and future works

This work succesfully obtain rapid diagnostic kit to detect TPO antibody with have more advantage over the commercial kit available in market. The further studies for clinical test to obtain specificity and sensitivity are needed to implement as well the kit stability test.

Acknowledgement

Authors would like to thanks to Directorate General of Innovation Strengthening, Ministry of Research, Technology and Higher Education the Republic of Indonesia for funding this research and publications through INSINAS Program of 2017.

References

[1] McGrogan A, Seaman HE, Wright JW, de Vries CS. The incidence of autoimmune thyroid disease: a systematic review of the literature. Clinical Endocrinology, 69(5):687-696, 2008.
[2] Vanderpump MP. The epidemiology of thyroid disease. British Medical Bulletin, 99:39-51, 2011.
[3] Agency for Health Research and Development, Ministry of Health Republic Indonesia. Helath Fundamental Research, 83-90. 2013
[4] Männistö T, Vääräsmäki M, Pouta A, Hartikainen AL, Ruokonen A, Surcel HM, Bloigu A, Järvelin MR, Suvanto-Luukkonen E. Perinatal outcome of children born to mothers with thyroid dysfunction or antibodies: a prospective population-based cohort study. The Journal of Clinical Endocrinology and Metabolism, 94(3):772-779. 2009.
[5] Päkkilä F, Männistö T, Surcel HM, Ruokonen A, Bloigu A, Pouta A, Hartikainen AL, Vääräsmäki M, Järvelin MR, Suvanto E. Maternal thyroid dysfunction during pregnancy and thyroid function of her child in adolescence. The Journal of Clinical Endocrinology and Metabolism, 98(3):965-972. 2013.
[6] Aggarawal N, Suri V, Singla R, Chopra S, Sikka P, Shah VN, and Bhansali A. Pregnancy outcome in hyperthyroidism: a case control study. Gynecologic and Obstetric Investigation, 77(2):94-99. 2014.
[7] He X, Wang P, Wang Z, He X, Xu D, Wang B. Thyroid antibodies and risk of preterm delivery: a meta-analysis of prospective cohort studies. European Journal of Endocrinology, 167(4):455-464. 2012.
[8] Singha, T.K., P. Gulati., A. Mohanty., Y Pal Khasa., R.K.Kapoor and S. Kumar. Efficient Genetic Approaches for Improvement of Plasmid Based Expression of Recombinant Protein In Escherichia coli: A Review, Process Biochemistry, 55:17-31. 2017.
[9] Fiqiyana, MA; Aulanni'am and A. Roosdiana. 2013. Characterization of Polyclonal Antibody Induced by Autoantibody TPO (Thyroid Peroxidase) From Autoimmune Thyroid Disease (AITD) Serum with ELISA and Western Blotting. J.Pur App. Chem. Res. 2(1) 19-26, 1 April 2013.
[10] Rista Nikmatu Rohmah, Soraya Widyasari, Aulanni'am, and Fatchiyah, Cloning and Expression of hGAD65 Gene in E.Coli BL21, Indonesian Journal of Biotechnology. Vol.18, No.1, pp.52-57, 2013.
[11] Aulanni'am, Ulhaq, ZSB., APW Marhendra, DK Wuragil, R Hermawan, and W. Riawan. Immunodot Technique for Early Detection Rheumatoid Arthritis Patient in Indonesia Based on Autoimmune Marker Matrix Metalloproteinase-3 (MMP-3). Journal of Applied Sciences Research, 8(9): 4795-4801, 2012.
[12] Aulanni'am. 2013. Detection Autoimmune Diseasese : Past , Present and Future. Invited Speaker in The 2nd Regional Symposium on Biosensors, Biodiagnostic and Biochip. Chiang Rai, December 11-13, 2013

IOP Conf. Series: Materials Science and Engineering **299** (2018) 012014 doi:10.1088/1757-899X/299/1/012014

Identification and Characterization of Bioactive Peptides of Fermented Goat Milk as a Sources of Antioxidant as a Therapeutic Natural Product

Chanif Mahdi[1], Handayu Untari[2] & Masdiana Cendrakasih Padaga[3]

[1]Department of Chemistry, Faculty of Sciences, Brawijaya University, Malang, Indonesia,
[2] Veterinary Medicinal Faculty, Brawijaya University, Malang, Indonesia
[3] Veterinary Medicinal Program, Brawijaya University, Malang, Indonesia

chanifmahdi@gmail.com

Abstract. The increasing of functional food is rising in line with public awareness for healthy food consumption. Provision of functional food source is developed through enhanced bioactive that has a regulatory function for body. Bioactive peptides in milk is known have variety of beneficial function of the body such as immunomodulator, immunostimulatory, anti-hypertension, anti-hyper cholesterol, as well as a variety of other beneficial function. The aim of this study is to obtain fermentation methods to product functional dairy product contain bioactive peptides and beneficial of fermented goat milk. The result of this study showed that goat milk fermented using 3 % commercial starter able to produce the best yoghurt than using local yoghurt starter. Analysis of protein content showed that the fermentation processing increased the amount of protein in goat milk sample. Using SDS- PAGE showed that the breakdown of protein into fraction of fermented goat milk greater than unfermented goat milk. The result of fractional protein was analyzed by LC MS/MS and showed that there were three kind bioactive sequences of bioactive peptides. Each of which consist of 16 amino acids that safely protected from gastrointestinal animal model that fed by dietary treatment of hypercholesterolemia.
Keywords: *Food functional; bioactive peptides; fermented goat milk; anti-hyper cholesterol*

1. Introduction

Nowadays, food protein is one of food component which interest for researcher and industries because protein is a source of amino acids and known has biology activities that influence functional and body system, both on whole protein or hydrolysis product. Hydrolysis product of food protein that has biologist function on regulation body function and healthy is known as bioactive peptides [1]. Bioactive peptides bonded on food protein precursor and will be released through hydrolysis process by proteolysis enzyme in gastrointestinal tract, or maybe by in-vitro using proteolysis enzyme that isolated from plant or microorganism, or by fermentation processing [2]. Several biological functions of bioactive peptides that have known are as an antioxidant, antimicrobial, anti-hypertension, anti-cholesterol, cyto-modulator and immunomodulator [3].

Bioactive peptides have been isolated and identified from several of food sources, such as instant milk, egg, cattle meat, chicken, fermented milk product, fish and soybean [4]. Milk is the main sources of bioactive peptides and has several biological function compared with throw foodstuff [2]. However, bioactive peptides are inactive form in protein binding condition. Its require hydrolysis to release bioactive peptides, so milk can be used as food sources with high content of bioactive peptide, thus milk can be used as food sources with high content of bioactive [4]. Hence, studying the effective methods to produce high bioactive content milk is important.

Fermentation using lactic acid bacteria as a famous method that is used to produce fermented milk is believed to give healthy benefits and cure several diseases ([2];[5]).Several results of research showed that bioactive peptides from fermented milk have many benefits for body regulation system and enhance body healthy [6]. Therefore, the researcher focus on developing and optimization fermentation method to product bioactive peptides in many states.

Goat milk as one of natural functional food has chemical characteristic that different from cattle milk, but same as human milk. Lactose and protein contained in goat milk are almost equal to cattle milk, but there are some differences on structure and immunological protein. Beside, goat milk contains several middle chain fatty acid and lipid globular that are relatively small compared to cattle milk. Therefore, it is important to do research use goat milk as source of bioactive peptides.

Nowadays, goat milk industry in Indonesia is very fast, It was important to develop processing technology to receive milk product processing that has economical value than fresh milk. Foods products with basic source of goat milk have been develop as a food medicinal product or biomedicine that is useful for health. Thus it is very important and urgent to explore bioactive peptides from goat milk through fermentation processing method.

The purpose of this study is to know the level concentration of lactic acid bacteria that is needed and how long fermentation processing needed to produce bioactive of goat milk. We hope with to find the optimal of fermentation through this study, it will be used to develop milk fermentation product with have better value if be compared with fresh milk.

2. Method

2.1 Starter Preparation

Starter was prepared according to a previously protocol by www.yogourmet.com: 100 mL of fresh goat milk was pasteurized at 72°C for 5 minute, then cooled to 40 °C – 45 °C. 0.5 gram frozen starter was solved in 5 mL of pasteurized milk (from 100 mL of Pasteurized milk). The starter was added into 100 mL pasteurized milk, then homogenized and incubated at 40°C – 45°C for 4 to 8 hours respectively until milk starter reached pH of 4.0 – 4.5.

2.2 Yogurt Preparation

The yogurt preparation method was conduct according to previously protocol by Position et al. (2005). 500 mL of goat fresh milk was pasteurized at 72°C for 5 minutes then cooled at 40°C – 45°C. After that, it was incubated with 3% and 5% milk starter solution (according to previously study), later homogenized and incubated at 40°C – 45°C for 4 to 8 hours until pH value of yogurt reaches 4.5 to 5.0 respectively. The product of yogurt was keep at 4 °C to 5 °C for the next analysis purpose

2.3 Protein Yogurt ≤ 3kD preparation

Hydrolyzed product of yogurt protein was prepared by solving 100 mg of dry sample frozen yogurt into 1 mL buffer ammonium bicarbonate solution (50 mM; pH 8.5), then it was sonicated (10 second, 40 Hz), and replicated 4 times. It was centrifuged of cold centrifugation (12,000 Xg; for 10 minutes at 4 °C). The supernatant (500 µL) was poured through in ultrafiltration membrane 3 kDa molecular weight cut of (MWCO). It was centrifuged in cold centrifugation for 15 minutes (12.,000 x g; 4 °C), and 100 µL buffer ammonium carbonate was added to avoid waste protein. The mixture was centrifuged with cold centrifugation for 7 mins (12,000 x g; 4 °C). The precipitation of solution considered as protein fraction ≤ 3 kDa, was collected for profile peptides identification immediately. The solution of protein fraction was frizzed at -20 °C for next laboratory analysis purpose.

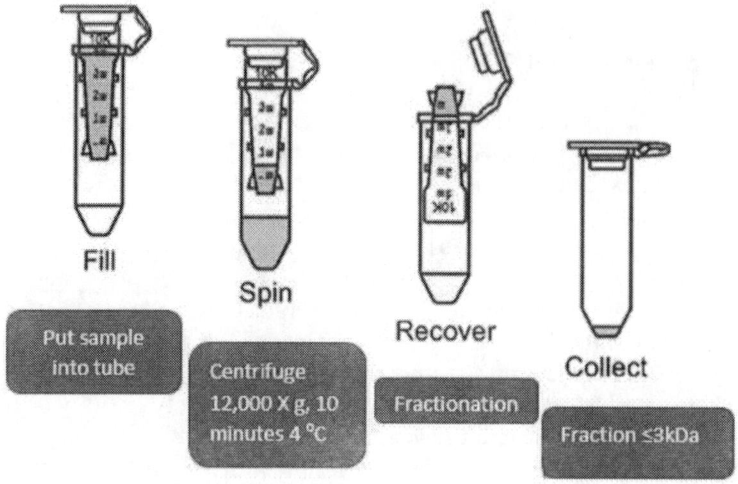

Figure 1. Hydrolysis result of protein fractionation processing (<3 kDa) used ultrafiltration method

2.4 Protein Determination by Kjehdahl Method

0.5 g of sample was weighted in Kjehdahl flask. The sample was transfer into the Installed digest tube using 10 mL concentrated H_2SO_4. Switch the digestion unit off and let the stand with the digestion tubes cool down in 400 °C. Destruction processing will be running as a programming. After raised color change from black to green color, the processing of digestion was stopped. The tube was waited for a moment (until cold room temperature reached), and continued to distillation and titration processing.

2.5 Peptides Identified by LC/-MS/MS

The peptides dry sample was solved in 5 % of acetonitrile and 0,1 % of formic acid in deionized water for LC-MS/MS analysis. LC- MS/MS analysis method was determined using LCQ Deka XP System MAX Thermo with Electrospray Ionization (ESI) (Thermo Scientific *Inc.* USA) (C 18 Bio-basic column, with 150 x 2,1 mm diameter, 5 µm particle size. LC- MS condition was gas flow rate 50 absorb prayer potential 4 kV, capillary potential 20 V, capillary temperature 300°C. Interval of MS Scan was between m/z 100 – m/z 1600 with flow rate 200 µL/min. Peptides separation was used with gradient linier gradually from 5% B solution to 70 B solution (Formic acid 0,1 % in acetonitrile) for 90 mins. Massa spectra data reading was used by Thermo– Xialibor/TM program (Thermo Scientific USA). Data of MS/MS were calculated by using Format File MGF with Mascot Distiller V2.3.2.0. (Matrix Saints, London United Kingdom), and continued with Blast by MGF file to Mascot search engine V 2.3 (Matrix Science, UK. Sequence peptides were identified by Based of peptides sequence in database.

3. Result and Discussion

3.1 Optimum Concentration, pH, and Fermentation Time of Both Local and Commercial Yoghurt

Table 1 Result of qualitative analyzed of yoghurt

	Treatment			
	A 3 %	A 5 %	B 3 %	B 5 %
pH	5.2	5,0	4.3	3.8
Viscosity consistency	Less	Less	Viscous	Viscous
Taste	Acid less	Acid less	Yoghurt	Acid
Aroma	Milk	Milk	Yoghurt	Acid

A : Local yoghurt starter B : Commercial yoghurt starter

Good quality yoghurt starter was able to produce with good viscosity needed. Table 1 shows that there are different of yoghurt quality produced by two kind of yoghurt starter in both 3 % and 5 %

concentration. Based on Table 1, showed that 3 % concentration of commercial yoghurt starter with pH 5.2 over 8 hours fermentation has a good quality. This results agree well with the previous experimental results that the best starter concentration was 2-3% ([2];[7]). Therefore, yoghurt starter with 3 % concentration was used for protein analysis.

3.2 Protein analysis

Table 2 Protein analysis of yoghurt of 3% concentration commercial starter

Sample	Protein content (% dry matter)
Milk	20.81
Yoghurt	22.80

Protein content analysis on goat milk and yoghurt were produced from 3% of starter yoghurt commercial showed to produce yoghurt with high quality, compared to yoghurt produced by local yoghurt starter. The purpose of protein content analysis was to investigated protein content in yoghurt and fresh goat milk. Protein content of yoghurt was 22.80 %, higher than protein content in fresh goat milk (unfermented goat milk) that was 20.81 % (Table 2). Many milk proteins possess specific biological properties that make potential components ingredients for healthy food. The increasing attention is being focused to physiological active peptides derived from milk protein. These peptides are inactive within the sequence of the source protein molecule and can be released by milk fermentation with proteolytic starter culture [2]. For this reason, it can be concluded that yoghurt of goat milk is potential to be developed as drinking product as a functional food.

3.3 Protein profile

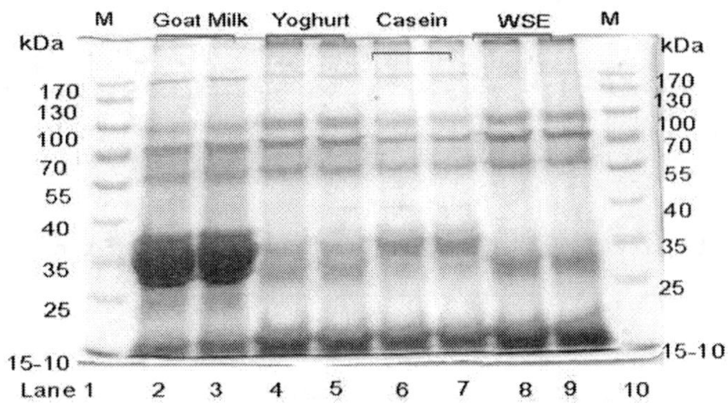

Figure 2. Figure of protein bands of yoghurt and fresh milk of goat

As can be seen in Figure 2, the fermentation treatment processing cause peptides of protein milk became hydrolyzed and denaturized. It was known from total band of peptides protein formed at SDS-PAGE. The electrophoresis processing showed the total of peptides protein bands of fermented milk more than peptides protein band of fresh milk. The hydrolyzed or denaturized of peptides protein fermented milk was reached optimally because of the presence of lactic acid bacteria. According to Korhonen and Pihlanto (2006) [2], that peptides are inactive within the sequence of the source protein molecule and can be released by (1) gastrointestinal digestion of milk, (2) fermentation of milk with proteolytic starter cultures or (3) hydrolysis by proteolytic enzymes.

Since there is difference variation of peptides bands total in fermented goat milk or yoghurt, the next research is suggested to determine peptides character that is useful to decrease cholesterol in blood. The peptides characterization is done by protein hydrolyzed fractionation method, that have molecule weight less than 3 kDa, that has potential of both as an antioxidant and anti hypercholesterol.

The functional properties of bioactive peptides are determined by the amino acid composition of the bioactive peptide, e.g peptides with the amino acid composition Val-Lys-Glu-Ala-Meta-Pro-Lys have antioxidant functions [1]. In general, the functional properties of bioactive peptides derived from milk can be divided into four groups that are the functional properties associated with the circulatory system (cardiovascular system), the nervous system (nervous system), the gastrointestinal system (immune system) and immune system (immune system) [2].

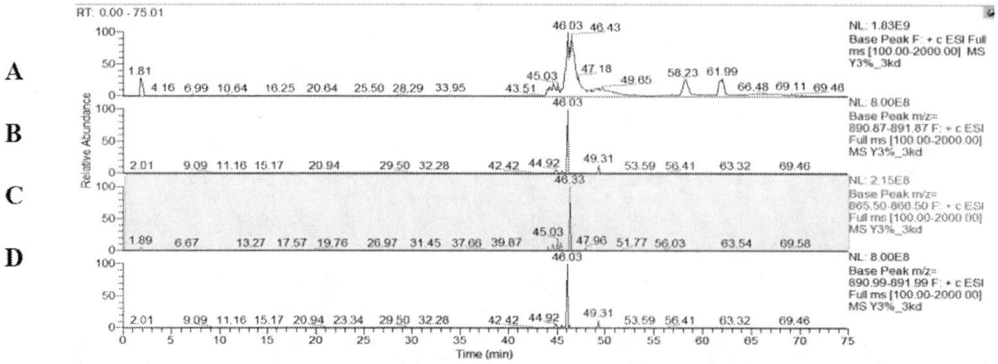

Figure 3. LC-MS/MS chromatogram of fermented milk yoghurt sample <3 kDa (A) LC-MS/MS chromatogram of fermented milk yoghurt (B) LC-MS/MS chromatogram of LYQEPVLGPVRGPFPI sequence (C) LC-MS/MS chromatogram of YQEPVLGPVRGPFPIL sequence (D) LC-MS/MS chromatogram of VQSWMHQPPQPLSPT sequence

The results of LC/MS chromatogram were showed at Figure 3. The fermented milk yoghurt was known by chromatogram A. The derivative sequence of fermented milk yoghurt amino acid from β-casein Capra hercus was showed by chromatogram B (LYQEPVLGPVRGPFPI), C (YQEPVLGPVRGFPIL), and D (VQSWMHQPPQPLSPT). These chromatograms had the highest peak in retention time 46.03; 46.33, and 46.03 respectively. This result in line with chromatogram A, suggesting that the three of sequence is derivate of amino acid from fermented milk yoghurt. Based on data in Table 3, it showed these three sequence had a low molecular mass. Hence it were keep from protease activities of gastrointestinal.

Table 3 Result of yoghurt sample peptides <3kDa identification sing Mascot Distiller and NCBI databse

LC-MS/MS Chromatogram	Identified Peptide Sequence	m/z	Molecule weight (Dalton)	Score	Retention time (min)
B	LYQEPVLGPVRGPFPI	891. 37	1780.90	50 (homology)	46.03
C	YQEPVLGPVRGPFPIL	891. 49	1780.90	52 (homology)	46.33
D	VQSWMHQPPQPLSPT	866.88	1731.84	51 (homology)	46.03

4. Conclusion

The best quality of yoghurt was produced by Canadian commercial yoghurt starter by 3 % concentration pH 5.2 over 8 h of fermentation. The content of dry matter protein fermented goat milk was 22,80 % higher than protein content of fresh goat milk. The result of isolation and characterization of peptides bioactive showed that there were 3 kinds of bioactive peptides sequences, each 16 amino acids protected from protease gastrointestinal enzymes. These peptide sequences had a function as antioxidant potential.

Acknowledgments

We are grateful to acknowledge Prof. Sumarno from Medical Faculty of Brawijaya University as a Professor of Biomedical Sciences for many helpful discussions and suggestions related to this work.

Moreover, Prof. Aulanni'am from Biochemistry Laboratory, Faculty of Sciences Brawijaya University for all support for this study

References

[1] Kitts, D.D. and K. Weiler. 2003. Bioactive proteins and peptides from food sources: Applications of bioprocesses used in isolation and recovery. *Curr. Pharm. Des.* **9** 1309

[2] Korhonen H. and A. Pihlanto. 2006. Bioactive peptides: Production and functionality. *International Dairy Journal,* **16** 945

[3] Hartmann, H. and H. Meisel. 2007. Food-derived peptides with biological activity: from research to food applications. *Curr. Opin. Biotechnol.* **18** 163

[4] Ebringer, L., M. Ferenčík, and J. Krajčovič. 2008. Beneficial health effects of milk and fermented dairy products – Review. *Folia Microbiol. (Praha)* **53** 378

[5] Donkor, O. N., A. Henriksson, T. K. Singh, T. Vasiljevic, and N. P. Shah. ACE-inhibitory activity of probiotic yoghurt. *Int. Dairy J.* **17** 1321

[6] Sarmadi. B. H. and A. Ismail. 2010. Antioxidative peptides from food proteins: A review, *Peptide.,* **31** 1949

[7] Herna'ndez-Ledesma, Amigo B, Ramos LM and Recio I. 2014 Angiotensin converting enzyme inhibitory activity in commercial fermented products: Formation of peptides under simulated gastrointestinal digestion. *J Agric Food Chem* **52** 1504

Green Chemistry Glucose Biosensor Development using *Etlingera elatior* Extract

A Fatoni[1,2], M D Anggraeni[3], Zusfahair[1], H Iqlima[1]

[1]Department of Chemistry, Faculty of Mathematics and Natural Sciences, Universitas Jenderal Soedirman, Purwokerto 53123, Indonesia
[2]Center for Maritime Biosciences Studies, Universitas Jenderal Soedirman, Purwokerto, Jawa Tengah 53123, Indonesia
[3]Department of Nursing, Faculty of Health Sciences, Universitas Jenderal Soedirman, Purwokerto, Indonesia
Email: aminfatoni@unsoed.ac.id

Abstract. Glucose biosensor development is one of the important strategies for early detection of diabetes mellitus disease. This study was aimed to explore the flower extract of *Etlingera elatior* for a green-analysis method of glucose biosensor. Flowers were extracted using ethanol: HCl and tested its performances as an indicator of glucose biosensor using glucose oxidase enzyme. The glucose oxidase react with glucose resulted hydrogen peroxide that would change the color of the flower extract. Furthermore, the extract was also studied including their stability to pH, oxidizing and reducing, temperature, and storage. The results showed that the *Etlingera elatior* extract had high correlation between color change and glucose concentration with regression equation of $y = -0.0005x + 0.4724$ and R^2 of 0.9965. The studied biosensor showed a wide linear range to detect glucose sample of 0 to 500 mM. The extract characterization showed a more stable in low pH (acid), reducing agent addition, heating treatment and storage.

1. Introduction

Diabetes mellitus (DM) is a group of carbohydrate metabolic disorder with a high blood glucose level. According to World Health Organization (WHO) (2016), there are 422 million people with diabetes worldwide, which was quadrupled since 1980 [1]. The DM complication lead to heart attack, blindness, kidney failure, lower limb amputation and stroke, which caused 1.5 million death in 2012. Indonesia is in the list of the 10 countries to have the highest number of diabetes, with diabetics of 8.4 million in 2000 and estimated to be 21.3 million diabetics in 2030 [1]. However, many people are not aware that they are living with pre-diabetes [2]. The diabetic patients are generally meet their physician in serious health complications. In this situation, it is very important to early detect diabetes.

The most popular in the diabetes early diagnostic is using a glucose biosensor, which is an analytical device in the combination of biological sensing elements and a transducer. Glucose biosensor was widely available on the market compare to other biosensor application. Improving the glucose biosensor remains interesting research topics to improve the performance [3], reduce the cost [4][5] or environment friendly method development in the analytical chemistry studies. The development of biosensors could be performed in the detection principles or the instrumentations.

On the other hand, the development of analytical technique should also consider the green chemistry for sustainable development. There are twelve principles in green analytical chemistry [6], which one of them would be applied in this study. The use of flower extract as biosensor colorimetric indicator was in the term from the ten list of "reagent obtained from renewable source should be preferred". Numerous study have been reported in the green analytical chemistry development such as solventless extraction [7], flower extract for nanoparticle preparation [8], eco-friendly enzyme [9] and non-toxic colorant [10]. In this study glucose detection was based on the reactions of glucose with glucose oxidase to produce gluconolactone and hydrogen peroxide (H_2O_2). The resulted hydrogen peroxide was then reacted with the *Etlingera elatior* flower extract to obtain the color change related to the glucose concentration. The performance of the *Etlingera elatior* flower extract stability were also studied.

2. Materials and Methods

2.1. Materials

Etlingera elatior was obtained from local market, which further was identified the taxonomy at the Biology Laboratorium of Universitas Jenderal Soedirman. Ethanol, hydrochloric acid, citric acid, acetic acid, sodium acetate, sodium dihydrogen phosphate, di-sodium hydrogen phosphate, hydrogen peroxide, and ascorbic acid were from E-Merck (Germany). Glucose Oxidase from Aspergillus niger (Type VII, lyophilized powder, ≥100,000 units/g, Sigma Aldrich)

2.2. Apparatus and measurements

Spectrophotometer (Shimadzu UV-1800) was used to measure the color change of the reactions. Temperature effect was used hotplate stirrer (Thermoline, Cimarec C), pH meter (Hanna instruments) was used to measure the pH of the solution. General laboratory instruments and glassware were also used in this research.

2.3. Etlingera elatior flower extraction

The *Etlingera elatior* flower was separated from their stalk, clean with water and grinded using blender. Solvent of mix ethanol 96% and HCl 1.5N (85:15) was then added to the flower (20 g flower / 100 mL solvent), mixed for 1 h and precipitated for 30 min. The filtrate was then separated using filter paper and keep in dark place at 4 ^{0}C when it was not in use. The flower extract was measured the absorbance using spectrophotometer, scanned at 400 – 700 nm. The extract was adjusted with the solvent to get the absorbance of about 0.8.

2.4. Effect of pH

Flower extract color was studied at various pH of 3.0 to 9.0 by diluting 10 mL of extract with 100 mL of buffer solution of the corresponding pH. The dilution of the extract was based on the previous procedure (2.1). Absorbance change with the pH of the buffer solvent were then recorded at 400 – 700 nm.

2.5. Effect of oxidizing and reducing agent

The study was performed at pH of 7.0, with the addition of 0.1 M hydrogen peroxide (0.1 mL/mL extract) as oxidizing agent and ascorbic acid (5 mg/mL) as reducing agent. The color change with the oxidizing or reducing agent was studied in various incubation time, before the extract oxidized or reduced to a colorless solution.

2.6. Effect of heat treatment

Extract of 10 mL was placed in well capped bottle to avoid the solvent to evaporate during heat treatment. The extract was then kept in the water bath at 40 ^{0}C and 60 ^{0}C for one and two hours. The color change was measured using spectrophotometer at 400-700 nm.

2.7. Effect of storage on extract decolorization

The effect of storage was performed by storing the concentrated filtrate at 4 ^{0}C in a dark bottle for a week. The color change was studied every day at 400-700 nm, by diluting the concentrated extract using pure water.

2.8. Glucose biosensor indicator study using Etlingera elatior extract

The performance of the flower extract as indicator of glucose biosensor have been tested using a standard glucose solution and glucose oxidase enzyme. Series glucose solution as samples have been prepare using phosphate buffer of pH 7.0. Five ml glucose solution was prepare in test tube, added 10 unit glucose oxidase enzyme and incubated in room temperature for 15 minutes. One ml flower extract was then added to the glucose – enzyme solution. The color change related to glucose concentration was measured using spectrophotometer at 400-700 nm.

3. Results and Discussion

3.1. Etlingera elatior flower extraction

The collected *Etlingera elatior* flower was confirmed that, the flower was *Etlingera elatior,* family of Zingiberaceae, with the local name of "kecombrang" (Fig 1).

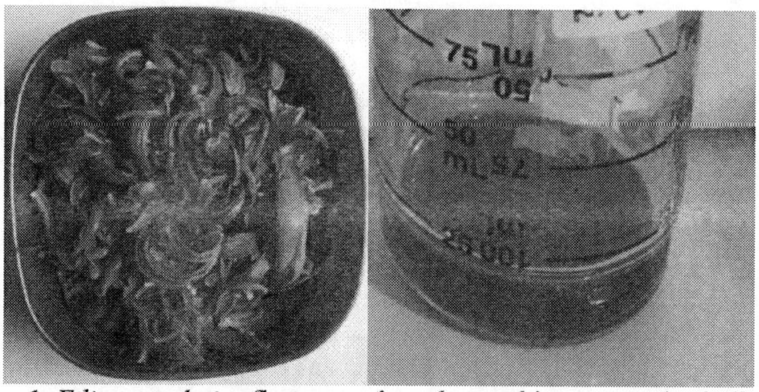

Figure 1. *Etlingera elatior* flower used as glucose biosensor color indicator

Measurement of flower extract was first diluted using the solvent for ten times to rich the good absorbance (0.2 to 0.8). The spectrum showed one peak at 523 nm. The visual color of the extract was purplish red, could be a 3-glycoside anthocyanidin, which was similar to the spectral characteristic of anthocyanin showed absorbance peak at 475-560 nm [11].

3.2. Effect of pH

Anthocyanin was the major compound in the flower extract as biosensor indicator studied. In the solution, anthocyanin could be transform into several ion form, depend on the pH of the solution. In this work, the extract color was studied in the range pH of 3.0 to 9.0 would cover acidic, neutral and alkaline condition. The result (**Fig. 2**) showed the decreasing of absorbance from pH 3.0 to 5.0 may due to the structure change of anthocyanin with the pH which could be in a blue quinoidal base, red flavylium cation or colorless carbinol pseudobase and calcone [12]. In the plant extract are consist of peroxidase and polyphenolase. These two enzyme were optimum at pH of 6-7 [13], catalyze the enzymatic browning which may involve in the increasing of absorbance at pH 5.0 to 6.0. The higher pH absorbance at the higher pH could also due to the structural change with the pH.

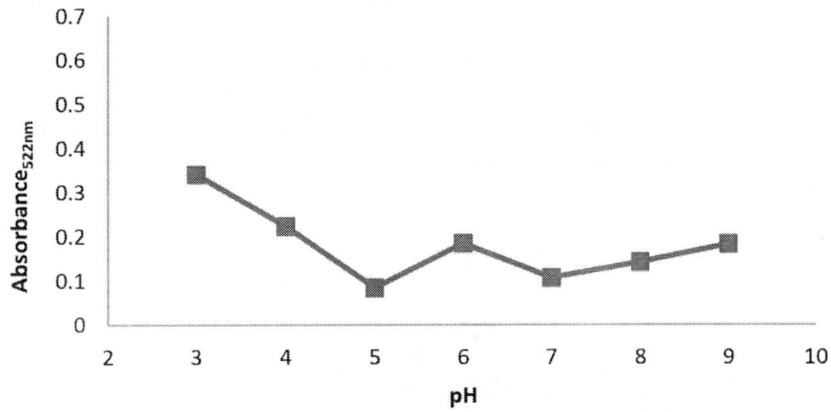

Figure 2. Effect of pH on the *Etlingera elatior* flower absorbance

3.3. Effect of oxidizing and reducing agent

This study was to measure the color change of the extract by the addition of ascorbic acid as an example of reducing agent and hydrogen peroxide as oxidizing agent. The result show that the addition of reducing agent did not effect to the color of flower extract up to 120 minutes incubation, whereas the addition of oxidizing agent of hydrogen peroxide was decrease the flower extract color with the incubation time (Fig. 3). The reducing of extract absorbance may due to the oxidizing agent attack the red color flavylium to lose their proton, resulted colorless solution [14], and the redox reaction of anthocyanin are irreversible [15].

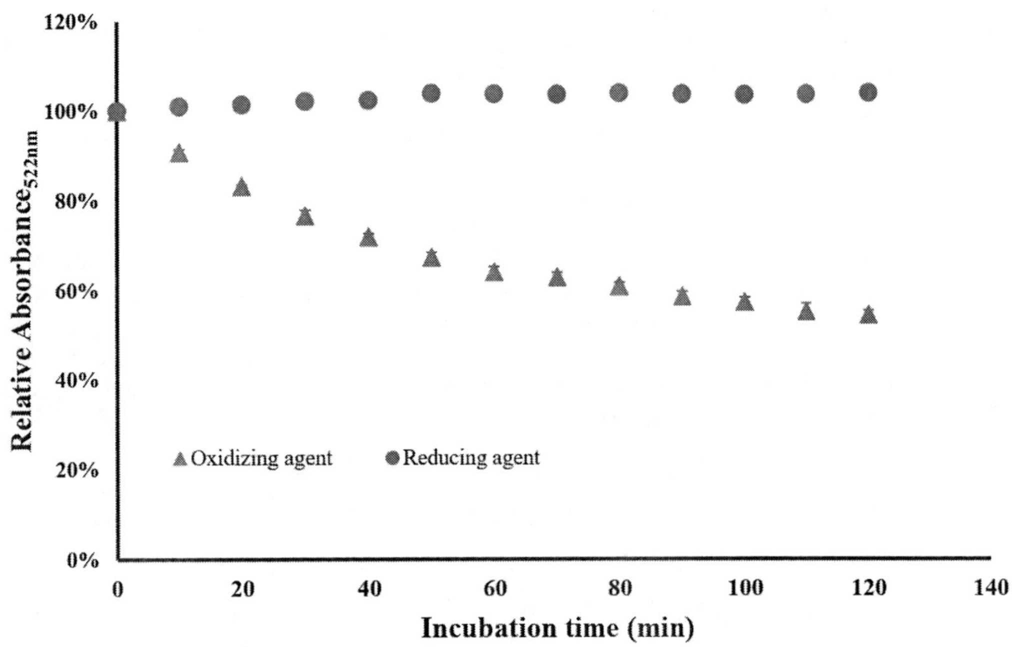

Figure 3. Effect of oxidizing (hydrogen peroxide) and reducing (ascorbic acid) agent on the *Etlingera elatior* flower absorbance with the incubation duration

3.4. Effect of heat treatment

Temperature is one of important factor of anthocyanin color stability, both during their using and storing. In this work, flower extract has been treated at 40 ^{0}C and 60 ^{0}C for one and two hours, in the closed dark containers. The result (**Fig. 4**) showed the higher temperature and longer incubation time would decrease the extract absorbance. The anthocyanin content in the plant extract would decreased with the increasing of temperature [16] and also the absorbance decreasing showed the anthocyanin degradation which was related to the increasing of the temperature [17][18].

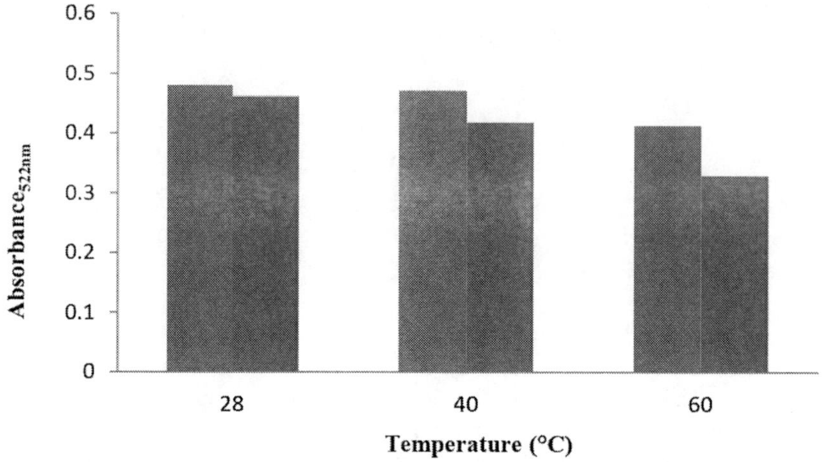

Figure 4. Effect of temperature and heat treatment duration (blue = 1 hr, Red = 2 hrs) to the extract absorbance

3.5. Effect of storage on extract color stability

Etlingera elatior extract contains of anthocyanin, pigments generally found in plants. The anthocyanin color and stability during storage are depend on the temperature and storage duration. In this work, the extract stability was studied in the room temperature and refrigerator temperature (4 ^{0}C) for one to seven days. The results showed that the extract remained their absorbance stable during storage study up to seven days (**Fig. 5**). The room temperature storage of seven days decrease the absorbance of the extract to 93%, whereas the refrigerator temperature storage, kept the absorbance of 100% on seven days storage duration. Thus, the extract should be kept in refrigerator (at 4 ^{0}C), in a closed dark container to prevent the color degradation. The decreasing of the color absorbance during storage may due to the intermolecular copigmentation [19]. The lower temperature (4 ^{0}C) kept the stability of the extract color compare to the room temperature, since the lower temperature would decrease the copigmentation reaction [12] and also the lower temperature would decrease the enzyme peroxidase and polyphenolase may present in the extract, thus the color was more stable.

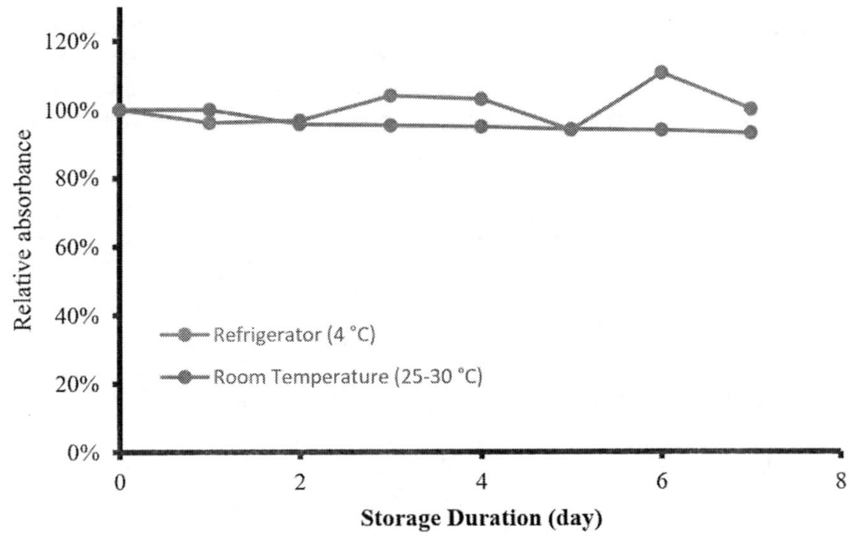

Figure 5. Storage duration and temperature effect to the *Etlingera elatior* extract stability showed high stability for up to seven days both for room temperature (red) and 4 ^{0}C (blue).

3.6. Glucose biosensor indicator study using Etlingera elatior extract

The glucose biosensor study was based on the color change of the *Etlingera elatior* extract with the resulted hydrogen as enzymatic reaction product of glucose and glucose oxidase. The higher glucose concentration, the higher hydrogen peroxide produced, would related to the reducing of extract color. This color change has been measured at 400-700 nm using spectrophotometer (**Fig. 6**). Furthermore, the absorbance change of the extract was linear to glucose concentration at 0 – 100 mM, with the regression equation of $y = -0.0005x + 0.4724$ and r^2 of 0.9965. The coefficient of determination (r^2) showed a high correlation between the glucose concertation and the color (absorbance) change, thus, the *Etlingera elatior* flower extract showed a great potential as a green reagent for glucose biosensor development, specially based on colorimetric detection by the glucose oxidase enzyme.

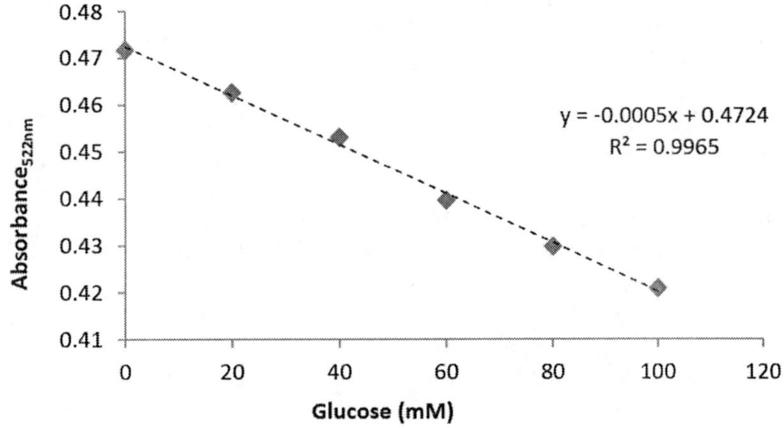

Figure 6. Calibration curve of glucose biosensor using *Etlingera elatior* flower extract

4. Conclusion

Etlingera elatior flower ethanol-HCl extract could be used as an indicator of glucose detection for a green chemistry approach of early glucose detection using colorimetric methods involving the use of glucose oxidase enzyme. The proposed method showed a good linearity (0 – 100 mM, R^2=0.9965) and high sensitivity (y = -0.0005x + 0.4724). Furthermore, the *Etlingera elatior* flower extract was also showed a high stability against pH, reductor, heating treatment and also stable during storage at 4 ^{0}C.

Acknowledgments

We would like to thank the Directorate General of Higher Education (DGHE, DIKTI) and the Jenderal Soedirman University for supporting this research through "Riset Unggulan Perguruan Tinggi" grant No 068/SP2H/LT/DRPM/IV/2017.

References

[1] Wild S, Roglic G, Green A, Sicree R and King H 2004 Global Prevalence of Diabetes: Estimates for the year 2000 and projections for 2030 *Diabetes Care* **27** 1047–53
[2] Troughton J, Jarvis J, Skinner C, Robertson N, Khunti K and Davies M 2008 Waiting for diabetes: Perceptions of people with pre-diabetes: A qualitative study *Patient Educ. Couns.* **72** 88–93
[3] Fatoni A, Numnuam A, Kanatharana P, Limbut W, Thammakhet C and Thavarungkul P 2013 A highly stable oxygen-independent glucose biosensor based on a chitosan-albumin cryogel incorporated with carbon nanotubes and ferrocene *Sensors Actuators, B Chem.* **185**
[4] Fatoni A, Anggraeni M D and Dwiasi D W 2016 Simple colorimetric glucose biosensor using chitosan cryogel supporting material *AIP Conference Proceedings* vol 1746
[5] Fatoni A, Dwiasi D W and Hermawan D 2016 Alginate cryogel based glucose biosensor *IOP Conference Series: Materials Science and Engineering* vol 107
[6] Gałuszka A, Migaszewski Z and Namieśnik J 2013 The 12 principles of green analytical chemistry and the SIGNIFICANCE mnemonic of green analytical practices *TrAC - Trends Anal. Chem.* **50** 78–84
[7] Tobiszewski M, Mechlińska A, Zygmunt B and Namieśnik J 2009 Green analytical chemistry in sample preparation for determination of trace organic pollutants *Trends Anal. Chem.* **28** 943–51
[8] Thovhogi N, Park E, Manikandan E, Maaza M and Gurib-Fakim A 2016 Physical properties of CdO nanoparticles synthesized by green chemistry via Hibiscus Sabdariffa flower extract *J. Alloys Compd.* **655** 314–20
[9] Riva S 2006 Laccases: blue enzymes for green chemistry *Trends Biotechnol.* **24** 219–26
[10] Sivakumar V, Anna J L, Vijayeeswarri J and Swaminathan G 2009 Ultrasound assisted enhancement in natural dye extraction from beetroot for industrial applications and natural dyeing of leather *Ultrason. Sonochem.* **16** 782–9
[11] Harborne J B 1998 *Phytochemical Methods; A Guide to Modern Techniques of Plant Analysis* vol 3
[12] Devi P S, Saravanakumar M and Mohandas S 2012 The effects of temperature and pH on stability of anthocyanins from red sorghum (Sorghum bicolor) bran *African J. Food Sci.* **6** 567–73
[13] Mizobutsi G P, Finger F L, Ribeiro R A, Puschmann R, Ludmila ;, De L, Neves M, Ferreira W and Mota D 2010 Peroxidase and polyphenoloxidase activities of litchi pericarp Effect of pH and temperature on peroxidase and polyphenoloxidase activities of litchi pericarp *Sci. Agric. (Piracicaba* **67** 213–7
[14] Welch C, Wu Q and Simon J 2008 Recent Advances in Anthocyanin Analysis and Characterization *Curr. Anal. Chem.* **4** 75–101
[15] Janeiro P and Brett A M O 2007 Redox behavior of anthocyanins present in Vitis vinifera L. *Electroanalysis* **19** 1779–86
[16] Boranbayeva T, Karadeniz F and Yilmaz E 2014 Effect of Storage on Anthocyanin Degradation in Black Mulberry Juice and Concentrates *Food Bioprocess Technol.* **7** 1894–902
[17] Hou Z, Qin P, Zhang Y, Cui S and Ren G 2013 Identification of anthocyanins isolated from black

rice (Oryza sativa L.) and their degradation kinetics *Food Res. Int.* **50** 691–7
[18] Sólyom K, Solá R, Cocero M J and Mato R B 2014 Thermal degradation of grape marc polyphenols *Food Chem.* **159** 361–6
[19] Eiro M J and Heinonen M 2002 Anthocyanin color behavior and stability during storage: Effect of intermolecular copigmentation *J. Agric. Food Chem.* **50** 7461–6

IOP Conf. Series: Materials Science and Engineering **299** (2018) 012016 doi:10.1088/1757-899X/299/1/012016

Stability Test of Partially Purified Bromelain from Pineapple (*Ananas comosus* (L.) Merr) Core Extract in Artificial Stomach Fluid

S Setiasih[1], A Adimas Ch. D.[1], V Dzikria[1], S Hudiyono[1]
[1]Departement of Chemistry, FMIPAUniversitas Indonesia, Kampus UI Depok, Depok 16424, Indonesia

E-mail: setiasih@ui.ac.id

Abstract. This study aimed to isolate and purify bromelain from pineapple core (*Ananascomosus* (L.) Merr) accompanied by a stability test of its enzyme activity in artificial gastric juice. Purification steps start with fractionation by a precipitation method were carried out stepwise using several concentration of ammonium sulfate salt, followed by dialysis prosess and ion exchange chromatography on DEAE-cellulose column. Each step of purification produced an increasing specific activity in enzyme fraction, starting with crude extract, respectively: 0.276 U/mg; 14.591 U/mg; and 16.05 U/mg. Bromelain fraction with the highest level of purity was obtained in 50-80% ammonium sulphate fraction after dialyzed in the amount of 58.15 times compared to the crude extract. Further purification of the enzyme by DEAE-cellulose column produced bromelain which had a purity level 160-fold compared to crude enzyme. The result of bromelain stability test in artificial stomach juice by milk clotting units assay bromelain fraction have proteolytic activity in clotting milk substrate. Exposing bromelain fraction in artificial stomach juice which gave the highest core bromelain proteolytic activity was achieved at estimated volume of 0.4-0.5 mL. Exposure in a period of reaction time to artificial stomach juice that contained pepsin showed relatively stable proteolytic activity in the first 4 hours.

1. Introduction

Bromelain is a protein found in the pineapple plants (*Ananas comosus*) of the family Bromeliaceae belonging to the group of protease enzymes.Bromelain is a protease that can hydrolyze proteins.The presence of proteolytic activity in bromelain is beneficial in the therapeutic field [1].

Proteases are the most well-known among industrial enzymes. The most well-known plant proteases are papain from *Carica papaya*, ficin from *Ficus spp.* and bromelain from pineapple (*Ananas comosus*). Bromelain is one of the protease enzyme found in pineapple [2]. Core bromelain is widely used in industri and medication. Based on literature, bromelain in digestive system can be absorbed without being degraded and without losing its activity [1].
In this research, core part of pineapple was used as source of bromelain. The parameters tested included proteolytic activity and specific activity after fractionation of bromelain using ammonium sulphate followed by dialysis and fractionation with cellulose DEAE column chromatography.After the purification steps were done, the proteolytic activity of the obtained bromelain fraction was tested based on the milk-clotting unit assay after bromelain was incubated in a medium of artificial gastric fluid.

2. Materials and Methods

Instruments used were UV-Vis spectrophotometer, shaker, buchner, refrigerated centrifuge, vortex, ice box, cuvette, pH meter, micropipette and tip, column, test vials, glass equipment (beaker glass etc), waterbath, thermometer, refrigerator, and blender.

Material and chemicals used were pineapple (core part) from Bogor, pepsin, phosphate buffer pH 7.0, Tris-HCl buffer (0.05 M, pH 8.0), casein 1%, HCl, NaOH, NaCl, TCA (trichloroacetate) 5%, BSA standard solution, Follin reagent 1 N, Lowy reagent, ice cubes, aquadest, aquademin, glacial acetic acid, anhydrous sodium phosphate, L-cystein, EDTA, aluminium foil, muslin, ringstand, and filtering paper.

2.1 Sample and Crude Extract Preparation.

Core part of the pineapple were separated, weighed, diced and juiced using blender while kept at 4°C. Then, it was filtered through muslin and centrifuged. Crude extract was stored in refrigerator when not in use. The next steps are testing the protein content and proteolytic activity of the crude extract and core pineapple juice.

2.2 Crude Extract Purification.

Cold filtrate was fractionated using ammonium sulfate at different concentration (0-20%; 20-50%; 50-80%). Protein precipitated produced (in form of pellets) by fractionation were separated through centrifuge, and resuspended in cold phosphate buffer solution pH 7.0. Each fraction, including residual fraction, were tested of its protein content and proteolytic activity [3].Ammonium sulfate fraction of bromelain with highest proteolytic activity were treated with dialysis to remove the effect of salt on bromelain's purity.

2.3 Purificationby ExchangeChromatography on DEAE CelluloseColumn

After dialysis, bromelain was purified further with ion exchange chromatography using DEAE-cellulose as a stationary phase. The sample was elutedwithTris-HClbuffer (pH 8.0; 0.05M) containing NaCl with concentration –rise gradually from 0.25 – 1.0 M. Gradient elution at a flow rate of 1.5 mL/min.

2.4 Bromelain ProteolyticActivity.

Proteolytic activity of each bromelain fractions were determined using Kunitz method, using casein as substrate [4,5].

2.5 Bromelain Specific Activity.

Enzyme specific activity was achieved by dividing proteolytic activity with protein content. Protein content was determined using Lowry method.

2.6 *Artificial Stomach Juice Preparation.*

Artificial stomach fluid was prepared by dissolving 2.0 g of NaCl and 3.2 g of purified pepsin(derived from porcine stomach mucosa, with an activity of 800 to 2500 units per mg of protein) in 7.0 mL of hydrochloric acid and water up to 1000 mL[6].

2.7 *Stability Test of Bromelain in Artificial Stomach Fluide.*

Bromelain fraction with highest activity were tested its stability in artificial stomach juice. Substrate used was buffered milk substrate based milk clotting units assay[7].

3. Results and Discussions

3.1 Crude Extract and Results of Fractionation with Ammonium Sulfate

From the pineapple core sample, core juice gave specific activity of 0.006 U/mg protein, while crude extract gave specific activity of 0.276 U/mg. Table 1 show the highest specific activity in bromelain faction can be found in 3rd fraction (50-80%), which is 14.591 U/mg with degree of purity 52.8 fold in comparison with crude extract, and yield of 0.97%.

Table 1 Results of Bromelain Activity Evaluation at Steps of Isolation and Fractionation using Ammonium Sulfate

Fraksi	V	Total		SA	SA	DoP	Y (%)
		PA	PC				
Core Juice	409,6	0,267	42,9	1,272	0,006	-	100
Crude Extract	359	6,85	24,74	7,318	0,276	46	57,67
F 0-20%	14,2	0,2	1,44	0,985	0,138	0,5	5,82
F 20-50%	17	1,9	1,42	11,323	1,331	4,8	5,73
F 50-80%	7,1	3,6	0,24	51,873	14,591	52,8	0,97
Residual Fraction	256	0,116	8,04	0,014	0,014	0,05	32,49

V –Volume (mL); PC –Protein Content (mg); PA –Proteolytic Activity (Unit); SA –Specific Activity (Unit/mg); DoP –Degree of Purity; Y(%) –Yield (%)

The result of fractionations is not fully purified, because the protein precipitated by ammonium sulfate is not only bromelain protein. Enzyme fraction with highest specific activity need to purified further, and to achieve higher degree of purity the fraction will be treated with dialysis.

3.2 Result of Dialyzed Fraction

After treating the ammonium sulfate fraction with dialysis, specific activity of the bromelain fraction is increased from 14.591 U/mg to 16.05 U/mg, with degree of purity of 58.15 fold in comparison with crude extract and yield of 0.6%. Specific activity of bromelain fraction before and after dialysis is shown at Table 2.

Table 2 Result of Dialyzed Bromelain Fraction

Fraksi 3	V	Total		SA	SA	DoP	Y (%)
		PA	PC				
Before dialisis	5	3,6	0,24	51,873	14,591	52,8	0,97
After dialisis	7	2,5	0,15	55,555	16,05	58,15	0,60

V –Volume (mL); PC –Protein Content (mg); PA –Proteolytic Activity (Unit); SA –Specific Activity (Unit/mg); DoP –Degree of Purity; Y(%) –Yield (%)

3.3 Result of Ion Exchange Chromatography

After treating the dialysis fraction with Ion exchange chormatography, specific activity of the bromelain fraction is increased from 55.55 U/mg to 157.61 U/mg,with degree of purity of 160-fold in comparison with crude extract. The ion exchange chromatogram is shown at Figure 1.

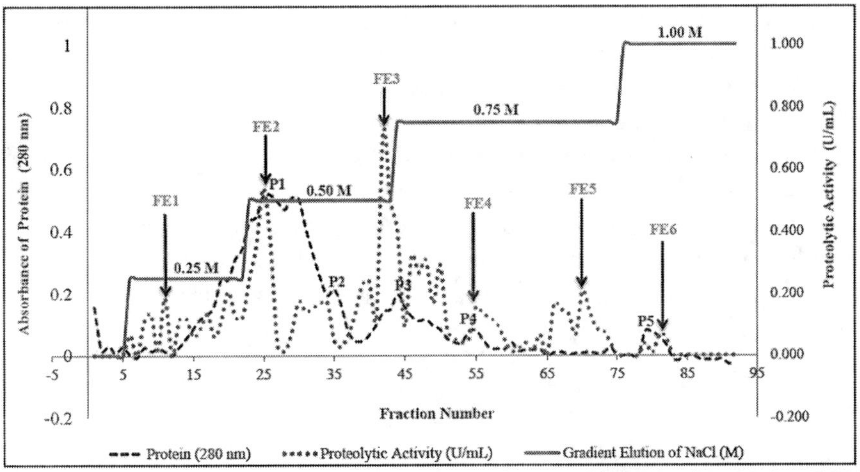

Figure 1. Chromatogram for 50-80% Ammonium Sulfat Fraction Using DEAE-CelluloseMatrix. Pr: The Peak of Protein at 280 nm, FE: The Peak of Protein with Proteolytic Activity

Separation conditions: column size 1.6 x 20 cm, matrix 135 mL volume flow rate was set at 1.5 mL/min and fraction were collected every 5 mL. The elution sycore using a combined method of linear gradient and step wise elution. The initial stage of the column was eluted with Tris-HCl buffer 0.05M pH 8.0; subsequent elution using a Tris-HCl buffer 0.05M pH 8.0 containing 0.25 M to 1.0 M NaCl.

In this process, the enzyme solution were diluted that the proteolytic activity was reduced, but the protein content was also reduced, resulting in an increase of specific activity. Bromelain fractions achieved at each step of purification gave an increasing factor of purity, as shown in Figure 2.

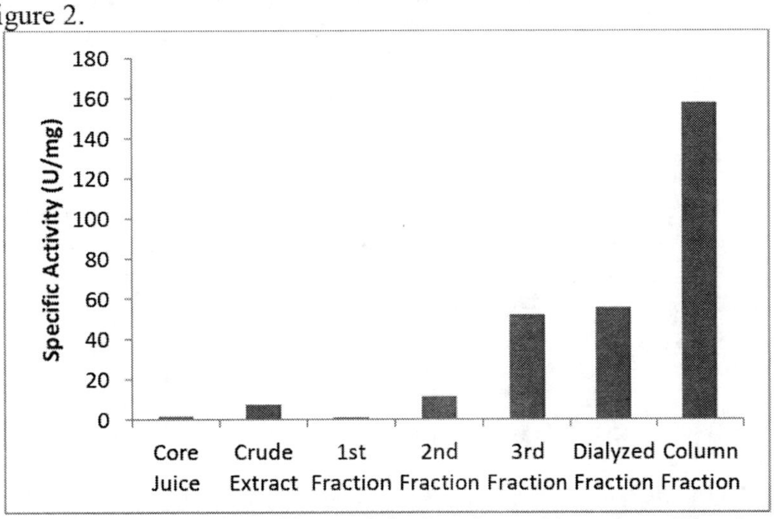

Figure 2. Specific activity of bromelain fraction with each purification step.

3.4. Results of Bromelain Stability Test in Artificial Stomach Fluid.

In this research, proteolytic activity of bromelain was tested in vitro in artificial stomach fluid. Bromelain will be 4ydrolyzin by stomach acid and other protease after consumption through oral pathway [6]. Simulated gastric fluid test solution is an artificial dissolution medium used to represent stomach acid. Stomach state simulated by this solution is fasted state. After the determined time for reaction with arficial stomach fluid has passed, the bromelain activity was tested using milk clotting unit assay. The assay was conducted using milk substrate that has been equilibrated to 40°C, where enzymatic activity in clotting milk substrate is at maximum capacity [7]. Due to artificial stomach fluid containing other protease (pepsin) which is capable of 5ydrolyzing milk protein, in this research a negative control containing only buffer is used. The stability test is conducted in three steps: (1) stability test of different bromelain fractions, (2) stability test of a bromelain fraction at different concentration, and (3) stability test of bromelain fraction at prolonged reaction time.

First, each bromelain fraction's stability was tested as a comparison, in this case the crude extract, 3rd ammonium sulfate fraction, and the dialyzed fraction. As a further comparison to the purified bromelain fraction, commercial bromelain's stability was also tested. Result of the test is shown in Figure 3.

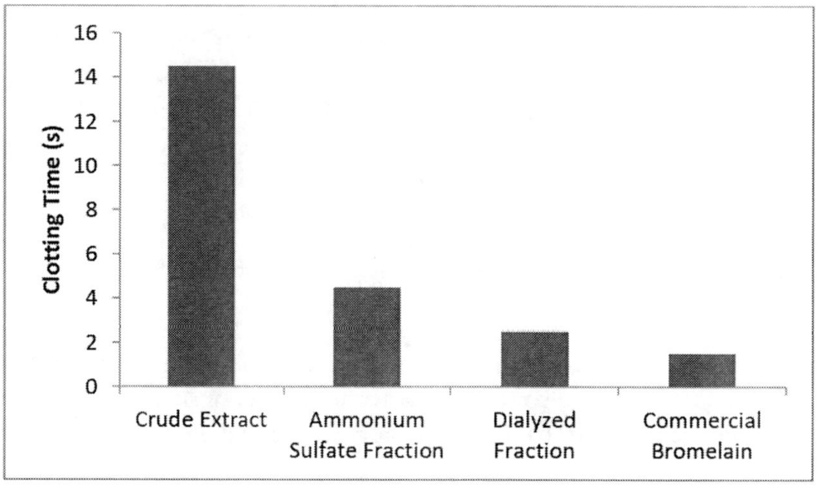

Figure 3. Clotting time of milk substrate by bromelain fractions after reaction with artificial stomach fluid.

Based on the clotting time of each sample, the bromelain fractions showed an existing clotting activity, with an increase of activity based on reduced clotting time needed by purer (dialyzed) fraction. The activity of dialyzed fraction is still relatively low when compared with commercial bromelain, which had higher degree of purity.

Second, a bromelain fraction's stability was tested at different concentration. The fraction tested was the dialyzed fraction. Result of this test is shown in Figure 4.

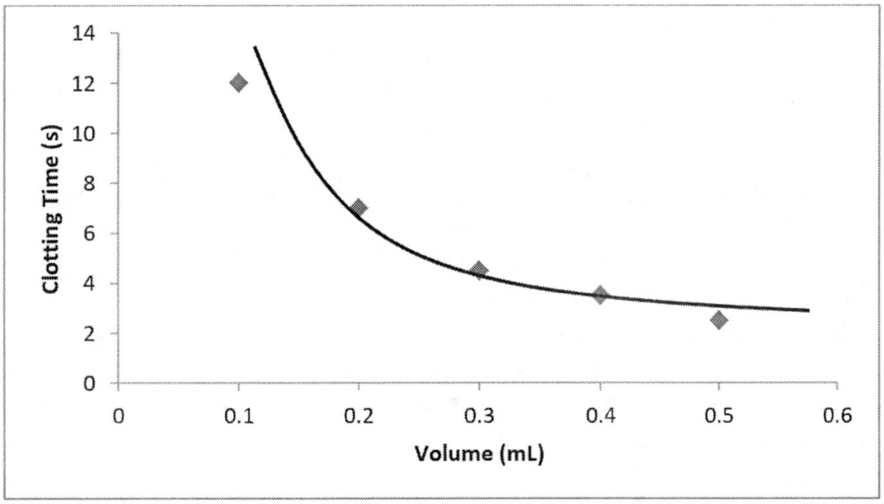

Figure 4. Milk clotting time by bromelain fraction at different concentration.

The time needed by higher volume of bromelain fraction to clot milk substrate is reduced. Higher bromelain concentration is capable of coagulating casein in milk at faster rate than lower concentration [2]. At increased concentration, as the volume approach 0.4 mL,the increase in proteolytic activity became insignificant.

Third, a bromelain fraction's stability was tested at a prolonged reaction time with artificial stomach juice. The volume of dialyzed fraction used was 0.5 mL. Result of this test is shown in Figure 5.

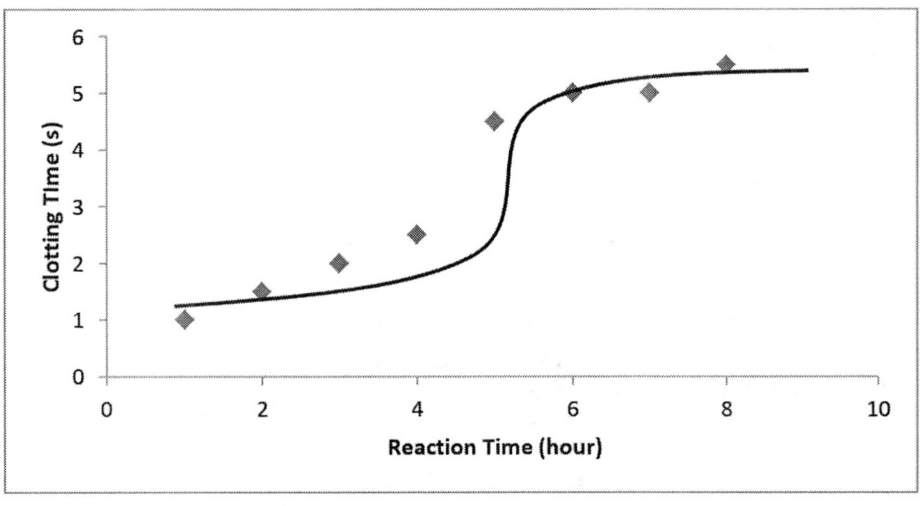

Figure 5. Stability of bromelain fraction in artificial stomach fluid after 8 hours of reaction.

Bromelain proteolytic activity at the first 4 hours is relatively stable, although slowly decrease as the reaction time goes on. After 4th reaction hour, the activity was reduced significantly. The reduced activity was due to deactivation and degradation of bromelain by gastric enzyme (pepsin) and low pH of stomach acid [2]. At 5th hour and onward, the reduction of bromelain activity became insignificant.

4. Conclusions

Bromelain was successfully purified through a series of purification steps, starting from isolation of crude extract, fractionation using ammonium sulfate, and dialysis. Further step of purification yielded purer bromelarin fraction, as shown by as shown by ion exchange chromatography fraction gave higher specific activity. Bomelainactivityinartificial stomach fluid showed that at higher degree of purity, bromelain fraction gave higher proteolytic activity. Bromelain also showed higher activity when higher concentration of bromelain fraction reacted with artificial stomach fluid, and relatively stable for a period of time. However, prolonged reaction time with artificial stomach fluids showed that bromelain would have significantly reduced activity, giving an insight on a relation between digestive system and bromelain (pineapple). For future study of bromelain stability in artificial stomach fluids, it's suggested to add another purification step to achieve bromelain fraction with higher degree of purity, like gel filtration column chromatography. It's also suggested to prepare another artificial stomach juice simulating stomach on fed state.

Acknowledgement

This research was funded by Biaya Operasional Perguruan Tinggi Negeri (BOPTN) DIKTI 2016.

References

[1] Bala M, Ismail N A, Mel M, Jami M S, Salleh M and Amid A 2012 Bromelain Production : Current Trends and Perspective. *Archives Des Sciences* **65** (11) 369–399

[2] Neta JLV, DaSilva LA, Lima AA, Santana JC, Leite NS, Ruzene DS, SilvaDP, DeSouza R R 2012 Bromelain Enzyme from Pineapple: In Vitro Activity Study under Different Micropropagation Conditions. *Appl Biochem Biotechnol 168 (2) 234-246*

[3] Setiasih S, Ilyas N M, Handayani S and Hudiyono S 2016 Proceeding of International Seminar: Collaborative Seminar of Chemistry and Industry (COSCI), Surabaya, 50-54

[4] Febriani K, Wahyuni I, Setiasih S and Hudiyono 2017 AIP Conference Proceedings 1862, 030095; doi: 10.1063/1.4991199

[5] Setiasih S, Wahyuni I and Hudiyono S 2016 Proceeding of International Conference on Biosciences & Medical Engineering (ICBME)s UTM-Johor Malaysia 163-170

[6] Shiew P S, Fang Y L, Adibah F and Majid A 2010 in Vitro Study of Bromelain Activity in Artificial Stomach Juice and Blood October

[7] Dash P, Maharana L and Ghosh G 2015 Proteolytic and Milk Clotting Activity of Fractionated Protein of Lagenaria siceraria. *Journal of Applied Pharmaceutical Science* **5** (12) 086–091

In Vivo antiplatelet activity aggregation assay of bromelain fractionate by ethanol from extract pineapple core (*Ananas comosus* [l.] merr.)

F F Musfiroh[1], S Setiasih[1], S Handayani[1], S Hudiyono[1] and N M Ilyas[1]
[1]Department of Chemistry, Faculty of Mathematics and Natural Sciences, Universitas Indonesia, Depok, West Java, Indonesia 16424

E-mail: Siswatisetiasih@gmail.com

Abstract. Processed fruit from pineapple is one of largest commodities tropical fruit production in Indonesia and will bring the waste from the skin and core .This study aims to isolate bromelain from the pineapple core (*Ananas comusus*) are purified by fractionation using ethanol and continued by activity test as an antiplatelets agent by in vivo method using white mice male ddy type with acetosal as positive control. Fractionation of crude enzyme bromelain with ethanol produces highest specific activity on ethanol 30-60% fraction (fraction 2) 3.107 Unit/mg and the protein content 61.25 mg with the degree of purity of 155 times compared to crude enzyme. Antiplatelet aggregation tests from in vivo method shows that faction of bromelain with doses 70 μg/KgBW, 140 μg/KgBW, and 210 μg/KgBW can increase a meaningful bleeding time. The highest percentage of increase shown by the isolate at a dose of 210 μg/KgBW in the amount of 515.10 ± 182.23%, when compared to aspirin (231.20 ± 140.66), the value is relatively higher.

Keywords: *Ananas comosus*, pineapple core, bromelain, purification, antiplatelet, in vivo

1. INTRODUCTION

Pineapple (*Ananas comosus*) is a commodity in the trade of tropical fruits which ranks second only to the bananas. Meanwhile Indonesia is the fifth-largest producer after Brazil, Thailand, the Philippines, and China [1]. According to Central Bureau of Statistics (2011), pineapple production in Indonesia increased from 2007 to 2009 is 1.395.566 tons, 1.433.133 tons and 1.558.196 tons.

Research enzyme bromelain in pineapple plants have been started since 1894 until today. Various ways of isolation has been conducted to get the enzyme bromelain with the best enzyme activity [2]. The protease enzymes useful in reducing the clumping of platelets, the formation of plaque in the arteries, and blood clots. All of these effects help in the treatment of cardiovascular disease [3]. Bromelain is widely used in therapeutic applications, among others as antitrombic, fibrinolytic, reversible inhibition of platelet aggregation [4]. Based on this background, research must be done test antiplatelet activity of extracts of pineapple fruit weevil.

The enzyme bromelain has been successfully demonstrated as an antiplatelet agent by in vitro method [5]. Hence, in this study wanted to know more comprehensive effects in the body through the test in vivo, using animal and clinical trials. This approach is employed to test the in vivo findings because it is more suitable to observe the overall effect on the living subject.

2. EXPERIMENTAL

2.1. Preparation of crude extract from pineapple core

Core was cut into pieces and weighed as much as 1000 gram, then blended and filtered using a filter cloth. Furthermore, core solution that has been obtained centrifuged at 6000 rpm for 15 minutes at 4°C. Supernatant was filtered and the filtrate taken as a crude extract of bromelain.

2.2. Bromelain purification from crude extract by ethanol precipitation

Bromelain crude extract obtained proceed to the purification process by precipitation method using ethanol. The concentration level was divided into three fractions, fraction 1 (0-30)%, fraction 2 (30-60)% and fraction 3 (60-80)%. Ethanol was added to the crude extract at 4°C. The extracts was kept for ± 12 hours. Extract was centrifuged with a speed of 6000 rpm for 15 minutes at 4°C. The precipitate obtained re-suspended using 0.2 M phosphate buffer pH 7. The filtrate was resumed to higher concentration levels.

2.3. Measurements of protein content and enzymatic activity

The protein content was measured by Lowry method. Standard was prepared using the protein BSA (Bovine Serum Albumine) with a concentration of 800 ppm. 0.1 mL sample was added 0.4 mL of 0.2 M phosphate buffer pH 7 and 5 mL reagent Lowry. Then incubated for 10 minutes at a temperature (4 ± 2)°C. Is then added and incubated 0.5 mL Folin Ciocalteu back for 30 minutes at a temperature (37 ± 2)°C. The solution was measured using a UV-Vis spectrophotometer at a wavelength of 595 nm [6,7,8].

Proteolytic activity was tested by the method Kunitz [9]. Made of control solution and sample. The sample solution is made by adding 0.9 mL 0.05 M tris buffer pH 8 at 0,1 mL sample. Then the substrate in the form of casein was added 1 mL 1% and incubated in a shaker for 30 minutes at a temperature (37 ± 2)°C. Further sampling was disabled with the addition of 3 mL of 10% TCA and incubated again in a shaker for 30 minutes at a temperature (4 ± 2)°C. The control solution was made equal to the sample, but added TCA 10% in advance so that the enzyme off, just after it added its substrate in the form of casein 1%.

2.4. Antiplatelet Aggregation Assays

Animal test that used in this study ddy strain male white mice were divided into five groups: positive control (acetosal), negative control (aquades), dose 1 (70 µg/Kgbw), dose 2 (140 µg/Kgbw), and dose 3 (210 µg/Kgbw).

2.5. Measurements of bleeding time

Bleeding time was measured by the tail of mice injured (cut 2mm), then the tail placed on filter paper to accommodate the blood coming out, note the time until the blood stopped flowing [10].

3. RESULTS AND DISCUSSION

3.1. Bromelain purification from crude extract by ethanol precipitation

The results of proteolytic activity from core solution and bromelain enzyme crude extract are shown in Table 1. The core solution and crude extract enzyme has proteolytic activity 0.967 Unit, and 1.222 Unit, and protein content of 84.11 mg and 61.15 mg of the total volume of solution. The value of the specific activity for both the solution amounted to 0.011 Units/mg and 0,020 Units/mg. From the data it showed an increase of proteolytic activity and specific activity of the enzyme after centrifugation of the solution hump despite the increase in value that occur less significant and purity level of enzyme was still very small. This is due to the enzyme crude extract still contains soluble proteins other than bromelain. Therefore, it is necessary to further purification stages to obtain a more pure enzyme fraction.

Table 1. Effect of ethanol concentration on bromelain precipitation

Sample	Total			Spesific activity (U/mg)
	Volume (mL)	Protein content (mg)	Proteolytic activity (U)	
Core solution	415	84.11	0.97	0.011
Crude Enzyme	362	61.15	1.22	0.020
Fraction 1 (0-30) %	15	2.33	1.33	0.571
Fraction 2 (30-60) %	46	3.30	10.27	3.107
Fraction 3 (60-80)%	24	3.45	1.57	0.454
Remaining filtrate	1440	7.44	0.00	0.000

Enzyme solution of each fraction were then determined their proteolytic activity and protein levels. Table 1 shows the value of proteolytic activity of each fraction obtained in the fractionation stage. The highest proteolytic activity was obtained in fraction 2 with a value of 10.267 Units and protein levels of 3.30 mg. From the data proteolytic activity and protein levels, the highest specific activity obtained in fraction 2 with a value of 3.107 units/mg with a purity level reached 155 times.

3.2. Antiplatelet aggregation assays

Lengthening of bleeding time showed a decrease in the activity of platelet aggregation induced by inhibition of the cyclooxygenase enzyme, thereby decreasing the synthesis of thromboxane A2. Thromboxane A2 is one of the mediators that influence the process of platelet activation and vasoconstriction when the process is mediated by platelet hemostasis [11].

The results of antiplatelet aggregation assay with parameter bleeding time showed an increased bleeding time of mice were significantly since the 7th day in all groups of test extracts compared to day 0. However, the increase is likely to fluctuate at each measurement (Figure 1). In the calculation, the highest percentage of increase shown by the isolate at a dose of 210 μg/KgBW in the amount of 515.10 ± 182.23%, when compared to aspirin (231.20 ± 140.66), the value is relatively higher. Each isolate of the enzyme indicates increasing time increase bleeding time.

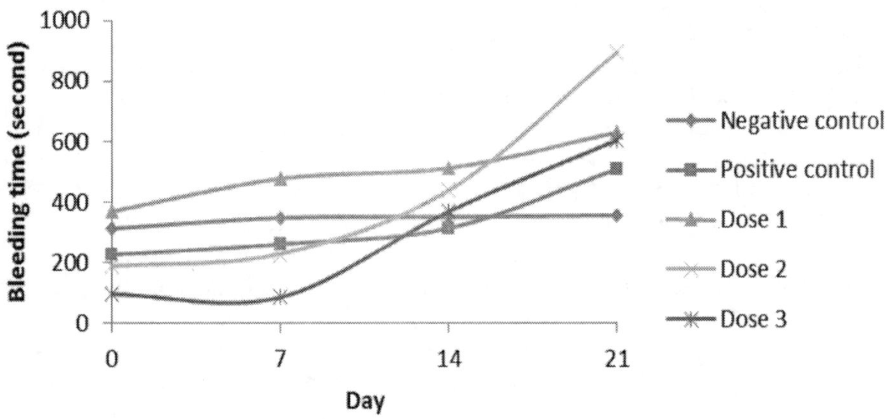

Figure 1. Bleeding time curve from isolate and control

Each fraction of enzyme indicates the increasing of bleeding time. Bleeding time of mice with various dose levels showed no significant difference from each fraction of enzyme.

4. CONCLUSIONS

Bromelain was successfully purified through a series of purification steps, starting from extraction of crude extract and precipitation of bromelain with ethanol. The result of bromelain after precipitation by ethanol gave higher specific activity than before precipitation. In vivo test of bromelain method by measuring bleeding time was divided into five groups of mice. All group of mice showed a positive response by increasing of the length time for each week.

5. ACKNOWLEDGEMENTS

This research was funded by Publikasi Internasional Terindeks Untuk Tugas Akhir Mahasiswa (PITTA) Universitas Indonesia 2017.

References

[1] Rugayah, Anggalia, I., Cahya, Y. (2012). Pengaruh Konsentrasi dan Cara Aplikasi Iba (Indole Butiric Acid) terhadap Pertumbuhan Bibit Nanas (Ananas Comosus [L.] Merr.) Asal Tunas Mahkota. Bandar Lampung: Jurusan Agroteknologi Fakultas Pertanian Universitas Lampung.

[2] Neta JLV, DaSilva LA, Lima AA, Santana JC, Leite NS, Ruzene DS, SilvaDP, DeSouza RR. (2012). Bromelain Enzyme from Pineapple: In Vitro Activity Study under Different Micropropagation Conditions. Appl Biochem Biotechnol; 168(2): 234-246.Another reference

[3] Bhattacharyya, B.K. (2008). Bromelain. India: Biotechnology and Molecular Biology. *East India Pharmaceutical Works Ltd. Vol 7(4).*

[4] Nawaz, Ali, Manzoor, Zoya, Mukhtar, Hamid, & Haq, Ikram. (2016). Bromelain: Methods of Extraction, Purification and Therapeutic Applications. *Journal of Brazilian Archives of Biology and Technology v.59: e16150010, Jan/Dec 2016.*

[5] Maurer, H.R., Metzig, Carola, Grabowska, Edyta, Eckert, Klaus, & Rehse, Klaus. (1999). Bromelain Proteases Reduce Human Platelet Aggregation in Vitro, Adhesion to Bovine Endothelial Cells and Thrombus Formation in Rat Vessels in Vivo. *Germany:Institute of Pharmacy, University ofBerlin. In vivo71-7-12.*

[6] Setiasih S, Ilyas N M, Handayani S and Hudiyono S. (2016). Proceeding of International Seminar: Collaborative Seminar of Chemistry and Industry (COSCI), Surabaya, 50-54.

[7] Febriani K, Wahyuni I, Setiasih S and Hudiyono S. (2017). AIP Conference Proceedings 1862, 030095; doi: 10.1063/1.4991199.

[8] Setiasih S, Wahyuni I and Hudiyono S. (2016). Proceeding of International Conference on Biosciences & Medical Engineering (ICBME), UTM-Johor Malaysia, 163-170.

[9] Soares, Paulo. A. G., Vaz, Antonio. F. M., Corela, Maria. T. S., Jr, Adalberto. Pessoa, Carneiro-da-Cunha, Maria. G. (2012). Purification of Bromelain from Bromelain Wastes by Precipitation Ethanol. *Separation and Purification Technology 98; 389–395.*

[10] Vogel, H. G. 2008. Drug *Discovery and Evaluation, Pharmacological Assay* (3rd Ed). Berlin: Springer.

[11] Anderson, H. Vernon., McNatt, Janice, Clubb, Fred. J., Herman, Michael ., Maffrand, Jean-Pierre., DeClerck, Fred., Ahn, Chul., Buja, L. Maximilian., & Willerson, James. T. (2001). Platelet Inhibition Reduces Cyclic Flow Variations and Neointimal Proliferation in Normal and Hypercholesterolemic-Atherosclerotic Canine Coronary Arteries. 2331-2337. http://doi.org/10.1161/hc4401.098434

IOP Conf. Series: Materials Science and Engineering **299** (2018) 012018 doi:10.1088/1757-899X/299/1/012018

The Potency of Ethanolic Extract of *Sauropus androgynus* (L.) Merr Leaves as Therapeutic herbal of Rats *(Rattus norvegicus)* Peptic Ulcer Model Induced by Aspirin

A Roosdiana[1*], S A Yudandi[2], A Erika[2]
1Faculty of Mathematics and Natural Sciences, University of Brawijaya, Malang, East Java, Indonesia
[2]Faculty of Veterinary Medicine, University of Brawijaya, Malang, East Java, Indonesia

E-mail: aroos@ub.ac.id

Abstract. Peptic ulcer is an erosion of the mucosa gaster and duodenum. Aspirin can increase the activity of ROS in duodenum leading to be oxidative stress condition, followed by the increase expression of TNF-α and changes in the duodenum histopathology. Ethanolic extract from *Sauropus androgynus* (L.) Merr leaves contains flavonoid compound that act as antioxidants. The research aimed to study the therapeutic effect of *Sauropus androgynus* (L.) Merr leaves towards the expression of TNF-α and the duodenum histopathology. The study used five groups: the negative control group, the positive control group (peptic ulcer), and three therapeutic groups. Peptic ulcer induced by aspirin 200 mg/kg BW given orally once a day for five days. The therapeutic group (dose 16.2 mg/200 g BW, 32.4 mg/200 g BW, 48.6 mg/200 g BW) administrated orally once a day for fourteen days. Expression of TNF-α were analyzed by ANOVA followed by Tukey test (α=5%) and histopathology description of duodenum were analyzed descriptive qualitative. The result showed that treatment of the ethanolic extract from *Sauropus androgynus* (L.) Merr leaves at dose 48.6 mg/ 200 g BW was the effective dosage. The conclusion of this research indicate that ethanolic extract from *Sauropus androgynus* (L.) Merr leaves can be used as alternative therapy peptic ulcer towards decrease the expression of TNF-α and repair duodenum histopathology such as structure of epithelial duodenum, vile and decrease of inflammatory cell and vascular vasodilatation

1. Introduction

Peptic ulcer or ulcuspepticum are ulcer that generally occur in gastric and duodenum as a result of imbalance between aggressive and defensive factors. The aggressive factor includes acid, pepsin and *Helicobacter pylori,* while the defensive factor are gastric mucus and bicarbonate secretion, prostaglandin, nitric oxide and innate resistance of mucosal cells. The pectic ulcer is indicated by a gnawing or burning sensation shortly after meal for gastric ulcer and 2-3 hours afterward for duodenal ulcer. In a gastric ulcer, usually, the acid secretion is normal or low, whereas in duodenal ulcer, acid secretion is high in half of patients but normal in the rest. Gastric and duodenal ulcers can be induced by a variety offactors such as stress, smoking, nutritional deficiencies and noxious agents including non-steroidal anti-inflammatory drugs (NSAID)[1].

Aspirin is a drug of NSAID that affects bothcyclooxigenase (COX -1 and COX-2 activity). Aspirin can inhibit COX-2 activity to reduce inflammation and pain while inhibition of COX -1 lowers the prostaglandin synthesis. The decrease of prostaglandin can reduce mucus secretion and vasoconstriction of capillary blood leading to disturbance of H^+ diffusion from gastric. The excessive gastric acid can lower bicarbonate and prevent proliferation of mucus epithelia that can injure the mucosa [2].

The increase of HCl production in gastric can trigger the excessive production of ROS, consequently, the duodenal tissue undergo stress oxidative condition [3]. The oxidative stress enhance oxidation process in normal cells that can severe to the body. This condition is responded by immune system through activation of transcription factor NF-κB with TLR signal in transduction pathway. The increase of NF-κB activation will be encountered by macrophage to produce pro-inflammatory cytokine (TNF-α) as inflammation indicator [4]. The inflammation can damage the duodenal mucosa lead to the change of duodenal histopathology.

Currently available medication for peptic ulcers include antacids (systemic and nonsystemic) and drugs which reduce acid secretion such as H_2 anti-histaminics, proton pump inhibitors, anticholinergics, prostaglandin analogues, ulcer protectives, ulcer healing drugs and anti-*H. pylori*drugs [5] . These drugs have decreased the morbidity rates, but produce many adverse effects including relapse of the disease, and are often expensive for the poor [6]. In light of the above, it is pertinent to study natural products from food/plants as potential anti-ulcer compounds. Due to less side effects compared to synthetic drugs, currently 80 % of the world population depends on plant-derived medicine for the first line of primary health care [7] . The flavonoid can act as antioxidant to neutralize free radicals, inhibit hydrolysis, oxidation and anti inflammation. The flavonoid in *Sauropus androgynus* (L.) Merr is 831.70 mg/100 g DW that comprised of kaempherol 805.48 mg/100 g DW [8]. *Sauropus androgynus* (L.) Merr is believed hitherto improve the flow of breast milk of breastfeeding mothers, however, this plant has not been reported as pectic ulcer medicine. Therefore, this research aimed to elaborate the potency of ethanolic extract of *Sauropus androgynus* (L.) Merr leaves as alternative therapeutic herbal for peptic ulcer.

2. Materials and Methods

2.1 Chemicals and instrumentation

Aspirin (Sigma-Aldrich), ethanol (Merck), liquid paraffin, NaCl, Hematoxylene-Eosin , Xylol (merck) ,H_2O_2,rat anti TNF-α , Rabbit anti-rat IgG – biotin, strepavidin horse raddish peroxydase (SAHRP), Bovine serum albumin (BSA) (Sigma), DAB (Sigma). Phospho buffer saline. Light microscope Olympus DP71®, rotatory evaporator, rat cages, a set of surgery apparatus. *Sauropus androgynus* (L.) Merr was purchased from UPT Materia Media, Batu, the rats were purchased from Institute of BioScience, UB, Malang.

2.2 Experimental animals and Design

Male Wistar rats (180-200 g) were purchased from Institute of BioScience, UB, Malang. The rats were acclimatized under controlled room temperature, fed standard diet and water *ad libitum* for 7 days prior to commencing experiment. The experiments for care and use of animal were approved by Ethical Clearance committee of Brawijaya University (No:673-KEP-UB).

In the whole study, a total of 20 rats were randomly divided into five groups (Groups 1-5). Groups 1 was set as normal control. The 16 aspirin induced- peptic ulcer were divided into 4 groups (Groups 2 to 5) of 4 rats in each, namely peptic ulcer control groups (Groups 2), receiving ethanolic extract of *Sauropus androgynus* (L.) Merr leaves at dose of 16.2 mg/200 g BW; 32.4 mg/200 g BW ; 48.6 mg/200 g BW (Groups 3,4,5).

The data of TNF-α were analyzed statistically using One Way Analysis of variance (ANOVA) followed by Tukey Test multiple comparation tests. The Level for statistical significance was set at a P value <0.05. The analysis used SPSS 23.0 for windows, while the duodenal histopathology was descriptively analyzed.

2.3 Induction of peptic ulcer in rats

The rats were administrated aspirin orally in dose 200mg/kg BW daily for 5 days. The administration of aspirin used sonde. Aspirin was dissolved in 1% CMC solution [9].

2.4. Preparation of ethanolic extract from Sauropus androgynus (L.) Merr leaves

Two hundreds g of *Sauropus androgynus* (L.) Merr leaves were macerated in 1.5 L of 80% ethanol, shake for 15 minutes at 50 rpm. Then, the mixture was left overnight. The liquid was separated from the *Sauropus androgynus*(L.) Merr dregs by using filter paper. The *Sauropus androgynus* (L.) Merr dregs were macerated 3 times. The liquid extract were concentrated by using rotary evaporator 60°C, the extract was further dried in vacuum oven, the result was thick liquid extract .

3. Results and Discussion

3.1 Therapeutic Effect of Ethanolic Extract of Sauropus androgynus (L.) Merr leaves towards TNF-α Levels of rats duodenum.

Naturally, TNF-α cytokine is produced in relative low level as immune component, which involved in apoptotic mechanism in body tissue such as duodenal organ. In the existence of antigen, macrophage, neutrophils, mast cell and lymphoid T will produce TNF-α cytokine in higher level to trigger or activate the nonspecific and specific immune system to responding inflammation [10].

Table 1. Profile of TNF-α Level in normal, peptic ulcer and therapeutic rats duodenum

Groups	TNF-α Level (%area)	Difference of TNF-α Level to negative control
Negative Control (normal)	11.98 ± 1.09^{a}	0.00
Positive control (peptic ulcer)	46.75 ± 1.58^{d}	290.23
Therapy 16.2 mg/200 g BW	32.12 ± 1.83^{c}	168.11
Therapy 32.4 mg/200 g BW	20.70 ± 1.27^{b}	72.79
Therapy 48.6 mg/200 g BW	18.50 ± 1.14^{ab}	54.42

As shown in Table 1, The expression of TNF-α in duodenum at pepticulcer were significantly higher compared to normal rats. The increase of TNF-α indicated the occurrence of inflammation in rats' duodenum which caused by the aspirin induction of 200 mg/kg BW. Aspirin inhibits Cyclooxygenase isoenzyme (COX-1) through irreversible acetylation COX-1 leading to prevent the synthesis of prostaglandin. The decrease of prostaglandin lowered the mucus production in digestive tract. The mucus function as duodenal defensive barrier will decrease that lead to increase of Reactive oxygen species (ROS). The excessive ROS production in cell cause activation of NF-κB and phosphorylation of NF-κB inhibitor that bring to induce the transcription of cytokine pro inflammatory (TNF-α)[11].

Therapy with ethanolic extract of *Sauropus androgynus* (L.) Merr leaves may reduce the elevated level of TNF-α. The level of TNF-α decreased with the increasing dosage of ethanolic extract of Sauropus *androgynus* (L.) Merr leaves. Statistical test results showed there were significant difference ($P<0.05$) between TNF-α levels of pectic ulcer and therapeutic peptic ulcer mice. It suggests that the kaempherolof *Sauropus androgynus* (L.) Merr enable to act as an antioxidant, especially as a hydroxyl radical's scavenger.

The kaempherol as an antioxidant function to neutralize ROS through releasing hydrogen atom of flavonoid hydroxyl group and produce unreactive stable radical with low energy. Therefore, the increase of ethanolic extract dosage declined ROS and oxidative stress in duodenum tissue. The decline of ROS can inhibit the activation of NF-κB to produce TNF-α. NF-κB remain bound to NF-κB inhibitor, and NF-κB cannot combine with response element which can trigger transcription and translation of pro inflammatory cytokine (TNF-α) as a result the lower pro inflammatory can reduce activity of inflammation cell [12].

3.2 Histopathology of duodenum tissue from control rats ,peptic ulcer rats, and the therapeutic rats

Free radicals are the result of normal product of cell metabolism and phagocytes produce large amounts of ROS *via* the NADPH oxidase system as part of their defensive mechanism against pathogens [13], however, under circumstances may occur the imbalance between ROS production and endogen antioxidants that lead to cell disfunction and cell damage. Epithelial and endothelial cell damage due to oxidative stress could cause damage to duodenum tissue and disfunction. The histology of duodenum tissue was observed to determine both the level of damage and organ repair

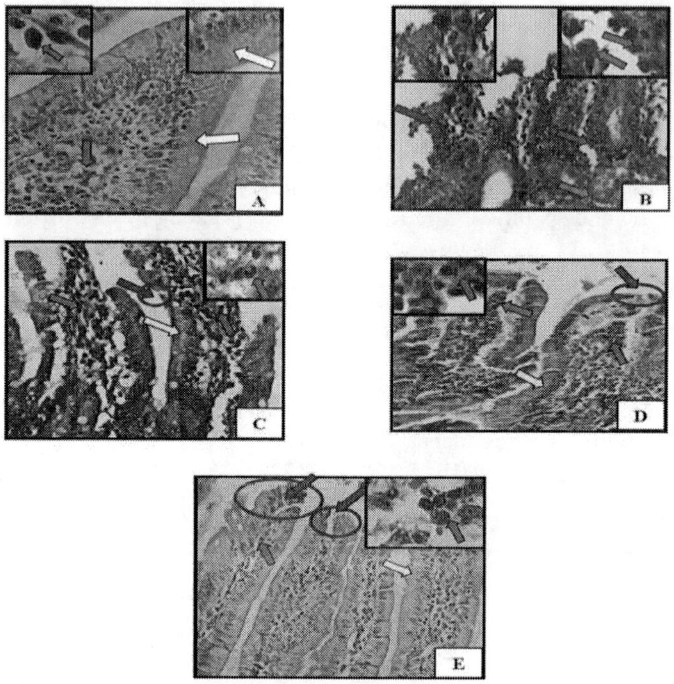

Figure 1. Histopathology of Rats Duodenum magnified 400x and 600x.**(A)**Negative Control (normal), **(B)**Positive control (peptic ulcer), **(C)**Therapy 16,2 mg/200 g BW, **(D)** Therapy 32.4 mg/200 g BW , **(E)**Therapy 48.6 mg/200 g BW. epithelia cell(⇧), vili epithelia cell erosion (⬆), neutrophil cell (⬆), epithelia cell regeneration(⇧),Blood vessel vasodilatations (⬆)

Epithelia vili structure in normal rat duodenum showed intact and compact, there are goblet cells and no inflammatory cell in lamina propia. Duodenum epithelial cell covered all mucosal layer by forming one cylindrical epithelial layer. In peptic ulcer rats showed erosion of cylindrical epithelial layer, inflammatory cell at lamina propia and increase blood vessel permeability. It indicates that aspirin induction has damaged to the epithelial cell of rats' duodenum.

After receiving therapy of the ethanolic extract from*Sauropus androgynus* (L.) Merr leaves the duodenum vili showed better that less erosion of cylindrical epithelial layer and infiltration of inflammatory cell, and occur proliferation of epithelial cells. The higher dose of the ethanolic extract from *Sauropus androgynus* (L.) Merr leaves therapeutic bring out a better repair of histology of duodenum tissues and the therapeutic dose ofethanolic extract from*Sauropus androgynus* (L.) Merr leaves 48.6 mg/kg BW in peptic ulcer rats can restore duodenum vili structure almost like normal rats duodenum . *Sauropus androgynus* (L.) Merr leaves extract can maintain the integrity of cell membranes by inhibited lipid peroxidation reaction resulting in lower (TNF-α) expression and inflammation.

4. Conclusions

Ethanolic extract from *Sauropus androgynus* (L.) Merr leaves can be used as alternatif therapy peptic ulcer towards decreasing the expression of TNF-α and repair duodenum histopathology such as structure of epithelial duodenum, vile and decrease of inflammatory cell and vascular vasodilatation with the effective dosage of 48.6 mg/ 200 g BW.

References

[1] Cullen D J, Hawkey G M and Greenwood D C 1997 Peptic ulcer bleeding in the elderly: relative roles of Helicobacter pylori and non-steroidal anti-inflammatory drugs. **Gut, 41 (4): 459–62.**

[2] Eifellia A R 2010 The effect of mungbean *Phaseolusradiatus*) administration towards the change of mucosal gastric structure histology of *Mus musculus induced by* Aspirin [Thesis]. Faculty of Medicine , UNS

[3] Fornai M., Antonioli L and Colluci R 2011 *Pathophysiology of Gastric Ulcer Development and Healing: Molecular Mechanisms and Novel Therapeutic Options*. Department of Internal Medicine, University of Pisa, Italy. Pp. 113-142

[4] Giugliano D, Ceriello A and Esposito K 2006 The effect of diet on inflammation. *Journal of the American College of Cardiology* 48 (4) : 677-685

[5] Rang and Dale 2003 The textbook of Pharmacology, An imprint of Elsevier., 5: 244-259.

[6] Srikanta BM, Siddaraju MN, Dharmesh SM: A novel phenol-bound pectic polysaccharide from Decalepishamiltonii with multi-step ulcer preventive activity. World J Gastroenterol 2007, 13(39): 5196-207.

[7] Samy R P and Gopalakrishnakone P 2007 Current status of herbal and their future perspectives. Nature Precedings pp: 1-13.

[8] Batari R 2007 Identification of flavonoid compounds on West Java indigenous vegetables, Bogor, Faculty of Agriculture Technology, IPB

[9] Acarturk G., Senol A., Akin M., Suteu R, Sahin O and Isler M 2014 The Protective Effect of Kefir in Aspirin-Induced Gastric Mucosal Damage : An Experimental Study. *ActaMedicaMediterranea.*30 : 875.

[10] Baratawidjaja K G and Rengganis J 2010. Basic Immunology : Cytokine , Balai Penerbit Faculty of Medicine, UI

[11] Campbell K J and. Perkins ND 2006 Regulation of NF-κB Function. *Biochem Soc Symp.* 73:165-180.

[12] Yilmaz H , Giizel Y. Onal Z, Altiparmak G, and Kocakaya S O 2011 4D-QSAR Study of p56 protein tyrosine kinase inhibitory activity of flavonoid derivates using MCET method. *Korean Chemical Journal* 32(12) : 4352-4360.

[13] Valko M, Leibfritz D, Moncol J, Cronin MTD, Mazur M, Telser J 2007Free radicals and antioxidants in normal physiological functions and human disease. Intl J Biochem Cell Biol 39:44–84

.

Enzymatic Synthesis of Glyserol-Coconut Oil Fatty Acid and Glycerol-Decanoic Acis Ester as Emulsifier and Antimicrobial Agents Using *Candida rugosa* Lipase EC 3.1.1.3

Sri Handayani, Ayu Tanissa Tamara Putri, Siswati Setiasih, Sumi Hudiyono*

Department of Chemistry, Faculty of Mathematics and Natural Sciences, Universitas Indonesia

* sumi.hudiyono@sci.ui.ac.id

Abstract. In this research, enzymatic esterification was carried out between glycerol and fatty acid from coconut oil and decanoic acid using n-hexane as solvent. In this reaction *Candida rugosa* lipase was used as biocatalyst. Optimization esterification reaction was carried out for parameter of the substrate ratio. The mmol ratio between fatty acid and glycerol were used are 1:1, 1:2, 1:3, and 1: 4. The highest conversion percentage obtained at the mole ratio of 1: 4 with the value of 78.5% for the glycerol-decanoic acid ester and 55.4% for the glycerol coconut oil fatty acid ester. Esterification products were characterized by FT-IR. The FT-IR spectrum showed that the ester bond was formed as indicated by the wave number 1750-1739 cm^{-1}. The esterification products were then examined by simple emulsion test and was proved to be an emulsifier. The glycerol-coconut oil fatty acid ester produced higher stability emulsion compare with glycerol decanoic ester. The antimicrobial activity assay using disc diffusion method showed that both glycerol-coconut oil fatty acid ester and glycerol-decanoic ester had the ability inhibiting the growth of *Propionibacterium acnes* and *Staphylococcus epidermidis*. Glycerol-decanoic ester shows higher antimicrobial activity than glycerol-coconut oil fatty acid ester.

1. INTRODUCTION

Coconut oil is widely used for food, industrial applications, health, and disease prevention [1]. This oil contains various fatty acids such as lauric acid, palmitic acid, decanoic acid (capric acid), myristic acid, etc [2]. Production of derivative products (oleochemicals) from coconut oil can provide added value and higher profits than just exporting crude coconut oil. In oleochemistry coconut oil can be converted into various products to get higher economic value, such as mono- and diglyceride. Mono- and diglycerides are highly valued oil diversification products with high market prospects. Those glycerides are needed in industry such as food and pharmaceutical, cosmetics, and cleaning products as surfactants or emulsifiers. Fatty acid monoesters are also known to have antibacterial activity against both gram-positive and gram-negative bacteria [3].

Monoglyceride (MAG) is an ester compound consisting of one glycerol molecule binding to one fatty acid molecule [4]. Fatty acid monoesters are also known to have the highest antibacterial activity against both gram-positive and gram-negative bacteria [3]. In addition, glycerol esters with short or medium chain fatty acids tend to have greater antimicrobial activity compared to glycerol esters with long-chain fatty acids [5].

The study of esterification using *Candida rugosa* lipase EC 3.1.1.3. between simple saccharides and fatty acids [2, 6] and trans and interesterification reactions of palm oil has been conducted [7, 8]. In this study, glycerol esters were synthesized by esterification reaction between glycerol and coconut oil fatty acid and oleic acid using *Candida rugosa* lipase as biocatalyst. In the presence of organic solvent and small amount of water, the lipase may act as biocatalyst in an esterification reaction [9]. The glycerol-fatty acid ester produced was then examined its emulsifier and antimicrobial activity.

Lipase is a hydrolase class of enzyme that catalyzes the triglyceride hydrolysis produces fatty acids and glycerol. On certain conditions, lipases can also catalyze the reverse reaction of hydrolysis, namely the esterification reaction. Esterification reactions catalyzed by lipase occur in the system with small amount of water (essential water) [10]. Moreover solvent hydrophobicity affects the amount of essential water needed by an enzyme for its catalytic activity. The use of non-polar solvent was chosen so that the lipase tended to catalyze the esterification reaction [11, 12].

In performing its function as a catalyst, an enzyme must be in a good condition to support its catalytic activity. A study on catalytic activity of enzymes in organic solvents showed that solvents with log P values between 2 and 4 can maintain enzyme activity and stability [13]. Based on earlier research, esterification was performed using n-hexane as solvent which has a log p value of 3.5 [2].

2. MATERIALS AND METHODS

2.1 Materials

Materials used in this study were *Candida rugosa* Lipase (2.45 U/mg) obtained from Sigma-Aldrich, coconut oil, commercial glycerol, decanoic acid, potassium hidroxyde, 96% ethanol, hydrogen chloride, sodium hydroxide, aquades, Na_2SO_4 anhydrate, n-hexane, phospate buffer pH 8, phenolphtalein, eosin, clyndamisin, DMSO 10%, nutrient broth and nutrient agar, *Propionibacterium acnes* and *Staphylococcus epidermidis*.

2.2 Methods

2.2.1 Hydrolysis of Coconut Oil

Hydrolysis of coconut oil was conducted by mixing 20 g of coconut oil and 100 mL KOH 1 M in ethanol and heated by reflux system for 1 hour at 62±2 °C. Coconut oil fatty acid was then extracted using n-hexane and continued by evaporation to remove the n-hexane [6].

2.2.2 Esterification

Esterification was carried out by mixing glycerol, coconut oil fatty acid, n-hexane as solvent, and *Candida rugosa* Lipase as a catalyst. The ratio of glycerol to fatty acid used were varied, *ie* 1:1, 2:1, 3:1, and 4:1 (mol / mol). The ratio solvent to substrate used was of 1: 1 (v / v substrates), while the enzyme used was 5% of the total mass of the substrate (w/w substrates). The mixture was then incubated in a horizontal incubator shaker at 200 rpm, 37 °C for 18 hours. The reaction was stopped terminated by heating at ± 80 °C. The mixture was then centrifugated at 3400 rpm for 15 minutes. Three layers were formed and then separated [2]. The same process were repeated performed using decanoic acid.

2.2.3 Determination of Conversion Percentage

The conversion percentage was determined by titrating the residual fatty acid in n-hexane using 0.1 N NaOH and phenolphthalein as indicator [2].

2.2.4 Characterization of Esterification Product using FT-IR

The esterification product, glycerol, fatty acid obtained from coconut oil hydrolysis, and decanoic acid were characterized using FT-IR.

2.2.5 Simple Emulsifier Test and Determination of Emulsion Type

Simple emulsifier test was performed by mixing 0.1 g esterification product, oil, and water according to Table 1. The mixtures were shaken and the emulsion stability was observed.

Table 1 Variations of Oil and Water in the Making of Emulsions

Type of emulsion	Water (mL)	Oil (drops)
Oil in Water	3	5
	Water (drops)	**Oil (mL)**
Water in Oil	5	3

The emulsion type determination was performed by mixing a drop of emulsion and eosin on an object glass. The mixture was then observed under microscope to determine whether the emulsion formed is oil in water (o/w) or water in oil (w/o) emulsion.

2.2.6 *Antimicrobial Activity Assay*
Antimicrobial activity assay was performed using disc diffusion method. An aliquot of 200 μL of *Propionibacterium acnes* suspension with cell density of 1 x 10^8 cells/mL was aseptically mixed by 20 mL of nutrient agar in a sterile petri dish. The medium was then allowed to harden. The sterile paper disc (6 mm in diameter) was placed on top of the medium and 4 μL sample was dropped onto this. The medium was then incubated at 37 °C for 24 hours. The positive control used was Clindamycin 0.5% and the negative control use was DMSO. The clear areas around the paper discs were measured. The same assay was performed using *Staphylococcus epidermidis*.

3. RESULT AND DISCUSSION
3.1. Hydrolysis of Coconut Oil
The hydrolysis reaction was carried out using sodium hydroxide in ethanol as catalyst. The function of ethanol is to lower the polarity difference between KOH and oil which in the end permits the reaction. These fatty acids obtained from this process were used for esterification. Yield percentage of coconut oil hydrolysis was 94.9%.

3.2.Determination of Conversion Percentage
The product obtained from esterification reaction formed an emulsion system. To break the emulsion system, the mixture was centrifuged. The three layers formed were separated and the top layer was then used for conversion percentage determination

The conversion percentage of esterification using various reactant mole ratios is shown in Figure 1. The conversion percentage increases with the increasing mole of glycerol. The amount of glycerol used was higher than fatty acids, thus the ester formed tends to be mono- or diglycerides.

The esterification reaction is an equilibrium reaction following the Le Chatelier principle. In order to improve the product of the reaction, excess reactants were used therefore the equilibrium will lead to the formation of product. The highest conversion percentage was achieved at mole ratio fatty acid to glycerol 1: 4 with the value of 55.4% using the coconut oil fatty acid and 78.5% using oleic acid as reactant.

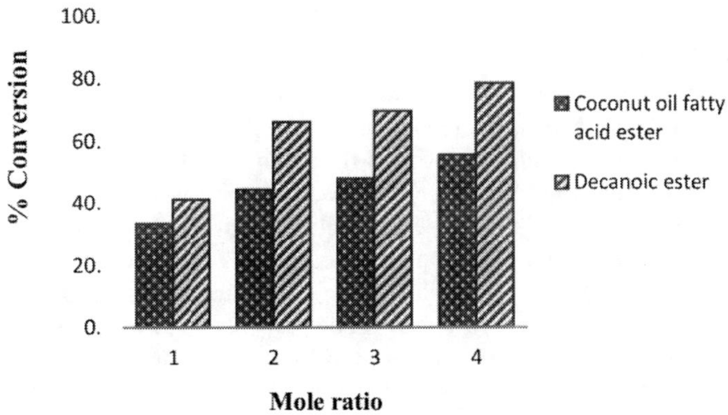

Figure 1. The Curve of Mole Variation vs Conversion Percentage

3.3 Characterization of Esterification Product using FT-IR

IR spectrum of Glycerol, fatty acids from coconut and palm oil, coconut oil fatty acid ester, Decanoic acid, and Decanoic ester are shown in Figure 2.

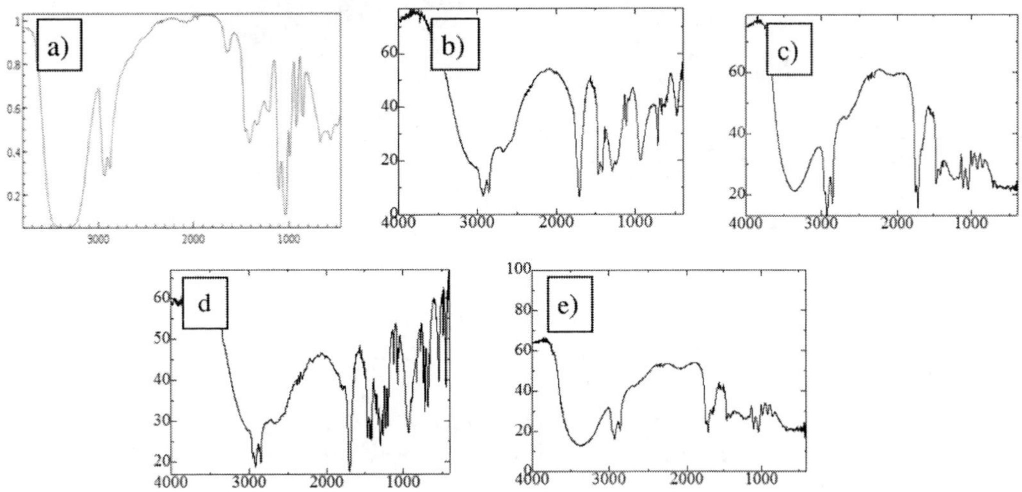

Figure 2. FTIR spectra of a. Glycerol, b. Coconut oil fatty acid, c. Coconut oil fatty acid ester, d. Decanoic acid, e. Decanoic ester

FT-IR spectrum of glycerol showed an absorption peak at wave number 3650-3200 cm^{-1} for -OH fuctional group, 2931 cm^{-1} for C-H group, and 2877 cm^{-1} for -CH_2 (Figure2.a). While FT-IR spectrum of coconut oil fatty acid showed there were absorption peaks at wave number 3000-2850 cm^{-1} indicating the presence of C-H functional group, 3000-2700 cm^{-1} which is a characteristic for CH_2 and CH_3 functional group, 1714 cm^{-1} for the C = O group (Figure 2.b and 2.d). Figure 2.c and 2.e showed that there are absorption peaks at wave number 3650-3200 cm^{-1} that indicate not all the O-H groups of glycerol were esterified. In addition there was an absorption peak at wave number 1748 cm^{-1} which indicated the presence of –C=O ester functional group. Therefore, it can be concluded that the ester product was successfully produced.

3.4 Simple Emulsifier Test

The simple emulsifier test showed that emulsion formed using glycerol-coconut oil fatty acid ester as emulsifier more stable than using decanoic ester. The emulsion system formed stable up to 24 hours.

The simple emulsion type test showed that glycerol-decanoic ester has the ability to form water in oil (w/o) and oil in water (o/w) emulsion. But the w/o type formed was more stable than o/w type. For glycerol-coconut oil fatty acid ester tends to show the ability to form w/o emulsion. Figure 4 shows the microscopic photograph of the emulsion. The emulsion formed was added with eosin, a red dye which is soluble in water, so the water phase will be red in color and easy to be observed and distinguished from the oil phase [2]. Both glycerol-decanoic and glycerol-coconut oil fatty acid ester have better ability to form a water in oil (w/o) emulsion type. This can be seen from the formation of pink droplets while the surrounding was yellow. The droplet formed was the water phase and the light yellow environment around pink droplets was oil.

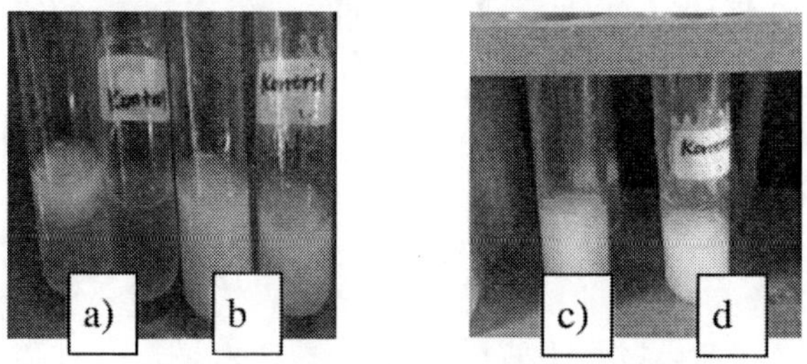

Figure 3 Simple emulsion test results
a) The oil in water (o/w) mixture + decanoic ester (left) and oil in water (o/w) mixture (right),
b) water mixture in oil (w/o) mixture + decanoic ester (left) and water in oil (w/o) mixture (right),
c) water in oil (o/w) mixture + coconut oil fatty acid ester, d) water in oil mixture

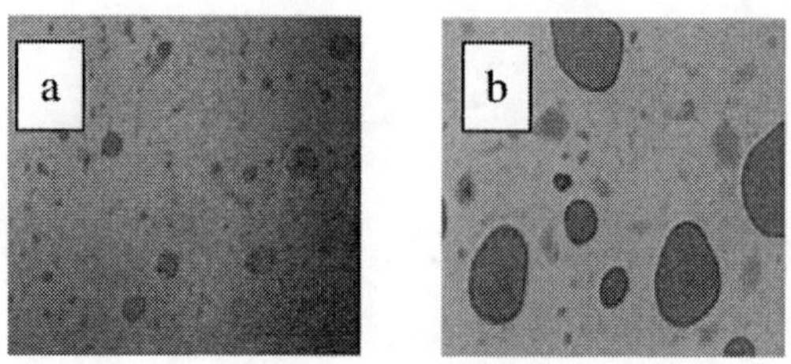

Figure 4 The Observed Emulsion Types with Microscopes
a. Water emulsion (w/o) using Glycerol-Decanoic ester
b. Water in oil emulsion (w/o) using Glycerol-Coconut Oil Fatty Acid ester

The type of emulsifier is influenced by the type and size of molecules of hydrophobic and hydrophilic part and discribed as HLB (Hydrophilic to Liophilic Balance) value. Emulsifiers with low HLB values (3-6) produce water-in-oil emulsions, whereas high HLB (8-18) produces an oil-in-water emulsion [14]. However, the glycerol-decanoic and glycerol-coconut oil fatty acid esters which tend to form monoester had a low HLB [15].

3.5 Antimicrobial Assay

Bacteria used in this study were *Propionibacterium acnes* and *Staphylococcus epidermidis*. *Staphylococcus epidermidis* is normal flora found in humans, while *Propionibacterium acnes* is acne-causing bacteria. Table 2 shows the antimicrobial activity of the ester products. Glycerol, n-hexane, and DMSO were also tested as control. Table 3 shows the effectiveness classification of antimicrobial substances [16].

The antimrobial assay showed that glycerol-decanoic and glycerol-coconut oil fatty acid esters had middle category against *P. acnes* at concentrations of 75%. At concentrations 50% Glycerol-decanoic ester showed showed middle antimicrobial activity against *S. epidermidis*, while glycerol-coconut oil fatty acid ester up to concebtration 75% showed weak activity against *S. epidermidis*.

Table 2 Diameter of the inhibitory zone of various compounds against *P. acne* and *S. epidermidis*

Sample	Concentration (%)	Inhibit zone diameter (mm)			
		P. acnes		*S. epidermidis*	
Glycerol-Decanoic Ester	25	14	weak	14	weak
	50	15	weak	16	medium
	75	16	medium	16	medium
	100	16	medium	17	medium
Glycerol-Coconut Oil Fatty Acid Ester	30	10	weak	8	not effective
	45	10	weak	8	not effective
	60	12	weak	10	weak
	75	16	medium	12	weak
Decanoic Acid	50	13	weak	14	weak
Coconut Oil Fatty Acid	50	10	weak	10	weak
Clyndamisin	0.5	16	medium	-	
DMSO	100	-		-	
Glycerol	100	-		-	
n-hexane	100	-		-	

Table 3 Classification of effectiveness of antimicrobial substances

Inhibit zone diameter	Response of growth barriers
>20 mm	Strong
16-20 mm	Medium
10-15 mm	Weak
<10 mm	Not effective

The antimicrobial activity of fatty acids is influenced by length of the carbon chain, unsaturation, and location of the unsaturated bond [5]. Short-to-medium chain fatty acids have antimicrobial inhibitory activity especially in the form of monoglycerides, whereas monoglycerides of long-chain saturated fatty acids do not act as antimicrobial agent [17]. The location and number of double bonds on C_{12}-C_{22} fatty acids also affect microbial activity compared to fatty acids with less than 12 carbon atoms. The addition of cis double bond can increase the antimicrobial activity of a straight chain fatty acid, while the configuration of the fatty acid structure geometry in the trans form is inactive as an antimicrobial agent [17].

Most of the fatty acids in coconut oil are fatty acids medium carbon chains, ie lauric, myristic, and decanoic acid having carbon chains ≤ 16 [2]. This caused the glycerol-coconut oil fatty acid ester produced in this research can act as antimicrobial agent.

4. CONCLUSION

Enzymatic synthesis of glycerol-decanoic and glycerol-coconut oil fatty acid esters using *Candida rugosa* lipase as biocatalyst have been successfully performed by showing the characteristic -C=O ester group at FT-IR spectra. Glycerol-decanoic and glycerol-coconut oil fatty acid esters have activity as emulsifier for water in oil (w/o) emulsion type. Both ester also showed the antimicrobial activity against *Propionibacterium acnes* and *Staphylococcus epidermidis*. Glycerol-decanoic ester shows higher antimicrobial activity than glycerol-coconut oil fatty acid ester.

ACKNOWLEDGEMENT

This work was funded by Hibah Kompetensi Publikasi Internasional Terindeks Untuk Tugas Akhir Mahasiswa (PITTA), Universitas Indonesia, Contract No. 1821/UN2.R3.1/PPM.00.01/2017.

REFERENCES

[1] Manisha, D.M. dan Shyamapada M. 2011. Coconut (*Cocos nucifera L. Arecaceae*): In Health Promotion and Disease Prevention *APJTM.*, 4 (3), 241-247.
[2] Handayani, S; I. Novianingsih; A. Barkah; S. Hudiyono. 2012. Enzymatic synthesis of sucrose polyester as food emulsifier compound. Makara Journal of Science, 141-148.
[3] Zhao, Lel, et al. 2015. In Vitro Antibakterial Activities and Mechanism of Sugar Fatty Acid Esters Againt Food- Related BacterJia. Journal Of Food Chemistry 187: 370-37.
[4] Ketaren, S. 1986. Introduction of Oil Technology and Food Fat. Penerbit UI Press. Jakarta.
[5] Kabara JJ; Swieczkowski DM., Conley AJ., Truant JP. 1972. Fatty Acids and Derivatives as Antimicrobial Agents. Antimicrob Agents Chemother 2(1):23–28.
[6] Hudiyono, S.; Risang Guritno; Rahayu Amanda; Sri Handayani; Ridla Bakri. 2015. Immobilized Lipase *Candida Rugosa* Ec 3.1.1.3 On Nanoparticles Fe_3O_4-Polydopamine As Catalyst Of Esterification Reaction Among Glucose To Palm Oil Fatty Acids. Conference Proceeding of The

3rd Asia-Pacific Conference on Life Science and Engineering (APCLSE). Chiang Mai, Thailand 18-20 November 2015.

[7] Anlisa, S.; Sri Handayani and Sumi Hudiyono. 2015. Effect of Solvents on Palm Oil Transesterification using Immobilized *Candida rugosa* Lipase on Fe_3O_4-Polydopamine Nanoparticles. International Symposium on Current Progres on Mathematics and Sience (ISCPMS) 2015. Depok, Indonesia 4-5 November 2015. AIP Conf. Proc. **1729**, 020044 (2016); http://dx.doi.org/10.1063/1.4946947.

[8] Kautsari,S.N.; Sri Handayani; Sumi Hudiyono. 2015. Interesterification of Palm Oil by using Immobilized Lipase *Candida rugosa* on Fe_3O_4-Polydopamine Nanoparticles and Isooctane Solvent. International Symposium on Current Progres on Mathematics and Sience (ISCPMS), Depok, Indonesia 4-5 November 2015. AIP Conf. Proc. 1729, 020044 (2016); http://dx.doi.org/10.1063/1.4946947

[9] Adamopoulos, Lambrini 2006 Understanding The Formation of Sugar Fatty Acid Esters. Faculty of North Carolina State University.

[10] Zaks, Aleksey; Klibanov, Alexander.1985. Enzyme-catalyzed processes in organic solvents. Vol. 82, pp. 3192-3196.USA 8

[11] Öztürk, Banu, 2001. Immobilization of Lipase from Candida rugosa on Hydrophobic and Hydrophilic Supports. ozmir Institute of Technology, Turkey.

[12] Zaks, A. and A.M. Klibanov. 1988. The Effect of Water on Enzyme Action in Organic Media. J. Biol.

[13] Yoo, I.S., S.J. Park and H.H. Yoon, 2007. Enzymatic Synthesis of Sugar Fatty Acid Esters. J. Ind. Eng. Chem., 13(1): 1-6.

[14] Luna, Prima; Andarwulan, Nuri. 2013. Potensi Produk Monoasilgliserol sebagai Emulsifier Nabati.Buletin Teknologi Pascapanen Pertanian Vol 9 (2), 2013 : 108-116.

[15] Christensen, Clemen. 2011. Emulsifiers used in food applications, focusing on the meat processing industry. Palsgaard Technical Paper.13.

[16] Greenwood. 1995. *Antibiotic susceptibility (sensitivity) test, antimicrobial and chemotherapy*. USA: Mc Graw Hill Company.

[17] Kabara JJ 1983 Medium-Chain Fatty Acids and Esters. Di dalam: Antimicrobials in Foods. Alfred L. B. dan P.M Davidson (eds). Mercel Dekker Inc., New York and Basel.

Depuration Study of Heavy Metal Lead (Pb) and Copper (Cu) in Green Mussels *Perna viridis* through Continues-discontinues and Acid Extraction Methods

Budiawan, Ridla Bakri, Intan Cahaya Dani, Sri Handayani, Rizki Ade Kurnia Putri, Riska Tamala

Chemistry Department, Faculty Mathematics and Sciences, Universitas Indonesia 16424
E-mail: dr.budiawan@gmail.com

Abstract. Green mussel or *Perna viridis* is filter feeder, which is very susceptible to heavy metals. It takes an effort to release heavy metal contents on the green shell, one of method that can be used to release heavy metal from green shell is depuration proccess. In this research, the depuration process was conducted by continues method of depuration, discontinues method by using various kind of water and acid extraction. The optimum time of continues depuration method is 1.5 hours, with circulation speed 250 L/h and result of Pb metal content decreased is equal to 30.048% and 29.748% for Cu. In the discontinues method, the optimum result was reached at 100oC by using PAM water as the media at 3 h immersion period with decrease of Pb metal content 35.001% and Cu metal content 39.015%. In the acid extraction method, the optimum condition was achieved by 11% acetic acid solvent with decreasing of Pb and Cu levels are 88.224% and 76.298%. For the determination of protein content, the decrease of protein content obtained by treatment with 11% acetic acid extract showed decrease of protein content 36.656% with Kjeldahl method.

1. Introduction

Based on the Decree of the Minister of Environment in 2004, the value of copper metal standard in seawater is 0.008 mg/L and if excess will be toxic and may cause disturbance to living organisms [1]. The content of Cu metal in green shells from the Jakarta bay is 0.0619-2.9065 ppm [2]. This value has not exceeded the maximum level of Ditjen POM No.03725/B/SK /1982 which is 20 ppm for copper [3]. The development of industrial area around the Jakarta bay, and the indication of metal pollution to the aquatic, it is necessary to monitor the accumulation of copper metal in the body of the shell. The prevention and efforts to reduce the level of heavy metal pollution in sea food products need to be conducted, one of the efforts is through the process of depuration. In this research the depuration process was conducted for lead (Pb) and copper (Cu). The depuration process is carried out by water flow recirculating method, discontinues water method and extraction (acetic acid, and citric acid). The water flow recirculating used sea water and PAM water as media depuration by variations of water speed and duration of flow. It was conducted by variations type of water, immersion time and variations of temperature. The extraction method was carried out using acetic acid solvent (8,9,10 and 11%), and citric acid (0.8, 1, 1.5 and 2%) at 30°C with stirring 120 rpm for 4 hours. In addition, the determination of protein content contained in green shells by Kjeldahl method to determine the protein content in the green shell before and after depuration due to depuration process using acid can cause the decreased of protein levels due to interaction between protein and metal disruption by the addition of other substances.

2. Experimental Method

The sample was obtained from Muara Kamal, North Jakarta. The green mussel samples hich have taken based on 3 parameters which are age, body length, and sampling location (only at one location). After the shellfish samples were taken, the samples were storage in the cooler box with temperature ± 4°C [4]. This research was conducted by several stages, namely determination of Pb and Cu levels of samples without treatment, depuration process, determination of metal content after depuration, and determination of protein content.

2.1 Acclimatization

The acclimatization process was conducted by immersing the green shellfish in sea water which has been filtered with filter paper and change every 12 hours to maintain green shell life.

2.2 Depuration with Continues Streaming Method (Water flow recirculating)

In this continues water method, a 35 x 30 x 30 cm aquarium was equipped with pumps and filters. The design of the aquariums have pump capacity 250 L/h and 500 L/hr. The variation of flow time were 6, 12, 18, and 24 hours, in which time the test sample were taken and measured the heavy metal content on the mussel meat using SSA instrumentation.

2.3 Depuration with Discontinues Method

This method was conducted by two types of water that are sea water and water PAM. Variation of immersion time were 1.5 h, 2 h and 3 h. After obtaining optimum result for variation of immersion time, the temperature variation was conducted in 80°C, 90°C, 100°C. In order to optimize the depuration process, it is stirred by 120 rpm.

2.4 Depuration with Acetic Acid Extraction

The extraction with acetic acid solvent was conducted by variation of concentration 8; 9; 10; 11% (v / v). The process of depuration was conducted by immersing the green shell in acetic acid solution. Further stirring 120 rpm at a temperature of 30°C for 4 hours. After the depuration process is complete, the green shells are rinsed and the metal content was measured with Atomic Absorption Spectrophotometer (SSA).

2.5 Depuration with Citrate Acid Extraction

The variation concentration of citric acid solution used was 0.8; 1; 1.5; 2% (w / v). The process of depuration is also done by immersing the mussels using citric acid solution with stirring 120 rpm at 30°C for 4 hours.

2.6 Determination of Protein content

Determination of protein content by Kjeldahl method was conducted by subtilize the dried sample meat. Furthermore, about 1 gram sample was put into a Kjeldahl flask and Selenium mixture was added to the flask. The concentrated H_2SO_4 solution was added and the mixture was heated until the solution was clear with the flask in a tilted state and then closed by short funnel. After the sample solution was cooled, the mixture was diluted in a 25 mL water. About 10 mL of sample solution was put from the solution. PP indicator and 30% NaOH were added to the solution until the solution was alkaline. After that, the solution was moved to the distillation apparatus. Then 10 mL of 5% H_3BO_3 solution and BCG-MM indicator on 10 mL erlenmeyer were prepared to accommodate the distillate. The distillation process ends when the distillate shaft changes color from red to green. The resulting destilates were then titrated with a standard 0.1 N HCl solution and recorded the required volume of HCl for the calculation of protein content, based on the formula:

$$Protein\ Content = \frac{(ml\ titration\ HCl\ x\ N\ HCl) x\ 14\ x\ 6{,}25}{1000\ x\ sample\ mass}\ x\ 100\%$$

3. Results and Discussion

3.1 Determination of metal in seawater samples

Measurement of metal content is carried out on seawater samples. It was obtained the results of Pb metal content in seawater samples before filtration was 0.109 mg/L while Cu content was 0.089 mg/L. From the above results it is known that the metal content of Pb and Cu has exceeded the standard quality determined by the Decree of the Minister of Environment in 2004 regarding the quality of sea water for marine biota, which is 0.008 mg /L for metal Pb and Cu. Cu metal is more stable to form complexes in sediments, based on Moore and Ramamoorthy (1984), the Cu complex bonds that occur in marine sediments are the most stable complexes, while the most complex in seawater, the lowest stability [5].

3.2 Depuration with Continues Method (Water Flow Recirculating)

In continues method, there are some variations of circulationvelocity with two pumps in capacity 250 L/h and 500 L/h. The purpose of this variation of circulation velocity is to know the effect of time contact in depuration media on the process of releasing heavy metals in the body of green shells. Here are the results of the measurement of the initial metal content of Pb and Cu:

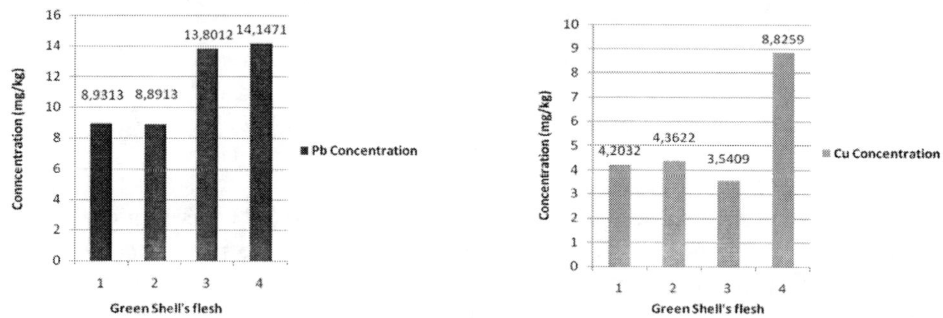

Figure 1 Pb and Cu content in green shells

Based on the picture above (Figure 4.1), the initial lead metal content is in the range of 8.9313-14,147 g/kg, with an average of 11.443 mg/kg. From this result, the accumulation of Pb in the body of the green shell has been categorized as dangerous. Meanwhile for Cu metal is in the range of 3,541-8,826 mg/kg, with a mean of 5.233 mg/kg. In this study, the depuration process was carried out within 24 hours in which the variations of water flow duration were 6, 12, 18 and 24 hours and variation of water circulation speed were 500 L/h and 250 L/hr. Here is the result of decreasing levels of Pb and Cu metal after depuration.

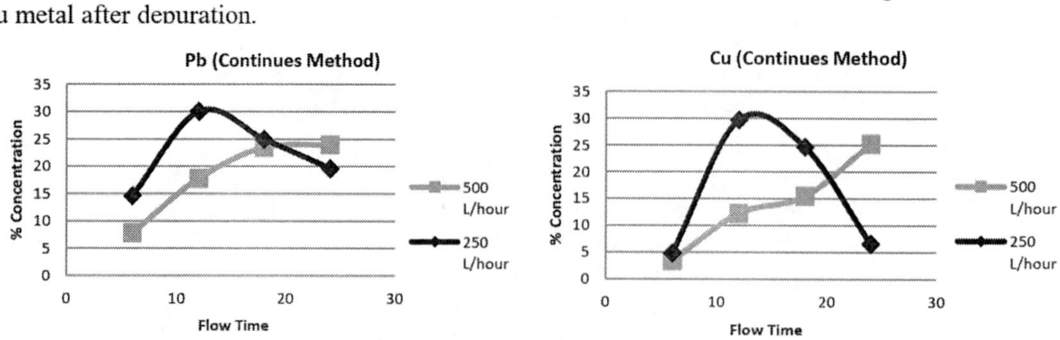

Figure 2 Percentage of Pb and Cu content decreases (continues method)

From the data above, by using the variation of circulation speed 500 L/h, the longer the drainage time, the higher percentage of lead and copper metal decrease. This is due to the water drainage assisting heavy metal excretion on the body of the green shell. The result of excretion in the form of faeces also contains metal, can be drawn with pump which has a stronger suction power and enter the filtration system. Optimal depuration are obtained at 24 hours with lead decrease of 23.971% and copper of 25.28%. This result is less efficient because it takes a long time to detoxify the metal.

At a circulation of 250 L/h, the highest percentage decrease in metal content at 12 hours with 30.047% decreases for lead metal and 29,748% decreases for copper metal. However, after 12 hours, the metal content has increased again, this may be due to the metal that has been excreted back into the shell body. From the graph above, when compared to the decreasing of heavy metal content at 500 L/hr and 250 L/h at 6-18 h, it was found that, both of metal decreasing at circulation of 500 L/h is smaller than circulation of 250 L/h,this is due to the existence of self-defence mechanisms by the green shell while in an environment that is less conducive to shell life such as heavy currents and extreme water conditions. One way to defend oneself is to close the shell and produce mucus [6].

This condition can be observed when the process of depuration is done with a faster circulation, shells are not fully open as in shells with a circulation of 250 L/hr. When the shell is not fully open, the heavy metal present in the mucus found in cleft of organ of the shell, is not eliminated properly.

3.3 Depuration of Discontinues Method

In this study, discontinue method was done on temperature variations of 80, 90 and 100°C and immersion time of 1.5 hours, 2 hours and 3 hours. Media used for immersion is PAM water and seawater that has been treated first. The average result of lead metal measurement is 11.0528 mg/kg and 4.9515 mg/kg for copper metal. The yield of heavy metals in the sample of clam meat after the depuration process using discontinues method with PAM water medium, for each of heavy metals, showed the optimum results at temperature 100°C with the immersion timewas3 hours. The decreases of protein content of Pb was 35.008% and copper was 39.016%.

From the results of sea water immersion, it was found that the highest decrease of heavy metal and copper was affected by treatment condition of temperature 100°C and immersion for 3 hours with 19.15% lead metal decay and 30.64% for copper metal. While the longer of the immersion time, obtained the decreases of heavy metals. Other metallic substances in seawater are likely to affect the process of releasing heavy metals in the body of the shell [7].

3.4 Depuration of Acid Extraction

The average of Pb metal content in initial shellfish was 9.684 mg/kg and for Cu metal 6.4141 mg / kg. In this study acetic acid concentration used for depuration is 8.9,10 and 11% (v/v). Based on the data, optimal results were obtained when using acetic acid 11% with decrease of Cu metal equal to 76,29% and decrease of metal Pb equal to 88,24% (Figure 4.4). Acetic acid can interfere with the interaction between metal and cysteine because acetic acid has the ability to protonate metal that binds to amino acids such as cysteine by donating H^+ ions and this results in bonds between amino acids and metals being able to break. According to Sari and Keman (2005), the bond between acetic acid and metals makes the metal lose its ionic properties, resulting in less toxicity [8].

The result of decrease of heavy metal content after depuration process with citric acid as in Figure 3. The percentage of the decrease in lead and copper metal content is equal to the increased concentration of citric acid. From the data above (Figure 4.3), the largest percentage of the decrease in both of metal content is on 2% citric acid with yield of 35,009% for Cu metal and 42,927% for lead metal. This is because citric acid is a polydentate ligand. In ligands that have two or more donor atoms, the donor atoms may bind to the same central atom forming a chelate complex [9]. The thing that makes citric acid can be metal because citric acid is able to donate 2 free electron pairs to the metal acting as Lewis acid.

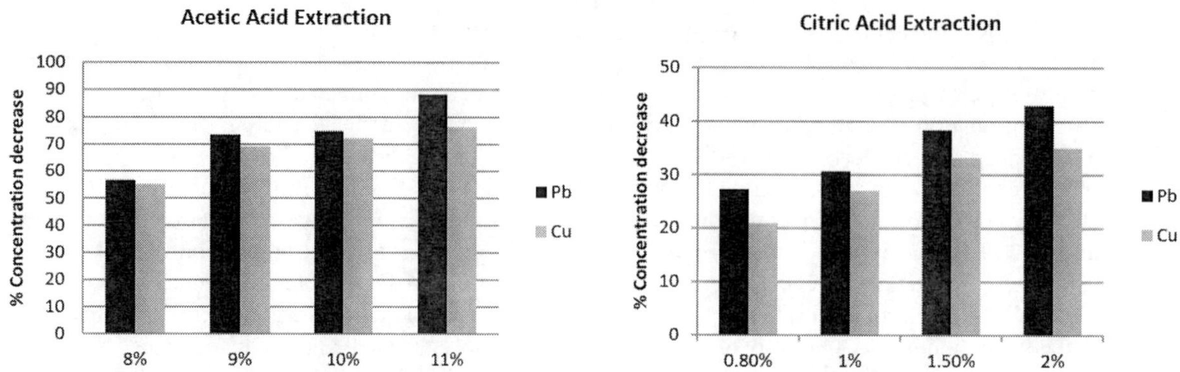

Figure 3 Percentage of Pb and Cu metal content decreases by Acetic Acid and Citric AcidExtraction

3.5 Determination of Protein content

In this study, the determination of protein content is only done on the depuration process which given the percentage of the most optimal metal content decreases. Here are the results of protein content in green mussel samples:

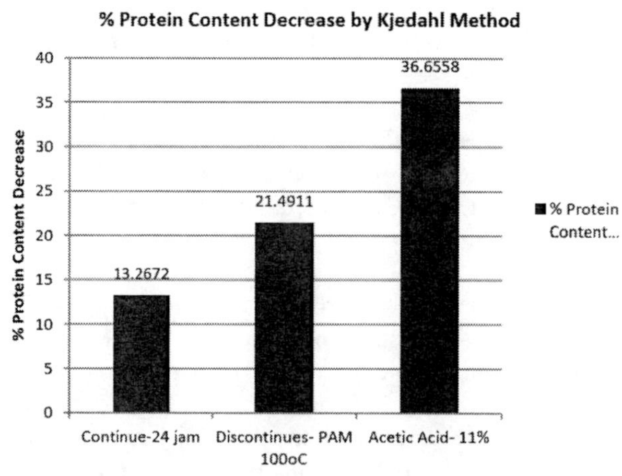

Figure 4 Percentage of Protein Content

From the data above (Figure 4), showed that the protein content of green shell after depuration with acetic acid 11% was 63.34%. The results of this Kjedahl method showed that there was a decrease of nitrogen content from clam meat samples after depuration treatment. The loss of these nitrogen levels can be affected by several factors. The first factor is stirring, according to Wu and Ling, mechanical treatment such as stirring can cause the isoelectric point to reach faster. Generally, denaturation speed is slower than stirring. The interaction between depuration media such as water and acid, while the depuration process takes place, also causes loss of nitrogen content in the sample of clam meat. The depuration process using acetic acid, causes the decreasing of metal content. Meanwhile, according to the protein content measurement, the increasing of acid can also alter the conformation of protein structure and protein damage [10].

4. Conclusion

From the depuration process of lead and copper as heavy metal in green shell (*Perna viridis*), it was obtained that optimum time of depuration by continues method is 1.5 h, with circulation velocity 250 L/h and obtained the decrease of Pb metal content as much as 30.048% and 29.748% for Cu. In discontinu method, optimum result was achieved at temperature 100°C and using PAM water as the media for 3 hours immersion period with decreasing Pb metal content of 35.001% and Cu metal content 39.015%. In the extraction method, the optimum condition was achieved in the extraction using 11% acetic acid with the decrease of Pb and Cu concentration of 88.243% and 76.298%. The depuration process by using water flow recirculating has a decreasion percentage of the protein content as much as 13.267% by Kjedahl method. The immersion treatment which was accompanied by heating on the discontinues method, has a percentage of the decrease in protein content as much as 21.491% by Kjedahl method. Acetic acid extraction showed a decrease in protein content for about 36.655% based on the Kjeldahl method.

References

[1] KLH. 2004 KeputusanMenteri Negara Kependudukan dan Lingkungan Hidup No Kep-02/MENKLH/1998 Tentang Pedoman Penetapan Baku Mutu Air Laut. Jakarta
[2] Inswiasri 1995 Kadar Logam Cu, Pb, Cd Dan Cr Dalam Ikan Segar dan Kerang. Jakarta: Badan Penelitian dan Pengembangan Kesehatan
[3] BPOM 2013 Peraturan Kepala Badan Pengawas Obat dan Makanan Republik Indonesia Nomor 8. Jakarta.
[4] Budiawan and Neera 2012 Biomonitoring Logam Berat In Diktat Biomonitoring (p. 18) Jakarta
[5] J.W.Moore S. Ramamoorthy 1984 Organic Chemicals in Natural Waters- Applied Monitoring and Impact Assessment.New York-Berlin-Heidelberg-Tokyo, Springer Verlag, 289 S., 81 Abb, DM 129
[6] Hughes, R. 1986 A functional biology of marine gastropods. London: Croom Helm.
[7] Hutagalung, H. P. 1991 Pencemaran Laut oleh Logam Berat dalam Status PencemaranLaut di Indonesia dan Teknik Pemantauannya. Jakarta: LIPI.
[8] Sari, F. I., Keman, & Soedjajadi. 2005 Efektifitas Larutan Asam Cuka untuk Menurunkan Kanduxngan Logam Berat Cadmium dalam Daging Kerang
[9] Effendy 2007 Perspektif Baru Kimia Koordinasi Jilid 1. Malang: Bayumedia Publishing.
[10] Purnomo T, M. 2007Analisis kandungan timbal (Pb) pada ikan bandeng (Chanoschanos Forsk.) di Tambak Kecamatan Gresik. Surabaya: Universitas Negeri Surabaya
[11] Power, A. R. 2004 First Occurrence of The Nonindigenous Green Mussel, Perna viridis in Coastal Georgia. 2004: United States: Journel of Shellfish Research 23
[12] Setyono, D. E. 2006 KARAKTERISTIK BIOLOGI DAN PRODUK KEKERANGAN LAUT. Oseana, Volume XXXI, Nomor 1, 1-7
[13] Skoog, D., Holler, F. J., & Crouch, S. R. 2007) Principles of Instrumental Analysis. USA: Thomson Brooks/Cole
[14] Soedarmadji, Haryono, & Suhardi. 1997 Prosedur Analisa Bahan Makanan dan Pertanian. Yogyakarta: Liberty
[15] TAN, W. (1975). Eggs and larva development in the green mussels, Mytilus viridis Linnaeus. The Velinger, 151-155
[16] WHO 1977 Environmental Health Criteria 2 : Copper. World Health Organization
[17] WHO 1977 Environmental Health Criteria 3: Lead. World Health Organization
[18] WHO 1977 International Programme on Chemical Safety Food Additives Series No. 12. Inchem
[19] Wibisono, M. S. 2005 Pengantar Ilmu Kelautan. Jakarta: PT. Grasindo
[20] Widowati, W. S. 2008 Efek Toksik Logam. Yogyakarta

The Proteomic Analysis of Pancreatic Exocrine Insufficiency Protein Marker in Type 2 Diabetes Mellitus Patients

Arie Srihardyastutie,* DW Soeatmadji, Fatchiyah,*** Aulanni'am***

**Chemistry Department, Faculty of Science, Brawijaya University*
***Medical Faculty of Brawijaya University*
****Biology Department, Faculty of Science, Brawijaya University*

arie_s@ub.ac.id / arishdyastutie@gmail.com

Abstract. Type 2 Diabetes Mellitus (T2D) is the vast majority case of diabetes. Patient with T2D is at higher risk for developing acute or chronic pancreatitis. Prolonged hyperglycemia results in damages to tissue, which also causes dysfunctions of some organ systems, including enzyme or hormone secretions. Commonly, dysfunction or insufficiency of pancreatic exocrine is evaluated by increasing activity of serum pancreatic enzyme, such as amylase and lipase. Although incidence of pancreatitis was found in Indonesian T2D, the pathogenic mechanism still unclear. The aim of this study was to characterize the marker protein that indicated the correlation of pancreatic exocrine insufficiency with progression of T2D. Proteomic analysis using LC-MS/MS was used in identification and characterization of protein marker which indicates insufficiency pancreatic exocrine. First step, protein profile was analyzed by SDS-PAGE methods using serum sample of T2D compared with normal or healthy control, as negative control, and pancreatitis patients, as positive control. Protein with 18 kDa was found as a candidate protein marker which indicated the pancreatic exocrine insufficiency in T2D. The further identification of that protein using LC-MS/MS showed 4 peptide fragments. *In silico* analysis of the peptide fragment indicated the correlation of pancreatic exocrine insufficiency with progression of T2D was METTL10 – methyltransferase like protein-10.
Keyword: T2D, pancreatic exocrine, protein, marker

1. Introduction

Type 2 Diabetes Mellitus (T2D) is the vast majority case of diabetes. The main pathophysiological feature of T2D is impaired insulin secretion and insulin resistance [1]. There are multiple defects insulin secretion and signal transduction in Type 2 DM which may effect in synthesis of enzymes and their release from pancreatic exocrine. Although the endocrine system seems to be involved in pathophysiological diabetes, exocrine functions is proposed to be involved as well, since exocrine system is close to endocrine system. Any process that diffusely injures the pancreas, such as pancreatitis, trauma, infection, and pancreatic carcinoma, mainly disturb the exocrine system, but the endocrine system could also be affected and causes diabetes. Diabetes due to that pancreatic exocrine disease is categorized in Type 3 DM, according to American Diabetes Association (ADA) 2012. Prevalence of pancreatitis has been increasing rapidly. Some research showed that alcoholism, consumption of healthcare resources, and using anti-diabetic therapies can increase the risk factor of pancreatitis. Our recent study has demonstrated the increasing incidence of pancreatitis in Indonesian T2D patients [2, 3].

Patient with T2D is at higher risk for developing acute or chronic pancreatitis. Prolonged hyperglycemia results in damages to tissue, which also causes dysfunctions of some organ systems, including enzyme or hormone secretions. Commonly, dysfunction or insufficiency of pancreatic exocrine is evaluated by increasing activity of serum pancreatic enzyme, such as amylase and lipase. Although incidence of pancreatitis was found in Indonesian T2D, the pathogenic mechanism still unclear. The aim of this study was characterization of the marker protein that indicated the correlation of pancreatic exocrine insufficiency with progression of T2D.

2. Research Methods

The study was approved by institutional Human ethics committee (Medical Faculty, Brawijaya University) and informed consent was obtained from all participant.

2.1. Subjects

This study included 67 subjects having general check-up in Central Laboratory of Dr. Saiful Anwar Hospital for a period of six months that meet the inclusion criteria. The inclusion criterion of them were males and females with type 2 diabetes mellitus with or without complications or any comorbid condition like hypertension, coronary artery disease, and also non diabetic healthy individual as the controls. All the subjects, including the controls, were fully informed about the study and their voluntary informed consents were taken. The subjects were categorized into four groups according to the criteria established by The ADA in 2012. Based on ADA criteria, the subjects were categorized into four groups including: healthy controls (n = 21), prediabetes (n = 10), diabetes (n=10), and prediabetes or diabetes with exocrine insufficiency (n=26).

2.2 Methods:

Clinical identification of normal or healthy control, prediabetes, and diabetes were classified according to ADA criteria. The diagnostic criteria of diabetes were assessed according to ADA, *i.e.* subjects with a fasting plasma glucose > 126 mg/dL and/or 2 h plasma glucose level > 200 mg/dL and/or HbA1c > 6.5 % were considered to have diabetes; subjects with a fasting plasma glucose 100 to 125 mg/dL (IFG) or 2 hour plasma glucose level 140 to 199 mg/dL (IGT) or HbA1c 5.7 to 6.4% were considered to have increased risk for diabetes (prediabetes); subjects with a fasting plasma glucose < 110 mg/dL or 2 hour plasma glucose level < 140 mg/dL or HbA1c < 5.6 % were regarded as a having normal glucose tolerance (NGT) [4]. The exocrine insufficiency was assessed according to increasing of amylase and or lipase level. The considered reference range of normal amylase was 30 - 110 U/L and lipase was 30 - 210 U/L. Fasting venous blood was collected from all of subjects, it was centrifuged (at 1500 g for 15 min). The separated plasma was used to assay the HbA1c. HbA1c was measured with ion-exchange high-performance liquid chromatography using an automated analyzer (BioRad D10). Amylase and lipase activity were measured by photometric enzymatic method. The amylase activity was assayed using BioAssay Systems' QuantiChromTM α- Amylase Assay Kit (DAMY-100). Lipase activity was assayed using BioAssay Systems' QuantiChromTM Lipase Assay Kit (DLPS-100).

2.3. Identification and characterization of protein marker of pancreatic exocrine defect in T2D

The protein profile of serum samples of all the groups (normal as negative control, patients with pancreatitis as positive control, pre-diabetes, T2D, and T2D with high amylase and/or lipase levels) were performed according to the methods described by Laemly with minor modification. The protein profiles were separated by 12.5% of separating gel with a 5% stacking gel. Then, the gel was stained with silver staining methods to visualize the protein profile.

Based on the protein band pattern resulted, band that was expressed in positive control (pancreatitis) but not in negative control (normal control) will be obtained 1 band. Otherwise, the same molecular weight of protein that express in positive control could be also express in T2D serum sample, thus, that band was suspected as a protein marker candidate of pancreatic exocrine defect.

Then the production process was conducted by SDS-PAGE with sample of serum pancreatitis patients that contained protein marker candidate on majority of electrophoresis gel wells. After the

IOP Conf. Series: Materials Science and Engineering **299** (2018) 012021 doi:10.1088/1757-899X/299/1/012021

silver nitrate staining was carried out, subsequent preparations were performed to isolate the target protein according to the Bringan method.

Protein samples were digested with trypsin, and peptides extracted according to standard techniques [5]. Peptides were analyzed by electrospray ionization mass spectrometry using the Ultimate 3000 Nano HPLC system (Dionex) coupled to a 4000 QTRAP mass spectrometer (Applied Biosystems). Tryptic peptides were loaded onto a C18 PepMap100, 3 µm (LC Packings) and separated with a linear gradient of water/acetonitrile/0.1% formic acid (v/v). Spectra were analysed to identify proteins of interest using Mascot sequence matching software (Matrix Science) with Ludwig NR database. Database: Ludwig NR Taxonomy: Homo sapiens (human).

3. Result and Discussion

3.1. Protein characters T2D sera

According to the previous study, we found that more than 50% T2D serum sample showed the pancreatic exocrine insufficiency condition [2]. The condition was showed by increasing level of serum amylase and or lipase. In order to explore protein marker for pancreatic exocrine insufficiency beside amylase and lipase marker analysis that commonly applied, electrophoresis SDS-PAGE was conducted as the first step of the exploration process. It was shown that 18 kDa peptide band appeared in positive controls and some T2D samples but was not obtained in negative control (Figure 1). The 18 kDa protein is suspected to be the candidate of expected protein marker.

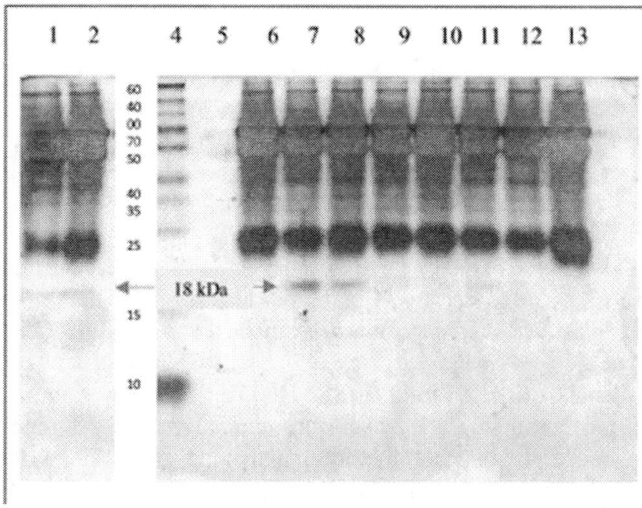

Notes:

1. positive control
2. positive control
3. -
4. Marker
5. -
6. Normal
7. T2D with high lipase level
8. T2D
9. T2D with high amilase and lipase level
10. T2D
11. T2D
12. prediabetic
13. T2D with normal amilase and high lipase level

Figure 1. The protein profile of serum sample of T2D patients, pre-diabetic, normal as negative control and pancreatitis patient as positive control, separated in 1D-GE with 12.5% polyacrylamide gel and visualized by silver staining method.

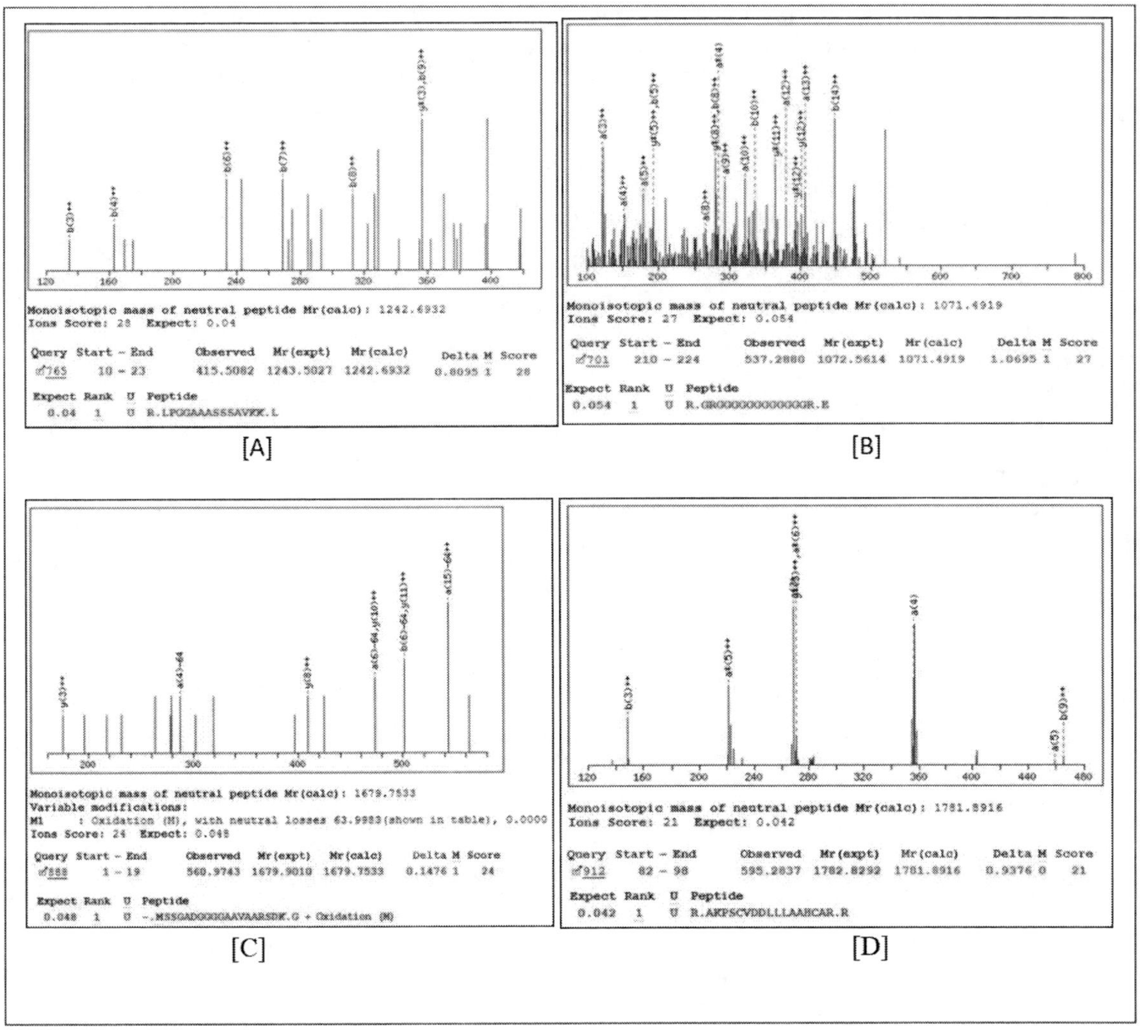

Figure 2. Spectra pattern and peptide fragmentation sequence of LC-MS/MS analysis of the 18 kDa protein band analyzed using *Mascot sequence matching software* (Matrix Science).

The protein 18 kDa was identified further using LC-MS/MS peptide analysis. Spectra pattern and peptide fragmentation sequence of LC-MS/MS analysis of the 18 kDa protein band analyzed using *Mascot sequence matching software* (Matrix Science) is presented in Figure 2. The further analysis of the peptide was identified as cDNA FLJ50670 that very similar to RNA polymerase I-specific transcription initiation factor RRN3, U2AF1 U2 small nuclear RNA auxillary factor 1 isoform, METTL10 Methyltransferase-like protein 10, and FAM19A3 isoform 2 of protein FAM19A3. The LC-MS/MS identification fragment was also confirmed by a Blast test to match the peptide sequences present with the peptide database in the National Centre for Biotechnology Information (NCBI) with a summary of the results as presented in **Table 1**. There are 2 peptides with high protein sequence coverage (10% and 18%) and approximating calculated molecular weight that estimated from electrophoregram data (18 kDa). Blast test results indicate that the protein has a similarity level of 100% with FAM19A3 protein isoform 2 of FAM19A3 and METTL10 protein methyltransferase-like protein 10 [8].

IOP Conf. Series: Materials Science and Engineering **299** (2018) 012021 doi:10.1088/1757-899X/299/1/012021

Table 1. Peptide identification result of protein marker candidate using LC-MS/MS

Character	cDNA FLI50670, highly similar to RNA polymerase I-specific transcription factor RRRN3	U2AF1 U2 small nuclear RN auxiliary factor 1 isoform	METLL10 methyltransferase-like protein 10	FAM19A3 isoform 2 of protein FAM19A3
Amino acid fragment from LC-MS/MS	R.LP GGAAASSSAV KK.L	R.GRGGGGGGGGG GGR.E	MSSGADGGGAAVAAR SDK.G	R.AKPSCVDDLLL AAHCAR.R
% identity with protein database	99	100	100	100
Calculated molecular weight	70364	27855	11340	18537
Predicted Pi	5.63	8.87	4.76	9.48
Protein sequence coverage	2	6	18	10
The number of amino acid sequence	621	240	104	169
Fragment location in the amino acid sequence	10-23	210-224	1-19	82-98

Protein FAM19A3 was predicted have MW 18,537 Da, pI (isoelectric point) of 9.48 Da [6, 7, 9-11]. According to Table 2, protein FAM19A3 was not expressed in pancreas. Protein FAM19A3 known as *Chemokine-like protein TAFA-3*. Protein FAM19A3 or TAFA protein that contain conserved cysteine residue that was dominant expressed in brain. The FAM19A3 protein was the FAM19A3 gene expression that demonstrated the highest associated with Type 1 DM [12, 13].

The other identified protein was METTL10 (Methyltransferase like protein 10). METTL10 was predicted have MW 11.340 Da, pI 4.76. *In silico* analysis of METTL10 have pI of 5.77, MW 31. 830.02 Da [6, 7, 9, 10, 14]. The differences between in-silico analysis and LC-MS/MS combined with Mascot software was supposed because there was conformational changing or protein sub unit splitting in electrophoretic process. According to the Table 2, METTL10 was highly expressed in almost all of body tissue, included pancreas and kidney. METTL10 was the enzyme that catalyzed methylation reaction of S-adenosyl-L-methionine (SAM or AdoMet). Product methylation of S-adenosyl-L-methionine (SAM or AdoMet) is S-adenosyl-L-homocysteine (AdoHcy) and other methylated biomolecules [15]. Increasing total concentration of Homocystein (tHcy) and methionine intermediate product (AdoMet and AdoHcy) sometimes correlate with kidney dysfunction and also cardiovascular disease that is common complication in T2D (15–17). Besides, overexpression of methyltransferase in methylation reaction was found in pancreas, especially in ductal, exocrine and islet cell, also found in early stage Pancreatic Carcinoma [18]. Pancreatic Carcinoma is one of the disease that indicated defect in exocrine pancreas. According to the Figure 1, protein 18 kDa that was showed in pancreatitis (positive control) but also found in some sample of T2D, so we can suppose that METTL10 (Methyltransferase like protein 10) was the protein marker candidate of insufficiency exocrine pancreas and correlated with progression of T2D.

Table 2. The local expression of Protein FAM19A3 dan METTL-10 in organ [6, 7]

Organ		Expression Level	
		FAM19A3	METTL-10
Liver and pancreas	Liver	++++	++
	Gallbladder		+++
	Pancreas		+++
GI Tract	Oral mucosa		
	salivary gland	+++	++
	eusophagus		++
	stomach		++++
	Duodenum		+++
	small intestine		+++
	Appendix		+++
	Colon		+++
	rectum		+++
Urinary Tract	kidney	++++	+++
	urinary bladder	++	+++
Man reproductive system	Testis	++++	++++
	Epididimis	+++	+++
	Prostat		++
	seminal vesicle	+++	++
breast and female reproductive system	breast		++
	vagina		++
	cervix, uternine	++	+++
	Endometrium	+++	+++
	Fallopian tube	+++	+++
	Ovarium		+++
	Placenta		+++
Skin and soft tissue	skin	++	++++
	adipose tissue		
	skeleton tissue		+++
	smooth muscle		+++
	soft tissue		+++
Blood and immune system	Bone marrow		
	Limphatic tissue		
	Tonsil		++
	Spleen		
CNS	cerebral cortex		+++
	Hippocampus		+++
	Lateral ventricle		
	Cerebellum		+++
Endocrine	thyroid tissue		+++
	parathyroid tissue		++
	adrenal tissue	++++	++
Respiratory system	nasopharynx		+++
	bronchus		+++
	lungs		+++
cardiovascular	heart		+++
+ = very weak			
++ = weak			
+++ = medium			
++++ = strong			

4. Conclusion

The current study has highlighted the proteomic analysis using LC-MS/MS to identify and characterize protein marker which indicated insufficiency pancreatic exocrine. A protein with 18 kDa has been suggested as a candidate protein that shown the pancreatic exocrine insufficiency in T2D. In addition, further identification of that protein using LC-MS/MS showed 4 peptide fragments. *In silico* analysis of the peptide fragment supported that pancreatic exocrine insufficiency correlated with progression of T2D.

References

1. DeFronzo R A 2004 *Med Clin North Am* **88** 787

2. Srihardyastutie A, Soeatmadji D W, Fatchiyah F, Aulanniam A. 2014 The 4th Annual Basic Science International Conference **2014** 33
3. Srihardyastutie A, Soeatmadji DW, Fatchiyah, Aulanni'am 2015 *Biomed. Res.* **26** 293
4. American Diabetes Association, 2013 Diabetes Care **Jan 1** S11
5. Bringans S, Eriksen S, Kendrick T, Gopalakrishnakone P, Livk A, Lock R, 2008 *Proteomics* 1081
6. Uhlen M, Oksvold P, Fagerberg L, Lundberg E, Jonasson K, Forsberg M, 2010 *Nat. Biotechnol.* **28** 1248
7. Uhlén M, Fagerberg L, Hallström B, Lindskog C, Oksvold P, Mardinoglu A, 2015 *Science* **2015** 80
8. Altschul SF, Madden TL, Schäffer AA, Zhang J, Zhang Z, Miller W 1997 *Nucleic Acids Res.* **25** 3389
9. Roy A, Kucukural A, Zhang Y. I-TASSER 2010 *Nat. Protoc.* **5** 725
10. Yang J, Yan R, Roy A, Xu D, Poisson J, Zhang Y 2015 *Nat. Methods.* **12** 7
11. Zhang Y. 2008 *BMC Bioinformatics.* **9** 40
12. American Diabetes Association 2012 *Input genes from type 1 diabetes-associated regions identified by GWAS* Published online at http://diabetes.diabetesjournals.org/lookup/suppl/
13. Tom Tang Y, Emtage P, Funk W, Hu T, Arterburn M, Park E, 2004 *Genomics* **83** 727
14. Zhang Y. 2008 *BMC Bioinformatics* **9** 40
15. Clarke S, Banfield K. 2001 *S-Adenosylmethionine-dependent Methyltransferases* In Carmel R, Jacobsen DW, editors. Homocystein in Health and Disease. United States of America: Cambridge University Press 63–78.
16. Herrmann W, Schorr H, Obeid R, Makowski J, Fowler B, Kuhlmann MK. D 2005 *Clin Chem.* **51** 891
17. Wijekoon EP, Hall B, Ratnam S, Brosnan ME, Zeisel SH, Brosnan JT. 2005 Diabetes **54** 3245
18. Donahue T, Tran L, Hill R, Li Y, Kovochich A, Calvopina J, 2012 Clin. Cancer Res. **18** 1352

IOP Conf. Series: Materials Science and Engineering **299** (2018) 012022 doi:10.1088/1757-899X/299/1/012022

The Effects of Root Extract *Ruellia tuberosa L* on Histopathology and Malondialdehyde Levels on the Liver of Diabetic Rats

Alfin Nur Laily Kurniawati,[1] Aulanni'am,[1,2] Arie Srihardyastutie,[1] Anna Safitri[1*]

[1]Departement of Chemistry, Faculty of Mathematics and Natural Sciences, Brawijaya University
[2]Faculty of Veterinary Medicine, Brawijaya University
Email: a.safitri@ub.ac.id

Abstract. The aim of this research is to study antidiabetic activity of root extract of *Ruellia tuberosa L* on rats (*Rattus novergicus*) induced by multiple-low dose streptozotocin as animal diabetic models. The parameters investigated were blood glucose levels, free radicals (MDA, malondialdehyde) levels and hepatic histopathology. The main materials used were n-hexane root extracts from *Ruellia tuberosa L*. Three groups of rats, including control group (group I), diabetic group (group II), and therapy group with *Ruellia tuberosa L* (group III), were used. Streptozotocin was given at multiple-low dose of 20 mg/kg of body weight for 5 times in 5 consecutive days *i.p.* to rats in groups II and III. The *Ruellia tuberosa L* extracts were then given orally for group III in the dose of 250 mg/kg of body weight per day for 3 weeks. Results of the current work showed that root extract *Ruellia tuberosa L* had lowered blood glucose levels on rats in group III by 60.3%, from 299.7±24.7 mg/dL up to 119.0±26.6 mg/dL. Moreover, the antidiabetic activity of *Ruellia tuberosa L* extracts also deduced from decrease of MDA levels in group III, from 3.5 ± 0.3 μg/mL up to 1.7 ± 0.4 μg/mL. The recovery of hepatic organ from treatment group has also been proven from the its histology profiles stained with hematoxylin-eosin.

1. Introduction

Diabetes mellitus type 1 is a metabolic disorder caused by the destruction of pancreatic β cells, resulted in reduced production of insulin hormones, that causes hyperglycemia [1]. Prevalence of diabetes worldwide is estimated at 171 million people in 2000, and is expected to increase to 366 million by 2030 [2]. In Indonesia, the number of people with diabetes mellitus reached 14 million in 2006, and is expected to reach 21.3 million people in 2030 [3].

Hyperglycemia in diabetes mellitus closely related with oxidative stress conditions [4] and increase of ROS (reactive oxygen species) production. Increasing ROS causes an increase in lipid peroxidation. ROS will degrade PUFA resulting in malondialdehyde compound [5]. Malondialdehyde compounds are one of the signs of lipid peroxidation due to excess ROS. The higher the amount of ROS in the body the higher the levels of malondialdehyde [6]. In addition, the presence of toxic substances in the body will cause damage to liver cells. The liver is an organ that plays role in the toxic detoxification process that enters the body thus it can easily be damaged. When the liver organ is constantly exposed to toxic substances, this will result in degeneration of liver function that causes damages to the liver cells. Damage to liver cells will be easily observed through microscopic histopathological observations.

Treatment of diabetes mellitus can be conducted with insulin or by admisteration of antidiabetic drugs such as sulphonylurea and biguarid derivatives. However, but such treatments can cause very serious side effects such as hypoglycemia, obesity, and renal dysfunction [8, 9]. In order to overcome these problems, it is necessary to search for alternative treatment from indigenous natural products from Indonesia. One of the plants that commonly used for diseases treatment is pletekan plants from the species of *Ruellia tuberosa L.* This plant has been proposed to have potency as anti-diuretic, antidiabetic, antipyretic, analgesic, antihypertensive, and antidotal agents [10]. Manikandan (2010) in his research reported that leaf of pletekan contains phenol compounds (0.36 mg/g), saponins (0.10 mg/g), glycosides (0.59 mg/g), flavonoids (0.75 mg/g) and vitamin (K, C) and carotenoid content [10]. The bioactive compounds of pletekan root obtained from GC-MS revealed 25 compounds including steroid form of stigmasterol 8.89%, sitosterol 3.99%, cholesterol 2.24% and triterpenoid form of Lupeol by 68, 14% [11]. Triterpenoid is the major bioactive compounds contained in the root pletekan can act as antioxidants [11], which can decrease ROS [12, 13].

The part of pletekan that already widely utilized to overcome hyperglycemia are leaves and stems. Rahmi *et al* studied of hypoglycemic effects purple gold leaf extract (*Ruellia tuberosa L.*) on Wistar rat mentioned that leaf water extract of pletekan can be lowered glucose levels in rat blood [14]. Utilization root pletekan actually already exists but still traditional and there has been no scientific research on the effects of hypoglycemia. Based on the description above, this study has been conducted to determine the effect of root pletekan root to decrease the malondialdehyde level and improve the histopathology on the liver of diabetic rat induced by MLD-STZ.

2. Materials and Methods

2.1 Materials

This research used the animal model of male Wistar white rat obtained from Institute of Biosciences Brawijaya University, Malang. The experimental animals have received a certificate of ethics from Research Committee of Universitas Brawijaya No: 624-KEP-UB-2016. The materials used in this research were root of *Ruellia tuberosa L.* obtained from Materia Medica, Batu City, East Java. Other materials used were distilled water, conc. H_2SO_4, citric acid solution, sodium citrate solution, citrate buffer solution pH 4.5, streptozotocin (Bioworld), PBS solution (Phosphate Buffer Saline), 1% PBS-azide solution, 10% trichloroacetic acid (TCA), HCl, 1% Na-thio, NaOH solution, ethanol, xylol, paraffin, hematoxyline eosin (HE), and 10% paraformaldehyde (PFA).

2.2 Methods

2.2.1 Preparation of Diabetic Rats

A week prior to the experiment, rats were adapted in the animal house. Rats were divided into three groups: (I) control group; (II) diabetic group; and (III) treatment group, 6 rats in each group. Rats were maintained in the animal house, Biosains Laboratory, Brawijaya University. A 100 mg of streptozotocin 100 was dissolved in 3 mL of citrate buffer at pH 4.5. A dose of 20 mg/kg of body weight was injected *i. p.* to rats in groups II and III, for 5 consecutive days, while rats in group I (control) were injected with PBS only [15-17]. Rats in group III were treated with root extracts at a dose of 250 mg/kg body weight per day for each rat for 21 consecutive days.

2.2.2 Extraction of Pletekan Root

Root powder was macerated with n-hexane, the volume was 7.5× the weight of the powder Maceration solution was stirred every 1 h in the first 5 h, and then allowed to stand for up to 48 h. The extract obtained was decanted to separate from pletekan powder. Solvent in the extract was evaporated with a rotary evaporator at 400 °C, 90 rpm, until the concentrated extract obtained.

2.2.3 Treatment with Root Extracts of Pletekan for Group III

Rats in group III were treated with pletekan root extracts at a dose of 250 mg/kg body weight per day for each rat for 21 consecutive days. Changes in blood glucose levels were monitored per week for during the treatment. At the end of the assay, rats were sacrificed, and the hepatic organ was collected for further analysis.

2.2.4 Measurement of Malondialdehyde (MDA) Levels
Measurement of MDA levels was conducted by spectrophotometry method, using TBA reagents. The liver homogenates from three groups group 1, II, and III were prepared based on previous method [18]. The liver supernatants were measured their absorbance using a UV-Vis spectrophotometer at the 530 nm.

2.2.5 Liver Histopathological observation
Histopathologic preparations of the rats' liver were made using the HE staining method. Histopathologic features of the liver were visualized using the Olympus BX51 microscope with 400× magnification to observe the degeneration of lipid sand damage to hepatocyte cells. Hepatic histopathology images were captured using a digital camera.

2.2.6 Statistical Analysis
The Student's *t*-test was used for statistical analysis of differences between two groups of samples, control and treatment, or between each treatment. In all cases, any result with a p threshold < 0.05 was considered to be significant.

3 Results and Discussion
Injection of streptozotocin with a dose of 20 mg/kg body weight 5 consecutive days caused rats suffered from DM type 1. Table 1 present the blood glucose levels on rats in all groups in the last week of the assay (week 3).

Table 1. Profiles of blood glucose levels on rats from group I, II, and III, after 21 d of treatment with pletekan root extract

Group	Blood glucose levels (mg/dL)*
I	83.17±7.46^{a}
II	299.67±24.71^{b}
III	119.3±24.63^{c}

*different notations show a significantly different effect in each group (p <0.05)

Based on the data on Table 1, it can be seen that blood glucose levels in group I was at a range of normal rats, 83.17 ± 7.468 mg/dL, while blood glucose levels in rats injected by MLD-STZ were 299.67 ± 24.695 mg/dL. These show an increase of 260.31% when compared with control rats. The treatment of pletekan root extract with the dose of 250 mg/kg body weight decreased blood glucose level to 119 ± 24,633 mg/dL, with a percentage of decrease was 60.29%. Decreased blood glucose levels in the treatment group is suggested as a result of improvement of pancreatic β cells, hence, insulin secretion into the bloodstream will improve the process of glucose uptake.

Next is determination of root pletekan extract on the hepatic MDA levels. Malondialdehyde is one of the end products of cell membrane lipid peroxidation by excess ROS, therefore, MDA is used as one of the markers of ROS levels in the body [16, 19]. Measurement of MDA levels (Table 2) shows that MDA levels in diabetic rats group were 3.48 ± 0.27 µg/mL, this value increased 136.58% when compared with the negative control group with MDA levels at 1.47 ± 0.32 µg/mL. In the treatment group MDA level was 1.69± 0.37, or decreased by 51.55%.

Table 2. MDA levels on rats from group I, II, and III, after 21 d of treatment with pletekan root extract

Group	MDA levels (μg/dL)*
I	1.47±0.32[a]
II	3.48±0.27[b]
III	1.69 ±0.37[a]

*different notations show a significantly different effect in each group (p <0.05)

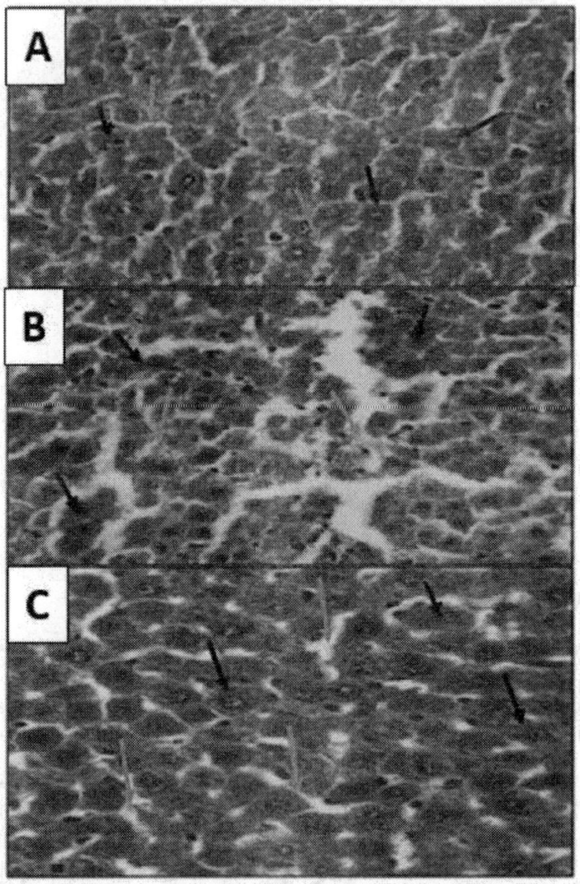

Figure 1. Comparison of histopathologic images of rats from group I, II, and III, after 21 d of treatment with pletekan root extract, with HE staining (400× magnification); (A) group I; (B) group II; (C) group III, black arrow(↑) indicates the hepatocyte cells, the blue arrow (↑) indicates a sinusoid.

Intraperitoneal injection of MLD-STZ resulted in toxic substances circulated into the body and going directly into the circulatory system. The liver plays a role in the detoxification process of toxic substances in the body is very susceptible to damage. The liver cells that are continuously exposed to toxic compounds such as STZ as diabetogenic agents can trigger an increase in the production of excessive free radicals resulting in oxidative stress. Excessive amounts of free radicals will increase the peroxidation of cell membrane lipid peroxidation resulting in increased levels of MDA. It is therefore, the MDA levels on the diabetic conditions generally is higher than in the normal state.

MLD-STZ injections and pletekan root extract therapy have an effect on the histopathologic picture of diabetic mouse liver. Changes in the histopathologic picture of the liver can be seen in Figure 1. Histopathologic changes of hepatic features can be seen from hepatocyte cells (hepatic cells). The hepatocyte cell condition of the negative control group shows that normal hepatocyte cells are arranged radial, the cell nucleus is clearly blue, the space between hepatocytes or sinusoids is clear and the absence of fat accumulation in the liver. The hepatocyte cells of diabetic rats show that the cell nucleus were damaged and black. In addition, there is dilation of sinusoids and suppress the cell nucleus to peripheral hepatocytes. Improvement in histopathologic features of the liver can be seen in the treatment group. Histopathologic improvement is evident from the presence of normal hepatocyte cells but still appears to be part of hepatocyte cells whose nuclei are damaged. In addition, the widening of the sinusoid begins to improve, thus, the space between hepatocyte cells is narrower.

Decreases in blood glucose and in MDA levels, improvements in hepatic profiles of diabetic rats treated with pletekan root extract can be associated with the content of triterpenoid in the composition of pletekan root extracts. Preliminary phytochemical test has been conducted, and result indicated that the root extracts of pletekan contained triterpenoid. These were deduced from the colour changed from orange after to red-purple when the sample solution was tested with Lieberman-Burchard reagent (result not shown). This suggests that triterpenoids in pletekan roots extract begin to suppress the presence of free radicals in order not to bind to the lipid component of PUFA in hepatocyte cells by removing the H atoms in the hydroxyl group (-OH), as a result, the free radicals will bind to atoms with high electronegativity. Moreover, triterpenoids will balance the amount of oxidants and antioxidants in the body that play the role of protecting cells from free radicals.

4. CONCLUSION

The current study demonstrated that administration of pletekan root extract decreased blood glucose level in diabetic rats, decreased malondialdehyde levels, and improved histopathological profiles on the liver. The triterpenoid contained in the pletekan root extracts may be responsible for the antidiabetic actions. Further research is needed to investigate molecular mechanisms of triterpenoid acting as free radical scavengers.

ACKNOWLEDGMENT

This research was supported by the DPP/SPP FMIPA Brawijaya University 2017 (Grants Number 18/UN10.F09.01/PN/2017).

REFERENCES

[1] Tibrani M M 2009 *Prosiding Seminar Nasional Penelitian, Pendidikan dan Penerapan MIPA*, Fakultas MIPA Universitas Negeri Yogyakarta Yogyakarta
[2] Wild S G Roglic A Green R Sicree and H King 2004 *Diabetes care* **27** 5
[3] BPOM 2009 Informasi Produk Terapeutik 1
[4] Rahmawati G Rachmawati F N Winarsi H 2014 *Scr. Biol.* **1** 19
[5] Konyalioglu S Durmaz G Yalcin A 2007 *Cell Biochem. Funct.* **25** 259
[6] Takahashi H Tran P O T Leroy E Harmon J S Tanaka Y Robertson R P 2004 *J. Biol. Chem.* **279**, 37316
[7] Wiyono P. 2003 *B.I. Ked.* **35** 55
[8] BPOM 2004 *DTCA (Direct to Consumer Advertising) dan Pengaruhnya* **5** 1
[9] Nathan M 2008 *Diabetes Care* **31** 173
[10] Manikandan A 2010 *Biomed. Sci. Res* **2** 190
[11] Rajendra K N Vasantha Mohan V R 2014 *Orig. Artic.* **2** 209
[12] Arivazhagan P Thilakavathy T 2000 *J. Nutr. Biochem.* **2863** 122
[13] Favier A Cadet J Kalyanaraman B Fontecave M Pierre J L 1995 *Analysis of Free Radicals in Biological Systems* Deutsche Bibliothek Basel

[14] Rahmi A N Sutjiatmo A B Vikasari S N 2014 *J. Ilm. Farm.* **2** 50
[15] Shofia V Aulanni'am Mahdi C 2013 *Student Chem. J.* **1** 119
[16] Dewi D R Aulanni'am Roosdiana A S. 2013 *Student Chem. J* **2** 351
[17] Bell D S H 2012 *Diabet. Obes. Metab.* **15** 291
[18] Aulanni'am Roosdiana A Rahmah N L 2012 *J. Life Sci.* **6** 144
[19] Sahin N Orhan C Tuzcu M Sahin K Kucuk O 2008 *Poult Sci.* **87** 276

Biosorption of Cr(VI) in Aqueous Solution using Microorganisms: Comparison of the Use of *Rhizopus oryzae*, *Bacillus firmus*, and *Trichoderma viride*

Anna Safitri[1*], Putri Mahardini[1], Sasangka Prasetyawan[1], Anna Roosdiana[1]

[1]Department of Chemistry, Brawijaya University, Malang

Email: a.safitri@ub.ac.id

Abstract. In this work, the study of biosorption of Cr(VI) from aqueous solution was conducted using *Rhizopus oryzae*, *Bacillus firmus,* and *Trichoderma viride* as microorganisms that can absorb Cr(VI). The research is focused on determination of optimum conditions including pH, the number of *R. oryzae*, *B. firmus*, and *T. viride* (inoculums), and initial concentrations of Cr(VI) used. Optimum pH was obtained at pH 5, 4.5 and 6, for biosorption of Cr(VI) with *R. oryzae*, *B. firmus*, and *T. viride*, respectively, in the capacity of 45.3%, 24.5%, and 90.3%. The highest amount of Cr(VI) adsorbed for biosorption with *R. oryzae*, *B. firmus*, and *T. viride*, were 55.4%, 18.5%, and 74.5%, respectively, using 6-mL inoculums. The equilibrium concentrations achieved for *R. oryzae*, *B. firmus*, and *T. viride* were 60 mg/mL, 40 mg/mL, and 40 mg/mL, with the amount of Cr(VI) adsorbed were 32.4%, 28.2%, and 89.3%, respectively. The adsorption capacity for *R. oryzae*, *B. firmus*, and *T. viride* were 45.3 mg/1x10^6 colonies, 36.2 mg/1x10^6 cells, and 77.8 mg/1x10^6 colonies, respectively. Overall, the biosorbents effectivity order in the biosorption process of Cr(VI) are *T. viride* > *R. oryzae* > *B. firmus*.

1. Introduction

Heavy metal pollution caused by both natural processes and human activities is serious problem in the environment. The main consequences of heavy metal pollution is contamination of water system. At high concentrations, heavy metal ions poisons aquatic life in the waters. These metal ions are non-degradable and are persistent in the environment [1]. One of toxic metals ion is Cr(VI). Cr(VI) shows elevated toxicity, being related to nasal irritation and ulceration, chronic bronchitis and emphysema, cause liver and kidney damage, lung and skin cancer, in addition to damaging DNA by its highly genotoxic crosslink to DNA [2, 3]. Inhaling and/or ingesting Cr(VI) compounds has been shown to be cancerous due to its easy permeation of the cell membrane and elevated oxidizing potential, causing lung and stomach cancer [3]. Therefore, the removal or reduction of Cr(VI), in particular in the waters, is essential.

Heavy metals such as Cr(VI) have been removed from industrial effluents by the application of conventional metal removal techniques such as reverse osmosis, solvent extraction, ion exchange and chemical precipitation [4, 5]. These methods are encountered with certain major disadvantages such as high energy requirements, incomplete metal removal and generation of a large quantity of toxic waste sludge. Consequently, the search for alternative solutions has become of eminent importance, either alone or coupled with another conventional method already used to remove Cr(VI).

The potential for growth and adaptation of microorganisms when faced with adverse conditions and their versatility in biotechnological applications linked to modern biotechnology, has qualified microorganism biosorption agents known as biosorbents [6]. The use of biosorbents in the removal of Cr(VI) is promising when compared to the conventional techniques mentioned above.

In the current work, the biosorption of Cr(VI) in aqueous solution has been conducted using microorganisms, *R. oryzae*, *B. firmus*, and *T. viride*. The use of microorganism for adsorption process of metal ions have been conducted before in some studies [7-10]. To date, the microorganisms used for biosorption process are mostly bacteria [7, 10] and fungi [8, 9]. Results of these studies indicated that microorganisms had high adsorption capacity and high effectivity to adsorb the metal ions [7-10]. The proposed mechanism of action of the biosorption of metal ions by microorganism is by adsorption in the cell membranes or cell walls (reviewed in [11]). The positively charged ions of metals present in the solution are attracted to the cell and adsorbed on a negatively charged surface of the cell. In addition, the biosorption process of metal ions should consider a number physicochemical factors including pH, contact time, concentration, or temperature, when living microorganism are applied [12, 13]. Therefore, in this study, the biosorption of Cr(VI) using *R. oryzae*, *B. firmus*, and *T. viride* are conducted under the influences pH, the amount of biosorbents, and initial concentrations of Cr(VI) used.

2. Materials and Methods

2.1 Materials

Culture of *R. oryzae* and *B. firmus* were obtained from Department of Food Science and Technology, Brawijaya University, Malang, while culture of *T. viride* was obtained from the microbial stock collection of Food and Nutrition Laboratory, Gadjah Mada University, Yogyakarta. All reagents of analytical or higher purity grade were purchased from Merck or Sigma-Aldrich and were used as received: HCl (37% aqueous solution), HNO_3 (trace pure, 65% w/w in H_2O), NaOH (99.9%), H_2SO_4 (98%, *d* = 1.84 g/mL), ethanol (96%), $K_2Cr_2O_7$, $MgCl_2{\cdot}2H_2O$, KH_2PO_4, Na_2HPO_4, NaCl, $NaKC_4O_6H_4$, KCl, NH_4Cl, CH_3COOH (100%; *d* =1.05 g/mL), $NaCH_3COON$, pepton, nutrient agar and Mueller Hinton Broth for microbiology. Water was purified using distillation technique. The growth media for all three species were potatoes dextrose agar (PDA). The determination of Cr(VI) concentrations before and after biosorption process was conducted using UV-Vis spectrophotometer (1601-Shimadzu). The functional groups change in the biosorbent before and after biosorption process was determined using FTIR spectrophotometer (8400S/Shimadzu).

2.2 Methods

A loop full of *R. oryzae*, *B. firmus*, and *T. viride* was grown in the sterile PDA for 6 days at 37 °C. Next, bacteria and spores grown in the PDA was dissolved in 1 mL of sterile bi-distilled water. These were cultivated in 100 mL liquid growth medium in 250-mL Erlenmeyer flasks on a rotary shaker at room temperature, for 36 h at 150 rpm. The suspension was the microorganisms' inoculums.

Chromium(VI) solutions with concentration of Cr(VI) of 10 mg/mL were added into liquid medium, these were contacted with cultures of *R. oryzae*, *B. firmus*, and *T. viride*, separately. These mixtures were then incubated at room temperature for 24 h, with continuously shaken on an orbital shaker at 350 rpm. The mixture was filtered and the resulting supernatants were taken in 1 mL and placed in a beaker for the next steps. A solution of 0.1 M H_2SO_4 at pH 1 and 2 mL of 1,5-diphenylcarbazide were added to supernatants, and allowed to stand at room temperature for 5-10 min. The absorbance of the solutions was measured by UV-Vis spectrophotometer at a wavelength of 540 nm. Similar steps were repeated with variations in the pH: 4, 5.5, 5, 6, 6.5, and 7; in the number of inoculums: 2, 3, 5 and 6 mL; and in the concentrations of Cr(VI): 20, 40, 60, 80, and 100 mg/mL. At the end of each biosorption process, the filtrate was separated and the final Cr(VI) concentrations in the solution were determined using AAS spectrometer (AA-6200/Shimadzu), and Cr(VI) adsorbed (%) was calculated. The adsorption capacity was calculated in the optimum condition of pH, the number of inoculum, and concentration. The adsorption capacity (q) was calculated using the following formula:

$$q = \frac{(C_0 - C_t)V}{\text{adsorbent mass (the number of microorganims)}}$$

where: C_o = initial concentration of Cr(VI) solution (mg/mL); C_t = concentration of Cr(VI) after biosorption process (mg/mL); and V = volume of solution (mL). The precipitates from biosorption process were collected and analyzed using FTIR spectrophotometer at a range of 4000-400 cm^{-1} (8400S/Shimadzu).

3. Results and Discussion

Figure 1 shows relationship between pH variations and the percentage of Cr(VI) adsorbed using three different microorganisms. The Cr(VI) uptake was highly affected by the initial pH of aqueous metal solution. In the range of pH 4-7, *R. oryzae*, *B. firmus*, and *T. viride*, showed slightly differences in their optimum pH in adsorbing Cr(VI). As shown in Figure 1, the optimum pH obtained for three species were different, and the biosorption capacity for Cr(VI) were also varied among those species. The highest biosorption capacity for Cr(VI) was obtained using *T. viride* at pH 6 with 90.3%, followed by *R. oryzae* at pH 5 with 45.3%, and *B. firmus* at pH 4.5 with 24.5%.

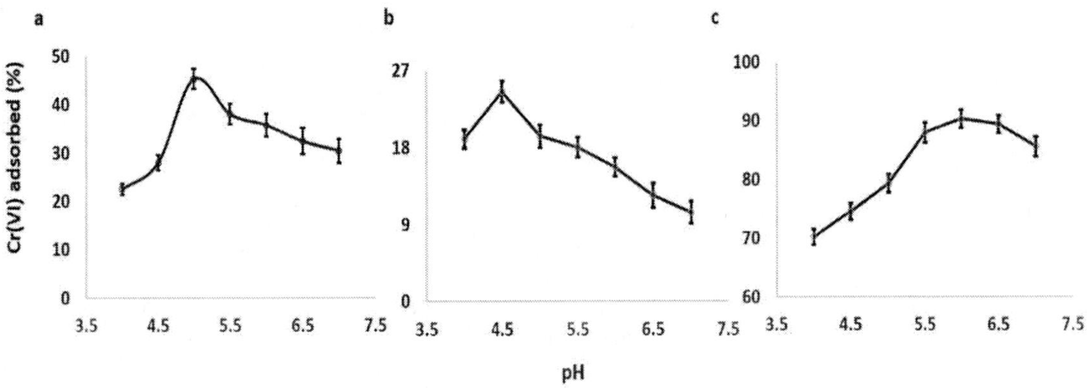

Figure 1. Result of biosorption of Cr(VI) using: (a) *R. oryzae*, (b) *B. firmus*, and (c) *T. viride*, under the influence of pH. The values are means and standard deviation of three replicates experiment.

The pH of the medium affects both the surface metal binding sites and metal chemistry in solution [11]. In solution, Cr(VI) exist in the form of its anions such as $HCrO_4^-$, CrO_4^{2-}, and $Cr_2O_7^{2-}$ [14]. Since the biosorption process was performed at acidic to neutral pH (4-7), these caused the H^+ ions in the solution were increased. As a result, Lewis base groups in the cell walls of fungi and bacteria, *i. e.* $-NH_2$ group was protonated to NH_3^+, making it possible to interact with protonated Lewis base groups and forming ion pair complexes with Cr(VI) species. Several researchers have investigated the effect of pH on biosorption of Cr(VI) by using different kinds of microorganism and have reported identical results [15, 16].

In the current study, the differences in optimum pH and the variation in the amount of Cr(VI) adsorbed for *B. firmus*, *R. oryzae*, and *T. viride* indicates the nature of the species used influence the biosorption capacity of Cr(VI). The fungal cell walls as in the case for *R. oryzae* and *T. viride* are mainly composed of polysaccharides β-1,3 glucan and β-1,6 glucan, whereas that of the Gram-positive bacteria such as *B. firmus* is constituted of a thick layer composed of peptidoglycan [11]. It is shown in this study that the capacity of *R. oryzae* and *T. viride* were greater than *B. firmus* in adsorbing Cr(VI), under the influence of pH.

The percentage of Cr(VI) removal from aqueous solution was found to increase concomitantly with increments in the number of microorganisms as shown in Figure 2. These indicate that the binding sites on the cell walls of fungi or in the cell membranes of bacteria is also increased, as a result, biosorption of Cr(VI) increased. At the highest amount of inoculum used (6 mL) the percentage values of Cr(VI) adsorbed by *B. firmus* was 18.5%, while *T. viride* showed the highest at 74.5%, and *R. oryzae* was at 55.4%. However, in these conditions, the adsorption capacity has not been achieved since the trend from all graphs (Figure 2a-2c) show that increasing inoculum will increase the amount of Cr(VI) adsorbed. Nonetheless, trend obtained in Figure 2 is in agreement with previous result, where the *T. viride* had the highest capacity in adsorbing Cr(VI), followed by *R. oryzae*, and *B. firmus*.

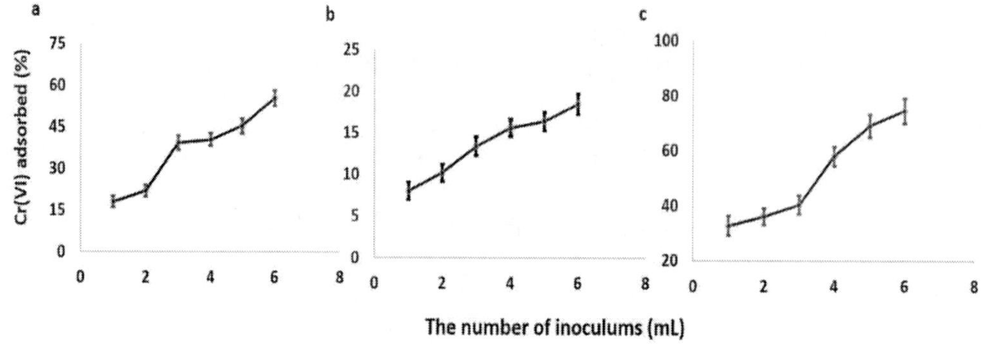

Figure 2. Result of biosorption of Cr(VI) using: (a) *R. oryzae*, (b) *B. firmus*, and (c) *T. viride*, under the influence of the number of inoculums (mL). The values are means and standard deviation of three replicates experiment.

The adsorption process of Cr(VI) is not only dependent on the availability of active sites for metal binding but also on the initial concentrations of Cr(VI). Effect of initial Cr(VI) ions concentration on its biosorption by *R. oryzae, and B. firmus*, and *T. viride* was investigated by incubating those species with previous optimum conditions that obtained before (pH 5, 4.5, or 6, for *R. oryzae*, and *B. firmus*, and *T. viride*, respectively, using 6-mL inoculum), with 100 mL of Cr(VI) solutions, concentration ranging from 20 to 100 mg/mL. Results are presented in Figure 3.

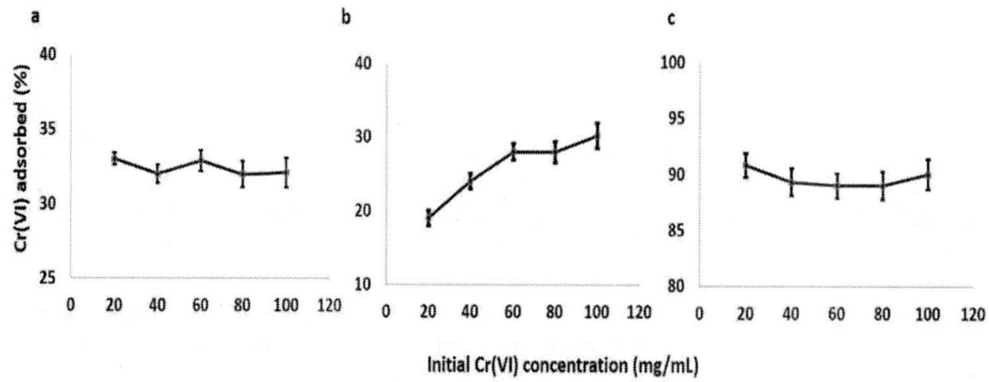

Figure 3. Result of biosorption of Cr(VI) using: (a) *R. oryzae*, (b) *B. firmus*, and (c) *T. viride*, under the influence of initial Cr(VI) concentration (mg/mL). The values are means and standard deviation of three replicates experiment.

From Figure 3, the uptake of Cr(VI) using *R. oryzae* and *T. viride*, showed similar trend at concentration of Cr(VI) ranging from 20-100 mg/mL. The equilibrium concentration obtained for *R. oryzae* and *T. viride* was at 40 mg/mL, in the percentages of 32.4% and 89.3%, respectively. For *B. firmus*, the concentration at equilibrium state achieved at concentration of 60 mg/mL, with 28.4% of

Cr(VI) adsorbed. The influence of concentration in the biosorption can be explained that at the beginning of the process, active sites at the cell walls of *T. viride* and *R. oryzae* or cell membrane of *B. firmus*, adsorbed metal ions fast. This may be due the unoccupied of the negatively active sites on the surface of cell walls, or may be due to the higher collision between the metal ions and negatively functional groups [14]. At high equilibrium concentration, uptake of Cr(VI) ions owing to surface binding was negligible due to saturation of biosorbent-binding sites. The slow increase in biosorption capacity at high concentrations could be related to the different concentration gradient between the solution and the inside of the microbial cells [15]. In these final conditions for three different biosorbent, the adsorption capacity for Cr(VI) was calculated. The calculated of the number of microorganisms *for R. oryzae, B. firmus,* and *T. viride* were 1.56 x 10^5 CFU/mL, 1.23 x 10^5 cells/mL, and 1.37 x 10^5 CFU/mL, respectively. Therefore, the adsorption capacity of *T. viride* toward Cr(VI) was 77.8 mg/1 x 10^6 colonies *R. oryzae* was 45.3 mg/1x10^6 colonies, and *B. firmus* was 36.2 mg/1x10^6 cells.

Again, the biosorption result of Cr(VI) under the influence of concentration show that *T. viride* had the highest capacity to adsorb Cr(VI), followed by *R. oryzae*, while *B. firmus* had the lowest capacity. Metal uptake mechanisms by various biosorbents depend on the cellular surface of the microbes. Generally, all microorganisms have a negative charge on their cell surface due to presence of anionic structures, that enable them to bind to metal ions. These negative charged groups includes alcohol, amine, carboxyl, ester, phoshoryl, sulfonate, sulfhydryl, and thiol group [17]. However, analysis of cell wall component, which also vary among different microbes, helps in assessing metal uptake by different microorganisms. In Gram-positive bacteria, such as *B. firmus*, peptidoglycan layer, which contains alanine, glutamic acid, polymer of glycerol and teichoic acid are the active sites involved in metal binding process [11]. Metals are attached to these ligands on cell surfaces, which displace essential metals from their binding sites [18].

The rigid cell walls of fungi, such as *R. oryzae* and *T. viride*, consist of chitin, inorganic ions, lipids, nitrogen-containing polysaccharide, polyphosphates, and proteins [19]. The surface of their cell wall acts as a ligand for binding metal ions, resulting in the removal of metals. Furthermore, fungi can also absorb heavy metals into their mycelium and spores [19, 20]. Therefore, it is suggested that *R. oryzae* and *T. viride* had higher capacity in adsorbing Cr(VI) compared to *B. firmus*, due to the mycelium and spores also contributed to biosorption process of Cr(VI). The differences between *R. oryzae* and *T. viride* show that microbial has different biosorptive abilities and effectivities, which also varies considerably within each group and each species.

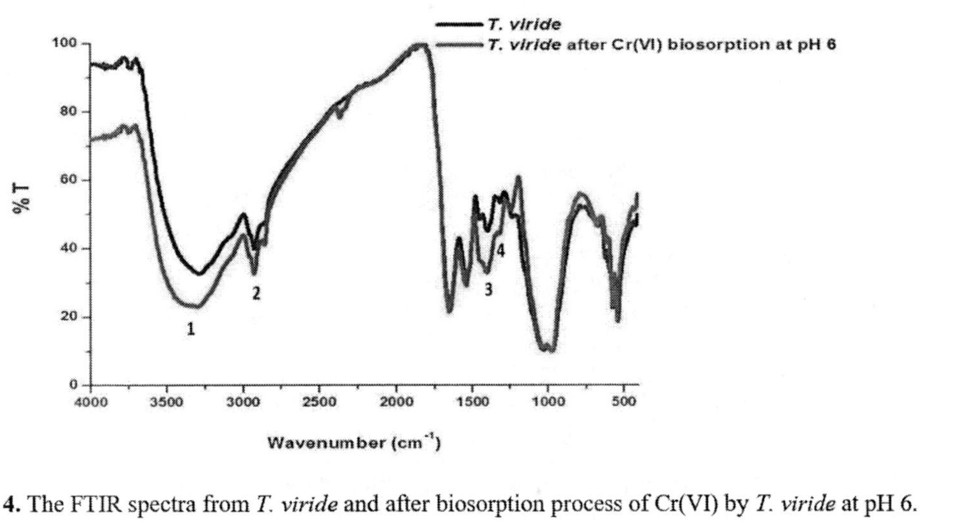

Figure 4. The FTIR spectra from *T. viride* and after biosorption process of Cr(VI) by *T. viride* at pH 6.

The functional groups in the biosorbents contributed to biosorption process of Cr(VI) separated from their filtrates and have been characterized using FTIR spectrophotometry. Figure 4 shows the FTIR results from *T. viride* before and after biosorption process in the pH 6. The *T. viride* was chosen to represent the biosorbent in Cr(VI) uptake, since it has the highest adsorption capacity among three different biosorbents. Some changes in FTIR bands were observed before and after biosorption process. The changes and relative intensities in the FTIR peaks are tabulated in Table 1. As shown in Figure 4, there were four significant areas affected due to biosorption process of Cr(VI). Those areas were 3450-3100 cm^{-1}, 2950-2850 cm^{-1}, 1470-1450 cm^{-1}, and 1180-1160 cm^{-1}. The band assignment for those are N–H bending from amine group, C–H stretching from alkane, another C–H stretching from alkane, and C–N bending from amine group, respectively [21, 22]. In addition, in the finger printing regions (1000-500 cm^{-1}) also showed some changes, these regions are generally phosphate groups and alkanes bands from nucleic acids (DNA). All those bands showed decreases in intensity relative to *T. viride* bands in the same regions. This is possible since *T. viride* as biosorbent is also microorganisms, that contains biomolecules, including proteins, lipids, carbohydrates, and nucleic acids.

Table 1. FTIR changes in intensity of bands of interests relative to *T. viride* bands after biosorption of Cr(VI) at pH 6

No	Wavenumber (cm^{-1})	Assignment	Intensity changes*
1	3450-3100	N–H amine	↓
2	2950-2850	C–H alkane	↓
3	1470-1450	C-H alkane	↓
4	1180-1160	N–H amine	↓

*The changes are relative compared to the intensity in the related FTIR bands of *T. viride*

4. Conclusion

The biosorption process of Cr(VI) by three different microorganisms, *R. oryzae, B. firmus,* and *T. viride* has been conducted successfully. The optimum pH obtained for *R. oryzae, B. firmus,* and *T. viride* were 5, 4.5 and 6 with the Cr(VI) adsorbed 45.3%, 24.5%, and 90.3%, respectively. Using 6-mL inoculum, the biosorption capacity for Cr(VI) were 55.4%, 18.5%, and 74.5%, with *R. oryzae, B. firmus,* and *T. viride*, respectively. The concentration in the equilibrium state for *B. firmus* and *T. viride* achieved at 40 mg/mL with Cr(VI) adsorbed were 28.2% and 89.3%, respectively, while *R. oryzae* was at 60 mg/mL (32.4%). Therefore, the order in the biosorption process of Cr(VI) using microorganisms as adsorbents are *T. viride* > *B. firmus* > *R. oryzae*, with adsorption capacity of 77.8 mg/1x10^6 colonies > 45.3 mg/1x10^6 cells > 36.2 mg/1x10^6 colonies, respectively.

References

[1] Vazquez C Errasquin E L 2003 *Chemos.* **50** 137
[2] Jasso-Chávez R Pacheco-Rosales A Lira-Silva E Gallardo-Pérez J C García N Moreno-Sánchez R 2017 *Aquat. Toxicol.* **100** 329
[3] Cefalu W T Hu F B 2004 *Diabet. Care* **27** 2741
[4] Mutongo F Kuipa O Kuipa P K 2014 *Bioinorg. Chem. Appl.* **2014** 1
[5] Ab Rahman N N N Shahadat M Mohd Omar F Chew A W Omar Ab Kadir M 2015 *Desalin. Water Treat.* **2015 1**
[6] Wang J Chen C 2009 *Biotechnol. Adv.* **27** 195
[7] Thatoi H Das S Mishra J Rath B P Das N 2014 *J. Environ. Manag.* **146** 383
[8] Sujatha P Kalarani V Kumar B N 2013 *J. Chem.* **2013** *open source* 7 pages
[9] Hemambika B Rani M J Kannan V R 2011 *J. Ecol. Nat. Environ.* **3** 168
[10] Samuel J Pulimi M Paul M L Maurya A Chandrasekaran N Mukherjee A 2013 *Biores. Technol.* **128** 423
[11] Ayangbenro A S Babalola O O 2017 *Int. J. Environ. Res. Pub. Health* **14** 16 pages
[12] Kumar D Gaur J P 2011 *Bioresour. Technol.* **102** 2529
[13] Kumar D Rai J Gaur J P 2012 *Bioresour. Technol* **104** 202

[14] B M Tissue 2013 *Basic of Analytical Chemistry and Chemical Equilibria*, John Wiley & Sons, Inc., 426 pages
[15] Bishnoi N R Kumar R Bishnoi K 2007 *Ind. J. Exp. Technol.* **45** 657
[16] Hillhor M R Figuiredo H Tavares T Gavrilescu M 2017 *Process Saf. Environ. Prot.* **108** 44
[17] Gavrilescu M 2004 *Eng. Life Sci.* **4** 219
[18] Akhtar K Akhtar M W Khalid A M 2007 *Water Res.* **41** 1366
[19] Gupta V K Nayak A Agarwal S 2015 *Environ. Eng. Res.* **20** 1
[20] Mishra A Malik A 2013 *Crit. Rev. Environ. Sci. Technol.* **43** 1162
[21] Munro K L Bambery K R Carter E A Puskar L. Tobin M J Wood B R Dillon C T 2013 *Vib. Spectrosc.* **53** 39
[22] Barth A 2007 *Biochim. Biophys. Acta* **1767** 1073

In vitro study of DNA Adduct 8-OHdG Formation by using Bisphenol A in Calf Thymus DNA and 2'-Deoxyguanosine

Budiawan, Intan Cahaya Dani, Ridla Bakri, Sri Handayani, Evi Ratna Dewi

Chemistry Department, Faculty Mathematics and Sciences, Universitas Indonesia 16424

E-mail: dr.budiawan@gmail.com

Abstract. The in vitro study of DNA Adduct 8-OHdG Formation due to BisphenolA (BPA) as xenobiotics has been conducted by using calf thymus DNA and 2'deoxyguanosine. The method of study was conducted by incubating calf thymus DNA and 2'dG with compounds trigger to radicals in the variation of pH (7.4 and 8.4), temperature (37°C and 60°C), and BPA concentrations (2 ppm and 10 ppm). To represent the work of CYP 450 enzyme in metabolic process of xenobiotics in the body and the effect of metal presence to the formation of radicals that can lead to 8-OHdG formation, we used iron(II) solution and also fenton reagent (Fe(II) and H_2O_2). The DNA used has 1.8 purity ratio (checked at λ260/ λ280 by using Spectrophotometry UV-Vis). The results by using HPLC method showed that BPA could interact with DNA and DNA base (represent as calf thymus and 2'dG) and potentially induced 8-OHdG formation. The presence of iron(II) metal and Fenton reagent also induced the higher 8-OHdG formation. The higher of pH, temperature and concentrations also lead to 8-OHdG formation (ranger between 4 – 70 ppb).

1. Introduction

Plastic is a material that is easy to find in the market, because it is not easily broken and relatively cheaper, making it a popular choice. The most common ingredient for plastic is polycarbonate (PC). Polycarbonate plastic is used for various food and beverage packagings, baby bottles, food packages, medical equipments, dental sealants (thin layer of plastic used to cover the surface of teeth), CD, DVD, spectacle lenses and sports equipments as well as several paper coating (ATM receipts and cash counter receipts) [1]. The main ingredient for making polycarbonate plastic is 2,2-bis (4-hydroxyphenyl) propane, also known as Bisphenol A (BPA). Aside from being the main material for polycarbonate (PC), BPA is also used as an epoxy resin material, mainly as inner coating on metal packaging products that serves to prevent corrosion, such as in food and beverage cans and water supply pipes [2].

BPA is carcinogenic - it induces cancer - because it interacts with DNA and causes what is known as DNA adducts due to mutations in DNA that can trigger continuous, out-of-control cell formation or division [3]. Carcinogenic compounds can trigger oxidative stress and contribute to the formation of reactive oxygen species (ROS). Upon entering the body, carcinogenic compounds will undergo a process called detoxification. It repairs damaged DNA through the mechanism of Excision Base Repair (BER) which can excrete cut-off damaged DNA. The DNA, which is damaged and truncated by BER, is found in the form of 8-hydroxy-2'-deoxyguanosine (8-OHdG) [4].

2. Experimental Method

2.1. Chemicals
Calf thymus DNA, 2'-deoxyguanosine monohydrate, 8-OHdG, aquabidest, Bisphenol A, $FeSO_4$, H_2O_2, phosphate buffer, acetate buffer, hydrochloric acid, sodium hydroxide, micrococcus nuclease enzyme (MN) and spleen phosphodiesterase (SPDE) enzyme, methanol, DMSO.

2.2 In vitro study of 2'-deoxyguanosine, BPA (2 ppm and 10 ppm), Fe(II), and H_2O_2
100 μL of 2'-deoxyguanosine (600 ppb and 3 ppm) in 0.1 M phosphate buffer solution with pH of 7.4 and 8.4 is added with 100 μL of BPA solution (6 ppm and 30 ppm) and 50 μL of H_2O_2 (12 ppm dan 60 ppm). The sample is then incubated over a varying period of 3 and 9 hours at a varying temperature of 37°C and 60°C.

2.2. In Vitro Studies with Calf Thymus DNA, BPA, Fe(II) and H_2O_2
A total of ± 10 μg (in 100 μL) of calf thymus DNA (100 μg / mL) in 0.1 M of phosphate buffer (pH 7.4 and 8.4) is incubated with 10 μg (in 100 μL) of BPA solution, 249 μg of Fe (II) (in 50 μl) at 37 ° C over a period of 6 hours.

2.3. Hydrolysis of Calf Thymus DNA Enzymatically
Mixture of Calf Thymus DNA and Bisphenol A, the mixture of Calf Thymus DNA, Bisphenol A, $FeCl_2$ and H_2O_2 and the mixture of Calf Thymus DNA , $FeCl_2$ and H_2O_2 were centrifuged and the filtrate hydrolyzed using micrococcus nuclease enzyme (MN) and spleenphosphodiesterase (SPDE) enzyme (enzyme ratio 0.02 units: 0.002 unit) with a total enzyme mix of 100 μL. Added with 133 μL sodium succinate 10 mM pH 6 and calcium chloride 5 mM, then incubated for 3 and 9 hours at 37 ° C and 60 ° C, then analyzed using HPLC.

3. Results and Discussion

3.1. Determination of DNA Adduct 8-OHdG with 2'Deoxiguanosine Chemicals
Carcinogenic compounds, can cause DNA damage due to its contribution to the formation of reactive oxygen species (ROS). If the mechanism of DNA repair in the body is slower than the rate of DNA damage, mutations will occur, eventually leading to the onset of cancer [5]. The hydroxyl radical (OH) formed by the ROS mechanism can attack the guanine base in DNA to form an 8-OHdG DNA adduct. 8-OHdG is a DNA damage produced by the addition of hydroxyl radical at C-8 guanine position to the DNA [6].

Table 1 8-OHdG Concentration (ppb) in the variation of reaction

Variation of reaction, Incubation time 3 hours, BPA 2 ppm	Concentration 8-OHdG (ppb)			
	Temperature 37° C		Temperature 60° C	
	pH 7,4	pH 8,4	pH 7,4	pH 8,4
dG + BPA + Fe(II) + H2O2	10.849	12.950	11.219	12.953

In the reaction of 2'-deoxiguanosine with BPA 2 ppm for 3 hours, the variation of pH and temperature was conducted. The results showed the increasion of 8-OHdG concentration as the increase of pH and temperature. The highest amount of 8-OHdG was obtained at temperature 60°C and pH 8.4 as much as 12.953 ppb.

3.2. Result of 8-OHdG DNA Adduct Formation with Calf thymus DNA

3.2.1. Incubation of Calf thymus DNA with BPA

The formation of DNA adducts from incubation with this HPLC-analyzed BPA compound detects the formation of 8 OHdG at retention time of 9.386 (Fig. 3.1) with a concentration of 8 OHdG of 2.580 ppb, whereas at retention time 6,817 it's the peak of deoxyguanosine (dG). This result is based on dG standard ad 8 OHdG which was previously analyzed under similar equipment condition.
The formation of 8 OHdG DNA Adduct in the sample indicates that BPA may contribute to the reactive oxygen species (ROS) that are hydroxyl radicals which bind with deoxyguanosine to form 8 OHdG [7].

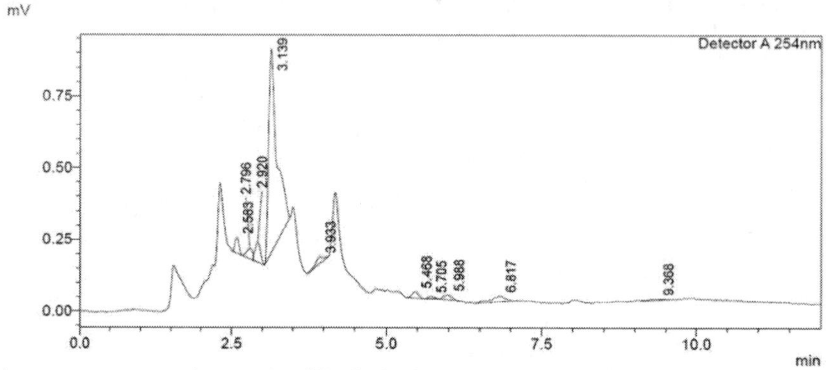

Figure 1 HPLC chromatogram on the result of hydrolysis between the calf thymus DNA and BPA

3.2.2. Incubation of Calf thymus DNA with BPA, Fe(II) and H202

Test result is the formation of 8 OHdG at retention time of 10.132. HPLC chromatogram of incubated calf thymus DNA at pH 7.4 and 37°C temperature shows 8-2 OHdG yield of 7.272 ppb, whereas peak deoxyguanosine appears at retention time of 7.030 (Fig. 3.2)

Fenton reagents (Fe (II) metal and H2O2) in this variation are used to determine the effect of fenton reagents in the formation of 8 OHdG. Based on the results, the amount of 8 OHdG is high. The reactive oxygen speciesproduced from this fenton reagent bound to dG forms 8-OHdG through the following mechanism:

$$Fe(II)+H_2O_2 \longrightarrow Fe(III) + HO^{\bullet} + OH^{-}$$

In the human body, the fenton reaction takes place through the mechanism of P450 cytochrome [8]. The mechanism of the fenton reaction itself is the reduction of hydrogen peroxide by transitional metal ions, producing reactive hydroxyl radicals and oxidized metal ions [9].

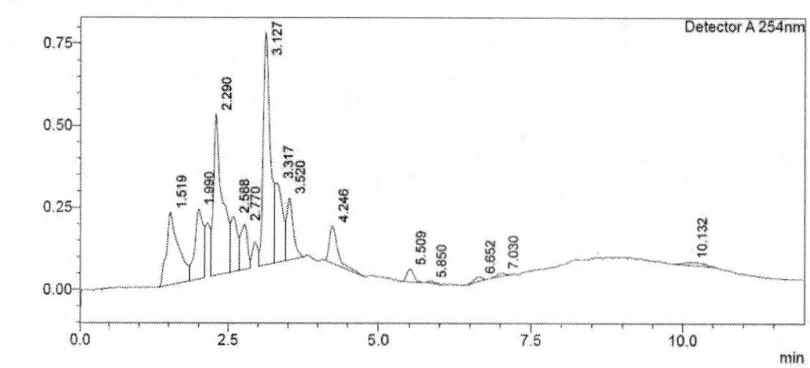

Figure 2 HPLC chromatogram on the result of hydrolysis between the calf thymus DNA and BPA, Fe(II) and H202

4. Conclusion

Most of the samples have an 8-OHdG yield trend that will increase as the temperature, pH, concentration increase and incubation time become longer. The reaction between dG and BPA may increase the concentration of 8-OHdG when in the presence of fenton reagents (Fe (II) and H_2O_2 in the reaction. Incubation of calf thymus DNA with free radical contributor compound that is BPA with addition of fenton reagent can produce 8 OHdG equal to 7.272 ppb.

References

[1] Gupta, R. C. 2011 *Reproductive and Developmental Toxicology*. London: Elsevier. 51

[2] European Food Safety Authority (EFSA) 2015 Scientific Opinion on the risks to public health related to the presence of bisphenol A (BPA) in foodstuffs: Executive summary. *EFSAJournal*.

[3] Vineis P, Husgafvel-Pursiainen K. 2005 Carcinogenesis **26** 1846

[4] Cooke, M. S., Evans, M. D., Dizdaroglu, M., and Lunec, J 2003 The FASEB Journal **1195**

[5] Best, Benjamin P 2009 Rejuven. Res. **12** 199

[6] Sharma,Pallavi.,Jha,Ambuj Bhushan.,Dubey, Rama Shanker and Pessarakli, Mohammad. 2012 J. Bot. **2012**

[7] Sakuma, S., Nakanishi, M., Morinaga, K., Fujitake, M., Wada, S.-i., & Fujimoto, Y. (2010). Bisphenol A 3,4-quinone induces the conversion of xanthine dehydrogenase into oxidase in vitro. *Food and Chemical Toxicology*, 2217-2222.

[8] Berka K, Hendrychová T, Anzenbacher P, Otyepka M 2011 J. Phys. Chem. A **115** 11248

[9] Nixon Ogendi Mwebi 2005 Fenton & Fenton-Like Reactions: The Nature Of Oxidizing Intermediates Involved *Unpublished doctoral dissertation* Faculty of the Graduate School of the University of Maryland, College Park

Fucoidan cytotoxicity against human breast cancer T47D cell line increases with higher level of sulfate ester group

Endang Saepudin[*1], **Fildzah Alfita Qosthalani**[1], **Ellya Sinurat**[2]
[1]*Department of Chemistry, Faculty of Mathematics And Natural Sciences, Universitas Indonesia*
[2]*Research and Development Center of Marine and Fisheries Product Processing and Biotechnology, Jalan KS. Tubun Petamburan VI, Slipi, Central Jakarta 10260, Indonesia*

[*]E-mail: endang.saefudin@sci.ui.ac.id

Abstract.
The anticancer activity of different sulfate ester group content in different molecular weight was examined. The anticancer activity was achieved in vitro on human breast cancer T47D cell line. Fucoidan with lower molecular weight (5.79 kDa) tends to have lower sulfate ester group content (8.69%) and resulted in higher IC_{50} value (184.22 µg/mL). While fucoidan with higher molecular weight (785.12 kDa) tends to have higher sulfate level (18.63%) and achieved lower IC_{50} value (75.69 µg/mL). The result showed that in order to maintain fucoidan cytotoxic activity against human breast cancer T47D cell line, the sulfate content should be remain high.
Keywords: fucoidan, sulfate ester group, human breast cancer

1. Introduction

Marine macroalgae have been exceptionally known to produce diverse compounds and cover a wide range of biological activities. Bioactive compounds act as chemical weapons against other organisms and/or as responses to physical stress [1]. Fucoidan is a sulfated polysaccharide extracted from brown seaweed and can be regarded as a bioactive compound to various biological activities including antibacterial, antioxidant, antiviral, and antitumor. However, the biological activity behavioral is not ubiquitous, depending on its structure [2].

Many studies suggested that molecular weight is one of the most important factors determining the biological activities of fucoidan. High molecular weight fucoidan may cause low solubility and processability, thereby hampering their penetration into the cell to perform a function [3]. Conversely, low molecular weight fucoidan (LMWF) has more biological actions than native fucoidan [4]–[6] Acid hydrolysis process is usually performed to break the glycosidic linkage and minimize the molecular size. However, acid hydrolysis can also remove the sulfate ester group. While the study on anticoagulant and antiproliferative effects in LMWF reveal that sulfate ester group content is necessary for their biological actions [7].

In this work, we attempted to determine the role of sulfate ester group and molecular weight on the anticancer effect of fucoidan extracted from *Sargassum binderi* Sonder. To that purpose, we prepared one fucoidan with hydrolysis treatment and one without hydrolysis treatment. The anticancer activity on both fucoidans was examined on human breast cancer T47D cell line.

2. Material and Method

2.1. Material

The raw material used in this work was brown seaweed *Sargassum binderi* Sonder which collected from Lampung, Indonesia. Chemical reagents used were mostly analytical grades and some technical grades.

2.2. Extraction, purification, and hydrolysis of fucoidan

Dried seaweed *S. binderi* (100 g) was grinded into small pieces and macerated into a mixture solvent MeOH – CH_2Cl_2 – H_2O (4:2:1) (500 mL) at room temperature for 3 hours to remove the lipophilic pigment (chlorophyll, fucoxanthin, carotene) and fats. The *S. binderi* then treated with 0.1 N HCl (1:20) (w/v) (2 L) at pH 4 and stirred for 6 hours at room temperature. The filtrate then collected and centrifuged at 5.000 rpm for 15 minutes. The filtrate was neutralized with 0.5 M NaOH, added with 2% $CaCl_2$ solution and incubated for 3 hours then centrifuged. The filtrate was collected and added to ethanol (1:2) and incubated overnight. The precipitate in the ethanol solution was separated by 5000 rpm centrifugation for 15 min at 15°C. The precipitate was dried with freeze-dry to obtain crude fucoidan. Fucoidan was purified using dialysis membrane (Thermo Scientific Slide-A-Lyzer 3,5 kDa) in a 0.5 M NaCl solution for 24 hours and continued with aquabidest for 24 hours. The purified fucoidan (50 mg) was hydrolyzed in 1 M TFA solution (5 mL) for 1.5 hours at 121°C in an autoclave. Finally, the fucoidan solution was neutralized with NaOH at the same concentration and then freeze-dried to obtain a hydrolyzed fucoidan.

2.3. Characterization of hydrolyzed and non-hydrolyzed fucoidan

2.3.1. Determination of total carbohydrate

The determination of total carbohydrate content was performed by Phenol - Sulfuric acid method. Each hydrolyzed and non- hydrolyzed fucoidan samples (5 mg) was dissolved in aquadest (1 mL) and added to concentrated H_2SO_4 (2.5 mL) and incubated in an ice bath for 30 minutes. Then, 5% phenol (0.5 mL) was added to the sample and also incubated in the cold ice tub for 30 minutes. Fucose, mannose, and xylose were used as standards in various concentrations (200-1000 ppm). Sample, standard, and blank were measured at 480 nm and 490 nm with a UV-VIS spectrophotometer.

2.3.2. Determination of sulfate ester group content

The determination of sulfate ester content in fucoidan was performed with $BaCl_2$ – Gelatin method. In 3.5 N HCl (2 mL), each hydrolyzed and non- hydrolyzed fucoidan samples (6 mg) were heated for 17-18 hours at 105-110°C. Then, each sample (10 μL) was placed in the 96-well plate, added with 3% TCA (190 μL) and 0.5% $BaCl_2$ – gelatin (50 μL). The plate was left for 20 minutes until the barium sulfate ready to be scanned. Na_2SO_4 was used as a standard with various concentrations (100 – 500 ppm) while aquadest used as a blank. The samples, standards and the blank were scanned with Microplate Reader Spectrophotometer at $\lambda = 360$ nm.

2.3.3. Determination of weight average molecular weight

The average molecular weights of hydrolyzed and non- hydrolyzed fucoidan were measured by the Size Exclusion Chromatography (SEC) method. Each fucoidan sample and pullulan standard (5 mg) were dissolved separately in aquabidest (1 mL) and left overnight in cold storage. Then, sample and standard were eluted on TSK-Gel G6000 PW column using RID detector with flow rate 0.8 mL/min for 25 minutes and water pure grade HPLC as the eluent.

2.4. Anticancer activity assay

2.4.1. Cell line and culture

The human breast cancer T47D cell line was cultured in RPMI medium supplemented with 10 % FBS, 2 % penicillin-streptomycin, and 0.5% fungison in an incubator (ESCO CelCulture) with a humidified 4.2% CO_2 at 37 °C.

2.4.2. Growth inhibition assay

The effect of hydrolyzed and non-hydrolyzed fucoidan on the inhibition of cancer cells was assessed in vitro by MTT assay [8]. The hydrolyzed and non-hydrolyzed fucoidan were respectively dissolved in aquabidest, DMSO, and medium growth into different dosages (non-hydrolyzed fucoidan: 50; 100; 200; 400; and 800 μg/mL, while hydrolyzed fucoidan: 20; 40; 80; 160; and 320 μg/mL). Three types of control were also prepared, which are cancer cell control, media control, and sample control.

Briefly, human breast cancer T47D cells lines were incubated in 96-well plate (1.5×10^4 cells/well) until reach a confluent phase (12 – 24 hours). After incubation, each fucoidan samples (100 μL) were added to the medium growth and incubated for 24 hours in a CO_2 incubator. Cells were then added to MTT reagent and incubated for 4 hours. The reaction stopped with the addition of SDS 10% and incubated for 12 hours. The cell growth was observed with microscope inverted and detected with Spectrophotometer Microplate Reader at wavelength 570 nm. The inhibition rate was calculated according to the formula:

$$\text{Inhibition rate (\%)} = \frac{(C-D)-(A-B)}{(C-D)} \text{ X } 100\%$$

where each absorbance stands for: A = fucoidan samples; B = control sample; C = control cells; D = control media. The IC_{50} value was calculated in probit statistic.

3. Results and Discussion

3.1. *Extraction, purification, and hydrolysis of fucoidan*

First, fucoidan extraction by acid method was carried out. The yield of crude fucoidan from brown seaweed *Sargassum binderi* Sonder was 6.64%. It is essential to be noted that the yield of crude fucoidan may vary from each species and the extraction techniques can also significantly improve the yield [9]. There are various methods in fucoidan extraction, includes dilute acid, hot water, or calcium chloride. Acid extraction was used as the major steps in fucoidan extraction [10]. Though it causes degradation and affects the integrity of overall fucoidan make-up, the advantage of using acid extraction will help the precipitation of alginic acid from alginates. The protons or hydroxide ions will interfere the hydrogen bonds between the various polysaccharides in brown seaweed (alginate, laminarin, and fucoidan) and then release them into the solution [11].

In order to remove the unnecessary compounds other than fucoidan, purification was conducted using dialysis membrane (3.5 kDa MWCO) with NaCl solution. After purification, the overall fucoidan yield decrease to 1.67%. Dialysis process separate small impurities attached to fucoidan by diffusion rate in a semipermeable membrane. The following process then split into two major in which one underwent hydrolysis and one without hydrolysis. In this work, mild acid hydrolysis was used to depolymerize fucoidan into low molecular weight fucoidan (LMWF). Though mild acid hydrolysis works in a nonspecific method [12], a reduction in molecular size is promising.

3.2. *Characterization of hydrolyzed and non-hydrolyzed fucoidan*

General characterization of fucoidan can be seen by reviewing the carbohydrate, molecular weight, and sulfate content [13]. Schweiger first suggested that fucoidan is a heteropolymer type of sulfated polysaccharide. Thus, fucoidan characterization on each species of brown seaweed needs to be clearly determined in order to see the effect to its bioactivity [14]. When fucoidan was hydrolyzed in mild acid solution, most of the chemical and physical characteristics in fucoidan have shown some modification [15]. From Table 1, we see that all aspect in fucoidan shows some decrement after hydrolysis.

Table 1. Characterization of hydrolyzed and non-hydrolyzed fucoidan

Fucoidan	Carbohydrate (%)	Sulfate (%)	MW (kDa)
Non-Hydrolyzed	82,79	18,63	785,12
Hydrolyzed	55,86	8,69	5,79

As a sulfated polysaccharide, the total carbohydrate in fucoidan is closely related to the amount of sulfate attached in carbohydrate. From the data showed in Table 1, the non-hydrolyzed fucoidan gave higher carbohydrate content than modified fucoidan (hydrolyzed). Mild acid hydrolysis at high temprature also reduces the sulfate ester group bond. The sulfate content in Table 1 showed that hydrolyzed fucoidan has lower sulfate content than the non-hydrolyzed. It confirmed that hydrolysis affect the total carbohydrate and therefore decrese the sulfate content.

The initial goal of hydrolysis is to depolymerize fucoidan into low molecule. Apart from the other effect of hydrolysis in fucoidan, the decreasement in molecular weight can help better penetration into cells and initiate higher bioactivity. In Table 1, the molecular weight of hydrolyzed fucoidan was successfully modified became low molecular weight fucoidan (LMWF). It proved that mild acid hydrolysis is suit to depolymerize fucoidan molecule, yet did cause a significant change in sulfate and carbohydrate content.

3.3. Cytotoxic fucoidan of human breast T47D cancer cell lines

To evaluate the biological activity in fucoidan against human breast cancer T47D cell line, cytotoxic activity was conducted with MTT assay. The cytotoxic of fucoidan from *Sargassum binderi* Sonder was compared over the hydrolysis treatment and to standard fucoidan commercial from *Fucus vesiculosus*. From Table 2, the result showed that standard fucoidan exhibit the highest cytotoxic activity against human breast cancer. Although the non-hydrolyzed fucoidan has slightly lower IC_{50} than the standard fucoidan, it can be rounded that non-hydrolyzed fucoidan was also achieved high cytotoxic activity. While the hydrolyzed fucoidan was obtained the lowest cytotoxic activity among the other fucoidan (Figure 1).

Table 2. IC_{50} of standard fucoidan from *Fucus vesiculosus*, hydrolyzed and non-hydrolyzed fucoidan from *Sargassum binderi* Sonder against human breast cancer

Fucoidan	Standard	Non-Hydrolyzed	Hydrolyzed
IC_{50} (µg/mL)	72.13	75.69	184.22

There are two main factors to increase the bioactivity of fucoidan, which is sulfate ester group content and molecular weight. Sulfate ester group carries a negative charge in the molecule which can facilitate the formation of fucoidan-protein complexes involved in cell proliferation [16]. On the other hand, low molecular weight fucoidan will increase the solubility and penetration into the cell. Generally, the optimum bioactivity of fucoidan has molecular weight range at 50 – 100 kDa, while large molecular weight (>850 kDa) showed a low activity [17]. However, the optimum molecular weight range for anticancer activity have not yet been reported to date.

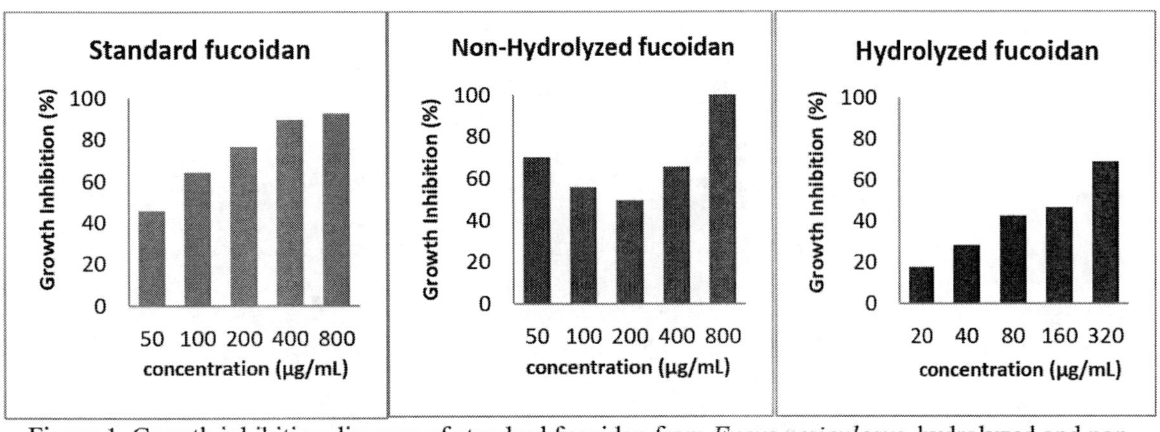

Figure 1. Growth inhibition diagram of standard fucoidan from *Fucus vesiculosus*, hydrolyzed and non-hydrolyzed fucoidan from *Sargassum binderi* Sonder against human breast cancer

The result of the cytotoxic test also implies that sulfate ester group content has played more important role than molecular weight. Non-hydrolyzed fucoidan with higher sulfate content (18,63%) achieved higher cytotoxic activity (75.69 µg/mL) than the hydrolyzed fucoidan (184.22 µg/mL). Compared to the molecular weight of non-hydrolyzed fucoidan (785,12 kDa), the hydrolyzed fucoidan has successfully achieved low molecular size (5,79 kDa), yet the sulfate content was not high enough to inhibit the cancer cell. Hence, in order to maintain fucoidan cytotoxic activity against human breast cancer T47D cell line, the sulfate content should be remain high.

4. Conclusion

This study describes the role of molecular weight and sulfate ester group to increase fucoidan cytotoxicity against human breast cancer T47D cell line. It was found that higher sulfate content is more significant to inhibit cancer activity rather than to lower the molecular weight of fucoidan. Hence, in order to maintain fucoidan cytotoxic activity against human breast cancer T47D cell line, the sulfate content should be remain high.

Acknowledgements

This research was financially supported by a grant from Universitas Indonesia under program of PITTA 2017.

References

[1] M. Hayes, *Marine Bioactive Compounds*. 2012 Springer New York Dordrecht Heidelberg London
[2] J. Y. Kwak, 2014 *Mar. Drugs* **12** 851
[3] J. Il Choi and H. J. Kim 2013 *Carbohydr. Polym.* **97** 358
[4] K. Matsubara, C. Xue, X. Zhao, M. Mori, T. Sugawara, and T. Hirata 2005 *Int. J. Mol. Med.*, **15** 695
[5] C. Yang *et al.* 2008 *Int. J. Biol. Macromol.* **43** 433
[6] S. K. Kim, 2015 *Handb. Anticancer Drugs from Mar. Orig* 1–805
[7] F. Haroun-Bouhedja, M. Ellouali, C. Sinquin, and C. Boisson-Vidal 2000 *Thromb. Res.* **100** 453
[8] A. Isnansetyo, F. Nor, L. Lutfia, M. Nursid, R. A. Susidarti, and A. Isnansetyo 2017 **9** 14
[9] J. Venkatesan, S. Anil, and S. K. Kim, 2017 Seaweed Polysaccharides: Isolation, Biological and Biomedical Applications. Elsevier Science
[10] M. Ale and A. Meyer 2013 *RSC Adv* **3** 8131
[11] T. Hahn, S. Lang, R. Ulber, and K. Muffler 2012 *Process Biochem.* **47** 1691
[12] V. H. Pomin, A. P. Valente, M. S. Pereira, and P. A. S. Mourão 2005 *Glycobiology* **15**, 137
[13] J. H. Fitton, "Fucoidans : Healthful Saccharides from the Sea 2006 *Glycosci. Nutr. Sci. Site* **2006** 1
[14] E. Sinurat 2011 Isolasi Dan Karakterisasi Serta Uji Aktivitas Fukoidan Sebagai Anti Koagulan Dari Rumput Laut Coklat (Sargassum crassifolium) Universitas Indonesia
[15] S. K. Kim, 2013 *Marine Nutraceuticals: Prospects and Perspectives* CRC Press
[16] F. Atashrazm, R. M. Lowenthal, G. M. Woods, A. F. Holloway, and J. L. Dickinson 2015 *Mar. Drugs* **13** 2327
[17] S. Dumitriu, 1996 Polysaccharides in Medicinal Applications. Taylor & Francis

IOP Conf. Series: Materials Science and Engineering **299** (2018) 012026 doi:10.1088/1757-899X/299/1/012026

Glucanase and Chitinase from Some Isolates of Endophytic Fungus *Trichoderma spp.*

Sasangka Prasetyawan,[1] Lilik Sulistyowati,[2] Aulanni'am[1]
[1]Faculty of Mathematics and Natural Sciences, Brawijaya University
2Faculty of Agriculture, Brawijaya University

Email: sasangka@ub.ac.id

ABSTRACT. Endophytic fungi are those fungi that are able to grow in plant tissue without causing symptoms of disease. It is thought that these fungi may confer on the host plants degree of resistance to parasitic invasion. Endophytic fungi have been isolated from stem tissue and these fungi are known to be antagonistic to pathogenic fungi. These endophytes produce chitinase and β-1,3-glucanase enzymes. Based on the fact that chitin and β-1,3-glucan are the main skeletal polysaccharides of the cell walls of fungal patogen. The aim of this research is to do potential test on some of isolates of Trichoderma's endophytic (L-1,L-2, Is-1, Is-2 and Is-7) in the chitinase and β-1,3-glucanase activity in effort to determine endophytic which be chossen to be gene resource for the next research. The gene will be transformed to citrus plant japanese citroen in effort to make citrus plant transgenic resistance to phytopatogenic invasion. The result of this research is endofit namely L-1 is the most potential endophytic fungi with chitinase activities is 4,8 10^{-2} Unit and glucanase 24,2. 10^{12} Unit. The addition of chitin and cell wall of phytophtora causes chitinase activity significantly increase, and also addition of laminarin and cell wall of phytophtora makes glucanase activity increase.

1. Introduction

Endophytic fungi are fungi that can grow in plant tissues of both stems, roots and leaves of plants [1] without causing symptoms of the disease even increasing the degree of resistance of host plants to parasitic attacks [2]. The relationship between endophytic microbes and their host plants is a form of symbiotic mutualism and ensured endophytic microbes will increase the degree of resistance of host plants to parasitic attacks [3]. Currently, these endophytes receive great attention because they can produce bioactive compounds such as antibiotics, enzymes because of their great activity in killing pathogenic microbes. Besides being able to produce antimicrobial compounds, endophytic microbes are also capable of producing potentially anticancer, antimalarial, anti HIV, antioxidant and so on. Liu et al [4] has successfully isolated 39 endophytes contained in *Artemisia annua*. Lu et al [1], from his research found that bioactive compounds isolated from endolitic microbes *Colletotrichum sp*. In *Artemisia annua* has the ability to inhibit the growth of the pathogenic fungi *Gaeumannomyces graminis* var. Tritici, *Rhizoctonia cerealis*, *Helminthosporium sativum* and *Phytophthora capisici*.

Trichoderma spp is one of the endophytes of fungi that grow on the root surface of various plants and soil. Trichoderma can inhibit the growth of pathogenic fungi by producing enzymes capable of degrading cell walls such as chitinase, β-1,3-glucanase, proteases, mannanases and other hydrolysed enzymes [7-9]. It has long been known that in addition to producing antibiotics Trichoderma also acts as a parasite for other fungi [10]. Trichoderma has the ability to compete with other microorganisms in terms of obtaining nutrients. It is also said that Trichoderma is able to degrade pectinases and other enzymes that have an important role for the life of pathogenic fungi [10]. β -1,3-glucan are the main framework of polysaccharides of the cell wall of pathogens. It is possible that chitinase and β-1,3-glucanase are enzymes involved in breaking the cell wall during the antagonistic action of the endophytes. A set of chitinases and gene glucanases have been isolated from *T. virens* and structural features and their expression has been deactivated [2]. Endophytic fungi have been successfully isolated from the citrus trunk tissue and these fungi are known to be antagonistic to pathogenic fungi. These endophytes produce chitinase and glucanase enzymes. Based on the fact that chitin and *Trichoderma harzianum* is a fungus that can produce chitinase [11, 12] so it can function as a controlling plant diseases. Chitinase is an extraeluler enzyme produced by fungi and bacteria [12] and plays an important role in the breakdown of chitin.

A number of endophytic fungi have been successfully isolated from citrus trunk tissues and rice stalks that are tested as antagonists against citrus plant pathogenic fungi. These endophytes are known to produce chitinase enzymes and 1,3-β glucanase enzymes. It is key that the enzyme is involved in the cell wall termination of phytopathogenic pathogens during the antagonist action of the endophytes. Such enzymes may act directly by degrading the cell wall of pathogens, and indirectly by releasing fragments of cell walls that may act as elicitors of a defense reaction such as the occurrence of phytoalexin accumulation [13].

2. MATERIALS AND METHODS

2.1 Materials

Endophytic isolates were obtained from a collection of endophytic isolates labeled L-1, L-2, Is-1, Is-2 and Is-3 from the biotechnology HPT laboratory of the Faculty of Agriculture Universitas Brawijaya, Malang. L-1 and L-2 are derived from orange stems while Is1, Is2 and Is 7 are derived from rice stem tissue. Chitin as the enzyme substrate khitinase and laminarin as glucanase substrate. In order to obtain the enzyme chitinase and glucanase is done by inoculating Trichoderma isolate on PDA medium (Potato Dextrose Agar). Then incubated for 5 days at 28 °C. After 5 days, the PDA media was centrifuged at 5000 rpm, 4 °C for 10 minutes. The supernatant (enzyme solution) obtained was then used for the enzyme activity of khitinase and glucanase. The chitinase enzyme activity test was conducted by mixing a 0.25 mL enzyme solution with 0.25 1% colloidal chitin solution in 50 mM phosphate buffer pH 6.7, incubated in a water bath temperature of 45 °C for 6 hours, after which the solution plus the Somogyi - Nelson and is read with a 550 nm. Glucanase enzyme activity test was done by mixing 0.25 ml of enzyme solution with 0.25 mL 1% laminarin substrate in Tris -HCl pH 8 buffer solution, incubated at 37 °C for 6 hours, plus Somogyi-Nelson reagent and quantized with spectrophotometer at 550 nm. One unit of activity is defined as the amount of enzyme liberated by 1 μmol GlcNAc/min under certain conditions.

3. RESULTS AND DISCUSSION

3.1 Activity test of chitinase enzyme from Trichoderma spp.

The results of chitinase enzyme activity test of several species of *Trichoderma spp* presented in Figure 1. Endophytes *Trichoderma spp.* L-1 has higher enzyme activity than Trichoderma spp at 4.8 x 10^{-2} units. On the other hand, the endophytes of Trichoderma Is-7 have the lowest enzyme activity, at 3.2 x 10^{-2} unit.

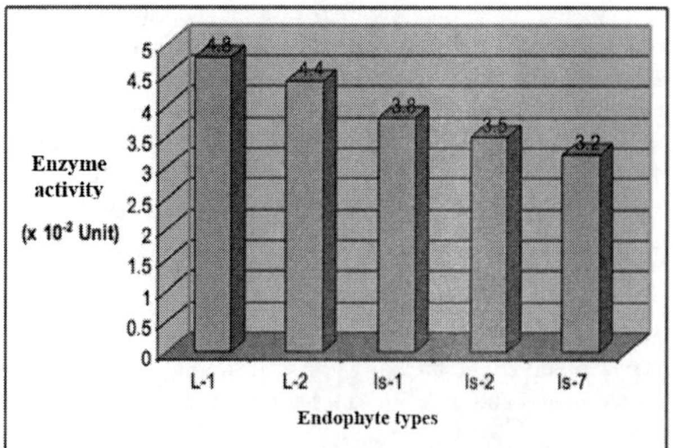

Figure 1. The relationship between endophytic type and chitinase enzyme activity

Trichoderma spp. is a fungus that can produce chitinase. Khitinase is an extracellular enzyme produced by fungi and bacteria [12]. In this study *Trichoderma spp.* grown with chitosan substrate. The difference of enzyme activity on the five types of *Trichoderma spp.* have been determined because of the different species of *Trichoderma spp.* According to Nugroho [13], the ability of *Trichoderma spp.* to produce enzymes varies greatly between strains, which may be due to differences in the gene that codes it. From the previous test it was found that the highest activity of several isolates tested, Trichoderma L-1 endophytic fungus had the highest enzyme activity of chitinase. In the next stage will be tested how the activity of enzyme chitinase when Trichoderma L-1 grown on the media with the addition of chitin. For this purpose, the concentration of chitin added to the medium is as follows: 0; 0.2; 0.4; 0.6; 0.8 and 1%, the results are presented in Figure 2.

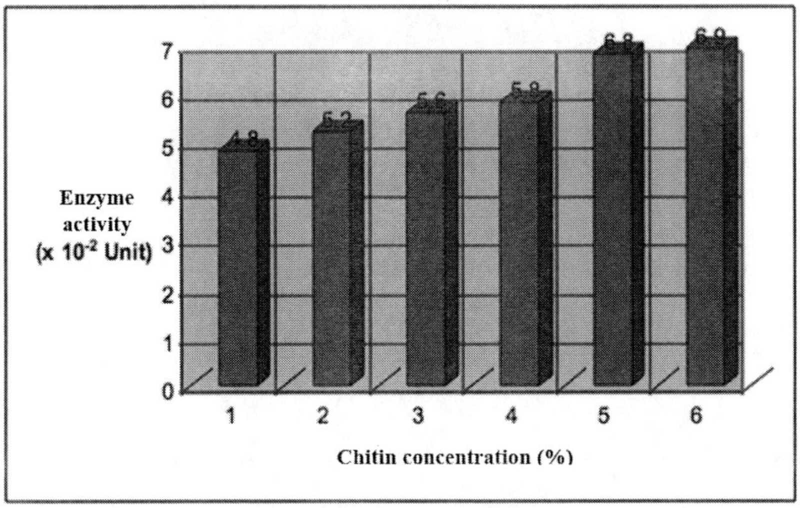

Figure 2. The relationship between the concentration of chitin on the activity of chitinase

From the experiment it was found that higher concentration of chitin added to the medium would increase the activity of chitinase. Chitin is a polymer composed of N-acetyl glucosamine. Chitin is a constituent of fungal cell walls and bacteria [14, 15]. The activity of chitinase will increase when the bean and pea plants are induced by pathogenic fungi.

3.2 Test of chitinase activity exposed to Phytophthora cell wall

The results of chitinase activity test on Phytophthora cell wall are presented in Figure 3. There is a significant relationship between cell wall concentration and chitinase activity. The higher concentration of Phytophthora cell wall added to the media, will increase the activity of chitinase. At 1% concentration, chitinase activity increased 3.2 x 10^{-2} unit higher than control.

Phytophthora spp. is the cause of fruit rot disease, stem cancer, leaf blight, seed blight, and wilting of water shoots. Chitin is a constituent of fungal cell walls and bacteria [14]. Boller et al. [3] states that the activity of khitinase enzymes will increase when the bean and pea plants are induced by pathogenic fungi. In this study, the increase in enzyme activity is suspected as a natural reaction mechanism performed by the enzyme chitinase because of the fungus pathogen, in this case is the cell wall *Phytophthora spp.* added to the media.

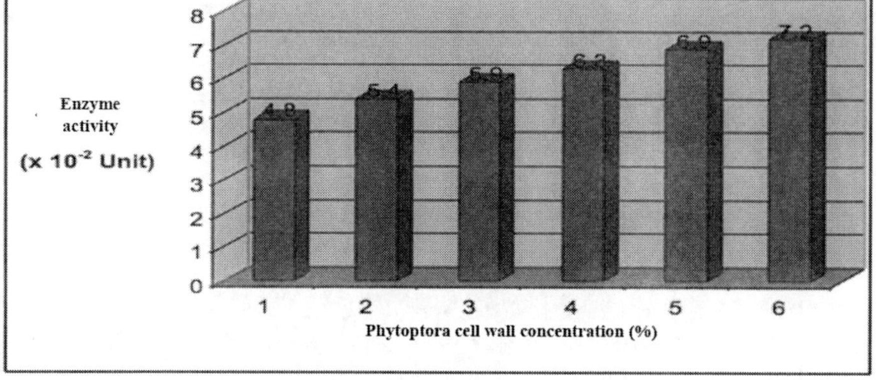

Figure 3. The relationship between Phytophthora cell wall concentrations and chitinase activity

3.3 Activity test of glucanase enzyme from Trichoderma fungus

The results of the glucanase activity showed that Trichoderma L-1 had the highest enzyme activity compared with other types of Trichoderma. (Figure 4). The glucanase activity on L-1, L-2, Is-1, Is-2 and Is-7 was 24.2 x 10^{-2} units, 22.4 x 10^{-2} units, 18.7 x 10^{-2} units, 18.9 x 10^{-2} Units and 18.7 x 10^{-2} units, respectively.

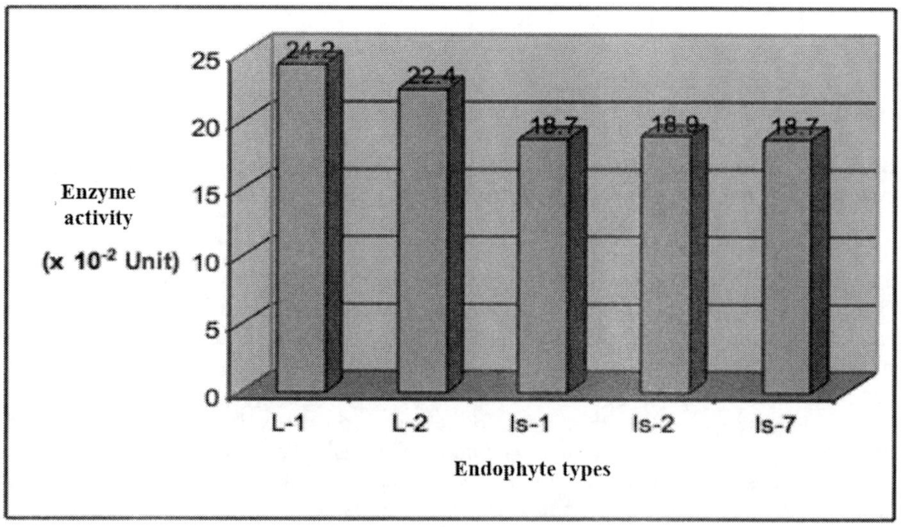

Figure 4. Relationship between types *Trichoderma spp* with their activity

In addition to produce chitinase, *Trichoderma spp.* is a fungus that can produce glucanase. Glucanase is an enzyme that hydrolyzes β-1,3-glucan, which is a constituent of fungal cell and bacterial cells [14]. The difference of enzyme activity on the five types of *Trichoderma spp.* tested because of the different species of *Trichoderma spp.*

IOP Conf. Series: Materials Science and Engineering **299** (2018) 012026 doi:10.1088/1757-899X/299/1/012026

3.4. The results of glucanase activity test with laminarin which is added to the PDA medium
To test the activity of glucanase used laminarin substrate added to PDA media. From the experiment it was found that the higher concentration of laminarin added to the medium would increase the activity of the glucanase (Figure 5).

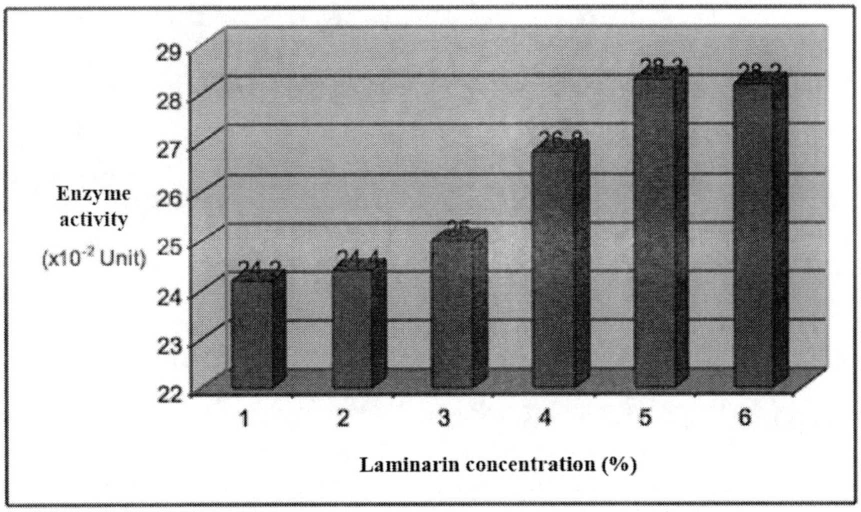

Figure 5. Relationship between laminarin concentration on glucanse activity

In addition to chitin, the fungus cell wall is also composed of β-glucans as a component of the cell skeleton and α-glucans as components that strengthen cell. From the results of his research Thrane *et al.* [15] found that the activity of β-1-3-glucanase increased when he was grown together with the presence of the antagonistic fungus Pitheum ultimum, which is the antagonistic fungus of *T. harzianum*.

3.5 Test of glucanase enzyme activity exposed to Phytophthora cell wall
The results of the test of glucanase activity on Phytophthora cell wall are presented in Figure 6. There is a significant relationship between cell wall concentration and glucanase activity. The higher concentration of Phytophtora cell wall added to the media, will increase the activity of glucanase. At the concentration of 1% cause glucanase activity increased 7 x 10^{-2} Unit higher than the control.

In this study an increase in enzyme activity is thought to be a natural reaction mechanism performed by the enzyme glucanase due to the presence of pathogenic fungi, in this case is the cell wall of *Phytophthora spp.* added to the media. Suzuki, et al. (1995) found that tobacco cell culture given chitinase and β-1,3-glucanase was able to suppress the growth of *P. infestans*. Noronho and Ulhoa [16] from the research they get their activity β-1,3-glucanase in culture containing the cell wall of *Rhizoctonia solani, Sclerotium rolfsii* and *Pythium spp.* Sivan and Chet in [5] found that the production of laminarinase (β-1,3-glucanase) enzyme by *T. harzianum* was induced by the cell wall of the target fungus added in the medium.

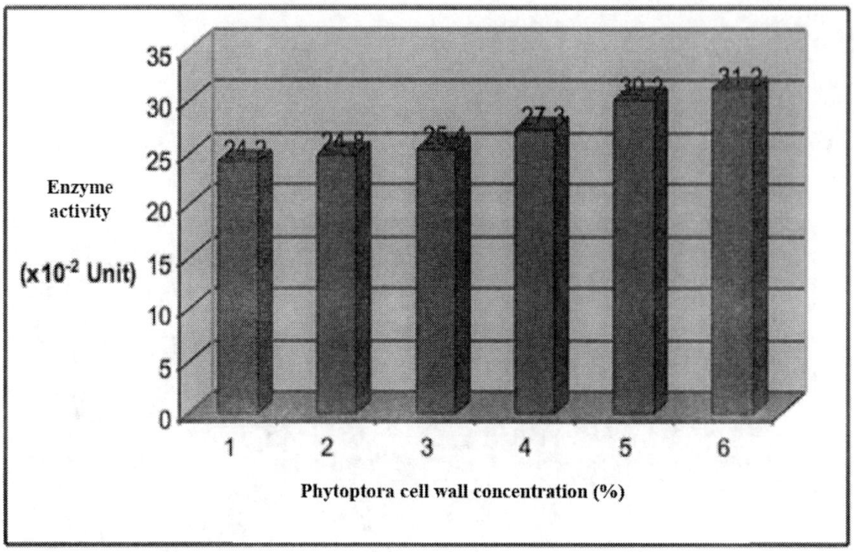

Figure 6. Relationship addition of Phytophthora cell wall concentration with glucanase activity

4. CONCLUSION:

From the research results can be concluded that Trichoderma fungus L-1 is the most potential endophytes with the highest chitinase and glucanase activity. The addition of chitin, laminarin or cell wall Phytophthora spp. causing an increase in enzyme activity of chitinase and glucanase. The addition of chitin, laminarin or cell wall causes increased activity of chitinase and glucanase.

5. REFERENCES

[1] Lu, H., Zou, W.X., Meng, J.C., Hu, J., and Tan, R.X. 2000. New bioactive metabolites produced by Colletotrichum sp., an endophytic in *Artemisia annua*. *Plant Science* **151** 67 – 73

[2] Kim, D.J,, Baek, J.M.,Uribe, P.,Kenerley, C.M and Cook, D.R. 2000 Cloning and characterization of multiple glycosyl hydrolasegenes from *Trichoderma virens*, *Curr.genet*. **40** 374-384

[3] Bolar, J.P., Norelli, J.L.,Wong, K.H.,Hayes, C.K., Harman, G.E. and aldwinkle H.H., 2000 Expression of endochitinase from *Trichoderma harzianum* apple increases resistance to apple scab ang reduce vigor *Phytopathology* **90** 72-77

[4] Liu C.H., W. X. Zou, H. Lu, R. X. Tan. 2001 Antifungal activity of *Artemisia annua* endophyte cultures against phytopathogenic fungi. *Journal of Biotechnology* **88** 277–282

[5] Chen, C.E.M. Bauske, G.Musson, R.Rodriguez-Kabana, J.W. Kloepper. 1995. Biological Control of Fusarium Wilt on Cotton by Use of Endophytic Bacteria. *Biological Control* **5** 83 -91

[6] Coombs J.T., C.M.M.Franco, and R. Loria. 2003. Complete sequencing and analysis of pEN2701, a novel 13-kb plasmid from an endophytic Streptomyces sp. Plasmid **49** 86 – 92

[7] Cotes, A., Thonart, P., and Lepoivre, P. 1994. Relationship between the protective activities of several strains of *Trichoderma* against damping-off agents and their ability to produce hydrolytic enzyme activities in soil or in synthetic media. Med. Fac. Landbouww. Univ. Gent. **59** 931-941.

[8] Goldman, G. H., Hayes, C., and Harman, G. E. 1994. Molecular and cellular biology of biocontrol by *Trichoderma harzianum*. *Trends Bio-technol*. **12** 478-482.

[9] Inbar, J., and Chet, I. 1995. The role of recognition in the induction of specific chitinases during mycoparasitism by *Trichoderma harzianum*. *Microbiology* **141** 2823-2829.

[10] Harman, G.E., Howell, C.R., Viterbo, a., Chet , I., and Lorito, M. 2004. Trichoderma Species — Opportunistic, Avirulent Plant Symbionts. *Microbiology* **2** 43-45

[11] Ulhoa,J.Cirano, and J.F.Pegerdy. 1991,regulation of Chitinase Synthesis in Trichoderma harzianum, *J. Gen. Microbol.* **137** 2163-2170

[12] Tsujibo,H.,Y. Yoshida,K.Miyamoto,T. Hasegawa, and Y.Inamori,1992 Purification and Properties of Two Types of Chitinases Produced by Alkalophilic Actinomycete, *Bisci Biotech Biochem.* **56** 1304-1305

[13] Nugroho, T.T., Ali M., Ginting, C., Wahyuningsih, Dahliaty A., Devi, S., dan Sukmarisa., Y., 2003. Isolasi dan Karakterisasi sebagian Kitinase *Trichoderma viride* TNJ63. *Jurnal Natur Indonesia* **5** 101-106

[14] Chlan, C.A. and Robert P.Bourgeois. 2001. Class I chitinases in cotton (*Gossypium hirsutum*): characterization, expression and purification. *Plant Science* **162** 143-154

[15] Thrane, C., Tronsmo, A., and Jensen, D.F. 1997. Endo-1,3-β-glucanase and cellulase from *Trichoderma harzianum*: purification and partial characterization, induction of and biological activity against plant pathogenic *Pythium* spp. *European Journal of Plant Pathology* **103** 331-344

IOP Conf. Series: Materials Science and Engineering **299** (2018) 012027 doi:10.1088/1757-899X/299/1/012027

Depigmentation and Characterization of Fucoidan from Brown Seaweed *Sargassum binderi* Sonder

Endang Saepudin[*1], Ellya Sinurat[2], Ira Azmi Suryabrata[1]
[1]Department of Chemistry, Faculty of Mathematics and Natural Sciences, Universitas Indonesia
[2]Research and Development Center of Marine and Fisheries Product Processing and Biotechnology, Jalan KS. Tubun Petamburan VI, Slipi, Central Jakarta 10260, Indonesia

[*]Corresponding email: endang.saefudin@sci.ui.ac.id

Abstract. Fucoidan has many uses in the field of pharmacology, therefore it is necessary to improve the quality of fucoidan by increasing its purity. The objective of this study was to remove brown pigment from seaweed and observe the effect of the result to the activity of isolated fucoidan. In this study, the pigment was removed by organic solvents in the maceration step. The pigment removal using ethanol was found to give a better result than that of the solvent mixture (methanol: chloroform: water) from previous study, indicated by the appearance of fucoidan color. The result showed fucoidan has a better color, total carbohydrate was 89.23% and total sulphate 18.74%.
Keywords: Characterization, Depigmentation, Fucoidan, Organic solvent

1. Introduction

Fucoidan is a polysaccharide composed of L-fucose with sulfate ester, found in brown seaweeds, *Sargassum binderi* Sonder. For over decades, research regarding fucoidan has developed rapidly, since its massive number of biological activities, such as anticoagulant, antivirus, antiinflammation, and antitumor were published [3]. Based on the previous study [8] [9], isolated fucoidan had blackish brown color. Meanwhile, the standard color of fucoidan is white yellowish to light brown. That blackish brown color indicates the presence of brown seaweed pigments, such as fucoxanthin, chlorophyll a and c, beta carotene, and violaxanthin, trapped in fucoidan during the isolation [6] [17]. Therefore, in this study, the pigment will be removed by edible organic solvents in the maceration step, which is ethanol. In earlier experiment, the maceration step had been done using organic solvents, such as methanol, chloroform, petroleum benzene [5] [10] [11] [15]. It is believed that edible solvent can give a better quality of fucoidan, safe for consumption, and environmentally friendly.

Depigmentation of brown seaweed *Sargassum binderi* Sonder using maceration method by edible organic solvent, ethanol, has been done in this research. The isolated fucoidan was characterized by FTIR, and the total carbohydrate, and total sulfate were also analyzed. The objective of this research is to discover whether ethanol can be used for depigmentation of brown seaweed and to see if the quality of treated fucoidan can be improved.

2. Material and Method

2.1. *Material*

The raw material used in this work was brown seaweed *Sargassum binderi* Sonder which was collected from Lampung, Indonesia. Chemical reagents used were mostly analytical grades and some technical grades.

2.2 *Depigmentation of Brown Seaweed Sargassum binderi Sonder*
Sample (50 g) was chopped, then macerated in Ethanol (95%) with the ratio of 1: 8 (w/v), with variations of time 24 h, 72 h, 120 h, and 168 h, respectively at room temperature.
2.3 *Isolation of Fucoidan from Brown Seaweed* Sargassum binderi *Sonder*
Brown seaweed (50 g) was soaked with 0.1 N HCl (1:10) (w/v) and stirred for 6 h at room temperature. The mixture was filtered using 500 mesh planktonet, the filtrate was collected and neutralized with NaOH (3 M). $CaCl_2$ (2%) was added to the filtrate and stirred. It was left at room temperature for 4 hours. The mixture was centrifuged (5000 rpm) for 15 min at 5° C. The residue was removed and then ethanol (1:2) was added to the filtrate. The mixture was allowed to stand for 24 hours. The precipitate was separated by centrifugation (5000 rpm) for 15 min at 5° C. The precipitate was dissolved in water, and the solution was dialyzed in 0.5 M NaCl and aquabidest to obtain extract fucoidan (F) (Sinurat 2011).
2.4 *Characterization of Fucoidan from Brown Seaweed* Sargassum binderi *Sonder*
2.4.1 Observation of Color Changes
Determination of fucoidan color was based on visual observation of the initial treatment of brown seaweed maceration against the standard fucoidan.
2.4.2 Identification of Functional Groups of Fucoidan Result of Isolation by FTIR
Sample (2 mg) was added with KBr to gain a mass of 200 mg. The mixture was grinded to homogeneous and fine powder then pressed to form a KBr-sample pellet and analyzed it with FTIR spectrophotometer. The FTIR used is *Perkin Elmer precisely Spectrum One FT-IR Spectrometer.*
2.4.3 Determination of Total Carbohydrates
Total carbohydrate was analyzed using the phenol-sulfate colorimetric method. L-fucose as a standard was diluted and concentrated into 20, 40, 80, 160, 320, and 640 μg/mL. Meanwhile, fucoidan sample (200 μL) was put into a reaction tube containing phenol 5% (0.5 mL) and H_2SO_4 18 M (2.5 mL). Each tube was shaken and incubated for 30 min. The absorbance was measured on a spectrophotometer at a wavelength of 480 nm [14].
2.4.4 Determination of Total Sulfate
Total sulfate on polysaccharides was done using $BaCl_2$-gelatin method (Dodgson 1961).
Gelatin (2 g) was diluted in hot water (60-70°C, 400 mL) and the resulting semi-gelatinous solution was left overnight at 4° C. $BaCl_2$ (2 g) dissolved in 400 mL semi-gelatinous solution to made 0,5% $BaCl_2$-gelatine and it was left for 2-3 hours before use. Fucoidan (2 mg) was dissolved in aquadest (2 mL). TCA 3% (30 mL) and $BaCl_2$-gelatine 0.5% (10 mL) was added and stirred. As standard, we used Na_2SO_4 in 10, 20, 30, 40, and 50 μg/mL concentration. The solution was allowed to stand for 15 min and the absorbance was measured at $\lambda = 360$ nm.

3. Results and Discussion

3.1. *Characterization of Fucoidan*
3.1.1. *Observation of Color Changes*
The result of maceration (**Table 1**) shows that maceration in ethanol resulted in fucoidan with light brown color. This result clearly indicates that ethanol was able to dissolve the pigments better than the mixture of methanol:chloroform:water. The solubility of pigments such as chlorophyll and fucoxanthin in ethanol is probably due to hydrogen bonding interaction. This interaction may also apply to the pigments in methanol or water. However, the presence of chloroform in the mixture may partially block this hydrogen bonding, resulting a darker fucoidan.

Table 1. Characteristics of Fucoidan from: a). Methanol: Chloroform: Water (4:2:1); b). Ethanol

Fucoidan	Color Changes	Total sulfate (SO_4^{2-})	Total Carbohydrate
Methanol : Chloroform : Water (4:2:1)[a)]	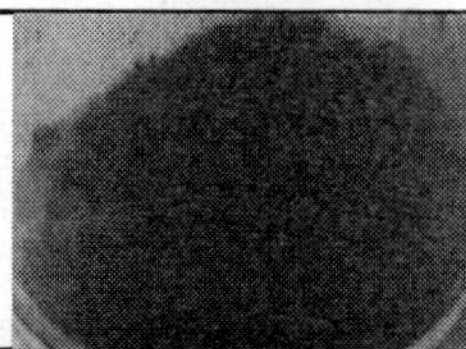	18,63 %	82,79 %
Ethanol[b)]		18,74 %	89,23 %

[a] Ref. [7]
[b] Data of this study

3.1.2. Total Carbohydrate and Total Sulphate
Fucoidan, which is a sulfated hetero-polysaccharide, shows variation in total carbohydrate and total sulfate content, depending on pre-treatment, isolation method, and how it is stored. Carbohydrate and sulfate in fucoidan were measured to determine the component and purity of fucoidan. Sulfate content has been shown to have a role for the activity of fucoidan. The higher sulfate content, the more active fucoidan. Related to sulfate activity, in the previous study, fucoidan without depigmentation has been proved to have good activity against breast cancer cell T47D [8].

Both solvent that used as pigment remover produced total carbohydrates and total sulfates around the common percentage, where the total carbohydrate around 35-60% [4] [9] [13] and total sulfate around 15-30 [2] [7] [12]. Based on Table 1, maceration with ethanol proved to give higher total carbohydrate and sulphate, 89.23% and 18.74%, respectively.

3.2. *Identification of Functional Groups of Fucoidan Result of Isolation by FTIR*
The FTIR spectrum of the sample shows some characteristics of fucoidan. Wave number at 1260-1250 cm^{-1} is the characteristic sulfate ester functional groups; Absorption at 3600-3400 cm^{-1} indicates the presence of –OH (hydroxyl); around 850 cm^{-1} to 820 cm^{-1} indicating position of sulfate, axial or equatorial; and in 1650-1600 cm^{-1} indicating carbonyl functional groups from uronic acid [16].

Peaks at fingerprint area confirm the results above. Bending vibration occurred at 1260-1250 cm^{-1} with medium intensity indicating the presence of sulfate ester. Then, the wavenumber of 825-819 cm^{-1} indicates the location of the sulfate bonded at the equatorial position. The results indicated that the samples were confirmed as fucoidan (Figure 1).

IOP Conf. Series: Materials Science and Engineering **299** (2018) 012027 doi:10.1088/1757-899X/299/1/012027

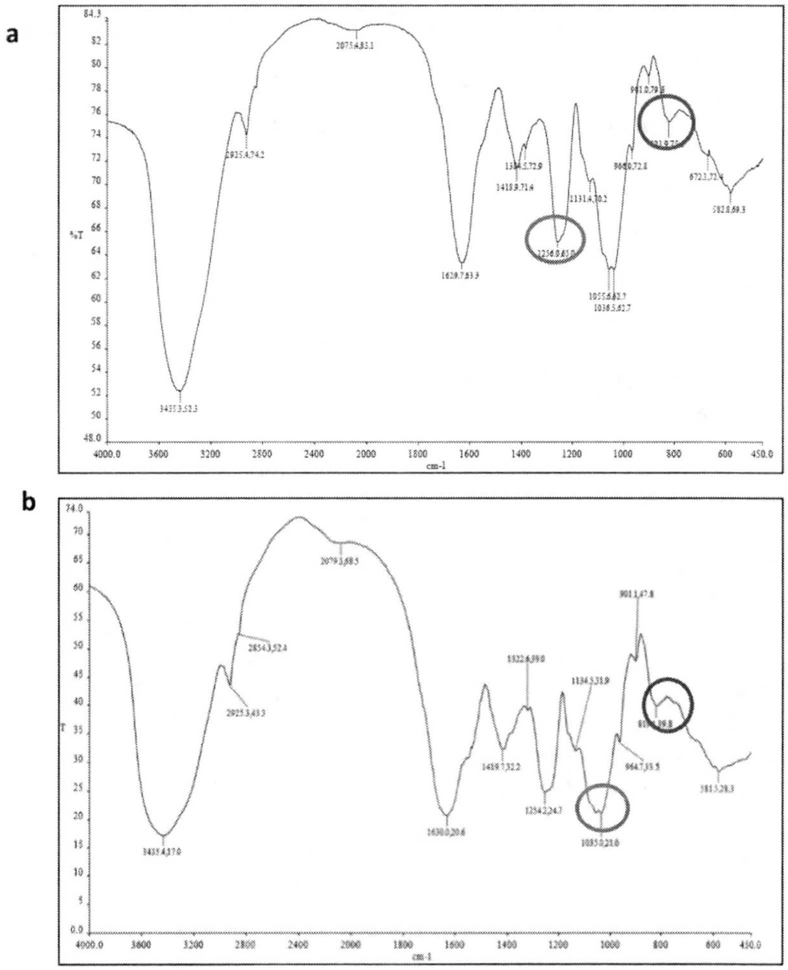

Figure 1. FTIR result: (a) Maceration with ethanol; (b) Maceration with methanol: chloroform: water

4. Conclusion

Macerated-pigment-removal with ethanol from brown seaweed, *Sargassum binderi* Sonder indicated lighter brown color as result. Based on the characterization of fucoidan, we can conclude that pigment removal gave fucoidan better appearance, higher total carbohydrate and total sulphate. Further research about the activity of fucoidan to support characteristics result is needed.

Acknowledgement

This work was supported by Research and Development Center of Marine and Fisheries Product Processing and Biotechnology, Ministry of Marine Affairs and fisheries and funded through the Grant PITTA from the Directorate of Research and Community Service Universitas Indonesia.

References

[1] Dodgson, K.S., 1961. Determination of Inorganic Sulphate in Studies on the Enzymic and Non Enzymic Hydrolysis of Carbohydrate and Other Sulphate Esters 312–319

[2] Hifney, A.F. et al., 2016. Industrial optimization of fucoidan extraction from Sargassum sp. and its potential antioxidant and emulsifying activities. *Food hydrocolloids* **54** 77–88

[3] Li, B. et al., 2008. Fucoidan: Structure and Bioactivity 1671–1695

[4] Lim, S.J. et al., 2016. Characterisation of fucoidan extracted from Malaysian Sargassum binderi **209** 267–273

[5] Lim, S.J. et al., 2014. Isolation and antioxidant capacity of fucoidan from selected Malaysian

seaweeds. *Food Hydrocolloids* **42** 280–288

[6] Nurhayati, Nurbaity, S. & Kusumawati, R., 2010. Pengaruh Kondisi Bahan Baku Rumput Laut Sargassum sp. terhadap Karakteristik Spektra UV-Vis Pigmen Klorofil dan Karoten. *Seminar Nasional Kelautan VI*

[7] Palanisamy, S., Vinosha, M. & Marudhupandi, T., 2017. Isolation of fucoidan from Sargassum polycystum brown algae : Structural characterization , in vitro antioxidant and anticancer activity **102** 405–412

[8] Qosthalani, F.A., 2017. *Pengaruh Hidrolisis Fukoidan dari Rumput Laut Cokelat Sargassum binderi Sonder Terhadap Aktivitas Antikanker Sel T47D.*

[9] Sinurat, E., 2011. *Isolasi dan karakterisasi serta uji aktivitas fukoidan sebagai antikoagulan dari rumput laut coklat Sargassum crassifolium*. Universitas Indonesia.

[10] Sinurat, E. & Marraskuranto, E., 2011. Fucoidan from brown seaweed and its bioactivity. *Squalen*, **7** 131–137

[11] Sinurat, E., Peranginangin, R. & Saepudin, E., 2015. Purification and Characterization of Fucoidan From The Brown Seaweed Sargassum binderi sonder. *Squalen Bulletin of Marine & Fisheries Postharvest & Biotechnology* **10** 79–87.

[12] Thuy, T.T.T. et al., 2012. Isolation, Characterization and Complexation with Bovine Serum Albumin **24** 3367–3370

[13] Wang, C.Y. et al., 2015. Antioxidant activity and growth inhibition of human colon cancer cells by crude and purified fucoidan preparations extracted from Sargassum cristaefolium. *Journal of Food and Drug Analysis* **23** 766–777

[14] Wang, C.Y. & Chen, Y.C., 2016. Extraction and characterization of fucoidan from six brown macroalgae. *Journal of Marine Science and Technology (Taiwan)* **24** 319–328

[15] Warkoyo & Saati, E.A., 2011. The Solvent Effectiveness on Extraction Process. *Makara Journal of Technology* **15** 5–8

[16] Widjanarko, S.B. & Soehono, L.A., 2014. Extraction Optimization by Response Surface Methodology and Characterization of Fucoidanfrom Brown Seaweed Sargassum polycystum **6** 195–205

[17] Yip, W.H. et al., 2014. Characterisation and Stability of Pigments Extracted from Sargassum binderi Obtained from Semporna, Sabah. **43** 1345–1354

Synthesis and Characterization of Hydroxyapatite- Collagen-Chitosan (HA/Col/Chi) Composite Coated on Ti6Al4V

Charlena[1,*], Ahmad Bikharudin[2] & Setyanto Tri Wahyudi[3]
[1]Department of Chemistry, Bogor Agricultural University, Bogor-16680, Indonesia
[2]Department of Chemistry, Gadjah Mada University, Yogyakarta-55281, Indonesia
[3]Department of Physic, Bogor Agricultural University, Bogor-16680, Indonesia
*Email: charlena.ipb@gmail.com

Abstract. HA-collagen-chitosan (HA/col/chi) composite is developed to increase bioactivity adhesiveness between the metal and the material composite and to improve corrosion resistance. The Ti6Al4V alloy was coated by soaking in HA/col/chi composite at room temperature and then allowed to stand for 5, 6, and 7 days. Diffraction pattern analysis of the coated Ti6Al4V alloy showed that the dominant phase were HA and Ti6Al4V alloy. Corrosion resistance test in media by using 0.9% NaCl showed the corrosion rate at the level of 0.3567 mpy, which was better than that of the uncoated Ti6Al4V alloy (0.4152 mpy). In vitro cytocompatibility assay on endothelial cell of calf pulmonary artery endothelium (CPAE) (ATCC-CCL 209) showed there was no toxicity in the cell culture with the percent inhibition of 33.33% after 72 hours of incubation.
Keywords: *chitosan, collagen, cytocompatibility, hydroxyapatite, Ti6Al4V*

1. Introduction

The use of titanium (Ti) and its alloys as bone plate substrates continues to evolve in accordance with the increasing need for human bone components. Titanium and its alloys are commonly used as orthopedic implants and that have high strength, light weight and high corrosive resistance compared to cobalt and stainless steel based alloys [1]. However, metal implants are susceptible to corrosion caused by body fluids, metal ions release may cause side effects on nearby tissues, bioactive properties on the surface are poorly for the formation of bonds with natural bones and tissues [2,3]. Therefore, biocompatible, fixation and life time properties of implantable materials able to enhanced by coating biomaterials on metal surfaces with a composite such as a HA/collagen composite. Collagen fibrils and HA are the main components of bone whose composites are more promising to produce preferable bone tissue reactions [4]. HA/collagen composite able to inhibit pathogenic bacteria present during the implantation process [5].

HA/collagen composite when implanted in the human body shown osteoconductive properties preferable than the monolithic hydroxyapatite and results in a similar bone matrix classification [6,7]. In this study hydroxyapatite was synthesized from the utilization of waste of snail (*Bellamya javanica*) shells through the wet method (precipitation). While collagen was extracted from waste white snapper (*Later calcarifer*) fish [8]. The development of HA/collagen/chitosan composite to increase the mechanical properties and strength of HA coated on Ti6Al4V (Ti grade 5). The addition of chitosan to increase the quality of the material, maintaining the hydroxyapatite position remains on the metal surface. Chitosan is natural polymers that biodegradable, non-toxic, and biocompatible [3]. Ti6Al4V alloy soaking in HA/col/chi composite at room temperature for 5, 6, and 7 days. The aim of this study is to increase biocompatibility and bioactivity properties between metal material and composite from HA/col/chi composite coatings on Ti6Al4V.

2. Material and Methods

The materials are used Ti6Al4V alloy from Baoji Xinglian Titanium Metal Co.Ltd, snail (*Bellamya javanica*) shells obtained from Pasar Anyar (Bogor, Indonesia), $CaCO_3$ standard from Wako, $(NH_4)_2HPO_4$ from Merck, collagen fish (*Lates calcarifer*) scales [8], chitosan (shrimp skin) obtained

IOP Conf. Series: Materials Science and Engineering **299** (2018) 012028 doi:10.1088/1757-899X/299/1/012028

from Department of Aquatics Products Technology, Bogor Agricultural University (Bogor, Indonesia).

2.1 Synthesis of HA

HA was synthesized from 2 types of precursors, suspension of 0.5 M $Ca(OH)_2$ as calcium precursor from extract snail (*Bellamya javanica*) shells and 0.3 M $(NH_4)_2.HPO_4$ as phosphate precursor. The solution of 0.3 M $(NH_4)_2.HPO_4$ was dropped on suspension of 0.5 M $Ca(OH)_2$ at a rate of 1 ml/min, heated conditions kept around 40±2 °C, and stirred for 1 hour. The pH conditions were monitored by pH meter and 1 M NH_4OH solution was used for pH adjustment into final pH kept around 10. The mixture was decanted for 24 hours at room temperature, sonicated for 6 hours, centrifugated for 15 minutes with a speed of 4500 rpm and filtrated. The precipitate was dried at 80 °C for 8 hours and heated at 800 °C for 2 hours.

2.2 Preparation of HA/Col/Chi Composite Coating On Ti6Al4V

As much as 2 ml of collagen 2%, 2 ml of chitosan 2% mixed with 2 gram of HA powder was dissolved in 20 ml of the ethanol solution and stirred for 1 hour at room temperature. Ti6Al4V alloy with a thickness of 5 mm and a diameter of 12 mm smoothed with paper grinding (1200 grit) and immersed in 96% ethanol for 24 hours, sonicated for 1 hour in 0.1 M NaOH solution and dried for 24 hours at room temperature. Ti6Al4V immersed in a composite HA/col/chi solution at room temperature with variation of immersion time (5, 6, and 7 days) and then dried at room temperature. HA/col/chi coatings was characterized by SEM, XRD, corrosion resistance test (galvanostatic/potentiostatic), in vitro assay in NaCl 0.9% solution and calf pulmonary artery endothelium (CPAE) (ATCC-CCL 209).

2.3 In vitro Test

2.3.1 Release of Ca^{2+} in NaCl 0.9%

HA/col/chi coatings immersed in NaCl 0.9% solution for 10 days. Immersion results were taken daily into 0, 1, 3, 5, 7, and 10 each 10 ml. Then that measured calcium content by using AAS (λ= 422.7 nm).

2.3.2 Cytotoxicity in calf pulmonary artery endothelium (CPAE) (ATCC-CCL 209) Cells [9]

The cells were grown using 6 well plates with 2×10^4 cells/wall and incubated for 24 hours at 37 °C with 5% CO_2. Meanwhile, samples of Ti6Al4V coated sterilized with gamma-ray irradiation doses of 15 kGy. Then sample was placed on a cultured CPAE cell (24 hours of cell culture). Endothelial cell culture without treatment was used as a negative control. Samples in the culture medium were then incubated for 72 hours at 37 °C with 5% CO_2. The number of cells is determined using a hemocytometer. Cell viability was examined with tripan blue dye and determined percent of its inhibition.

3. Results and Discussion

3.1 Phase Structure Analysis

Phase structure of HA/col/chi coatings was characterized by X-ray diffraction (XRD) shown in Figure 1. XRD results compared with JCPDS No. 00-009-0432 (HA) and JCPDS No. 00-044-1294 (Ti6Al4V). The diffraction peak of HA occurs shifts after coating. As shown in Figure 1b, a sharp peak and indicating the presence of HA at 2θ= 25.65° (002), 31.79° (211), 32.89° (300), 33.88° (202), and 49.34° (213). While the diffraction peak of Ti6Al4V appeared at 2θ= 35.18° (100), 38.29° (002), 40.20° (101), 53.01° (102), 76.82° (112), and 77.88° (201). The diffraction peak not show any peak diffraction of chitosan, collagen or other compounds. The diffraction peak in chitosan will appear when chitosan contained in the composite more than 30% [10]. The degree of crystallinity of HA/col/chi coatings was obtained 73.35 %.

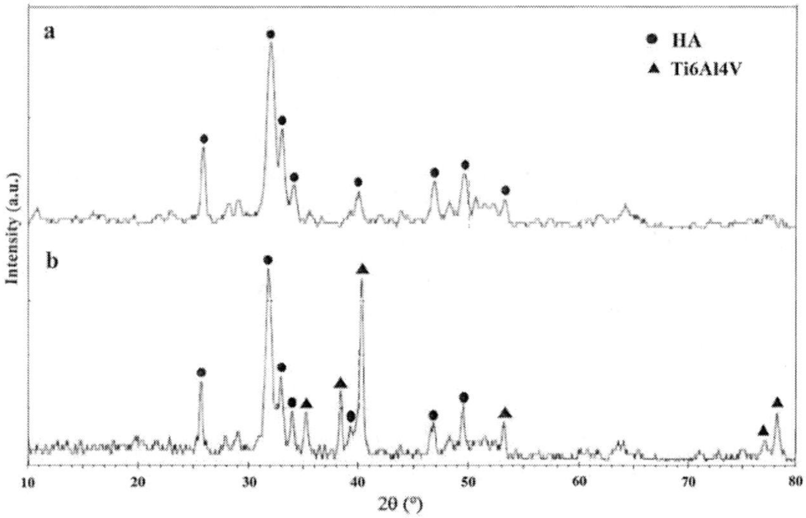

Figure 1 X-ray diffraction patterns: (a) HA; (b) HA/col/chi coatings

3.2 Surface Morphologies

The surface morphology of HA/col/chi coatings observed using by SEM shown in Figure 2. The image shows the presence of granules showing of HA and its distribution less homogeneous. Also shows the shape of the chunks that distributed less evenly. In surface morphology, the presence of fibers, ie; collagen fibers (Figure 2c).

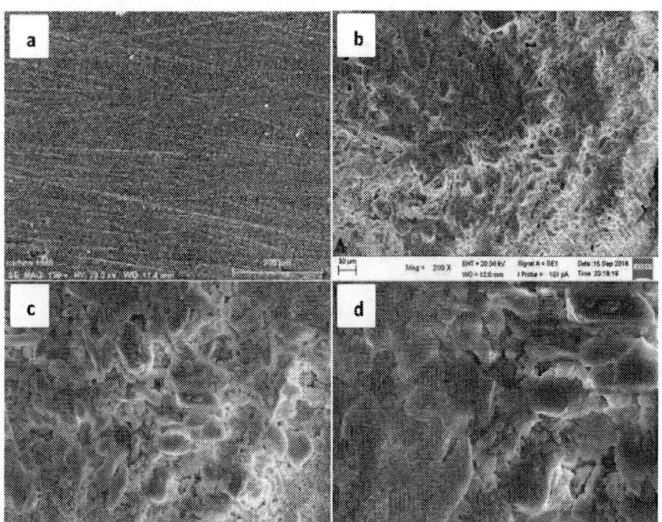

Figure 2 The surface morphology of Ti6Al4V (a); HA/col/chi coatings magnification 200x (b); 500x (c); 1000x (d).

3.3 Measurement of Ca^{2+} Release in 0.9% NaCl Solution

HA/col/chi coatings was carried out measurement of release of Ca^{2+} by soaked sample in 0.9% NaCl solution for 10 days. The average concentration release of Ca^{2+} of shown in Figure 3 using by AAS was obtained 624.46 ppm and shown an increase Ca2+ concentration from day 1 to day 10. According Sharma, *et al.* in [11], calcium concentration increased due to ion exchange in SBF solution between the surface of the HA-coated and due to differences in chemical potential between HA with the metal release and increased Ca^{2+} concentration. According Gu, *et al.* in [12], reported the concentration of Ca^{2+} release in SBF, HA-coated for 56 days soaking obtained 54.4% and increased concentration on day 0 until day 14, and decreased concentration of Ca^{2+} after day 14 to day 56. The longer the immersion time causes the concentration of calcium ions to decrease exponentially and indicate continuous precipitation [13].

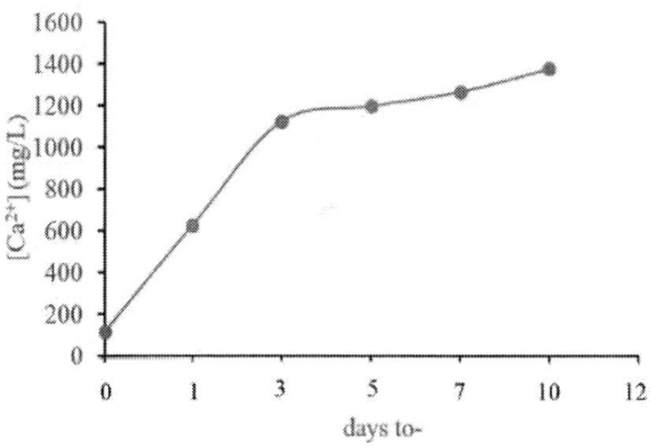

Figure 3 Concentration of Ca^{2+} release for 10 days

3.4 Corrosion test

The corrosion rate on uncoated Ti6Al4V obtained 0.4152 mpy (0.0105 mmpy) with I_{corr} 0.61 (μA/Cm²) and coated Ti6Al4V decreased obtained 0.3567 mpy (0.0090 mmpy) with I_{corr} 0.52 (μA /Cm²). The resultant corrosion rate under the permissible corrosion rate for implanted material and based on european standards of 0.457 Mpy (0.0116 mmpy). Inspection curves (Figure 4) of Ti6Al4V coated indicates a general tendency for potential values to be more stable and indicates excellent value. This indicates the formation of a major passive layer comprising TiO_2 and an additional oxide such as Al_2O_3 and V-oxide. Aluminum is capable of forming the Al_2O_3 passive oxide layer on Ti6Al4V alloys.

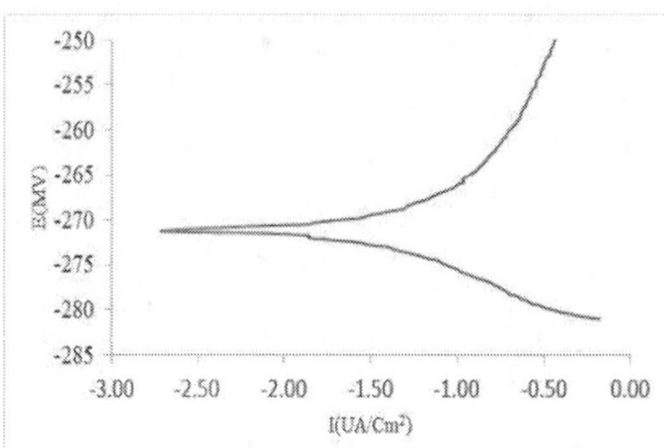

Figure 4 Polarization curve of Ti6Al4V coated

3.5 In vitro cytocompatibility assay on endothelial cell

The cytocompatibility test aims to determine the viability of cells when there is direct contact with the sample and was performed in vitro on endothelial cell culture media. Endothelial cells are the main cells involved in the formation of blood vessels. HA/col/chi coatings as implants that accelerate the healing process should not cause cytotoxicity in endothelial cells. The morphology of endothelial cells in HA/col/chi coatings was observed for 24 hours (Figure 5c) and 72 hours after incubation (Figure 5d), that shown endothelial cells able to alive and adapted in the presence of samples. The results of the cytocompatibility test obtained percent inhibition which showed inhibition of cell growth due to exposure by the sample. According Matsuura, *et al.* in [14], the samples with percentage inhibitions exceeds 50% is toxic. The percent inhibition of HA/col/chi coatings obtained 33.33% that classified as non-toxic and qualifies as a biocompatible implant.

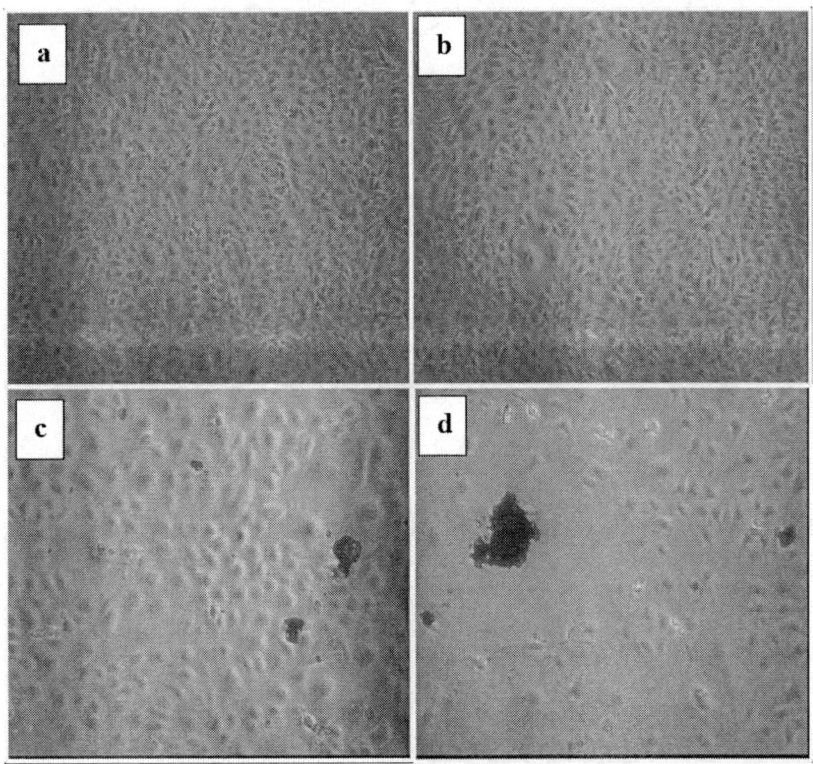

Figure 5 Morphology of endothelial cells of 80x magnification on 24 h control (a); 72 h control (b); HA/col/chi coated Ti6Al4V for 24 h (c); HA/col/chi coated Ti6Al4V for 72 h (d)

4. Conclusion

Ti6Al4V alloy with HA/col/chi composite coatings was successfully performed with the best coated at immersion for 7 days. The corrosion resistance properties Ti6Al4V coated has better than Ti6Al4V uncoated. Based on the corrosion resistance and in vitro test of in vitro endothelial cells, Ti6Al4V coatings is biocompatible and non-toxic for bone implant applications.

References

[1] Elias, CN., Lima, JHC., Valiev, R., Meyers, MA 2008 *Biol. Mat. Sci.* **60** 46
[2] Yuhua, Li., Chao, Yang., Haidong, Zhao., Shengguan, Qu., Xiaoqiang, Li., Yuanyuan, Li. 2014 *Materials* **7** 1709-1800
[3] Kim, P., Wakai, S., Matsuo, S., Moriyama, T., Kirino, T. 1998 *J. Neurosurgery* **88** 21
[4] Haiguang, Zhao., Lie, Ma., Changyou, Gao., Jiacong, Shen. 2008 *Polym. Adv. Technol.* **19** 1590
[5] Wahl, DA., Czernuszka, J T 2006 *Eur. Cells Mat.* **28** 43
[6] Serre, CM., Papillard, M., Chavassieux, P., Boivin, G 1993 *Biomaterials* **14** 97
[7] Wang, RZ., Cui, FZ., Lu, HB., Wen, HB., Ma, CL., Li, HD 1995 *J. Mat. Sci. Lett* **14** 490
[8] Charlena, Bikharudin, A., Wahyudi, ST., Erizal 2017 *Rasayan J. Chem.* **10** 766
[9] Miki, M, Morita, M., 2015 *Mat. Transact.* **56** 1087
[10] Marist, AI., 2011 Undergraduate Thesis., Department of Chemistry., Bogor Agricultural University, Bogor
[11] Sharma, S., Son, VP., Bellare, JR., 2009 *J. Mat. Sci.* **20** 1427
[12] Gu, YW., Khor, KA., Pan, D., Cheang, P. 2003 *Biomaterials* **24** 1603
[13] Chang, E., Chang, WJ., Wang, BC., Yang, CY 1997 *J. Mat. Sci.* **8** 201
[14] Matsuura, T., Hosokawa, R., Okamoto, K., Kimoto, T., Akagawa, Y 2000 *Biomaterials* **21** 1121

IOP Conf. Series: Materials Science and Engineering **299** (2018) 012029 doi:10.1088/1757-899X/299/1/012029

Molecular Self-Assembly of Group 11 Pyrazolate Complexes as Phosphorescent Chemosensors for Detection of Benzene

N F Ghazalli[1], L Yuliati[2,3,4] and H O Lintang[2,3*,4]
[1]Department of Chemistry, Faculty of Science, Universiti Teknologi Malaysia, 81310 UTM, Johor Bahru, Johor, Malaysia
[2]Centre for Sustainable Nanomaterials, Ibnu Sina Institute for Scientific and Industrial Research, Universiti Teknologi Malaysia, 81310 UTM, Johor Bahru, Johor, Malaysia
[3]Ma Chung Research Center for Photosynthetic Pigments, Universitas Ma Chung, Malang 65151, East Java, Indonesia
[4]Department of Chemistry, Faculty of Science and Technology, Universitas Ma Chung, Malang 65151, East Java, Indonesia

E-mail: hendrik.lintang@machung.ac.id

Abstract. We highlight the systematic study on vapochromic sensing of aromatic vapors such as benzene using phosphorescent trinuclear pyrazolate complexes (**2**) with supramolecular assembly of a weak intermolecular metal-metal interaction consisting of 4-(3,5-dimethoxybenzyl)-3,5-dimethyl pyrazole ligand (**1**) and group 11 metal ions (Cu(I), Ag(I), Au(I)). The resulting chemosensor **2(Cu)** revealed positive response to benzene vapors in 5 mins by blue-shifting its emission band in 44 nm (from 616 to 572 nm) and emitted bright orange to green, where this change cannot be recovered even with external stimuli. Comparing to **2(Ag)** with longer metal-metal distance (473 nm) with same sensing time and quenching in 37%, **2(Au)** gave quenching in 81% from its original intensity at 612 nm with reusability in 82% without external stimuli and emitted less emissive of red-orange from its original color. The shifting phenomenon in **2(Cu)** suggests diffusion of benzene vapors to inside molecules for formation of intermolecular interaction with Cu(I)-Cu(I) interaction while quenching phenomenon in **2(Au)** suggests diffusion of benzene vapors to between the Au(I)-Au(I) interaction. These results indicate that suitable molecular structure of ligand and metal ion in pyrazolate complex is important for designing chemosensor in the detection of benzene vapors.

1. Introduction

Vapor sensor materials have been developed using phosphorescent metal complexes with characteristics of luminescent properties. These materials have received more attention as one of chemical sensors (chemosensors) that display selective, reversible and naked-eye luminescent changes upon exposure to specific volatile organic compounds (VOCs). Moreover, it has been proved to be one of the most effective, cheap and convenient methods and materials in the 21st century for the detection of harmful vapors [1-2]. In particular, phosphorescent d^6, d^8 and d^{10} metal compounds with extended π or coordination donor systems have been widely developed as vapochromic chemosensors with several advantages from their photophysical properties such as sensitivity of emission, Stokes shifts, excitation of single photon and longer lifetime [3-5]. Considering luminescent changes such as shifting, quenching, positive response and photoinduced energy transfer, phosphorescent metal complexes from those metal ions have showed great attention as vapochromic chemosensors with high tunability and rigidity of supramolecular metal-solvent, metal-metal, π–π, hydrogen bonding and host-guest interactions [3, 6-8].

Development of chemosensors with high sensitivity and selectivity has received much attention for sensing benzene vapors and its homologues [9] such as benzene, ethylbenzene, trimethylbenzene, toluene and xylenes. These vapors in our environment such as workplaces and home [10-11] can cause many diseases from their high volatility and harmful properties to human inhalation or dermal absorption such as leukemia, myeloma and lymphoma [1]. However, they are less attention to develop chemosensors based on their optical properties with high sensitivity and selectively. On the other hands, many researchers have been focused on the use of expensive and complicated instrumentations for development of sensing methods using gas chromatography [12], fiber optic [13-14], diode laser IR spectroscopy [15], electronic nose sensor array [16] and metal oxide semiconductor films [17].

Recently, many reports have showed the study of chemosensors for the detection of benzene vapors using metal complexes such as Ru(II) [18], Zn(II) [19], Au(I) [20], Ag(I) [21], Pt(II) [22] and Cu(I) [24]. Since the sensing sites have been reported based on the formation of crystal voids without any effective interaction with the chromophore as functional groups, low sensing capabilities were found so far. Hence, rational design of chromophores of chemosensor molecules is really important to provide stronger interactions between sensing sites and vapors (analytes). For example, Che *et al.* have used diimine platinum(II) bis(s-acetylide) complexes [25] to trigger luminescent color and emission changes upon interaction with VOCs due to hydrogen bonding and Pt–Pt interactions, leading to different crystals packing [26-27]. Considering the phosphorescent properties of pyrazolate complexes from the weak metal-metal interactions as supramolecular columnar assembly in nanostructures [28-30] and nanocomposites [31-35], we have reported the utilization of trinuclear copper(1) pyrazolate complexes for chemosensors of ethanol [36-37] using different kinds of pyrazole ligands bearing short alkyl side-chains. We found that chemosensors from 4-(3,5-dimethoxybenzyl)-3,5-dimethyl pyrazole ligand provided the best sensing capability. On the other hand, Kishimura have reported that this chemosensor gave positive response upon exposure to benzene vapors with irreversible changes [38]. Since group 11 metal ions have been synthesized as pyrazolate complexes with different metal-metal distances [30], it is interesting challenge to synthesize different pyrazolate complexes **(2)** from group 11 metal ions and 4-(3,5-dimethoxybenzyl)-3,5-dimethyl pyrazole ligand **(1)** as chemosensors for the detection of benzene vapors. Indeed, we observed that chemosensor **2(Au)** was the best sensing material for the detection of benzene vapors by quenching of its emission intensity up to 81% (reversible changes) compared to chemosensor **2(Cu)** with shifting phenomenon after prolong the exposure time. These results suggest that benzene vapors will probably diffuse to inside molecules to cut-off the metal-metal interaction for light-emitting capability and to benzyl ring for the formation of π–π interaction.

2. Experimental

2.1 Instrumentations

Fourier Transform Infrared Spectroscopy (FT–IR) spectra was measured using Thermo Scientific on a model ofNicolet iS50 using potassium bromide (KBr) for preparation of a pellet with sample. Mass spectra were recorded using JEOL on a model of JMS–T100LP (AccuTOF LC–plus). Sample in dichloromethane (DCM) and methanol (MeOH) or acetone for direct injection (10 μL/min, 2000 V) was ionized using electrospray ionization (ESI) method. Absorption spectra were measured using Shimadzu Diffuse Reflectance (DR) Ultraviolet-Visible (UV-Vis) spectrophotometer (UV-2600). Luminescent properties were monitored on a JASCO model of FP-8500 spectrofluorophotometer. The photography images were taken on a Panasonic Lumix of DMC–FZ38 digital camera under exposure to a UV lamp with a model of Vilber Lourmat (8 watt) at daylight or in the dark room. Photon technology international (PTI) with a model of QuantaMaster 400 containing of near infrared (NIR) PMT detector model 914 under liquid N_2 cooling system and a Xenon lamp (75 W) was used as steady state photoluminescent spectrophotometer for the determination of emission and excitation maxima in the lifetime measurement.

2.2 Synthesis of Group 11 Trinuclear Metal Pyrazolate Complexes

Trinuclear copper(I) pyrazolate complexe **2(Cu)** was prepared according to our previous reports [28,29]. As shown in Figure 1, at room temperature, the mixture of 4-(3,5-dimethoxybenzyl)-3,5-dimethyl pyrazole ligand (**1**) and tetrakis(acetonitrile)copper(I)hexafluorophosphate ($[Cu(MeCN)_4]PF_6$) in dry tetrahydrofuran (THF) was reacted for 5 mins using a Schlenk technique and followed by adding the distilled dry triethyl amine (Et_3N) under an inert condition. After overnight reaction, THF was removed from the reaction mixture under reduce pressure and the remaining residues were dried-off to isolate the desired complexes. In–situ recrystallization was carried out using dy DCM and MeOH at room temperature to give solid powder of **2(Cu)** after dried under reduce of vacuum pressure in 85% yields. The same procedure was repeated for synthesis other metal complexes using silver(I)hexafluorophosphate ($AgPF_6$; 1.02 g, 4.06 mmol) and chloro(dimethylsulfide) gold(I) ($[Au(SMe_2)]Cl$; 1.19 g, 4.06 mmol) to give **2(Ag)** and **2(Au)** as white powder in 75% and 95% yields.

2.3 Evaluation of Sensing Capability

Chemosensors **2(Cu, Ag, Au)** in 22 mg were used for evaluation of sensing capability by directly expose of benzene vapors (200 μL) from 0 to 5 mins. The study was carried out by arranging cell holder for spectrofluorophotometer containing of the chemosensors into the sealed beaker for desired exposure time. The sensing capability was evaluated using spectrofluorophotometer and photographs.

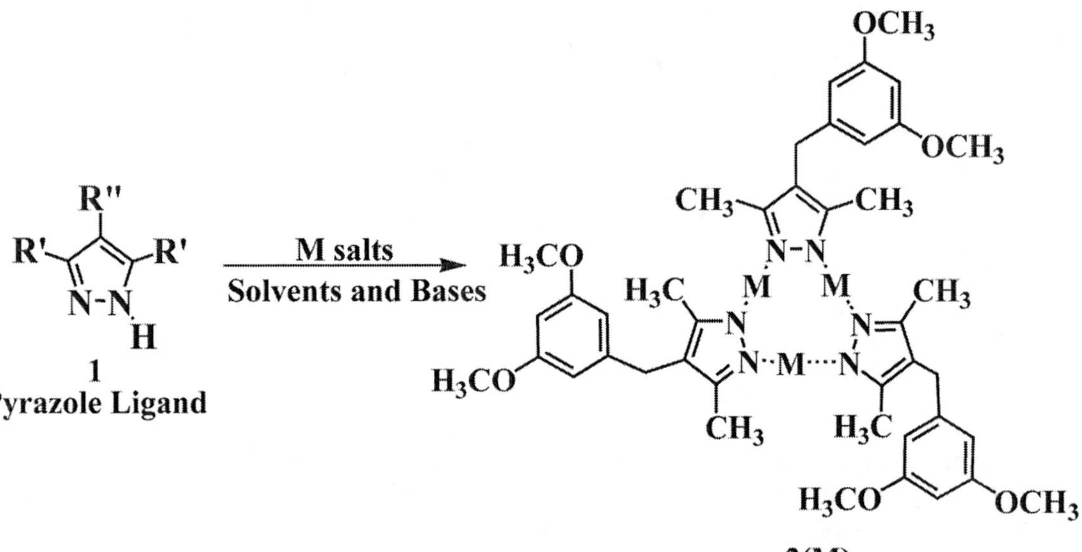

1
Pyrazole Ligand

2(M)
Trinuclear Group 11 Metal Pyrazolate Complexes

Figure 1. Schematic and molecular structures of trinuclear pyrazolate complexes **2(M)** from pyrazole ligand **1** with their metal ions (**M** salts; $[Cu(MeCN)_4]PF_6$ salt for **Cu(I)**, $AgPF_6$ salt for **Ag(I)** and $[Au(SMe_2)]Cl$ salt for **Au(I)**).

3. Results and Discussion

3.1 Structure Elucidation of Group 11 Trinuclear Metal Pyrazolate Complexes

In order to confirm the synthesis of complexes **2(Cu, Ag, Au)**, FT-IR and mass spectroscopy measurements were carried out to observe characteristic vibration bands of the functional groups and molecular weight compared to the ligand **1**. Figure 2(i) shows FT-IR spectrum of ligand **1** with characteristic vibration bands at 3201-3058 and 1750-1595 cm^{-1} for N–H stretching and bending from

1° or 2° amine groups. These vibration bands of **1** did not observe in that of **2(Cu, Ag, Au)** (Figure 2(ii-iv)) due to the formation of N–Cu–N with vibration bands at 560-832 cm^{-1}. As the results, **2(Cu, Ag, Au)** still preserved C–H, C–N and N–N stretchings of **1** with vibration bands at 2835, 1371-1289 and 1289-1151 cm^{-1} for **2(Cu)**, 2834, 1339-1291 and 1205-1156 cm^{-1} for **2(Ag)** and 2991, 1376-1289 and 1207-1155 cm^{-1} for **2(Au)**, respectively. The mass spectra showed that the observed molecular weights were 927.60, 1057.12 and 1326.29 Da for $[M+H]^+$ with molecular formula of $C_{42}H_{51}M_3N_6O_6$ (M is group 11 metal ions). Moreover, monoisotopic patterns and observed molecular weights for all complexes were almost similar to the predicted ones. Hence, all complexes were successfully synthesized.

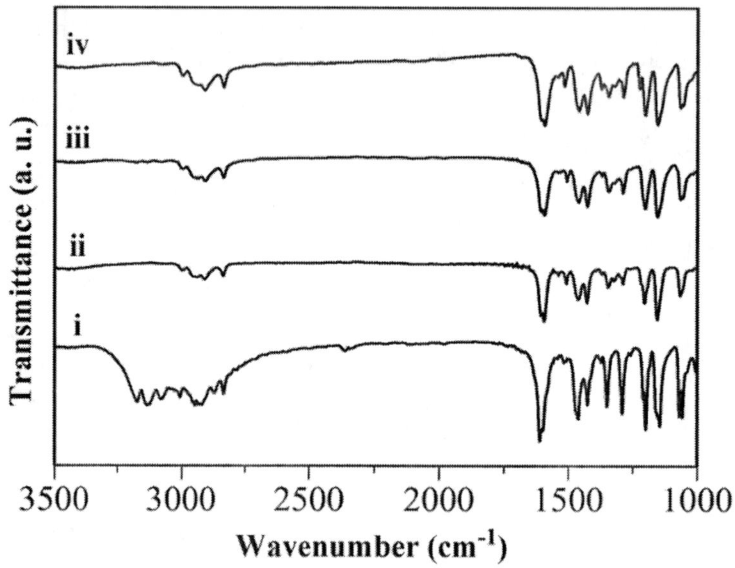

Figure 2. FT-IR spectra of (i) pyrazole ligands **1** and (ii) trinuclear pyrazolate complexes **2(Cu)**, (iii) **2(Ag)** and (iv) **2(Au)**.

3.2 Optical Properties of Group 11 Trinuclear Metal Pyrazolate Complexes

The luminescent properties of d^{10} complexes have been found to show phosphorescent properties with unique molecular geometry. The absorption spectra were measured to find the maximum energy for excitation of electrons from ground to excited state in the emission measurement. By using the peak top of the absorption spectra at 270, 273 and 280 for **2(Cu, Ag, Au)**, respectively as excitation wavelength (λ_{ext}) as shown in Figure 3a, emission spectra were monitored to give a peak centered at 616, 473 and 612 nm for **2(Cu, Ag, Au)** as shown in Figure 3b. Monitoring at maxima emission wavelength (λ_{em}), it provided the excitation spectra for the maximum absorption of light with a peak centered at 274 (**2(Cu)**), 280 (**2(Ag)**), and 280 nm (**2(Au)**), which were closed to the peak top of absorption of the luminophores (Figure 3a). By exposure to a UV lamp (λ_{ext} = 254 nm) in the dark room, the complexes **2(Cu, Ag, Au)** as greenish, gray and white powders at daylight (Figure 3c(i-iii)) emitted green, orange and red color emissions (Figure 3c(iv-vi)). These photograph images and emission spectra were in good agreement with visualized color emissions. Moreover, the calculated Stokes shifts ($\Delta\lambda$) from the difference between wavelength of emission and excitation maxima was 338, 193 and 332 nm with lifetime (τ) of 8.16±1.8, 7.84±0.7 and 8.91±0.7 μs at room temperature, indicating the characteristics of phosphorescent compounds from the weak Cu(I)-(Cu(I) interaction [28-38].

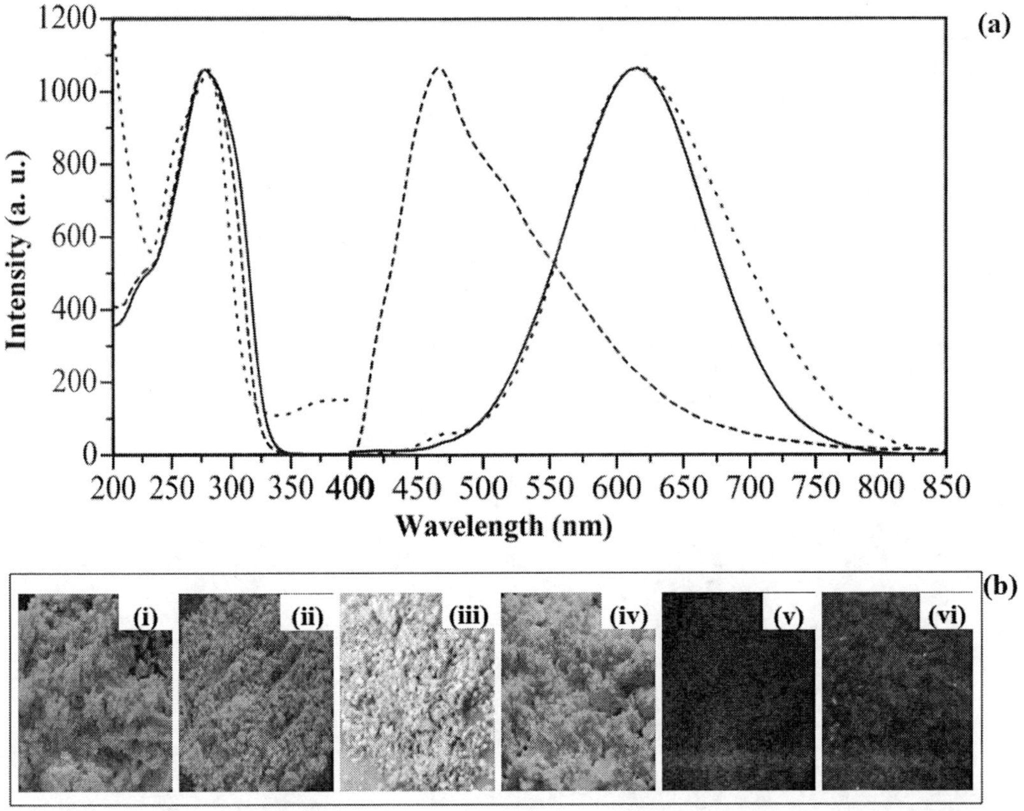

Figure 3. (a) Normalized excitation and emission spectra of **2(Cu)** (solid line), **2(Ag)** (dash line) and **2(Au)** (dot line) as well as (b) their photography images of chemosensors **2(Cu)** (i), **2(Ag)** (ii) and **2(Au)** (iii) at daylight and **2(Cu)** (iv), **2(Ag)** (v) and **2(Au)** (vi) in the dark room upon exposure to a UV lamp (λ_{ext} = 254 nm).

3.3 Evaluation of Sensing Capability

Sensing capability of chemosensor group 11 pyrazolate complexes was evaluated using spectrofluorophotometer and photography images for the vapochromic chemosensors of benzene vapors in the closed system. Upon exposure to benzene vapors from 0 to 5 mins, chemosensor **2(Cu)** showed blue-shifting of its emission band centered at 616 to 572 nm (Figure 4a) with a Stokes shift ($\Delta\lambda$) in 44 nm (Figure 4b). This phenomenon was in good agreement with their color changes from orange to orange-green (Figure 5a). This shifting was less than the sensing of ethanol with the same chemosensor due to the number of stacking for benzene vapors to benzyl of the pyrazole ring (3 arms) is lesser than that of hydrogen bonding interaction of ethanol to the methoxy of the benzyl (6 methoxy groups) [37]. Moreover, the same phenomenon was also observed for the reusability testing as reported by Kishimura [38]. Interestingly, when chemosensor **2(Ag)** was utilized for the same VOC, the emission band centered at 473 nm was quenched (Figure 4c) in 37% (Figure 4d) with color changes from dark green to almost less emissive with increasing sensing time to 5 mins as shown in Figure 5b(i-vi). The metal-metal distance of Ag(I)-Ag(I) is longer that Cu(I)-Cu(I) interaction so that certain number of benzene vapors will only allow to be diffused into the metal-metal.

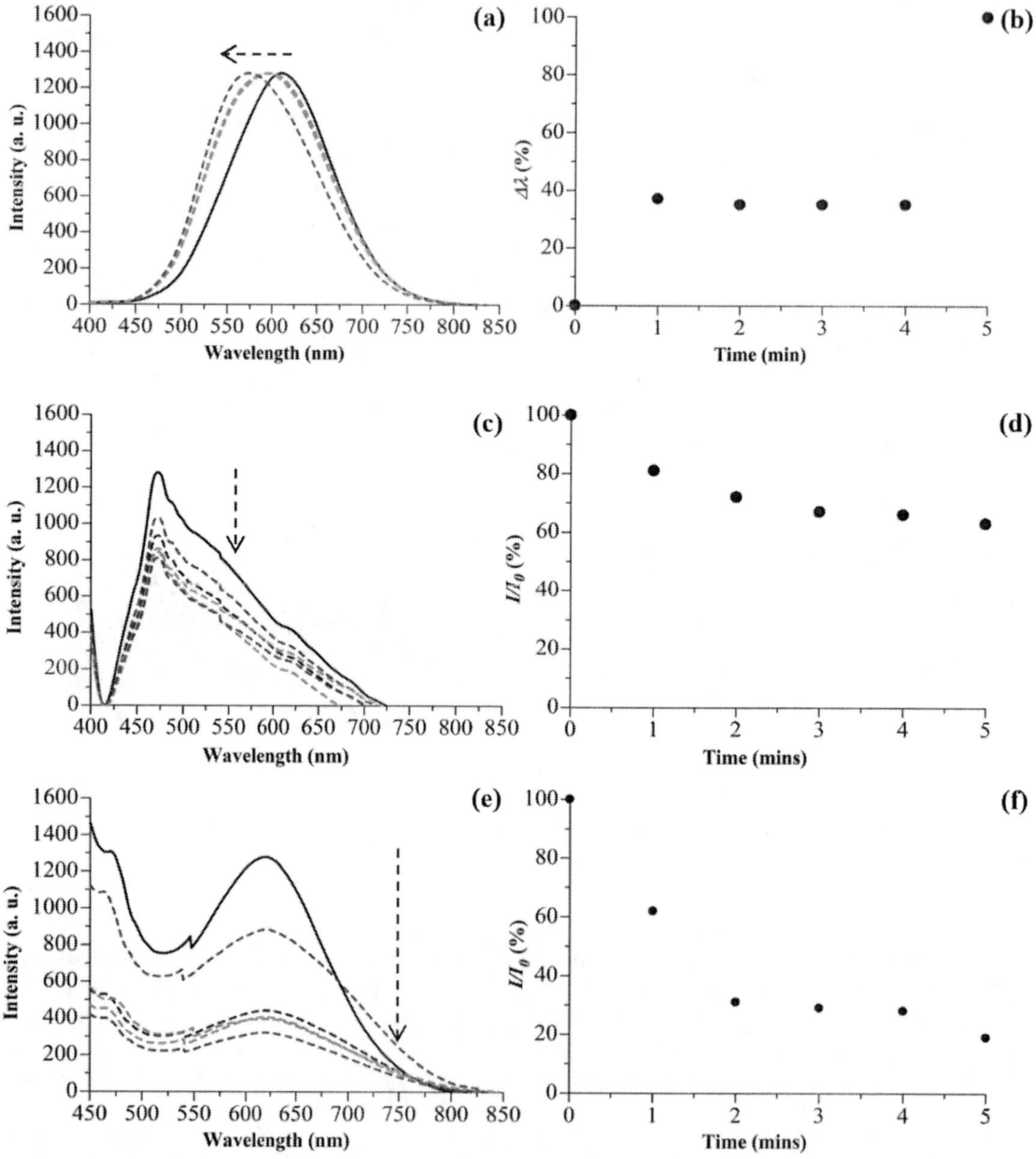

Figure 4. Emission spectral changes of (a) chemosensors **2(Cu)**, (c) **2(Ag)** and (e) **2(Au)** upon exposure to benzene vapors from 0 (black solid line) to 5 mins (red, blue, green, orange and pink dash line, respectively) and their sensing capabilities for (b) chemosensors **2(Cu)**, (d) **2(Ag)** and (f) **2(Au)**.

Figure 4e shows quenching phenomenon in an emission intensity of chemosensor **2(Au)** upon exposure to benzene vapors within 5 mins. In sharp contrast to chemosensors **2(Cu)** and **2(Ag)**, this chemosensor showed about 81% quenching from its original peak top (Figure 4f). Moreover, this positive response emitted less emissive of red-orange from its original color as shown in Figure 5c(i-vi) to support the emission spectral changes. Such quenching using this chemosensor has not yet been

IOP Conf. Series: Materials Science and Engineering **299** (2018) 012029 doi:10.1088/1757-899X/299/1/012029

reported so far. The shorter Au(I)-Au(I) interaction with more broader near-infrared emission peak compared to chemosensor **2(Cu)** will tend to highly encourage benzene vapors to be diffused and stacked to between two complex molecules so that the metal-metal interaction will be cut-off. Such diffusion will also possible to stack benzene molecule with benzyl ring of the pyrazole ring with appearance of small emission peak at 460 nm from aggregation induce emission. Considering the reusability up to 82% without using external stimuli in 10 mins as shown in Figure 5(d), chemosensor **2(Au)** provided the best sensing capability for the detection of benzene vapors compared to the others.

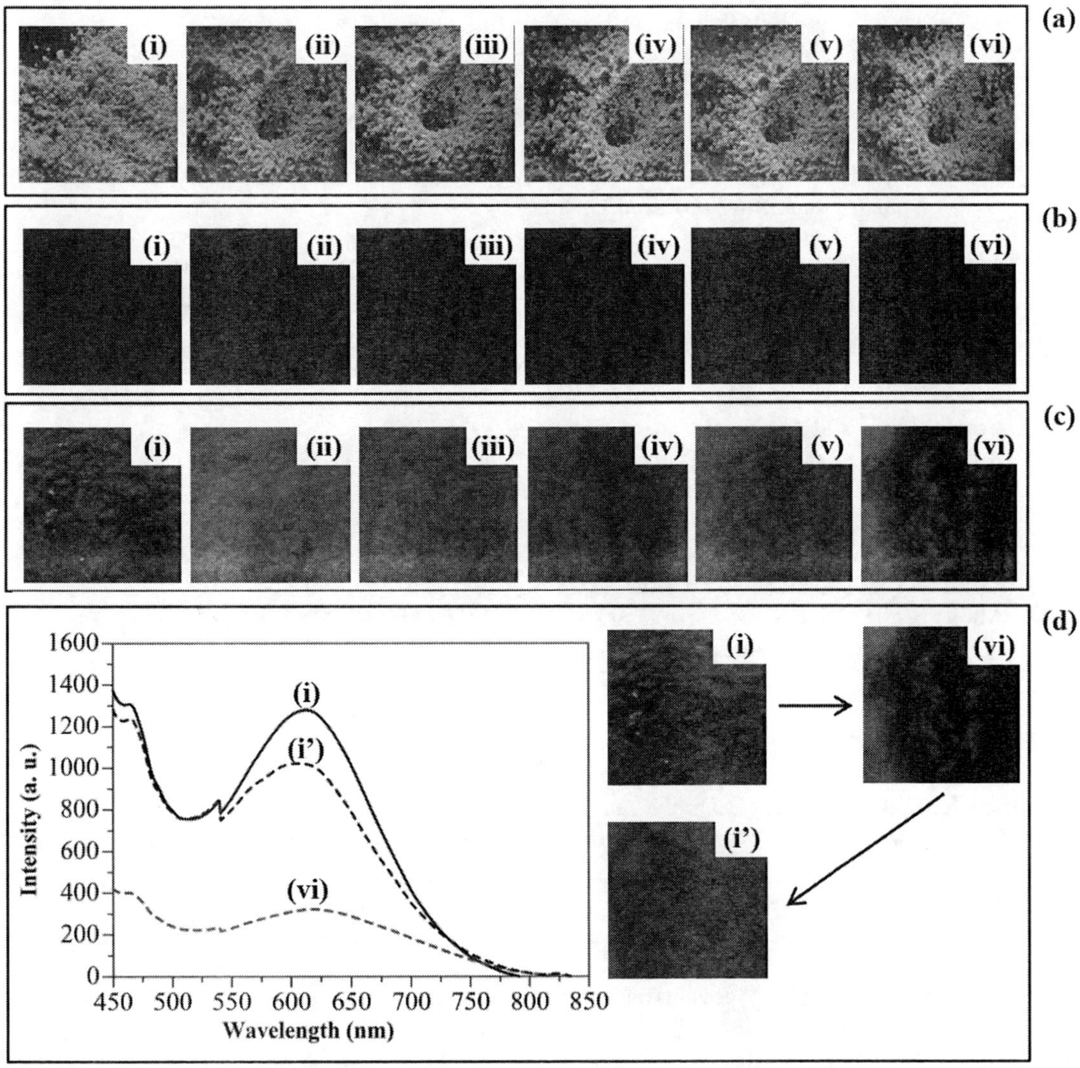

Figure 5. The photography images of (a) chemosensors **2(Cu)**, (b) **2(Ag)** and (c) **2(Au)** upon exposure to benzene vapors for (i) 0, (ii) 1, (iii) 2, (iv) 3, (v) 4 and (vi) 5 mins and (d) emission spectral and color changes from (i) 1 to (vi) 5 mins and (i') after standing for reusability testing in 10 mins. The photographs were taken at room temperature on exposure to a UV lamp with wavelength excitation of 254 nm in the dark room.

4. Conclusion

In conclusion, we have successfully synthesized trinuclear group 11 pyrazolate complexes **2(Cu, Ag, Au)** from 4-(3,5-dimethoxybenzyl)-3,5-dimethyl pyrazole ligand **1** in 85%, 75% and 95% yields. All complexes showed characteristics of phosphorescent properties with emission bands centered at 616, 473 and 612 nm, large Stokes shifts and luminescent lifetime decays in the range of microseconds, which were in good agreement with their colors emissions from green, orange and red, respectively. Upon exposure to benzene vapors, chemosensors **2(Au)** showed the best sensing capability with quenching in intensity of its emission center up to 81% and the reusability in 82% without external stimuli compared to chemosensors **2(Ag)** and **2(Cu)**. These results demonstrated that benzene vapors will able to be diffused and then inserted between the Au(I)-Au(I) interaction to cut-off the interaction for light-emitting capability. By using molecular design of ligand and metal ion in the synthesis of complexes, the phosphorescent chemosensors with high sensing capability can be prepared for the detection of VOCs in next future.

References

[1] Ni J, Kang J-J, Wang H-H, Gai X-Q, Zhang X-X, Jia T, Xu L, Pan Y-Z and Zhang J-J 2015 RSC Adv. **5** 65613
[2] Subramanian E, Mercy Leela Jeyarani B, Murugan C and Pathinettam Padiyan D 2016 *Journal of Chemistry and Material Research* **5** 129
[3] Yu L, Song H, Tang Y, Zhang L and Lv Y 2014 *Sens. Actuators B Chem.* **203** 726
[4] Dubar A D F, Brittle S, Richardson T H, Hutchinson J and Hunter C A 2010 *J. Phys. Chem. B* **114** 11697
[5] Zhao Q, Li F and Huang C 2010 *Chem. Soc. Rev.* **39** 3007
[6] Ma D-L, Ma V P-Y, Chan D S-H, Leung K-H, He H-Z and Leung C-H 2012 *Coord. Chem. Rev* **256** 3087
[7] You L, Zha D and Anslyn E V 2015 *Chem. Rev.* **115** 7840
[8] Xu Z, Chen X, Kim H N and Yoon J 2010 *Chem. Soc. Rev.* **39** 127
[9] Hatanaka S, Ono T and Hisaeda Y 2016 *Chem. Eur. J.* **22** 10346
[10] Dodson R E, Levy J I, Spengler J D, Shine J P and Bennet D H 2008 *Atmos. Environ.* **42** 1569
[11] Sekhar P K and Subramaniyam K, 2014 *ECS Electrochemistry Letters* **3** B1-B4
[12] Sarafraz-Yazdi A and Yekkebashi A 2014 *New J. Chem.* **38** 4486
[13] Allouch A, le Calve S and Serra C A *2013 Sens. Actuators, B* **182** 446
[14] Elosua C, Matias I R, Bariain C and Arregui F J 2006 *Sens.* **6** 1440.
[15] Jeffers J D, Roller C B, Namjou K, . Evans M A, McSpadden L, Grego J and McCann P J 2004 *Anal. Chem.* **76** 424
[16] Bakker E 2004 *Anal. Chem.* **76** 3285
[17] Zhang F, Wang X, Dong J, Qin N and Xu J 2013 *Sens. Actuators,B* **186** 126
[18] McGee K A, Marquardt B J and Mann K R 2008 *Inorg. Chem.* **47** 9143
[19] Pang J, Marcotte E J P, Seward C, Brown R S and Wang S N 2001 *Angew. Chem., Int. Ed.* **40** 4042
[20] Jiang H, Wang Y, Ye Q, Zou G and Zhang Q 2010 *Sens. Actuators B Chem.* **143** 789
[21] Rawashdeh-Omary M A, Omary M A, Fackler Jr. J P, Galassi R, Pietroni B P and Burini A 2001 *J. Am. Chem. Soc.***123** 9689
[22] Rawashdeh-Omary M A, Rashdan M D, Dharanipathi S, Elbjeirami O, Ramesh P and Dias H V R 2011 *Chem. Commun.* **47** 1160
[23] Drew S M, Smith L I, McGee K A and Mann K R 2009 *Chem. Mater.* **21** 3117
[24] Hettiarachchi C V, Rawashdeh-Omary M A, Korir D, Kohistani J, Yousufuddin M and Dias H V R 2013 *Inorg. Chem.* **52** 13576
[25] Chan C W, Cheng L K and Che C M 1994 *Coord. Chem. Rev.***132** 87
[26] Ni J, Zhang X, Wu Y H, Zhang L Y and Chen Z N 2011 *Chem.–Eur. J.* **17** 1171
[27] Zhang Y, Zhang H Y, Mu X Y, Lai S W, Xu B, Tian W J, Wang Y and Che C M 2010 Chem. Commun. **46** 7727
[28] Enomoto M, Kishimura A and Aida T 2001 *J. Am. Chem. Soc.* **123** 5608
[29] Kishimura A, Yamashita T and Aida T 2005 *Nature Mater.* **4** 546

[30] Omary M A, Rawashdeh-Omary M A, Gonser M W A, Elbjeirami O, Grimes T and Cundari T R 2005 *Inorg. Chem.* **44** 8200
[31] Lintang H O, Kinbara K, Yamashita T and Aida T 2010 *Int. Conf. on ESciNano 2010 – Proceedings 2010, IEEE proceedings (Kuala Lumpur, Malaysia, 2010),* Article number 5700970
[32] Lintang H O, Kinbara K, Yamashita T and Aida T 2012 *Int. Conf. on ESciNano 2012 – Proceedings 2012, IEEE proceedings (Johor, Malaysia, 2012),* Article number 6149684
[33] Lintang H O, Yuliati Y and Endud S 2014 *Mater. Res. Innov.* **18**, S6-444
[34] Jalani M A, Yuliati Y, Endud S and Lintang H O 2014 *Adv. Mater. Res.* **925**, 233
[35] Jalani M A, Yuliati Y and Lintang H O 2014 *Jurnal Teknologi* **70**, 131
[36] Ghazalli N F, Yuliati L, Endud S, Shamsuddin M and Lintang H O 2014 *Adv. Mater. Res.* **970** 44
[37] Lintang H O, Ghazalli N F and Yuliati L 2017 *Indones. J. Chem.* **17(2)** 191
[38] Kishimura A, Ph.D. Thesis, Submitted to The University of Tokyo, 2005.

Acknowledgements
The authors thank Centre for Sustainable Nanomaterials, Universiti Teknologi Malaysia (UTM), Malaysia for characterization facilities. The authors also thank consultancy project from RGS Corporation Sdn. Bhd. through Uni-Technologies Sdn. Bhd. with vote no. 1079 for the financial supports.

IOP Conf. Series: Materials Science and Engineering **299** (2018) 012030 doi:10.1088/1757-899X/299/1/012030

Tempe Waste Water Degradation Using TiO_2-N/Bentonite alginate Granule Photocatalyst with Ultraviolet Light Irradiation

Aldila Khoirun Nisaa',[1] Sri Wardhani,[1]* Danar Purwonugroho,[1] Darjito[1]

[1]Chemistry Department, University of Brawijaya, Malang, Indonesia

Email: wardhani@ub.ac.id

Abstract. Tempe waste water stew has high ammonia concentration which causes odor due to polluting by anaerobic decay. Free ammonia in the waste has exceeded the limit, thus endangering the aquatic environment. This research aims to determine the activity of photocatalyst granule TiO_2-N/bentonite-alginate as decomposers of compounds in the photodegradation process. Photodegradation is the decomposition process of compounds by semiconductors with light. Results expected includes the photocatalyst activity of TiO_2-N/bentonite-alginate granule produced by ultraviolet rays is known based on the effect of dopant N concentration on the catalyst and the effect of photocatalytic ratio toward tempe waste water. Methods proposed in this research are activation of bentonite using H_2SO_4 0.8 M, TiO_2-N synthesize by sonication method with urea as the source of N, then TiO_2-N impregnation into bentonite. Photocatalyst in granule form synthesized with alginate was then dripped with syringe pump into 3% (w/v) $CaCl_2$. The photocatalyst characterization will be performed using XRD. The optimum tempe waste water degradation at the concentration of TiO_2-N 0.4 (g/g) bentonite is 53.66%. The ratio of photocatalyst and tempe waste water, optimum at 150 mg of photocatalyst with 25 mL of waste equal to 53.66%.
Keyword: photodegradation, photocatalyst, tempe waste water, TiO_2-N/bentonite-alginate

1. Introduction

Soybean consumption in Indonesia is 50% for tempe production, 40% of tofu production, and 10% in other products such as tauco, soy sauce, and others. Currently, average tempe consumption every person per year in Indonesia is estimated to reach about 6.45 kg [1]. One potential alternative method for simultaneously processing organic wastes is photocatalytic photodegradation [2]. Photodegradation is the process of decomposing organic compounds into simpler compounds making them safer for the environment. The decomposition uses the aid of photon energy [3]. Photocatalytic reactions require active catalysts when exposed to light [4]. One of the catalysts that can be active under the sun is TiO_2. TiO_2 has a band gap energy (Eg = 3.2 eV) which can be activated by UV light from UV lamps [7].

The dopant element will form a new catalyst matrix with a smaller gap energy and the equivalent of visible light energy. TiO_2 has a weak adsorption power in degrading the target compound. The dispersed TiO_2 properties in each part of the solution cause the contact of TiO_2 with less than optimal pollutant. TiO_2 can be modified by expanding on a mineral having a high adsorption capacity so that the semiconductor adsorption power is increased [5]. The widely used carrier is bentonite because bentonite is a silica-alumina that has a large pore and surface area, can absorb both organic and inorganic substances, as a catalyst for various reactions, and as a cation exchange [6].

Photocatalyst TiO_2 can be made with a granular form to facilitate the photodegradation process. This is because when the photocatalyst is in the form of a spherical bead, the photocatalyst does not mix with the waste that causes the formation of suspension during the photodegradation process. Making granules is need and can be done with the addition of alginate.

Based on the above background, this study aims to determine the effect of TiO_2-N concentration on TiO_2-N/bentonite-alginate, the ratio of TiO_2-N/bentonite-alginate granule to tempe waste water, and the effect of degradation on the long-exposure of TiO_2-N/bentonite-alginate with ultraviolet light

2. Procedures

2.1. Preparation of Activated Bentonite

Bentonite was grinded with a porcelain mortar and sieved with a size of 200 mesh. The bentonite solid that passes on the 200 mesh sieve is used for further research. The resultant bentonite was weighed 50 g, soaked with 500 mL of H_2SO_4 0.8 M in erlenmeyer 1000 mL and covered with aluminum foil, then shook at 9 rpm for 3 h. The soaking samples were washed with distilled water until SO_4^{2-} free. This can be known by the universal pH that shows the pH according to the aquadest. The samples were dried in an oven at a temperature of 110 °C for 2 h and then the bentonite solution was calcined 5 h at a temperature of 500 °C.

2.2. Preparation of Photocatalyst TiO_2-N

Preparation of TiO_2-N was made with a mole ratio of (20:3) by mixing 2.4 g of TiO_2, 0.27 g of urea and 5 mL of aquademineralization in erlenmeyer, then the mixture was sonificated for 30 minutes and heated over hot plate heaters until the evaporated. Subsequently the suspension was transferred in the oven at 110 °C and calcined at a temperature of 300 °C for 2 h.

2.3. Development of TiO_2-N in Bentonite

Preparation of TiO_2-N impregnated on bentonite was done by mixing TiO_2-N variations 0.9; 1.2; 1.5; and 1.8 g with 3 g of activated bentonite and 10 mL of 96 % ethanol in 50 mL beaker, then stirred with magnetic stirrer for 5 h. The formed TiO_2-N/bentonite was dried in an oven at 120 °C for 5 h. And then it was calcined at 500 °C for 5 h.

2.4. Synthesis of TiO_2-N/bentonite-alginate granule

The TiO_2-N/bentonite photocatalyst is impregnated with sodium alginate gel to prepare granular-shaped adsorbents in a calcium chloride solution. 5 g of sodium alginate was dissolved in 200 mL of distilled water, stirred using a magnetic stirrer while heated at 40 °C to form a gel phase. TiO_2-N/ bentonite of 20 g is mashed using the mortar and then mixed into the alginate gel. The suspension then injected into 3% (w/v) calcium chloride solution with syringe 50 mL/h. Complex granules TiO_2-N/bentonit-alginate that was formed, then separated and dried at 70 °C for 3 h.

2.5. Effect of TiO_2-N Concentration on TiO_2-N/bentonite-alginate Photocatalyst

A total of 25 mL of tempe waste water was prepared in 4 cups, each added a photocatalyst contained 3 g of bentonite and TiO_2-N of 0.9; 1.2; 1.5; and 1.8 g. Then illuminated with sunlight and UV for 5 h. Measurement of COD value was done on tempe waste water before degradation for 5 h. The photodegradation process was conducted twice.

2.6. Effect of TiO_2-N/bentonite-alginate Photocatalyst Ratios and Tempe Liquid Waste Industries Against Liquid Waste Degradation of Tempe Industry

The concentration of TiO_2-N in the TiO_2-N/bentonite-alginate photocatalyst provides the highest reduction in COD values used to degrade tempe waste water with the ratio between the mass of the photocatalyst and the volume of the liquid waste. Mass ratio photocatalyst with volume of tempe waste water is 4:1 (100 mg photocatalyst:25 mL tempe waste); 2:1 (100 mg photocatalyst:50 mL of tempe waste); 3:1 (150 mg photocatalyst:50 mL of tempe waste); and 6:1 (150 mg photocatalyst:25 mL of tempe waste); as well as done twice. The COD degradation of the waste in a variety of ratio was measured.

2.7. Test of Chemical Oxygen Demand
The degraded tempe waste water was diluted 1 mL in 250 mL, then 25 mL plated was added to the Erlenmeyer, and added HgSO 40.5 g and 5 mL concentrated H_2SO_4 are also mixed. Next, the solution plus $K_2Cr_2O_7$ as much as 25 mL was added. Erlenmeyer containing the mixture of the solution is arranged in the reflux apparatus with the condenser using water flows. Then the solution in the erlenmeyer was added with concentrated H_2SO_4 of 32.5 mL through the condenser and rinsed with a small amount of distilled water. The solution was refluxed for an hour at a temperature of 265 °C. The reflux solution is awaited to cooldown and followed by adding 5 drops of Ferroin indicator. The solution was then subjected to a solution of FAS and recorded titration volume. Titration and reflux were performed twice. The COD test was carried out for the test results of the tempe wastewater activity of the tempe industry treated by the variation of the concentration of TiO_2-N on TiO_2-N/bentonite-alginate photocatalyst and the mass ratio of TiO_2-N/bentonite-alginate photocatalyst and the volume of tempe waste.

3. Result
3.1. Various types of photocatalyst activity
In the various types of photocatalyst activity three tempe waste water treated, the first without additional photocatalyst, the second tempe waste water with TiO_2-N/bentonite powder photocatalyst, and the third tempe waste water with TiO_2-N/bentonite-alginate granule photocatalyst.

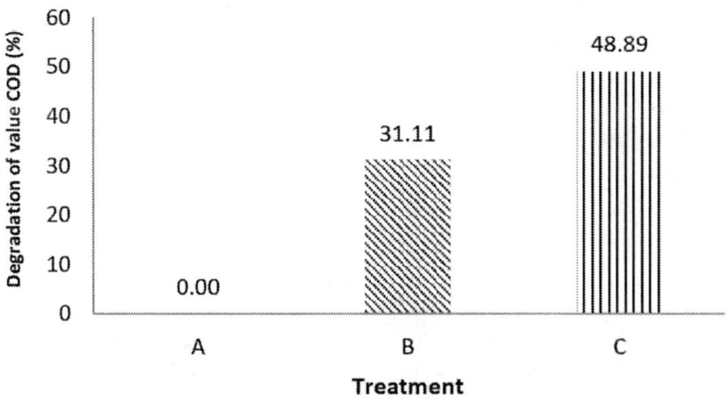

Figure 1. Relation of various type of photocatalyst with tempe waste water degradation
A: tempe waste water without photocatalyst
B: tempe waste water with TiO_2-N/bentonite powder photocatalyst
C : tempe waste water with TiO_2-N/bentonite-alginate granule

Activity test was conducted to determine the effectiveness of photocatalyst form in decreasing the COD value of tempe waste water. The volume of the tempe waste water has used 25 mL of filtered to reduce sediment and photocatalyst weighing 0.15 g. Irradiation did for 5 h. In the test results can be seen in Figure 1, the addition of photocatalysts gives the effect of the COD value of tempe waste water. Tempe waste water without the addition of photocatalyst did not decrease the COD value, the addition of TiO_2-N/bentonite powder photocatalyst gave a decrease of 31.11% the COD value, while the addition of TiO_2-N/bentonite-alginate granule photocatalyst showed a 48.89% decrease for the the COD value. The photocatalyst shape affects the photodegradation process. Photocatalyst in the form of granule works better than powder form. This is because ultraviolet light cannot affect the photocatalyst particles evenly so that the photodegradation process does not run optimally. In addition, photocatalyst in powder form can only photodegradation process.

Photocatalyst added to absorb the energies of photons from ultraviolet light. The addition of N dopant can decrease the energy of the TiO_2 catalyst band gap so that it can absorb more photon energy to produce hydroxyl radicals. The small band gap energy absorbs the energy of ultraviolet light photons larger [8]. TiO_2 when exposed to ultraviolet light will produce electrons (e^-) in the conduction band and hole (h^+) in the valence band. Hole when reacting with H_2O produces a radical ·OH that can degrade organic substances. The inclusion of bentonite causes the photocatalyst to have multiple capabilities, in addition to being able to degrade also as an adsorbent.

3.2. Effect of TiO_2-N Concentration on TiO_2-N/bentonite-alginate Photocatalyst

The concentration of catalyst applied to the carrier was varied to four, ie 0.9; 1.2; 1.5; and 1.8 gr into 3 gr of bentonite.

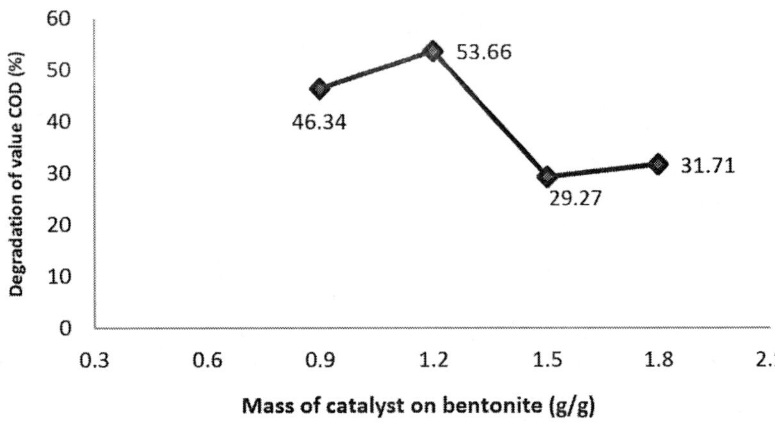

Figure 2. Relation of mass of photocatalyst with tempe waste water degradation

Based on Figure 2 shows the effect of TiO_2-N concentration on the degradation of tempe waste water. The addition of TiO_2-N in photocatalysts can increase the degradation of tempe waste water, with an optimum value of TiO_2-N 1.2 g degrading to 53.66%. In addition, 0.9 g of degradation reached 46.34%, 1.2 g decreased by 29.27% degradation, while 1.8 g of degradation increased to 31.71%. The addition of TiO_2-N as a catalyst affects tempe waste water degradation, since the number of catalysts will have an impact on photodegradation.

The optimum concentration of TiO_2-N in per gram bentonite is 0.4 gr. This happen because the addition of excess TiO_2-N will reduce the adsorption capacity of the photocatalyst, so the adsorption process will decrease. Photocatalysts that have a balanced adsorption and photocatalytic ability provide better degradation results than photocatalysts with only one process. The composition between the catalyst and the carrier shall be suitable for obtaining an optimum photocatalyst [9]. In addition, excess TiO_2-N in the photocatalyst will cover the pores of the adsobent so that the photocatalyst surface area will shrink and decrease the photocatalyst activity [10].

3.3. Effect of TiO_2-N/bentonite-alginate Photocatalyst Ratios and Tempe Liquid Waste Industries Against Liquid Waste Degradation of Tempe Industry

The ratio of photocatalyst and tempe waste water was used two variations. The first variation of tempe waste water volume is 25 and 50 mL. Variations in both photocatalyst mass of 100 and 150 mg. Based on Figure 3 shows the ratio of the mass of photocatalyst and tempe waste water volume to degradation which gives the different result on the parameters of photocatalyst and tempe waste water volume. The tempe waste water volume parameters use 25 mL and 50 mL, while the photocatalyst mass parameters use 100 mg and 150 mg. At 25 mL volume, 100 mg photocatalysts degrade 39.02%, while the 150 mg mass was able to degrade up to 53.66%. Volume 50 mL, 100 mg photocatalyst mass degrades 43.90%, for 150 mg photocatalyst mass degrades 48.78%.

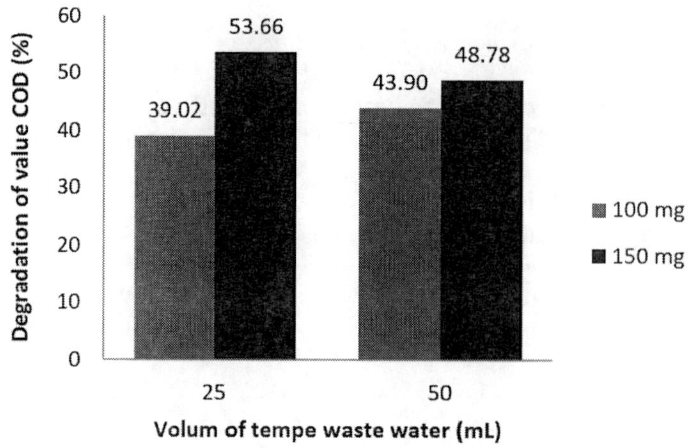

Figure 3. Relation of ratio of photocatalyst mass and with tempe waste water degradation

The optimum degradation of each volume at photocatalyst mass 150 mg, optimum degradation value 53.66% with ratio 150 mg photocatalyst and 25 mL tempe waste water volume. The addition of an increasingly high photocatalyst at low tempe waste water volume will provide the best degradation. This is because the higher number of photocatalysts will decompose and optimally adsorb the organic substances in tempe waste water volume. Ultraviolet light will affect and activate photocatalyst that are based glass at low volume so that photodegradation runs optimum. The amount of photocatalyst used is quite large, 150 mg per 25 mL. It is expected that more photocatalyst, degradation and adsorption process run better so that tempe waste water discharged into the waters will not pollute the environment.

3.4. X-Ray Diffraction Characterization

The photocatalyst characterization was performed using XRD spectrophotometer to determine the photocatalyst crystallinity. In addition, to find out the N dopant added to the photocatalyst. Characterization results were analyzed by comparison with JCPDS (Journal Committee of Powder Diffraction Standard) characterization for TiO_2 and bentonite. Characterization was performed on natural bentonite, activated bentonite, TiO_2, TiO_2-N, and TiO_2-N/Bentonite.

Natural bentonite and activated bentonite were characterized to determine the effect of the activation process on the bentonite crystal structure. The characterization results shown in Figure 4, there is no change in the bentonite crystal structure, but there are changes in some peaks. Activated bentonite showed an increase in intensity at some peaks compared to natural bentonite. This is because the activation process has removed the free oxide impurities but did not alter the bentonite structure. Thus the crystal structure is more dominant and the adsorption capacity of bentonite increases.

Characterization of TiO_2 and TiO_2-N was performed to determine the dominant phase in TiO_2 used and the effect of N dopant on TiO_2-N. Diffractogram showed different results for TiO_2 and TiO_2-N at peak height and 2θ, as in Figure 5. Based on the comparison of 2θ samples of TiO_2 with JCPDS TiO_2 anatase in Table 1, there are several similarities of properties, so the TiO_2 crystals are anatase. Some peaks that have high intensity on the TiO_2 decrease and the 2θ per value shift becomes greater in TiO_2-N. This is due to the presence of an N dopant which replaces the position of the O atom on the TiO_2 bond.

TiO_2-N is impregnated with activated bentonite, the success of impregnation can be known by XRD characterization. Figure 6 is a TiO_2-N diffractogram, bentonite after activation, and TiO_2-N/bentonite. The TiO_2-N/bentonite diffractogram is a combination of the diffraction patterns of TiO_2-N and bentonite. Some of the peaks of TiO_2-N and bentonite diffractogram were changed in the TiO_2-N/bentonite diffractogram showing the lattice changes due to the impregnation process. The shift of the 2θ value and the peak change of the diffractogram are shown in Table 2.

Table 1. The 2θ value of TiO_2 diffractogram

JCPDS of TiO_2 rutil No. 00-021-1276		Sample of TiO_2		JCPDS of TiO_2 anastase No. 00-021-1272	
2θ	Int (%)	2θ	Int (%)	2θ	Int (%)
		25.3143	35.25	25.281	100
27,477	100	27,4991	17.86	-	-
-	-	29,4360	100	-	-
36,086	50	36,1064	13.26	-	-
-	-	36.9502	1.64	36.947	10
-	-	37.8186	7.36	37.801	20
-	-	38.5818	2.04	38.576	10
41,226	25	41,2607	3.46	-	-
-	-	48.050	10.42	48.050	35
-	-	53.9306	6.01	53.891	20
54,323	60	54.3252	10.43	-	-
-	-	55.0573	5.11	55.062	20
56,642	20	56.6369	4.21	-	-
-	-	-	-	62.690	14
69,010	20	-	-	-	-
-	-	-	-	75.032	10

Table 2. The 2θ value of photocatalyst diffractogram

TiO_2		TiO_2-N		Bentonite		TiO_2-N/Bentonite		Phase
2θ	Int (%)	2θ	Int (%)	2θ	Int (%)	2θ	Int (%)	
-	-	-	-	22.2796	100	-	-	Montmorillonite
25.3143	35.25	25.4316	78.10	-	-	25.5434	70.11	Anatase
-	-	-	-	26.9632	61.70	-	-	Montmorillonite
27,4991	17.86	27.5839	23.33	-	-	27.6563	30.66	Anatase
29,4360	100	29.5833	100	-	-	29.6202	100	Anasase
36,1064	13.26	36.1405	18.45	36.3805	31.92	36.2187	31.23	Montmorillonite
37.8186	7.36	37.9470	18.84	-	-	37.9869	14.65	Anatase
48.050	10.42	48.1315	26.04	-	-	48.2130	15.03	Anatase
-	-	-	-	51.0653	16.75	-	-	Montmorillonite
54.3252	10.43	54.4227	17.22	-	-	54.5028	14.29	Anatase
55.0573	5.11	55.1444	15.76	-	-	55.2999	9.19	Anatase

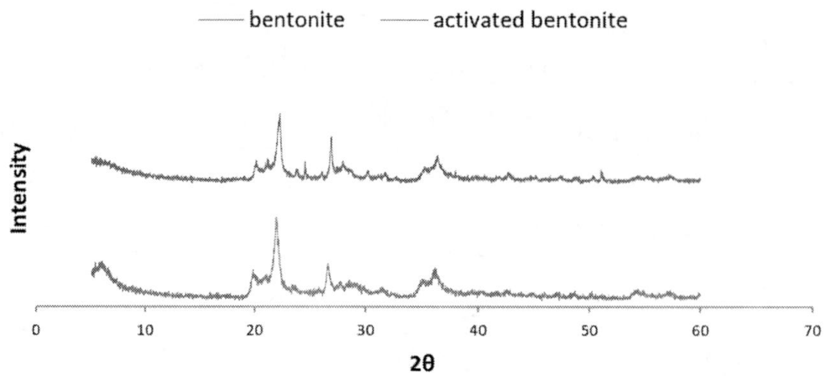

Figure 4. Diffractogram of Bentonite

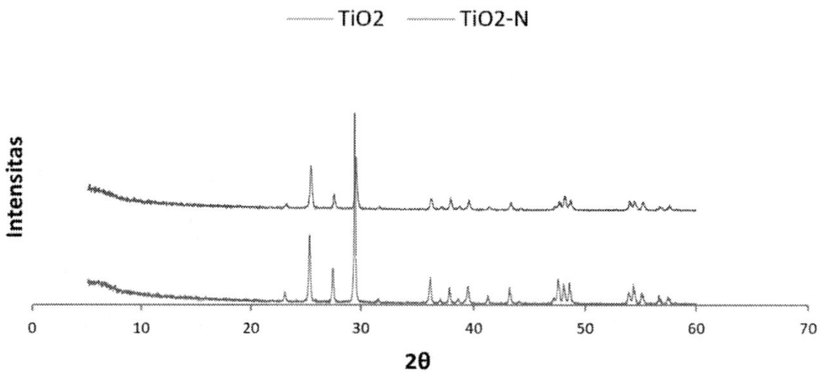

Figure 5. Diffractogram of TiO_2 and TiO_2-N

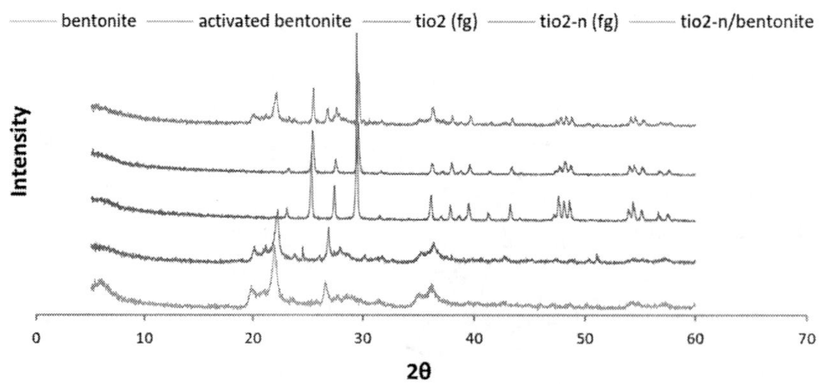

Figure 6. Diffractogram of Bentonite, TiO_2, TiO_2-N, and TiO_2-N/bentonite

4. Conclusion

Based on the result of research, degradation of tempe waste water with the optimum result using photocatalyst in form of granule TiO_2-N/bentonite-alginate. The optimum concentration of TiO_2-N 0.4 g/g bentonite to degrade the tempe waste water with the ratio of photocatalyst and tempe waste water is 1:6 (150 mg:25 mL). The characterization results show that TiO_2 used has similarity properties with

TiO_2 anatase, it can also be seen from the comparison of JCPDS anatase. Addition of dopant N into the photocatalyst decreases the intensity of TiO_2 due to substitution of N atoms and is expected to decrease the band gap energy of TiO_2.

References

[1] PUSIDO BSN, 2012, Tempe : Persembahan Indonesia untuk Dunia, Badan Standardisasi Nasional, Jakarta.

[2] Fatimah, I., 2006, Titanium Oxide Dispersed on Natural Zeolite (TiO_2/Zeolite) and Its Application for Congo Red Photodegradation, Indo. J. Chem., Vol. 6, No. 1, 38-42.

[3] Septiana, R., 2014, Pengaruh Penambahan Ion Logam Cu(II) terhadap Dekolorisasi Zat Warna Methylene Blue dan Methyl Orange oleh Komposit TiO_2 Kitosan, Skripsi, Fakultas Sains dan Teknologi, Universitas Islam Negeri Sunan Kalijaga, Yogyakarta.

[4] Abdullah, M. Arutanti, O. Isnaeni, V.A. Fitria, I. Amalia. Maturi. Aliah, H. dan Khairurijjal., 2011, *Pengolahan Air Limbah dengan Material Struktur Nanometer*, Jurnal Seminar Kontribusi Fisika, INV05

[5] Naimah, S., 2014, Degradasi Zat Warna pada Limbah Cair Industri Tekstil dengan Metode Fotokatalitik Menggunakan Nanokomposit TiO_2-Zeolit, J. Kimia dan Kemaasan, Vol. 36, No. 2, 215-224.

[6] W.T. Tsai, H.P. Chen, M.F. Hsieh, H.F. Sun, S.F. Chien, 2002, *"Regeneration of Spent Bleaching Earth by Pyrolysis in a Rotary Furnace"*, Journal of Analytical and Applied Pyrolisis, 63, hal. 157-170.

[7] Stamate, Marius and Gabriel Lazar, 2007, *Application Of Titanium Dioxide Photocatalysis To Create Self-Cleaning Materials,* MOCM 13 – Volume 3–ROMANIAN TECHNICAL SCIENCES ACADEMY – 2007.

[8] Wismayanti, D.A., Diantariani, N.P., dan Santi, S.R., 2015, Pembuatan Komposit ZnO-Arang Aktif sebagai Fotokatalis Untuk Mendegradasi Zat Warna Metilen Biru, *Jurnal Kimia*, Vol. 9, No. 1, 109-116.

[9] Zuhriah S, 2011, Degradasi Zat Warna Methyl Orange menggunakan Fotoakatalis ZnO-Zeolit dengan Sinar **UV,** Skripsi, Fakultas Matematika dan Ilmu Pengetahuan, Universitas Brawijaya, Malang

[10] Riyani, K., Setyaningtyas, T., Andreas., R.(2008). Pengolahan Limbah Logam Berat Industri Tekstil Menggunakan Fotokatalis TiO_2/Arang Aktif. Molekul. 3.(1) : 40 – 47

IOP Conf. Series: Materials Science and Engineering **299** (2018) 012031 doi:10.1088/1757-899X/299/1/012031

Mg^{2+} Doped into Electro-synthesized HKUST-1 and Their Initial Hydrogen Sorption Properties

W W Lestari[1*], D Ni'maturrohmah[1], U S F Arrozi[2] & H Suwarno[3]

[1]Research Group of Porous Materials for Sustainability, Department of Chemistry, Faculty of Mathematics and Natural Sciences, Sebelas Maret University, Jl.Ir Sutami No. 36A, Kentingan-Jebres, Surakarta, 57126, Central Java, Indonesia
[2]Department of Inorganic Chemistry, Dresden University of Technology Mommsenstrasse 6, 01069 Dresden, Germany
[3]Center for Nuclear Fuel Technology, BATAN, Serpong, South Tangerang, 15314, Indonesia

*Email: witri@mipa.uns.ac.id

Abstract. The hydrogen storage materials are essentially play important roles in supporting the utilization of hydrogen as a promising alternative energy. Several innovative materials have been proposed and intensively investigated in this regard, including Metal-Organic Framework (MOFs). MOFs type HKUST-1 [$Cu_3(BTC)_2$] (BTC = benzene-tri-carboxylate) is the most explored materials in hydrogen storage. In this research, HKUST-1 was electro-synthesized under 15 volt for 1.5 h. This material was ex-situ modified with magnesium(II) ion with variation: 3, 5 and 10 wt% to add attractive sites for hydrogen to form Mg^{2+}@HKUST-1. The final materials were characterized by XRD, FTIR, SEM-EDX, and SAA. Hydrogen sorption measurement was conducted using Sievert system at 30 and 80 °C with pressure from 0.2 to 1.5 bar in 10 minutes for each condition. According to XRD analysis, the basic structure of Mg^{2+}@HKUST-1 was remaining stable. In contrary, SEM analysis showed that HKUST-1 morphology was changed after modification with Mg^{2+}. In addition, the surface area of materials significantly increased from 372.112 to 757.617m^2/g, based on SAA analysis. The presence of Mg^{2+} in the HKUST-1 increased the hydrogen sorption capacity up to 0.475 wt% at 1.4 bar at 30 °C and 0.256 wt% at 80 °C (1.4 bar).

1. Introduction

Metal-organic frameworks (MOFs) have been more attractive to be investigated due to their interesting characteristics, such as highly flexible in structural design in producing various pores and coordination sites, high crystallinity, various chemical and physical properties. Metal ions or metal oxide clusters and organic ligands combination is a way to design the various type of framework [1]. The pore size and surface area of MOFs and the occupancy of unsaturated metal center in their structure, especially in the node part, provide its active site and ability to store gases with non-covalent interaction, like-Van Der Waals, electrostatic and dipol-dipol interactions [2].HKUST-1 is a well-known type of MOFs consisting of Cu^{2+} as nodes and trimesic acid [benzene-1, 3, 5-tricarboxylate] as linker in the cube lattice (Fm-3m) with paddle wheel structure of secondary building unit. HKUST-1 has two types of pore size 9 and 3.5 Å [3]. The hydrogen storage and capture is highly dependent on the inorganic and organic inherent structure molecules in three dimensional structure[4]. In the framework of HKUST-1, Cu^{2+} metal ion forms dimers, each Cu^{2+} coordinated with 4 oxygen atoms from water molecule and BTC ligand. The ability to bind water molecules of Cu^{2+} ions coordination sphere indicates that there are still available coordination in Cu^{2+} [5] which can be used to interact with hydrogen after the activation process[6].

So far, HKUST-1 for hydrogen storage is obtained from solvothermal synthesis [6, 7, 8]. This method is not efficient because the reaction requires high energy and longer reaction time with low yields and the process must be conditioned at high temperature and pressure. Furthermore, the use of metal salts containing nitrate or chloride in the solvothermal synthesis produces unsafe side reaction, therefore it is not environmentally friendly [9]. Electrochemical synthesis HKUST-1 pioneered by Mueller *et al* [10], is simple and safe because it is conducted at ambient temperature. In the electrochemical synthesis, electrode plates will be oxidized and the ligand will be deprotonated and reduction reaction take place to form hydrogen gas and MOF itself [11]. This method requires an electrolyte solution for ion transfer and voltage and electric current for energy source [12].

In order to enhance the hydrogen sorption capacity, HKUST-1 need to be modified with electropositive metal. The metal ions investigated for HKUST modification so far are Pd^{2+} [13] Li^{+} and Na^{+} [14] and Pt^{2+} [15]. The use of Pd^{2+} can increase hydrogen storage capacity up almost twice than unmodified HKUST-1 [13]. In-situ Na^{+} @HKUST-1 and Li^{+}@HKUST-1 did not change the basic structure, but Na^{+} ion can block the HKUST-1 pores [14]. The loading of Pt^{2+} ion nanoparticles also did not change HKUST-1 structure. However, the Pt^{2+} nanoparticles can decrease the pore volume and surface area of HKUST-1 [15]. Meanwhile, Mg^{2+} doping into HKUST-1 have never been studied. Based on previous literature, Mg^{2+} is an electropositive metal and it can be modified into MOF due to its low binding energy with the material [16]. In this research, we have studied hydrogen storage capacity on HKUST-1 prepared under electrochemical method and ex-situ modified with Mg^{2+}.

2. Materials and Methods

2.1 *Materials*

Cu plate (99%), H_3BTC (95%), $MgCl_2{\cdot}6H_2O$ (95%), Tetrabutylammoniumtetrafluoroborate (TBATFB) (99%) were purchased from Sigma Aldrich Germany, Ethanol (96%) was commercially obtained from Merck, deionized water was obtained from Bratachem Indonesia.

2.2 *Synthesis of Mg^{2+}@HKUST-1*

HKUST-1 was synthesized via electrochemical method using *Regulated DC (Power Supply)* ATTEN TPR 3005T-3C according to reported procedures [17] then doped with Mg^{2+} trough ex-situ modification. $MgCl_2{\cdot}6H_2O$ was used as Mg^{2+} source doped into HKUST-1 with variation of 3, 5 and 10 wt% Mg^{2+} of 1 g of HKUST-1. The mixture of $MgCl_2{\cdot}6H_2O$ and HKUST-1 were stirred in 24 mL solvent (deionized water : ethanol, ratio 1:1) for 15 minutes then placed into Teflon vessels covered with stainless-steel autoclaves and heated to 393 K in an oven MEMMERT (type 67 UN30) for 12 h. The final product was cooled, filtered and dried for 24 h at ambient temperature.

2.3 *Characterization*

The X-ray powder diffraction (PXRD) data were obtained on Rigaku Miniflex 600 Benchop, using Cu-Kα radiation (λ = 1.5406 Å), with the current and voltage under 30 mA and 40 kV (2θ = 5-50°) at a scan rate of 1°/min. The morphology of the final materials were performed using scanning electron microscopy (SEM) type: inspect S50-FEI. Infra-Red spectra were collected by using Shimadzu IR Prestige-21 in KBr pellets in the range of 400-4500 cm^{-1}. The nitrogen sorption isotherm at 77 K were obtained from a Quadrasorp evo (Quantachrome instruments).

2.4 Hydrogen sorption measurement
General procedures: the material (0.5 g) was heated under vacuum condition at 473 K for 1 h. Hydrogen storage measurement was carried out using Sievert system as published by Suwarno *et al* [18] at 30 and 80 °C with pressure from 0.2-1.5 bar, in each pressure the measurement was conducted for 15 minutes.

3. Results and Discussion

Diffractogram of Mg^{2+}@HKUST-1 compared with HKUST-1 are presented in Figure 1. The diffraction pattern of the obtained materials have three main peaks at 2θ: 6.73° (2 0 0); 9.47° (2 2 0) and 11.65° (2 2 2) which are confirmed through similarities with the simulated pattern created from CCDC number.112954..Moreover, there are no peaks identified at 2θ: 36.43°; 35.5° and 38.7° indicating the absence of Cu_2O and CuO [19] and [20]. The XRD peaks of Mg^{2+}@HKUST-1 are similar with unmodified HKUST-1. It may be assigned that structure of HKUST-1 was remaining stable after being doped with Mg^{2+}. In addition, Mg^{2+} ions are not chemically bonded with the HKUST-1 atoms, which can be inferred from the absence Mg^{2+} peaks detected in the diffractogram. Based on previous report [14], the electropositive metals ion doped into HKUST-1 do not chemically change its main structure.

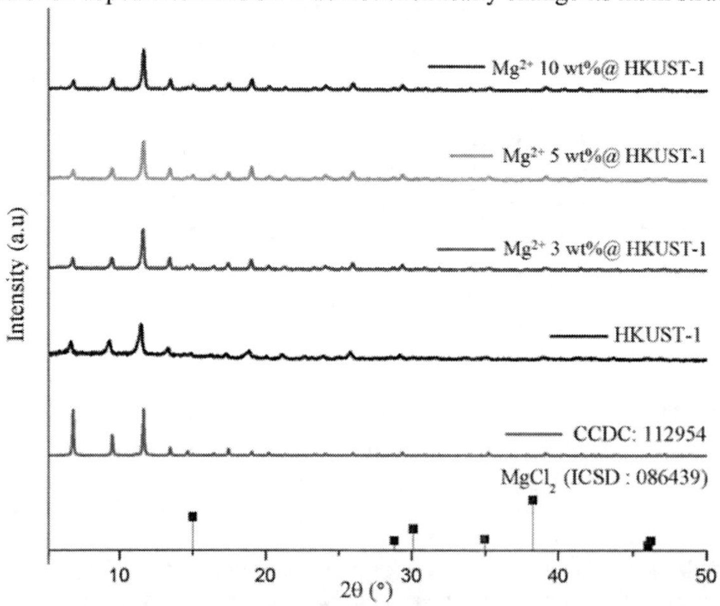

Figure 1. Diffractogram of doped HKUST-1 with Mg^{2+} and undoped HKUST-1 compared to simulated pattern #CCDC 112954

IR spectra of the final products (Figure 2) revealed that Mg^{2+} doped into HKUST-1 does not change the basic structure. The band identified at around 1715 cm^{-1} shows C=O stretching vibration present in the H_3BTC. Ligand deprotonation after complexation with Cu^{2+} causes C=O shift to 1665 cm^{-1}. As a result, the band at 714 cm^{-1} can be assigned to stretching vibration of Cu-O revealed O^{2-} ion is coordinated with Cu^{2+} [12]. The broad absorption at 3100-3600 cm^{-1} shows the existence of coordinated water molecules in the structure. The peak at 3000-3100 and 1452 cm^{-1} indicates vibrational stretching of C-H aromatic and C=C aromatic of the ligand. In addition, the band at 714 cm^{-1} can be assigned to stretching vibration of

Cu-O, revealed O^{2-} ion is coordinated with Cu^{2+} from the paddle wheel moiety in the secondary building unit in HKUST-1 [12].

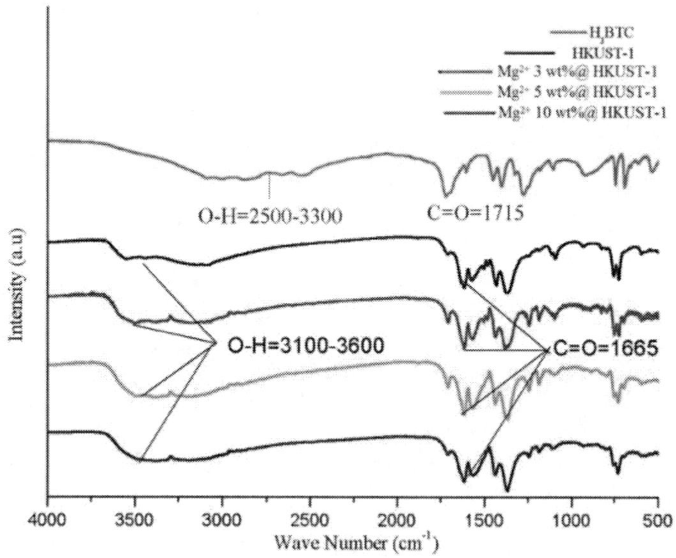

Figure 2. FTIR spectra of doped HKUST-1 with Mg^{2+} and undoped HKUST-1 compared to H_3BTC ligand

Nitrogen sorption isotherm (Figure 3) indicates that the materials could be classified as microporous materials, before and after they modified with Mg^{2+}. It was concluded from IUPAC that the sorption type-1 with small hysteresis loop. Doping of Mg^{2+} decreases pore volume and surface area of HKUST-1 (Table 1). This phenomenon is probably due to blocking of Mg^{2+} into HKUST-1 pores. After doped with Mg^{2+}, the pore volume changed with inhomogeneous pore distribution. It is assumed that distribution of Mg^{2+} particles in the HKUST-1 is not well spread.

Table 1. Pore Volume and Surface Area of doped HKUST-1 with Mg^{2+} and undoped HKUST-1

Material	Pore Volume (cc STP g^{-1})	Langmuir Surface Area (m^2/g)
HKUST-1	0.066	372.112
Mg^{2+} 3 wt%@ HKUST-1	0.042	111.271
Mg^{2+} 5 wt%@ HKUST-1	0.036	33.255
Mg^{2+} 10 wt%@ HKUST-1	0.057	757.617

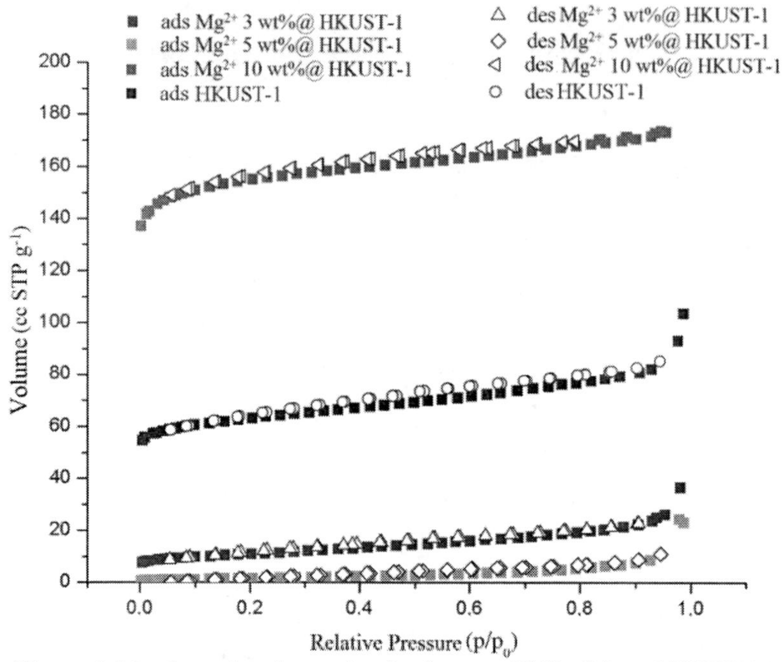

Figure 3. N_2 adsorption-desorption isotherm at 77 K of doped HKUST-1 with Mg^{2+} and undoped HKUST-1

Based on SEM-EDX analysis (Figure 4), HKUST-1 morphology was changed after doped with Mg^{2+}. Based on Table 2, it is observed that the content of Mg elements were not exactly detected according to theoretical calculation. It may because Mg^{2+} lost during stirring or filtering and it was not measured at the well spread particles position. Moreover, identified Cl^- ions show that doping method was not optimum. Mg^{2+} and Cl^- ions size are too small, 0.072 nm [21] and 0.18 nm [22] which may diffused into HKUST-1 pores.

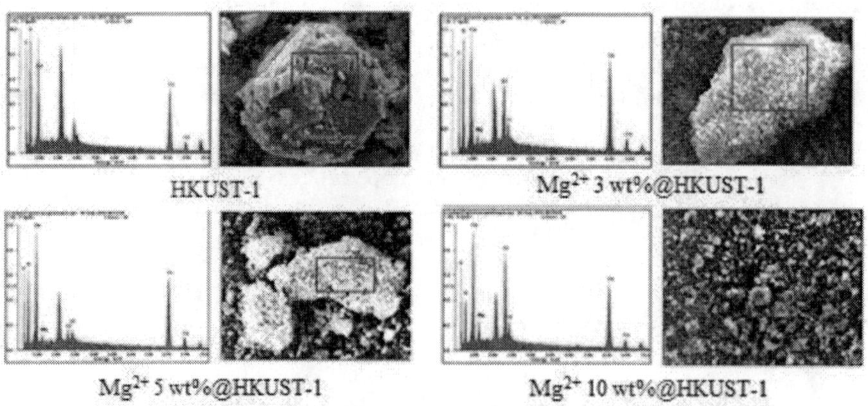

Figure 4. SEM-EDX images of doped HKUST-1 with Mg^{2+} and undoped HKUST-1

Table 2. Elements Composition of doped HKUST-1 with Mg^{2+} and undoped HKUST-1

The material	Element (wt%)				
	C	O	Cu	Mg	Cl
HKUST-1	34.43	27.26	38.31	-	-
Mg^{2+} 3 wt%@HKUST-1	30.84	21.47	41.94	1.69	4.05
Mg^{2+} 5 wt%@HKUST-1	35.03	13.43	49.43	2.12	-
Mg^{2+} 10 wt%@HKUST-1	37.48	13.05	38.06	3.52	7.89

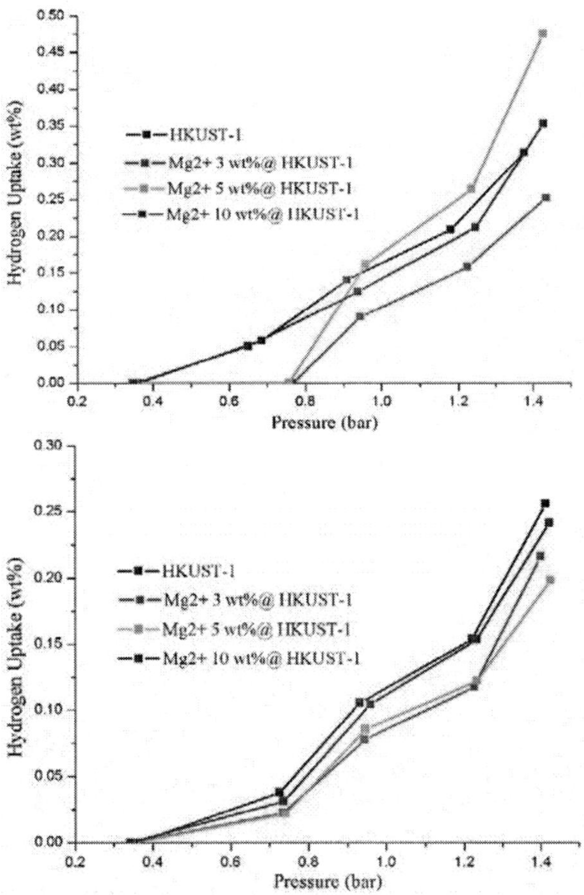

Figure 5. Hydrogen adsorption capacity of Mg^{2+}@HKUST-1 under conditional temperature (top) 30 °C (bottom) 80 °C

Figure 5 shows the hydrogen storage capacity of Mg^{2+}@HKUST-1 compared with HKUST-1 under operational temperature 30 and 80 °C. The optimum storage capacity of Mg^{2+}@HKUST-1 is up to 0.475 wt% (measured at 30 °C, 1.4 bar, obtained from Mg^{2+} 5 wt%@HKUST-1. Additionally, the optimum storage capacity under operational temperature 80 °C is up to 0.256 wt% at 1.4 bar, obtained from undoped HKUST-1 with Mg^{2+}. The result shows that at ambient temperature, the hydrogen storage capacity was higher than at high temperature [6]. Agglomeration of HKUST-1 particles causes doped Mg^{2+} to be spread unevenly and blocking in the pores. Finally, doping Mg^{2+} can't increase the hydrogen storage capacity of HKUST-1. This study is in line with previous research conducted by Lim et al [23].

4. Conclusion

Ex-situ modification of Mg^{2+} into electro-synthesized HKUST-1 was successfully carried out and applied for hydrogen sorption. The result indicates that presence of Mg^{2+} ions do not change the HKUST-1 structure but increases HKUST-1 surface area. After doped with Mg^{2+}, hydrogen storage capacity was increased up to 0.475 wt% at temperature 30 °C (1.4 bar). Further investigation on hydrogen gas interactions with HKUST-1 and Mg^{2+} doped into HKUST-1 should be studied by Inelastic Neutron Scattaring (INS). TG-DTA study could also be performed in order to assess the adsorption enthalpy, so that interaction strength of the hydrogen with the materials at low and high temperature can be discussed intensively.

References

[1] Tranchemontagne D J, Mendoza-Cortés J L, O'Keeffe M & Yaghi O M, 2009 *Chemical Society Reviews* **38** 1257-1283.
[2] Broom D P, 2011 *Springer Science & Business Media.*
[3] Chui S S Y, Lo S M F, Charmant J P, Orpen A G & Williams I D, 1999 *Science* **283** 1148-1150.
[4] Panella B & Hirscher M, 2005 *Advanced Materials* **17** 538-541.
[5] Prestipino C, Regli L, Vitillo J G, Bonino F, Damin A, Lamberti C, Zecchina A, Solari P L, Kongshaug K O & Bordiga S, 2006 *Chemistry of Materials* **18** 1337-1346.
[6] Lin K S, Adhikari A K, Ku C N, Chiang C L & Kuo H, 2012 *International Journal of Hydrogen Energy* **37** 13865-13871
[7] Lee J, Li J & Jagiello J, 2005 *Journal of Solid State Chemistry* **178** 2527-2532.
[8] Rowsell J L & Yaghi O M, 2006 *Journal of the American Chemical Society* **128** 1304-1315.
[9] Lestari W W, Adreane M, & Suwarno H, 2017 *J. Math. Fund. Sci* **49** 213-224.
[10] Mueller U, Schubert M, Teich F, Puetter H, Schierle-Arndt K & Pastre J, *2006 Journal of Materials Chemistry* **16** 626-636.
[11] Lestari W W, Winarni I D and Rahmawati F, 2017 *IOP Conference Series: Materials Science and Engineering*, **172**
[12] Kumar R S, Kumar S S & Kulandainathan M A, 2013 *Microporous and Mesoporous Materials* **168** 57-64.
[13] Zlotea C, Campesi R, Cuevas F, Leroy E, Dibandjo P, Volkringer C, Loiseau T, Férey G & Latroche M, 2010 *Journal of the American Chemical Society* **132** 2991-2997.
[14] Anbia M. & Faryadras M, 2015 *Journal of Nanostructure in Chemistry* **5** 357-364,
[15] Wang L, Stuckert N R, Chen H & Yang R T, 2011 *The Journal of Physical Chemistry* C **115** 4793-4799
[16] Srinivasu K & Ghosh S K, 2011 *The Journal of Physical Chemistry* C **115** 16984-16991.
[17] Lestari W W, Adreane M, Purnawan C, Fansuri H, Widiastuti N & Rahardjo S B, 2016 *IOP Conf. Series: Materials Science and Engineering* 107.

[18] Suwarno H, Wiryolukito S & Dewa M D K, 2015 *Jurnal Teknologi Bahan Nuklir* **10** 96-103
[19] Schlichte K, Kratzke T & Kaskel S, 2004 *Microporous and Mesoporous Materials* **73** 81-88.
[20] Maiti S, Pramanik A, Manju U & Mahanty S, 2016 *Microporous and Mesoporous Materials* **226** 353-359.
[21] Yamaga M, Imai T & Kodama N, 2000 *Journal of luminescence* **87** 992-994.
[22] Sitharaman B, Kissell K R, Hartman K B, Tran L A, Baikalov A, Rusakova I, Sun Y, Khant H A, Ludtke S J, Chiu W & Laus S 2005 *Chemical Communications* **31** 3915-3917.
[23] Lim D W, Yoon J W, Ryu K Y & Suh M P, 2012 *Angewandte Chemie* **124** 9952-9955.

Coordination Polymer of M(II)-Pyrazinamide (M = Co, Cd) with Double End-to-End Thiocyanate Bridge

Yuniar Ponco Prananto

Department of Chemistry, Faculty of Mathematics and Natural Sciences, Brawijaya University, Jl. Veteran 01 Malang, Indonesia 65145

E-mail: prananto@ub.ac.id

Abstract. Pyrazinamide (pza, $C_4N_2H_3$-$CONH_2$) is a good ligand for coordination polymer. Their transition metal complexes are known to have antibacterial activities, magnetic properties, *etc*. Coordination polymers of M(II)-pyrazinamide with thiocyanate (M = Co (**a**), Cd (**b**)), prepared using bench-top layering technique with M(II):pza:SCN ratio of 1:2:2, is successfully crystallised at room temperature. Single crystal XRD was used to determine the crystal structure. Infrared and melting point determination were also performed. Crystal structure of both complexes, solved in Triclinic *P*-1, show that each octahedral metal centre is connected to two adjacent metal centres by double end-to-end thiocyanate bridge forming a 1D polymeric structure with M···M distances of 5.524 Å (in **a**) and 5.887 Å (in **b**). Two monodentate pyrazinamide ligands occupy the rest of the coordination sites on the metal centre in a *trans* relationship. Only in complex **a**, one lattice pyrazinamide molecule is involved in the asymmetric unit. Crystal packing of both **a** and **b** are also displaying non-covalent networks as a result of hydrogen-bonding involving the pyrazine ring, amide and carbonyl groups between adjacent chains and $\pi\cdots\pi$ interactions (only occurred in **a**). In addition, the observed melting points of both **a** and **b** are relatively close to each other (around 180°C), and ATR-IR spectra support the presence of the bridging thiocyanate and terminal pyrazinamide.

1. Introduction

Pyrazinamide or pyrazine-2-carboxamide (pza, $C_4N_2H_3$-$CONH_2$) is a known compound for decades but it was started to be widely used in the 1970s mostly for medical uses, *i.e.* for tuberculosis treatment [1,2]. In inorganic chemistry, pyrazinamide is also a good ligand for the construction of metal complexes, clusters, and/or coordination polymers due to their versatility to coordinate to the metal centre in mono, bi, or polydentate fashions (Figure 1) [3-5]. Transition metal complexes of pyrazinamide, with or without co-ligand, have been explored since then and their properties, such as antibacterial activity [6], electroluminescence [7], and antiferromagnetic behaviour [8], were reported.

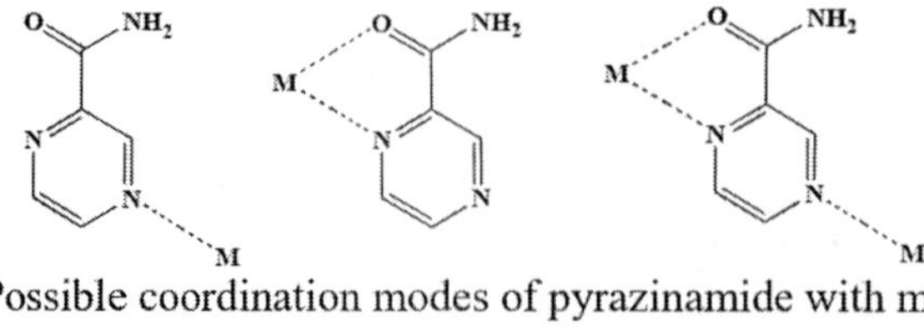

Figure 1. Possible coordination modes of pyrazinamide with metal ion (M)

Metal complexes and/or coordination polymers of M(II)-pza (M = transition metal ions) have been reported many times and to the best of our knowledge, M(II)-pza coordination polymers with double end-to-end thiocyanate bridge were never been published. There are, however, several analogue complexes published with formula of poly-[M(μ-NCS)$_2$(X)$_2$] (X = monodentate terminal ligand), such as poly-[Cd(μ-NCS)$_2$(pia)$_2$] [4], poly-[Cd(μ-NCS)$_2$(3pica)$_2$] [9], poly-[M(μ-NCS)$_2$(4pica)$_2$] (M = Co, Cd) [9], poly-[M(μ-NCS)$_2$(py)$_2$] (M = Mn [10], Fe[11], Co [11], Cd [12]) and poly-[M(μ-NCS)$_2$(4pi)$_2$] (M = Cu [13], Cd [14]), poly-[Cu(μ-NCS)$_2$(ambzim)$_2$] [15], poly-[Cd(μ-NCS)$_2$(3ampy)$_2$] [16], [Cd(μ-NCS)$_2$(val)$_2$]·H_2O [16], *etc.* The source of the thiocyanate could be from NaSCN, KSCN, or NH_4SCN. In addition, some of these coordination polymers also shows interesting magnetic properties [9, 10, 15].

In this paper, we reported the synthesis and crystal structure of {[Co(μ-NCS)$_2$(pza)$_2$]·pza}$_n$ (**a**) and [Cd(μ-NCS)$_2$(pza)$_2$]$_n$ (**b**). The melting point and ATR-IR analyses of the complexes are also discussed. Synthesis of M(II)-pza-thiocyanate (M = Mn, Ni) compounds using similar procedure only gives amorphous products and were not included in this paper. Improved procedure to produce better XRD quality-crystalline products are ongoing.

2. Materials and Methods

2.1. *Materials and Measurements*

Sigma–Aldrich chemicals and solvent (pyrazinamide, NaSCN, $Co(NO_3)_2·6H_2O$, $Cd(ClO_4)_2·6H_2O$, and methanol) were used as received. Solid-state IR spectra were recorded by using an Agilent ATR spectroscope at 4000-600 cm^{-1}. Melting points were measured with an Electrothermal Melting Point Apparatus in open capillaries. Low-temperature single-crystal X-ray diffraction experiments were performed using a Bruker Apex II KAPPA CCD with Mo-Ka radiation (λ 0.71073 Å) equipped with an Oxford Instruments nitrogen gas cryostream. Single crystals were mounted on nylon loops in viscous hydrocarbon oil. Crystals were quench-cooled to 123(2) K. Analysis of diffraction data collected with the Bruker Apex II KAPPA CCD was performed by using SAINT within the APEX2[17] software package. Empirical absorption corrections were applied to all data by using SADABS.[18] The structures were solved using SHELXS[19] and refined using SHELXL-13[19] within the graphical interface Olex2.[20] All hydrogen atoms were refined using a riding model, while all non-hydrogen atoms were refined using an anisotropic model.

2.2. *Synthesis of the Complexes*

2.2.1. {[Co(μ-NCS)$_2$(pza)$_2$]·pza}$_n$ (a). A 3-mL methanolic solution of pyrazinamide (50 mg, 0.406 mmol) was carefully layered on top of a 5-mL aqueous solution of $Co(NO_3)_2·6H_2O$ (34.6 mg, 0.203 mmol) and NaSCN (32.9 mg, 0.406 mmol), separated by 5 mL buffer layer of H_2O:CH_3OH (1:1). Two weeks later, a clear homogenous pink solution was formed. Upon slow evaporation at room temperature for 2 months, the solution gives red block crystals. The crystals were then isolated, washed with cold water and then air-dried for overnight (yield = 35 mg, 32%).

2.2.2. [Cd(μ-NCS)$_2$(pza)$_2$]$_n$ (b). A 3-mL methanolic solution of pyrazinamide (50 mg, 0.406 mmol) was carefully layered on top of a 5-mL aqueous solution of $Cd(ClO_4)_2·6H_2O$ (85.1 mg, 0.203 mmol) and NaSCN (32.9 mg, 0.406 mmol), separated by 5 mL buffer layer of H_2O:CH_3OH (1:1). A month later, long block colourless crystals were yielded. The crystals were then isolated, washed with cold water and then air-dried for overnight (yield = 45 mg, 47%).

3. Result and Discussion

3.1. *Crystal Structure*

Crystal structures of both complexes, {[Co(μ-NCS)$_2$(pza)$_2$]·pza}$_n$ (**a**) and [Cd(μ-NCS)$_2$(pza)$_2$]$_n$ (**b**), were solved in Triclinic *P*-1 space group with unit cell volumes of 1077.52(10) Å^3 and 420.62(8) Å^3, respectively. The unit cell volume of **a** is nearly three times larger than that of **b** since there is one lattice pyrazinamide molecule, which hydrogen-bonded to one of the pyrazinamide ligand, involved in the asymmetric unit of **a** (Figure 2). Coordination environment around the metal centre in **a** and **b** is similar,

IOP Conf. Series: Materials Science and Engineering **299** (2018) 012032 doi:10.1088/1757-899X/299/1/012032

in which each metal centre is surrounded by two monodentate pyrazinamide in a *trans* relationship and four bridging thiocyanate ligands forming an octahedral neutral complex. The pyrazinamide ligands are coordinated through one of the pyrazine nitrogen atom (N5 and N8 in **a**; N3 in **b**). Moreover, two adjacent metal centres are connected by two end-to-end thiocyanate ions with M···M distances of 5.524 Å (in **a**) and 5.887 Å (in **b**), thus the complexes form 1D coordination polymers with a metal centre coordination core of MN_4S_2 (Figure 3). The M···M distance in this work is slightly shorter to those observed in poly-[M(μ-NCS)$_2$(4pica)$_2$] (M = Co, Cd) [9], 5.581 Å in Co(II) and 5.926 Å in Cd(II) complexes, which is likely due to different size, hydrogen bonding, flexibility or steric effect of the terminal monodentate ligand. Two different M···M distances may also be observed in a 1D M(II)-double thiocyanate chain as reported in poly-[Cd(μ-NCS)$_2$(3ampy)$_2$] with Cd···Cd distances of 5.7329(7) Å and 6.0354(8) Å [16].

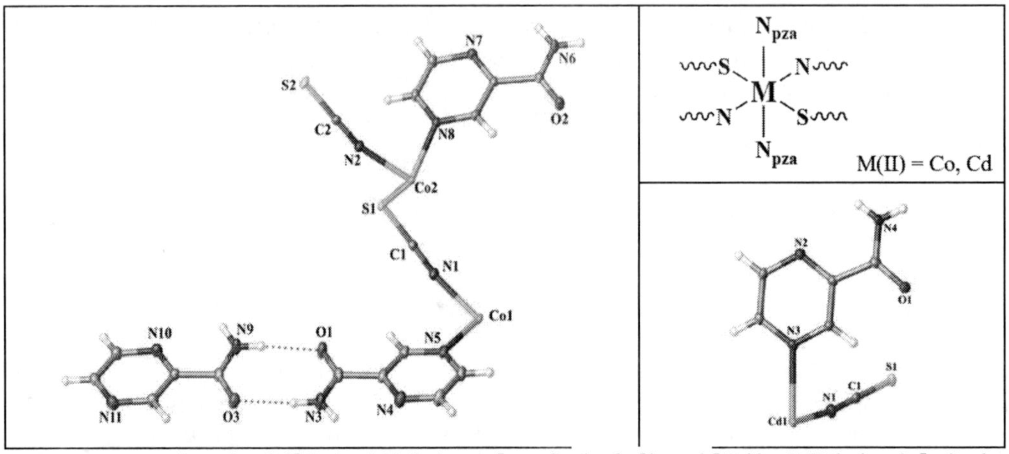

Figure 2. Asymmetric unit of {[Co(μ-SCN)$_2$(pza)$_2$]·pza}$_n$ (**a**, left) and [Cd(μ-SCN)$_2$(pza)$_2$]$_n$ (**b**, bottom right) shown with thermal ellipsoids at 50% probability except hydrogen atoms (note: cyan = cobalt (**a**) or cadmium (**b**), blue = nitrogen, red = oxygen, grey = carbon, yellow = sulphur, white = hydrogen); coordination environment around the metal centre (top right)

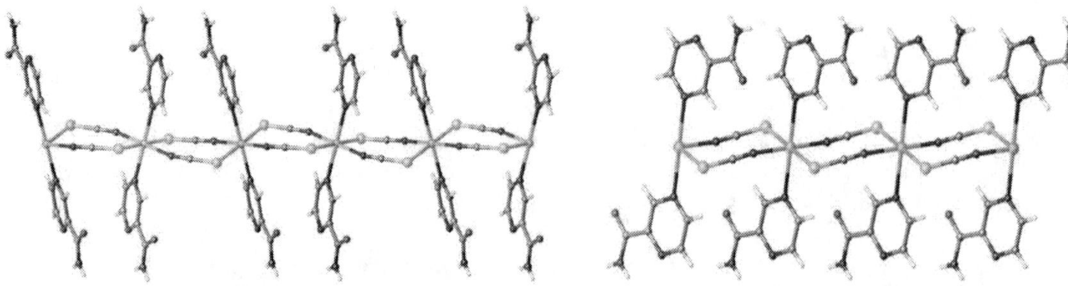

Figure 3. Both **a** and **b** forms 1D chains due to double end-to-end thiocyanate bridges with M···M distances of 5.524 Å (in **a**, left) and 5.887 Å (in **b,** right); note: cyan = cobalt (**a**) or cadmium (**b**), blue = nitrogen, red = oxygen, grey = carbon, yellow = sulphur, white = hydrogen.

Selected bond lengths and angles around the metal centres in the complex **a** and complex **b** are presented in Table 1, whereas crystallographic data of both complexes are presented in Table 3. As expected, the coordination bond length around the metal centre in complex **b** is longer than in complex **a** due to Cd(II) has a bigger size than Co(II). The Cd-N/S bond lengths are ranged from 2.2701(14) to 2.6964(5) Å, whereas the Co-N/S bond lengths are ranged from 2.0370(14) to 2.5306(4) Å. The longest coordination bonds in both **a** and **b** occur between the metal ion and the thiocyanate sulphur atoms.

Table 1. Selected bond lengths and angles around the metal centres in **a** and **b**

Complexes	Bond Lengths (Å)		Angles (°)			
a	Co1-N1	2.0370(14)	S2#c-Co1-S2#b	180.0	S1#b-Co2-S1	180.0
	Co1-N1#a	2.0370(14)	N5#a-Co1-S2#b	89.68(4)	N2#b-Co2-S1	94.68(4)
	Co1-N5	2.2052(15)	N5-Co1-S2#c	89.68(4)	N2-Co2-S1#b	94.69(4)
	Co1-N5#a	2.2052(15)	N5-Co1-S2#b	90.32(4)	N2-Co2-S1	85.32(4)
	Co1-S2#b	2.5352(4)	N5#a-Co1-S2#c	90.32(4)	N2#b-Co2-S1#b	85.32(4)
	Co1-S2#c	2.5352(4)	N5-Co1-N5#a	179.99(14)	N2-Co2-N2#b	180.0
	Co2-N2	2.0441(14)	N1-Co1- S2#c	84.93(4)	N2#b-Co2-N8	91.88(5)
	Co2-N2#b	2.0441(14)	N1#a-Co1-S2#b	84.93(4)	N2-Co2-N8#b	91.88(5)
	Co2-N8	2.1903(14)	N1-Co1-S2#b	95.07(4)	N2-Co2-N8	88.12(5)
	Co2-N8#b	2.1903(14)	N1#a-Co1-S2#c	95.07(4)	N2#b-Co2-N8#b	88.12(5)
	Co2-S1	2.5306(4)	N1-Co1-N5	87.54(6)	N8#b-Co2-S1	89.60(4)
	Co2-S1#b	2.5306(4)	N1#a-Co1-N5#a	87.54(6)	N8-Co2-S1#b	89.60(4)
			N1-Co1-N5#a	92.46(6)	N8-Co2-S1	90.40(4)
	#a = -x, 2-y, 1-z		N1#a-Co1-N5	92.46(6)	N8#b-Co2-S1#b	90.41(4)
	#b = -x, 1-y, 1-z		N1#a-Co1-N1	180.0	N8#-Co2-N8	180.0
	#c = +x, 1+y, +z					
b	Cd1-N1	2.2701(14)	S1#e-Cd1-S1#f	180.0	N1#d-Cd1-S1#e	88.15(4)
	Cd1-N1#d	2.2700(14)	N3-Cd1-S1#f	89.77(3)	N1-Cd1-S1#f	88.15(4)
	Cd1-N3	2.4314(14)	N3#d-Cd1-S1#e	89.77(3)	N1-Cd1-N3	87.71(5)
	Cd1-N3#d	2.4313(14)	N3#d-Cd1-S1#f	90.23(3)	N1#d-Cd1-N3#d	87.71(5)
	Cd1-S1#e	2.6964(5)	N3-Cd1-S1#e	90.23(3)	N1-Cd1-N3#d	92.29(5)
	Cd1-S1#f	2.6964(5)	N3#d-Cd1-N3	180.0	N1#d-Cd1-N3	92.29(5)
			N1-Cd1-S1#e	91.85(4)	N1#d-Cd1-N1	180.0
	#d = 2-x, 1-y, 1-z		N1#d-Cd1-S1#f	91.85(4)		
	#e = 1-x, 1-y, 1-z					
	#f= 1+x, +y, +z					

As reported in polymeric double end-to-end thiocyanate complexes [9-14], the linear thiocyanate anions bind to the metals have M–N–S angles close to 180° (at the N-bound end) and M–S–N angles close to 90° (at the S-bound end). Accordingly, the M–N–S and M–S–N angles in **a** are 163.06° (Co1-N1-S1), 162.21° (Co2-N2-S2), and 99.85° (Co2-S2-N2), while the M–N–S and M–S–N angles in **b** are 163.18° (Cd1-N1-S1), and 99.66° (Cd1-S1-N1). In addition, the double end-to-end thiocyanate bridge in **b** creates eight-membered $Cd_2(SCN)_2$ rings with a chair conformation, whereas in the **a**, the $Co_2(SCN)_2$ rings form a flat conformation which probably due to the M···M distance in **a** is shorter than in **b**. The lattice pza molecules in **a** which hydrogen bonded to one of the pza ligands may also affect this conformation.

In both **a** and **b**, several hydrogen-bond motifs are observed (Table 2). Each lattice pza molecule in **a** is hydrogen bonded to two adjacent pza ligand from the same Co-$(SCN)_2$-Co chain. The Co-$(SCN)_2$-Co chain is also hydrogen bonded to two neighbouring chains through the amine and carbonyl groups of the pza ligand resulting 2D hydrogen-bonded sheets with a (Co–pza)···(pza–Co) distance of 14.734 Å. The crystal packing of **a** then grows into a 3D hydrogen-bonded π···π stacked networks (Figure 4) due to π···π interactions between two closest pyrazine rings with a plane centroid-to-plane centroid of 3.582 Å. Unlike in **a** where lattice pza is observed, only hydrogen bonds between the pza ligands occurred in **b**. Each pza ligand is hydrogen-bonded to two neighbouring pza ligand in intramolecular and intermolecular fashion. Each chain in **b** is also hydrogen bonded to two adjacent chains, thus 2D hydrogen-bonded sheets are formed with a (Cd–pza)···(pza–Cd) distance of 11.984 Å. Although there are no π···π interactions between two closest pyrazine rings observed, crystal packing of **b** also forms non-covalent networks due to hydrogen bonding formed between the sheets (Figure 4).

Table 2. Hydrogen bond data of **a** and **b**

Complex	D	H	A	D-H (Å)	H···A (Å)	D···A (Å)	DHA (°)
a	N3	H3A	O3	0.86	2.03	2.887(2)	172.5
	N9	H9A	O1	0.86	2.03	2.887(2)	173.2
	N6	H6A	O2#g	0.86	2.05	2.905(2)	178.3
	N6	H6A	N10#e	0.86	2.36	3.044(2)	136.8
	N9	H9B	N7#e	0.86	2.54	3.274(2)	143.6
b	N4	H4A	O1#h	0.86	2.01	2.8623(18)	168.4
	N4	H4B	N2#i	0.86	2.48	3.177(2)	138.9

Note: #e = 1-x, 1-y, -1-z; #g = -1-x, 1-y, -z; #h = -x, 1-y, -z; #i = 1-x, 2-y, -z

Table 3. Crystallographic data of {[Co(pza)$_2$(μ-SCN)$_2$]·pza}$_n$ (**a**) and [Cd(pza)$_2$(μ-SCN)$_2$]$_n$ (**b**)

Complexes	a	b
Chemical formula	$C_{17}H_{15}CoN_{11}O_3S_2$	$C_{12}H_{10}CdN_8O_2S_2$
Formula Mass	544.45	474.80
Crystal system	Triclinic	Triclinic
Space group	*P-1*	*P-1*
a/Å	6.8711(4)	5.8871(6)
b/Å	11.0487(6)	6.3004(7)
c/Å	14.6891(8)	11.8296(13)
α/°	94.949(2)	96.219(4)
β/°	103.144(2)	102.848(4)
γ/°	93.867(2)	96.821(4)
Unit cell volume/Å^3	1077.52(10)	420.62(8)
Temperature/K	123(2)	123(2)
Z	2	1
μ (mm^{-1})	1.038	1.571
Reflections measured	15710	11887
Independent reflections	5122	2107
Observed reflections ($I > 2\sigma(I)$)	4365	2075
R_{int}	0.0319	0.0383
Final R_1 values (obs. data)	0.0274	0.0180
Final $wR(F^2)$ values (obs. data)	0.0706	0.0411
Final R_1 values (all data)	0.0353	0.0184
Final $wR(F^2)$ values (all data)	0.0750	0.0412
Goodness of fit on F^2	1.066	1.084

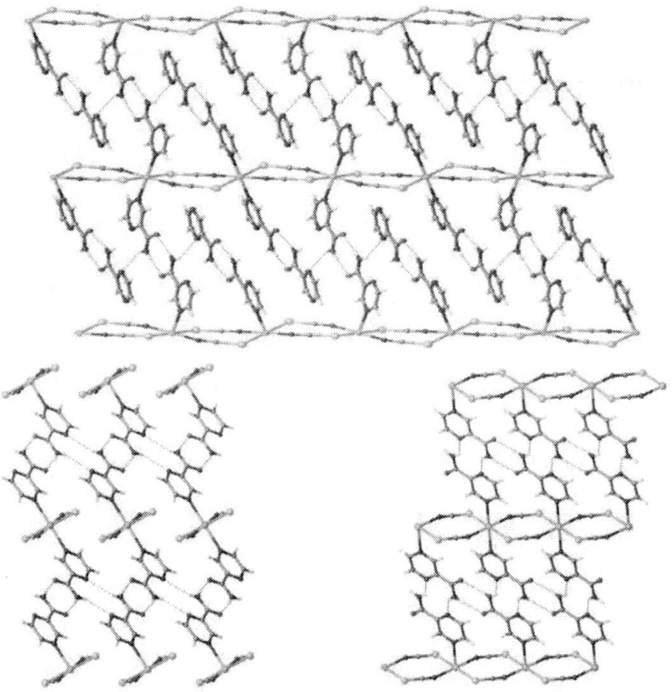

Figure 4. Crystal packing of **a** and **b** displays 3D non-covalent networks, viewed along *a* axis in **a** (top) and along *a* (bottom left) and *b* axis in **b** (bottom right).

3.2. *ATR-IR and MP analyses*

In general, the ATR-IR spectra of both complexes shows characteristic $N_{pyrazine}$-coordinated pyrazinamide and bridging thiocyanate absorptions as expected which consistent to the X-ray structure analysis (Figure 5). Due to non-coordinated (free) pza molecules in the lattice, there are more bands observed in the complex **a** than in complex **b**.

The asymmetric and symmetric stretching mode of NH_2 in **a** appeared at 3457 cm^{-1} (free pza), 3408 cm^{-1} (coordinated pza), and 3287 cm^{-1} (both free and coordinated pza), respectively; while in **b**, they are appeared at 3416 and 3285 cm^{-1}, respectively. These bands are also supported by a sharp peak at 1606 cm^{-1} (free pza in **a**), 1590 cm^{-1} (coordinated pza in **a**), and 1591 cm^{-1} (in **b**), which correspond to scissoring of the NH_2 group and medium peak at 1590 cm^{-1} (both in **a** and **b**), which belong to the bending of the NH_2 group. A very sharp peak at 1706 cm^{-1} (free pza in **a**), 1686 cm^{-1} (coordinated pza in **a**), and 1686 cm^{-1} (in **b**), which belongs to a C=O stretch, suggest that the C=O of pza ligand is not coordinated to the metal ions. Coordinated C=O is commonly observed in a lower wavenumber (around 1650-1500 cm^{-1}). This also means that the pza ligands in both **a** and **b** are coordinated to the metal ion in a monodentate fashion through the pyrazine nitrogen atom, as already confirmed by the X-ray structure analysis. The presence of pza in both complexes are also supported by medium peaks around 3100-3000 cm^{-1} which correspond to the C–H stretch.

In general, the bridging thiocyanate group (M–NCS–M) has the highest ν(CN) wavenumber (2165–2065 cm^{-1}) compare to the ν(CN) wavenumber of terminal thiocyanate groups (M–NCS = 2010–2050 cm^{-1}; M–SCN = 2130–2085 cm^{-1}), while the ν(CS) wavenumber of M–NCS (870–820 cm^{-1}, weak) is higher than that of M–NCS–M (800 – 750 cm^{-1}) and M–SCN (760 – 700 cm^{-1}, broad), respectively, although other factors such as mass, charge, and size of the central atom influences the frequencies of thiocyanate ligand [21]. In these works, the ν(CN) band is observed at 2118 cm^{-1} in **a** and 2115 cm^{-1} in

b, while the *ν*(CS) band appears at 783 cm^{-1} in **a** and 808 cm^{-1} in **b**. The slight difference of *ν*(CN) and *ν*(CS) frequencies in the complex **a** and complex **b** is due to the thiocyanate bridge is coordinated to two different metal centre, which is cobalt(II) in **a** and cadmium(II) in **b**. This observed *ν*(CN) and *ν*(CS) bands in the complex **a** and complex **b** are in line with the bridging thiocyanate ligand.

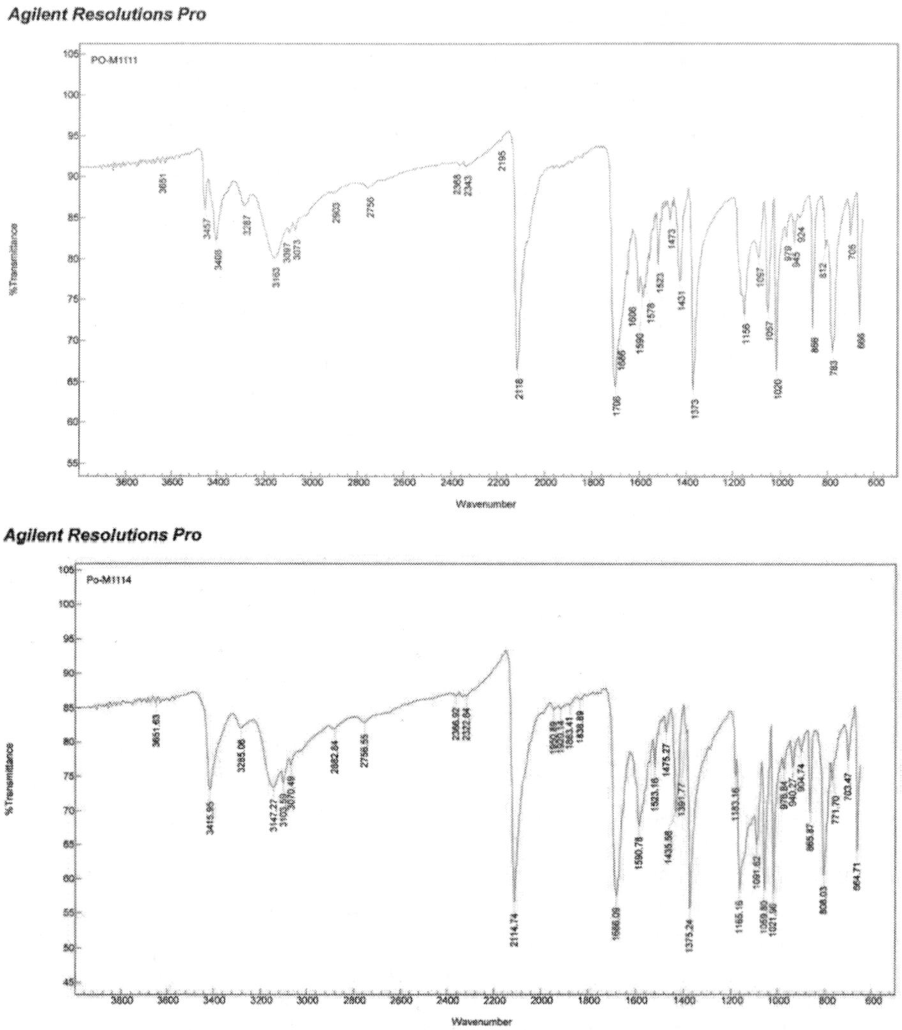

Figure 5. ATR-IR spectra of complex **a** (top) and **b** (bottom)

The melting point of the complex was also determined for further characterisation. The observed melting points of both complexes are relatively close to each other (180°C for **a** and 185°C for **b**). Upon heating, complex **a** colour is gradually changed from red to pink to orange, and finally at 180°C, complex **a** was changed from dark orange crystal to dark blue liquid, indicating the phase change of the complex at high temperature. Although no further chemical analysis was conducted on the dark blue liquid, this indicates the formation of tetrahedral cobalt(II) complex as commonly observed in cobalt(II) complexes. In complex **b**, there was no phase change observed throughout the heating as cadmium(II) complexes are usually more stable than cobalt(II) complexes. The crystal remains colourless upon heating and then turn into a light yellowish liquid at 185°C.

4. Conclusion

Two 1D coordination polymers of $\{[Co(\mu\text{-}NCS)_2(pza)_2]\cdot pza\}_n$ and $[Cd(\mu\text{-}NCS)_2(pza)_2]_n$ were made and structurally characterised. In both complexes, double end-to-end thiocyanate ions bridges between two adjacent metal centre and the pza ligand only coordinate to the octahedral metal ion in a monodentate fashion via pyrazine nitrogen atom forming a metal centre coordination core of MN_4S_2. Crystal packing of both **a** and **b** are also displaying non-covalent networks as a result of hydrogen-bonding involving the pyrazine ring, amide and carbonyl groups between adjacent chains and $\pi\cdots\pi$ interactions (only occurred in **a**). In addition, the observed melting points of both complexes are relatively close to each other (around 180°C) and ATR-IR spectra support the presence of the bridging thiocyanate and terminal pyrazinamide.

Acknowledgments
Author thanks Chemistry Department of Brawijaya University, Indonesia for financial support and acknowledges Prof. Stuart Batten and his group member (Monash University, Australia), Assoc. Prof. Jack Clegg (The University of Queensland, Australia), and Australia Award Scholarship–DFAT Australia.

References

[1]. Murray J F, Schraufnagel D E and Hopewell P C 2015 *Annals. Am. Thor. Soc.* **12** (12) 1749
[2]. Zumla A, Nahid P and Cole S T 2013 *Nat. Rev. Drug Disc.* **12** 388.
[3]. Brandi-Blanco M P, Gonzalez-Perez J M, Choquesillo-Lazarte D, Carballo R, Castineiras A and Niclos-Gutierrez J 2003 *Inorg. Chem. Commun.* **6** 270.
[4]. Dakovic M, Popovic Z, Giester G and Rajic-Linaric M 2008 *Polyhedron* **27** 210.
[5]. Shirvan S A and Dezfuli S H 2012 *Acta Crystallo.g* **E68** m1082.
[6]. Chiniforoshan H, Radani Z S, Tabrizi L, Tavakol H, Sabzalian M R, Mohammadnezhad G, Görls H and Plass W 2015, *J. Mol. Struct.* **1081** 237.
[7]. Tabrizi L, Chiniforoshan H and McArdle P 2015, *J. Coord. Chem.* **68** (11) 1936.
[8]. Eckberg R P and Hatfield W E 1975 *J. Chem. Soc. Dalton Trans.* 616.
[9]. Prananto Y P, Urbatsch A, Moubaraki B, Murray K S, Turner K S, Deacon G B and Batten S R 2017 *Aust. J. Chem.* **70** (5) 516.
[10]. Boeckmann J and Nather C 2012 *Polyhedron* **31** 587.
[11]. Porai-Koshits M A and Tishchenko G N 1960 *Kristallog.* **4** 239.
[12]. Wohlert S, Boeckmann J, Jess I and Nather C 2012 *Cryst. Eng. Commun.* **14** 5412.
[13]. Koziskova Z, Kozisek J and Kabesova M 1990 *Polyhedron* **9** 1029.
[14]. Taniguchi M, Shimoi M and Ouchi A 1986 *Bull. Chem. Soc. Japan* **59** 2299.
[15]. Machura B, Switlicka A, Mrozinski J, Kalinska B and Kruszynski R 2013 *Polyhedron* **52** 1276.
[16]. Saber M R, Abu-Youssef M A M, Goher M A S, Sabra B A, Hafez A K, Badr A M A and Mautner F A 2012 *J. Mol. Struct.* **1008** 17.
[17]. Bruker AXS Ltd 2005 *APEX2* Madison Wisconsin USA.
[18]. Sheldrick G M 1996 *SADABS* University of Göettingen Germany.
[19]. (a) Sheldrick G M 2008 *Acta Crystallog.* **A64** 112; (b) Sheldrick G M 1997 *SHELXS97 Program for the Solution of Crystal Structures* University of Göttingen Germany; (c) Sheldrick G M 1997 *SHELXL97 Program for the Refinement of Crystal Structures* University of Göttingen Germany.
[20]. Dolomanov O V, Bourhis L J, Gildea R J, Howard, J A K and Puschmann H 2009 *J. App. Crystallog.* **42** 339.
[21]. Norbury A H 1975 *Adv. Inorg. Chem. Radiochem.* **17** 231.

Synthesis of Composit From Bamboo Fiber, Zeolite and Epoxy for Room Separation

Dhany Raihan Muhammad,[1] Kris Tri Basuki,[1] Bangun Wasito,[1] Suroso[2]
[1]Nuclear Technochemistry Study Program, Polytechnic of Nuclear Technology – National Nuclear Energy Agency, Jalan Babarsari KP6101 YKBB Yogyakarta 55281
[2]Technophysics Study Program, Polytechnic of Nuclear Technology – National Nuclear Energy Agency, Jalan Babarsari KP6101 YKBB Yogyakarta 55281

Email: dhany_raihan@yahoo.com.sg

Abstract This research aims is to search a subtitute of the asbestos for the separator rontgen room using bamboo fiber filled with zeolite; which harden using epoxy it is all caused because the hazard of the asbestos to the human body. Bamboo stem degenerated using NaOH (20%) to get the bamboo fiber. Bamboo fiber added with CS2 (10 mL) to form xanthate cellulose. Xanthate Cellulose mixed with filler zeolite and harden of epoxy, layer by layer until getting the right width. The variant of the mass composition is 3: 0:1; 3: 0.25:0.75; 3 :0.5:0.5; 3: 0.75:0.25; 3: 1:0, and the variant of the temperature 28 °C; 40 °C; 60 °C; 80 °C; and 100 °C. The sample tested using microscopic method, impact test with Charpy method, corrosivity method, Electricity conduct method, thermal conduct method, and radiation resistance or attenuation method. The result shown the optimum composition of the composite it is at the variant 3 :0.5:0.5, with the optimum temperature is 40°C with the density of the sample is 1.5789 g/cm^3. Impact resistance of the sample is 44 Joule. The Radiation resistance is 0.46, with the thermal conductivity of the sample is 0.016 Kkal/ m.s.c. it shown that the sample is isolator. From the result is shown that the sample is can be a substitute for asbestos as material of the separator in the Rontgen room. Key word: Composite, Epoxy, Bamboo, Zeolite, Xanthate cellulose

1. INTRODUCTION

Current technological developments are beginning to lead to environmental problems in order to reduce the negative impacts of synthesis materials. Such technologies include the manufacture of composites that utilize natural fibers. The development of natural materials can slowly replace other materials to meet human needs. Material from nature is expected to be a solution because it is easy to obtain, cheaper, environmentally friendly and can be degraded by nature.

Generally, iron and asbestos are used as the basic material of space divider, but there are many studies that indicate a weakness in their use. Iron weakness is heavy, expensive, and quite difficult to degrade by nature while asbestos produces harmful asbestos dust.

Bamboo can be used as wall barrier material, has strong rod, flexible, straight, and light so easy to be processed for various products. The addition of zeolite filler and epoxy hardener is expected to produce a strong and light composite. Bamboo is a low cost biomass (poor man's timber). However, as most of bamboo is produced by smallholder farmers, collection costs and logistical costs may be high [1]. The researchers used natural fiber composite as a superior product in accordance with its specialties. Although not fully shifting synthetic fibers, the use of environmentally friendly natural fibers is a wise move to save the environment [2].

Application of bamboo fiber composite as a radiation wall boundary material needs to be tested radiation attenuation value. Attenuation of radiation is the ability of the material to withstand the intensity of radiation

2. EXPERIMENT

2.1 Chemicals and instrumentation

Chemicals used for this research it is, Epoxy, NaOH, HCl, CS2 and Bamboo cellulose. Instrumentation applied for analysis it this research it is Impact test, Microscope, Multimeter, Thermometer, Geiger Counter

2.2 Synthesis of Bamboo fiber

Making bamboo fiber is done by cutting bamboo fiber and softening the bamboo fiber using NaOH, after that Grind the bamboo fiber and for the last it is to form the fiber cellulose xanthate using CS2.

2.3 Synthesis of zeolite powder

Making zeolite powder is by pounding zeolite with mortar until it becomes smooth. To add the zeolite surface area value so it is expected to be more easily grooved and blend with the composite later, with the hope of increasingly zeolite surface area, the smaller the zeolite volume so that it can more easily be absorbed and bound by bamboo fiber and epoxy hardener.

2.4 Synthesis of the composite

Preparation of composites in this study is to combine bamboo fiber, with epoxy mixed zeolite filler with fiber ratio: zeolite: epoxy which will be varied with composite making method of Hand layup. The process of manufacture by this method is by pouring resin by hand into woven-shaped fibers, cloth, then give pressure as well as leveling it using rollers or brushes. The process is repeated until the desired thickness is reached.

2.5 Destructive test

Destructive test is a test method of the material is tested until the material becomes damaged and will be compared with the standard material to be replaced later. Here are all kinds of broken tests performed. Impact test was testing the load to the sample, to know the form of fault and the ability to withstand momentum. Corrosion test was test immersing samples in HCl 5N per time to know the rate of material corrosity. Microscope test was the test looks at the sample surface in a micro-way to see the shape of the sample arrangement.

2.6 Non-destructive test

Non-destructive test is a method of testing the material not being damaged and the state is maintained at its original state. Here are the various nondestructive tests performed. Test of conductor resistance was test by giving current to see the value of sample resistance to electric current. Thermal conductivity test was test through the provision of heat flow to determine the value of the ability of materials in conducting heat. Test of radiation resistance was test to determine the ability of the material to withstand radiation through the giving of an intensity from a previously known source.

3. RESULT AND DISCUSSION

3.1 Radiation resistance test

If known, the initial intensity in the sample is obtained on the test data that has been done to determine the value of the blank and the initial intensity, performed using gs detector with the Cs-137 source for 60 seconds. Table 1 is the data table of the blank test results with the initial intensity.

Table 1. Starter Intensity and Blanko

Data	Radiation resistance (CPM)
Io	2809.33 ± 134.82
Blanko	46.00 ± 15.10

The result of research for variation of composition, got sample test data calculated by using data of initial intensity and blank using formula as follows.

$$I = Io \times e^{-\mu t} \quad (1)$$

With (I) is intensity after being resist by the sample (CPS), (Io) is background Intensity (CPS), (μ) is attenuation, and (t) is thickness (m).

3.2 Attenuation test for Variety of mass composition
Based on the formula is found Table 2 is the data attenuation value of variations in mass composition.

Table 2 . Variety of mass composition Counting data

Sample (bamboo(g))	CPM	Atenuation
Bamboo 0,25	1841,83	0,43
Bamboo 0,5	1797,17	0,46
Bamboo 0,75	1749,75	0,48
Bamboo 1	1712,33	0,51
Cement	1760,67	0,48

Based on Table 2, the attenuation value of bamboo compound 0.75 g and the bamboo composition of 1 g has been obtained with good attenuation that can equalize and exceed the cement. This suggests a connection between the presence of cellulose that absorbs gamma rays.

3.3 Attenuation test for Variety of Temperature
Based on the formula is found Table 3 is the data attenuation value of variations in temperature composition.

Table 3. Variety of temperature composition Counting data

Sample (Temperature °C)	Count (CPM)	Attenuation
Temperature 28°C	1797.17	0.46
Temperature 40 °C	1788.42	0.46
Temperature 60 °C	1805.17	0.45
Temperature 80 °C	1712.50	0.51
Temperature 100°C	1832.17	0.44
Cement	1760.67	0.48

Based on Table 3, the attenuation value of temperature 80°C obtained attenuation over cement. This is because of the relation between temperature gain and material density, so it is found that the best temperature to be used in composite manufacture is at 80°C.

3.4 Corrosivity Test
After the test results have been done that is by soaking the sample with the desitas average of the test material is 1.5789 into HCL 5N within a certain time then obtained data results shown by the graph in Figure 1.

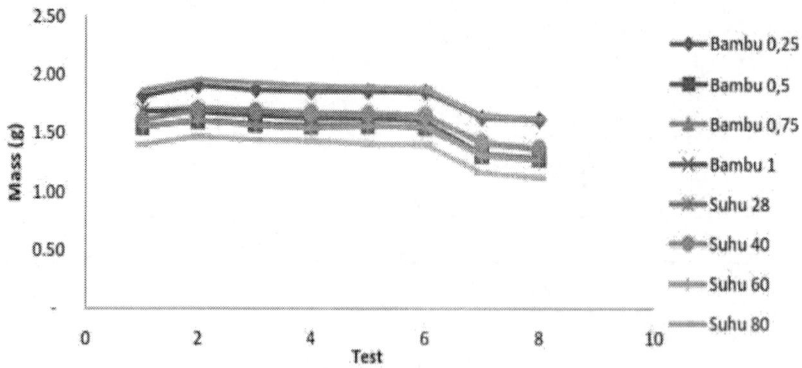

Figure 1. Total Data Korositas

IOP Conf. Series: Materials Science and Engineering **299** (2018) 012033 doi:10.1088/1757-899X/299/1/012033

In Figure 2, the color of the line indicates the type of sample, the sample used in the test is the Bamboo sample with the temperature composition variation in °C. The X axis in Figure 12 shows the test number in unit time. The Y axis in Figure 12 shows the mass value of the sample in grams.

Based on the graph in Figure 1 can be seen that the phenomenon that occurs in all samples is the same outline, so to further simplify can be taken one data that will represent the discussion of all the data sample of other test materials, for that will be taken one graph data on the variation of bamboo 0.25 g to represent the results of the test data.

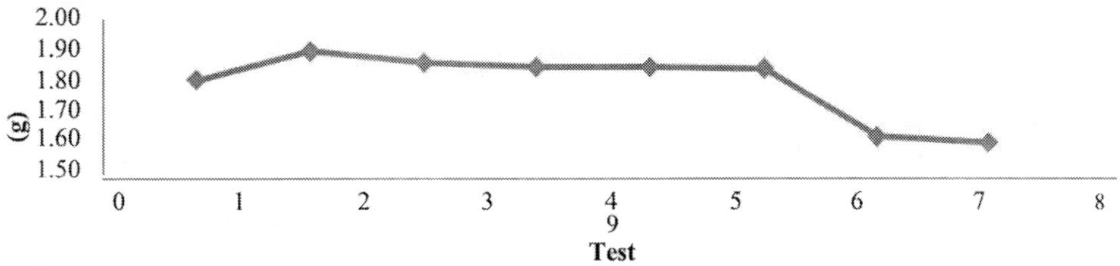

Figure 2. Corrosivity Variation of Bamboo composite 0,25

From Figure 2 it can be seen that the increase of mass that occurs at the time after the sample of the test material soaked is not in accordance with the existing theoretical basis. This is due to the presence of pores either macro or micro in the sample, these pores are significantly absorbed, so that the mass is the mass of the reduced sample plus the mass of the absorbed HCl into the sample, resulting in an very significant mass increase. Based on these results can be concluded that the calculation of the value of corosity rate is cannot be done.

3.5 Impact test (charpy method)

From the results of tests that have been done then obtained the test results of test material samples using Impact Charpy method.

3.6 Impact test for Variation of Composition

The result of composition variation test using Impact Charpy method is shown in Figure 3 as follows. In the impact test, aluminum is used as a comparison on the results of data. Seen in the bamboo variation test sample 0.25 g showed an increase in the ability to withstand the impact of the aluminum value. This proves that with the addition of zeolite filler can add value of mechanical ability of more and more zeolite material tied to bamboo as filler material that can reduce the effect of momentum on the sample of test material.

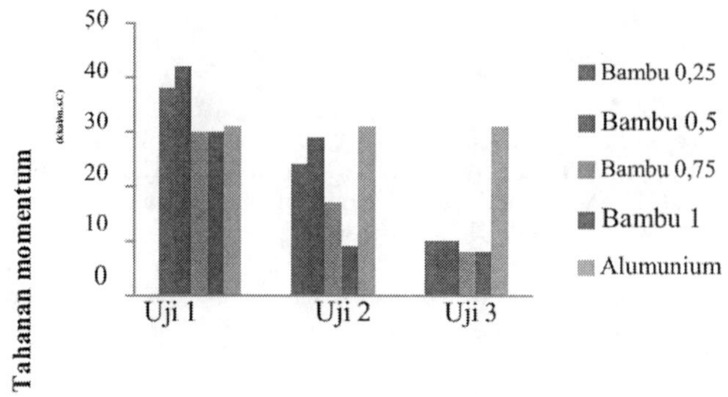

Figure 3. Impact test Composition Variation Result

3.7 Impact test for Variation of Temperature

The result of temperature variation test using Impact Charpy method is shown in Figure 4.

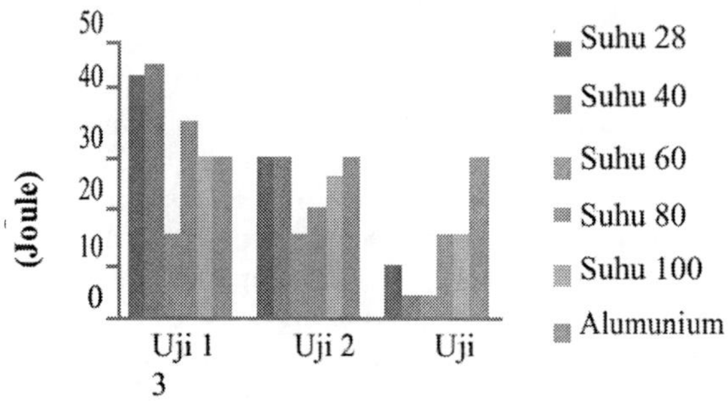

Figure 4. Result Impact test for Variation of Temperature

Based on the data in Figure 4 It can be seen that there is a capability retention in the impact resistance of the test material sample with the increasing temperature of the material making, this is due to the influence of the speed of composite formation with the higher the temperature used the faster the composite dries but on the other side will form many small particles so that it will be more easily broken than by using its optimum temperature at 40°C.

3.8 Electrical Conductivity Test

The electrical conductivity test found that the test result data using amperemeter is shown in Table 4.

Table 4. Electric conductivity test

Sample (g)	Arus (A)	Sampel (^{O}C)	Arus (A)
Bamboo 0,25	0	Temperature 28	0
Bamboo 0,5	0	Temperature 40	0
Bamboo 0,75	0	Temperature 60	0
Bamboo 1	0	Temperature 80	0
		Temperature 100	0

Based on Table 4 shows the manufacture composite is a type of insulator material which is an electrical resistor, so no current is read by ampere meter. Therefore, this composite is well used to substitute materials that usually interact directly as a barrier between humans with electrical materials.

3.9 Microscopic Test

Based on the results of microscopic observations obtained results as follows.

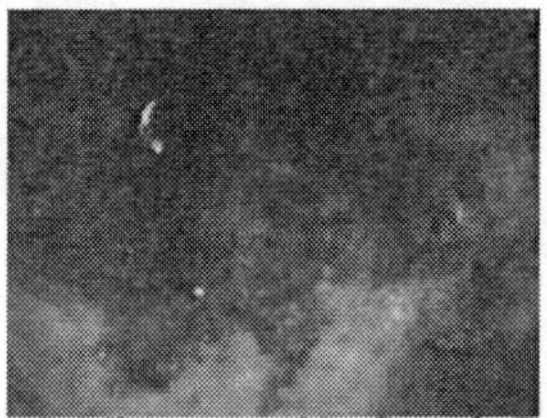

Figure 5. Microscope 20x Sample Bamboo 0,25 g

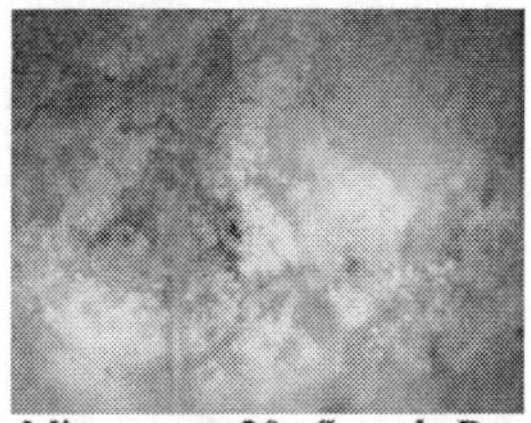

Figure 6. Microscope 20x Sample Bamboo 0,5 g

In Figure 5 it can be seen that the micro form of the sample there are silica granules that break down and form groups individually. In figure 5 it can be proven that the bonds do not occur micros but only occur physically and there are porosity values that affect the properties of the material. It is proven by the separation between each shiny material that means it did not create the chemical bond, only the physical material that wrap each other

Figure 6 shows the presence of opaque cellulose. This is because the three-dimensional cellulose form so difficult to get the focus of the image on cellulose. The image proves the presence of cellulose bonding which helps add properties to the material in the sample.

3.10 Thermal Conductivity Test

Sought value Thermal conductivity of the material by using Alumunium comparator to know the value of Q so that it can search the thermal conductivity value of materials using the formula.

$$Q = K A (\Delta T/\Delta x) \qquad (10)$$

Where Q = heat flow rate (kkal/m s c); K = thermal conductivity; A = cross section (m^3); ΔT = temperature changes (□C); and ΔT = length (m); K for aluminium is 0.05 kcal/m s c.

3.11 Thermal Conductivity Test composition variety

Here is the data from the test results on the sample of the test material with the variation of the composition. The results of the test data are shown in Table 5.

Table 5. Change of tempertature on composition variety sample **s**ample

Bamboo (g)	K.heat flow rate (kkal/m.s.C)
0,25	0,018 ± 0,005
0,5	0,018 ± 0,003
0,75	0,016 ± 0,002
1	0,019 ± 0,004

From Table 5, it can be seen the optimum point of thermal conductivity value for composition variation is bamboo 1 composition without zeolite, with its thermal conductivity point is 0,019 kcal/ms.

3.12 Thermal conductivity test temperature variety

Here is the data from the test results on the sample of the test material with the variation of the composition. The results of the test data are shown in Table 6.

Table 6. Change of tempertature on temperature variety sample Sample

Temperature (^{o}C)	K. heat flow rate (kkal/m.s.C)
28	0,018 ± 0,003
40	0,016 ± 0,002
60	0,016 ± 0,002
80	0,016 ± 0,003
100	0,016 ± 0,002

From Table 6 data it can be seen that the optimum point of thermal conductivity value for temperature variation is at temperature 28°C with its thermal conductivity point is 0.018 kcal / m.s. Based on the calculation that has been done the sample of test material will be compared with conductivity value at cement which is the main material which is expected to be replaced by composite.

Table 7. Result of equation for thermal conductivity

Sample		k.Thermal (kkal/m.s.C)
Bamboo (g)	Bamboo 0,25	0,018± 0,005
	Bamboo 0,5	0,018 ± 0,003
	Bamboo 0,75	0,016± 0,002
	Bamboo 1	0,019± 0,004
Temperature (°C)	Temperature 28	0,018 ±0,003
	Temperature 40	0,016 ± 0,002
	Temperature 60	0,016 ±0,002
	Temperature 80	0,016 ± 0,003
	Temperature 100	0,016 ± 0,002
Data	Zeolit	0,016 ± 0,001
	Paint	0,015 ± 0,002
	Cement	0,043
	Alumunium	0,050

Based on the results of the data in Table 7 it can be seen that the sample of the test material in the form of bamboo fiber composite filler zeolite epoxy hardener has a thermal conductivity value that is still far below the cement and aluminum so it can be concluded that the test material sample is a good heat retardant material and resembles asbestos resistance. So with the similarity of composite materials with the properties of asbestos, composite materials are considered to replace asbestos materials due to its safety in fire retention due to the high value of heat resistance of materials that resemble asbestos.

From the result all of the data it will be gain the characteristhic of the material it is stronger than asbestos, the attenuation number is greater than asbestos, and included as isolator material. But on the other side the material it is only occur with physically bond.

4.CONCLUSION

Based on the results of research that has been done can be concluded that. The more the composition of bamboo in the composite will decrease the value of ability to impact resistance or momentum, and more zeolite will add value of mechanical ability of composite material. Increasing the amount of fiber in the composite composition has the effect of an increase in attenuation value. Based on these results, the optimum point in this experiment is the comparison of epoxy: bamboo: zeolite at point 3: 0.5: 0.5 with the value of mechanical resistance over aluminum, heat resistance approaching asbestos, and having attenuation resembling cement plaster.

Temperature variations affect the mechanical strength, the higher the temperature, the smaller the composite ability to withstand impact with rising temperature values will decrease the strength of the material holding impact. The optimum temperature in the manufacture of bamboo fiber composite, zeolite and epoxy is at 40°C. The thermal conductivity generated by the sample of the test material shows that the composite sample has a good enough resistance value near the asbestos, meaning that the test sample includes an insulator material which is not good enough to conduct heat.

The optimum variation in the manufacture of bamboo fiber composite, zeolite, and epoxy is by comparison of epoxy: bamboo: zeolite is at the point of comparison of 3: 0,5: 0,5 and at temperature 40°C. Based on the advantages of this composite material can be concluded this composite material can be used as a substitute for asbestos based on the properties obtained have many advantages compared to asbestos materials that will be replaced. The material properties at the optimum point it is the strength of the impact resistance is 44 Joule, with the value of radiation attenuation is 0.46, and the thermal conductivity is 0.016 Kcal / m.s.c which mean the material it is included into the insulator material and the density of the material is 1.5789 g/cm^3.

ACKNOWLEDGMENT

Acknowledgment will be granted to prof. dr. Kris Tri basuki, Ir. Bangun wasito, and Suroso M.Sc as his excellency and guidance to make me able to finish this project.

IOP Conf. Series: Materials Science and Engineering **299** (2018) 012033 doi:10.1088/1757-899X/299/1/012033

REFERENCES

[1] Poppens R,Dam J V, Elbersen W 2013 Analyzing the potential of bamboo feedstock for the biobased econom Netherland.

[2] Jamasri, Opportunities and Challenges of Development of Natural Fiber Composites in, Proceedings of National Seminar on Machinery and Industry (SNMI4) 2008 Tarumanagara University Jakarta, pp. 1-13 Indonesia

[3] Filgueiras, T. S.; Gonçalves, A. P. S. A 2009 The Journal of the American Bamboo Society **18** pp 7-18

[4] Widya F, Hermiati E 2006 *Fiber morphology analysis and physical chemical properties on six bamboo types as pulp and paper raw materials* Bogor Indonesia

[5] Bowman, B.S 2003 *Microporous Mesoporous Materials* **6** pp 43-56

[6] Guan, H., Bestland, E., Zhu, C., Zhu, H., Albertsdottir, D., Hutson, J., Simmons, C.T., Ginic - Markovic, M., Tao, X. and Ellis, A.V 2010 *Journal of Hazardous Materials* **183** pp 616-621

[7] Idris A.A, Firmanti A, Purwito 1994 *Prosiding Seminar Strategi penelitian Bambu di Indonesia* Serpong: PUSPITEK.

[8] Meijer R; Nazarpour K 2015 *Prosthetic* Falkrik Great Britain

[9] Mehta, Bhavin, V; Setlock, Robert, J. ;Sindelar,Betty; Bantle, John 2006 *Prosthetic Bone Implant*, Ohio, United State

[10] Oishi M. 2012 *Bone Prosthetic Material*, Tokyo, Japan.

[11] Wahyudi T 2015 Extraction of Fiber from bambu (Gigantochloa Apus) for raw material of creative industry

Low Cost $CaTiO_3$ Perovskite Synthesized from Scallop (*Anadara granosa*) Shell as Antibacterial Ceramic Material

Is Fatimah*, Rico Nur Ilahi, Rismayanti Pratami

[1]*Program Studi Kimia, FMIPA, Universitas Islam Indonesia, Kampus Terpadu UII, Jl. Kaliurang Km 14, Sleman, Yogyakarta, Indonesia 55581,*

email: isfatimah@uii.ac.id

Abstract. Research on perovskite $CaTiO_3$ synthesis from scallop (*Anadara granosa*) shell and its test as material for antibacterial ceramic application have been conducted. The synthesis was performed by calcium extraction from the scallop shell followed by solid-solid reaction of obtained calcium with TiO_2. Physicochemical character of the perovskite wasstudied by measurement of crystallinity using x-ray diffraction (XRD), diffuse-reflectance UV Visible spectrophotometry, scanning electrone microscope-energy dispersive x-ray (SEM-EDX) and Fourier-Transform InfraRed. Considering the future application of the perovskite as antibacterial agent, laboratory test of the peroskite as material in antibacterial ceramic preparation was also conducted. Result of research indicated that perovskite formation was obtained and the material demonstrated photocatalytic activity as identified by band gap energy (Eg) value. The significant activity was also reflected by the antibacterial action of formed ceramic.

1. Introduction

Antibacterial materials paid intensive attention recently due to the world's need on hygiene and sanitation. Due to this trend, investigation on new and low cost and economist antibacterial materials is widely studied. Titanium dioxide-based materials get more attention in this purpose since it is known as photoactive and having high band gap energy responsible for bacteria disinfection. As composite of TiO_2, $CaTiO_3$ perovskite material is a kind of material have been attracting significant attention due to their rich physical and chemical properties such as stability especially for ceramic application[1,2]. Perovskite ABO_3 is a new inorganic nonmetallic material which has unique physical and chemical properties; A is usually rare earth or alkaline earth elements ion; B is the transition element ions; A and B can be partially replaced by other radius similar metal ions to maintain its crystal structure essentially the same, so in theory, it is the ideal sample to study the surface of the catalyst and catalytic properties. Perovskite has stable crystal structure, unique electromagnetic properties and high oxidation-reduction, hydrogenolysis, isomerization, electrocatalysis and other activity. With the interstitially titanium having semiconductor properties, perovskite is also widely used for photocatalytic applications. Within this scheme, the properties of material can be implemented for photocatalytic disinfection of some bacteria[3,4].

Basic principle of the synthesis of $CaTiO_3$ perovskite material is the solid-solid reaction of CaO or $Ca(OH)_2$ with TiO_2 at a certain molar composition[5]. Calcium source itself can be derived by low cost and/or animals wastes such as molusca shells. Beside of the advantageous as the low cost calcium source, the utilization of molusca shells is also an alternative of waste utilization. By high potency of biogenic calcium from some snail animals, this research aimed to synthesize $CaTiO_3$ perovskite material from scallop (*Anadara granosa*) shell and its laboratory-scale testing as antibacterial agent in the production of ceramic material. Refer to some investigations that the character of material is driven by the composition of Ca and Ti in the structure, investigation on the effect of Ca: Ti ration on the physicochemical character of prepared $CaTiO_3$ perovskite material.

2. Materials And Methods

2.1. Materials

Materials used in this research were Scallop shell obtained from traditional market waste in Sleman District, Yogyakarta, Indonesia, kaolinite obtained from Sukabumi, West Java, TiO_2 purchased from Merck, Germany. Method: Scallop shell was calcined at 1000°C for 2 hrs in order to get CaO from the shell. The obtained powder was then mixed with TiO_2 at the Ca: Ti molar ratio of 1:20; 1:30 and 1:50 and grinded together before was calcined at 400°C for 4 hrs. The obtained sample was encoded as $CaTiO_3$ for further analysis using X-Ray diffraction, diffuse reflectance-UV Visible spectrophotometry and scanning electrone microscope-energy dispersive x-ray (SEM-EDX). Ni filtered-Cu Kα radiation was used for XRD analysis using Philips Benchtop instrument and the step size of the analysis was 0.4°/min. JASCO V760 UV-DRS instrument was employed for determination of diffuse reflectance spectra and SEM-JEOL was utilized for surface profile and EDX analysis.

2.2. Ceramic formation

$CaTiO_3$ material was applied as antibacterial agent in ceramic formation. The preparation of ceramic material was conducted by mixing kaolinite clay, ball mill clay and gypsum together with $CaTiO_3$ in the certain weight ratio. The mixture was molded and the surface coating of ceramic was conducted by smearing $CaTiO_3$-kaolinite smeary mixture on the mold compounded by using water until the stable dough of the mixture was obtained. The stable dough was then calcined at 1200°C for 4 hrs. Steps of the ceramic preparation is presented in Figure 1.

(a) (b) (c)

Figure 1. (a) preparation of antibacterial agent using $CaTiO_3$ (b) surface coating of ceramic (c) stable molded composite

Antibacterial activity of $CaTiO_3$-coated ceramic was tested in water containing coliform. Varied treatment of the water was:

(a) No treatment
(b) Water sample + UV
(c) Water sample + Coated Ceramic + UV
(d) Water sample + Uncoated Ceramic + UV

The coliform disinfection was calculated by following equation:

$$\% \, desinfection = \frac{[initial\ coliform - coliform\ after\ treatment]}{[initial\ coliform]} x\ 100\% \ ...... (1)$$

Total coliform was analyzed by using most probable number (MPN) method.

3. Results and Discussion

Figure 2 shows the comparison on XRD pattern of derived CaO from scallop shell and preapared $CaTiO_3$. Derived calcined scallop shell exhibits the reflections at 2θ: 18.97 and 37.39° that are corresponding to the presence of CaO and the reflection at 17.87° and 34.18° as indication of $Ca(OH)_2$. The data representing the incomplete conversion of Ca-containing in the scallop shell during the calcination. Furthermore, the prepared $CaTiO_3$ express the formation of TiO_2 in composite form with Ca in the form of $CaTiO_3$ by some reflections as indication of anatase formation [6 ,7].

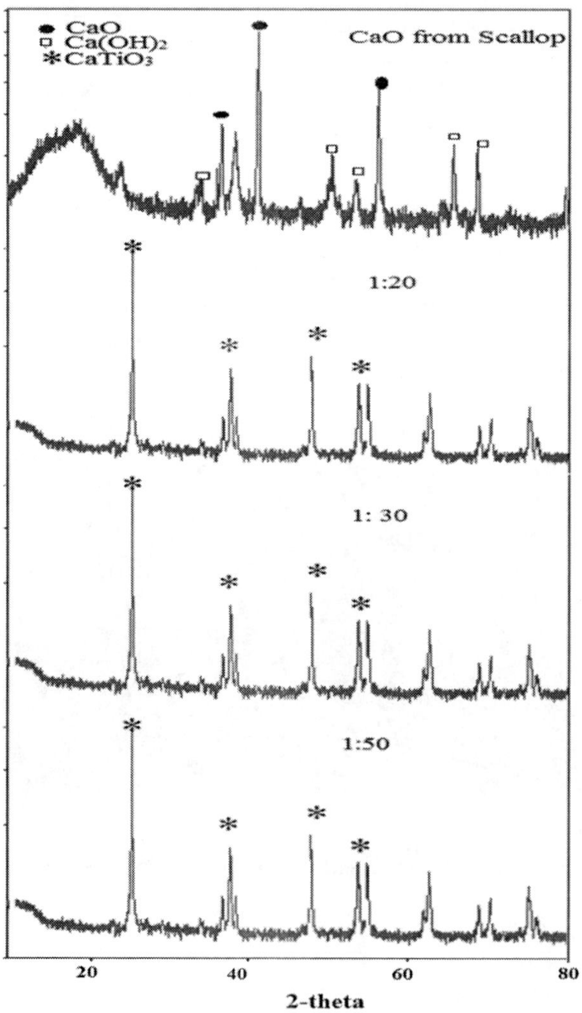

Figure 2. XRD pattern of CaO from Scallop and CaTiO3 from varied Ca:Ti ratio

From the SEM-Profile of materials (Figure 3), it can be seen that there is no significant difference of the surface over perovskite synthesis.

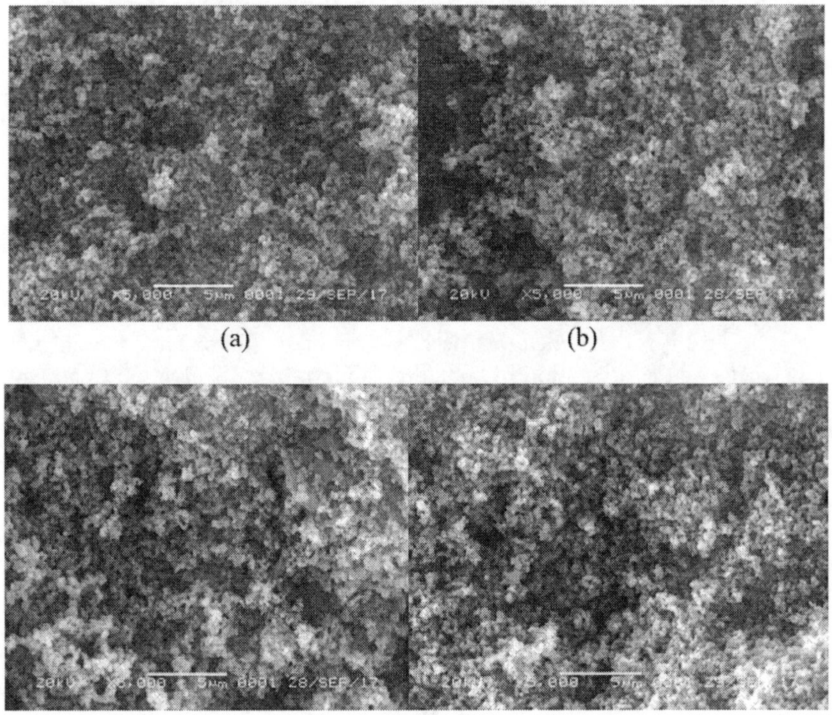

Figure 3. SEM profile of (a) CaO from Scallop (b-d) $CaTiO_3$ from varied Ca:Ti of 1:20, 1:30 and 1:50

The potency of prepared $CaTiO_3$ as photoactive material is hown by the DRS spectra (Figure 4). The calculation of band gap energy (Eg) from Kulbeka-Munk equation (f[R]) gives the value of 3.2 eV. This value is fit with the band gap energy range of TiO_2 material responsible for a semiconductor material[8].

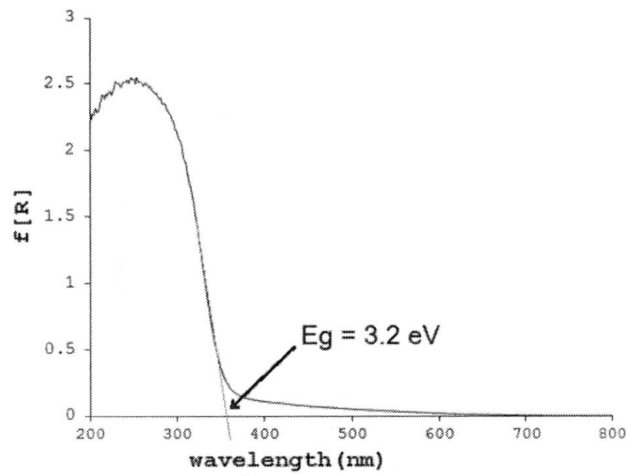

Figure 4. UV-DRS spectra of $CaTiO_3$

Antibacterial activity of the material is expressed by the percentage of total coliform disinfection from Table 1.

Table 1. Antibacterial activity test of coated ceramic

Treatment	**Total Coliform disinfection (%)**
Water sample	0± 0.0001
Water sample + UV	0± 0.0001
Water sample + Coated Ceramic (1:20) + UV	99± 0.04
Water sample + Coated Ceramic (1:30) + UV	99± 0.04
Water sample + Coated Ceramic (1:50) + UV	99± 0.04
Water sample + Uncoated Ceramic + UV	90± 0.04

From the data in Table 1 it is concluded that derived $CaTiO_3$ exhibited antibacterial activity as the total disinfection increases by the utilization of coated ceramic in the disinfection treatment while the UV-treatment gives no disinfection activity. From the varied Ca:Ti ratio, it is found that the ratio does not give effect to the disinfection activity means that similar activity found for Ca:Ti range of 20-50.

4. Conclusion

Based on the results, it can be concluded that $CaTiO_3$ was successfully prepared from scallop shell. The derived $CaTiO_3$ shows activity in coliform disinfection as utilized in the formation of antibacterial ceramic.

Acknowledgments
The authors gratefully acknowledge the financial support of Chemistry Department, Universitas Islam Indonesia.

References

[1] Heydari M, Fazaeli R, Yousefi M. Preparation of Perovskite Nanocomposites and Photochemical Degradation Kinetics of Acid Yellow 199 2012;15:7–15.
[2] Lertpanyapornchai B, Ngamcharussrivichai C. Mesostructured Sr and Ti mixed oxides as heterogeneous base catalysts for transesterification of palm kernel oil with methanol. Chemical Engineering Journal 2015;264:789–96. doi:10.1016/j.cej.2014.12.031.
[3] Dong W, Song B, Meng W, Zhao G, Han G. A simple solvothermal process to synthesize $CaTiO_3$ microspheres and its photocatalytic properties. Applied Surface Science 2015;349:272–8. doi:10.1016/j.apsusc.2015.05.006.
[4] Han C, Liu J, Yang W, Wu Q, Yang H, Xue X. Enhancement of photocatalytic activity of CaTiO3 through HNO3 acidification. Journal of Photochemistry and Photobiology A: Chemistry 2016;322–323:1–9. doi:10.1016/j.jphotochem.2016.02.012.
[5] Kaliaraj GS, Kirubaharan K, Pradhaban G, Kuppusami P, Vishwakarma V. Isolation and characterization of biogenic calcium carbonate/phosphate from oral bacteria and their adhesion studies on YSZ-coated titanium substrate for dental implant application. Bulletin of Materials Science 2016;39:385–9.
[6] Linggawati A, Anita S, Amri TA. Preparation and Characterization of Calcium Oxide Heterogeneous Catalyst Derived from AnadaraGranosaShell for Biodiesel Synthesis 2016;2016:1–8. doi:10.18502/keg.v1i1.494.
[7] Yang H, Han C, Xue X. Photocatalytic activity of Fe-doped CaTiO3 under UV-visible light. Journal of Environmental Sciences (China) 2014;26:1489–95. doi:10.1016/j.jes.2014.05.015.
[8] Gaikwad SS, Borhade A V., Gaikwad VB. A green chemistry approach for synthesis of CaTiO3 photocatalyst: Its effects on degradation of methylene blue, phytotoxicity and microbial study. Der Pharma Chemica 2012;4:184–93.

IOP Conf. Series: Materials Science and Engineering **299** (2018) 012035 doi:10.1088/1757-899X/299/1/012035

Synthesis of Hydroxyapatite using Precipitated Calcium Carbonate (PCC) from Limestones

Sri Wardhani*, Noor Isnaini Azkiya, Rachmat Triandi Tjahjanto

Department of Chemistry, Faculty of Mathematics and Natural Sciences, Brawijaya University, Jl. Veteran 01 Malang, Indonesia 65145

*E-mail: wardhani@ub.ac.id

Abstract. Hydroxyapatite (HAp) is a material that widely applied in bone and teeth implant due to its biocompatibility and bioactivity. This material can be prepared from PCC by precipitation method using CaO and H_3PO_4 in ethanol. In this work, variations of phosphoric acid amount and aging time were investigated. The synthesized HAp was characterized by FT-IR, AAS, UV-Vis Spectrophotometer, PSA, SEM, and powder XRD. The results showed that the high concentration of calcium in PCC gives better yields in which PCC obtained from carbonation method has higher yield than that of caustic soda method. The determination of optimum phosphoric acid addition based on targeted Ca/P ratio (1.67) from HAp was obtained on the addition of 0.1271 mol phosphoric acid with Ca/P ratio of 1.66. The aging time gave significant effect to the particle size of synthesised HAp. The smallest particle size was obtained in aging time for 48 hours as high as 49.25 μm. FTIR spectra of the synthesized HAp show the presence of hydroxyl (-OH) group at 3438.8 cm^{-1}, PO_4^{3-} at 557.39 and 1035.7 cm^{-1}, and CaO at 1413.72 cm^{-1}. The synthesized HAp forms agglomeration solid based on the SEM analysis. The powder XRD data shows three highest peaks at 2θ *i.e.* 27.8296; 31.1037; and 34.3578 which corresponds to β-TCP (tricalcium phosphate) in accordance with JCPDS no.09-0169. The characteristic 2θ peak of hydroxyapatite with low intensity is observed from the synthesized HAp refer to the JCPDS data no. 09-0432.

1. Introduction

Bone and teeth are the important role in the human's body. The injury of their tissues may cause disabled structure and deficiency of body's function. The efforts to repair the tissue's structure can be conducted by addition and substitution of tissue so called implant [1,2]. The materials widely applied for bone and dental implant such as metal, ceramic, polymer, composite, and carbon [3].

Ceramic is one of material that capable of implant due to its high compatibility and inert. There are two type of ceramic, they are bioactive and non-reactive ceramic. The bioactive ceramic often utilized for implant is hydroxyapatite due to its high suitability in the body and bioactive [3]. Hydroxyapatite (HAp) is the mineral apatite ($Ca_{10}(PO_4)_6(OH)_2$) which have Ca/P ratio 1.67. 65% of its mineral fraction in the human's bone comprised of hydroxyapatite [4]. This mineral can be applied for bone reconstruction and teeth implant [5].

The production of HAp requires sufficient high calcium source such as egg-shell [6], gypsum [2], blood clams [7], mammal's bone, clamshells, and coral [8]. However, these materials are not easily found in the nature.

The calcium source in the huge amount can be obtained from limestone. Lailiyah, et.al[8] reported the natural limestone which located at Tuban Regency, East Java, Indonesia has the high purity and calcium fraction until 98%. To synthesize HAp, the limestone treated to be Precipitated Calcium Carbonate (PCC) by caustic soda and carbonation method. PCC is the derived product from limestone which re-carbonated by producing $CaCO_3$ [9].

This paper reports the synthesis hydroxyapatite by wet-precipitation method using variation of Ca amount, phosphate amount, and aging time. The advantages of this method include easier manufacturing in low temperature [10], cost-effective, and simple [6]. Wahdah, *et. al.* had successfully synthesized HAp from cow-bone by wet-precipitation at 60C in ethanol media and H_3PO_4 as phosphate source [11].

2. Materials and Methods

2.1. Materials and Measurements

The chemicals used in this work include ethanol (96%), H_3PO_4 (85%), ammonium molybdate tetrahydrate, ammonium metavanadate, HNO_3 (65%), HCl (37%), and distilled water. To calculate the Ca/P ratio, the P amount was determined by UV Vis spectrophotometry whereas the Ca amount was measured using atomic adsorption spectroscopy. FTIR, powder XRD and SEM were also used to characterized the synthesized HAp.

2.2. Preparation of Precipitated Calcium Carbonate (PCC)

Precipitated Calcium Carbonate (PCC) was obtained from caustic soda and carbonation method and both were calcined at 900°C for 20 mins, cooled down to room temperature and then grinded using mortar. In caustic soda method, the limestone is firstly calcined at high temperature into CaO and then added with water to form $Ca(OH)_2$. The PCC was obtained by reacting the $Ca(OH)_2$ solution with sodium carbonate solution [12,13]. Unlike in caustic soda method, the PCC synthesized using carbonation method was obtained by adding CO_2 gas into the $Ca(OH)_2$ solution. The precipitated was formed when at pH 7 [14].

2.3. Synthesis of Hydroxyapatite (HAp)

2.3.1. Effect of Phosphate Amount

Two different solution are prepared in a beaker glass by dissolving 10 g of CaO in 50 ml ethanol (solution A) and phosphoric acid (0.0871; 0.1071; 0.1270 moles) with 50 ml ethanol (solution B), heating the solution might be needed to dissolve the mixture. Solution B is then inserted into an intravenous bottle to be passed through in a line into solution A. Solution B is added into solution A, then mixed and heated at 60C. After solution B was run out, the mixture was continued to be heated at 60C for 1 h. The mixture was then allowed to stand at room temperature for 24 hrs. After that, the mixture was stirred and heated again at 60C for 30 min which eventually filtered off. The precipitate was dried in an oven at 100C for 1 h, then heated in a furnace at 800C for 5 hrs [11]. The dry solid that was obtained as HAp was weighed and characterized.

2.3.2. Effect of Aging Time

Procedure 2.2.1 was repeated using phosphoric acid amount (mol) that gives the highest yield and done in a variety of aging time (12 and 48 hrs). The synthesized HAp obtained from the highest yield in this procedure was characterised using FT-IR, powder-XRD, and SEM.

3. Result and Discussion

3.1. Preparation of PCC

PCC synthesized by caustic soda method has Ca amount of 31.5%, on the other hand, the PCC obtained from carbonation process has Ca amount of 34.1%. As presented in Table 1, the Ca amount influenced the yield. The highest yield was obtained by carbonation method which was 79.4%. This may due to higher Ca amount in the PCC from carbonation than in PCC from caustic soda method.

Based on this result, the PCC from carbonation was used as the starting material on the synthesis of hydroxyapatite. The hydroxyapatite was synthesized by mixing calcined CaO and phosphoric acid (H_3PO_4) as phosphate source in ethanol media. The synthesized HAp result was represented in Figure 1.

Table 1. Preparation of PCC

	Caustic soda	Carbonation
Ca amount (%)	31.5	34.1
HAp Mass (g)	13.7	14.2
Yield (%)	76.4	79.4

Figure 1. HAp (calcined at 800°C) synthesized from PCC obtained using carbonation method

3.2 Effect of Phosphate Amount

Synthesis of hydroxyapatite from PCC was conducted by adding various mole of H_3PO_4 as the source of phosphate. This mol variation of phosphate was aimed to investigate the exact ratio of Ca/P and hydroxyapatite. Petit, *et. al.* reported that hydroxyapatite had Ca/P ratio as high as 1.67 [4]. The Ca/P ratio of synthesized HAp obtained from various amount of phosphate is given in Table 2.

Table 2. Ca/P ratio in various H_3PO_4 amount

H_3PO_4 amount (mole)	Ca amount (%)	P amount (%)	Ca/P ratio
0.0871	52.74	21.29	2.48
0.1071	42.67	23.27	1.83
0.1271	41.77	25.22	1.66

From several variations of H_3PO_4, the Ca/P ratio which has nearest value to 1.67 was obtained at 0.1271 mol. The increase of Ca/P ratio can enhance the strength of HAp. This H_3PO_4 mol that give the best Ca/P ratio was used in the aging time experiments.

3.3 Effect of Aging Time

The effect of aging time toward the particle size of HAp was conducted in various aging time of 12, 24, and 48 hours. The synthesized HAp was characterized by Particle Size Analyzer (PSA) to determine the particle distribution on each aging time. The particle size of synthesized HAp is shown in Table 3 and their particle distribution is displayed in Figure 2.

Based on the Table 3, it is found that the aging time influences the particle size of the hydroxyapatite. The smallest particle size was obtained at aging time 48 hours. Longer aging time may result in a second precipitation phase. The precipitate size was smoother in a lower temperature. However, it may cause roughing of particle which was the small particle tend to dissolved and the large particle getting bigger [15].

Table 3. Particle size of synthesized HAp in various aging time

Aging time (hours)	Diameter (μm)		
	10%	50%	90%
12	15.36	63.53	117.41
24	15.71	40.67	111.03
48	11.93	27.48	49.25

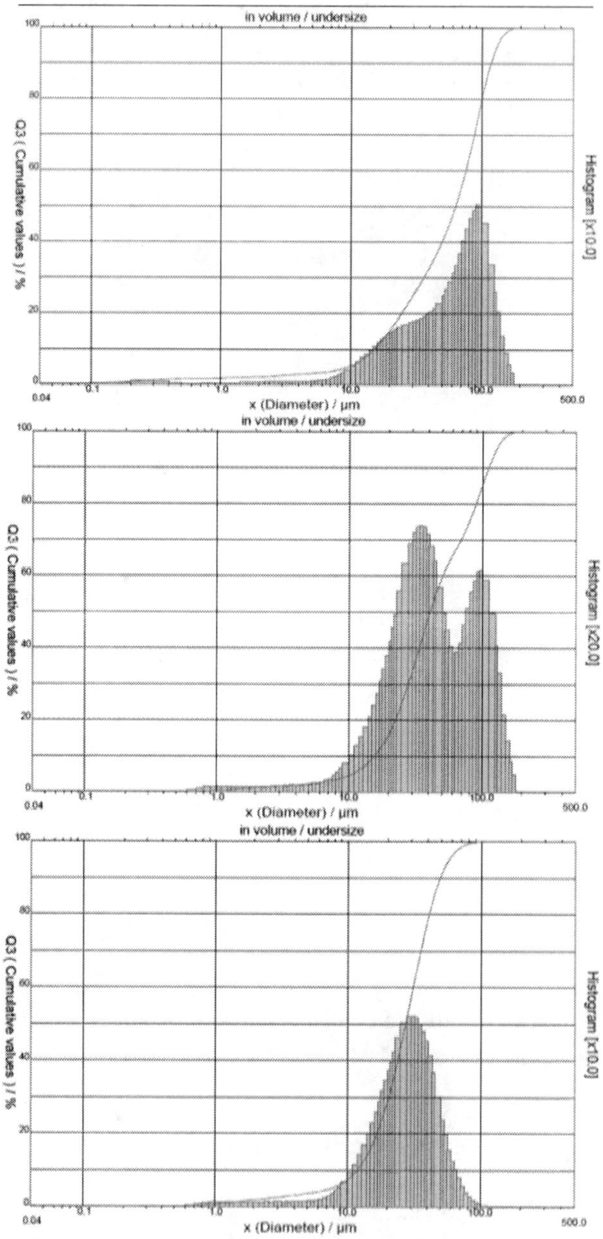

Figure 2. Particle distribution of the synthesized HAp in various aging time (top = 12 h, middle = 24 h, bottom = 48 h).

IOP Conf. Series: Materials Science and Engineering **299** (2018) 012035 doi:10.1088/1757-899X/299/1/012035

3.4 Characterisation of the Synthesized HAp

The synthesized HAp was characterized with FTIR, SEM, and powder XRD. The IR spectra of both PCC and synthesized HAp is presented in Figure 3 for comparison within a range of 4000-450 cm^{-1}. The IR spectra of PCC indicates the carbonate group of $CaCO_3$ polymorph calcite at 2981, 2871, 2515, 1799, 1421, 875, and 711 cm^{-1}. Moreover, the polymorph vaterite appeared at 2362 cm^{-1}. The FT-IR spectra of synthesized HAp in various aging time showed the stretching vibration of PO_4^{3-} at 1031.63 and 972.06 cm^{-1}. The band at 3452.34 cm^{-1} assigned to the hydroxyl (-OH) group whereas the band at 1413.72 cm^{-1} is correspond to Ca-O. The comparison between FTIR spectra of the synthesized HAp and the standard and the reference hydroxyapatite is presented in Table 4. The synthesized HAp shows some similarities with the standard and reference HAp spectra including the presence of PO_4^{3-}, CaO, and OH bands.

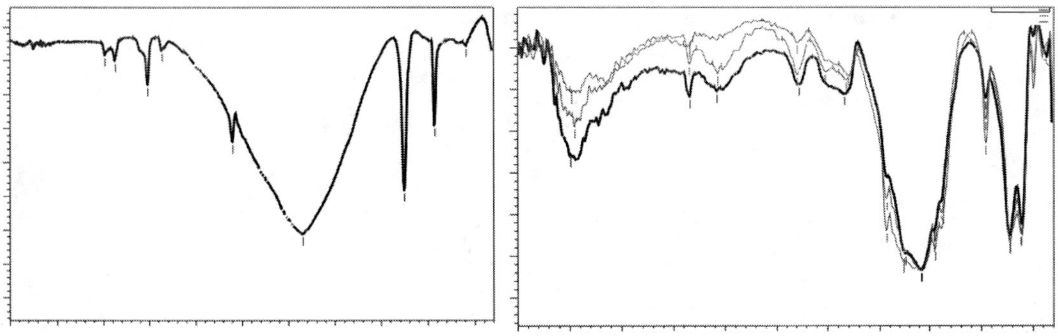

Figure 3. FTIR spectra of PCC (left) and synthesized HAp (right) in various aging time (black = 12 hrs, red = 24 hrs, blue 48 hrs).

Table 4. Comparison of FT-IR spectra between the synthesized HAp with the standard and the reference HAp

Functional Groups	Wavenumber (cm^{-1}) 12 h	24 h	48 h	Standard HAp	Reference HAp [16]
	561.6	553.5	557.4	565.1	560-610
				601.7	
	611.4	611.4	611.4	-	-
PO_4^{3-}	-	-	-	960.5	960
	975.9	972.1	974.0	-	-
	1041.5	1400.2	1035.7	1035.7	
	1118.6	1647.2	1128.3		1000-1150
CaO	1417.6	1400.2	1413.7		
	1635.5	1647.1	1510.2	1639.5	1400-1700
			1645.2		
OH	3492.9	3483.2	3452.3	3438.8	-

Characterization of PCC and the synthesized HAp using SEM was conducted to study and to compare the size and morphology before and after the synthesis. The SEM images of PCC and synthesized HAp is shown in Figure 4. The PCC has a dominant cubic structure and known to be in the calcite polymorph [17]. In contrast, the synthesized hydroxyapatite is agglomerated in a smooth morphology, which is similar to that of reported by Hui in 2010 [18].

In order to investigate the crystallinity, the synthesized HAp was analyzed using powder-XRD which then compare it with the JCPDS data based on the 2θ value. The powder spectra of PCC and the synthesized HAp is presented in Figure 5. The powder XRD pattern of PCC has 5 identical 2θ peaks with the JCPDS data of $CaCO_3$ no. 05-0586 which correspond to calcite polymorph. Unlike the PCC,

IOP Conf. Series: Materials Science and Engineering **299** (2018) 012035 doi:10.1088/1757-899X/299/1/012035

the powder XRD pattern of the synthesized HAp has different characteristic 2θ peaks compared to the JCPDS data of HAp no 09-0432. Instead, it has close similarity to the JCPDS of β-TCP (tricalcium phosphate). There was, however, 2θ peaks that correspond to the hydroxyapatite in lower intensity than in the JCPDS data of HAp which indicate the presence of HAp in a small amount. This result reveals that the bulk sample of the synthesized HAp is not pure and contains β-TCP hence intensive purification is needed for further investigation. The impurity occurred may be due to the solution pH was not suitable for synthesis of HAp, in which in this work the pH effect was not investigated yet. According to Kannan, *et.al.* [19], the pure HAp can be synthesized at pH 10. At pH 9, it may produce the mixture of β-TCP and HAp, whereas in pH 8, only β-TCP is obtained.

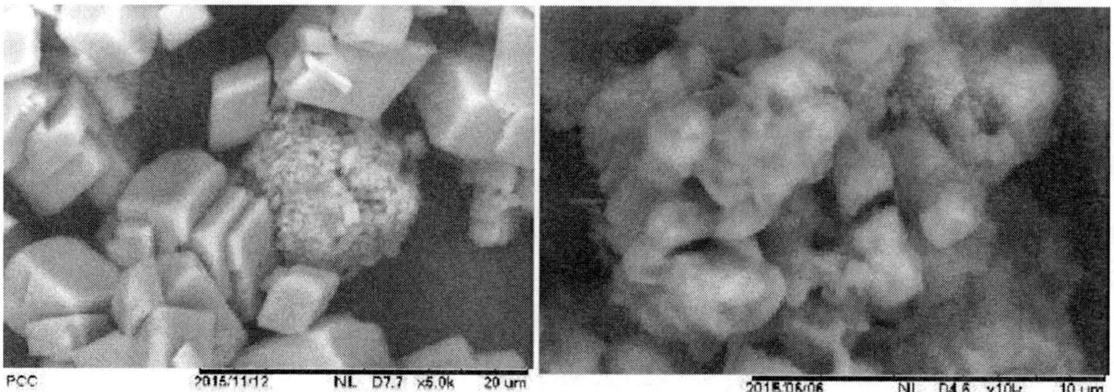

(A) (B)

Figure 4. SEM images of (A) PCC and (B) synthesized HAp

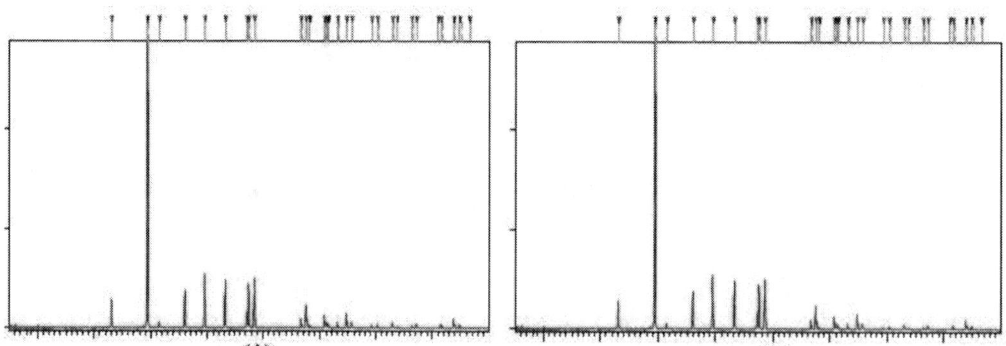

(A) (B)

Figure 5. Powder XRD data of PCC (A) and the synthesized HAp (B)

4. Conclusion

Synthesis of HAp was conducted from PCC as starting material which obtained from limestone by precipitation method. The high concentration of calcium in PCC gives better yields in which PCC obtained from carbonation method has higher yield than that of caustic soda method. The determination of optimum phosphoric acid addition based on targeted Ca/P ratio (1.67) from HAp was obtained on the addition of 0.1271 mol phosphoric acid with Ca/P ratio of 1.66. The aging time gave significant effect to the particle size of synthesised HAp. The smallest particle size was obtained in aging time for 48 hours as high as 49.25 μm. However, the synthesized HAp obtained using this method was agglomerated in a smooth morphology and based on powder XRD pattern, the product not pure and contained the mixture of HAp and β-TCP.

Acknowledgments
The authors would like to thank Brawijaya University, Indonesian Directorate of Higher Education, and Indonesian Ministry of Research and Technology and High Education for their financial support.

References

[1]. Dahlan K, Sari Y W and Soejoko D S 2006 *J. Sains. Mater Indones.* 221.
[2]. Putri E D, Prasetia F, Azkiya N I, Rosiana A and Wardhani S 2015 *Prosiding Seminar Nasional Kimia (Indonesian)* Jurusan Kimia FMIPA Universitas Negeri Surabaya, C108.
[3]. Iezzi G, Scarano A, Petrone G and Platelli A 2007 *J. Periodont.* **78** (5) 940.
[4]. Petit R 1999 *Eur. J. Orthop. Surg. & Amp. Traum.* **9** (2) 71.
[5]. Manurung R 1997 *J. Ked. Gigi (Indonesian)* Universitas Padjajaran **9** 28.
[6]. Gergely G, Weber F, Lukacs I, Toth A L, Horvath, Z E, Mihaly J and Balazsi C 2009 *Ceramics Internat.* **36** (2) 803.
[7]. Darwis D and Warastuti Y 2008. *J. Ilm. Apl. Iso. Rad.* (Indonesian) **4** (2) 143.
[8]. Lailiyah Q, Baqiya M A and Darminto 2001 *J. Sains POMITS* (Indonesian) **1** (1) 1.
[9]. Liu H S, Chin T S, Lai L S, Chiu, S Y, Chung, K H, Chang, C S and Lui, M T 1997 *Ceramic Internat.* **23** (1) 19.
[10]. Vahdettin B, Murat B and Elif B 2010 *Digest J. Nanomat. Biostruct.* **7** (3) 267.
[11]. Wahdah I, Wardhani S and Darjito 2014 *J. Ilmu Kimia* (Indonesian Student Journal), Universitas Brawijaya **1** (1) 92.
[12]. Brady J E and Holum J R 1992 *Chemistry The Study of Matter and its Changes* John Wiley & Sons New York.
[13]. Jamarun N, Rahmadanis and Arif S 2005 *J. Kimia Andalas* (Indonesian) **11** (1).
[14]. Renaudin G, Bertrand A, Dubois M, Gomes S, Chevalier P and Labrosse A 2008 *J. Phys. Chem. Sol.* **69** (7) 1603.
[15]. Jillavenkates A and Condrate RA 1998 *J. Mater. Sci.* **33** (16) 4111.
[16]. Pattanayak D K, Divya P, Upadhyay S, Prasad R C, Rao B T and Mohan T R R 2005 *Trends Biomater. Artif. Organs*, **18** (2) 23.
[17]. Sabriye P and Ozgul D O 2012 Inter. J. Biol. Ecol. Enviro. Sci. **1** (6).
[18]. Hui P, Meena S L, Singh G, Agrawal R D and Prakash S 2010 *J. Min. Mater. Charac. Eng.* **9** (8) 683.
[19]. Kannan S, Lemos A F and Ferreira J M F 2006 *Chem. Mater.* **18** (8) 2181.

Effect of metal cation ratio on chemical properties of $ZnFe_2O_4$/AC composite and adsorption of organic contaminant

Demara Meilia*, Mochamad Misbah Khunur, Tutik Setianingsih
Dept. of Chemistry, Brawijaya University, Malang

Email : tutiksetia@ub.ac.id

Abstract. Porous woody char is biochar prepared through pyrolisis. The biochar can be used as adsorbent. In this research, $ZnFe_2O_4$/AC composite was synthesized through imregnation of the woody biochar with $ZnFe_2O_4$ to study effect of mol ratio of Fe(III) and Zn(II) toward their physicochemistry and adsorption of drug wastewater. Paracetamol was used as adsorbate model. This research was conducted in several steps, including activation of the woody biochar using KOH activator at temperatur 500 °C for 15 min to produce the activated carbon, fungsionalization of the carbon using H_2SO_4 oxidator (6M) at temperature of 80 °C for 3 h, impregnation of the oxidized activated carbon with Zn-Fe-LDH (Layered Double Hydroxide) at various mol ratio of Fe(III) and Zn(III), including 1:2, 1:3 and 1:4 using NaOH solution (5M) for coprecipitation, and calcination of Zn-Fe-LDH/AC at 950 °C for 5 min to produce $ZnFe_2O_4$/AC. FTIR diffraction characterization indicated existence of M-O (M=Zn(II), Fe(III)) and OH functional groups. FTIR spectra showed increasing of bands connected to –OH by increasing of the ratio till the ratio was achieved at 1:4, then decreased again. The ratio mol showed effect on the adsorption of paracetamol. Profile of adsorption value was fit with changing of functional groups. The highest adsorption was achieved at the ratio of 1:4. After calcination it gave the adsorption value of 17,66 mg/g.
Key words : activated carbon, $ZnFe_2O_4$, Zn-Fe-LDH, adsorption

1.Introduction

Pharmacy industry is one of industries which produce dangerious wastewater for ecosystem, derived from production process, washing process, laboratory activity, and remain product due to failure of production process. The wastewater can be overcame by adsorption method carbon as adsorbent. Adsorption method is more advantageous than other methods due to economical,eficient, and easily applied [1].

Adsorption is a process in which atoms, ions, or molecules from substances such as gas, liquid, or solluble solid are concentrated on porous surface of an adsorbent [2]. There are various adsorbents, such as alumina, zeolite, activated carbon, biomass, polymers, and silica gel [3]. Activated carbon adsorbent has been applied and developed, due to its characteristics such as not poisonous, abundant, economical, and biodegradable. Activated carbon is the best adsorbent in adsorption system because it has large surface area and high adsorption peformance so that its usage can be optimal [4]. Surface area of activated carbon is about 300 – 3000 m^2/g. Activated carbon can be made from animal bones or plant biomass such as wood, rice husk, corn cob, coconut shell, hard wood, and coal [5].

Paracetamol is one of medicinal wastewater components. Activated carbon has been applied in adsorption of paracetamol and showed good performance [6]. Development of many reserachs have been conducted to improve performance of activated carbon as adsorbent. One of the is modofication of activated carbon with metal chloride salts such as $ZnCl_2$ and $FeCl_3$ [7] or ZnO and Fe_2O_3 as product of calcination [8]. Fe (III) and Zn (II) have different charges so that they have different affinities toward polar organic subtances in adsorption process. So that, this research is focussed on studying of Fe(III)/Zn(II) mol ratio influence on paracetamol adsorption.

2. Experiment

2.1. Preparation of actived carbon from wood char

Wood char from market (1 kg) was crushed and conditioned to obtain particle size of 30 -60 mesh. Mixture of KOH solution (4M, 100 mL) and wood char (10 g) was shaked for 2 h at room temperature. After decantation, the residue was dried at 105°C and calcined at 500°C for 10 minutes. The product was washed with HCl solution (1M) and aquadest, then it was dried at 105°C for 6 h.

2.2. Funcionalitation of actived carbon by oxidation

The actived carbon was supended in H_2SO_4 solution (6 M) at the carbon and solution ratio of 1:10 (g/mL) and heated at 800°C for 3 h. The product was washed by aquadest and dried at 70 °C for 1 h.

2.3. Preparation of composites

This procedures is combination of procedures applied in previous researchs [9, 10]. The actived carbon (0.5 g), $FeCl_3$ and $ZnCl_2$ solutions (each 0.2M and 50 mL) were mixed and shaked at 175 rpm for 1 h, then dropped with NaOH solution (5 M) under stirring to achieve pH of 7 condition. The synthesis was conducted at the Zn(II)/Fe(III) ratio of 1:2 to 1:4. Then, the mixture was heated at 80°C for 3 h. The mixture was filtered, washed, dried at 70°C for 24 h, and finally calcined to 950°C for 5 minutes to get $ZnFe_2O_4$/AC composite. The same procedure was applied without $FeCl_3$ solution to produce ZnO/AC and without $ZnCl_2$ solution to produce Fe_2O_3/AC.

2.4 Adsorption test

The prepared composites (0.1 g) and paracetamol solution (100 mg/L ; 25 mL) were mixed and shaked at 200 rpm for 24 h. After filtration, paracetamol concentration was analyzed by UV-Vis spectrophotometry analysis. The procedures were conducted in 3 times of repeatation.

3. Results and Discussions

3.1. Functional group characterization by FTIR spectrophotometry

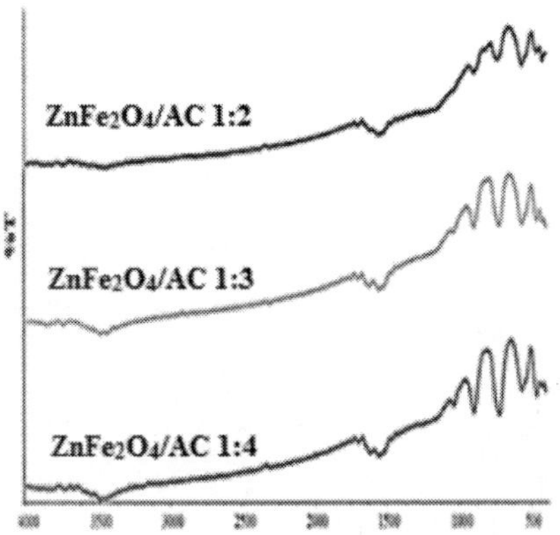

Figure 1. FTIR spectra of $ZnFe_2O_4$ composite at various Fe(III) and Zn(II) mol ratio

Characterization of the products by FTIR spectrophotometry was conducted to identify the change of wavelength shift and sharpness of bands of the composites' spectra, as shown at Figure 1. FTIR spectra in Figure 1 shows that no significant wavelength difference of the composite spectra at various Fe(III) and Zn(II) ratio. Spectra of $ZnFe_2O_4$/AC has bands at 3000-3500 cm^{-1}, connected to hydroxil group (ν_{O-H}). The spectra bands at 903.39; 905.31; and 905.31 cm^{-1} indicate presence of Cl^- ion. The bands at range of 400-900 cm^{-1} are connected to Fe(III) – O and Zn(II) – O. However, the cation ratio gives significant effect on intensity of the spectra. Intensities of the bands at range of 400-900 cm^{-1} were increased by increasing of Zn(II)/Fe(III) ratio from 1:2 to 1:4 due to improvement of metal content in the composite [11]. Besides that, band of hydroxil was also increased by increasing of the ratio which indicates increasing of hydroxil functional groups connected to phenolic groups on graphene layers of the activated carbon in the composite.

3.2. Characterization of crystal structure by XRD method

Characterization by XRD has been conducted to identify chemical structure of the composites. Figure 2 shows that the diffractogram of $ZnFe_2O_4$/AC composite contains of diffractogram peaks of the synthesized ZnO/AC and the synthesized Fe_2O_3/AC which indicates that the $ZnFe_2O_4$/AC composite also contains both structures.

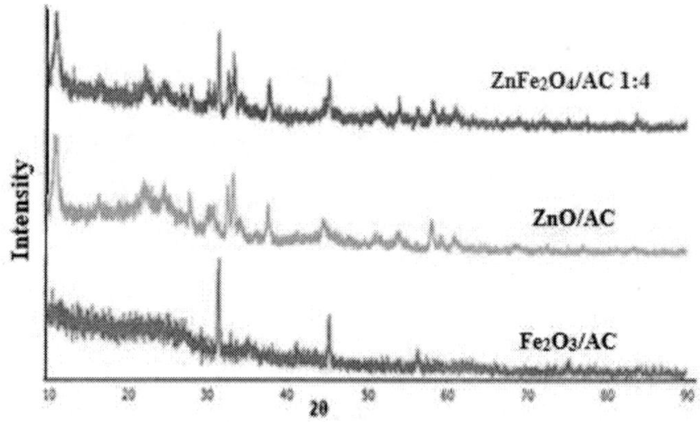

Figure 2. X-ray diffractogram of the synthesized $ZnFe_2O_4$/AC, ZnO/AC, and Fe_2O_3/AC

Table 1. Interpretation of the $ZnFe_2O_4$/AC, ZnO/AC, and Fe_2O_3/AC composite diffractograms

Synthesized $ZnFe_2O_4$/AC composite at 1:4 ratio Zn(II)/Fe(III)			Synthesized ZnO/AC			Synthesized Fe_2O_3/AC			$ZnFe_2O_4$ [12]
2Θ	d (Å)	I_R (%)	2Θ	d (Å)	I_R (%)	2Θ	d (Å)	I_R (%)	d-spacing (Å)
11,16	7,93	100,0	11,09	7,97	100,00	-	-	-	-
22,43	3,96	23,32	22,22	4,00	27,00	-	-	-	-
24,86	3,58	15,97	24,78	3,59	32,59	-	-	-	-
31,70	2,82	94,01	-	-	-	31,66	2,82	100,0	2,97 (220)
32,89	2,72	43,47	32,80	2,73	65,19	-	-	-	-
33,58	2,67	69,44	33,45	2,68	79,31	-	-	-	2,53 (311)
45,42	1,99	61,91	-	-	-	45,41	1,99	61,91	2,10 (400)
51,33	1,78	10,29	51,31	1,78	8,39	-	-	-	1,71 (422)
56,47	1,63	12,10	-	-	-	56,46	1,63	17,18	1,61 (511)
58,34	1,58	26,96	58,20	1,58	33,11	-	-	-	1,50 (440)
83,93	1,15	7,42	-	-	-	-	1,15	5,34	-

As interpreted in Table 1, the d-spacing data of the synthesized $ZnFe_2O_4$/AC diffractogram is fit with d-spacing data of the $ZnFe_2O_4$/AC diffractogram from the previous research [12]. It indicates that the synthesized composite at the 1:4 ratio contains $ZnFe_2O_4$ structure.

3.3. Adsorption test

Adsorption test has been performed and the data of adsorption was reported in Figure 3.

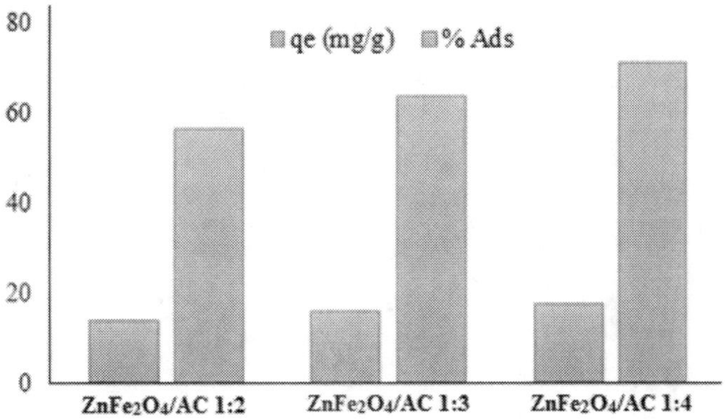

Figure 3. Adsorption values of paracetamol by $ZnFe_2O_4$/AC composite

Figure 3 shows increasing of adsorption by increasing of the Zn(II)/Fe(III) ratio. It is due to increasing of Lewis acid such as Zn(II) cations in the composites which improve affinity of the composite toward Lewis bases paracetamol structure such as O atoms at –OH and C=O groups and N atom at –NH group.

4.Conclusion

$ZnFe_2O_4$/AC has been synthesized at various cation ratios of Zn(II/Fe(III). Characterization by FTIR spectrophotometry indicates increasing of hydroxil and metal oxide contents by increasing of the cation ratio. Characterization by XRD indicates presence of $ZnFe_2O_4$, ZnO, and Fe_2O_3 in the composite. Adsorption test shows improvement of adsorption by improvement of the caion ratio.

Reference

[1] Oktem, Y.A. et al 2008 *Bioresource Technology* 99 **(5)** pp.1089-1096.
[2] Liu, Y., Y. Zheng, dan A. Wang. 2009 *Journal of Environmental Sciences* 22(4) pp 486-493
[3] Tripathi, A. dan M. R. Ranjan 2015 *Bioremed Biodeg* **6** p315
[4] Prabowo, A.L. 2009 Pembuatan Karbon Aktif dari Tongkol Jagung serta Aplikasinya untuk Adsorpsi Cu, Pb, dan Amonia Depok Universitas Indonesia
[5] Sembiring, T.M dan T.S Sinaga 2003 Arang Aktif (Pengenalan dan Proses Pembuatannya), Universitas Sumatera Utara, Medan.
[6] Dutta, M., Das, U., Mondal, S., Bhattachriya, S., Khatun, R., and Bagal, R. 2015 *International Journal of Environmental Sciences* 6(2) pp 270-281
[7] Park, S.J., Jang, Y.S., Shim, J.W., Ryu, S.K. 2003 *Journal of Colloid and Interface Science* **260** pp 259–264
[8] Pereira, M.F.R., Soares, S.F., Órfão, J.J.M., Figueiredo, J.L. 2003 *Carbon* **41** pp 811–821
[9] Hongo, Teruhisa., Takeshi Iemura, and Atsushi Yamazaki 2008 *Journal of the Cereamic Society of Japan* **116** pp 192-197
[10] Ahmed, A.A.A., Talib, Z.A., Hussein, M.Z., Flaifel, M.H., and Al-Hada, N.M., 2014 *Journal of Spectroscopy* 2014 Article ID 732163, 6 pages.
[11] Wan, J., Jiang, X., Li, H., and Chen, K. 2012 *Journal of Material Chemistry* **22** pp 13500-13505.

[12] Lemine, O.M., M.Bououdina, M.Sajieddine, A.M. Al-Saie., M.Shafi, A.Khatab, M.Al-Hilali, and M.Henini. 2011 *Physica B* 406 pp 1989-1994

IOP Conf. Series: Materials Science and Engineering **299** (2018) 012037 doi:10.1088/1757-899X/299/1/012037

Strategies for characterizing compositions of industrial pulp and paper sludge

Solmaz Aslanzadeh, Rahmat A. Kemal and Amadeus Y. Pribowo

Department of Bio Technology, School of Life Sciences, Indonesia International Institute for Life Sciences, Pulomas Barat Kav 88, Jakarta, 13210, Indonesia

Email: solmaz.aslanzadeh@i3l.ac.id

Abstract. The large quantities of waste sludge produced by the pulp and paper industry present significant environmental challenges. In order to minimize the amounts of waste, the pulp sludge should be utilized for productive applications. In order to find feasible solutions, the sludge need to be characterized. In this study, the potential of using acid pretreatment and ashing method to determine the chemical compositions of the sludge is investigated. This study shows that acid pretreatment could be used to dissolve and determine the composition of $CaCO_3$ in the pulp sludge. $CaCO_3$ removal also facilitates the measurement of fiber and ash (clay) contents by using the ashing method. The optimum acid concentration used to completely dissolve $CaCO_3$ was determined using a titration method. Using this method, the measurement of the chemical composition of the sludge sample revealed that it consisted primarily of $CaCO_3$ (55% w/w), clay (25%, w/w), and fibers (18%, w/w). Based on these chemical compositions, potential utilization for the sludge could be determined.

1. Introduction

In 2017, Indonesia is projected to produce 10 million tonnes of pulp and 17 million tonnes of paper [1]. Based on real pulp and paper production and sludge estimation, sludge production from pulp and paper production would be 0.20-0.27 million tonnes/year and 0.31-0.40 million tonnes /year [2]. This large quantity of pulp sludge causes environmental concerns. Currently, industrial pulp sludge is disposed of in landfills or used for land spreading, but this represents low-value use of the sludge [3]. Research has been done to produce value-added products from pulp sludge, including using sludge for the production of concrete block [4] chipboard [2], and cement composite [5]. Components in pulp sludge can also be separated and processed to produce other materials. Wajima and Ikegami [6] used paper sludge ash to synthesize zeolite. The valorization of pulp sludge depends on the compositions and physical properties of the sludge which in turn are affected by the raw materials, manufacturing process, and the types of final products [7, 8]. Pulp sludge is typically composed of water, organic materials such as cellulose fibres, and inorganic materials such as calcium carbonate ($CaCO_3$) and Kaolin clay [9]. Thus, a characterization step is required to determine the types of value-added products that can be produced from a given pulp sludge. This work analyses a strategy to characterize the composition of pulp and paper sludge and discusses potential applications of the sludge based on the its compositions.

2. Materials and Methods

2.1. Material

The pulp sludge used in this work was obtained from PT Bukit Muria Jaya in Karawang regency, Indonesia. According to the company, the sludge consisted mainly of fibres, $CaCO_3$, and trace amounts of chemical additives used as coagulant and flocculating agents. The material contained 70 % moisture content. Prior to characterisation, materials were stored at 4°C.

2.2. $CaCO_3$ Removal and Quantification

Sludge samples (10 g wet weight) were mixed with 150 ml of HCl at concentrations ranging from 0.1 - 1.2M in order to find the optimum concentration needed to dissolve all $CaCO_3$. This mixture was stirred for 24 h with a magnetic stirrer at room temperature. After stirring, vacuum filtration was applied to separate the solid from the liquid fraction. The liquid fraction was analysed for pH and $CaCO_3$ content. The pH of each filtrate was measured using a pH meter OHAUS ST3100-F. The $CaCO_3$ content in the original sludge sample was quantified by titrating excess HCl in the filtrate with NaOH and 3 drops of phenolphthalein as an indicator. Filtrates from 0.8 – 1.2 M of HCl treatment were titrated using 0.47 M of NaOH, and filtrates from 0.6 M of HCl treatment were titrated using 0.02 M of NaOH. Prior to titration, NaOH solutions were standardised using 0.25 M of potassium hydrogen phthalate. The amount of $CaCO_3$ was then calculated according to the Equations (1) - (3).

$$CaCO_{3(s)} + 2HCl_{(aq)} \rightarrow CaCl_{2(aq)} + CO_{2(g)} + H_2O_{(l)} + \text{excess } HCl_{(aq)} \quad (1)$$

$$\text{Excess } HCl_{(aq)} + NaOH_{(aq)} \rightarrow NaCl_{(aq)} + H_2O_{(l)} \quad (2)$$

$$CaCO_3 \text{ Mass } \% = \frac{Mass\ of\ \text{CaCO3}\ (g)}{mass\ of\ the\ sludge\ (g)} \times 100\% \quad (3)$$

2.3. Measurement of Total Solids (TS) and Volatile Solids (VS)

The solid fraction after the HCl treatment was washed with tap water until the samples reached pH 7, indicating that the acid had been completely washed out. Distilled water was used as the final washing. Compositions of TS and VS in the solid fraction were determined by ignition at 550°C± 25°C in a furnace [10]. A mass balance calculation was then carried out based on the ash and VS contents in the solid fraction after acid treatment. Samples were also analyzed for lignin and cellulose content. Lignin content was determined according to an NREL procedure [11]. Fibre samples were milled to 40 mesh fractions using Thomas Willey- mini cutting mills 1171H10. Milled fibres were subjected to acid hydrolysis with 72% H_2SO_4 in order to release the sugars (hemicellulose and cellulose fractions). Upon separation, the solid residue was analysed for acid-insoluble lignin after drying and ignition at 575°C. The supernatant was analysed for acid-soluble lignin using UV-VIS spectrophotometer optima SP-3000 Nano at 280 nm. All lignin analyses were carried out in triplicates. Cellulose content in the solid fraction was carried out according to the standard anthrone method described by Hodge and Hofreiter [12]. The samples were mixed with H_2SO_4 and anthrone reagent and then boiled. The mixture was cooled down and measured using UV-VIS spectrophotometer optima SP-3000 at 620 nm.

2.4. Fiber analysis

The fiber dimensions were analyzed by washing the sludge and separating the fibers manually. The fibers were then treated in a dilute HCl acid (0.1 M) to dissolve the residuals of the $CaCO_3$ in the sludge in order to separate the fibers. The fiber lengths were measured by Olympus CX 22 RFS1 microscope, with LED illumination, using a millimeter paper (size: 1x1mm) at the background.

IOP Conf. Series: Materials Science and Engineering **299** (2018) 012037 doi:10.1088/1757-899X/299/1/012037

3. Results

3.1. Optimization of the dissolution of $CaCO_3$ with HCl

Figure 1 shows the result of ignition loss expressed as a weight percentage (%wt) of the dry mass of insoluble residue after treatment with different concentrations of HCl. After washing and drying, the untreated sludge lost 2.63% in weight and had 97.3% insoluble residue. Increasing acid concentration from 0. 1 to 0.8 M resulted in a steady decrease in the dry weight of the solid residue after the pretreatment (from 90.00% to 41.89%). The decrease in weight after HCl treatment may indicate the dissolution of $CaCO_3$ from the pulp sample. An increase in HCl concentrations from 0.8 to 1.2 M did not result in any further change in the remaining percentage weight of solid residue after the treatment, which remained stable at 41.9 wt%. This indicated that there was an optimum HCl concentration for the dissolution of the total amount of $CaCO_3$ at ~0.8M. $CaCO_3$ dissolution after HCl treatment can also be confirmed by the ignition of the solid fraction at 550°C (Figure 2). The VS showed an increase from 33.1% for the untreated sludge to 74.1 wt % for the sample treated with 0.8 M HCl. Further increases in HCl concentration to 1.2 M did not significantly increase the VS content (75.1 wt%). Similarly, the ash content decreased from 66.6 wt% for the untreated sludge to approximately 25.5wt% for the 0.8M HCl and remained stable with further increases in HCl concentration beyond 0.8 M.

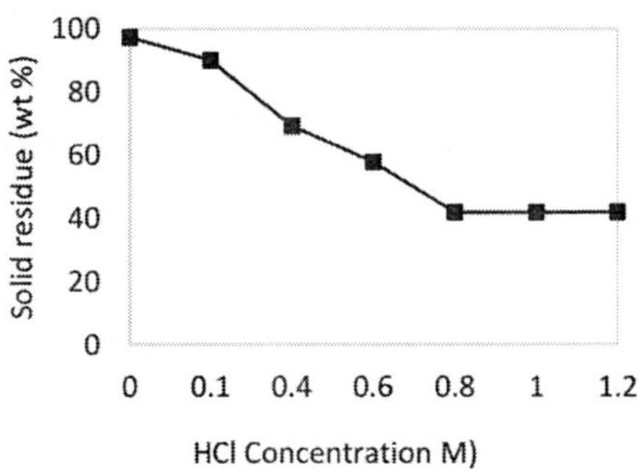

Figure 1. The percentage loss of the dry mass of the sludge solid residues after 24 h HCl treatment based on mass balance

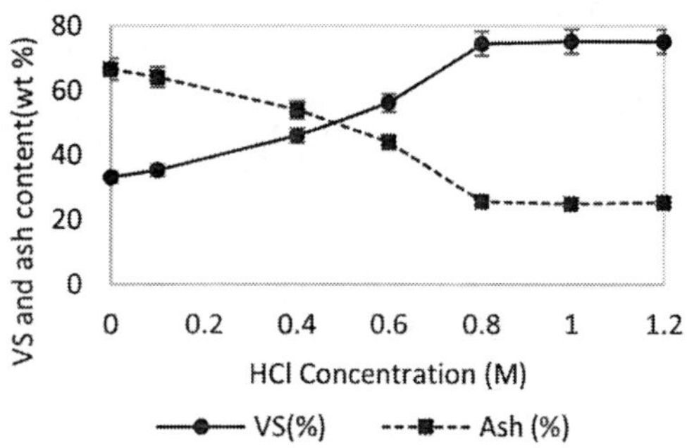

Figure 2. The loss of ignition of the solid residues after HCl treatment, expressed as volatile solid (VS %) and ash content (Ash%) of solid residues

In order to confirm that weight loss after acid treatment was due to $CaCO_3$ dissolution, the liquid fractions after treatments with HCl between 0.6 to 1.2M were analyzed using back titration with standardized NaOH. Figure 3 shows the calculated $CaCO_3$ content using back titration method· The result showed that 0.6M HCl did not result in complete $CaCO_3$ dissolution in the sludge, which is in accordance to with the result from previous method using mass balance of ignition loss of solid residue. The titration with 0.8M showed extensive fluctuation in the weight percent of $CaCO_3$. Repetition of the dissolution step with HCl after separation of the solid residual with acid concentration 0.8M for another 24 h did not significantly change the ash content of the solid residues, which remained stable at 24.5 % as shown in figure 4. The titration result on the other hand was more stable. This indicates that the HCl concentration at 0.8M was rather a transitionary concentration for the $CaCO_3$ dissolution rather than the optimum concentration for a complete $CaCO_3$ dissolution. Further increases in HCl concentrations to 1 and 1.2M stabilized the titration result to 58.3 wt% to 56.7 wt% respectively (Figure 3). This shows that the optimum HCl concentration for dissolution of total $CaCO_3$ in the sludge should be higher than 0.8M. This is also confirmed by the pH of the liquid fraction after the acid treatment (Figure 5) where pH decreased from 1.71 in the liquid residue treated with 0.8M HCl to less than 0.5 after treatment with 1.2M HCl. At higher HCl concentration all of the $CaCO_3$ were dissolved and the excess unreacted acid in sample resulted in the low pH.

IOP Conf. Series: Materials Science and Engineering **299** (2018) 012037 doi:10.1088/1757-899X/299/1/012037

3.2. Fiber and clay content analysis

The ash contents in the solid fraction after HCl treatments at concentrations between 0.8 to 1.2M were shown to be stable between 24- 25 wt% (Figure 4). However, a change in the ash color from white gray to pink was observed with the samples treated at HCl concentrations between 0.6 to 1.2M (Figure 8). This color alteration could be caused by a lack of $CaCO_3$ dissolution when the sludge samples were treated with HCl at concentrations below 0.8M. The remaining pink color ash could indicate the presence of clay in the pulp sludge.

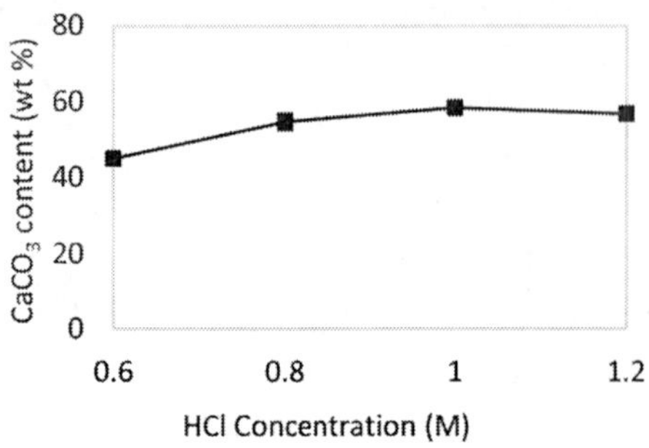

Figure 3. Determination of $CaCO_3$ content (wt %) with titrimetric method using standardized NaOH of the liquid filtrate after 24h HCl treatment at different concentration.

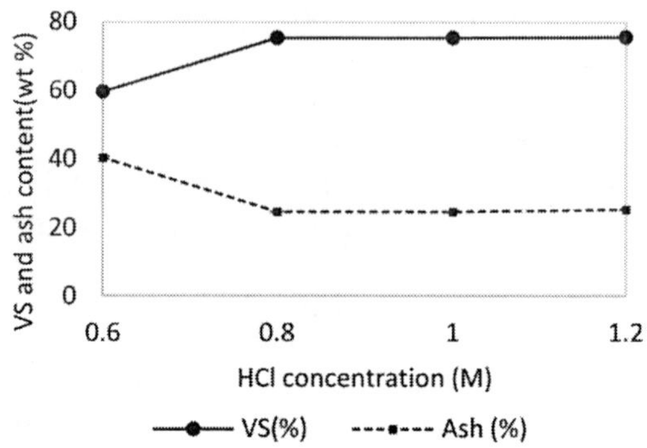

Figure 4. Titrimetric method - The loss of ignition of the solid fraction after HCl treatment, expressed as volatile solid (VS %) and ash content (Ash wt %) of solid residues

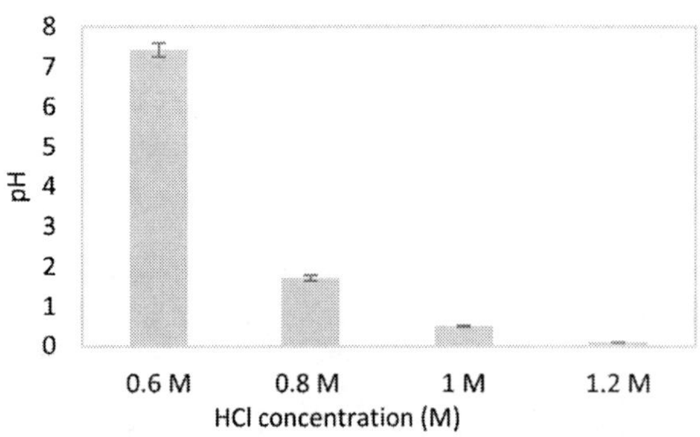

Figure 5. The pH of the liquid filtrate separated from the solid residue after HCl treatment at different concentrations

The ash content after combustion was used to indirectly determine the amount of organic matters (fiber) in the pulp sludge. In this case, the sample treated with 1.2 M HCl was used to ensure total $CaCO_3$ dissolution from the pulp sample. The VS from the solid fraction were estimated to be 18.5 wt. % (Table 1). The VS of the pure pulp (the raw material) was determined to be 99.8% (wt%). In other words, the pure pulp had negligible ash content. Based on this information, we could ascribe the 18.5% VS from the solid fraction to be the fiber content in the pulp sludge. However, as some fiber might have been lost during acid treatment and filtration process, the percentage weight of the fibers could be slightly higher than the estimated 18 wt%.

The fibers dimension in the sludge was measured to be approximately around 1 mm to 2 mm shown in Figure 6. The acid treatment caused partial degradation of the cellulose fibers, which was lost during filtration process. Figure 6 shows the untreated sludge (6b), and acid treated sludge with 24 h (6c) and 48 h hydrolysis (6d). The 48h-acid treatment caused significant fiber damage while the fibers were more intact after 24h-acid treatment. No significant difference was observed in the ash and VS contents after 24h and 48h-acid treatment: 24 wt% and 75 wt%, respectively (Figure 7).

The cellulose content of the pure fibers was measured to be around 94.6±5 (%). The lignin content of the pure fibers was measured to be around 6.32 ± 0.33. The acid soluble lignin fraction of the fibers was shown to negligible. This indicates that the cellulose content in the untreated sludge should be somewhat higher than the measured value for the pure fibers.

Table 1. The result of percentage weight (wt %) of the fibers in the sludge treated with 1.2 M HCl and the volatile solid (VS), cellulose, total lignin and acid soluble lignin content in pure fibers

Components	Content (%)
Fibers[a]	18.5±0.55
Volatile solid (VS)[b]	99.8±0.1
Cellulose	94.6±5.6
Total Lignin	6.32±0.33
Acid soluble lignin	0.044±0.0003

[a] Calculated value based on mass balance after ignition loss of the solid residues after 1.2 M HCl treatment
[b] Determined based on pure fibers used in the paper making process.

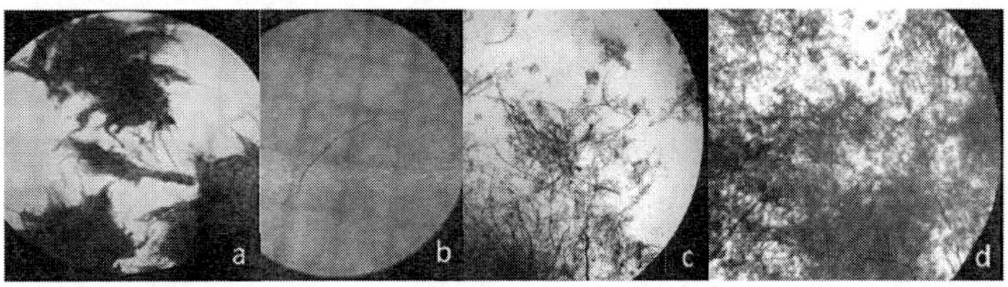

Figure 6. The fibers in the a) untreated sludge, b) Separated fibers from untreated sludge and oven dried at 7 °C, c) 24 h acid treatment with 0.8 M HCl and d) 48h acid treatment with 0.8 M HCl

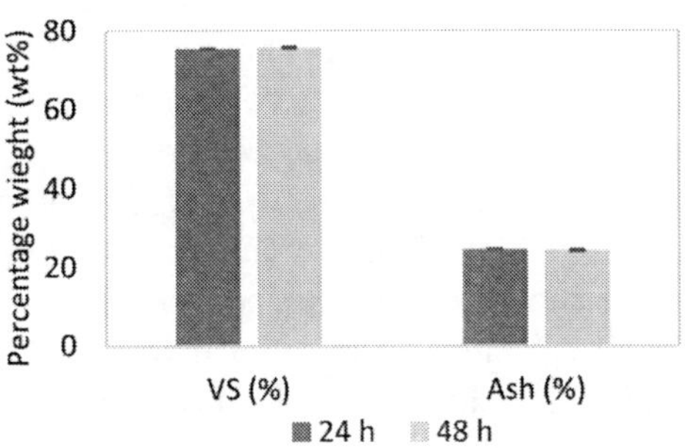

Figure 7. Weight percentage of volatile solid (VS) and ash content of the solid residues after 24 h and 48 h of treatment with 0.8 M HCl

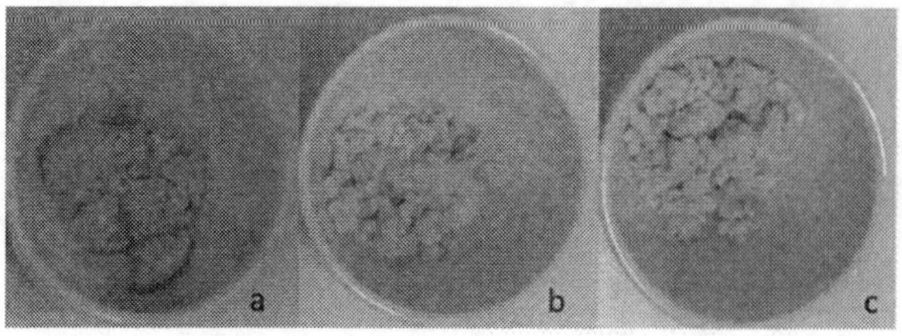

Figure 8. The colour of ash following ignition of solid residues at 550°C changed from white to pink as the solid residues were treated with increasing HCl concentrations: a) 0.6 M HCl, (white grey colour), b) 0.8 M HCl (pink color), c) 1.2 M HCl (pink colour)

4. Discussion

The composition of the primary sludge from paper making process was analyzed using a combination of loss on ignition (LOI) method and titrimetric method after optimization of HCl concentration for the total amount of $CaCO_3$ dissolution. The result shows that using these methods, compositions of $CaCO_3$, fiber, and ash in the pulp sludge could be determined. $CaCO_3$ composition was measured between 56-58 wt%. Even though the composition of the sludge vary widely depending on the process, the results of compositional analysis in this study is close to other previous studies [13, 14].

As the decomposition and volatilization of $CaCO_3$ usually occur above 600°C, combustion at 550°C would not affect $CaCO_3$ content [15]. Therefore, an acid treatment step was required prior to LOI analysis to ensure total $CaCO_3$ dissolution. The titrimetric method could be used to determine the optimum HCl concentration that would result in complete $CaCO_3$ removal. Following complete $CaCO_3$ dissolution, the clay and the fiber contents were measured using LOI to be at 24-25 wt % and about 18 wt%, respectively. This result implies that the simple methods described here can be used to determine the compositions of a typical pulp sludge.

Compositional analysis is important for determining potential utilization for the pulp sludge. $CaCO_3$ seems to the dominant part of our sludge sample (56-58 wt%), which could potentially be used as a soil amendment to balance the pH of acidic soil such as soil affected by coal mining operations [16]. The cellulosic fiber content in the sludge would also improve the ability of the soil to retain water [17].This direct application of paper sludge on land is the preferred method of utilizing paper mill sludge due to its cost-effectiveness.

Applications of waste sludge in the construction industry, especially in the production of bricks and cement, have also been recognized. Using sludge with 13% ash content in cement tiles has been shown to increase plasticity, workability, strength, and waterproof property of the cement[18]. Sludge with 20% to 30% ash content may be used to produce a facing brick for commercial use [19, 20]. Another potential use of sludge is as filler in bio plastic production as the $CaCO_3$ and the fibers work as a binding agent. Cellulose fibers are already known to improve the starch film performance [21] while $CaCO_3$ could improve the homogeneity of the plastic by providing good thermal conductivity.

The cellulose fibers in the sludge also have potential applications. Utilization of the cellulose fibers may require separation of the fiber from other components in the sludge. The method described here could be used to remove $CaCO_3$ from the sludge, increasing the proportion of the fiber in the treated sludge. The acid treatment to dissolve $CaCO_3$, however, may cause fiber damage, and thus the acid concentration needs to be optimized based on the requirements for the final application of the fiber. In addition, the economics of the process should be considered for such application especially when the cellulose fiber content is not very high.

Kaolin clay was another major fraction (24 wt%) in the sludge sample. The recovery of clay from paper mill sludge using wet air oxidation has been investigated earlier (28). However, it would be more cost effective and easier to separate the clay easily upon ignition after dissolution of the $CaCo_3$ as almost all the cellulose fibers are volatilized. The clay component can be potentially be recycled into the papermaking process, used as a masonry additive, or for any other purposes [22, 23]. Kaolin is a natural pigment and the Kaolin clay in the sludge could have potential use for pigment recovery or dye in various application, even though no studies could be found for such application [24, 25].

5. Conclusion

This paper has described an inexpensive and easy-to-do method to determine the chemical compositions of a given pulp sludge. By using an acid treatment followed by ignition of the solid residues, the mass balance of $CaCO_3$, fiber, and ash (clay) could be determined. Understanding the chemical compositions of a pulp sludge is necessary to determine the possibility of producing value-added products from the sludge.

References

[1] Handoyo 2016 The Production Capacity of Paper and Pulp will be increasing in 2017. Ministry of Industry of Republic of Indonesia)
[2] Hardiani H and Masriani R 2016 *Potensi Sludge dari Industri Kertas Sebagai Bahan Baku Chipboard* vol 9
[3] Monte M C, Fuente E, Blanco A and Negro C 2009 Waste management from pulp and paper production in the European Union *Waste Management* **29** 293-308
[4] Hardiani H and Sugesty S 2017 pemanfaatan limbah sludge industri kertas sigaret untuk bahan baku bata beton *2017* **44**
[5] Yadollahi R, Hamzeh Y, Ashori A, Pourmousa S, Jafari M and Rashedi K 2013 Reuse of waste sludge from papermaking process in cement composites *Polymer Engineering & Science* **53** 183-8
[6] Wajima T and Ikegami Y 2007 zeolite synthesis from paper sludge ash via acid leaching *Chemical Engineering Communications* **195** 305-15
[7] J. Jackson M and A. Line M 1997 *Organic composition of a pulp and paper mill sludge determined by FTIR, C-13 CP MAS NMR, and chemical extraction techniques* vol 45
[8] Battaglia A, Calace N, Nardi E, Petronio B M and Pietroletti M 2003 Paper mill sludge–soil mixture: kinetic and thermodynamic tests of cadmium and lead sorption capability *Microchemical journal* **75** 97-102
[9] Likon M and Trebše P 2012 *Industrial waste*: InTech)
[10] A. Sluiter B H, R. Ruiz, C. Scarlata, J. Sluiter, and D. Templeton 2005 Determination of ash in biomass *National renewable energy laboratory*
[11] A. Sluiter B H, R. Ruiz, C. Scarlata,J. Sluiter, D. Templeton, and D. Crocker 2008 Determination of structural carbohydrates and lignin in biomass *National renewable energy laboratory*
[12] Hodge J E and Hofreiter B T 1962 *Determination of reducing sugars and carbohydrates.* (New York,: Academic Press,)
[13] Klyosov A A, Philippidis G P and Monovoukas Y A 1999 Production of precipitated calcium carbonate from papermaking sludge and sludge-derived ash. Google Patents)
[14] Klyosov A A, Philippidis G P, James A M and Monovoukas Y A 1999 Method of producing glucose from papermaking sludge using concentrated or dilute acid hydrolysis. Google Patents)
[15] Halikia I, Zoumpoulakis L, Christodoulou E and Prattis D 2001 Kinetic study of the thermal decomposition of calcium carbonate by isothermal methods of analysis *European Journal of Mineral Processing and Environmental Protection* **1** 89-102

[16] Shimada H, Kusuma G, Hiroto K, Sasaoka T, Matsui K, Gautama R and Sulistianto B 2012 *Development of a new covering strategy in Indonesian coal mines to control acid mine drainage generation: A laboratory-scale result* vol 26

[17] Abdullah R, Ishak C F, Kadir W R and Bakar R A 2015 Characterization and Feasibility Assessment of Recycled Paper Mill Sludges for Land Application in Relation to the Environment *International Journal of Environmental Research and Public Health* **12** 9314-29

[18] Johnson O, Napiah M and Kamaruddin I 2014 Potential uses of waste sludge in construction industry: a review *Research Journal of Applied Sciences, Engineering and Technology* **8** 565-70

[19] Jain S 2015 *Utilization of Waste Paper Sludge in Construction Industry*

[20] Wiegand P S U, J.P. 1994 Alternative management of pulp and paper industry solid wastes *Tappi J* **77** 91-7

[21] Dufresne A and Vignon M R 1998 Improvement of Starch Film Performances Using Cellulose Microfibrils *Macromolecules* **31** 2693-6

[22] Scott G M and Smith A 1995 Sludge characteristics and disposal alternatives for the pulp and paper industry. In: *Tappi International Environmental Conference* TAPPI PRESS) pp 269-

[23] Verhoff C R 1996 Papermaking sludge recovery process and apparatus. Google Patents)

[24] Grönfors J 2010 Use of fillers in paper and paperboard grades, *Tampere University of Applied Sciences International Pulp and Paper Technology* **Final Thesis**

[25] Berube R R, Babiec J S, Jameson M P, Negele A R and Willis M J 1991 Calcined kaolin clay filler pigment for paper. Google Patents)

Acknowledgments

The authors would like to acknowledge PT Bukit Muria Jaya for financing this project. Authors would like to express their sincere appreciation to Dr. Irnayuli Sitepu for her advice and support on this work. We would like to thank Ms Ester Yusmarina for her assistance in the practical part of this work. We would like to thank our students, Zabrina Quiteria, Fahreza G. Avicena, and Khoe E. Sudibya for their help in the practical part of this work.

Determination of process parameters for curcumin – dextrose cocrystallization

Katherine[1,2], Denny Nugroho[2] and Asaf K Sugih[2]

[1] Department of Biotechnology, Indonesia International Institute for Life Sciences, Jalan Pulomas Barat Kavling 88, Jakarta, Indonesia
[2] Department of Chemical Engineering, Universitas Katolik Parahyangan, Jalan Ciumbuleuit 94, Bandung, Indonesia
Email: katherine.k@i3l.ac.id

Abstract. Curcumin is a polyphenol that could act as anti-oxidant and anti – inflammation agent. It is usually isolated from rhizome plants such as turmeric and temulawak. Despite its many favorable properties, curcumin is practically insoluble in water, thus limiting its application. In the present investigation, variables affecting preparation of curcumin-dextrose cocrystal were examined with the aim to increase the solubility of curcumin. The effect of different processing conditions, such as water to dextrose ratio, final heating temperature and water bath temperature to the formation of cocrystal, were studied and the yield and solubility of curcumin – dextrose cocrystal products were analyzed. The morphology of the cocrystals were also analyzed using SEM and fluorescence microscopy.. Curcumin – dextrose cocrystals showed a significant increase in solubility up to 25 mg curcumin per mL water compared to pure curcumin.

Keywords: *cocrystallization, curcumin, dextrose*

1. Introduction

Curcumin ((1,7-Bis-(4-hydroxy-3-methoxyphenyl)-hepta-1,6-diene-3dione) is a polyphenol extracted from turmeric (*Curcuma longa*) and temulawak (*Curcuma xanthorrhiza*). It is widely used as natural food colorants. In addition, curcumin has demonstrated several therapeutic effects in the treatment of cancer, HIV infection, and antioxidant [1]. However, the low solubility of curcumin in water limits its usage and bioavailability. In order to overcome this limitation, various methods has been suggested, such as using nanocarrier [2], high pressure homogenization [3] and cocrystallization with pyrogallol and recorcinol [4].

Cocrystallization is one interesting alternative method to improve curcumin solubility as the method is simple, shows high yield and six times increase in solubility [4]. However, the coformer used previously was considered toxic. Therefore, there is a need to find a safer coformer. In this study, dextrose was chosen as a coformer because it is a generally considered as safe (GRAS) compound and easily obtained. In addition, a simple method of cocrystallization employing solubilization of coformer by heating followed by cooling to achieve supersaturation was introduced. While the method has not been used for curcumin, the method has been successfully employed in encapsulation of cardamom oleoresin [5], yerba mate[6], honey [7], and orange peel oil [8] to stabilize the active ingredients.

2. Material and methods

2.1. *Material*
Curcumin and ethanol were obtained from Merck (Jakarta). Dextrose monohydrate was purchased from Bratachem (Bandung). All chemicals were of analytical grade.

2.2. *Methods*
2.2.1. *Determination of dextrose recrystallization process parameters*
Preparation of supersaturated dextrose solution is the initial step in the recrystallization process. Important factors investigated in this step were water to dextrose ratio, final heating and cooling temperature. Water of different volumes was added to 15 g of dextrose to obtain ratio of water to dextrose as shown in Table 1. The mixture was then heated with a laboratory hot-plate while stirring with magnetic stirrer. The temperature of the mixture was monitored continuously during heating process. The heating process was stopped when the desired final temperature was achieved (Table 1). Afterwards, the solution was removed from hot-plate and cooled inside a water bath. In order to control the cooling rate, the temperature of water bath was set according to Table 1 . Once the temperature reached 50°C, ethanol was added while the solution was continuously stirred. The solution was further transferred into a silicone mold and cooled inside a water bath until cocrystals were formed. Water and ethanol were finally removed from the cocrystals by drying in oven for 24 hours at 40°C.

Table 1 Recrystallization process parameters

Mass ratio of water to dextrose	0:15	1:15	2:15	3:15
Final heating temperature (°C)	90	100	110	120
Water bath temperature (°C)	30	35	40	-

2.2.2. *Cocrystallization of curcumin with dextrose*
Cocrystallization was performed by mixing ethanolic solution with sugar solution. 15 g of dextrose was dissolved in water (weight ratio of 1:15). The dextrose solution is then heated while stirring until 100 °C. Once the desired temperature was achieved, the dextrose solution was cooled while continuously stirred in a preheated water bath (35 °C). until its temperature decreased to 50°C. Curcumin solution (0.125 mg curcumin/mL ethanol) was subsequently added while stirring to minimize heat-induced curcumin degradation. The mixture was then poured into silicon mold and, placed in water bath to further decrease the temperature of the solution. The cocrystals formed upon cooling were transferred into evaporation cups, and finally dried in a vacuum oven for 24 hours at 40°C to evaporate the remaining solvent. Dried cocrystal products were stored in tightly closed container to protect them from moisture and light.

2.2.3. *Morphology analysis*
The morphology of curcumin dextrose cocrystal was analyzed using Scanning Electron Microscopy (JEOL JSM 6510 LA, Japan). Prior to analysis, the samples were spread on the surface of double – face carbon tape and coated with gold. In addition, bright-field and fluorescence microscopy images of the cocrystals were captured using fluorescence microscope type Nikon Eclipse E800 at 400x magnification.

2.2.4. Curcumin cotent determination

Curcumin content was analyzed by using UV – vis spectrophotometer (Genesys 20, Spectronic) at the peak absorbance wavelength for curcumin, i.e. 430 nm. At this wavelength, dextrose posed no interference to curcumin, because dextrose absorption occured at lower wavelength (260 - 270 nm)[9]. The ethanol – water mixture is chosen as solvent for its ability to dissolve curcumin at higher concentration. To determine curcumin content in the cocrystal, 300 mg cocrystal was dissolved in a known amount of water. A 10 mL aliquot was taken and centrifuged for five minutes at 6000 rpm, and 8 mL of ethanol was further added to 2 mL of the obtained supernatant. The absorbance of obtained mixture was then measured with the spectrophotometer, and curcumin concentration was determined using a callibration curve built from measuring the absorbancs of series of curcumin in 80% ethanol-water standard solutions. The yield of curcumin is defined as the actual amount of curcumin in the cocrystal divided by the initial amount of curcumin added.

2.2.5. Solubility of curcumin – dextrose cocrystal

The solubility of curcumin in cocrystal in water was measured by adding amount of cocrystal to 10 mL water while stirring until it reached saturation. After centrifugation at 6000 rpm, 2 mL of the supernatant was mixed with 8 mL of ethanol. The solution was then analyzed for its curcumin content with Vis spectrophotometer.

3. Result and Discussion

3.1. Dextrose recrystallization process parameters

Prior to cocrystallization of curcumin – dextrose, the crystallization process parameters were investigated to ensure consistent crystal formation. Crystallization process is affected by temperature at which the crystallization is initiated, water content and cooling rate [7].

Water content affects the formation of crystals. In cocrystallization of curcumin - dextrose, low water content is desired due to low solubility of pure curcumin in water. In addition, high water content decelerates crystal formation [10]. However, absence of water necessitates higher heating temperature to completely melt dextrose. Moreover, high temperature usage should be avoided due to tendency of dextrose to caramelize. By fixing final heating temperature to 100°C, it was found that mass ratio of water to dextrose mass is optimum at 1:15. Crystals were formed within 7 minutes. At mass ratio of water to dextrose 0:15, dextrose could not completely melt. Meanwhile, both mass ratio of water to dextrose 2:15 and 3:15 yielded highly viscous dextrose solution which could only crystallized after more than 24 hours.

The final heating temperature was varied from 90°C to 120°C. Not all dextrose particles were dissolved at 90°C. Meanwhile, when heated to 110°C and above, the color of dextrose solution changed to slightly yellow, which indicated the formation of caramel. At the optimum temperature of 100°C, dextrose was dissolved into transparent colorless liquid.

Solution supersaturation significantly determines the development of nucleation and crystal growth. One method to control supersaturation is by controlling cooling rate. A fast cooling rate favors nucleation over crystal growth as it increases supersaturation level, which leads to formation of large population of small size crystals. On the other hand, low cooling rate favors crystal growth over nucleation, resulting in formation of small population of large crystals [11]. In addition, cooling rate needs to be investigated to determine the time needed for addition of curcumin solution and homogenization of the mixture. Fast cooling rate will result in formation of cocrystals with inconsistent curcumin content. In this work, cooling rate was controlled by immersing dextrose solution container in the water bath. Setting the water bath to 30°C accelerated crystal formation, which caused crystals

IOP Conf. Series: Materials Science and Engineering **299** (2018) 012038 doi:10.1088/1757-899X/299/1/012038

formation prior to homogenization. Therefore, minimum water bath temperature for cocrystallization was set at 35°C.

3.2. Curcumin – dextrose cocrystal characterization

Figure 1 shows the morphology of curcumin – dextrose cocrystal. SEM image (Figure 1a.) shows that the cocrystal is a mixture of flakes and smaller irregular particles, with length of flakes about 10 µm and length to width ratio of 5. The size of the small irregular particles is about 1 µm. Some of the flake particles formed aggregates with the smaller particles trapped inside or attached to the surface of the flakes.

SEM image is incapable of differentiating dextrose from curcumin. In order to locate curcumin, fluorescence microscope was employed. Figure 1b shows the cocrystal under bright – field microscope while Figure 1c shows the same sample under fluorescent microscope. Under the bright-field microscope, the cocrystals appeared as an aggregate with dark color. Using mercury – vapor lamp as light source, some part of the crystals appeared bright green. The green color came from curcumin as curcumin has fluorescence property [12]. The image indicates that distribution of curcumin in cocrystal is not uniform.

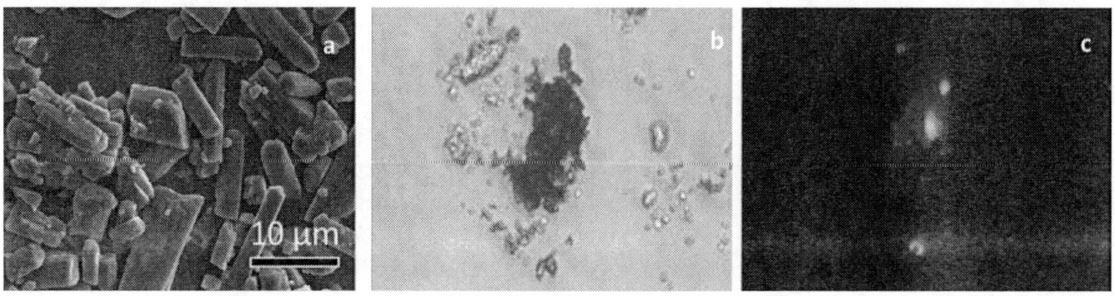

Figure 1. Curcumin – dextrose cocrystal observed using (a). SEM at 1500 x magnification, (b) bright-field microscope at 400x magnification and (c) fluorescence microscope at 400x magnification

The yield of curcumin was up to 90% in cocrystal. Pure curcumin (Figure 2a) could not be detected by spectrophotometer as it was insoluble. According to a study, the water solubility of curcumin was reported to be 0.6 µg/ml [13]. Initial trial to dissolve the crystal in water showed that the cocrystal is highly soluble (Figure 2c). Quantitative analysis indicates that the solubility of the cocrystals containing 0.2% curcumin is about 25 mg curcumin per mL of water. This shows a significant increase compared to pure curcumin. The curcumin dispersed in dextrose solution was also insoluble, which can be shown in Figure 2b. This confirms that curcumin solubility is not caused by presence of dextrose.

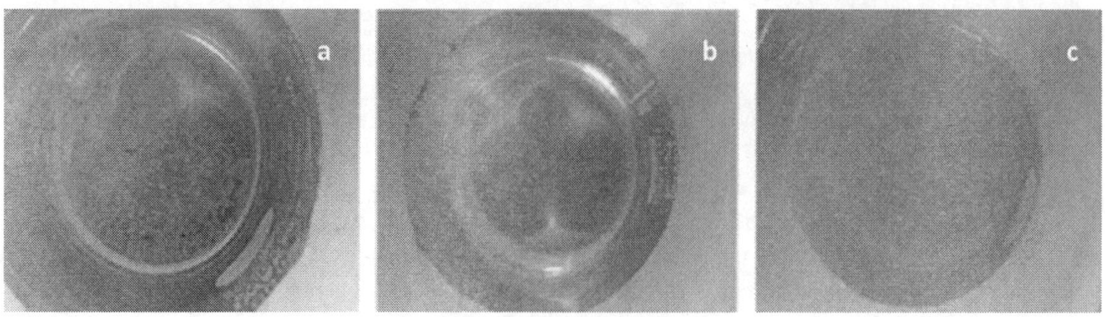

Figure 2. (a) curcumin dispersed in water, (b) curcumin dispersed in dextrose solution, (c) curcumin – dextrose cocrystal dispersed in water

4. Conclusion

This work demonstrates the preparation of highly – soluble curcumin – dextrose cocrystal. Parameters such as mass ratio of water to dextrose, final heating temperature and cooling temperature were identified. The optimum parameters for curcumin – dextrose cocrystallization were mass ratio of water to dextrose 1:15, final heating temperature of 100°C, and water bath temperature of 35°C. The curcumin – dextrose cocrystals consisted of a mixture of flakes and small irregular particles in which curcumin was distributed unevenly. The increase solubility of curcumin shows a promising potential of curcumin – dextrose cocrystals. Investigation on stability of curcumin – dextrose cocrystals, effect of curcumin concentration in the cocrystallization, and also the curcumin – dextrose interaction is currently carried out and will be reported in forthcoming papers .

5. Acknowledgement

This work was funded by Hibah Monodisplin Universitas Katolik Parahyangan. The authors would like to thank Universitas Katolik Parahyangan for the grant received. The author would also like to thank Dr. Solmaz Aslanzadeh for proofreading the manuscript.

6. References

[1] Sousdaleff M, Baesso M L, Neto A M, Nogueira A C, Marcolino V A and Matioli G 2013 Microencapsulation by Freeze-Drying of Potassium Norbixinate and Curcumin with Maltodextrin: Stability, Solubility, and Food Application *Journal of Agricultural and Food Chemistry* **61** 955-65

[2] Bisht S, Feldmann G, Soni S, Ravi R, Karikar C, Maitra A and Maitra A 2007 Polymeric nanoparticle-encapsulated curcumin ("nanocurcumin"): a novel strategy for human cancer therapy *Journal of Nanobiotechnology* **5** 3

[3] Donsi F, Wang Y, Li J and Huang Q 2010 Preparation of curcumin sub-micrometer dispersions by high-pressure homogenization *J Agric Food Chem* **58** 2848-53

[4] Sanphui P, Goud N R, Khandavilli U B R and Nangia A 2011 Fast Dissolving Curcumin Cocrystals *Crystal Growth & Design* **11** 4135-45

[5] R. Sardar B and S. Singhal R 2013 *Characterization of co-crystallized sucrose entrapped with cardamom oleoresin* vol 117

[6] López-Córdoba A, Deladino L, Agudelo Mesa L and Martino M 2014 *Yerba mate antioxidant powders obtained by co-crystallization: Stability during storage* vol 124

[7] Bhandari B, Datta N, D'Arcy B and B. Rintoul G 1998 *Cocrystallization of Honey with Sucrose* vol 31

[8] Beristain C I, Vazquez A, Garcia H S and Vernon-Carter E J 1996 Encapsulation of Orange Peel Oil by Co-crystallization *LWT - Food Science and Technology* **29** 645-7

[9] Kaijanen L, Paakkunainen M, Pietarinen S, Jernström E and Reinikainen. S-P 2015 Ultraviolet Detection of Monosaccharides: Multiple Wavelength Strategy to Evaluate Results after Capillary Zone Electrophoretic Separation *International Journal of Electrochemical Science* **10** 2950-61

[10] Parisi M, Rivallin M and Chianese A 2006 Prediction of Dextrose Nucleation Kinetics by the Growth Rate of Crystallites *Chemical Engineering & Technology* **29** 265-70

[11] Markande A, Nezzal A, Fitzpatrick J J and Aerts L 2009 Investigation of the Crystallization Kinetics of Dextrose Monohydrate Using In Situ Particle Size and Supersaturation Monitoring *Particulate Science and Technology* **27** 373-88

[12] Chignell C F, Bilskj P, Reszka K J, Motten A G, Sik R H and Dahl T A 1994 SPECTRAL AND PHOTOCHEMICAL PROPERTIES OF CURCUMIN *Photochemistry and Photobiology* **59** 295-302
[13] Kurien B T, Singh A, Matsumoto H and Scofield R H 2007 Improving the solubility and pharmacological efficacy of curcumin by heat treatment *Assay and drug development technologies* **5** 567-76

Synthesis and Characterization of Hydroxyapatite Powder by Wet Precipitation Method

S. E Cahyaningrum[1], N. Herdyastuty[1], B. Devina[1] and D. Supangat[1]
[1]Department of Chemistry, Faculty Mathematics and Natural Science
Universitas Negeri Surabaya

Email: saricahyaningrum@unesa.ac.id

Abstract. Hydroxyapatite is main inorganic component of the bone with formula $Ca_{10}(PO_4)_6(OH)_2$. Hydroxyapatite can be used as substituted bone biomaterial because biocompatible, non toxic, and osteoconductive. In this study, hydroxyapatite is synthesized using wet precipitation method from egg shell. The product was sintered at different temperatures of 800°C to 1000°C to improve its crystallinity. The hydroxyapatite was characterized by X-ray analysis, Scanning Electron Microscopy (SEM) and Fourier Transform Infrared Spectroscopy (FTIR) to reveal its phase content, morphology and types of bond present within it. The analytical results showed hydroxyapatite had range in crystallinity from 85.527 to 98.753%. The analytical functional groups showed that presence of functional groups such as OH, $(PO_4)_3^{2-}$, and CO_3^{2-} that indicated as hydroxyapatite. The result of characterization SEM indicated that hydroxyapatite without sintering and HAp sintering at 800 °C were irregular shape without pore. The best hydroxyapatite with temperature sintering at 900 °C showed oval shaped with pores without agglomerated.

1. Introduction

Hidroxyapatite is one of bioceramic was used as materials making of filler bone and theet [1]. Filler Bone of hidroxyiapatite can occupy by bone network because hidroxiapatite have looking like with bone composition. Hidroxyapatite have high biocompatibility and also have the character of osteoconduktive that is can stimulate growth of bone. Chemical formula of HAp is $Ca_{10}(PO_4)_6(OH)_2$ with ratio of molar Ca/P 1.67 [2]. Hydroxiapaptite has characteristic pore, bioactive, biocompatible, and osteoconductive [3-4]. Making of HAp can use the source of natural and synthetic calcium [5]. Source of synthetic calcium which is generally used for the synthesis of HAp is CaO, $Ca(NO_3)_2$, $Ca(OH)_2$, $CaCO_3$ and $CaCl_2$ [6-7]. Source of natural calcium which used for the synthesis of HAp generally have high calcium rate among others, coral [8], seashell [9], eggshell [10] and also limestone [3].

At this research, source of calcium which is used in HAp synthesis is eggshell. Exploiting of egg in society still limited for food, by exploiting its just flesh while egg shells are thrown. The waste of egg shell can contaminate of land, water ground and air. Waste of egg shell contain of calcium compound ($CaCO_3$) which is high enough, that is around 53-78% from weight of crab shell dry. This calcium rate height can be used as materials of HAp. With content of $CaCO_3$ big, eggshell can be used as CaO precursor by calcinated at temperature 1000°C. The aim of calcinations is to eliminate carbonate ion able to bother process of sythesis. HAp can be synthesed with few methods, among others, hydrothermal [11-12], sole gel [13-14], dry method [15] and wet precipitation method [5]. Sythesis HAp with wet precipitation method have many excellence, the side product only water, and possibility of contamination during processing is very low, so that in course of HAp yield is perity high level enough[16], moreover expense of which released is cheap relative. Other advantage of wet precipitation method is the reaction easy and do not contaminate environment.

2. Material and methods

2.1 Synthesis Hidroxiapatite

Eggshells of duck were collected and cleaned from impurities in their surface, after that the egg shells of duck were calcinated at 900 °C to produce calcium oxide. Synthesis hidroxyapatite was done by reacting calcium precursor and phosphate with comparison of concentration of molar Ca/P 1.67. Solution of $Ca(OH)_2$ 0.5 M was added by phosphoric acid 0.3 M with wise drop methods later then trap closed chemical glass to use alumunium foil to produced suspension. The suspension was added with NaOH 1 M until pH 10. The solution was aging over night to produce HAp powder. The HAp powder was cleaned with aquademin until neutral, and then was added with HNO_3 6 M and sintering in furnace with variation temperature are 800, 900 and 1000 °C for 2 hours. The crystal HAp was storaged in furnace until cold and is weighed with analitycal balance until constant in mass. The HAp crystal was characterized with some instrument. The phase composition of the HAp crystal was studied by X-ray diffractometry. The functional group was studied with Fourier transform infrared spectroscopy (FTIR) and scanning electron microscopy (SEM) to examined the surface of morphological characteristics.

3. Results and Discussion

CaO was produced from calcination eggshell of duck that reacted with H_2O to produce $Ca(OH)_2$ solution as equation:

$$CaO\ (s) + H_2O\ (l) \rightarrow Ca(OH)_2\ (aq) \qquad ...(1)$$

The $Ca(OH)_2$ was added with H_3PO_4 solution with *wise drop* methods at 60 °C until pH 7. Synthesis of HAp under 60 °C produce amorf structure, so that the synthesis of Hap must be done at 60 °C (Wang, 2010). Hap that produce in this research was characterized with some intrumentations Analysis crystalinity with XRD was showed in picture 1.

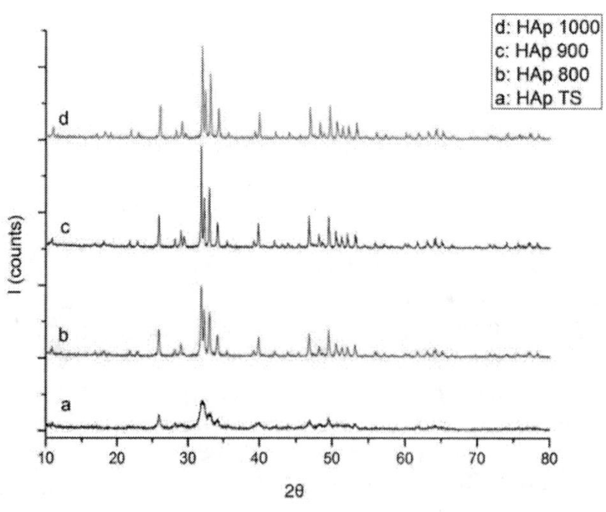

Figure 1. X-ray diffraction patterns of hydroxyapatite synthesized HAp-TS; HAp 800; HAp 900 and HAp 1000

Figure 1 shows X-ray diffraction of HAp-TS, HAp-800, HAp-900, and HAp-1000. At HAp-TS sampel have 2θ = 32.045°; 32.272°; and 33.116°. HAp-800 has highest top at 2θ= 31.810°; 32.197°; and 32.949°; HAp-900 have highest top at 2θ= 31.792°; 32.197°; and 32.933°. HAp-1000 have highest top at 31.811°; 32.214°; and 32.951°. Highest top of HAp-TS represent highest top of carbonate apatite matching with JCPDS no. 35-0180 that property of type carbonate apatite of A (AKA) and of JCPDS no. 19-0272 that property of type carbonate apatite of B (AKB). Highest topat HAp-800, HAp-900, and HAp-1000 represent highest top of HAp matching with JCPDS no. 09-0432 that property of HAp.

Table 1. Composition of HAp.

Sample	HAp (% b/b)	Apatite carbonate (% b/b)	Tetracalcium ciclo Decaphospate 16 Hydrate (% b/b)
HAp-TS	50.4	49.6	-
HAp-800	72.0	28.0	-
HAp-900	82.7	17.3	-
HAp-1000	99.1	0.9	-

Table 1 shows that increase of temperature of sintering, perity of HAP carbonate apatite composition and excelsior smaller. Composition of HAp at HAp-TS, HAp-800, HAp-900, and HAp-1000 are 50.4; 72.00; 82.70; and 99. 10% b/b. HAp owning highest perity that is HAp-1000, if compared to HAp-TS show composition which look like that is with difference 0.7 %b/b. At HAp TS there are carbonate apatite phase is bigger. According to ISO-13779:2008, HAp as implant have minimize 50% b/b HAP fase. Pursuant to quantitative analysis of XRD, all HAp are without and with treatment of sintering have larger one mass from 50 % w/w.

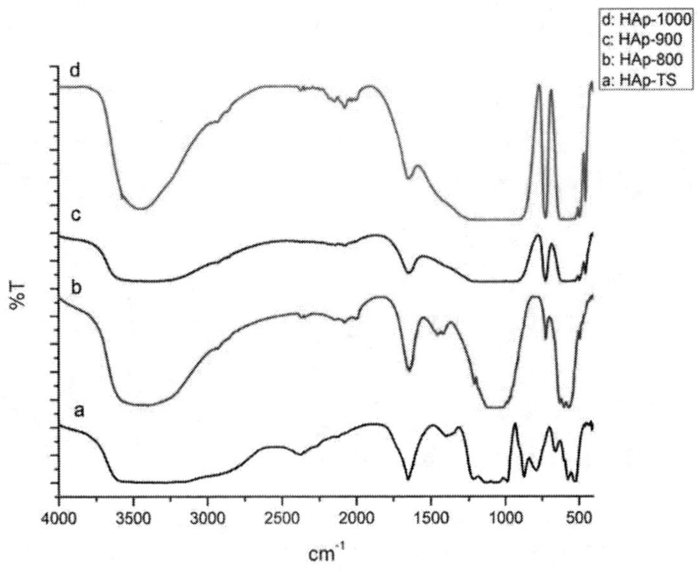

Figure 2 FTIR spectra of HAp –TS; HAp- 800; HAp- 900 and HAp-1000

The results of FTIR spectra in different sintering temperatures are shown in the Fig 2. This spectrum (Figure 2) that informs about the hydroxyl (OH) functional group by absorption band at 600, 700, 800 and 900°C, respectively at around 3400-3571 cm^{-1}. FTIR spectra showed that absorption from the PO_4 3- on 1150 – 1000 cm^{-1} and known as the characteristics of bands for hydroxyapatite. The absorption bands of functional groups are presence at around 1069, 1011, 1043, and 1016 cm^{-1}. The phosphate functional ($PO4^{3-}$) at HAp with stretching vibrasi there are at1000 - 1150 medium and at 960 cm^{-1} and also bending vibrasi perceived at 560 - 610 cm^{-1}. The identifying the existence of carbonate in crystal perceived from FTIR spectra for the sampel of HAp and carbonate apatite.

Table 2 Crystalinity of HAp

Sample	β (rad)	Fraction		Crystalinity (%)
		Amorf	Crystal	
HAp-TS	0.033	2.970	19.074	86.527
HAp-8	0.022	1.760	52.800	96.774
HAp-9	0.019	1.140	57.570	98.058
HAp-10	0.017	1.020	80.750	98.753

Crystalinity HAp-TS, HAp-800, HAp-900, HAp-1000 respectively that are 86.527; 96.774; 98.058; and 98.753%. The smallest crystalinity is HAp-TS. The so small HAp-TS crystalinity for no process sintering at sample. At HAp which was not sintering, slimmer particles are atom formation and crystal more is not regular compared to HAp which is sintering.

(C) (D)

Figure 3 SEM images of hydroxyapatite: HAp-TS (A), HAp-800 (B), HAp-900, (C), HAp-1000 (D), 5000x.

The surface morphology of the HAp crystal was investigated by SEM. The result of characterization SEM indicated that hydroxyapatite without sintering and HAp sintering at 800 °C were irregular shape without pore. The best hydroxyapatite with temperature sintering at 900 °C showed oval shaped with pores without agglomerated. Crystal size measure of HAp as implant have to span certain so that earning application as implant. Smallest size measure of application able to reside in at scale of nano (Chandrasekar, al et 2013). Biggest size measure according to standard of HAp-Bank Tissu RS Dokter Sutomo, Surabaya that is size measure 1193.48 m.

4. Conclusion

The results of this research hydroxyapatite can be synthesised from eggshell of duck with wet precipitation methods. Temperature of sintering has effect in degree and phase crytalination of HAp.

The FTIR spectra analysis showed that HAp have some functional group there are –OH, $-(PO_4)_3^{2-}$, dan $-CO_3^{2-}$. Analysis with SEM morphology showed that hidroxypatite in the form of block with spread over pore to flatten.

Acknowledgment
The researcher wish thanks to DRPM DIKTI for support this research with Hibah Penelitian Berbasis Kompetensi Program 2017 No. Contract: 0001.88/UN38.11-P/LT/2017

References

[1] Nayak, A K 2010 *Int. J. ChemTech Res.* **2** 903-907
[2] Oshida, Yoshiki., Wang, C. Sun., Ou, and K. Liang. 2015 *Hydroxyapatite Synthesis and Applications*. New York: Momentum Press.
[3] N. Jamarun, Z. Azharman, ZILFA and U. Septiani 2016 *Oriental Journal of Chemistry* **23** 2095-2099
[4] Krishnamurithy G 2013 *J. Heal. & Trans. Med* **16** 1-6
[5] M. Ansari, S.M. Naghib, F.Moztarzadeh. 2011 J. *Ceramics-Silikáty*. **55** 123-126
[6] S. Adzila, I. Sopyan, M. Hamdi, 2011 *Int. Fed.Med.& Biol.Eng.* **35** 97-101
[7] V. Sergey, Dorozhkin 2009 *Materials* **2** 1975-2045
[8] U. Ripamonti, J. Crooks, L. Khoali, L. Roden, 2009 *Biomaterials* **30** 1428–1439
[9] K.S. Vecchio, X. Zhang, J.B. Massie, M. Wang, C.W. Kim 2007 *Acta Biomaterialia* **3** 910–918
[10] A. A Baba, I.T.,Oduwole, F.O Salami, F.A. Adekola, and S.E Adeboye 2013 *Life Journal of Science* **15** 435-443
[11] S. Y. Bircan, I. Naruse, K. Matsumoto, K. Kitagawa 2013 *Journal of Sustainable Bioenergy Systems* **3** 74-78
[12] R. Puntharoda, C. Sankrama, N. Chantarameec, P. Pookmaneea, and K. J. Haller 2013 *Journal of Ceramic Processing Research* **14** 198-201.
[13] K. Suchanek, A. Bartkowiak, A. Gdowik, M. Perzanowski, S. Kąc, B. Szaraniec, M. Suchanek, and M. Marszałek 2015 *Materials Science and Engineering*:C **51** 57-63
[14] D. Sidane, H. Ramma, A. Beljebbar, S.C Gangloff, D. Chicot, F. Velard, H. Khiddeine and A.M. Kerdjoudj. 2017 *Materials Science and Engineering C* **53** 650-658
[15] B.T. Reves, J. A. Jennings, J D. Bumgardner and W. O. Haggard 2012 *J. Funct. Biomater.* **3** 114-130
[16] W. Michał, D.Ewa, and C.Tomasz 2015 *Colloid and polymer science* **293** 1561-1568

Effect of Evaporation Time on Separation Performance of Polysulfone/Cellulose Acetate (PSF/CA) Membrane

Intan Syahbanu[1], Bambang Piluharto[2], Syahrul Khairi[3], Sudarko[2]

[1]Departement of Chemistry, Faculty of Mathematics and Natural Science, Universitas Tanjungpura
[2]Departement of Chemistry, Faculty of Mathematics and Natural Science, Universitas Jember
[3]Departement of Chemical Engineering, Faculty of Technique, Universitas Tanjungpura

Email : intan.syahbanu@chemistry.untan.ac.id; intansyahbanu@gmail.com

Abstract. Polysulfone and cellulose acetate are common material in separation. In this research, polysulfone/cellulose actetate (PSF/CA) blend membrane was prepared. The aim of this research was to study effect of evaporation time in casting of PSF/CA membrane and its performance in filtration. CA was obtained by acetylation process of bacterial cellulose (BC) from fermentation of coconut water. Fourier Transform Infra Red (FTIR) Spectroscopy was used to examine functional groups of BC, CA and commercial cellulose acetate. Subtitution of acetyl groups determined by titration method. Blend membranes were prepared through phase inversion technique in which composition of PSF/PEG/CA/NMP(%w) was 15/5/5/75. Polyethyleneglycol (PEG) and N-methyl-2-pyrrolidone (NMP) were act as pore forming agent and solvent, respectively. Variation of evaporation times were used as parameter to examine water uptake, flux, and morphology of PSF/CA blend membranes. FTIR spectra of CA show characteristic peak of acetyl group at 1220 cm^{-1}indicated that BC was acetylated succesfully. Degree of subtitution of BCA was found at 2.62. Highest water flux was performed at 2 bar obtained at 106.31 $L.m^{-2}.h^{-1}$ at 0 minute variation, and decrease as increasing evaporation time. Morphology of PSF/BCA blend membranes were investigated by Scanning Electron Microscopy (SEM) showed that porous asymetric membrane were formed.

1. Introduction

Ultrafiltration technology has been interesting due to effectivities in separation performance. Polysulfone is common material for ultrafiltration. It has high durability and stability under mechanical, thermal and chemical treatment [1]. Unfortunately, it strongly hydrophobic, thus susceptible to fouling. Fouling in membrane was a major problem in environmental application, which cause lower water flux and membrane life, thus higher energy in operation [2]. Many research has been taken in order to improve performace of polysulfone, they are impregnating PSF with nanoparticles such as silver, TiO_2, ZrO_2, SiO_2[3][4] or blending with hydrophilic polymers such as nanocellulose [5] and cellulose acetate [6].

Blend polysulfone to another hydrophilic polymer is a simple way to improve its performance. Cellulose acetate known as good hydrophilic ultrafiltration materials, but has low stability in chemical treatment. Adding cellulose acetate in small portion to polysulfone, improve its properties, and reduce fouling properties. Cellulose acetate is able to obtained from natural resources such as refined fiber, cellulosic waste, deinked newspaper, cotton linter pulp and bacterial cellulose [7][8][9][10]. Bacterial cellulose is one of cellulose material with high crystallinity, and purity. They bring advantages in reduce step in isolation of cellulose, such as delignification and bleaching [10].

Polymer blend membranes can be prepared by phase inversion technique. Polymer solution is poured into glass plate then immerse into water in coagulation bath [6][11][12][13]. There are many factors affecting characteristic of blend membranes such as coagulation bath temperature [11][13], post treatment after solidification [12], and evaporation time [14] before immersion into coagulation bath.Variation on evaporation time has different characteristic due to evaporation of solvent before immersion into non solvent.

In this research, PSF/CA blend membranes are prepared by phase inversion technique. In the present work, cellulose acetate from renewable resources are used. Bacterial cellulose are used as the source for cellulose acetate (CA). Evaporation time is used as parameter to study characteristic of membrane and separation performance on filtration of oil/water emulsion.

2. Materials and Methods

Food grade sugar and coconut water were obtained from traditional market. *Acetobacter xylinum* was obtained from biology laboratory of Universitas Negeri Semarang, glacial acetic acid, sodium hydroxide, n-methyl-2-pyrrolidone (NMP), polyethyleneglycol-400 (PEG-400), sulfuric acid, acetic anhydride were obtained from Merck. Polysulfone (MW~35.000) and commercial cellulose acetate (MW~30.000) were obtained from Sigma Aldrich.

2.1. Preparation of Cellulose Acetate

Bacterial cellulose acetate was prepared by acetylation of bacterial cellulose. Bacterial cellulose was produced by fermentation of *Acetobacter xylinum* in coconut water medium [10]. Acetylation of bacterial cellulose was using acetic anhydride as precursor and sulfuric acid as catalyst [7][8]. Brown solution was obtained after acetylation at 60 °C for 4 hours. The final product was filtered and washed until neutral pH and then dried at 50 °C. The degree of subtitution was determined by titration method [8]. Functional groups of bacterial cellulose and bacterial cellulose acetate were investigated using Fourier Transform Infra Red (FTIR) Spectroscopy. Thermal characteristic of cellulose acetate was examined by Differential Scanning Calorimetry (DSC 8230- Rigaku) analysis by determine Transition Glass Temperature (T_g), Melting Temperature (T_m), Entalphi of fusion on crystalline region (ΔH_f) and crystallinity index (%X_c) which determined by equation (1)[15]. Fusion entalphi of perfectly crystalline region of cellulose triacetate (ΔH_{100}) was 58.8 $J.g^{-1}$ taken from[16].Thermal analysis were carried out from 30°C to 350 °C with heating rate at 5 °C/minutes. Furthermore, cellulose acetate was used for membrane preparation.

$$X_c(\%) = \left(\frac{\Delta H_f}{\Delta H_{100}}\right) x100 \dots\dots\dots\dots(1)$$

2.2 Preparation of PSF/CA blend membranes

Blend membranes was prepared through phase inversion technique. Polysulfone (PSF) was blended with Cellulose Acetate (CA) using Polyethyleneglycol-400 (PEG) and N-Methyl-2-Pyrrolidone (NMP) as an aditive and solvent, respectively. Composition of PSF/BCA/PEG/NMP were 15/5/5/75 (%w). Evaporation times (0; 0.5 ; and 1 minutes) were use as parameter to study separation performance of blend membranes. Morphology of blend membranes was examined by Scanning Electron Microscopy (SEM) at Politeknik Negeri Jember. Porosity of PSF/CA membranes were calculated by equation (2), with density of water (ρ) was 998 kg/m^3, membrane area (A) and membrane thickness (l). Weight of dry (w_{dry}) and wet (w_{wet}) membranes were obtained by immersion dry membrane on water for 24 hours. Surface of the membrane was wiped out to remove excess water, then weight as wet weight [19].

$$Porosity\ (\%) = \frac{w_{wet}-w_{dry}}{\rho\ .A.l} \text{ x } 100 \ldots\ldots\ldots\ldots(2)$$

2.3 Separation Performance of PSF/CA blend membranes

Blend membranes performace was examined using dead-end ultrafiltration cell at 200 kPa in pressure[17]. Water flux (J) was determined by permeation of pure water flux (eq.3), where V was defined as volume of water passed through membrane (L), A was membrane area (m^2) and Δt was sampling time for permeation of pure water.

$$J = \frac{V}{A.\Delta t} \quad \ldots\ldots\ldots(3)$$

Separation performace of blend membrane was studied by filtration of dichloromethane in water emulsion. Dichloromethane in water was prepared as previous research [18]. Fouling character was evaluated by Flux Recovery Ratio (FRR), Total Fouling (R_t), Reversible Fouling (R_r) and Irreversible Fouling (R_{ir}). FRR was determined by equation (4), which Jw_1 was pure water flux and Jw_2 was recovered water flux after filtration.Total, Reversible, and Irreversible Fouling were determined by equation (5), (6) and (7), respectively, with J_p was permeate flux[2][19].

$$FRR\ (\%) = \left(\frac{Jw_1}{Jw_2}\right) x100 \ldots\ldots\ldots\ldots(4)$$

$$R_t(\%) = \left(\frac{J_{w1} - J_p}{J_{w1}}\right) x\ 100 \ \ldots\ldots\ldots(5)$$

$$R_r(\%) = \left(\frac{J_{w2} - J_p}{J_{w1}}\right) x100 \ldots\ldots\ldots(6)$$

$$R_{ir}(\%) = \left(\frac{J_{w1} - J_{w2}}{J_{w1}}\right) x100 \ldots\ldots\ldots(7)$$

3. Results and Discussion

3.1Characterization of Cellulose Acetate

Cellulose acetate has successfully synthesized from bacterial cellulose (BC). BC was obtained after fermentation of *Acetobacter xylinum* on coconut water. White pellicle was obtained after six days of fermentation. It has large amount of hydroxyl groups which interact each other through intermolecular hydrogen interaction.Therefore, BC was difficult to dissolve in various solvent.

Fourier Transform Infra Red (FTIR) spectra (Fig.1a) showed thatBC had large amount of hydroxyl groups indicated by broad spectra at 3336.07 cm^{-1}, hence, BC was naturally hydrophilic which indicated by sharp peak of H-O-H bending from absorbed water at 1649.69 cm^{-1}. Acetylation process by acetic anhydride had substituted hydroxyl groups of BC with acetyl groups, which indicated by reduction intensity of–OH hydroxyl grups at 3573.36 cm^{-1} leads reduced intensity of H-O-H bending at 1640.46 cm^{-1}. Presence of acetylgroups were showed by characteristic peak of C-O acetyl at 1223.35 cm^{-1} and increased peak at 1742.18 cm^{-1} indicating stretching of C=O esters. Shift peak at 887 cm^{-1} to 903.67 cm^{-1} showed vibration of β-glucosidic linkages between sugar unit. Detailed of FTIR spectra interpretation were showed at Table 1.

Table 1 FTIR spectra of bacterial cellulose and cellulose acetate

Wavenumber(cm^{-1})		Functional Groups [8][15]
Bacterial Cellulose	**Cellulose Acetate**	
3336.07	3573.36	O-H *stretching*
	2954.67	C-H *stretching*
2922.55		
1752.04	1742.18	C=O *stretching* of ester
1639.69	1640.45	H-O-H bending of absorbed water
1330.74	1372.23	C-H deformation
-	1223.35	C-O stretching of acetyl group
1058.96	1041.85	C-O asymmetric stretching of primary alcohol
887.00	903.67	β-glucosidic linkages between sugar unit

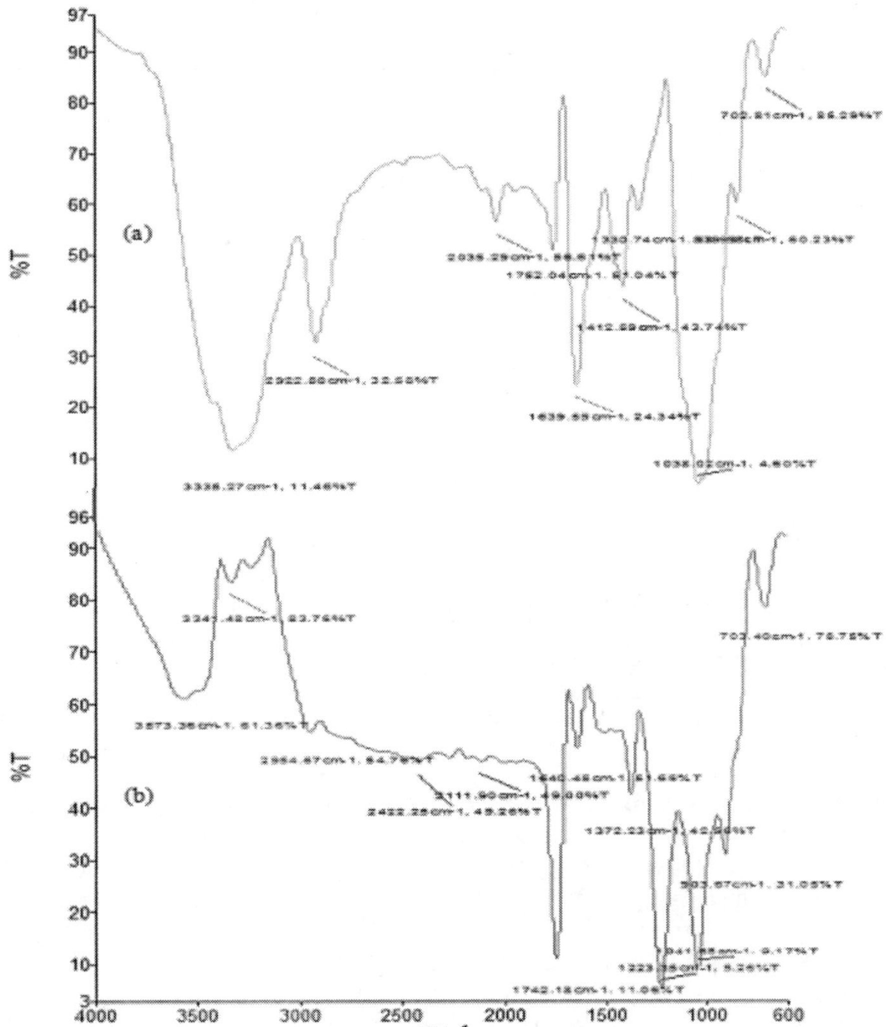

Figure 1 FTIR Spectra of (a) Bacterial Cellulose (b) Cellulose Acetate

IOP Conf. Series: Materials Science and Engineering **299** (2018) 012040 doi:10.1088/1757-899X/299/1/012040

Degree of subtitution (DS) is a value that indicate average number of hydroxyl groups on glycosidic rings of cellulose subtituted by acetyl groups [8]. DS value is affected on character and solubility of cellulose acetate. In this research, DS value and percentage of acetyl content were found at 2.62 and 41.413%, respestively. In comparison with commercial cellulose acetate (*Sigma Aldrich*), DS value and acetyl content were 2.17 and 36.943%. It was concluded acetylation process at 4 hours results cellululose diacetate. This result similar to previous research which variation of acetylation time (0.5 to 24 h), were found cellulose diacetate [15].

Thermal characteristic of cellulose acetate was examined by DSC. Figure 2 was DSC curve of commercial and synthetic of cellulose acetate.Thermal behaviour of synthetic cellulose acetate showed higher transition glass (T_g) and melting temperature (T_m) than commercial cellulose acetate (Table 2). It might cause by lower DS value and acetyl content of commercial cellulose acetate. However, commercial cellulose acetate had higher fusion entalphi and crystallinity index. Both value were lower for synthetic cellulose acetate because of acetylation process had used high concentration of acid, hence, formation of crystalline region were influenced. Overall, there were no significant differences between both cellulose acetate.

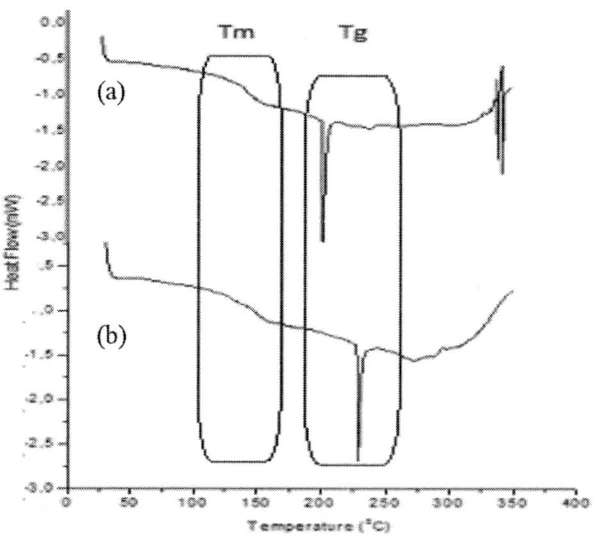

Figure 2 DSC curve of Cellulose Acetate

Table 2 Thermal Characteristic of commercial and synthetic cellulose acetate

Cellulose Acetate	Tg (°C)	Tm (°C)	ΔH_f (J/g)	Xc (%)
Commercial	134.5	203.5	23.68	40.27
Synthetic	150.2	228.8	15.43	26.24

3.2. Characterization of Blend Membrane PSF/CA

Blend membranes PSF/CA were prepared by phase inversion technique using water as non solvent agent. SEM images (Fig.3) of blend membranes PSF/CA showed porous structure for all variation of evaporation times. PEG-400 as additive in dope solution act as pore forming agent [6]. Evaporation times in phase inversion technique affected to the pore formation and membrane structure. Finger-like structure of pores were found on membranes at evaporation time 0 minutes (Fig 3a). Longer evaporation time induce shrinkages of the pore (Fig. 3b and 3c). It can be explained by kinetic aspect on membrane formation by phase inversion technique. The kinetic aspect was related to mass transfer during coagulation or solvent-non solvent exchange. Mass transfer was conducted by different concentration between solvent (NMP) and non solvent (water). When polymers solution had contact with non solvent, liquid-liquid demixing was occured. In the case of instaneous demixing, membrane pores were formed immediately induced porous membrane structure. On the other hand, delayed demixing was conducted membranes with more dense structure, due to slow pores formation [14]. Therefore, in the case effect of evaporation time, different porosity of PSF/CA membrane were obtained.

Porosity of PSF/CA membranes were decrease as increasing evaporation time (Fig.4). Membrane with no evaporation time had higher porosity by instaneous demixing process. However, membranes with 0.5 and 1 minutes evaporation time introduce to delayed demixing, hence, give lower porosity due to evaporation of solvent before immersion on water. When solvent had evaporated, concentration and viscosity of polymer solutions were incerase, thus, solvent-non solvent exchange process was being restricted. Therefore, lower porosity were obtained for 0.5 and 1 minutes evaporation time.

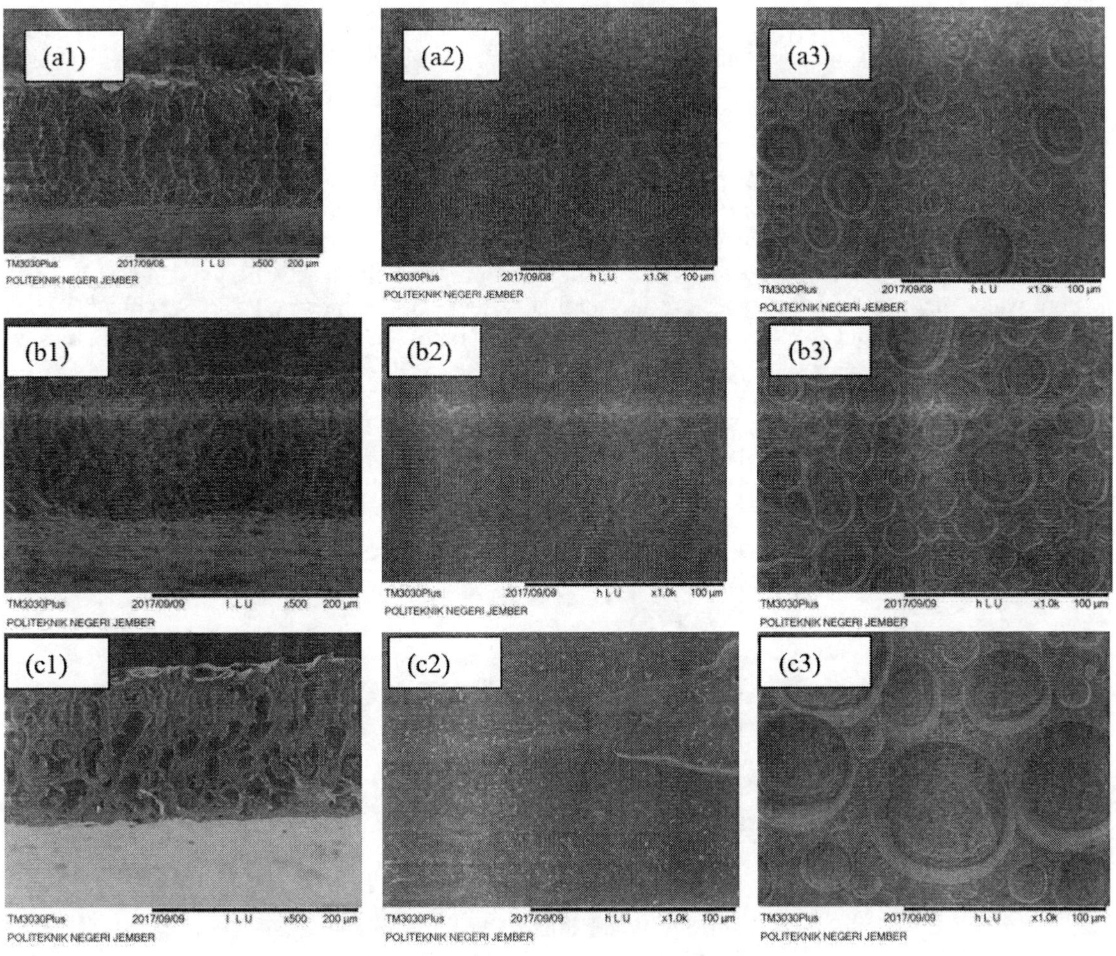

Figure 3 SEM Images of PSF/CA membranes on cross-sectional (1), surface (2) and bottom(3), respectively, at (a) no evaporation time (b) 0.5 min and (c) 1 min evaporation time

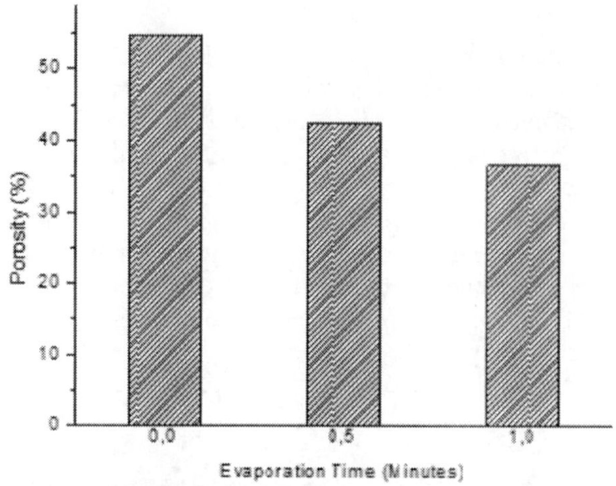

Figure 4 Porosity of PSF/CA membranes

3.3 Separation Performance of PSF/CA Blend Membranes

Pure water flux value of PSF/CA membranes with different coagulation time os shown on Fig.5. Adding small portion of CA on polymer solution had increase hydrophilic properties of the membranes. Increasing evaporation time has lowering water flux due to reduction of porosity. Moreover, delayed demixing (on 0.5 and 1 min) could produce denser pores structure, hence, reducing water flux. PEG-400 has play important role on pore formation. Pore network gave higher change for pure water run into the membranes produce higher water flux. Membrane PSF/CA without PEG has low water flux, about 9.8 $L.m^{-2}.h^{-1}$ [6]. It was 5-10 times lower than PSF/CA membranes with addition PEG-400.

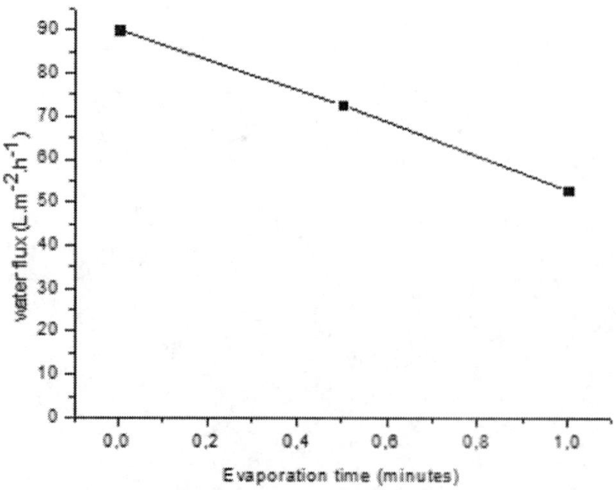

Figure 5 Water Flux on PSF/CA membranes

Separation performance of PSF/CA membranes were studied using surfactant stabillized dichloromethane/water emulsion.Water and permeate flux for membranes with different evaporation time depicted on Fig.6. Membrane with no evaporation time had higher flux and decrease for the next variation.Pores structure and porosity were play important role as describe in previous. The flux on membranes decrease sharply when water (J_{w1}) had replaced by permeates (J_p). It caused by fouling process had started on membrane and clog the pores. The flux remain increase for the second filtration of water (J_{w2}) after cleaning process by water, indicate PSF/CA membranes had ability to recover the performance which determined by Flux Recovery Ratio (FRR) value.

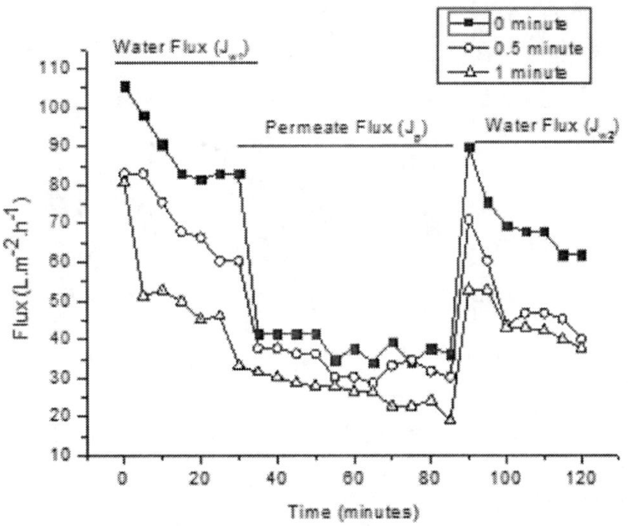

Figure 6 Water and permeate fluks on PSF/CA membranes

Flux recovery ratio (FRR) study show membranes with 0.5 minutes evaporation time had lower FRR (Fig.7). Fouling on membrane surfaces or pores has contribute to the recovery flux (J_{w2}). There were reversible (R_r) and irreversible (R_{ir}) fouling. Highest irreversible fouling was found on PSF/CA membrane with 0.5 minutes evaporation time (Fig.8), hence, decrease FRR value. Fig.8 was depict study of fouling resistance on PSF/CA membranes.

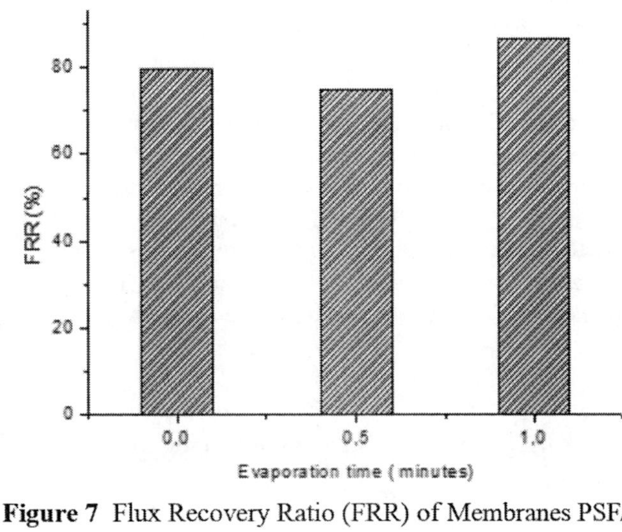

Figure 7 Flux Recovery Ratio (FRR) of Membranes PSF/CA

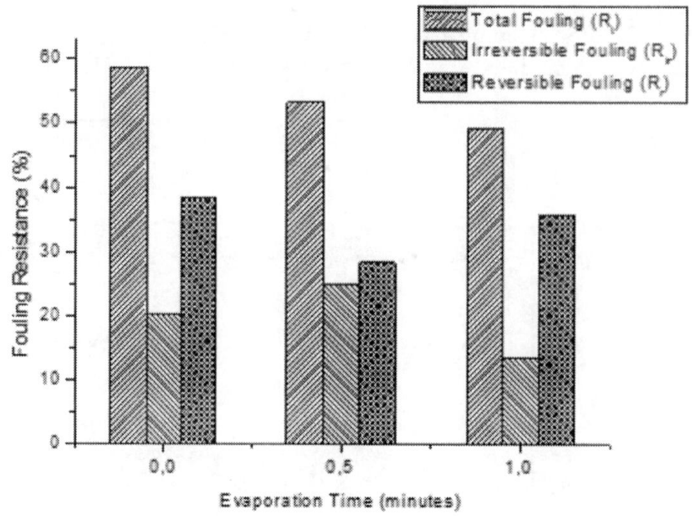

Figure 8 Fouling Resistance of PSF/CA membranes

Fouling resistance study on PSF/CA membranes show that increase on evaporation time had decrease the total fouling (R_t). The highest total fouling (R_t) was up to 58% for 0 minutes evaporation time. All of PSF/CA membranes show higher reversible fouling (R_r) that irreversible (R_{ir}) one. It might be caused by physical interaction between emulsion with membrane pores, therefore, it could be replaced from the pores by cleaning process with water. Irreversible fouling (R_{ir}) in this research did not exceed 25%, obtained by 0.5 minutes evaporation time, due to penetration and entrapping emulsion on membrane pores. As a results, blend membrane PSF/CA had good antifouling properties.

4. Conclusion

PSF/CA blend membranes were prepared with using bacterial cellulose as the source for CA. Characteristic of CA from bacterial cellulose were evaluated and show similar properties with commercial cellulose acetate (*Sigma Aldrich*). Evaporation time (0; 0.5; 1 min) before immersion were used as parameter to study characteristic and performance of PSF/CA blend membranes. SEM images show porous structure of membranes which influenced by presence of PEG-400. Porosity of PSF/CA membranes were decrease as increasing evaporation time. Membrane performance on filtration show highest flux obtained by 0 minutes evaporation time. FRR study show all membranes has recover the second flux up to 75%. However, membrane on 0.5 min evaporation time show lowest FRR due to irreversible fouling by entrapped emulsion on its pores. The highest total fouling (R_t) was 58%, but all of membranes show higher reversible fouling (R_r) than irreversible (R_{ir}) one. In conclusion, PSF/CA based on bacterial cellulose has good antifouling properties..

Acknowledgment

The authors thanks to Ministry of Research, Technology and Higher Education for financial support to this research through PEKERTI Program.Furthermore, authors thanks to Mita Yuspitasari, Afaf Sri Hartini and Ardita Rusmaningsih for being a good teamwork.

References

[1] Liu C, Mao H, Zhu J, Zhang S 2017 *J.Seppur.***179** 215
[2] Manawi Y, Kochkodan V, Mohammad A W, Atieh M A 2017 *J.memsci.***529** 95
[3] Zodrow K, Brunet L, Mahendra S, Li D, Zhang A, Li Q, Alvarez PJJ 2009 *J. Wat. Res.***43** 715
[4] Richards HL, Baker PGL, Iwuoha E 2012 *J.SEMAT.* **2** 183

[5] Zhong L, Ding Z, Li B, Zhang L 2015 *BioResources.* **10(2)** 2936
[6] Shivakumar M. Susithra L, Mohan DR, Rangarajan R 2006 *J. Macromol. Sci.***43** 1541
[7] Fei Z, Huang S, Yin J, Xu F, Zhang Y 2015 *J. Polym. Environ.***2** 383
[8] Nedjma S, Djidjelli H, Boukerrou A, Benachour D, Chibani N 2012 *J. Appl. Polym. Sci.* **12** 4795
[9] Luo P, Xin C, Ma X, Liang X, Jiao Z, Xu S, Zhang J 2013 *BioResources***8(4)** 5635
[10] Radiman C and Yuliani G 2008 *Polym. Int.* **52** 502
[11] Saljoughi E, Amirilargani M, Mohammadi T 2010*Desalination***262** 72
[12] Mansourizadeh A and Asad A J 2014 *J. Polym. Res.* **21** 375
[13] Peng J, Su Y, Chen W, Shi Q, Jiang Z 2010 *Ind. Eng. Chem. Res.* **49** 4858
[14] Holda A K, Aernouts B, Saeys W, Vankelecom IFJ 2013 *J. Mem. Sci.* **442** 196
[15] Barud HS, Junior AMA, Santos DB, de Assuncao RMN, Meireles CS, Cerqueira DA, Filho GR, Ribeiro CA, Messadeq Y, Ribeiro SJL, *Thermochimica Acta* 2008 **471** 61
[16] Cerqueira DA, Filho GR, Assuncao RMN 2006 *Polymer Bulletin***56** 475
[17] Jalali A, Shockravi A, Vatanpour V, Hajibeygi M 2016 *J. Micomeso.* **228** 1
[18] Chen PC and Xu ZK 2013 *Scientific Reports***3** 2776 *doi : 10 1038/srep 02776*
[19] Golpour M and Pakizeh M 2017 *J. Seppur.* **183** 237

Synthesis and Characterization of Hydrophobic Silica Thin Layer Derived from Methyltrimethoxysilane (MTMS)

Adi Darmawan*, **Riyadini Utari, Riza Eka Saputra, Suhartana, Yayuk Astuti**

Department of Chemistry, Diponegoro University, 50275 Semarang, Indonesia

*E-mail: adidarmawan@live.undip.ac.id

Abstract. This study investigated the synthesis and characterization of MTMS hydrophobic silica prepared by sol-gel method. In principle, silica xerogels and silica thin layer were obtained by reacting MTMS in ethanol solvent in some pH variations. The MTMS solution was used to modify the surface of the ceramic plate by dipcoating method to further be calcined at two different temperatures of 350°C and 500°C. The silica xerogels were analysed by FTIR, TGA-DSC and GSA to determine functional group characteristics, thermal properties and pore morphology respectively. Meanwhile, the silica thin layers were analysed their hydrophobic properties using water contact angle measurement and surface roughness determination using SEM. The results showed that the higher the pH used in the MTMS solution, the higher the resulting contact angle. The highest contact angle was obtained at pH 8.12 which reached 94.7° and 79.5 ° for silica thin layer calcined at 350°C and 500°C, respectively. The TGA results indicated that the methyl group survived up to 400°C and disappeared at 500°C which had implications on silica thin layer hydrophobic nature. GSA result exhibited that the silica xerogel had a close structure with a very low pore volume. While the SEM-EDX results displayed that the silica thin layer prepared at acidic pH had smoother surface morphology and became rough when prepared at an alkaline pH.
Keywords: silica thin layer, hydrophobic silica, methyltrimethoxysilane

1. Introduction

In recent years, silica-based materials have become very attractive because of their unique properties. The magnitude of the possibility to functionalize the silica material, on the surface, the pores and the structure, makes silica applicable to a wide range of applications. The industry's most promising application is the need for thin layers with various structural parameters such as thickness, size and hydrophobicity [1]. Thin layers made from organic silane functioned as precursor were used extensively to modify surfaces hence they had a wide spectrum of applications for example, for coating automotive parts, medical purposes and coating of diagnostic devices. In addition, silane was also commercially available and easy to use [2]. Silane is a substance that can react with an inorganic substrate such as glass or ceramic to form a stable covalent bond thus as to alter the physical interaction of the coated substrate [3].

The various silanes can be used as starting materials for the manufacture of thin silica coatings, some of which are tetraethoxysilane (TEOS) [4], trimethyletoxysilane (TMES) [5], phenylthriethoxysilane (FTES) [6], tetramethoxysilane (TMOS)) [7], ethyltriethoxysilane (ETOS) [8], trivinylmethoxysilane (TVMS) [9], trimethylchlorosilane (TMCS) [10] and hexamethoxydisilane [9]. From various sources of silica precursor, TEOS is the most widely used [11]. Several studies [12-15] have reported that the use of TEOS resulted in the surface of silica becoming hydrophilic. This indicates that TEOS has a strong hydrophilic character.

On the other hand, the hydrophobic layer is of great concern because of its remarkable nature and has a wide range of potential applications in industry and domestic such as for oil-water separation membranes, anti-dirty surfaces, and coatings on cars [16]. The hydrophobic surface displays a water contact angle greater than 150° consequently it is water resistant. This contact angle is influenced by Si-O-Si, Si-C and C-H functional groups. The more the number of those functional groups, the greater the hydrophobicity properties [17]. To obtain materials that have hydrophobic properties it is necessary to have alternative compounds having larger hydrophobic group contributor.

Methyl trimethoxy silane (MTMS) is a silica precursor having a methyl group which is hydrophobic. Hence, if a surface coated by the MTMS it will be non-polar and more hydrophobic. Several related studies of the hydrophobic silica thin layer of methyl trimethoxy silane (MTMS) have been performed [16, 18-20]. The silica thin layer was prepared by dipcoating method in methanol solvent with volume ratio of MTMS:methanol:oxalic acid:NH_4OH of 1: 10: 1: 1 produced a water contact angle of 168° [16]. The aerogel-glass fibres ceramic matrix composite was also successfully produced by sol gel and freeze drying technique using MTMS precursor with superior properties. The composite had a great hydrophobicity and the contact angle up to 131° was achieved [18].

On the other hand, the pH of the sol-gel reaction has been revealed to have considerable influence on the silica polymerization process thus affecting the characters of the resulting silica polymer [21, 22]. The pH affects the hydrolysis and condensation reaction rate, the structure and amount of silica formed. Whilst, the reaction rate of hydrolysis and condensation had the most important effect on the final structure [23]. However, as far as we know, there is no in-depth study of the relationship between pH to silica thin layer hydrophobicity.

In this research, synthesis and characterization of hydrophobic silica thin layer derived from methyl trimethoxysilane (MTMS) through sol-gel process and dip coating were conducted. The effect of pH variation and the calcination temperature on the hydrophobic properties of silica thin layer obtained was assessed using water contact angle measurements, FTIR, TGA-DSC, GSA and SEM-EDX.

2. Experimental Section

2.1. Materials

Materials used were methyl trimethoxy silane (MTMS) 98% Merck Sigma Aldrich (CAS 1185-55-3), ethanol 99.99% Merck Sigma Aldrich (CAS 64-17-5), glacial acetic acid 17.4 M 100% Merck Sigma Aldrich (CAS 64-19-7), ammonia 12.6 M 25% Merck Sigma Aldrich (CAS 7664-41-7), distilled water and a commercial ceramic plate.

2.2. Synthesis and characterization of MTMS xerogel

The MTMS solution was prepared using ethanol solvent with a mole ratio of MTMS: ethanol = 50: 1. Five pH variations were used wherein the ethanol solvent pHs were arranged first by adding acetic acid or ammonia before mixing with the MTMS. The ammonia or acid compositions added are presented in Table 1. Furthermore 3.386 mL of MTMS was added to 70 mL of ethanol solution and stirred using magnetic stirrer for 30 min. The MTMS solution was then dried at room temperature for 2 days, after that the uncalcined xerogels were characterized by the gravimetric analysis performed on a Shimadzu thermo gravimetric analyser TGA-50 using air flow rate of 60 mL min^{-1} and heating rate of 2°C min^{-1}. Furthermore, the MTMS xerogels were calcined at 350°C and 500°C for 30 min with a ramp rate of 2°C min^{-1}. The MTMS silica xerogels in powder form were characterised using a Shimadzu IRAffinity-1 Fourier-transform infrared (FTIR) spectrometer with a Pike MIRacle attenuated total reflectance accessory (ATR-FTIR) at wavelength range 600-4000 cm^{-1}. Samples were degassed under vacuum for more than 6 h at 150 °C before nitrogen adsorption analysis was performed at 77 K and 1 bar using Micromeritic TriStar 3020 instrument. The specific surface area was determined by the Brunauer, Emmett and Teller (BET) method and total pore volume was taken from the last point of the isotherm. Average pore diameter (4V/A by BET) was used to determine the average pore sizes.

Table 1. The concentration of acid or base added in ethanol solution

	pH before adding MTMS	pH after the addition of MTMS
1 M acetic acid in ethanol solution	3.44	2.91
0:25 M acetic acid in ethanol solution	4:52	3.48
0.025 M acetic acid in ethanol solution	5.82	4.24
Liquid ethanol without addition of acid or base	8.27	5.49
0.1 M ammonia in ethanol solution	9.78	8.12

2.3. Hydrophobic silica layer formation on glass surface

The ceramic surface modification was carried out by a dipcoating process in which the ceramic was immersed in the MTMS solution by dipping and withdrawal rate of 10 cm min^{-1} until the entire surface was submerged for 60 seconds. The ceramics were removed from the MTMS solution slowly and dried at room temperature, then calcined at 350°C and 500°C for 30 min at a heating rate of 2°C min^{-1}. The process of dipcoating and calcination was repeated 4 times hence the layer formed was not too thin to prevent the surface rupture. The hydrophobic silica layers formed on the ceramic surface were measured their water contact angle by a water dropping on their surface and then photographed using Samsung Galaxy A5 camera (2016). The θ/2 method is generally used for measurement of contact angle. The contact angle was determined by the equation given below. Where radius r and height h of the droplet.

$$\tan\theta_1 = \frac{h}{r} \rightarrow \theta = 2\ arctan\frac{h}{r}$$

The contact angle was obtained from the angle of straight lines connecting the left and right ends and the apex of the droplet against a solid surface as shown in Fig. 1.

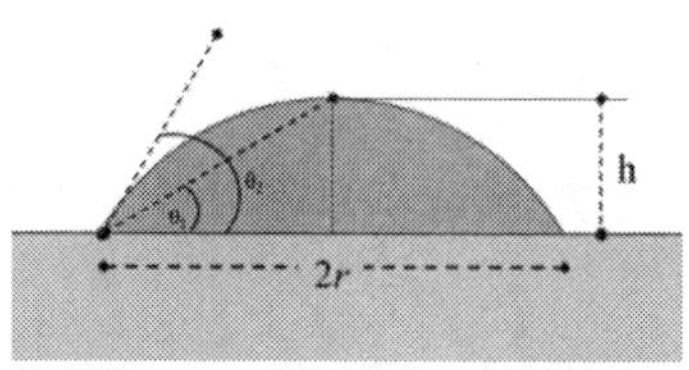

Figure 1. Water contact angle measurement

The surface morphology of the ceramic surface was analysed using SEM-EDX 6510 (LA) operated at 20 kV acceleration voltage with dispersive Si (Li) X-ray energy spectrometer. The EDX spectrum was obtained using a JED-2300 system at 20 kV and a distance of 10 mm. The EDX spectra to see the morphology of a thin layer of hydrophobic silica MTMS were analysed by integrated JEOL Analysis Station Software (Version 3.8)

3. Results and Discussion

3.1. The water contact angle on MTMS ceramic plates

The water contact angle analysis aims to determine the hydrophobicity of the MTMS silica thin layer on the ceramic surfaces prepared at different pH and calcination temperatures. On each surface, the water was dropped 3 times on different sides to see the contact angle uniformity on the MTMS silica layer formed. Then the average contact angle was obtained by calculating the average value of three different side water contact angles. Water contact angles at various surfaces are shown in Fig. 2.

350 °C

500 °C

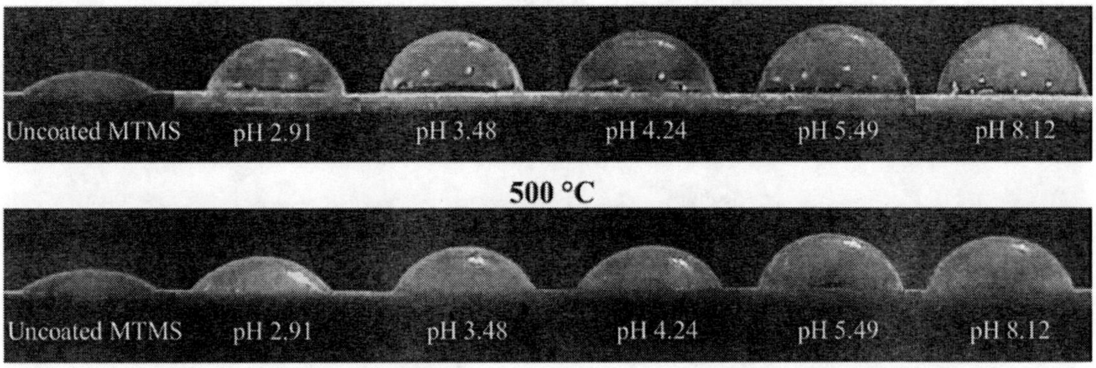

Figure 2. The water droplets on the surface of MTMS silica thin films

Figure 2 shows the water droplets on the thin layer of the ceramic surface without MTMS coating and MTMS coated ceramic surfaces prepared at various pHs and calcination temperatures of 350°C and 500°C. It can be seen that the MTMS silica thin layer had a great influence on the surface hydrophobicity. In ceramic without MTMS coating, the water droplets formed were very wide and spread on the ceramic surface, the water contact angle on the ceramic surface was 37°, while after the addition of MTMS, the water droplets were much more rounded which directed the ceramic surface was more hydrophobic. The higher the pH the greater water contact angle whereas the contact angles of the MTMS silica thin films calcined at 350°C were always higher than that of 500°C for all pH variations.

To find the relationship between the contacts angles and pH, the a correlation graph was made as presented in Figure 3. Figure 3 presents the relationship between the pH of the MTMS solution to the water contact angles generated by the MTMS silica thin layers, which were calcined at two different calcination temperatures. The higher the pH the greater the contact angle, both on the silica calcined at 350°C or 500°C. The MTMS silica thin layers calcined 350°C had water contact angles greater than that calcined at 500°C since at 500°C the methyl group had been oxidized henceforth the surface hydrophobicity became considerably reduced. These results are supported by the TGA-DSC results presented in Fig. 4 and Fig. 5. According to Mahadik et. al [24], the silica layers began losing their hydrophobicity above 550° C and becoming superhydrophilic at 600°C due to the oxidation of methyl groups and other organic groups. According to Budunoglu et. al [10], the silica thin layer could retain the hydrophobic properties up to 500°C and at higher temperatures the contact angle decreased and the superhydrophilic layer with contact angle <5°C was obtained at a temperature of 600°C. Meanwhile, according to [25], xerogels lost the hydrophobic properties above 325°C, this was due to the fact that at 325°C, the Si-CH_3 group, which was the hydrophobic driving group, underwent oxidation causing the drop of hydrophobicity. Figure 3 has an error margin of 5% and indicates a high linearity in which the regression at 500°C is 0.998 and the regression at 350°C is 0.99. These results reveal that pH is linearly related to the water contact angle.

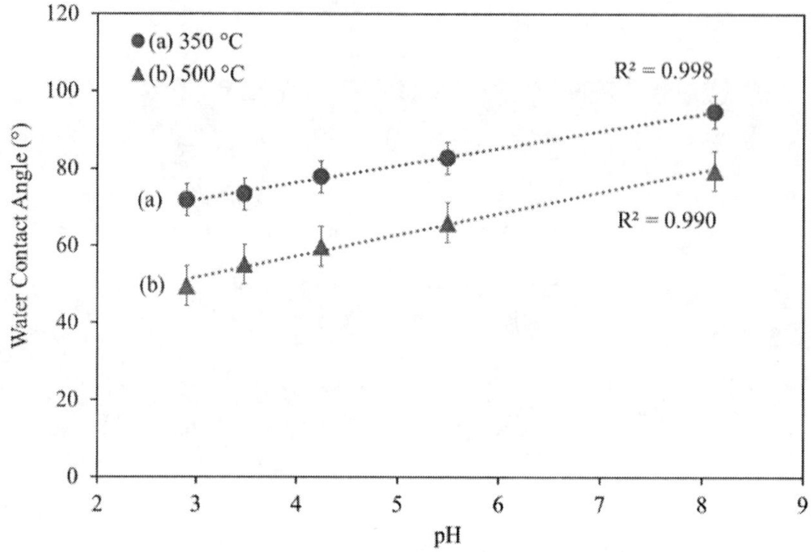

Figure 3. The relationship between the pH of the MTMS solution and the water contact angles

According to Uhlmann [26], acidic pH generated a long, branched and weak branch of a silica, thereby causing silica thin layer surfaces to become smoother. On the other hands, the alkaline pH resulted in the formation of highly branched and rough silica. The more rough the surface morphology, the higher the surface hydrophobicity of the material [27]. On the hydrophobic surface, surface roughness increases the water contact angle, therefore the surface rises resisting water. Whereas on a hydrophilic surface, the surface becomes softer hence the contact angle decreases and the surface more easily draws water [28]. This explains why the higher (the more alkaline) the pH the greater the contact angle formed.

3.2. Thermal Properties

The analysis of mass changes due to the thermal process was carried out using TGA to identify temperature region at which the dominant weight loss occurred on the MTMS silica xerogel. TGA test results are shown in Fig. 4 and 5.

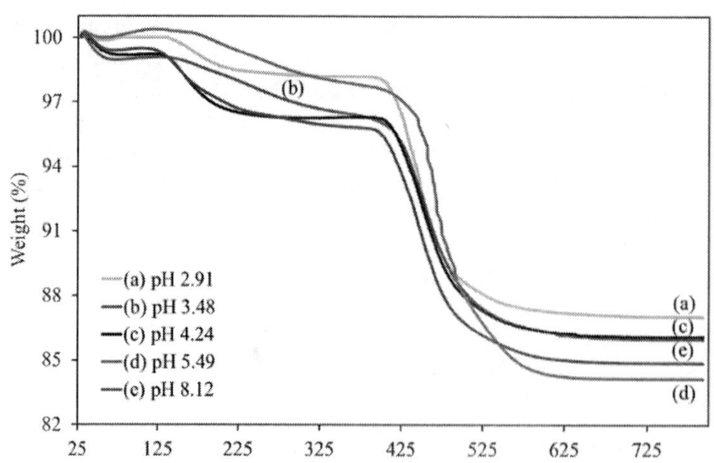

Figure 4. The mass loss of MTMS silica thin layer at various pHs

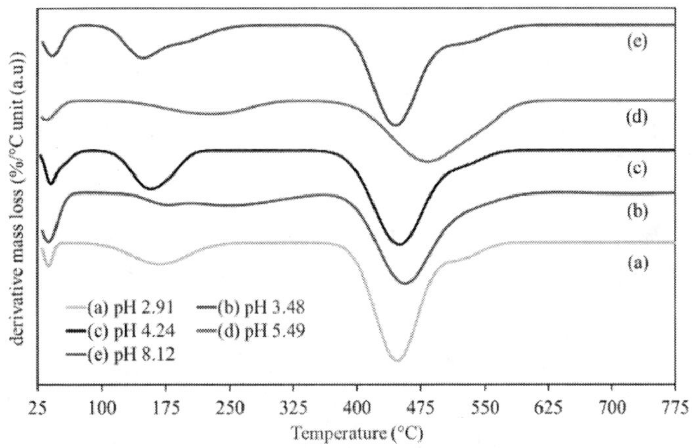

Figure 5. The first derivative of the TGA curves (the DTG curves) at various pHs

Fig. 4 presents a change in the weight of the MTMS silica xerogel as temperature upsurges. Overall, the TGA curves provide the same patterns for the five pH variations with a similar percentage of weight reduction. Fig. 4 and then clarified by Fig. 5 which is the first derivative of the TGA curve display that the process of weight change could be divided into 4 sections i.e. (i) 30°-125°C, (ii) 125°-325°C, and (iii) 325°-550°C (iv) 550°-800°C. At the section (i) there was a mass reduction caused by the loss of ethanol solvent and other volatile organic materials. The average weight loss was small which less than 2%. This result indicates that most of the ethanol has evaporated during the drying process at room temperature, as well as acetic acid or ammonia. The solvent and organic compound which evaporated in this first section were the solvent and volatile material trapped in the MTMS silica xerogel pores that have not been vanished during the drying at room temperature.

At section (ii) there was a reduction in weight at temperature of 125°-225°C and characterized by a peak at around 175°C in Fig. 5. This might be due to the further condensation process of the unreacted Si-OH group and formation of the siloxane bond (Si-O-Si) which also released the H_2O molecules as the by-product. In this phase, there was shrinkage and compaction of the silica polymers structure. Mass decline may also be due to the decomposition of a small fraction of methyl groups from MTMS and unreacted ethoxy groups. This is in line with [29] which suggested that a weight reduction of the sample heated on air atmosphere at a temperature of 150°C-280°C corresponded to the decomposition of an ethoxy or methoxy group.

In section (iii) there was a sharp mass decline observed at 375-550°C caused by the release of methyl (-CH_3) groups from the MTMS xerogel structure because the oxidation reaction formed CO_2 (equation 1) and the oxidation of Si-C group became Si-O which further formed a siloxane group. This result is in line with surface hydrophobicity data (Fig. 3) which indicating that the silica thin layer calcined at 500°C had water contact angles which were always lower than that of 350°C. According to Bhagat et. al. [30], this methyl group was responsible for the silica hydrophobicity. The sharp decline in temperature 375-550°C was also accompanied by an exothermic peak on the DSC curve at the same temperature (Fig. 6) due to the decomposition of the -CH_3 group and oxidation to CO_2 [31].

$$C_xH_y + O_2 \rightarrow CO_2 + H_2O \qquad (1)$$

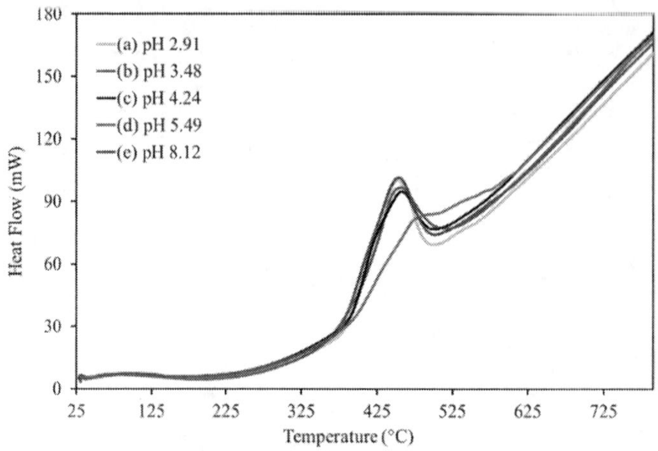

Figure 6. DSC curves for MTMS silica xerogel at various pH condition

Fig. 6 shows changes in heat flow as temperature increased. At temperatures below 250°C, the curves tend to be flat with slight escalations marked by low and wide peaks at 100°C due to the release of water molecules. This indicated that the silica weight reduction process caused by the solvent evaporation and the following condensation reaction did not release heat. Starting at 250°C there was a sharp rise noticed by an exothermic peak at a temperature of 450°C caused by the oxidation of the methyl group to CO_2 and Si-O. After a temperature of 525°C, heat flow continued to rise which might be due to shrinkage continuation of the silica polymer structure even though there was no change in weight occurred at that temperature. Based on the TGA-DSC data, it can be concluded that the MTMS hydrophobic silica xerogels were able to maintain their hydrophobic properties up to 370°C. The calcination above 370°C reduced the hydrophobic properties because the methyl group as a hydrophobic nature guide had been oxidized. However, the incidence of DSC curves for all pH variations indicated that pH gave little effect on the thermal properties of the MTMS silica xerogel.

3.3. Functional group on MTMS silica thin layer

The identification of the bond type and functional group presented in the MTMS silica xerogel was performed by FTIR analysis. The FTIR spectra of the synthesized MTMS silica xerogel are shown in Fig. 7. Fig. 7 confirms that the pH variation and the calcination temperature affected the position, shape and intensity of the MTMS silica xerogel absorbance. The emerging peaks are characteristic for silica polymers located at wave numbers between 1200-750 cm^{-1}, i.e. (i) wave numbers of 1150-1000 cm^{-1} indicated the presence of the symmetric and asymmetric vibrations of Si-O-Si [32], (ii) the shoulder region at wave number of 955-835 cm^{-1} in the form of weak absorption overlapping with strong adsorption was attributed to the Si-OH vibration, (iii) the wave number of 865-750 cm^{-1} was assigned to Si-C group which was easily recognized because of the strong and sharp absorption peaks [33], and (iv) a small peak of 1260 cm^{-1} corresponded to the symmetrical vibration of the C-H group [34].

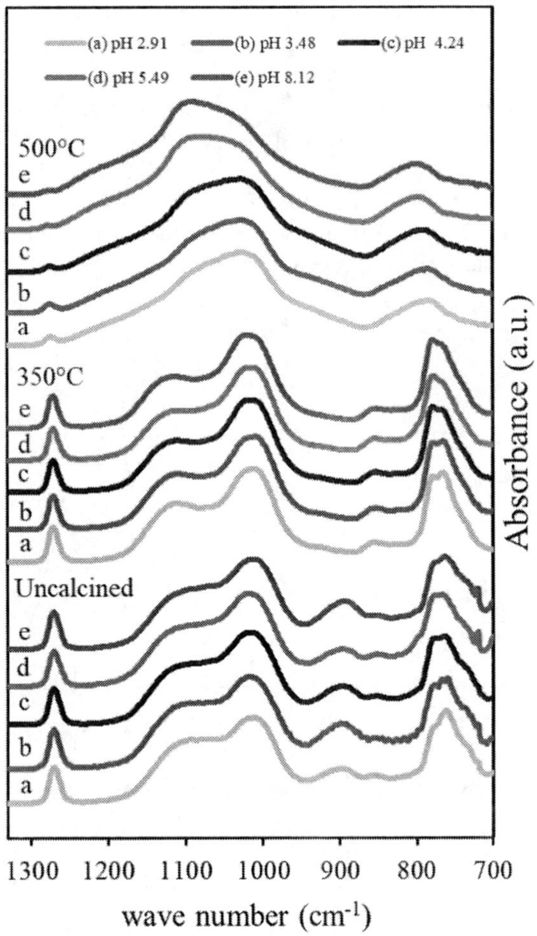

Figure 7. FTIR spectra of MTMS silica xerogel on various pH conditions

According to Mahadik et. al. [31], the difference in calcination temperature resulted in higher Si-OH contents at higher temperatures. The peak intensity associated with -OH increased with temperature step up. At Fig. 7 and the wave number of 955-835 cm^{-1}, the MTMS silica xerogels calcined at 500°C had shoulders higher than that of 350°C. This result implied that the Si-OH group content was higher for silica xerogels calcined at 500°C.

Meanwhile, the intensity of the alkyl group (-CH_3) decreased after thermal decomposition at a temperature of 352°C, therefore at a calcination temperature of 500°C the peak at the wave number of 865-750 cm^{-1} was not seen [31]. The MTMS silica xerogel calcined at 350°C had a Si-C peak at 865-750 cm^{-1} while that of 500°C did not appear at all which indicated that at temperature of 500°C the Si-C group had been much reduced due to high calcination temperature. The same thing happened at the C-H peak at 1260 cm^{-1} where the peak height was considerably reduced at a temperature of 500°C. The reduced peaks at the wave number of 900 cm^{-1} appearing on uncalcined silica xerogels, then weakened at calcination temperature of 350°C and completely lost at 500°C were possibly due to burning methyl groups of unreacted MTMS.

For further analysis of the FTIR spectrum, the deconvolution was performed to obtain information about the almost imperceptible differences in the FTIR spectra. In principle, deconvolution develops the experimental spectra into a derived spectrum in which the derived spectra corresponded to the type of functional group to be measured, such as Si-OH, Si-C and Si-O-Si. Quantification of the absorption peaks associated with silanol and siloxane groups may provide an overview of the hydrolysis and condensation processes in sol-gel reactions [35]. To perform a quantitative analysis, the deconvolution of the peak was used to identify the various vibrations in the overlapping region. All spectra were normalized using the intense peak at 1080 cm^{-1} (Si-O-Si stretching vibration). This peak was baseline resolved and observed to be the least varying peak. Deconvolution of the FTIR spectra was performed using the Fityk software with Gaussian peak during curve fitting. The same number of peaks was used in the entire spectral peak fitting. The half width at half maximum (HWHM) was fixed for each peak, whilst the peak position was allowed to change slightly to realize qualified fitting. Then, the ratios of Si-OH area to Si-O-Si area and Si-C area to Si-O-Si area were calculated to obtain Si-OH/Si-O-Si ratio and C-H/Si-O-Si ratio. The fityk analysis yielded the deconvolution peak of the MTMS spectra as shown in Fig. 8 with the deconvolution peak description shown in Table 2.

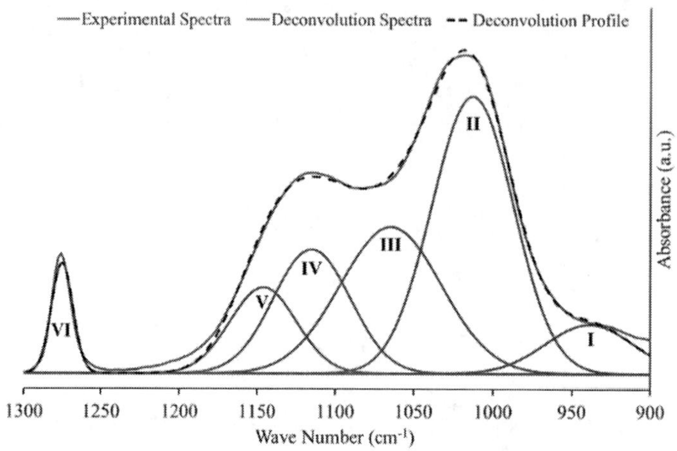

Figure 8. Deconvolution of the FTIR spectra of silica xerogels MTMS at pH 2.91 and calcined at 350°C in the region between 1300 and 900 cm^{-1}

Table 2. Description of the deconvolution peaks of the MTMS in Fig. 8

Deconvolution peak	Wave number (cm^{-1})	Vibrational mode	References
I	960 cm^{-1}	Si-OH	[36, 37]
II	1030 cm^{-1}	Linier Si-O-Si	[38, 39]
III	1080 cm^{-1}	Cyclic Si-O-Si	[39-41]
IV	1150 cm^{-1}	Si-O	[40]
V	~1220 cm^{-1}	longitudinal optic mode Si-O-Si	[38, 39],
VI	1260 cm^{-1}	C-H	[34]

Fig. 9 and 10 present the ratios of peak area between Si-OH/Si-O-Si and C-H/Si-O-Si as functions of pH. Based on Fig. 9, it can be seen that the ratios of the area of Si-OH/Si-O-Si of samples calcined at 350°C were lower than that of 500°C for all pH variations. These results indicated that the relative amount of Si-OH groups on silica calcined at 350°C was less than that of 500°C. On the other hand, the higher the pH, the lower Si-OH/Si-O-Si ratio which may be caused by the Si-OH group binding together and forming the siloxane (Si-O-Si) group. Hydrophobicity is determined by the Si-OH group because Si-OH is easy to interact with water because of the polar OH group [31]. The relatively low Si-OH/Si-O-Si ratio resulted in the interaction between silica thin layer surfaces and water also weak. This result explained why the MTMS silica thin layer calcined at 350°C was more hydrophobic than that of 500°C.

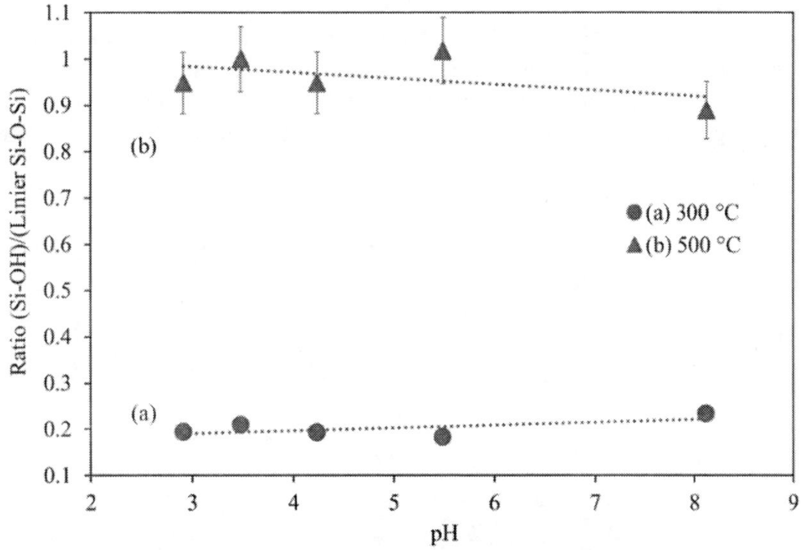

Figure 9. The (Si-OH)/(linier Si-O-Si) ratio for MTMS silica xerogel as function of pH

Fig. 10 shows the C-H/Si-O-Si ratio as a function of pH. The ratio of the area of C-H/Si-O-Si from the silica xerogel calcined at 350°C is higher than that of 500°C for all pH variations. This proved that the relative amount of C-H/Si-O-Si in silica xerogel calcined at 350°C was greater than that of 500°C. The hydrophobic properties are determined by C-H groups derived from methyl groups from MTMS in which the C-H group has weak interaction properties with water [31]. Although, on the silica xerogels calcined at 500°C, the amount of C-H group was considerably reduced because of the oxidation process of the C-H group becoming CO_2. The loss of the C-H group caused the declining of hydrophobic properties of the silica thin layers.

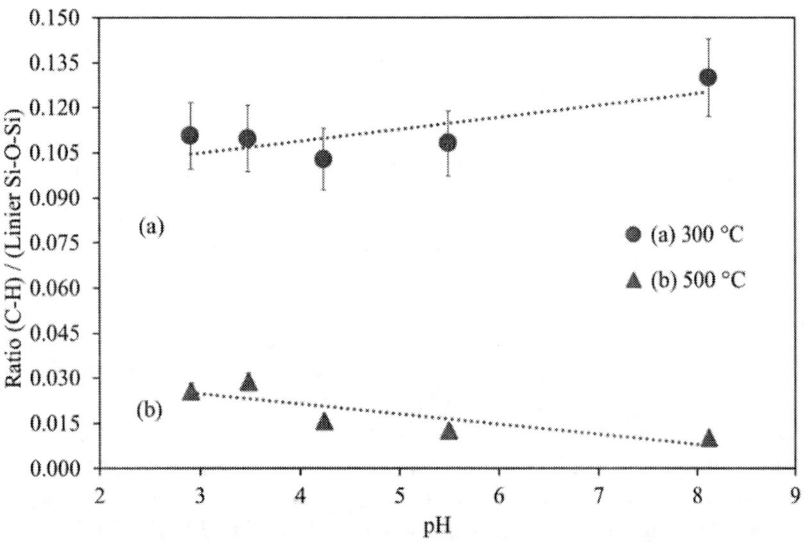

Figure 10. The (C-H)/(linier Si-O-Si) ratio for MTMS silica xerogel as function of pH

3.4. Gas Sorption Analysis

Fig. 11 shows the N_2 adsorption isotherm of the MTMS silica xerogel in which all specified type IV isotherms (as defined by the IUPAC classification), indicating mesoporous network characters. The

MTMS silica xerogel exhibited a low adsorption quantity both in the samples calcined 350°C and 500°C. However, the adsorbed quantities at 350°C were higher than that of 500°C indicating that the higher the calcination temperature, the denser the silica xerogel structure. Apparently, the presence of methyl groups from MTMS had a huge effect on pore structure, since silica xerogel obtained from TMOS (all bonded C binding to methoxy groups) produced large pore volume and surface area [42]. Possibly, the CH_3 group was deposited on the outer surface and inside the cavity thus covering the silica xerogel pores.

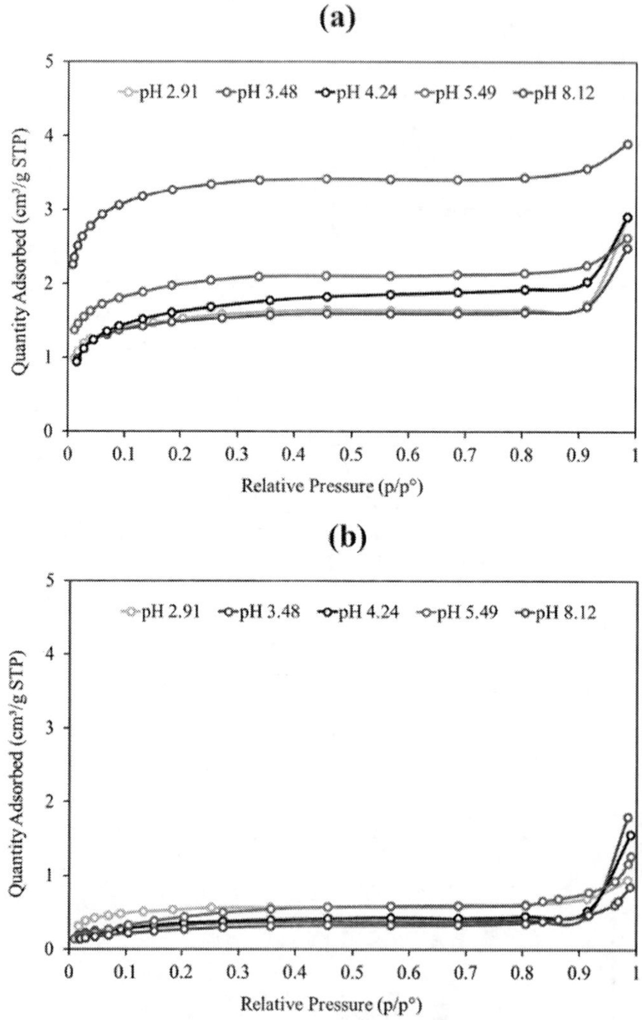

Figure 11. Nitrogen adsorption isotherms of MTMS silica xerogel samples calcined at (a) 350°C and (b) 500°C

Fig. 12 shows the BET area, pore diameter and pore volume of the MTMS silica xerogel. In general, pore diameter and pore volume was unchange with pH variation even though BET area increased. This results suggested that the increase in pH was related to the BET surface area while the pH had no effect on pore diameter and volume.

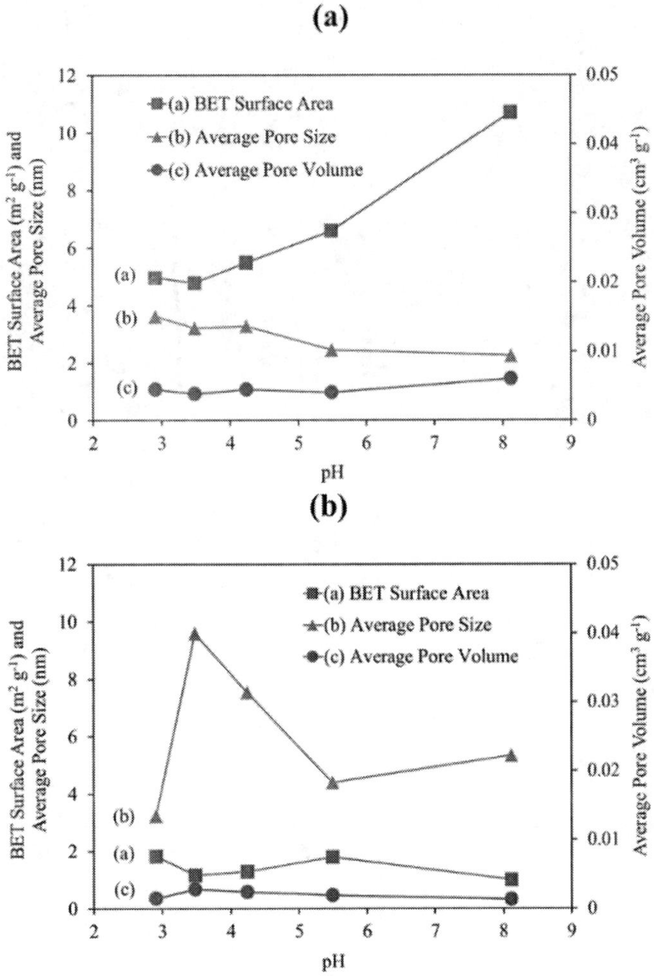

Figure 12. BET surface area, average pore size, and average pore volume of MTMS samples calcined at (a) 350 °C and (b) 500 °C

3.5. Morphology and C/O composition (% mass) in MTMS silica thin layer

Analysis of surface morphology and elemental composition of the silica thin layer were performed with SEM-EDX and presented in Fig. 13 and Table 3.

Fig. 13 presents SEM images showing the morphology of two MTMS silica thin layer samples synthesized at the most acidic pH of 2.91 and the most alkaline pH of 8.12. Both were calcined at 350°C with magnification of 500 times. Based on Fig. 12, it can be seen the difference of the resulting silica surface where in the acidic solution which the resulting surface was smoother than that of produced in an alkaline solution. The alkaline pH generated a rougher surface which tended to be more hydrophobic. According to Mahadik et. al. [24], the surface roughness was responsible for the hydrophobic behaviour of the silica layer which reduced the contact of the surface area between the water droplets and the surface of the coating.

IOP Conf. Series: Materials Science and Engineering **299** (2018) 012041 doi:10.1088/1757-899X/299/1/012041

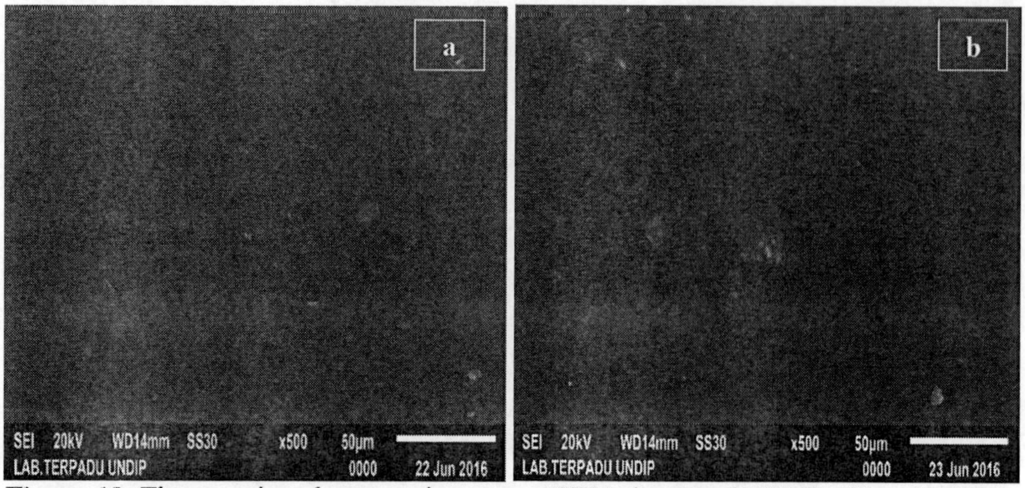

Figure 13. The scanning electron microscope (SEM) image of MTMS silica thin layer calcined at 350°C for pH of (a) 2.91 and (b) 8.12

Table 3. EDX results of carbon and oxygen element

pH	Element	%Mass	C/O (%Mass)
2,91	C	16.19	0,30
	O	52.96	
8,12	C	17,89	0,35
	O	51.32	

Table 3 presents the percentage ratio of carbon/oxygen mass obtained from the EDX analysis. It was found that the ratio of carbon to oxygen at pH 2.91 was lower than that of pH 8.12 which were 0.30 and 0.35 respectively. The carbon element tends to be hydrophobic while due to its higher electronegativity, the oxygen element is more hydrophilic. Thus, an upsurge in C/O ratio rose the hydrophobic properties of silica thin layer surfaces [43]. A larger C/O ratio in alkaline pH indicates that at base pH the ceramic surface was more hydrophobic because it contained more carbon than that of acid pH conditions. The high carbon content on the MTMS silica thin layer surface induced the surface became more hydrophobic. A larger C/O ratio at an alkaline pH indicated that under alkaline pH conditions, the silica surface was more hydrophobic because it contained more carbon than the acid pH conditions.

4. Conclusions

A synthesis and characterization study of xerogel and thin silica coatings prepared by MTMS precursors on the surface of the ceramic plate was conducted. The pH greatly influenced the properties of the MTMS silica thin layer, where the higher the pH used in the sol-gel process, the higher the water contact angle which indicated the amplified surface hydrophobicity of silica thin layer. FTIR analysis showed that there was a correlation between the hydrophobic properties of the silica surface and the content of Si-OH and C-H groups. The higher the pH, the lower the Si-OH group as the C-H group increased. The TGA analysis results displayed that the MTMS silica thin layer maintained the hydrophobic properties up to 350°C. The increase in pH was also proportional to the surface area and the roughness of the silica thin layer surface.

Acknowledgments

Adi Darmawan and Yayuk Astuti gratefully acknowledge financial support from Ministry of Research Technology and Higher Education, Republic of Indonesia via the Penelitian Kompetensi (Hikom) with the grant number of 186-01/UN7.5.1/PG/2016. Adi Darmawan specially thanks for FIM2lab at the University of Queensland for their support for all experimental work.

References

[1] Laskowski L, Laskowska M, Jelonkiewicz J, Dulski M, Wojtyniak M, Fitta M and Balanda M 2016 SBA-15 mesoporous silica free-standing thin films containing copper ions bounded via propyl phosphonate units - preparation and characterization *Journal of Solid State Chemistry* **241** 143-51

[2] Dietrich P M, Glamsch S, Ehlert C, Lippitz A, Kulak N and Unger W E S 2016 Synchrotron-radiation XPS analysis of ultra-thin silane films: Specifying the organic silicon *Applied Surface Science* **363** 406-11

[3] Kavale M S, Mahadik D B, Parale V G, Wagh P B, Gupta S C, Rao A V and Barshilia H C 2011 Optically transparent, superhydrophobic methyltrimethoxysilane based silica coatings without silylating reagent *Applied Surface Science* **258** 158-62

[4] Yan H, Yuanhao W and Hongxing Y 2015 TEOS/silane coupling agent composed double layers structure: A novel super-hydrophilic coating with controllable water contact angle value *Applied Energy*

[5] Cui S, Liu Y, Fan M-h, Cooper A T, Lin B-l, Liu X-y, Han G-f and Shen X-d 2011 Temperature dependent microstructure of MTES modified hydrophobic silica aerogels *Materials Letters* **65** 606-9

[6] Purcar V, Stamatin I, Cinteza O, Petcu C, Raditoiu V, Ghiurea M, Miclaus T and Andronie A 2012 Fabrication of hydrophobic and antireflective coatings based on hybrid silica films by sol–gel process *Surface and coatings technology* **206** 4449-54

[7] Latthe S S, Nadargi D Y and Rao A V 2009 TMOS based water repellent silica thin films by co-precursor method using TMES as a hydrophobic agent *Applied Surface Science* **255** 3600-4

[8] Ramezani M, Vaezi M R and Kazemzadeh A 2015 The influence of the hydrophobic agent, catalyst, solvent and water content on the wetting properties of the silica films prepared by one-step sol–gel method *Applied Surface Science* **326** 99-106

[9] Spirk S, Findenig G, Doliska A, Reichel V E, Swanson N L, Kargl R, Ribitsch V and Stana-Kleinschek K 2013 Chitosan–silane sol–gel hybrid thin films with controllable layer thickness and morphology *Carbohydrate Polymers* **93** 285-90

[10] Budunoglu H, Yildirim A, Guler M O and Bayindir M 2011 Highly transparent, flexible, and thermally stable superhydrophobic ORMOSIL aerogel thin films *ACS applied materials & interfaces* **3** 539-45

[11] Ferrero F and Periolatto M 2013 Application of fluorinated compounds to cotton fabrics via sol–gel *Applied Surface Science* **275** 201-7

[12] Yan J, Shamim T, Chou S K, Li H, Yan H, Yuanhao W and Hongxing Y 2015 Clean, Efficient and Affordable Energy for a Sustainable Future: The 7th International Conference on Applied Energy (ICAE2015)TEOS/Silane-Coupling Agent Composed Double Layers Structure: A Novel Super-hydrophilic Surface *Energy Procedia* **75** 349-54

[13] Burunkaya E, Kesmez Ö, Kiraz N, Çamurlu H E, Asiltürk M and Arpaç E 2012 Hydrophilic hybrid SiO_2: polyethylene glycol–3-isocyanato-propyltrietoxysilane nanocomposite films *Thin Solid Films* **522** 238-46

[14] Li H, Liang T, Lai X, Su X, Zhang L and Zeng X 2018 Vapor-liquid interfacial reaction to fabricate superhydrophilic and underwater superoleophobic thiol-ene/silica hybrid decorated fabric for oil/water separation *Applied Surface Science* **427** 92-101

[15] Yang X, Zhu L, Chen Y, Bao B, Xu J and Zhou W 2016 Controlled hydrophilic/hydrophobic property of silica films by manipulating the hydrolysis and condensation of tetraethoxysilane *Applied Surface Science* **376** 1-9

[16] Mahadik S A, parale V, Vhatkara R S, Mahadik D B, Kavale M S, Wagh P B, Gupta S and Gurav J 2013 Superhydrophobic silica coating by dip coating method *Applied Surface Science* **277** 67-72
[17] Latthe S S, Imai H, Ganesan V and Rao A V 2010 Porous superhydrophobic silica films by sol–gel process *Microporous and Mesoporous Materials* **130** 115-21
[18] Ul Haq E, Zaidi S F A, Zubair M, Abdul Karim M R, Padmanabhan S K and Licciulli A 2017 Hydrophobic silica aerogel glass-fibre composite with higher strength and thermal insulation based on methyltrimethoxysilane (MTMS) precursor *Energy and Buildings* **151** 494-500
[19] Wankhede R G, Morey S, Khanna A S and Birbilis N 2013 Development of water-repellent organic–inorganic hybrid sol–gel coatings on aluminum using short chain perfluoro polymer emulsion *Applied Surface Science* **283** 1051-9
[20] Filho C M C, Matias T, Durães L and Valente A J M 2017 Efficient simultaneous removal of petroleum hydrocarbon pollutants by a hydrophobic silica aerogel-like material *Colloids and Surfaces A: Physicochemical and Engineering Aspects* **520** 550-60
[21] Yang R T 2003 *Adsorbents: fundamentals and applications*: John Wiley & Sons)
[22] Brinker C J and Scherer G W 1985 Sol → gel → glass: I. Gelation and gel structure *Journal of Non-Crystalline Solids* **70** 301-22
[23] Milea C A, Bogatu C and Duta A 2011 The influence of parameters in silica sol-gel process *Bulletin of TheTransilvania University of Brasov* **4** 53
[24] Mahadik S A, Kavale M S, Mukherjee S K and Rao A V 2010 Transparent Superhydrophobic silica coatings on glass by sol–gel method *Applied Surface Science* **257** 333-9
[25] Rao A P, Rao A V and Pajonk G M 2007 Hydrophobic and physical properties of the ambient pressure dried silica aerogels with sodium silicate precursor using various surface modification agents *Applied Surface Science* **253** 6032-40
[26] Uhlmann D 2011 Robust Microporous Cobalt Oxide Doped Silica Membranes for High Temperature Industrial Coal Gasification Syngas Separation. In: *School of Chemical Engineering,* (Brisbane, Queensland: The University of Queensland)
[27] Nakajima A, Hashimoto K and Watanabe T 2001 Recent studies on super-hydrophobic films *Monatshefte für Chemie/Chemical Monthly* **132** 31-41
[28] Sanjay S L, Annaso B G, Chavan S M and Rajiv S V 2012 Recent progress in preparation of superhydrophobic surfaces: a review *Journal of Surface Engineered Materials and Advanced Technology* **2012**
[29] Jitianu A, Lammers K, Arbuckle-Kiel G A and Klein L C 2011 Thermal analysis of organically modified siloxane melting gels *Journal of thermal analysis and calorimetry* **107** 1039-45
[30] Bhagat S D, Oh C-S, Kim Y-H, Ahn Y-S and Yeo J-G 2007 Methyltrimethoxysilane based monolithic silica aerogels via ambient pressure drying *Microporous and Mesoporous Materials* **100** 350-5
[31] Mahadik S A, Mahadik D B, Parale V G, Wagh P B, Gupta S and Rao A V 2012 Recoverable and thermally stable superhydrophobic silica coating *Journal of sol-gel science and technology* **62** 490-4
[32] Durães L, Maia A and Portugal A 2015 Effect of additives on the properties of silica based aerogels synthesized from methyltrimethoxysilane (MTMS) *The Journal of Supercritical Fluids* **106** 85-92
[33] Amoriello S, Bianco A, Eusebio L and Gronchi P 2011 Evolution of two acid steps sol–gel phases by FTIR *Journal of sol-gel science and technology* **58** 209-17
[34] Homma H, Kuroyagi T, Izumi K, Mirley C, Ronzello J and Boggs S 1999 Diffusion of low molecular weight siloxane from bulk to surface *IEEE Transactions on Dielectrics and Electrical Insulation* **6** 370-5
[35] Wang S, Wang D K, Smart S and da Costa J C D 2015 Ternary Phase-Separation Investigation of Sol-Gel Derived Silica from Ethyl Silicate 40 *Scientific reports* **5**

[36] Lam H C and Lam Y W 1977 Physical and conduction measurements of spin-on oxide *Thin Solid Films* **41** 43-56
[37] Capeletti L B, Baibich I M, Butler I S and dos Santos J H Z 2014 Infrared and Raman spectroscopic characterization of some organic substituted hybrid silicas *Spectrochimica Acta Part A: Molecular and Biomolecular Spectroscopy* **133** 619-25
[38] Young S K 1999 Organic-inorganic composite materials: Molecular tailoring of structure-property relationships
[39] Ratner B D and Castner D G 2013 *Surface Modification of Polymeric Biomaterials*: Springer Science & Business Media)
[40] Farmer V t and Russell J 1964 The infra-red spectra of layer silicates *Spectrochimica Acta* **20** 1149-73
[41] Achyuta A K H, White A J, Pryce Lewis H G and Murthy S K 2009 Incorporation of Linear Spacer Molecules in Vapor Deposited Silicone Polymer Thin Films *Macromolecules* **42** 1970-8
[42] Vollet D R, Nunes L M, Donatti D A, Ibañez Ruiz A and Maceti H 2008 Structural characteristics of silica sonogels prepared with different proportions of TEOS and TMOS *Journal of Non-Crystalline Solids* **354** 1467-74
[43] Latthe S S, Imai H, Ganesan V and Rao A V 2009 Superhydrophobic silica films by sol–gel co-precursor method *Applied Surface Science* **256** 217-22

Micro Structure and Hardness Analysis of Brass Metal Welded

Lukman Faris N[1], Muljadi[1], Djuhana[2]

[1]Research Center for Physics - LIPI
Kawasan Puspiptek, Serpong, Tangerang Selatan 15314, Indonesia
[2]Mechanical Engineering Department – UNPAM
Tangerang Selatan 15314, Indonesia

Email: lukm008@lipi.go.id

Abstract. Brass metals are widely used for plumbing fittings. High tensile brasses are more highly alloyed and find uses in marine engineering. The welding of brass metal has been done by using electrical weld machine (SMAW). The microstructure of brass metal welded was observed by optical microscope. The result can see that the microstructure has been changed due to heat from welding. The microstructure of original brass metal is seen a fine laminar stucture, but the microstructure at HAZ appears bigger grains and some area at HAZ is seen coarser microstructure. The microstructure at weld zone can be seen that it was found some of agglomeration of materials from reaction between brass metal and electrode coating wire. According the hardness measurement, it is found highest hardness value about 301.92 HV at weld zone, and hardness value at base metal is 177.84 HV

1. Introduction

Brass is sample of copper zinc alloy, it easy to cast, with excellent machinable and good resistance to corrosion in air and fresh water. They are widely used for plumbing fittings. High tensile brasses are more highly alloyed and find uses in marine engineering [1]. Welding is a process in which materials of the same basic type or class are joined together trough the formation of primary (and occasionally and secondary) atomic or molecular level bonds under the combined action of heat and pressure [2]

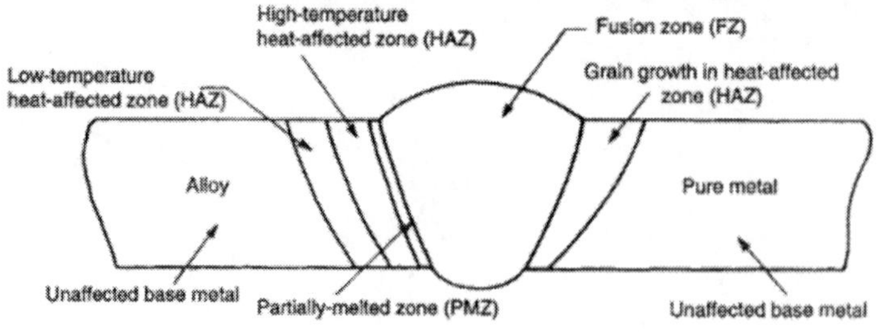

Figure 1. The American Welding Society Classification of Welding and Allied Processes [3]

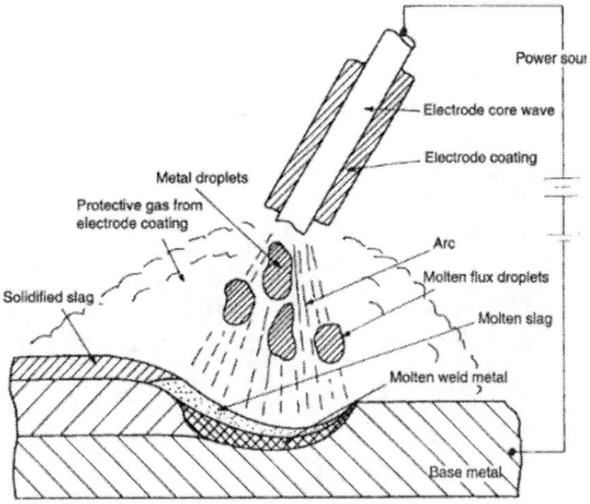

Figure 2. Schematic Ilustration of the shielded metal arc welding (SMAW) process [3, 4].

SMAW can operate with DC power sources (with electrode positive and negative) or AC power sources, depending of coating design. Typically current range from 50- 300 Amperes, largely based on electrode core wire diameter, at 10-30 volts. Resulting in 1-10 kg (2-20 lb) per hour deposition rate.[3, 4].

2. Experimental Method

The welded samples have bean prepared by using brass metal and welding process was use SMAW machine. The flow diagrams of research can be seen at figure 3.

Brass Metal
↓
Cutting the sample
↓
Welding
Preparation sample to test
↓
Microstructure analize and Hardness test
↓
Sample Brass Metal

Figure 3. Schematic Diagram for sample test

The characterization of welded samples ware done include: micro structure analysis by using optical microscope and Vickers hardness by using micro hardness tester. There are three zones for micro structure and hardness measurement, such as : base metal, HAZ, and weld zone.

3. Result and Discussion

The result of measurement of hardness vickers can be seen at table 1. The hardness value is different each zone, where base metal has hardness about 177.84 HV, HAZ has hardness about 131.41 HV and welding zone has hardness about 301.92 HV. The hardness at welding zone is highest due to heat effected from welding.

Table 1. Hardness Vickers value of Brass Welding.

Welding Zone	Weld	HAZ	Base Metal
HV	301.92	131.41	177.84

The result of measurement of micro structure can be seen at figure 4-6, the observation was done by using optical microscope with magnification 40x.

Figure 4. Base metal of brass with 40x magnification.

The micro structure of base metal of brass is seen that fine laminar grains are found according the literature that brass metal has generally a laminar grains [3].

Figure 5. Heat affected zone of brass

The micro structure at HAZ for Brass metal has been changed it is due to heating from welding effect as seen at figure 5. It was found some bigger grains than grains at base metal brass. The grains growth effect at HAZ can decrease hardness value. According the reference that effect of welding can trend decrease hardness from base metal to HAZ [3].

Figure 6. Welding zone of brass

The micro structure at welding zone for brass metal has been changed it is due to heating and melting from welding effect as seen at figure 6. It was found some bigger grains than grains HAZ brass. The grains growth effect at welding zone can increase hardness value. According the measurement of micro structure shows that the hardness value is enough high about twice compare that hardness value at base metal. It indicates that the material at weld zone becomes brittle, that is not good for welding result[5]

4. Conclusion

The joining technique of brass metal by using welding machine has been done, the result is indicated that hardness value and microstructure has been changed significantly due to heat from welding effect. The joing brass metal by using welding can causes the material at weld zone become brittle, properly it is not used joining of brass by using welding.

5. Acknowledgment

Acknowledgments are recommended to Research Center for Physics - LIPI, helping the authors in many ways.

References

[1] John B R 1999 *Foseco Non-Ferrous Foundrymans Handbook* pp 225-251
[2] Mogaddam S M, Parvizi R, Haddad-Sabzevar M, Davoodi A 2011 J.matdes 01.015 2749-2755
[3] Robert M W Jr 2004 *Joining of Materials and Structures From Pragmatic Process to Enabling Technology* pp 285-348
[4] Kazi J, Zaid S, Tatha S M, Yasir M, Akib D 2015 IJMER ISSN: 2249-6645 Vol 5 Iss2 22
[5] *Buehler Hardness Convertion Charts Razi Metallurgical Research Center* p 7

IOP Conf. Series: Materials Science and Engineering **299** (2018) 012043 doi:10.1088/1757-899X/299/1/012043

The characterization of ceramic alumina prepared by using additive glass beads

Suprapedi[b)], Muljadi[a)], Priyo Sardjono

Research Center for Physics-Indonesian Institute of Scienses, Kawasan Puspiptek Serpong Tangerang Selatan
a)mulj004@lipi.go.id
b)supr002@lipi.go.id

Abstract. The ceramic alumina has been made by using additive glass bead (5 and 10 % wt.). There are two kinds of materials, such as : gamma Alumina and glass bead. Synthesis of alumina was done by ball milling for 24 hours, then the mixed powder was dried in drying oven at 100 °C for 6 hours. Furthermore, the dried powder was mixed by using 2 % of PVA and continued with compacted to form a pellet with pressure of 50 MPA. The next step is sintering process with variation temperature of 1150, 1200, 1250, 1300 and 1400 °C and holding time for 2 hours. The characterization conducted are consist of test density, hardness, shrinkage, and microstructure. The results show that ceramic alumina with addition of 10 % wt. glass bead has the higher value of density, hardness and shrinkage than addition of 5% wt. glass bead. The highest characterization of ceramic alumina with addition 10 % glass bead was achieved at sintering temperature of 1400 °C with density 3.68 g/cm^3, hardness vickers 780.40 Hv and shrinkage 15.23 %. The XRD results show that it was founds a corrundum (alpha Alumina) as dominant phase and mullite as minor phase.

1. Introduction

Alumina is one of the most important ceramic oxides and has a wide range of uses, including high-temperature applications (refractory materials), electrical insulator and microelectronics. Alumina is also important from the point of view that it has many meta stable polymorphs such as $\varkappa$ -, γ-, δ-, θ-Al_2O_3 and stable form α Al_2O_3 [1]. Among the several meta stable polymorphs of alumina (transition alumina phases), γ-alumina is important due to its application in industry as catalyst, adsorbents, coatings, and soft abrasives. A stable form of α-Al_2O_3 are widely used in blast finishing and surface preparation due to their hardness, cost and longevity [1].

Alumina is the most widely used oxide ceramics because of its hardness, good corrosion resistance, high insulation and ease of processing [1,2]. Furthermore, its thermal expansion coefficient is nearer to those of metals than that of structural nitrogen ceramics, such as silicon nitride (λ Iron = 15 x10^{-6} °C^1, λ Al_2O_3 = 9 x10^{-6} °C^{-1}, λSi_3N_4=3 x10^{-6} °C^{-1}) [3]. Therefore, if the mechanical properties of alumina are further improved, it is reasonable to expect that alumina could be widely used for engine part [3]. There are some steps for manufacturing of ceramic product, sintering process is one of the most important and crucial part in fabricating ceramic component. In fact, most of the ceramic component that fabricated via plastic forming and powder route for the variety applications must undergo this sintering steps [4]. Ceramic alumina has a very high melting point of about 2040°C, therefore to produce alumina ceramic product with maximum density requires sintering temperature near the melting point, which is about 1700 - 1800°C [5,6]. There are two general approaches to enhance sintering process or to lower the sintering temperature. The first is to improve powder processing, that is, to use fine starting powders and to eliminate agglomerates in the green pre forms such as by chemical routes [7].

The second approach is to use sintering aids or additives [7]. Additives in solid solutions can enhance diffusion and hence sintering by increasing defect, while additives forming a liquid phase can facilitate particle rearrangement and solution / reprecipitation [8,9]. Some of additives of sintering were used commonly such as: SiO_2, Borax, MgO, TiO_2,Y_2O_3 and so on [10]. M. Sathiyakumar et all have reported that beneficial effect of small addition of MgO in alumina powder in achieving full density and even translucency by sintering and prevention of abnormal grain growth in the final stage of densification [11]. Glass beads are made of silica and other minerals melted at a elevated temperature to form a thick, viscous liquid. The liquid is molded into the desired shape and hardens as it cools. Glass beads are made in many sizes and shapes [12]. Commonly, beads are manufactured by winding molten glass around a long iron rod. Different shapes can be created while the bead is still semisoft. For instance, square and oval beads can be manufactured by pressing the bead while it is still hot against a flat surface. Shapes that are more intricate are manufactured with molds. Major oxides in most glasses comprise Na_2O, MgO, SiO_2, Al_2O_3, K_2O, and CaO, whereas other oxides are usually minor and trace compositions of the glass. Glass beads are generally glassy materials such as glass windows that have an amorphous structure with the main content of silica (SiO_2), glass beads containing alkali elements such as Ca, Na and K. Glass beads are formed in form a granule with a size of about 0.1 - 0.3 mm. Generally, the composition of glass bead consists of 18 ± 2% Na_2O, 67 ± 3% SiO_2 and 9 ± 2% CaO [12, 13]. In this study we used a glass bead as an alumina sintering additive. The purpose of this research is to finf out the influence of addition of a glass bead additive composition to the sintering of alumina ceramic.

2. Experimental works

The starting material was a high-purity (>99.99%) γ - alumina powder with an average particle size of 10 μm and Glass bead as additive. The additive composition expressed in weight percent was 5 % and 10 %. Both of raw materials were milled by ball milling for 24 hours with using pure Al_2O_3 ball and water as milling media. Then the slurry was dried at 100°C and the dried powder was hand ground using mortar and pestle, then the fine powder was mixed with 2 % wt. of binder polyvinyl alcohol (PVA). Pellet samples with 12 mm in diameter were prepared by pressing under pressure 50 MPa using stainless steel die. The pellets were then dried for 24 hours under room temperature and sintering were conducted at variation temperature namely 1150, 1200, 1250, 1300 and 1400°C for 2 hours with holding time and heating rate of 10 °C/min in the programmable electrical furnace. The characterization of sintered samples were done namely : measurement of bulk density, firing shrinkage, vickers hardness and crystal structure by using X-Ray Diffractometer.

3. Results and discussion

Figure 1 shows relationship of bulk density and firing shrinkage to sintering temperature with different of percentages of glass beads. The results show that the pellet sintered for 2 hours with 10 % wt.of glass bead has a higher bulk density value compared to the pellet with 5 % wt.of glass bead. This can be explained that sample with the higher of percentage of glass beads will increases the quantity of melts of glass bead and can enter between the grain so as to cover the existing cavities, consequently the density tends to increase. Likewise with increasing sintering temperature, the value of bulk density tends to increase. It is seen in Figure 1 that the rate of increase in density from 1300 °C to 1400 °C tends to be constant for sample with 10 % glass beads, indicating that the sintering process has ended. But it is different in the sample with 5% glass beads, where the bulk density value is still increasing with increasing sintering temperature. This indicates that the sintering process is still continuing at higher temperatures.

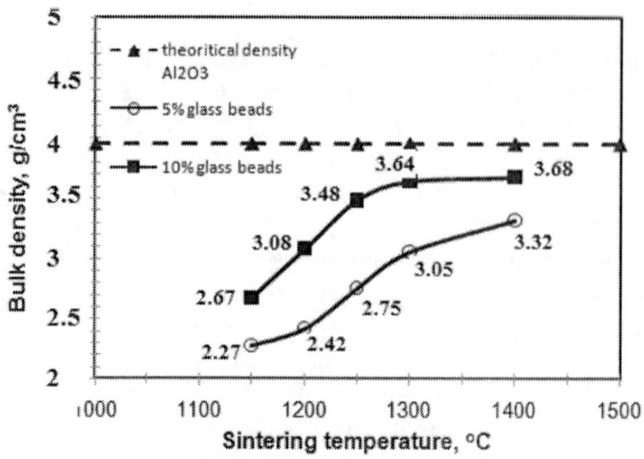

Figure 1. The curve of relationship between bulk density and temperature sintering with different of composition of glass beads

The highest density of 3.68 g/cm^3 was achieved in samples with 10% glass beads and at sintering temperature of 1400°C. When compared with theoretical density value of Al_2O_3 that is 3.95 g/cm^3 [11] to the experimental results, it shows that samples with additive 10% glass beads have a density level of 93.16% with a sintering temperature of 1400°C. These results indicate that the glass beads can accelerate and promotes the densification during sintering stage.
Figure 2 shows relationship between firing shrinkage and temperature sintering with different composition additive of glass beads.

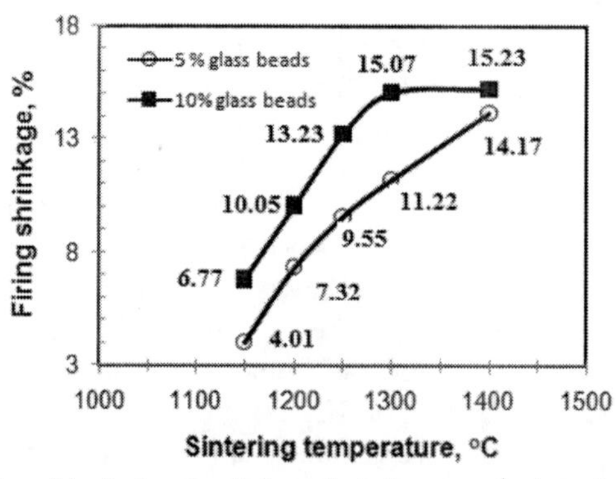

Figure 2. The curve of relationship between firing shrinkage and sintering temperature with different of composition of glass beads.

Figure 2 shows that the firing shrinkage for both samples increase as the firing temperature is increased, the firing shrinkage value of sample with 10% glass beads is higher than sample with 5 % glass beads. The optimum of sintering process is achieved when bulk density and firing shrinkage are a maximum value, tending to be constant. The results of measurement of vickers hardness can be seen at figure 3. Sample after sintering below 1250°C cannot be measured because the surface of the samples was stell roughness and contains more pores. It is seen that the Vickers hardness increases with sintering temperatures and composition of glass beads.

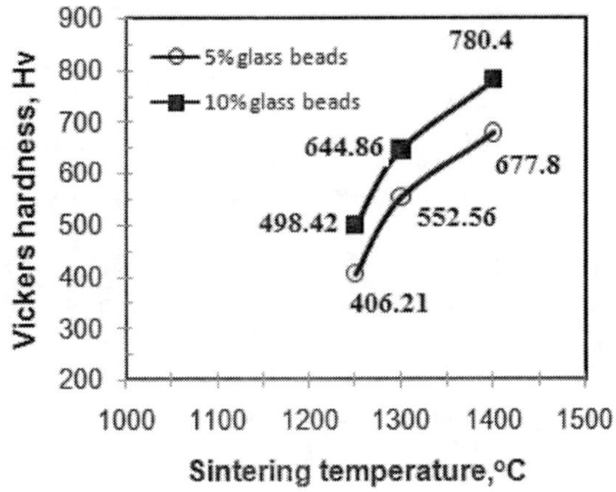

Figure 3. The curve of relationship between vickers hardness and temperature sintering with different of composition of glass beads.

The hardness value of sample with 10 % glass beads is higher than sample with 5 % glass beads, it is due to densification of samples, where sample with 10 % glass beads has highest density value. The maximum value of vickers hardness is achieved about 780.4 Hv (7.65GPa), theoretically vickers hardness of alumina ceramic (99.0% Al_2O_3) is about 10 GPa [11]. Figure 4 shows the results of XRD measurement for sample after sintering at temperature 1400 °C with different composition of glass beads.

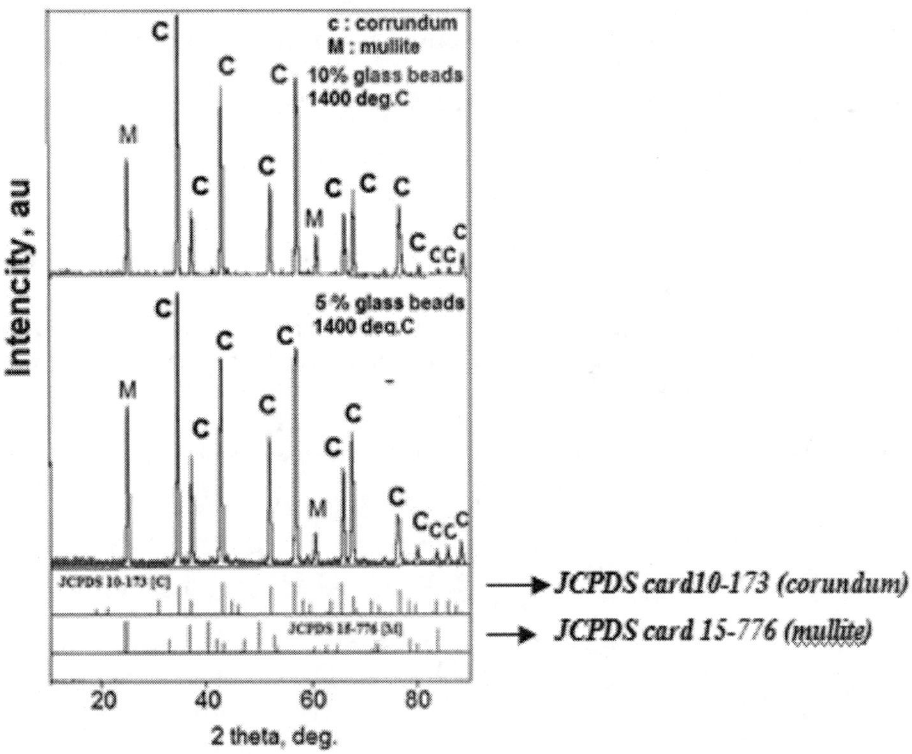

Figure 4. XRD patterns of sintered at 1400°C samples with 5 % and 10 % of glass beads

The analysis the crystal structure of sintered samples were conducted by using X-ray Diffractometer-Rigaku. Based on the matching of the diffractogram peaks with JCPDS card10-173 (corundum) and JCPDS card 15-776 (mullite), the XRD peaks as shown in figure 4. have two phases : corundum (α-Al_2O_3) as major phase and mullite ($3Al_2O_3.2SiO_2$) as minor phase. The formation of mullite phase is due to reaction between Al_2O_3 with glass beads which glass beads contains a silica or SiO_2.

4. Conclussions

The use of glass beads additives containing the main components of silica can have a significant effect on the decreasing sintering temperature of alumina ceramic. Alumina ceramics can be successfully sintered with a densification level of 93.16 % at sintering temperature 1400°C and with the addition of 10 % wt.of glass beads additives. The properties of ceramic alumina at this condition are bulk density value = 3.68 g/cm^3, firing shrinkage value = 15.23 %, vickers hardness value = 780.4 Hv (7.65 GPa). The sintered 1400°C of ceramic alumina with additive 5 % and 10 % of glass beads contain two phases which are corundum as major phase and mullite as minor phase.

Acknowledgments

The authors would like to thank to operators of Vickers hardness and XRD at the Research Center for Physics-LIPI.

References

[1] Karen Davis, *Material Review: Alumina (Al_2O_3)*, School of Doctoral Studies (European Union) Journal – 2010
[2] Singha Roy R, Manoj Mitra and Basu D 2005 *Trends Biomater. Artif. Organs*, **18** 2
[3] Chaouki Sadik, Iz-Eddine, Abderrahman 2014 *Journal of Asian Ceramic Societies* **2** 310
[4] Zawati Harun, Nur Farhani Ismail, Nur Azam Badarulzaman 2012 *Advanced Materials Research* Online: **488-489** 335
[5] Pontikes Y, Rathossi C, Nikolopoulos P, Angelopoulos G.N, Jayaseelan D.D, Lee W.E 2009 *Ceramics International* **35** 40
[6] Chaouki Sadik, Iz-Eddine El Amrani, Abderrahman Albizane, 2011 *Journal of Asian Ceramic Societies* **2** 83
[7] Bussarin Ksapabutr, Erdogan Gulari, Sujitra Wongkasemjit 2004 *Colloids and Surfaces A:* Physicochem. Eng. Aspects **233** 145
[8] Liang A. Xue and I-Wei Chen 2011 *J. Am. Cerom. Soc.,* **74** 11991
[9] Igor tubòa, Peter Vín, Mart Viljus, Anton Trník 2014 *Materials and technology* **48** 331
[10] Liu, Z. Xie W 2014 *Science of Sintering*, **46** 3
[11] Sathiyakumar M, Gnanam F D, 2003 *Journal of Materials Processing Technology* **133** 282
[12] Behzad Mehdikhani and Gholam Hossein Borhani, 2013 *International Letters of Chemistry, Physics and Astronomy* **14** 58
[13] Peter Robertshaw, Sonja Magnavita, Marilee Wood, Erik Melchiorre, Rachel Popelka-Filcoff & Michael D. Glascock, 2009 *Journal of African Archaeology*, Monograph **series 2** 105

Effect of Heating Time to Density, Hardness, and Resistivity Againt Fungus of Yellow Bamboo (*Bambusa Vulgaris* Var Schard. Vitata)

A.H. Dawam Abdullah[1], Saukani Nasution[2]
[1]Research Unit for Clean Technology, Indonesian Institute of Sciences, Bandung, Indonesia
[2]Department of Physics, UIN Sunan Gunung Djati, Bandung, Indonesia

Corresponding author: dawamdullah@yahoo.com

Abstract. It has been studied the influence of heating process to the density, hardness and resistivity againt fungus of Yellow bamboo (*Bambusa Vulgaris* Var Schard. Vitata*)*. Structure of bamboo is identified by optical microscope. Bamboo is heated at 100 °C with variation of heating time: 0, 1, 2, 3 and 4 hours. Density of bamboo is measured by using pycnometer, whereas hardness is measured by using a Rockwell Hardness Type- R (HRR). Resistivity againt fungus is observed by naked eye. OM image shows that bamboo consist of hollow structures wherein the water is stored. The diameter of vascular at inner part (~ 150 μm) is quite larger than outer part (~50 μm). Heating time of one hour caused density decrease by about 39%, whereas heating four hours will decrease by 52%. The longer heating time, the hardness of bamboo is increased. For the longitudinal side, two hours heating increased hardness up to 31%. For tangential side, four hours heating time increased hardness until 82% . Furthermore, heating time 3-4 hours will persist bamboo from fungus attack more than 14 days.
Keywords: bamboo, density, hardness, fungus

1. Introduction

Bamboo is one of the plant that easily to be cultivated in Indonesia and has been widely used for various purposes like traditional houses, furniture, or musical instrument. The price of bamboo is relatively cheap compare with previous commodities such as wood, because the quantity of bamboo is so abundant. It's due to growth-time of bamboo is fairly quick, in a matter of 4 years bamboo is ready to be harvested. Moreover, time processing to become finished goods is relatively short, so it made bamboo more attracted in industrial manufacture.

Yellow bamboo *(Bambusa Vulgaris Var Schard. Vitata) is* one type of bamboo that has a cylindrical shape, hard-walled culm, light, and good mecahnical strength. Yellow bamboo is easy to breed because it has fast growth. Some potential of yellow bamboo are as decoration in yard and furniture. However, the properties of Yellow bamboo mostly will be determined by its water content [1] Generally, the less water content in the bamboo, the better bamboo properties [2].

Bamboo culm consist of cells, fibers and parenchyma with a percentage of 10%, 40%, and 50% respectively [2]. Beside that, bamboo's culm also cantain an empty space called vascular. Chemitry of bamboo culm consist of lignin, pentose, and holoselulosa with a percentage of 20-25%, 30%, and 50-70% [3]. Combination of these structure and chemistry caused water likely to be stored in bamboo's culm (hidrophilic). Those maybe caused a drawback in mechanical properties of bamboo. Moreover, high water content could accelerate growth of fungus that can initiate the destruction of bamboo culm. So this researched is aimed to enhance mechanical properties of bamboo by address the heating for reducing water content where evaluation of density, hardness, and fungus growth has not been done before.

2. Materials and methods

2.1 Materials

Yellow Bamboo that was studied is obtained from Bandung area, West Java, Indonesia (Figure 1). The age of bamboo is around 3 years. Sample for experiment is taken from the bottom part.

Figure 1 Yellow Bamboo

2.2 Morphology analysis

Bamboo's morphology is identified by using optical microscope (MeijiTechno MT6000). Observations were made on the cross section of bamboo. Bamboo is being cut and sliced bamboo is observed. Identification of bamboo is conducted in the 3 area the inner, middle, and outside areas.

2.3 Density test

Bamboo was cut with dimension of 2 x 0.5 x 0,5 cm. Then bamboo was heated 100°C by oven with five variations time: 0, 1, 2, 3, and 4 hours. For each bamboo, the density was measured by using pycnometer with principle similar to the Archimides. Density test for each specimen is performed three times.

2.4 Hardness test

Yellow bamboo is being cut into two categories: 3 x 3 x 3 cm and 3 x 3 x 2 cm as presented in Fig.2. The former category is for tangential side measurement, and the rest is for longitudinal side measurement. After cut then bamboo is heated with variation: 0, 1, 2, 3, and 4 hours. And Then bamboo is measured by using rockwell hardness tester type R (HRR). Hardness for each sample is measured five times repetition.

2.5 Resistivity againts fungus

Slicing of bamboo by heating time variation is obeserved by naked eye. Initial time for fungus growth is recorded. All sequence of experimet including morfology analysis, density measurement, hardness test and resistivity against fungi are presented in Figure 2.

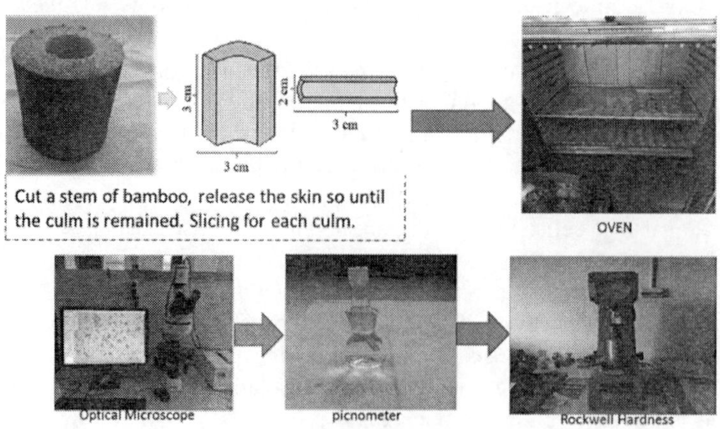

Figure 2. Sequence of Expriment

3 Result and discussion

3.1 Morfology of Bamboo

Morfology of bamboo is presented in Figure 3. Figure 3 shows a cross section of bamboo culm. The vascular of bamboo is marked by white circular area which are represent the hollow structure in culm. The area surrounding vascular consist of fiber and parenchyma which is represented by orange color area [4]. The image shows that diameter of vascular at inner part (~ 150 μm) is quite larger than outer part (~50 μm). The vascular is the most water storaged because it is a water channel from bottom to top bamboo. So Inner part will the more water than outer part so that the inner side of bamboo will be softer than outer side.

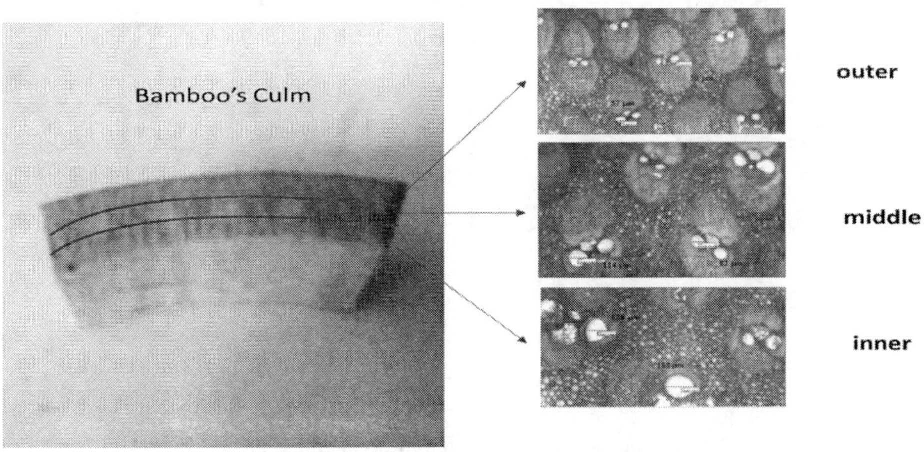

Figure 3 Morfology of Bamboo

3.2 Bamboo's Density

The density of bamboo corresponding to heating time is presented in Figure 4. It could be seen that the bamboo density decreases with increasing heating time. Heating for one hour caused the bamboo density decreased by about 39%. Heating up to 4 hours caused bamboo density decreased by 52%. This condition reflected that the highest evaporation occurs in the first hour periode. So it could be estimated that evaporation in the first hour is occured in the vascular area where the water is mostly stored. However, after one hour, water evaporated mostly occured from the area which contain low water such as parenchyma and fiber area.

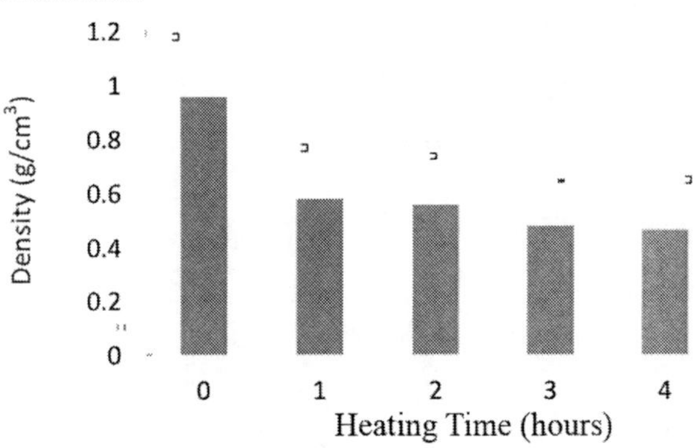

Figure 4 Density vs Heating time

3.3 Hardness of bamboo

In the Figures 5 and 6 both hardness in the longitudinal and tangential sides show the same tendency. The longer heating time, the hardness of bamboo is increased. For the longitudinal side, by heating time for two hours, the hardness increased by 31%, whereas for additional two hours (heating time 4 hours) the hardness only increased by additional 10%. However, slightly different is shown in the measurement of hardness in the tangential plane. The hardness indicates continuous increase up until up to 82% upon 4 hours heating time. Principally, the longer heating time, the less water content in the bamboo, so it will make the bambo's harder [5]. Certainly, beside the water content the properties of bamboo is also determined by the ingredients contents [6].

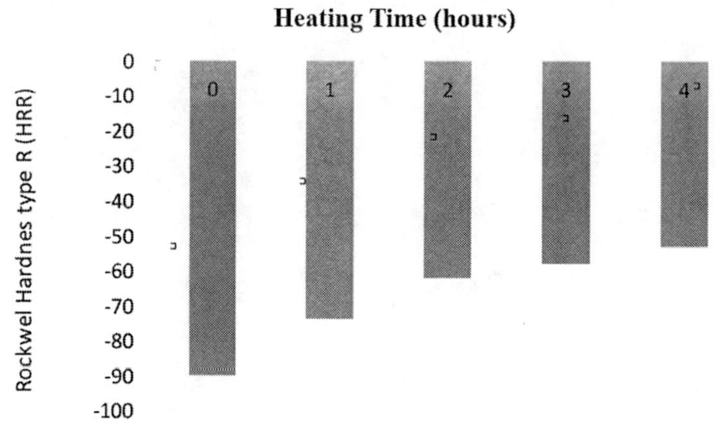

Figure 5 Hardness vs Heating time (longitudinal side)

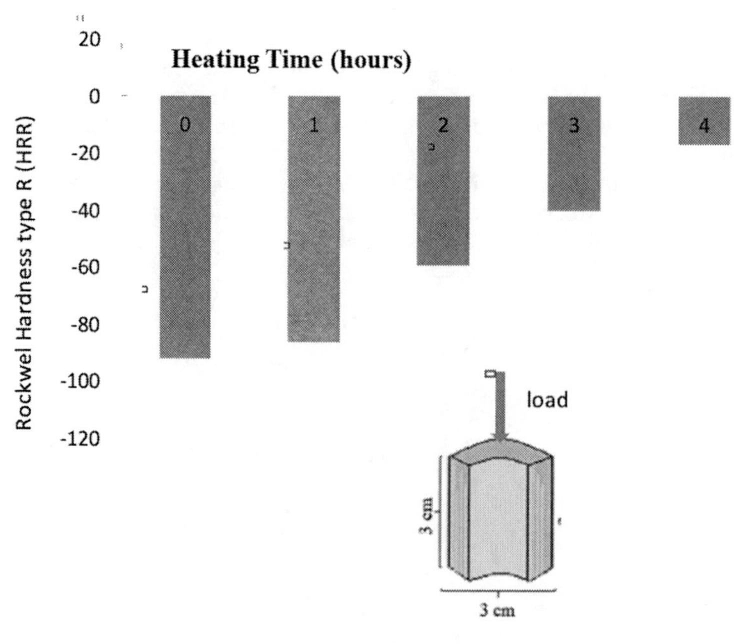

Figure 6 Hardness vs Heating time (tangential side)

3.4 Resistivity against fungi

The resistivity of bamboo againts fungus is resumed in Table 1.

Table 1. Resistivity against fungus of bamboo

Heating time (hours)	**0**	**1**	**2**	**3**	**4**
	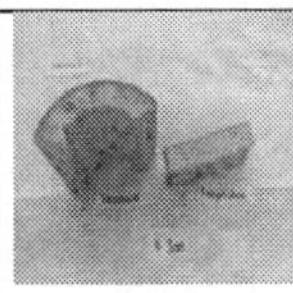	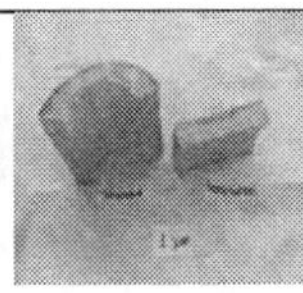	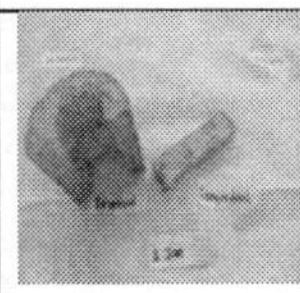	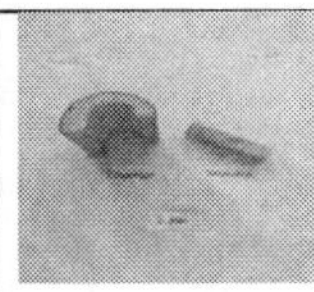	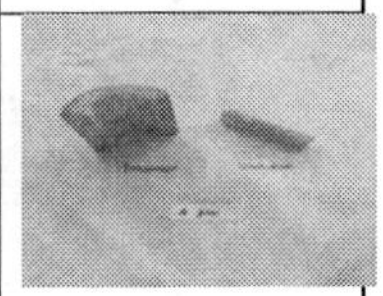
Fungus appeared (day)	3	3	6	>14	>14

Table 1 shows that bamboo with 0 and 1 hour heating time will be attacked by fungus after 3 days. Bamboo with longer heating time (2 hours) are able to survive from fungus attack until 6 days. While bambo with heating time of 3 and 4 hours can survive from fungus attack up to more than 14 days. The more heating time causes water content in the bamboo is decreased. This respectively causes the growth of fungi is low. In contrarily, the short heating time still result a high water content which results in relatively rapid growth of fungi. Thus, heating time up to 3 hours is quite optimum in keeping the bamboo from the fungus attack.

4. Conclusions

The bamboo' culm structure contains many cavities (hollow) caused a lot of water stored. An efforts to reduce the water content of bamboo is successfully done with heating at 100°C. Through evaluation of bamboo's density, water decreases significantly after 1 hour heating. However , in order to improve mechanical properties (hardness), the heating process should be extend up to 2-3 hours. While, for enhancing the resistivity againt fungus attack, bamboo should be heated until 3-4 hours.

References

[1] Fahrina, R. 2014 Pemanfaatan Bambu Betung Bangka Sebagai Pengganti Tulangan Balok Beton Bertulangkan Bambu. *Fropil,* **2** 59

[2] Dransfield, S., & Widjaja, E. A. 995 *Dendrocalamus Asper in Bamboos*. Bogor: Plant Resources of South-East Asia7 Prosea Fondation.

[3] Husnil, Y. A. 2009 *Perlakuan Gelombang Mikro dan Hidrolisis Enzimatik Pada Bambu Untuk Pembuatan Bioetanol*. Jakarta: Fakultas Teknik UI.

[4] Praptoyo, H., & Yogasara, A. 2012 Sifat Anatomi Bambu Ampel (Bambusa Vulgaris Schrad) Pada Arah Aksial Dan Radikal. *Seminar Nasional Mapeki XV*, (pp. 26-29). Makassar.

[5] Jr, A. S., Torres, S. M., Barbosa, N. P., Ortiz, S. R., & Ghavami, K. 2012 Impregnation of Bambusa Vulgaris with Polymeric Resins. *Key Engineering Materials*, **517** 1662

[6] Syafi'i, L. I. 1984 *Pengujian Sifat-sifat Fisis dan Mekanik* . Bogor: Fakultas Kehutanan IPB.

Acknowledgements

We would like to thank the ministry of research and technology for its support so that it can disseminate this research through the project INSINAS fiscal year 2017.

Effects of Sodium and Chlorine Doping on Optical Properties of Germanene: Density Functional Theory Calculation

Mauludi Ariesto Pamungkas, Istiroyah, Muhammad Fikri Salim, Dina Nurul Afifah

Lab Simulasi dan Pemodelan, Jurusan Fisika, Universitas Brawijaya

m_ariesto@ub.ac.id

Abstract.Germanene is a two-dimensional material which is derived from germanium elements and arranged in a honeycomb-like hexagonal. This study aims to determine the optical properties of pristine germanene and effect of doping Sodium (Na) and Chlorine (Cl) atom on germanene. These elements are chosen owing to their ability to create strong bonds with germanene and easily available in nature. This research uses Density Functional Theory approach to measure the effects of sodium and chlorine doping on germanene's optical properties. The results showed that pristine germanene had higher optical and energy than the doped with sodium and chlorine. Cl-doped germanene had similar graphical shape with pristine germanene

1. Introduction

Since the discovery of graphene, interest on two dimension materials are continuously increasing. Germanene and silicene attracted special interest due to their similarities with graphene[1]. Even, compared to graphene, germanene and silicene are more likely to be implemented in current silicon based semiconductor technology. Today, applications of germanene are not restricted as electronic materials anymore but become more widely such as gas sensors[2], [3] and photo- catalyst supports[4].

Band structure is one of features of electronic structure which has close relation with optical properties. Electronic structure, including band structure of pristine germanene have been intensively studied. Similar to graphene, pristine germanene has zero band gap[5]. This semimetalic property hinder germanene to be applied in semiconductor industry.

Controlable band gap is the key for an electronic material like germanene to widen its applications. There are several ways to engineering band gap, these include hydrogenation[6][7], halogenation[8], adsorption of gas [2],metal elements[9], alkali and alkali earth [10] and organic molecules[11], applying an external electrical field[12], [13] and applying uniaxial[14] and biaxial strain[12]. Band Structure of a material has close relation [15].

Comparison of optical properties of pristine graphene, silicene, and germanene have been investigated previously using density functional theory. Refractive index of germanene is relatively larger than that of graphene. While, reflectivity of germanene is lower than that of graphene [16], [17]. Eventhough electronic structure and optical properties of pristine germanene have been studied insensively, electronic structure and optical properties of doped germanene is still very rare investigated. Therefore, this work is aimed to study the effects of Na and Cl adatoms on the optical properties of germanene.

2. Methods

In this work, we focus on the optical properties of Na-doped germanene and Cl-doped germanene using the *ab initio* pseudopotential density functional method. Pseudopotentials were expanded within a plane

IOP Conf. Series: Materials Science and Engineering **299** (2018) 012045 doi:10.1088/1757-899X/299/1/012045

wave basis set with an energy cutoff of 40 eV. Integrations in Brillouin zone were performed using special *k* points generated with a 12 × 12 × 1 Monkhorst–Pack mesh mesh parameters grid. Exchange and correlation effects were described by Perdew-Burke-Eruzerhof in generalized gradient approximation. Calculations of optical properties are based on norm-conserving pseudopotentials available in ABINIT. Na/Cl adatoms is placed on germanene with three possible sites i.e : 1) Bridge site (B-site), between two Ge/As atoms 2) Hollow site (H-site), in the centre of pentagon shape of silicene, and 3) Top site (T-site), above one Ge/As atom.

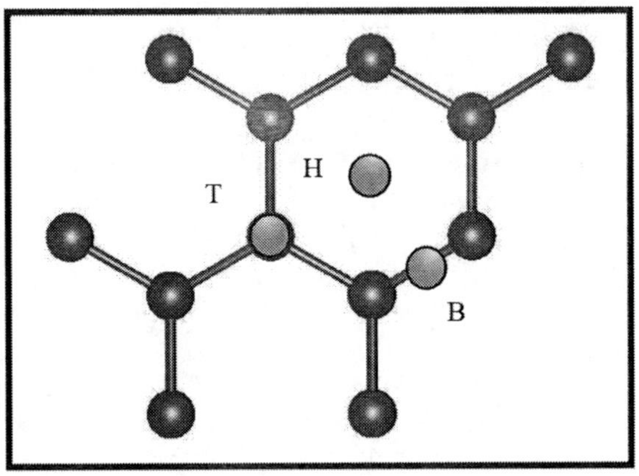

Figure 1. Three possible positions of Na/Cl adatom on germanene structure: Bridge (B), Hollow (H), and Top (T).

3. Result and discussion

In this calculations, the most favorable sites was determined by calculating the binding energy for adsorption. The system with the lowest binding energy is the most favorable one.

$$E\text{ binding} = E\text{ system} - (E\text{ pristine germanene} + E\text{ adatom})$$

Based on binding energy calculation, Na-adatom at B-site is the most favorable position. T-site is the most favorable position of Cl adatom in germanene. Therefore, next we focus on optical properties of germanene with Na-adatom in B-site and with Cl-adatom in T-site.

In order to observe the effect of Na adatom on the optical properties of germanene, here we compare optical properties of pristine germanene and germanene with Na-adatom in B-site. Figure 1, Figure 2, and Figure 3 show density of states and band structure of pristine germanene, Na-doped germanene, Cl-doped germanene respectively.

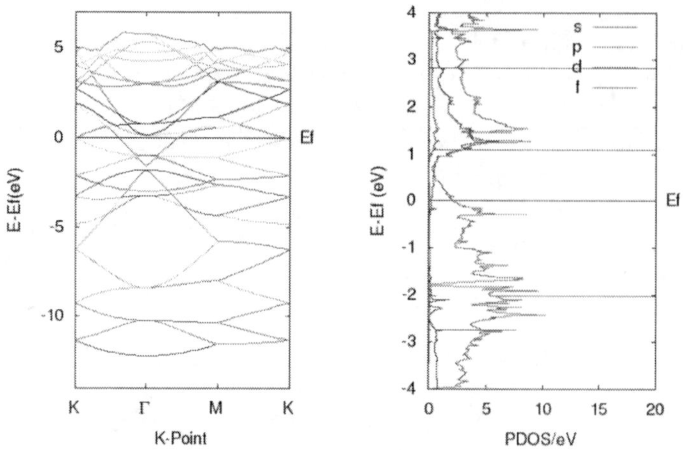

Figure 2. Band Structure and Desity of State of pristine germanene

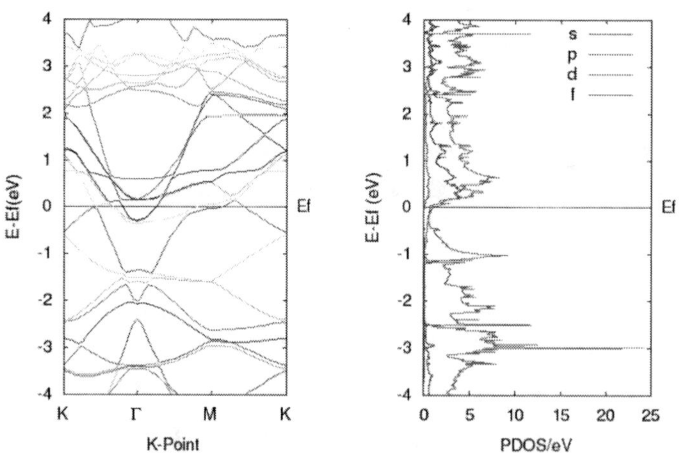

Figure 3. Band Structure and Desity of State of Na-doped germanene

Eventhough band structure of pristine germanene, Na-doped germanene, and Cl-doped germanene have different shape, all of them are conducting materials due to overlap of their energy levels near fermi energy. The Density of states spectra of the three systems also have similarities in term of orbital contribution. For all three system, p orbitals are dominant.

One of the most important optical constants of a material is its refractive index, which in general depends on the wavelength of the electromagnetic wave, through a relationship called *dispersion*. In addition to dispersion, an electromagnetic wave propagating through a lossy medium experiences attenuation, which means it loses its energy, due to various loss mechanisms such as the generation of phonons (lattice waves), photogeneration, free carrier absorption, scattering, In materials where an electromagnetic wave can lose its energy during its propagation, the refractive index becomes complex. The real part is usually the refractive index, *n*, and the imaginary part is called the *extinction coefficient*, *K*. The refractive index of an optical or dielectric medium, *n*, is the ratio of the velocity of light *c* in vacuum to its velocity *v* in the medium; *n* □□*c*/*v*.

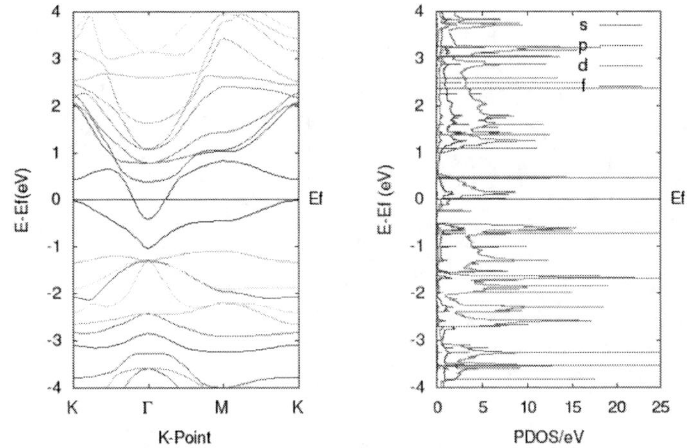

Figure 4. Band Structure and Desity of State of Cl-doped germanene

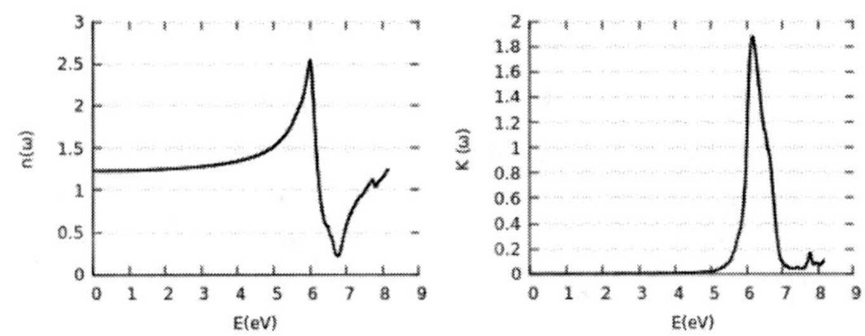

Figure 5. Real and imaginary part of refractive index of pristine germanene

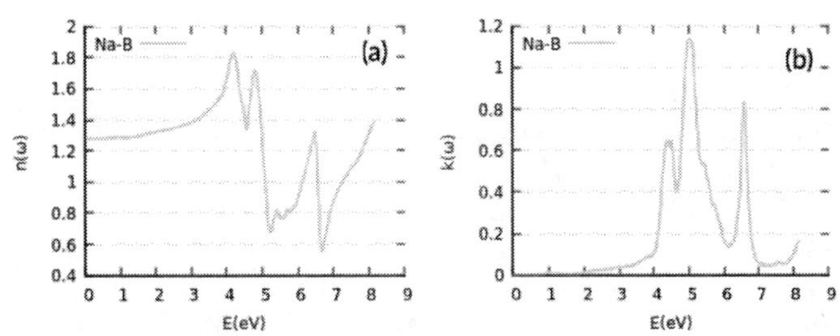

Figure 6 Real and imaginary part of refractive index of germanene with Na-adatom

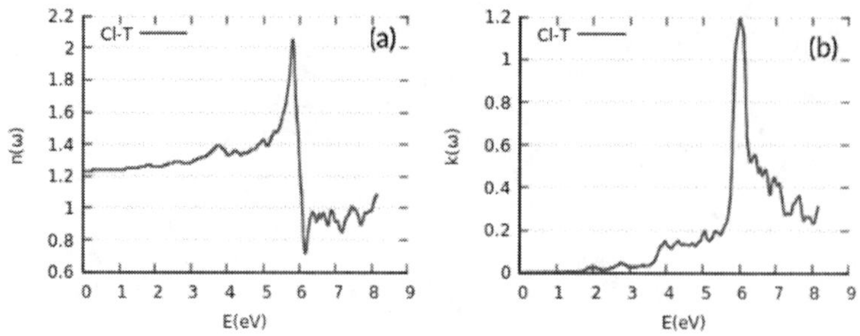

Figure 7. Real and imaginary part of refractive index of germanene with Cl-adatom

By adding Na adatom (Figure 6), both real part and imaginary part of refractive index decrease. The decrease of real part of refractive index indicates normal dispersion, while decrease of imaginary part of refractive index which represents extinction coefficient, indicates decrease of absorpsion. Peaks of real part and imaginary are observed at lower energy compared to pristine germanene. It indicates that Na adatom induces a new state such that it requieres lower energy to absorb. While, Cl adatom (Figure 7) hardly does not change refractive index of germanene.

4. Conclusion

The all three systems i.e pristine germanene, Na-doped germanene, and Cl-doped have different shapes but are the same in terms of having no band gap. Their Density of States shows that for all systems, p-orbital are dominant. Na adatom shifts refffractive index, while Cl adatom seems has no effect on refractive index There is split of reflection and absorption index of Na-doped germanene.

Acknowledgement
The authors thank to physics department of Universitas Brawijaya for their supports.

References

[1] Jomehpour Zaveh, S., Roknabadi, M. R., Morshedloo, T. & Modarresi, M. *Superlattices Microstruct.* **91** (2016) 383
[2] Pang, Q. *et al. Phys. E Low-dimensional Syst. Nanostructures* **88** (2017) 237
[3] Bianco, E. *et al. ACS Nano* **7** (2013) 4414
[4] Rupp, C. J., Chakraborty, S., Anversa, J., Baierle, R. J. & Ahuja, R. *ACS Appl. Mater. Interfaces* **8** (2016) 1536
[5] Acun, A. *et al. J. Phys. Condens. Matter* **27** (2015) 443002
[6] Trivedi, S., Srivastava, A. & Kurchania, R. *J. Comput. Theor. Nanosci.* **11** (2014) 781
[7] Houssa, M. *et al. Appl. Phys. Lett.* **98** (2011) 2011
[8] Ma, Y., Dai, Y., Niu, C. & Huang, B. *J. Mater. Chem.* **22** (2012) 12587
[9] Pang, Q. *et al. Appl. Surf. Sci.* **314** (2014) 15
[10] Li, S. *et al. Phys. Chem. Chem. Phys.* **16** (2014) 15968
[11] Rubio-Pereda, P. & Takeuchi, N. *J. Phys. Chem. C* **119** (2015) 27995
[12] Yan, J.-A., Gao, S.-P., Stein, R. & Coard, G. *Phys. Rev. B* **91** (2015) 245403
[13] Ni, Z. *et al. Nano Lett.* **12** (2012) 113
[14] Bayani, A. H., Dideban, D. & Moezi, N. *Superlattices Microstruct.* **100** (2016) 198
[15] Van Den Broek, B. *et al. Appl. Surf. Sci.* **291** (2014) 104

[16] Matthes, L., Pulci, O., Küfner, S., Furthmüller, J. & Matthes, L. *New J. Phys.* **16** (2014) 105007
[17] John, R. & Merlin, B. *J. Phys. Chem. Solids* **110** (2017) 307

IOP Conf. Series: Materials Science and Engineering **299** (2018) 012046 doi:10.1088/1757-899X/299/1/012046

Evaluation of the suitability of tin slag in cementitious materials: Mechanical properties and Leaching behaviour

Andi Rustandi[1], Fuad Wafa' Nawawi[1], Yudha Pratesa[1], Agung Cahyadi[1]

[1] Department of Metallurgy and Materials Engineering, Universitas Indonesia
Corresponding email: fuad.wafa@ui.ac.id

Abstract. Tin slag, a by-product of tin production has been used in cementitious application. The present investigation focuses on the suitability of tin slag as primary component in cement and as component that substitute some amount of Portland Cement. The tin slags studied were taken from Bangka, Indonesia. The main contents of the tin slag are SiO_2, Al_2O_3, and Fe_2O_3 according to the XRF investigation. The aim of this article was to study the mechanical behaviour (compressive strength), microstructure and leaching behaviour of tin slag blended cement.
This study used air-cooled tin slag that had been passed through 400# sieve to replace Portland Cement with ratio 0, 10, 20, 30, 40 by weight. Cement pastes and tin slag blended cement pastes were prepared by using water/cement ratio (W/C) of 0.40 by weight and hydrated for various curing ages of 3, 7, 14 days
The microstructure of the raw tin slag was investigated using Scanning Electron Microscope (SEM). The phase composition of each cement paste was investigated using X-ray Diffraction (XRD). The aim of the leachability test was to investigate the environmental impacts of tin slag blended cement product in the range 4-8 pH by using static pH-dependent leaching test.
The result show that the increase of the tin slag content decreasing the mortar compressive strength at early ages. The use of tin slag in cement provide economic benefits for all related industries.

1. Introduction

Portland cement has a big impact in the large CO_2 emissions in the world, that are generated during the production of Portland cement. This production also need big energy and natural resources. Considering the process of cement clinker production, half of the carbon emission originates from the dissociation of limestone during clinkering, whereas the other half is due to fuel combustion [3]. Utilization of metallurgical slags can be solution for energy and environmental problems, in the production of Portland cement.

An example of metallurgical slags is tin slag. Like other metallurgy slags, tin slag has the potential to be used in cementitious material, due to the similarity between the oxides that are present in both Portland clinker and Tin slag. The annual production of tin in Indonesia is between 70.000 and 96.000 tonnes (Kemenperin 2016). This big amount of production tin slag in Indonesia also generates big quantity of tin slag. At present the tin slag is stored at landfill, and no solution to utilize it.

Tin slags can be treated as a cementitious material due to its silicate calcium minerals. However, the hydration activity of tin slag is much lower than that of Portland cement due to their mineralogy properties and relatively high content of inert components.

In this study, Air cooled tin slag (ACTS) was used as component to replace Portland cement. Air cooled Tin Slag, is a by-product in tin smelter. Molten tin slag, which is discharged from the furnace, is cooled by air quenching. The slag was solidified slowly and forms slag lumps. Air cooled tin slag has partly glassy and partly crystalline phase. Air cooled tin slag composed primarily of SiO_2, Al_2O_3, Fe_2O_3 and CaO. The main objective of this research was to investigate the mechanical properties and leaching behaviour of tin slag as pozzolanic materials in Portland cement to form Portland pozzolan cement.

2. Materials and methods

This research used OPC and Air-cooled Tin Slag (ACTS). Commercially available ordinary Portland Cement (OPC) was used as the principal binder and ACTS as pozzolanic materials. This research used ACTS with particle size under 0.037 mm, coresponding to ASTM C 595 (Standard Specification for Blended Hydraulic Cement). Table 1 shows the chemical composition of OPC and tin slag which were evaluated by X-ray fluorescence (XRF).

The microstructure of the raw tin slag and hydrated cement paste was investigated using Scanning Electron Microscope (SEM). The phase composition of cement paste (OPC, S10, and S20) was investigated using X-ray Diffraction (XRD).

Table 1. OPC and Slag chemical composition

	CaO	SiO_2	Al_2O_3	Fe_2O_3	SO_3	TiO_2	P_2O_5	MgO	SnO_2	Na_2O
OPC	62.34	23.83	5.96	2.56	3.23	-	-	2.92	-	0.46
Slag	5.65	39.66	13.18	8.77	-	6.84	5.26	1.91	4.00	3.75

2.1 Compressive strength

The test used ASTM C109 as a standard. Specimens were cast in 50 mm cube molds with water/cement (W/B) ratio of 0.4. These samples were first cured in room temperature at 95% relative humidity for 24 h and later demolded. The cement cube was kept in a water until testing time (3, 7, and 14 days). Specimens were removed from the bath and tested at laboratory temperature.

2.2 Leaching test

The leaching test used CEN/TS 14429 pH dependence test standard. The test was carried out using deionised water at Liquid/solid ratio (L/S) = 20: 1 L/kg, with range of pH test $4 \leq pH \geq 8$ for 48 hours. Acid (1M HNO_3) or base (1M NaOH) was added to added to adjust the pH values [5]. An experiment at the natural pH was included in all cases without addition of acid or base. Leachates were filtered through a 0.45 µm pore size nitrocellulose filtration membrane and divided into subsamples for measuring Cr, Mn, and Pb concentrations. The leachate solutions was analysed with ICP to identify Cr, Mn, and Pb concentration

Table 2. Samples mix design

Samples	Cement (wt%)	Tin Slag (wt%)	W/B ratio
OPC	100	-	0.4
TS-10	90	10	0.4
TS-20	80	20	0.4
TS-30	70	30	0.4
TS-40	60	40	0.4

3. Result and Discussion

3.1 Air Cooled Tin Slag Characterization

The main constituents of ACTS determined by XRF are SiO_2, Al_2O_3, CaO, Fe_2O_3 and TiO_2. Figure 1 presents the microstructure of Air Cooled Tin Slag (ACTS) investigated with Scanning Electron Microscopy (SEM). Through the SEM investigation, is observed the presence of particle with rough surface texture, with smooth and very low porosity surfaces.

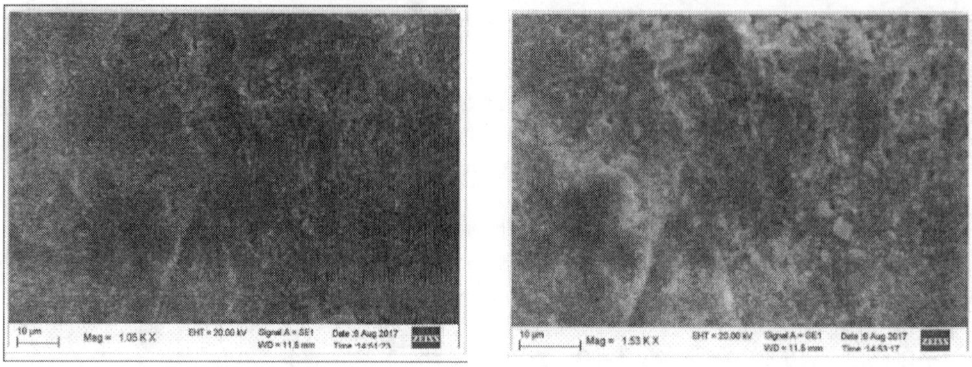

(a) (b)

Figure 1. SEM image of Air Cooled Tin Slag (a) magnification 1000 x (b) magnification 1500 x.

Figure 2 shows the XRD pattern of ACTS slag. ACTS slag has amorphous phase as dominant phase. Mineralogical phases that can be identified from this XRD pattern: Quartz (SiO_2), Hematite (Fe_2O_3), Cassiterite (SnO_2), and Corundum (Al_2O_3)

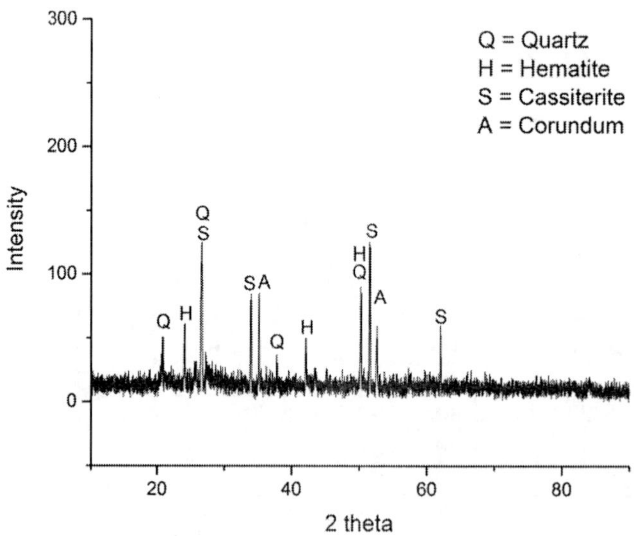

Figure 2. XRD pattern of the Air-Cooled Tin Slag

3.2 XRD Analysis

Figure 3 shows the XRD graphic of 7 days curing samples. According to the result shown in Figure 3, portlandite, calcite, ettringite, and CSH were observed in all the pastes products. Sample with replacement of 10% and 20% ACTS show lower portlandite and Calcium Silicate Hydrate (CSH) than in OPC. This was attributed to the fact that replacement of OPC with ACTS decreased the calcium oxide in slag-cement blended paste. The lower of calcium oxide also has impact in the lower amount of tricalcium silicate (C_3S) in the slag-cement blended paste. C_3S can react with water to form CSH and calcium hydroxide or portlandite according to the reaction:

$$2(3CaO.SiO_2) + 6H_2O = 3\,CaO.2SiO_2.3H_2O + 3Ca(OH)_2$$

Tricalcium Silicate Calcium Silicate Hydrate Calcium Hydroxide

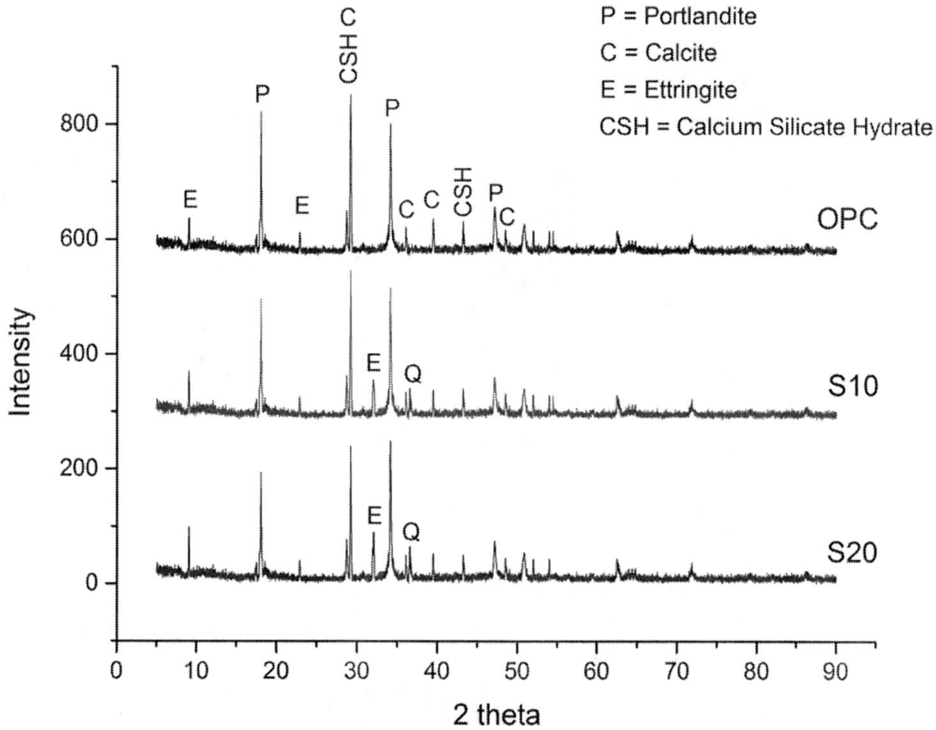

Figure 3. XRD pattern of cement hydrated for 7d

XRD patterns also show that samples S10 and S20 have higher ettringite peaks than OPC. Increasing amount of ACTS in cement blended paste can produced higher ettringite due to increasing ticalcium aluminate (C_4F) in cement paste. The ettringite (Calcium trisulfoaluminate hydrate) transformation from C_4F corresponds to the reaction:

$$3CaO.Al_2O_3 + 26H_2O + 3CaSO_4.2H_2O = 3CaO.Al_2O_3.3CaSO_4.32H_2O$$

Tricalcium aluminate gypsum Calcium trisulfoaluminate hydrate

The calcite content shows no difference with or without ACTS addition in cement paste. In contrast, the characteristic peak of quartz phase shows a clear difference with or without ACTS addition. The quartz's peak was higher with the increasing ACTS in cement blended paste, which is probably due to higher amount of quartz in ACTS than in OPC

3.3 Compressive Strength

Figure 4 presents the compressive strength of mortar with different ACTS replacement ratio at W/B ratio 0.4. The early compressive strength of mortar containing ACTS is significantly lower than OPC. The compressive strength of mortar, decrease with the increasing of ACTS replacement. At 3 days curing, comparing to the reference mortar (OPC), mixtures with the lowest replacement (TS10), presented a decrease 42% of compressive strength. The largest decrease of compressive strength occurred in sample with the most replacement content (TS40), reaching 68.9% of OPC compressive strength. OPC compressive strength can reached 19.79 MPa at 3 days curing, while slag-cement paste just only reach 11.44, 8.81, 7.59 and 6.15 MPa for 10, 20, 30, and 40 of ACTS replacement in OPC, respectively. At 7 days, the difference value of compressive strength between PC and slag-cement composite decrease. The lowest replacement (S10) gave a decrease 40.7% and the largest replacement (S40) gave a decrease 68.5%. The 14 days curing samples show the lowest difference value of compressive strength of OPC and slag-cement paste, with 37.1% difference for 10% ACTS replacement in OPC.

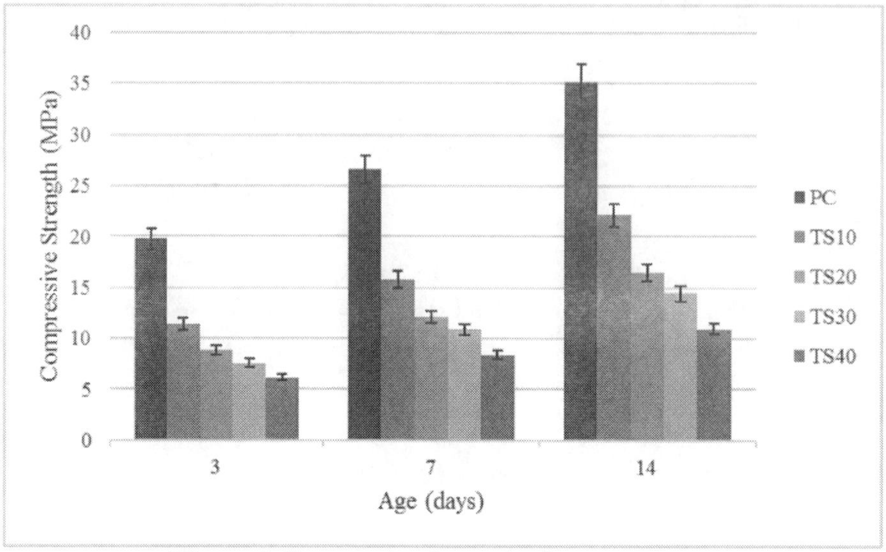

Figure 4. Compressive Strength of cement mortar

Decrease of compressive strength of mortar samples, with the increase of tin slag content is due to several reasons: 1). Cement dilution by less active component. 2). Increase of water demand (due to less water absorption of tin slag) and easily workable mixtures (less viscous mixtures than OPC), 3). The content of calcium hydroxide ($Ca(OH)_2$) and chemically bounded water ($W_{c.bw}$) in hardened cement paste decrease with increasing slag content. The lower compressive strength of slag-cement also corresponding to the lower amount of CSH and Portlandite as a hydration product as detected by XRD. Portlandite or calcium hydroxide give slightly strength to cement paste. Decrease of mechanical properties also attributed to higher levels of iron oxides, zinc oxides, and traces of sulphur in the chemical composition of the slag.

3.4 Leaching Behaviour

The pH dependence test CEN/TS 14429, can assess the metal matter release at different pH values [5]. The evaluation of pH dependence of leaching test is one of the important tools in the assessment of the expected long-term leaching behaviour of materials in different scenarios of utilisation or disposal [8].

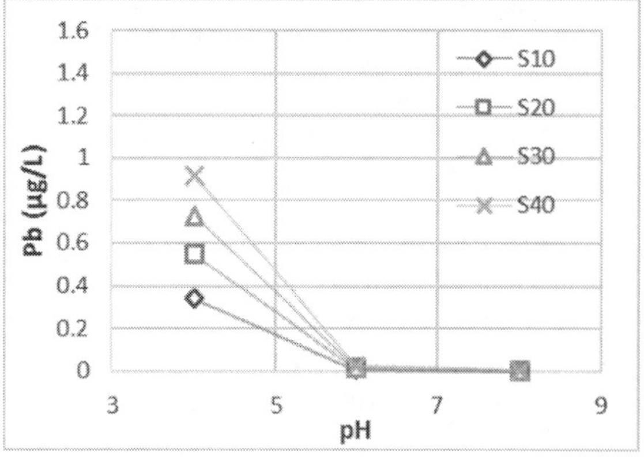

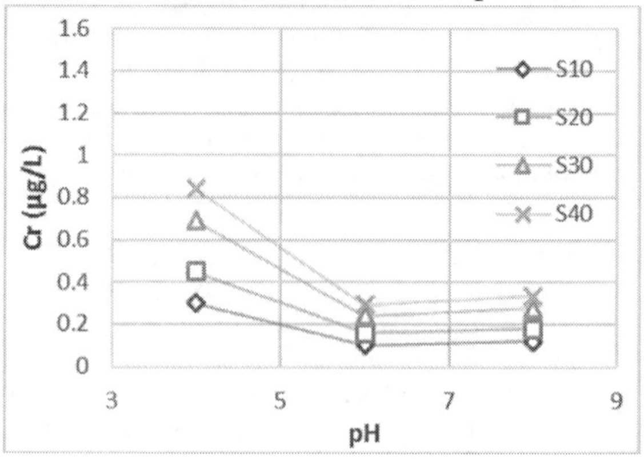

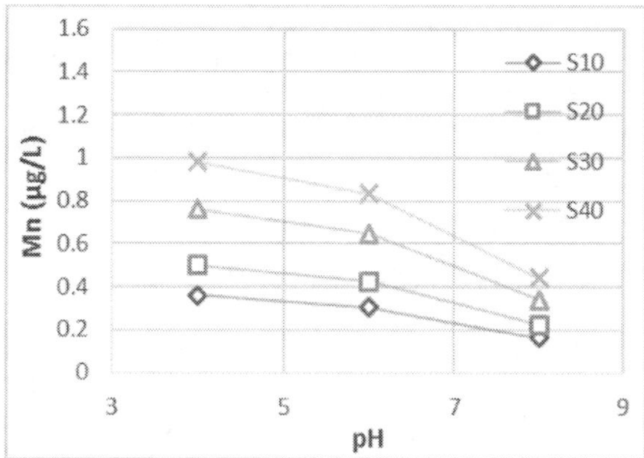

Figure 5. Pb, Cr, and Mn concentration in leachates solution

Figure 5 shows the concentration of Pb, Cr, and Mn in leachate solution. Concentration of Pb, Cr, and Mn increasing with the increasing amount of ACTS replacement in cement paste, due to higher amount of Pb, Cr and Mn, than in OPC. Concentration of Cr, Pb, and Mn in aqueous leachates of cement composite increased with decreasing pH. The highest Cr concentration occurred in S40 leachate samples with 0.84 µg/L in pH = 4. Pb and Mn also show same leaching behaviour with Cr. The highest concentration of Pb and Mn in leachate solution occurred in S40 leachate sample with pH test = 4 (0.92 µg/L and 0.98 µg/L). The trace elements Pb, Cr, and Mn in slag-cement composite tend to soluble in acid solution. This characteristic due to dissolving of some $Ca(OH)_2$ and C-S-H in acidic solution [15, 16].

4. Conclusions

ACTS is a by-product of tin production that has major constituent SiO_2, Al_2O_3, CaO, Fe_2O_3, and TiO_2. Portlandite, C-S-H, Calcite, and Ettringite are the major phase that detected in both OPC and slag-cement composite according to the XRD evaluation. Slag-cement composite also shows quartz phase that not detected in OPC. Replacement of OPC with ACTS decreasing the compressive strength of the cement mortar. The trace elements Pb, Cr, and Mn in slag-cement composite tend to soluble in acid environment, due to dissolving of $Ca(OH)_2$ C-S-H.

Acknowledgment
This research was supported by DRPM Universitas Indonesia with the PITTA program.

References

[1] İ. Alp, H. Deveci, H. Süngün. 2008 Utilization of flotation wastes of copper slag as raw material in cement production. *J. Hazard. Mat.* **159** 390-395
[2] P.E. Tsakiridis, G.D. Papadimitriou, S. Tsivilis, C. Koroneos. 2008 Utilization of steel slag for Portland cement clinker production. *J. Hazard. Mat* **152** 805-811
[3] Iacobescu, R., Angelopoulos, G., Jones, P., Blanpain, B. and Pontikes, Y. 2016 Ladle metallurgy stainless steel slag as a raw material in Ordinary Portland Cement production: a possibility for industrial symbiosis. *Journal of Cleaner Production* **112** 872-881
[4] Qiang, W., Mengxiao, S. and Jun, Y. 2016 Influence of classified steel slag with particle sizes smaller than 20 µm on the properties of cement and concrete. *Construction and Building Materials* **123** 601-610
[5] Martín-Torre, M., Payán, M., Galán, B., Coz, A. and Viguri, J. 2014 The Use of Leaching Tests to Assess Metal Release from Contaminated Marine Sediment under CO_2 Leakages from CCS. *Energy Procedia* **51** 40-47

[6] Bhatty, J.I., Miller, F.M., Kosmatka, S.H. 2004 Innovations in Portland Cement Manufacturing, first ed. Portland Cement Association, Skokie, Illinois, USA
[7] Shi, Y., Chen, H., Wang, J. and Feng, Q. 2015 Preliminary investigation on the pozzolanic activity of superfine steel slag. *Construction and Building Materials* 82 227-234
[8] Van der Sloot, H.A., Heasman, L., Quevauviller P., 1997. Harmonisation of leaching / extraction tests. Elsevier Science B. V., Amsterdam
[9] B. Das, S. Prakash, P.S.R. Reddy, V.N. Misra. 2007 An overview of utilization of slag and sludge from steel industries. *Resour. Conserv. Recycl.* **50** 40-57
[10] M. Malhotra, P.K. Mehta. 996 Pozzolanic and Cementitious Materials. Gordon and Breach Science, Amsterdam
[11] A.S. Reddy, R.K. Pradhan, S. Chandra. 2006 Utilization of Basic Oxygen Furnace (BOF) slag in the production of a hydraulic cement binder. *Int. J. Miner. Process.* **79** 98-105
[12] G. Habert, N. Choupay, J.M. Montel, D. Guillaume, G. Escadeillas. 2008 Effects of the secondary minerals of the natural pozzolans on their pozzolanic activity. *Cem. Concr. Res.* **38** 963-975
[13] M. Schneider, M. Romer, M. Tschudin, H. Bolio. 2011 Sustainable cement production—present and future. *Cem. Concr. Res.* **41** 642-650
[14] K.L. Lin, W.C. Chang, D.F. Lin. 2008 Pozzolanic characteristics of pulverized incinerator bottom ash slag. *Constr. Build. Mat.* **22** 324-329
[15] X.D. Li, C.S. Poon, I.M.C. Lo, D.W. Kirk. 2001 Heavy metal speciation and leaching behaviors in cement based solidified/stabilized waste materials. *J. Hazard. Mat.* **82** 215-230
[16] D.H. Moon, D. Dermatas. 2006 An evaluation of lead leachability from stabilized/solidified soils under modified semi-dynamic leaching conditions. *Eng. Geol* **85** 67-74

IOP Conf. Series: Materials Science and Engineering **299** (2018) 012047 doi:10.1088/1757-899X/299/1/012047

Corrosion Behavior of Carbon Steel in Concrete Material Composed of Tin Slag Waste in Aqueous Chloride Solution

Andi Rustandi[1], Agung Cahyadi[1*], Sonia Taruli Siallagan[1], Fuad Wafa' Nawawi[1], Yudha Pratesa[1]
[1]Metallurgy and Materials Engineering Department, Faculty of Engineering, Universitas Indonesia, Depok, Indonesia

*Corresponding email: agung.cahyadi@ui.ac.id

Abstract. Tin slag is a byproduct of tin ore smelting process which is rarely utilized. The main purpose of this work is to investigate the use of tin slag for concrete cement material application compared to the industrial Ordinary Portland Cement (OPC). Tin slag composition was characterized by XRD and XRF analysis. The characterization results showed the similar chemical composition of tin slag and OPC. It also revealed the semi crystalline structure of tin slag sample. Several electrochemical tests were performed to evaluate corrosion behavior of tin slag, OPC and various mixed composition of both materials and the addition of CaO. The corrosion behavior of OPC and tin slag were evaluated by using Cyclic Polarization, Electrochemical Impedance Spectroscopy (EIS) and Electrochemical Frequency Modulation (EFM) methods. Aqueous sodium chloride (NaCl) solution with 3.5% w.t concentration which similar to seawater was used as the electrolyte in this work. The steel specimen used as the reinforce bar (rebar) material of the concrete was carbon steel AISI 1045. The rebar was embedded in the concrete cement which composed of OPC and the various composition of tin slag including slag without addition of CaO and slag mixed with addition of 50 % CaO. The electrochemical tests results revealed that tin slag affected its corrosion behavior which becoming more active and increasing the corrosion rate as well as decreasing the electrochemical impedance.

1. Introduction

Tin slag is byproduct of tin ore smelting process which is rarely utilized. The smelting process of tin ore produces slag-1 and slag-2. Slag-1 contains 20-30% tin, and slag-2/final slag which produced from smelting slag-1 contains only 2-3% tin [1]. The final slag (slag-2) is commonly discarded as waste material and usually it will be kept and controlled by the authorized environmental agency. Based on the XRD and XRF characterization analysis, it shows that the oxides composition of slag-2 (final slag) is quite similar to the Ordinary Portland Cement (OPC).

In addition, the similar characteristic between final slag and OPC is water absorption value. Water absorption value is the enhancement of weight percentage of a material to theoriginal weight expressed as the absorption value (in percent). The value of final tin slag was measured at 1.97% [2]. This value is quite moderate when compared with other slag, even close to the water absorption value for OPC, which is about 1.08%. This means the hydration reaction between water and final slag can happen quite well. This is based on the study that the average absorbance value of cement should not be more than 5% with none being more than 7% [3].

Concrete is material that widely use for an infrastructure,that has constraint with damage caused by corrosion, especially by salinity in coastal area. Salinity is one of the major sources of damage in building on Coastal Zone [4]. The presences of aggressive chloride in coastal area can initiate corrosion in concrete reinforcement. The previous studies have suggested that there is a shoreline effect on the presence of destructive sea salt very quickly at the first 100 meters from the shore [5]. In fact, the wind can carry chlorides in the air up to distances beyond 3 km [6]. Whereas the concrete structure is generally expected to survive without repair and maintenance in a long time (up to 100 years). Therefore, concrete materials constructed as infrastructure in coastal areas should have good corrosion resistance to the presence of aggressive ions such as chloride so that innovation is required on the raw material of cement material in steel reinforcement to address this problem.

The chemistry in concrete involves hydration reaction by addition of water. Lime stone as main sources of CaO is used to form hydration product and bind other constituent including final tin slag. The addition of limestone has a purpose to increase the CaO content in final tin slag which only contain small amount of CaO. The important reaction on hydration reaction in concrete is given below:

$$2(3CaO.SiO_2) + 6H_2O \rightarrow 3CaO_2.SiO_2.3H_2O + 3Ca(OH)_2$$

$$(2CaO.SiO_2) + 4H_2O \rightarrow 3CaO_2.2SiO_2.3H_2O + 3Ca(OH)_2$$

This work focused on the aspect of corrosion behavior of OPC and final slag with CaO and without CaO addition in 3.5 % aqueous chloride solution (NaCl) which is evaluated by using electrochemical measurement testing methods which are: Cyclic Polarization, Electrochemical Impendence Spectroscopy (EIS) and Electrochemical Frequency Modulation (EFM) methods respectively.

2. Materials and Methods

Final tin slag used in this study comes from Bangka Island, Indonesia. Based on XRF characterization, the major component in final tin slag was SiO_2 and the other constituent component are shown in Table. 1.

Table 1. Chemical Composition of Final Tin Slag (%)

SiO_2	Al_2O_3	Fe_2O_3	TiO_2	CaO	P_2O_3	Na_2O	SnO_2	ZrO_2
39.7	13.2	8.70	6.84	5.64	5.26	4.30	4.10	3.75

In addition, this experiment used AISI 1045 as reinforced steel that will be embedded in concrete material which composed of tin slag waste. The composition of reinforced steel is shown in Table 2.

Table 2. Chemical Composition of AISI 1045

Chemical Composition (%)						
C	Si	Mn	P	Cr	Ni	Cu
0.40	0.23	0.67	0.011	0.06	0.01	0.02

The resistivity and pH of the various composition materials: OPC, final tin slag without CaO addition and tin slag material mixed with 50 % CaO addition were measured to provide preliminary corrosivity data prior to performing electrochemical testing. The cement and tin slag sample materials were put in the soil box in wet condition by adding some distilled water then measured by resistivity meter to obtain resistivity data followed by simply measured by using pH meter.

Cylindrical concrete specimens used for the electrochemical testing were prepared with diameter 100 mm and length 140 mm which steel bar was embedded in symmetrical position. The exposed area of 700 mm^2 was made prior to the measurements testing (Fig. 1). All of the concrete specimens were exposed in NaCl 3.5% at room temperature.

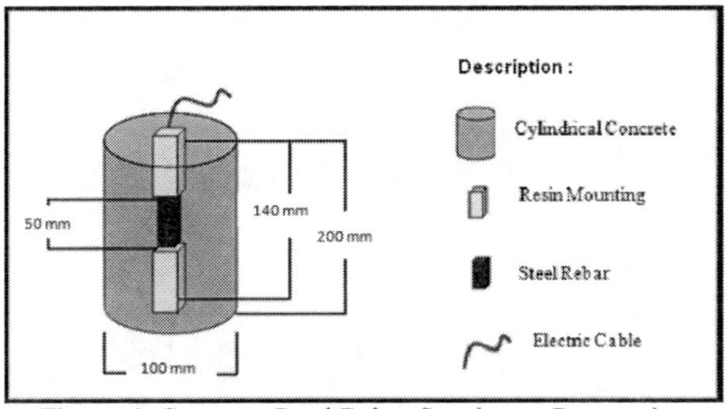

Figure 1. Concrete Steel Rebar Specimens Preparation

The electrochemical measurements testing were conducted with Cyclic polarization, EIS and EFM methods by using Gamry Instruments Potentiostat PCI4G750 completed with Global Software. The measurements apparatus consists of reference and counter electrodes with steel rebar testing specimen was treated as working electrode as shown in Fig. 2.

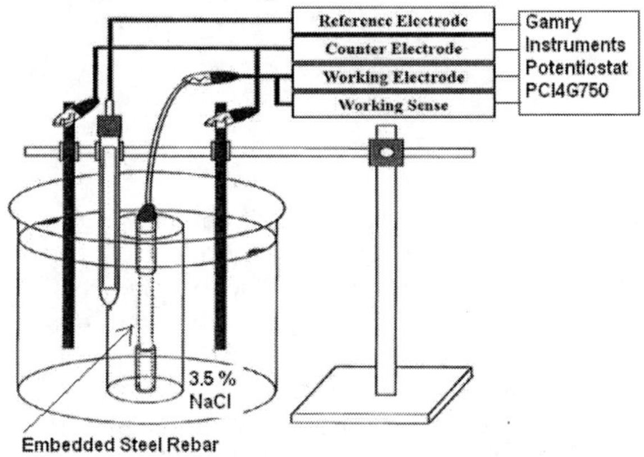

Figure 2. Electrochemical Measurement Testing Apparatus

3. Results and Discussion

3.1 Resistivity and pH Measurement

Prior to electrochemical testing, the resistivity and pH of OPC and tin slag in wet condition were measured as preliminary corrosivity data. The testing results can be seen in Fig. 3. which shows that the resistivity and pH increases with increasing CaO content in the mixture.

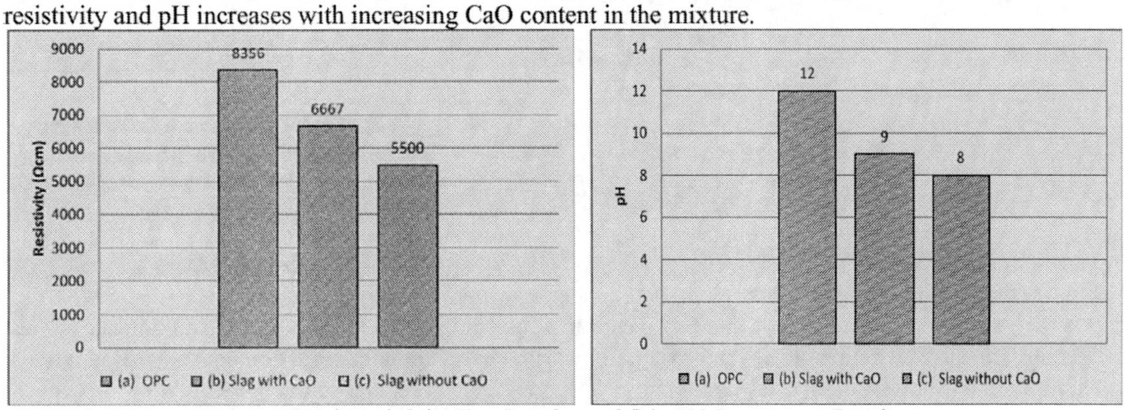

Figure 3. (a.) Resistivity Test Results. and (b.) pH Measurement Results

The effect of CaO content on pH that is the value increases with increasing CaO content as resulted from basic or alkaline property of CaO compound and will further hydrated to form $Ca(OH)_2$ followed by curing process.

3.2 Cyclic Polarization Testing

The results of cyclic polarization measurement are shown in Fig. 4.

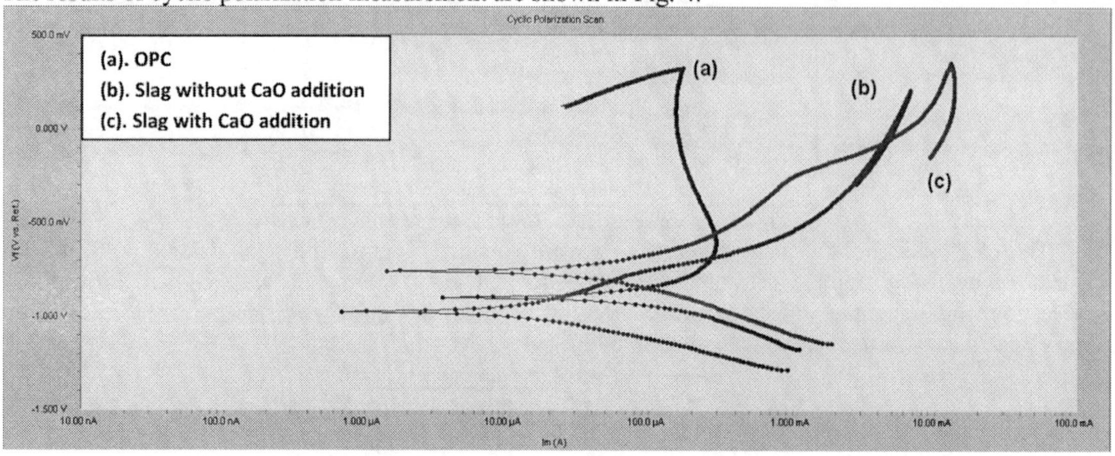

Figure 4. Cyclic Polarization Curves for OPC, Slag with CaO addition , and Slag without CaO addition

The Cyclic Polarization testing of steel rebar which embedded in OPC shows that the significant passive condition behavior takes place without the occurrence of pitting breakdown. Steel rebar embedded in tin slag without CaO addition shows the active condition of corrosion behavior. There is pitting breakdown encountered by steel bar embedded in tin slag with CaO addition caused by the aggressive ion chloride (penetration) in aqueous 3.5 % NaCl solution which contains dissolved oxygen.

3.3 Electrochemical Impedance Spectroscopy (EIS)Test

The results from EIS test can be shown in Fig.5 which represented by Nyquist plot curves.

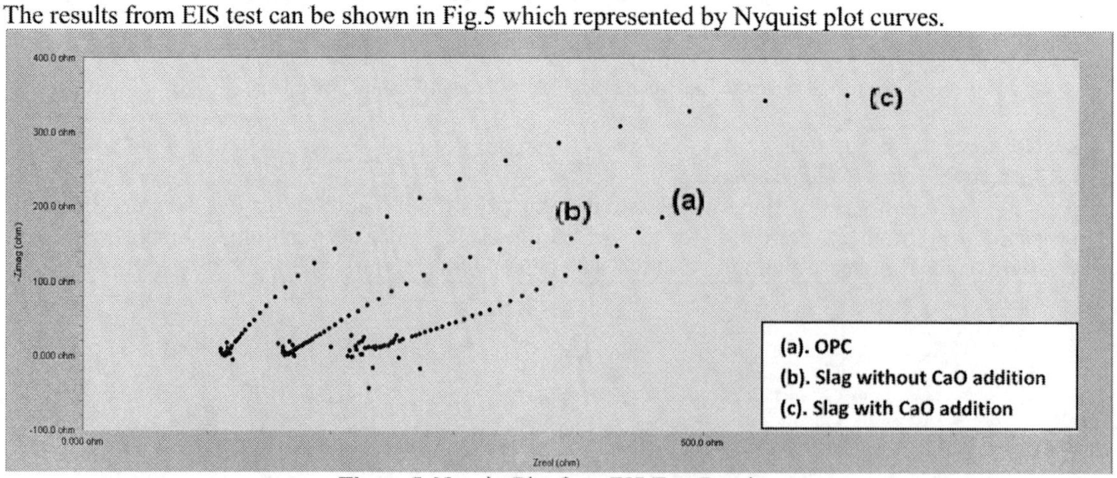

Figure 5. Nyquist Plot from EIS Test Results

As can be shown by Nyquist plots, OPC material which represented by (a) curve shows with nearly 45° line slope indicates diffusion and passive corrosion behavior. Slag without CaO addition shows nearly semi-circle curve which indicates active condition with charge transfer corrosion behavior. Slag with CaO addition also shows nearly semi-circle curve which indicates relatively in active condition or controlled by charge transfer corrosion behavior but has higher magnitude of electrochemical impedance which compared to the impedance magnitude of slag without CaO addition.

3.4 Electrochemical Frequency Modulation (EFM) Test

The results of EFM test can be shown in Fig. 6 (a), 9 (b) and 9 (c) curves which represent the materials: OPC, slag without CaO and slag with CaO respectively.

(a) (b) (c)

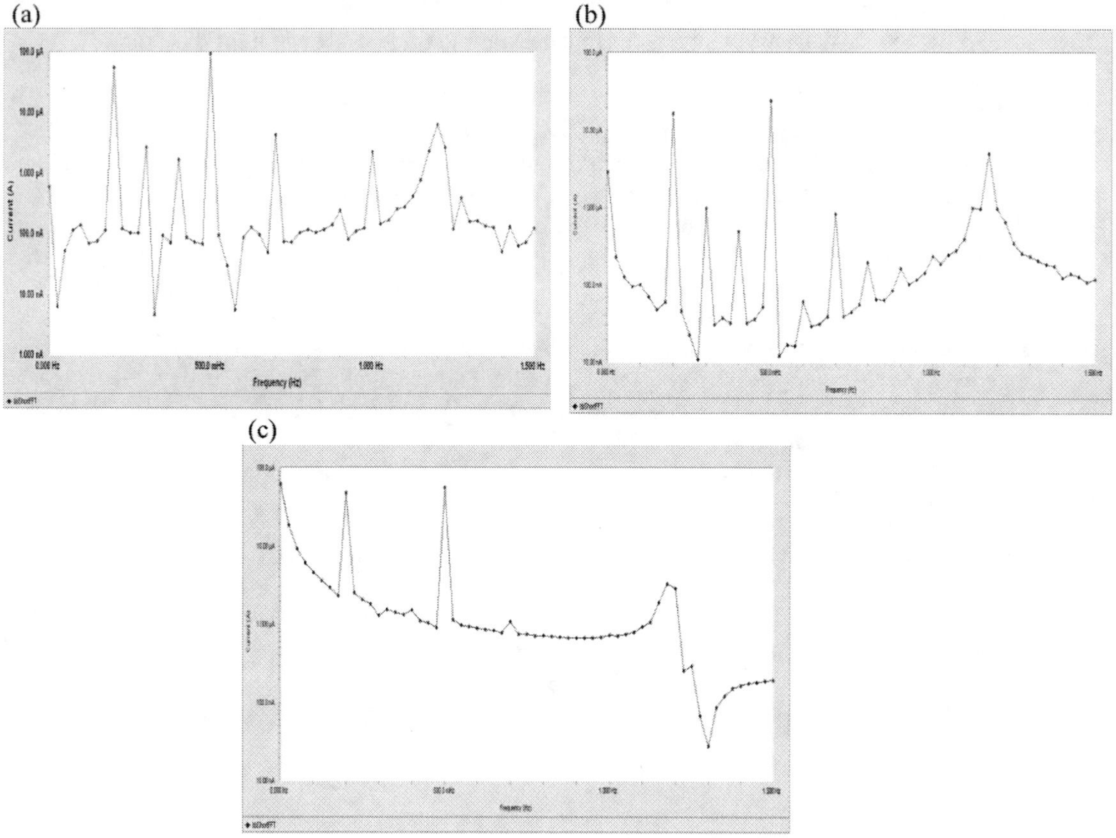

Figure 6. EFM Inter-modulation Spectrum for **(a)** OPC, **(b)** Slag without CaO addition **(c)** Slag with CaO addition.

Resulted from the above EFM curves, the following corrosion rates results analysis can be shown in Table 3.

Table 3. Corrosion Rates Resulted From EFM Testing

	OPC	Slag with CaO	Slag without CaO
Current (μA)	92.77	44.99	88.13
Corrosion Rate (mpy)	1.70	13.70	26.85

The corrosion rates decreased with increasing CaO content and influenced by the cement mixture composition like the presence of specific oxide compounds which cause the occurrence of passivity corrosion behavior and contribution of curing process especially in OPC concrete material.

4. Conclusions

The following conclusions can be made from this work is the corrosion rates decreased with increasing CaO content due to the occurrence of passivity corrosion behavior and the magnitude of impedance increased. Furthermore, the addition of approximately 50% CaO with final tin slag was required to improve its corrosion property so that it can be used for concrete cement application properly.

5. Acknowledgement

The authors wish to thanks DRPM-Universitas Indonesia for the financial support of this work.

6. References

[1] Permana S, Soedarsono J W, Rustandi A and et.al, *Other Oxides Pre-removed from Bangka Tin Slag to Produce a High Grade Tantalum and Niobium Oxides Concentrate,* IOP Conference Series: Materials Science and Engineering. Vol. 012006, p10, (2016).

[2] Wang, George C. The Utilization Of Slag In Civil Infrastructure Construction,1st ed. Elsevier Science, (2016).

[3] Shah, Darshan S., and Jayeshkumar Pitroda. An Experimental Study On Durability And Water Absorption Properties Of Pervious Concrete. The Master Builder, p50-60, (2016).

[4] S. Muthulingam, and B.N. Rao., *Non-uniform time-to-corrosion initiation in steel reinforced concrete under chloride environment.*, Corrosion Science 304–315, p82,(2014).

[5] M.A. Mustafa, K.M. Yusof, *Atmospheric Chloride Penetration into Concrete in Semi tropical Marine Environment*, Cem. Concr. Res. 24, p661–670,(1994).

[6] A. Neville, *Chloride Attack of Reinforced-Concrete – An Overview*, Mater. Struct. 28, p63–70, (1995).

IOP Conf. Series: Materials Science and Engineering **299** (2018) 012048 doi:10.1088/1757-899X/299/1/012048

Characteristic of Low Temperature Carburized Austenitic Stainless Steel

Istiroyah[1], M A Pamungkas, G Saroja, M Ghufron, and A M Juwono

Physics Department, Faculty of Natural Science, Brawijaya University, Indonesia

istie@ub.ac.id

Abstract. Low temperature carburizing process has been carried out on austenitic stainless steel (ASS) type AISI 316L, that contain chromium in above 12 at%. Therefore, conventional heat treatment processes that are usually carried out at high temperatures are not applicable. The sensitization process due to chromium migration from the grain boundary will lead to stress corrosion crack and decrease the corrosion resistance of the steel. In this study, the carburizing process was carried out at low temperatures below 500 ° C. Surface morphology and mechanical properties of carburized specimens were investigated using optical microscopy, non destructive profilometer, and Vicker microhardness. The surface roughness analysis show the carburising process improves the roughness of ASS surface. This improvement is due to the adsorption of carbon atoms on the surface of the specimen. Likewise, the hardness test results indicate the carburising process increases the hardness of ASS.

1. Introduction

Austenitic stainless steel has has a good resistance to corrosion, so, is widely used in the manufacturing industries and structural sectors. However, its mechanical properties, especially its hardness and resistance to wear are relatively low, hence, limiting its applications, mainly those related to friction [1-2]. Modifications of its surface, based on the diffusion process, have been developed to increase the hardness and resistance to wear of ASS, such as through nitriding and carburising [3-4]. Interstitial solid solution hardening is a technique that is considered effective without causing sensitization and stress corrosion cracking [5-6].

The solubility of carbon and nitrogen in Fe-C alloy, in the phase diagram is very low, and increases with increasing temperature. However, for the case of stainless austenitic, with a relatively high content of chromium will mobile at high temperatures, and tends to form precipitates. The presence of precipitates, either chromium precipitates, or iron precipitates, indeed increases the hardness of the steel. However, it decreases the corrosion resistance. Therefore, the process of interstitial diffusion at low temperature is an option treatment for modifying the surface of the stainless steel austenitic. Low temperature carburizing (LTC) is the process of improving the hardness of steel by increasing carbon content in the parent material. A specimen of steel resulted from carburising process is being harder on its surface. This will also increase corrosion resistance and resilience of fatique [7]. Meanwhile, from

[1] Corresponding author

This research financially support by BOPTN Faculty of Natural Science Brawijaya University contract number: 19/UN10.F09.01/PN/2017.

the investigation of the process of low temperature plasma carburizing for steel of AISI316, it was concluded that the hardness obtained from carburising process was due to the presence of supersaturated carbons [8].

Coconut shells and rice husks are a lot of waste around us. coconut shell and rice husk has a high enough carbon content that has the potential as a carbon sumbrant in the process of carburising.In present study, we investigate carburized austenitic stainless steels, produced by low temperature solid carburizing (LTC) with coconut chacoar and rice husk cacoal as carbon source. Vickers microhardness test, optical microscopy, and non-destructive profilometre are utilized to investigate the characteristic of carburized specimens.

2. Experiment

Austenitic stainless steel type AISI 316L was used as substrate. The specimen dimension is 2 x 2 cm^2 and 0.1 cm in thickness, respectively. The specimens were degreased to remove surface contaminant. Ultrasonic cleaner system with alkali solution was used fo cleaning and degreasing.

Coconut shell and rice husk after cleaning and dried, then carbonized at 600°C. then powdered and sieved using mesh 50 meshes. After that, specimens were dried. Finally, both specimens and carbon source were packed by aluminium foil, then inserted stainless steel carburizing pack, and heated in high temperature furnace. Carburizing was performed at 400°C for 8 hours.

Figure.1 Stainless steel carburizing pack

Furthermore, specimen analyzed by Vickers microhardness testing tools. Asahi vickers microhardnes operated with load 50 g. Olympus LOM utilized to analysis of specimen microstructure. Surface roughness evaluated by a Hommelwerke T2000 stylus profilometre.

3. Results and Discussion

Microstructure of untreated AISI 316L austenitic stainless steel and carburized specimens are shown in figure.2. Untreated specimens are dominated by ferrite phase that appear as white grains. The size of the ferrite phase grains is unchanged on LTC carburized specimens. This behavior indicates there is no grain growth and grain coalesence. The carbon distribution on the surface of the carburized specimen shows that the carburized specimen with coconut shell charcoal carbon source is distributed more homogenly than carburized specimens with rice husk carbon source, as can be seen in fig. 2b and 2c. This is related to the characteristics of the two carbon sources. Coconut shell charcoal has a

65.6% carbon content while rice husk charcoal has a bonded carbon content of 41.3%. In addition, rice husk has enough high silica content. This will affect the properties of adsorption and diffusion of carbon into steel matrix.

The adsorbed carbon on the surface increases the surface roughness of the specimens as can be seen in Table 1. The average roughness values of the specimens increased from 0.26 μm (untreated specimens) up to0.46 μm and 0.31 μm for carburized specimens with coconut shell charcoal and charcoal rice husk, respectively.

The carbon diffusion into the steel matrix is the diffusion of interstitial [9], where the carbon will occupy the interstitial site position. Furthermore, the diffused carbon has an opportunity to become a solid solution or to precipitate with Fe / Cr. It depends on the condition of the carburizing process. Generally, in LTC, carbon diffuse to form a solid solution. At low temperature, both Fe and Cr are not mobile so no precipitate formation occurs.

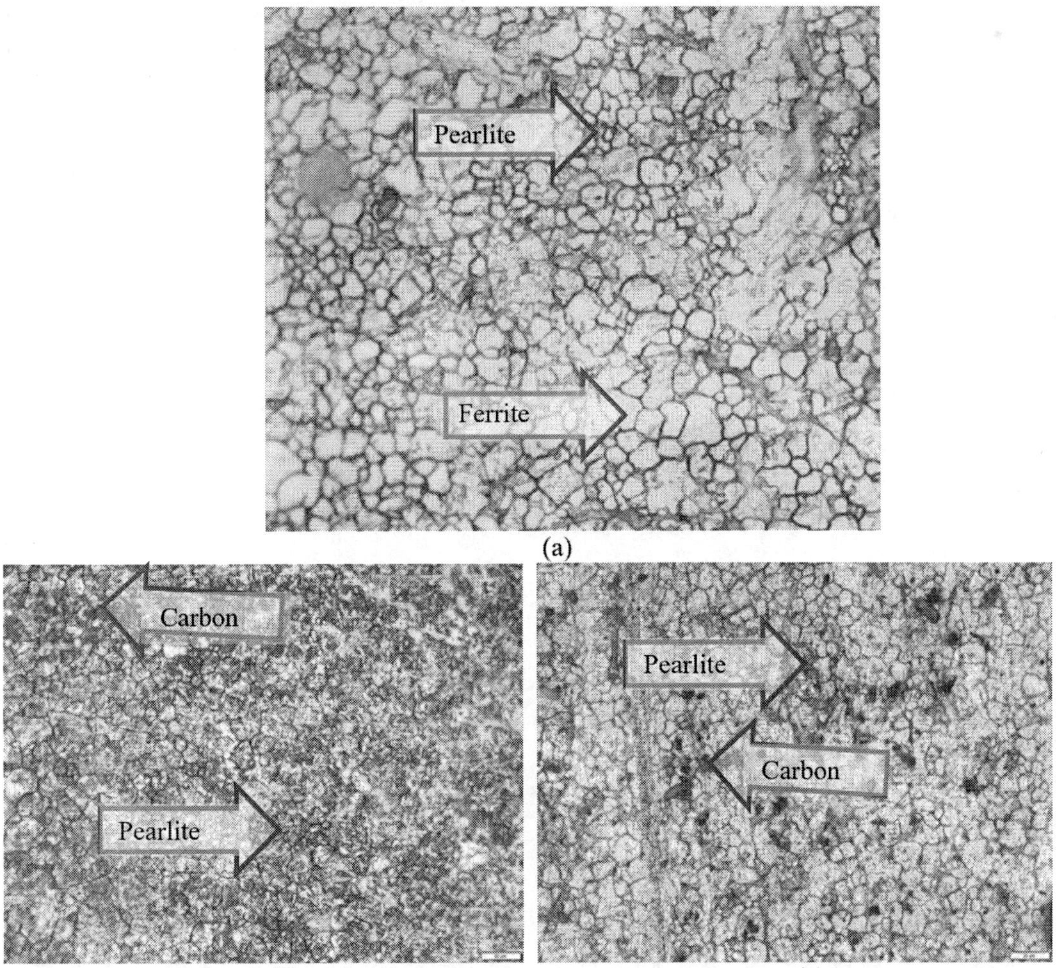

(b) (c)

Figure 2. Microstructure of untreated AISI 316L (a), carburized (LTC) AISI 316L at 400 °C for 8h with coconut charcoal (b), and carburized (LTC) AISI 316L at 400 °C for 8h with rice-husk charcoal (c).

The presence of carbon in the steel matrix will affect steel hardness. Figure 3 shows the results of measuring the hardness of specimens before and after carburizing process. Hardness of carburized specimens increased up from 26% to 40% compared to untreated specimens.

Table 1. Surface roughness of austenitic stainless steel specimens

Treatment condition	Ra (micrometre)
Untreated	0.26
LTC coconut charcoal	0.46
LTC rice-husk charcoal	0.34

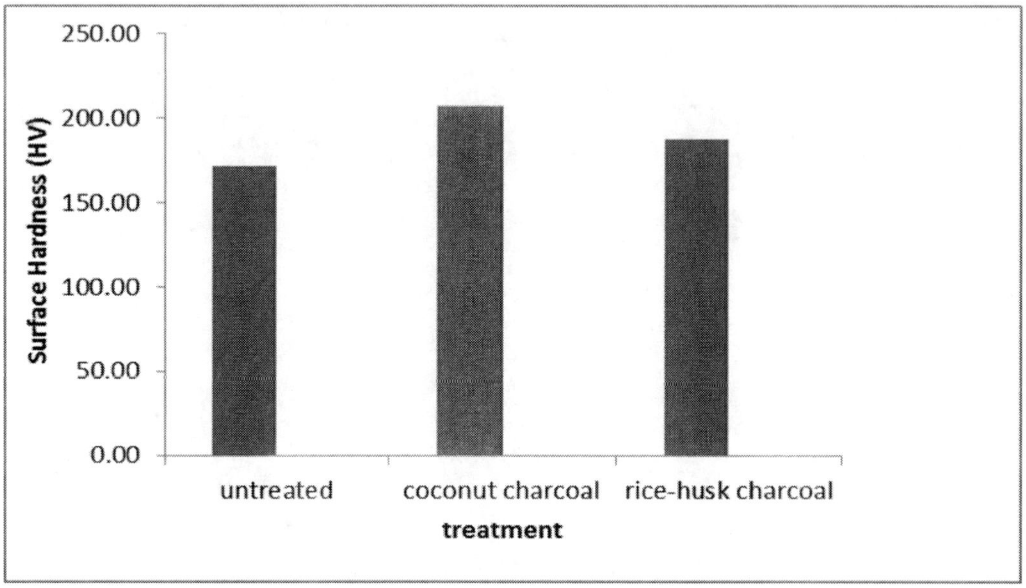

Figure 3. Surface hardness of AISI 316L specimens

4. Conclusion

Low temperature carburizing (LTC) on austenitic stainless steel does not affect grain size which indicates no grain growth and grain coalesence. The adsorbed carbon on the specimen surface increased roughness and hardness of the specimen compared to untreated specimen.

References

[1] Oliviera, A.M., Riofano, R.M.M, Casteletti, L.C., Tremiliosi, G.F., and Bento, C.A.S. 2003. Effect of the temperature of plasma nitriding in AISI 316L austenitic stainless steel. *Revista Brassileira de Aplicacoes de vacuo*, **22(2):** 63-66..

[2] Ahmadi, P., and Czerwiec, T., 2008. Some Consideration on Double Layer structure in plasma assisted nitriding of austenitic stainless steel. *IUST International Journal of Engineering Science*, **19**: 51–59.

[3] Yuksel Sarıkaya, and Muserref Onal, 2012, High temperature carburizing of a stainless steel with uranium carbide, *Journal of Alloys and Compounds* **542** : 253-256.

[4] Istiroyah, I.N.G Wardana, and D.J Santjojo, 2014, Comparison of AISI 316L Plasma Nitriding Behavior at Low and Medium Temperature, *Applied Mechanics and Materials* , **Vol. 493** pp

755-760.

[5] A. Kosmac, Surface hardening of stainless steels, *Mater. Appl. Ser.* 20 (2015) 1-17.

[6] S.R. Collins, P.C. Williams, S.V. Marx, A.H. Heuer, F. Ernst, and H. Kahn, 2014, Low-temperature carburization of austenitic stainless steels, in: J. Dossett, G. Totten (Eds.), Heat Treating of Irons and Steels, **vol. 4**, ASM International, Materials Park, Ohio, . 451-460

[7] Suh B-S,and LeeW-J., 1997, Surface hardening of AISI316L stainless steel using plasma carburizing. *ThinSolidFilms* ;295:185–92.

[8] TsujikawaM,NoguchiS,YamauchiN,UedaN,SoneT. Effect of molybdenum on hardness of low-temperature plasma carburized austenitic stainless steel. *Surf. Coat. Technol.* 2007;201:5102–7.

[9] Budinski, G. & Budinski., K. (1999). Engineering materials-properties and selection (6th ed.). New Jersey: Prentice Hall International.

IOP Conf. Series: Materials Science and Engineering **299** (2018) 012049 doi:10.1088/1757-899X/299/1/012049

The Effect of Various Acids to the Gelation Process to the Silica Gel Characteristic Using Organic Silica

NA Rahman[1], W Widiyastuti[2], D Sigit[3], M Ajiza[4], W Sujana[5]

[1]Dept. of Chemical Engineering, National Institute of Technology Malang
[2]Dept. of Chemical Engineering, Sepuluh Nopember Institute of Technology Surabaya
[3]Dept. of Chemistry, Malang State University, Malang
[4]Dept. of Geodetic Engineering, National Institute of Technology Malang
[5]Dept. of Geodetic Engineering, National Institute of Technology Malang

E-mail: nanik_ar29@yahoo.com

Abstract. Bagasse ash is solid waste of cane sugar industry which contain of silica more than 51%. Some previous study of silica gel from bagasse ash have been conducted often and been applied. This study concerns about the effect of various acid used in the process of gelation to the characteristic of silica gel produced. Then, this silica gel will be used as adsorbent. As that, the silica gel must fulfill the requirements of adsorbent, as have good pores characteristics, fit in mesoporous size so that adsorbent diffusion process is not disturbed. A fitted pores size of silica gel can be prepared by managing acid concentration used. The effect of acid, organic acid (tartaric acid) and inorganic acid (hydrochloric acid), is investigated in detail. The acid is added into sodium silicate solution in that the gel is formed, the pores structures can be investigated with BET, the crystal form is analyzed with XRD and the pore structure is analyzed visually with SEM. By managing the acid concentration added, it gets the effect of acid to the pore structure of silica gel. The bigger concentration is, the bigger the pore's size of silica gel produced.

1. Introduction

Bagasse, the solid waste of cane sugar industry cause environmental pollution if it is not handled well. All this time, bagasse is utilized as animal feed, brick or landfill. Bagasse is also utilized as boiler fuel which the result of the burning in a form of bagasse ash is containing of high level of silica and is potential as source of silica. Material exploration consisting of silica is regularly conducted, such as rice husk [1,2,3] bagasse ash [4,5,6] and geothermal sludge [7]. Silica gel prepared from bagasse ash has been conducted and applied for wastewater purifier [4]. Many methods developed to get expected characterization based on the needs, e.g., as adsorbent. This research is aiming to develop the technique for expanding surface area and pore size by adding organic and inorganic acids as the controller of pore structure and external morphology in the synthesized mesoporous silica from bagasse ash. Acid added to the gel-forming process gives effect as a template. The pores are formed when the acid enters the silica matrix and is trapped inside. Strong acids provide the effect of accelerating the condensation polymerization so that the formation of silica webs has not been uniformly formed. The otherwise, the polymerization condensation process proceeds more slowly when a weak acid is used. The pores are formed when the acid is released from the silica matrix in the process of washing the gel [8].

2. Material and Method

2.1. Materials

Raw material used on this study is bagasse ash from PG. Kebon Agung, Malang, East Java, Indonesia. All chemical used to prepare mesoporous of silica gel from bagasse ash were reagent grade and used without further purification. The sodium hydroxide (NaOH, Merck)), tartaric acid ($C_4H_6O_6$, Sigma)), hydrochloric acid (HCl, Merk). Demineralized water was used during all synthesis and treatment process.

2.2. Synthesis of Silica Gel

The process of sodium silicate made from bagasse ash according to the process that has improved before [4]. Briefly, silica in bagasse ash was extracted with 2M NaOH solution on its boiling point under stirring for 1 hour. After that, the mixture was cooled until room temperature. Then, solution was filtered through an ashless filter Whatman No. 41 (Whatman Plc, Kent, England). The filtrate solution was sodium silicate. Sodium silicate solution will be used as raw material for silica gel fabrication. HCl solution 2M is added into sodium silicate solution and the gel is formed. Then,aging at 40 °C for 18 hours. Later, the gel is washed using demineralized water untilneutral and then dried in temperature100 °C for 24 hours. Mesoporous silica obtained and characterizing the pores.

2.3. Characterization

N_2 adsorption-desorption analysis was done using a Nova 1200e apparatus (Quantachrome) under liquid N_2 temperature (77 K). The samples were degassed at 150 °C under vacuum for 4 hours. The specific surface area was calculated by the BET method. The pore size distribution analysis was done by the BJH method using the desorption branch of the isotherm. The grafting of the functional groups of adsorbents were analyzed by Fourier transform infrared spectroscopy (FTIR, Shimadzu 8400s), using KBr pellets. X-ray diffraction (XRD) analysis was used to obtain information on the diffraction model and particle's crystallography.

3. Result and Discussion

Isotherm adsorpsi-desorpsi of nitrogen of silica gel prepared using HCl and $C_4H_6O_6$ is shown in Fig. 1. The both curve are identical and according to IUPAC classification type IV. The isotherm of silica prepared HCl has low slope and narrow hysteresis. This isotherm different if silica prepared with $C_4H_6O_6$, the hysteresis more wider. It indicated that $C_4H_6O_6$ has ability to increase the pore. So, adsorption capacity of nitrogen in silica with $C_4H_6O_6$ was higher than silica with HCl. It means that the silica gel which is fabricated by pH control has uniform pore size.

The typical pore size distributions of silica gel prepared by hydrochloric acid (HCl) and tartaric acid ($C_4H_6O_6$) are shown in Fig. 2. The distribution is narrow indicating that the pore size is nearly uniform. the pore size distribution curve on silica prepared with HCl indicates that many particles are 0.62 nm in size, which also means that the pores formed are smaller than those prepared with tartaric acid.

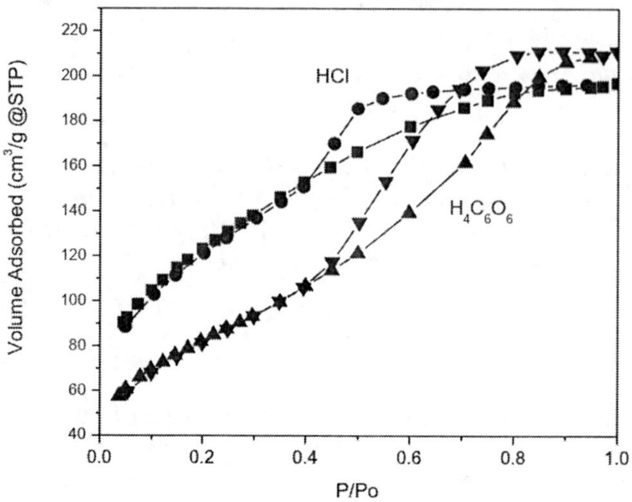

Figure 1. Isoterm adsorption-desorption N_2 in silica gel prepared from bagasse ash using HCl and $C_4H_6O_6$

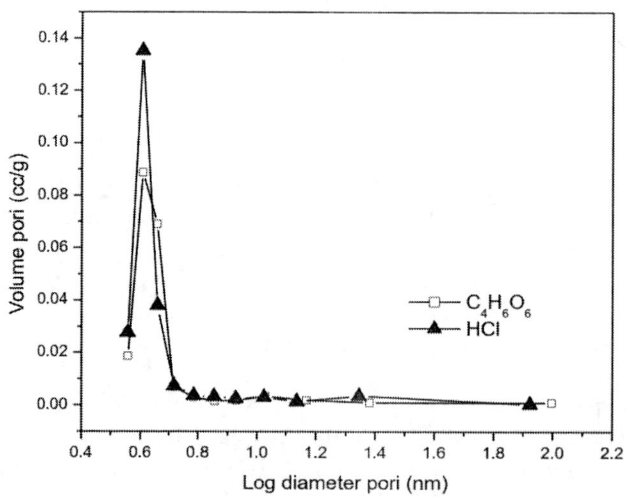

Figure. 2. Typical pore size distribution of porous silica prepared by various acids (HCl and $C_4H_6O_6$)

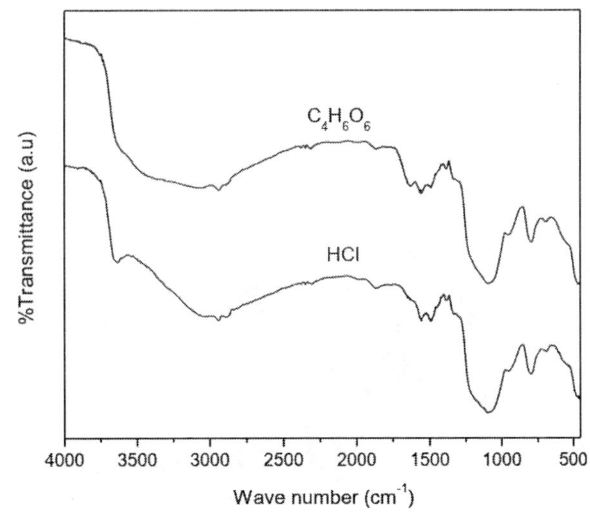

Figure 3. FTIR Spectrum of silica using HCl and $C_4H_6O_6$

Figure 3 shows the FTIR spectra of the sample prepared by various acid treatments. It showed that the curve of silica prepared using HCl and $C_6H_6O_6$ are identic..The peaks appearing in 500–1700 cm^{-1} are typical of silica gel (800, 960, 1090 and 1630 [9]. The results show that functional groups of silica products obtained from various acids are similar to each other. The peak at 805 cm^{-1} related to the Si–OH bond, and 1115 cm^{-1} related to the Si–O bond. The O–H stretching vibrations appear in the region of 3200–3700 cm^{-1}. The band resulted from silanol OH groups and water bound to the silica surface by hydrogen bonding.

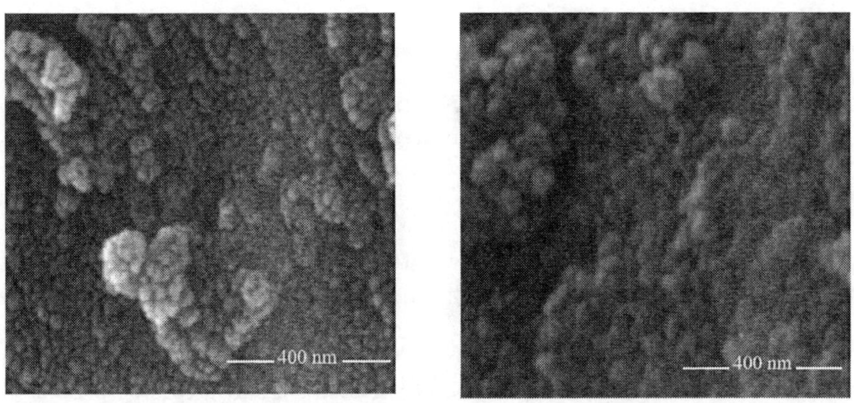

Figure 4. SEM image of silica gel prepared from bagasse ash using HCl (left) and $C_4H_6O_6$ (right)

Figure 5 shows SEM image of silica gel prepared from bagasse ash using HCl and $C_4H_6O_6$. This figure to further characterize the physical properties of the prepared silica gel. That, the silica prepared using HCl and $C_4H_6O_6$ were irregular in shape. In addition, XRD analysis indicated that the silica gel were amorphous, which was confirmed by the absence of any ordered crystalline structure, at 23 – 25 (2θ degree) show in Fig. 6.

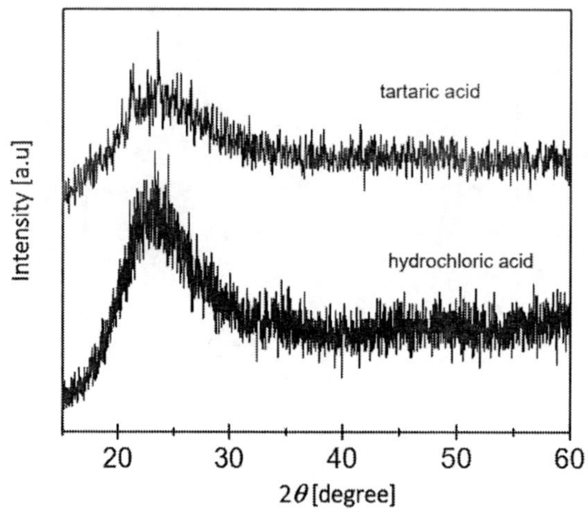

Figure 5. X-ray diffraction patterns of silica prepared from bagasse ash using HCl and $C_4H_6O_6$

4. Conclusion

Base on the research finding, it can be concluded that acid addition is significantly influenced to pore structure of silica. The organic acid (tartaric acid) gives more effect to extend pore size than the inorganic acid (hydrochloric acid).

References

[1] Kalapathy U, Proctor A, Shultz J 2000 *Bioresour. Technol.* **73** 257–262
[2] Kalapathy U, Proctor A, Shultz J 2002 *Bioresour. Technol.* **85** 285 – 289
[3] Kalapathy U, Proctor A, Shultz J 2000 *J. Chem. Technol. Biotechnol.* 72 99 – 106
[4] Affandi, S., Setyawan, H., Winardi, S., Purwanto, A., Balgis, R. 2009 *Adv. Powder Technol.*, **20**, 468-472
[5] Nazriati N, Setyawan H, Affandi S, Yuwana M, Winardi S 2014 *J. Non-Cryst. Solid.* **400**, 6 – 11
[6] Rahman, N.A, Widhiana, I., Juliastuti, S.R., Setyawan, H., 2015 *Colloid and Surface A : Physicochem* Eng. *Aspect* **476** 1-7
[7] Muljani S, Setyawan H, Wibawa G, Altway A 2014 *Adv. Powder Technol* **25** (5) 1593 - 1599
[8] Rahman NA, Setyawati, H, Anggorowati DA, Astuti S., Ajiza, M. 2016 Proc. Conf. on SNAST (Yogyakarta, Indonesia) Pengaruh Suhu Kalsinasi pada Penghilangan Template Terhadap Karakterisasi Silika periode IV 219 – 222
[9] Izutsu H, 1997 *J, Am. Ceram.Soc.* **80** 2581 – 2589

IOP Conf. Series: Materials Science and Engineering **299** (2018) 012050 doi:10.1088/1757-899X/299/1/012050

Enhancing the Hardness of Sintered SS 17-4PH Using Nitriding Process for Bracket Orthodontic Application

B Suharno[1*], S Supriadi[2], S T Ayuningtyas[1] and T Widjaya[1] and E R Baek[3]

[1] Metallurgy and Materials Engineering Department, University of Indonesia, Depok, Indonesia

[2] Mechanical Engineering Department, University of Indonesia, Depok, Indonesia

[3] School of Materials Science and Engineering, Yeungnam University, South Korea

E-mail : suharno@metal.ui.ac.id

Abstract. Brackets orthodontic create teeth movement by applying force from wire to bracket then transferred to teeth. However, emergence of friction between brackets and wires reduces load for teeth movement towards desired area. In order to overcome these problem, surface treatment like nitriding chosen as a process which could escalate efficiency of transferred force by improving material hardness since hard materials have low friction levels. This work investigated nitriding treatment to form nitride layer which affecting hardness of sintered SS 17-4PH. The nitride layers produced after nitriding process at various temperature i.e. 470°C, 500°C, 530°C with 8hr holding time under 50% NH_3 atmosphere. Optical metallography was conducted to compare microstructure of base and surface metal while the increasing of surface hardness then observed using vickers microhardness tester. Hardened surface layer was obtained after gaseous nitriding process because of nitride layer that contains Fe_4N, CrN and Fe-αN formed. Hardness layers can achieved value 1051 HV associated with varies thickness from 53 to 119 μm. The presence of a precipitation process occurring in conjunction with nitriding process can lead to a decrease in hardness due to nitrogen content diminishing in solid solution phase. This problem causes weakening of nitrogen expansion in martensite lattice.

1. Introduction

Orthodontic treatment performed to deal with malocclusion problem in human teeth. The principle of this instrument is based on provision of force on teeth through attachment bracket to teeth. Sliding method is commonly used as a way to create teeth movement by giving a force through arch wire which then proceed to the teeth with the use of orthodontic bracket [1]. This sliding method generate a frictional resistance between bracket and arch wire that lock bracket position and reduce destructed force to the teeth. Frictional resistance adopts up to 60% of provided forces which might cause the arch wire get caught and tilt the tooth [2, 3, 4].

Several factors have been involved in affecting the frictional resistance, such as bracket material [5], dimensions [6, 7] and arch wire material [8] have been investigated. Current research showed friction produced by sintered stainless steel has a friction rate of 40-45% less than cast brackets stainless steel since harder and smoother material has a low coefficient of friction [9, 10]. On behalf of this matter, the study was conducted to modifying material through increased hardness to reduce material friction.

In previous research, investment casting process used to manufacture bracket and the result shows rough surface that require further processing end [11]. It is necessary to study other manufacturing process for bracket manufacturing, namely metal injectin molding that produces sintered material.

Sintered stainless steel 17-4PH can achieve hardness up to 394.58 HV after sintering treatment at temperature of 1360 °C [12]. In the previous study, an aging treatment performed to enhance SS 174-PH hardness with a 10% enhancement [13]. Therefore, this work tries to investigate an alternative technique to engineer the bracket material.

One of the most effective ways to increase hardness several times larger than initial hardness is nitriding treatment because SS 17-4 PH contains nitride former such as chromium. The most practical type of nitriding process to modify the bracket orthodontic surface is gas nitriding due to its complicated shape with many hard surface to reach [14]. Gas nitriding is also capable to overcome limitations of plasma nitriding since this treatment consists of gas that can easily penetrate into the pores of sintered material and capable to forming a nitride layer inside the pores [15]. Certain things that should be concerned during gas nitriding of precipitation hardened (PH) stainless steels summarized as two aspects: first is a surface passivation due to high contents of chromium (surface activation is necessary), second is presence of the precipitation processes proceeding in the matrix during thermo-chemical treatment [14].

2. Experimental

2.1. *Material and Treatment*

Cubical specimen (5x5x5mm) made of 17-4PH stainless steel feedstock for MIM provided by Ryer Inc. and contained approximately 7 wt.% of binder with particle distribution and size d10%: 2.8μm, d50%: 6.9 μm , d90%:18.6 μm. The feedstock composition is shown in Tabel 1.

Table 1. Chemical composition of SS 17-4PH manufactured by Ryer, Inc (wt.%).

Fe	**C**	**Cu**	**Co**	**Cr**	**Mn**	**Mo**	**Nb**
74.26	0.04	3.94	0.03	16.44	0.16	0.02	0.29
Ni	**O**	**P**	**S**	**Si**			
4.04	0.35	0.02	-	0.39			

Feedstock was compacted through injection molding machine at 200°C with 89m/min injection speed and 5.2 kN/m^2 injection force. Green part samples debounded by submerging the components into the hexane solutions while heated at 50 °C with agitation to remove primary binder. The remain binder then thermally debound at 510°C using heat rate of 1 deg/min for 60 minutes in the Barkeley furnace [16]. Before sintering, purging treatment with argon gas is needed to remove oxygen inside the tube furnace. Sintering process performed by heating the specimen up to 1360°C with a rate 5 deg/min in vacuum condition for 1.5 hour and slowly cooled until room temperature under vacuum [12]. After sintering, samples grounded using #400 SiC sandpaper and shot blasting to remove the oxide layer from material surface. The nitriding process was conducted at vacuum furnace with atmosphere condition which consisted 75m^3/h N_2, 75m^3/h NH_3, and 5 m^3/h CO_2 and submitted to cooling down under N_2 atmospheric condition inside the treatment chamber. Samples were nitrided at 470, 500 and 530 °C, for a constant treatment time of 8 hours.

2.2. *Characterization*

Conventional metallographic procedure was applied for microstructural analysis. After nitriding process samples were cross-sectioned, mechanically polished and then they were slightly etched with Nital 2% with composition 2 ml of nitric acid and 98 ml of methanol. Microstucture was characterized using Zeiss Microscope to identify the white/compound zone and the diffusion zone. Thickness of the formed layers also measure with Labscope application under Zeiss Microscope.

For the surface hardness number, microhardness tester equipped with a vickers indenter was conducted with 0.01 kg load and 10 seconds dwell times. The result was the average value of three microhardness values at different positions of the treated layer.

X-ray diffraction analysis used to analyze formed phase on the nitride layer. The nitrided sample surfaces were exposed to x-ray without any mechanical surface preparation. This sample was characterized with a type PAN-Analytical using Cu Ka radiation. The scanning angle (2θ) comprises the range of 20-110°.

3. Results and Discussion

3.1. Effect of Gas Nitriding Process on Surface Microstructure of SS 17-4PH

Observation by optical microscopy in Figure 1 revealed cross-section microstructure of nitride layer in the surface of specimen. Nitride layers were produced at all nitriding temperature and indicates a boundary with the substrate. According to the different concentration of nitrogen content, microstructure can be distinguished into two zone which are compound/white layer and diffusion layer [17].

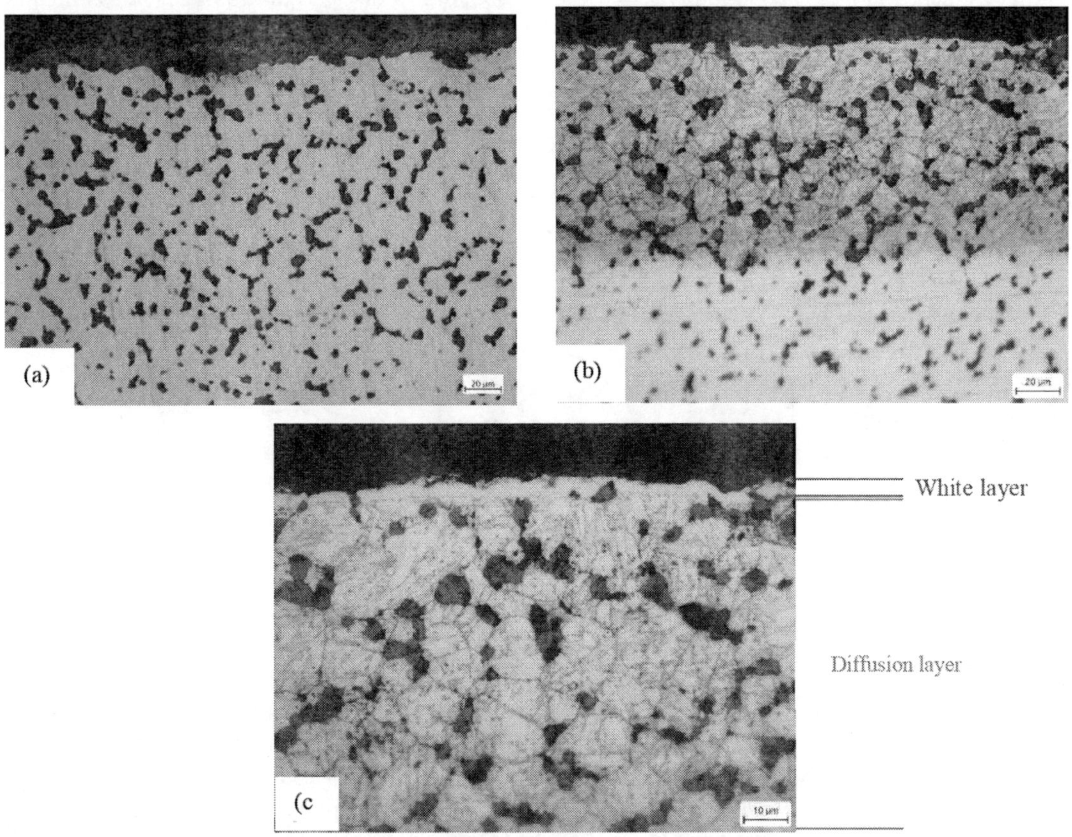

Figure 1. Microstructure of 17-4 PH stainless steel (a) untreated specimen (b) nitrided specimen at 530 °C at 500X magnification (c) nitrided specimen at 530 °C at 1000X magnification.

Compound layer is white nitride colour that comprises high concentration of nitrogen and configures a ceramic layer made of iron nitrides Fe_4N-γ`. As a metastable compound, ammonia was dissociated into active nitrogen-hydrogen gas mixture and brings up nitrogen potential or activity of nitrogen dissolved in iron. In conjunction with increasing active nitrogen-hydrogen gas mixture nitrogen, the solubility also increases until the nitrogen potential is high enough to reach the equilibrium of nitrogen in Fe_4N-γ` [18]. Formation of this ceramic layer then makes the hardness on the surface of the material increases since compound layer results in a tensile residual stress.

Another nitride layer formed is diffusion layer. This layer contains low concentration of nitrogen and defined as base microstructure hardened by interstitial atom in solid solution and/or by nitrides precipitation. Due to high concentration of chromium in stainless steel 17-4 PH, type of precipitate which is formed is CrN. Increasing hardness on this layer is caused by interstitial nitrogen in iron and volume expansion of nitrides precipitation that fabricates a compressive residual stress field in the diffusion layer [17].

3.2. Effect of Nitriding Temperature on Nitride Layer Thickness

Gas nitriding process on SS 17-4 PH indicated formation of two type nitride layer such as compound/white layer and diffusion layer. Based on the results presented in the graph below, it can be conclude that the thickness of those two types of layer was convinced on temperature process. Layer thickness owned by compound/white layer varies between 2–5 μm under the temperature range of 470–530°C while diffusion layer thickness varies between 53–119 μm. Overall data exhibit an increasing thickness of nitride layer along with increased temperature. This result occur because of the increase in diffusivity of the interstitial elements into base metal as the temperature increases. Temperature enhancement can raise activation energy of nitrogen atom to diffuse and this behavior is intended for atomic diffusion that is controlled by thermally activated treatment [19].

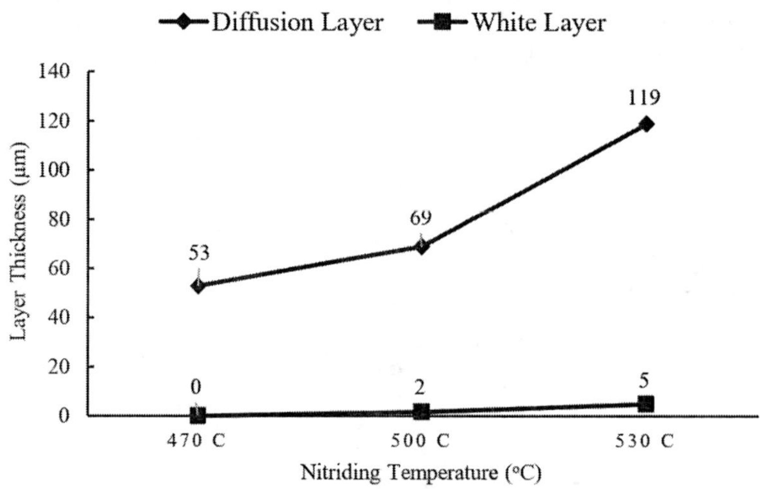

Figure 2. Effect of Nitriding Temperature on thickness of nitride layer.

On a research that was conducted by Cardoso [20], deployment of nitrogen during nitriding treatment ensued through volume (lattice) diffusion and through high diffusivity paths (e.g. grain boundaries and dislocations), which both of mechanism were performed depending on temperature process. Diffusion process that occurs through high diffusivity paths required low activation energy since this mechanism already has stored activation energy. Therefore, this mechanism remarkably takes place during low-temperature of nitriding treatment. On the other hand, nitriding treatment with high temperature process occurs through volume diffusion. It happens because of grain growth and defects annihilation due to temperature increment.

3.3. The Effect of Nitriding Temperature on Material Hardness

Graphic in Figure 3 showed microhardness number of nitride layer as a function of temperature. This figure indicated microhardness number between 1008 HV at temperature 470°C and 1051 HV at temperature 530°C. Material hardness was high on the surface and decreased along with the indentation of hardness towards the base metal since nitride layer only formed on the surface of material. The highest attainment for microhardness number was obtained while nitriding process was carried out at temperature 530 °C with 2.7 times greater than sintered SS 17-4 PH (395 HV). Therefore, interstitial nitrogen atom inside iron lattice, nitride compound precipitation and ceramic layer formation on the surface of material affect hardness enhancement in material.

Hardness enhancement was expected to reduce friction value on bracket surface as the result of study about relationship between hardness and friction on comparison of bracket SS 17-4 (investment casting) in contrast to Synergy® (RMO, USA) and Gemini® (3M, USA) from commercial bracket. Bracket Synergy® (RMO, USA) that had a hardness value of 265 HV conducts 1.37 friction coefficient, while SS 17-4PH (investment casting) with 295 HV generated 1.17 friction coefficient and Gemini® (3M, USA) with 312 HV produced 0.67 friction value [21]. Therefore, the enhancement of hardness up to 1051 HV is expected to reduce friction coefficient less than the friction values above.

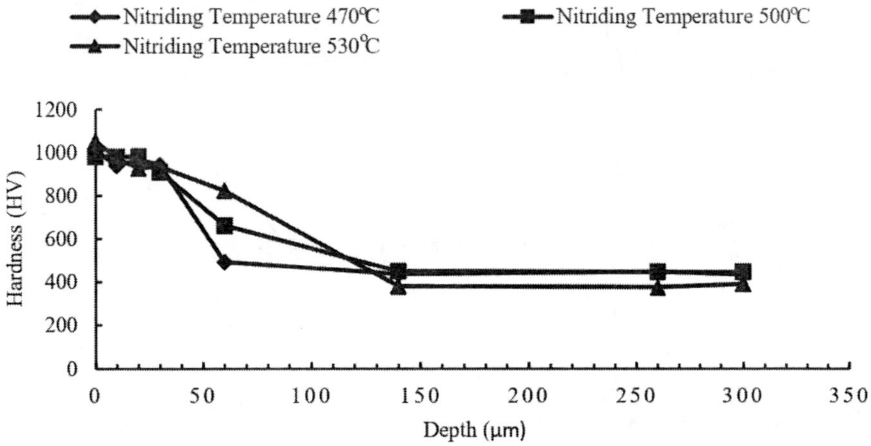

Figure 3. Microhardness profile of SS 17-4 PH from surface to base material.

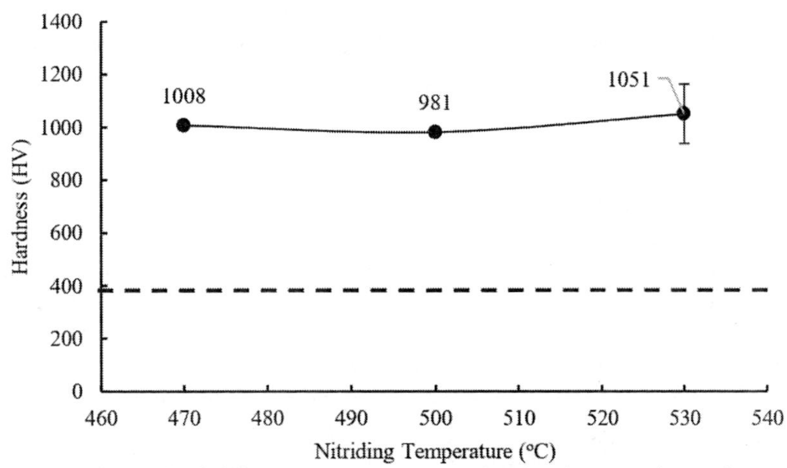

Figure 4. Effect of Nitriding Temperature on microhardness at the surface of SS 17-4PH

Based on Figure 4, hardness at the surface of material influenced by temperature process. This figure explained the decreasing hardness of material at temperature 500 °C. Hardness reduction on material occurred in relation to precipitation of chromium nitride, which reduces nitrogen content in solid solution phase. Therefore, nitrogen expansion in the martensite lattice has weakened. As illustrated by Wang [22], equilibrium solubility limit of nitrogen within martensite structure has been exceeded at temperature more than 450 °C and at the same temperature, mobility chromium also increases. Therefore, formation of CrN or Cr_2N is more desirable. The presence of precipitation process which happened in the matrix during nitriding treatment caused distinction on the nitriding kinetics and structure characteristic [23].

3.4. *Effect of Nitriding Temperature on Phase Formation inside Nitride Layer*

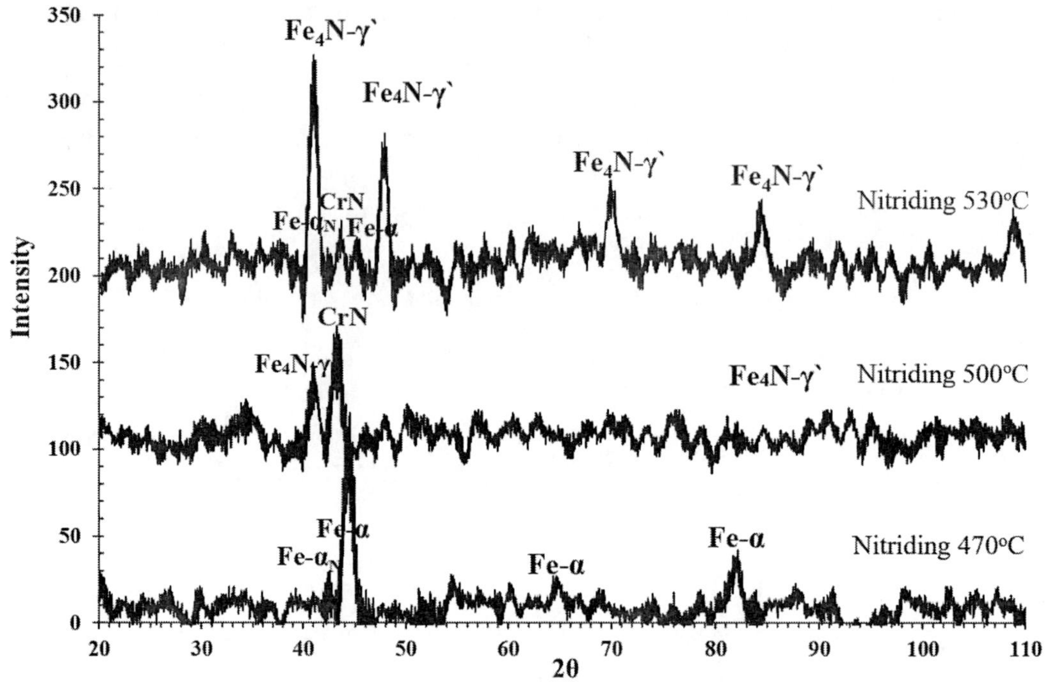

Figure. 5 Effect of Nitriding Temperature on Phase Formation inside Nitride Layer.

The X-ray diffraction results of stainless steel 17-4PH with three different nitriding temperatures were shown in Figure 5. The main phase that appears during nitriding temperature at 530 °C is Fe_4N-γ` with some CrN precipitates. The presence of these phases is appropriate with the OM micrograph shown in Figure 1. $Fe_4N$-γ` phase was established as white nitride layer that indicated a high nitrogen concentration area. High intensity of Fe_4N-γ` peak on this parameter specified the thickness of the white layer formed on this surface. Otherwise, this figure also showed decreasing intensity of $Fe_4N$- γ` peak at temperatures below 530 °C even disappearing at nitriding temperature 470 °C parallel to compound layer thinning. At nitriding temperature 470 °C nitrogen potential was not high enough to reach the equilibrium of nitrogen in Fe_4N- γ`. Therefore, compound layer was not formed on the surface. Figure 5 also showed the high peak of CrN formed at 500 C nitriding temperature so that the hardness of the material at this temperature decreasd as shown in Figure 4. The high peak of CrN showed that the amount of CrN precipitates formed at this temperature were more plentiful than other parameters.

4. Conclusion

From the study above, it can be concluded that the experimental results reveal that gas nitriding significantly improve the hardness of 17-4 PH up to 1051 HV which is 2,7 times greater than ordinary bracket (395 HV). A white layer made of iron nitrides (Fe_4N-γ') makes the hardness on material surface increase because it generates a tensile residual stress while diffusion layer strengthens material by interstitial atom in solid solution (Fe-α_N) and/or by the precipitates of chromium nitrides (CrN) which fabricate a compressive residual stress field in the diffusion layer. The surface hardness increases as an increasing of nitriding temperature and formed nitride layer up to 119 μm width at 530 °C. Numerous CrN precipitates which are formed at temperature 500 °C reduce material hardness because nitrogen content in solid solution phase disappear. Thus, the optimum parameter with the maximum hardness value and the thickest nitride layer is achieved at nitriding temperature 530°C.

References

[1] Guendeman G, Bednar J and Sandrik J 1991 *Am. J. Orthod. Dentofac. Orthop.* **100** pp 13–22
[2] Nanda R 2007 Biomechanics in clinical orthodontics (Philadelphia: WB Saunders)
[3] Hain M, Dhotpatkar A and Rock P 2003 *Am. J. Orthod. Dentofac. Orthop.* **123** pp 414–22
[4] Tecco S, Festa F, Caputi S, Traini T, Di lorio D and D'Attillio M 2005 *Angle Orthod* **75** pp 828–32
[5] Ketih O, Jones S and Davies E 1993 *Br. J. Orthod* **20** pp 109–15
[6] Frank C and Nokalai R 1980 *Am. J. Orthod.* **78** pp 593–609
[7] Peterson L, Spencer R, Andreasen G 1982 *Quintessence Int.* **13** pp 563–71
[8] Kusy R and Whitley 1990 J *Am. J. Orthod. Dentofac. Orthop.* **98** pp 300–12
[9] Voughan J, Duncanson M, Nanda R and Currier G *Am. J. Orthod. Dentofac. Orthop.* **107** pp 20–27
[10] Moore A and Tegart G 1952 *Proc. R. Soc. Lond. A Math. Phys. Sci.* **212** pp 452–58
[11] Supriadi S, Sitanggang T, Irawan B, Suharno B, Kiswanto G and Prasetyadi T 2015 *I J Tech* **4** pp 312–21
[12] Suharno B, Saputro H, Supriyadi S, Suharno L and Baek E 2017 *Solid State Phen.* **266** pp 231–37
[13] Suharno B, Supriyadi S, Togar M, Abdullah H, Prasetyadi T, Saputro H, Suharno L 2016 *Sem.Nas. Met. Mat IX* (Banten: SENAMM IX)
[14] Kochmanski J and Nowacki J 2008 *Surf. Coat. Technol.* **202** pp 4834–38
[15] Baranowska J and Borrás V 2011 *Deffect and Diffusion Forum* **312-15** pp 524–29
[16] Supriadi S, Suharno B, HIdayatullah R, Maulana G and Baek E 2017 *Solid State Phen.***266** pp 238–44
[17] Santulina E 2011 Effect of surface treatments on mechanical properties of low alloy sintered steel (Italy: University of Toreto)
[18] Shohoji N, Marcelo T and Katsura M 1990 *Solid State Ionics* **38** PP 187–94
[19] Scheuer C, Zanetti F, Cardoso R and Brunatto S 2016 *Cong. Br. Eng. Ci. Mat. 22°* (Brasil: CBECiMat 22°)
[20] Cardoso R, Scheuer C and Brunatto S 2016 *Enc. of Iron, Steel, and Their Alloys* Colás R and Totten G (Location: Taylor & Francis)
[21] Prasetyadi T 2017 Innovation of Orthodontic Brackets 17-4 PH Stainless Steel with Investment Casting Method (Depok: Universitas Indonesia)
[22] Kochmanski J and Nowacki J 2006 *Surf. Coat. Technol.* **200** pp 6558–62

Acknowledgement

The authors would like thanks to Directorate of Research and Community Engagement of Universitas Indonesia (DRPM) UI for financial support through HIBAH PITTA (747/UN2.R3.1/HKP.50.00/2017)

IOP Conf. Series: Materials Science and Engineering **299** (2018) 012051 doi:10.1088/1757-899X/299/1/012051

Effect of Shot Peening in Different Shot Distance and Shot Angle on Surface Morphology, Surface Roughness and Surface Hardness of 316L Biomaterial

Erich Umbu Kondi Maliwemu[1,2,*], **Viktor Malau**[1] **and Priyo Tri Iswanto**[1]

[1] Department of Mechanical and Industrial Engineering, Gadjah Mada University, Jl. Grafika 2 - Yogyakarta, Indonesia, 55281

[2] Department of Mechanical Engineering, Nusa Cendana University - Jl. Adi Sucipto, Penfui, Kupang, East Nusa Tenggara, Indonesia, 85100

*email: erich.umbu@mail.ugm.ac.id

Abstract. Shot peening is a mechanical surface treatment with a beneficial effect to generate compressive residual stress caused by plastic deformation on the surface of material. This plastic deformation can improve the surface characteristics of metallic materials, such as modification of surface morphology, surface roughness, and surface hardness. The objective of this study is to investigate the effect of shot peening in different shot distance and shot angle on surface morphology, surface roughness, and surface hardness of 316L biomaterial. Shot distance was varied at 6, 8, 10, and 12 cm and shot angle at 30, 60, and 90°, working pressure at 7 kg/cm^2, shot duration for 20 minutes, and using steel balls S-170 with diameter of 0.6 mm. The results present that the shot distance and shot angle of shot peening give the significant effect to improve the surface morphology, surface roughness, and surface hardness of 316 L biomaterial. Shot peening can increase the surface roughness by the increasing of shot distance and by the decreasing of shot angle. The nearest shot distance (6 cm) and the largest shot angle (90°) give the best results on the grain refinement with the surface roughness of 1.04 μm and surface hardness of 534 kg/mm^2.

1. Introduction

Metallic biomaterials continue to be used extensively for the fabrication of surgical implants. AISI 316L stainless steel is widely used in components of the oil and gas, food and beverage and chemical industries, as well as a biomaterial due to its outstanding corrosion resistance in aggressive environments [1, 2]. The biomaterial of 316L austenitic stainless steel is one of the most used materials for fracture fixation devices compared to the other mentioned alloys, due to its suitable mechanical properties and low cost [3]. The biomaterials of stainless steel 316L are self-protected by the spontaneous formation of a thin oxide film. The passive layer formed on surgical stainless steel. [4-7]. Shot-peening method is a cold-working process used to produce a layer containing huge compressive residual stresses, modifying the metal surface structure, and creating the ultrafine grain and nanostructure. It entails bombardment of the surface with shots inducing the sufficient force to create plastic deformation resulting in increased compressive residual stress on the metal surface [8-10]. The hardness increment resulted mainly from a significant increase in the strain-induced strain-hardening [11]. The increase in the peening coverage and the Almen intensity markedly improved the microhardness and compressive residual stresses in the surface layer, while the peening intensity depends on the shot size [12]. Shot peening effectively changed the topography and removed the grinding marks, increase the surface roughness and initially increase the surface roughness [13]. The shot peening surface material was constrained by the surrounding material and suffered reaction force because the target surface is stable and remains plane [14]. The surface roughness increases with increase in the peening time. The hardness of surface layer increase with increasing shot-peening time [15, 16]. It is possible to generate not only a compressive residual stress state and an increase of hardness in the surface layer of material, but also a nanocrystalline layer of material. Roughness is strongly increased and is somehow comparable to the different treatments that were tested. It induces severe plastic deformation that causes crystallite refinement in the surface layers [17]. Grain diameter was adjusted by changing bright annealing temperature after cold working. It could be realized the effect on the decreasing of grain diameter [18]. Consistent reduction in grain size and enhancement in dislocation density with increase in strain, indicate that no dynamic recovery occurred during the shot peening process [19]. The references give the support statement that the shot peening is an appropriate mechanical surface treatment to be applied on surface of 316L to achieve the objective of this research to investigate the effect of shot peening in relation with shot distance and shot angle on surface morphology, surface roughness, and surface hardness of 316L biomaterial.

2. Experimental work

The biomaterial of 316L stainless steel was used as the specimen of shot peening (20 x 25 x 3 mm) with chemical composition as shown in table 1. This chemical composition was obtained by spectrometry (Hilger spectrometer E-9 OA701).

Table 1. Chemical composition of 316L.

Element	C	Si	S	P	Mn	Ni	Cr	Mo	Ti	Nb	Fe
%	0,031	0,969	0,003	0,037	1,079	10,866	16,782	1,894	0,005	0,014	67,78

The specimens were polished by abrasive paper and metal polish. Steel balls S-170 (diameter of 0.6 mm) were used in this research. Air blast machine was used as shot peening machine with 7 kg/cm^2 pressure of air compressor for 20 minutes. Surface morphology was investigated by an optical microscope, surface roughness was investigated by surface roughness machine Surfcom 120A and surface hardness was investigated by Micro-Vickers indentation using Buchler Micromet. Shot peening was applied on the surface of 316L by shot distance variations of 6, 8, 10, and 12 cm, and shot angle of 30, 60, and 90° as shown in figure 1.

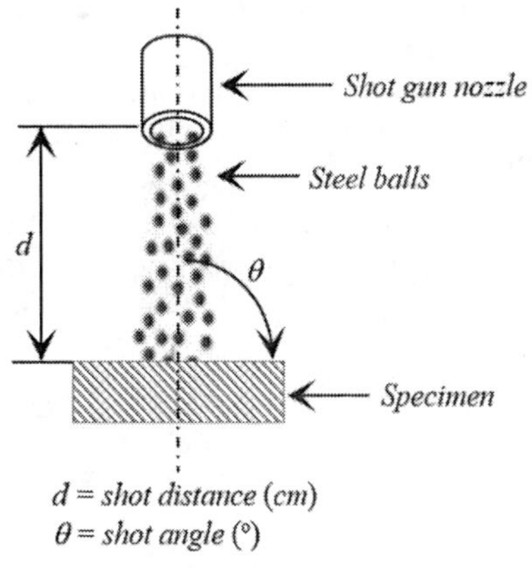

Figure 1. Scheme of shot peening.

The morphology modification on the surface of 316L was investigated by an optical microscope. The specimens were etched with HCl and HNO_3 and then specimen can be observed by an optical microscope. Surface roughness was investigated by contact stylus profilometer Surfcom 120A with setting parameter cut off of 0,800 mm, length of 5 mm, meas-mag of 2000 where the data can be used to get data of surface roughness (Ra). Surface hardness was investigated by micro-Vickers indentation using 100 grams indentation load for 10 seconds (ASTM standard E384). Hardness distribution for every specimen always takes in 10 points to get the hardness distribution be spread evenly on the surface of the specimen.

3. Result and discussion

3.1. Surface morphology

Shot peening gives a significant effect on surface modification in the form of plastic deformation characterized by the indentations formed on the surface of the specimen. The shot distances give a significant effect causing surface modification marked by the color change on the surface of the specimen, where the farther the shot distance, the color of the specimen becomes brighter as shown in figure 2, while the smaller the shot angle the specimen's color becomes brighter as shown in figure 3. This is due to the greater shot intensity at nearer distances, so the surface looks darker / gray.

Figure 2. Surface modification on the surface of specimens at shot angle of 90°; (a) Shot distance of 6 cm, (b) Shot distance of 8 cm, (c) Shot distance of 10 cm, (d) Shot distance of 12 cm.

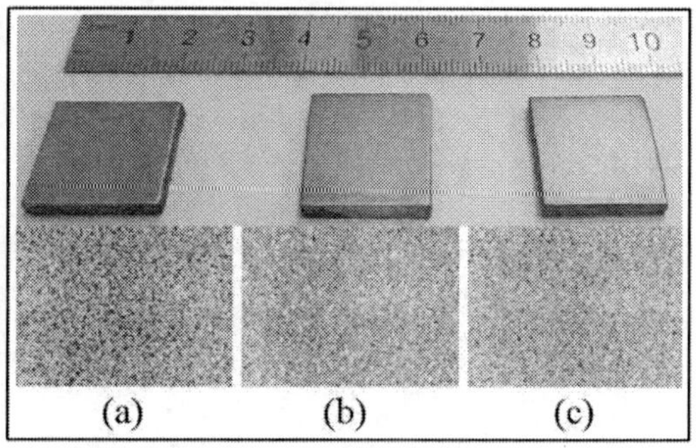

Figure 3. Surface modification on the surface of specimens shot distance of 6 cm; (a) Shot angle of 90°, (b) Shot angle of 60°, (c) Shot angle of 30°.

Figure 4 shows that the grain refinement occurs on the surface of specimen caused by multiple impacts of shot peening. Gariépy et al [18] report that shot peening gives an effect on the decreasing of grain diameter. The shot distances give a significant effect on the grain refinement on the surface of the specimen. The nearer of shot distance can create the smoother of the shape and size of the grain, because of the nearer of the shot distance, the greater of the shot intensity. The changes in shot angle also give an effect, the larger the shot angle the smoother the shape and size of the grain as shown in figure 5. According to Ranaware and Rathod [19], reduction in grain size and enhancement in dislocation density with increase in strain, indicate that no dynamic recovery occurred during the shot peening process. This grain refined will also increase the hardness properties of the material, due to the increasingly denser structure on the surface of the specimen.

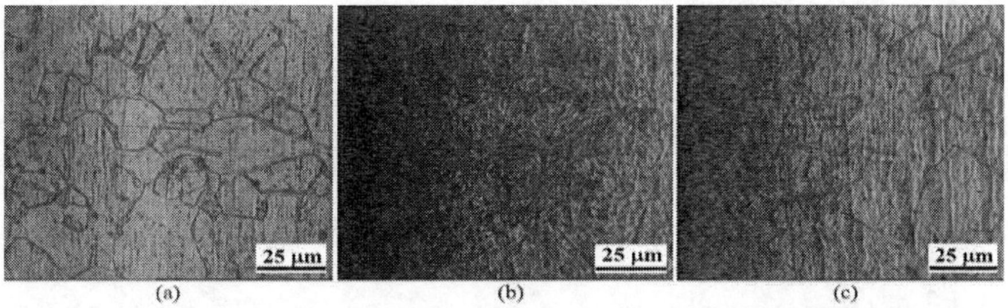

Figure 4. Cross-sectional microstructure of 316L at shot angle of 90°;
(a) Non-Shot Peening, (b) Shot distance of 6 cm, (c) Shot distance of 12 cm.

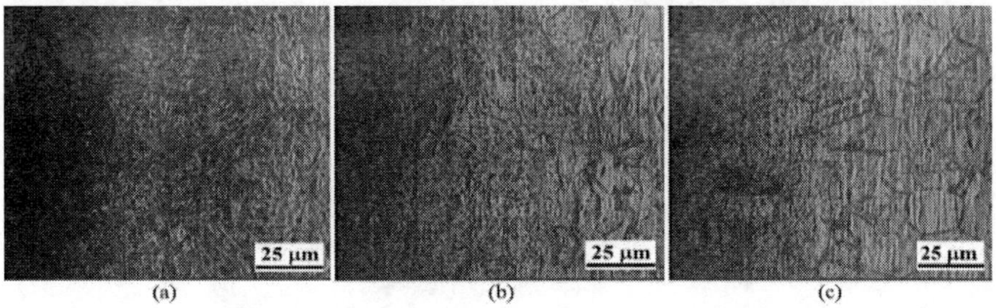

Figure 5. Cross-sectional microstructure of 316L at shot distance of 6 cm;
(a) Shot angle of 90°, (b) Shot angle of 60°, (c) Shot angle of 30°.

3.2. Surface roughness

The surface of the specimen after the shot peening has a modification in surface roughness, where the steel ball impact has deformed the surface to become rougher than non-shot peening. Libor et al [17] report that roughness is strongly increased because it induces severe plastic deformation that causes crystallite refinement in the surface layers. The roughness value of the non-shot peening specimen increases significantly (133%) after shot peening in the different shot distance 6, 8, 10 and 12 cm (0.46, 1.07, 1.10 and 1.16 µm) at shot angle of 90°, but decreases in increasing of shot angle 30°, 60° and 90° (1.09, 1.07, and 1,04 µm) at shot distance of 6 cm. The farther the shot distance, the rougher the surface of the specimen, while the larger the shot angle, the roughness decreases as shown in figure 6.

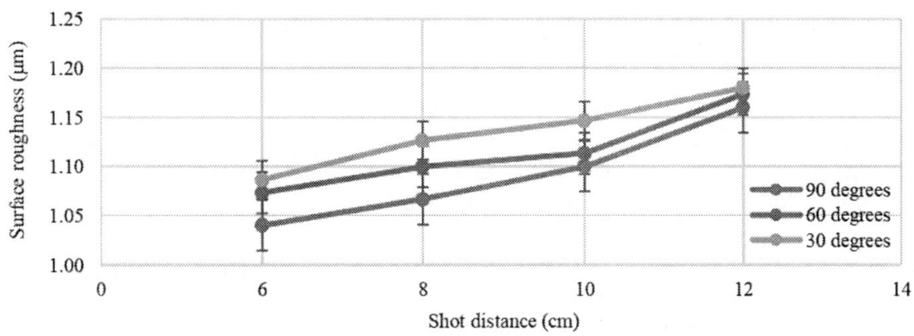

Figure 6. Effect of Shot Distance and Shot Angle of Shot Peening on Surface Roughness of 316L.

Nordin and Alfredsson [13] report that the surface roughness increased with increasing intensity of a given coverage. For a given intensity shot peening initially increase the surface roughness. This is due to the greater intensity of the shot at a distance of 6 cm so that the roughness decreases, while the shot intensity decreases at shot angle of 30° because the angle formed between the surface of the specimen with the shot direction is smaller so the tendency of the number of shot impact on the surface decrease.

3.3. Surface hardness

The shot distance and shot angle of shot peening also give the effect of enhancement in microhardness on the surface of the specimen. Nordin and Alfredsson [13] report that the largest hardness increases at the surface of the target. Hardness increases significantly (115%) at a distance of 6 cm with a shot angle of 90° (from 248 kg/mm^2 to 534 kg/mm^2). Hardness decreases at the farthest of the shot distance, respectively at 6, 8, 10 and 12 cm (534, 489, 441 and 391 kg/mm^2) at shot angle of 90°, but increases in increasing of shot angle at 30°, 60° and 90° (516 kg/mm^2, 532 kg/mm^2, 534 kg/mm^2) at shot distance of 6 cm. According to Natori et al [11], the shot peening gives the hardness increment resulted mainly from a significant increase in the strain-induced strain-hardening. Figure 7 shows the best distance is the nearest distance (6 cm) and the best angle is the largest angle (90°). The farther the shot distance causes the lower the shot intensity because the required distance of the steel balls to reach the specimen surface farther away and the smaller the shot angle, the hardness decreases. This corresponds to the influence of the area of contact between the surface of the specimen and the shot direction so that the smaller shot angle causes the affected area of specimen getting smaller too.

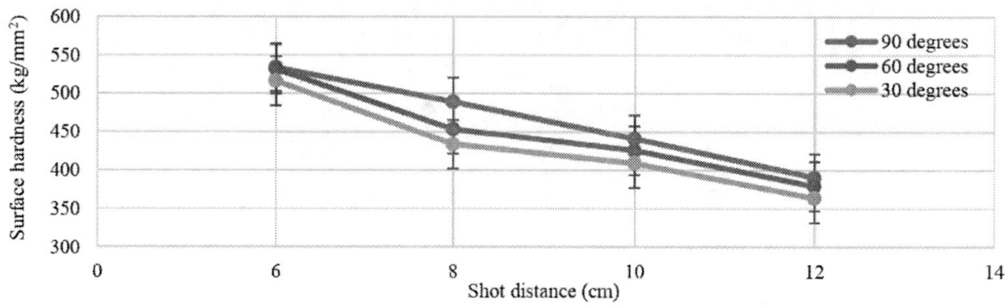

Figure 7. Effect of Shot Distance and Shot Angle of Shot Peening on Surface Hardness of 316L.

4. Conclusion

Shot peening presents the effect of the modification on surface morphology, surface roughness and surface hardness by different shot distance and shot angle. Shot peening can increase the surface roughness by the increasing of shot distance and by the decreasing of shot angle. Shot peening can increase the surface hardness by the decreasing of shot distance and by the increasing of shot angle. The nearest shot distance (6 cm) and the largest shot angle (90°) give the best results on the grain refinement with the surface roughness of 1.04 μm and surface hardness of 534 kg/mm^2.

References

[1] Pilliar RM 2009 *Biomedical Materials* 41–81
[2] Escalada L, Lutz J, Brühl SP, Fazio M, Márquez A, Mändl S, Manova D and Simison SN 2003 *Surface and Coatings Technology* **223** pp 41–46
[3] Sloten JV, Labey L, Audekercke RV and Perre GV der 1998 *Biomaterials* **19** pp 1455–1459
[4] Schultze JW and Lohrengel MM 2000 *Electrochimica Acta* **45** pp 2499–2513
[5] Schmuki P 2002 *Journal of Solid State Electrochemistry* **6** pp 145–164
[6] Milošev I, Metikoš-Huković M and Strehblow H-H 2000 *Biomaterials* **21** pp 2103–2113
[7] Milošev I and Strehblow H H 2000 *Journal of Biomedical Materials Research.*, **52** pp 404–412
[8] Umemoto M, Todaka Y and Tsuchiya K 2003 *Materials Transactions* **44** pp 1488–1493
[9] Bagherifard S, Fernández Pariente I, Ghelichi R and Guagliano M 2010 *Procedia Engineering* **2** pp 1683–1690
[10] Okada H, Tange A, Ando K 2006 *Fracture of Nano and Engineering Materials and Structures* pp 1315–1316
[11] Natori M, Song SM and Sugimoto KI 2013 *PRICM* pp 927–935
[12] Ahmed AA, Mhaede M, Wollmann M and Wagner L 2014 *Surface and Coatings Technology* **259** pp 448–455

[13] Nordin E and Alfredsson B 2017 *Experimental Techniques*, Springer Nature **41** pp 433–451
[14] Wang J, Han J, Li W, Yang Z, Li Z and Zhao Y 2016 *Journal of Wuhan University of Technology* **31** pp 1352–1362
[15] Luo S, Jie X, Zheng Q, Zhang L, Wu H and Huang X 2017 *Transactions of the Indian Institute of Metals* **70** pp 2175–2183
[16] Iswanto PT, Malau V, Priyambodo BH, Wibowo TN and Amin N 2017 *Materials Science Forum* **901**, 91–96
[17] Trško L, Guagliano M, Bokůvka O, Nový F, Jambor M and Florková Z 2017 *Journal of Materials Engineering and Performance* **26** pp 2784–2797
[18] Gariépy A, Perron C, Bocher P and Lévesque M 2012 *The Minerals, Metals & Materials Society*, pp 251–256
[19] Ranaware PG and Rathod MJ 2017 **33** pp 667–678

Acknowledgments
The authors are grateful to The Ministry of Research, Technology and Higher Education of The Republic of Indonesia that support this research by a Postgraduate Scholarship (BPPDN-DIKTI).

IOP Conf. Series: Materials Science and Engineering **299** (2018) 012052 doi:10.1088/1757-899X/299/1/012052

Synthesis and Characterization Pectin-Carboxymethyl Chitosan crosslinked PEGDE as biosorbent of Pb(II) ion

Budi Hastuti,[1] Dwi Siswanta,[2] Mudasir,[2] and Triyono[2]

[1]Department of Chemistry Education, Faculty of Faculty of Teacher Training and Education,Universitas Sebelas Maret, Jl. Ir. Sutami 36A Surakarta, 57126, Indonesia
[2]Department of Chemistry, Faculty of Mathematics and Natural Sciences Universitas Gadjah Mada, Sekip Utara, Yogyakarta 55281, Indonesia

*Corresponding author E-mail: budihastuti@staff.uns.ac.id

Abstract Pectin and chitosan are biodegradable polymers, potentially applied as a heavy metal adsorbents. Unfortunately both biosorbents pectin and chitosan have a weakness in acidic media. For this purpose required modified pectin and chitosan. The modified adsorben is intended to obtain a stable adsorbent and resistance under acid. The research was done by experimental method in laboratory. The stages of this research are the synthesis of carboxymethyl chitosan (CMC), synthesis of Pec-CMC-PEGDE film adsorbent, stabily test under acid, the characterization of active group using FTIR, stability characterization of Pec-CMC-PEGDE powder adsorbent using XRD, termo stability using DTA- TGA. The results of the research have shown that: pectin and CMC can be cross-linked using PEGDE crosslinking agent, the film adsorbent was stable under HCl 1 M, the film adsorbent have active group comprise of carboxylate and amine groups. The result of characterization using XRD, shows that the adsorbent is semi-crystalline. Base on termo stability, the film adsorbent Pec-CMC-PEGDE stable up to 600°C. The film can be applied as an adsobent of Pb (II) ion remediation. The optimum pH of pec-CMC-PEGDE in adsorbed of Pb(II) was reached at pH 5 with 99.99% absorbent adsorbed and of and adsorption capacity was 46.11 mg/ g.

1. Introduction

Waters contaminated by waste materials from industrial processes containing hazardous and non-biodegradable materials such as Cu, Ni, Zn, Cr, Hg, Cd, Pb will disrupt the dynamic relationship of the balance of nutrient component composition, organic matter and biomass which are very important for the stability of aquatic ecosystems. The existence of input materials that are toxic will reduce the potential of biological resources in addition to increasing cases of poisoning and public health disorders [1]. Thus it is necessary to have a simple technology capable of dealing with aquatic environmental problems, particularly with regard to methods for dealing with heavy metals contamination.

The adsorption is a simple method, economical, and naturally occurring, but highly effective method used to overcome wastewater problems in the aquatic environment. Methods for dealing with heavy metals by means of adsorption are generally based on the interaction of metals with functional groups present on the surface of the adsorbent through complex forming interactions and usually occur by involving groups such as -OH, -NH, -SH and -COOH.

Various biomaterials have been applied as adsorbents, and Pectin is one of those compounds which is applied as metals absorbent. Pectin is an anionic plant cell wall polysaccharide based on α-(1–4) linked D-galacturonic acid [2]. Pectin widely available in the middle lamella of plant cell walls. Pectin has many active groups. The main functional groups contained in the pectin are hydroxyl, carboxyl, amide and methoxy [3]. Lately the pectin was widely used in pharmaceuticals and cosmetics, in the food industry and as adsorbent to remove heavy metal ion [4,5]. Pectin contains a lot of active groups, that can be used as a source biosorbent. These functional groups can be used to bind heavy metals, especially hydroxyl groups [6]. Another biomaterials that capable of acting as biosorbents are well known are chitosan. Chitosan containing the active group -OH and -NH_2 or in neutral state -NH_3 is able to adsorb heavy metal ions well, either by ion formation mechanism and / or ion exchanger.

However, chitosan is easily soluble in acetic acid, it also partially soluble in dilute acid, such as HNO_3, HCl, $HClO_4$, and others, so that the direct use of chitosan as an adsorbent will be less effective. In a acidic conditions, the number of amine group which is an active group that capable to bind a metal ions will be reduced. So to solve this problem, chitosan chains must be crosslinked in order to improve the stability and hold up with acid condition. Several chitosan cross-linker agents suitable for chitosan such as Glutaraldehyde, epichlorhidrin chloromethyl oxirane, poly ethylene glycol (PEG), ethylene diamine diglycidyl ether (EDGE) and so on. Cross-linked chitosan is insoluble acids condition because it is very stable [7].

Chitosan can be modified into carboxymethyl chitosan (CMC). CMC is one derivative of chitin obtained by a carboxymethylation process of chitosan alkalis with monochlororoacetic acid. This CMC powder has the advantages of chitin and chitosan because CMC powder is easily soluble in water compared with chitin powder and chitosan powder, in addition it is biodegradable, biocompatible and non-toxic [8]. Carboxymethyl chitosan contains more active groups than chitosan so carboxymethyl chitosan can adsorb Pb (II) more than chitosan.

In this research, crosslinking between pectin and carboxymethyl chitosan was studied. The crosslinking agent to be used is Poly (ethylene glycol) Diglycidyl Ether (PEGDE). The presence of a cyclic ether group in both ends of PEGDE is expected to bind the -OH group of pectin and chitosan. This research synthesizes CMC-Pectin-PEGDE film to adsorb heavy metals Pb based on pH variation, contact time, and concentration.

2. Experimental section

2.1. Material

Chitosan with the deacetylation degree in the range of 75-85% and Poly ethylene glycol diglycidyl ether (PEGDE) were purchased from Sigma Aldrich (Germany). Chloroacetic acid, acetone, HCl 37 % v/v, HNO_3, $Pb(NO_3)_2$ were purchased from E. Merck (Germany).

2.2 Procedure

2.2.1. Preparation of adsorbent

Chitosan (3.0 g) was dispersed in 80 mL isopropanol (0.2 M) and stirred for 30 min. NaOH 40% (28 mL) was then added dropwise for 30 min and followed by dropwise addition of chloroacetic acid in isopropanol. The reaction mixture was stirred at room temperature for 12 h. The mixture was filtered and washed with ethanol and then dried on oven to obtain CMC. The next step, 0.2 g pectin and 0.2 g CMC was dispersed in 10 mL acetic acid 5% respectively and then 1.78 ml PEGDE in 25 mL ethanol was added and stirred for 2 h. The mixture was poured in a petri dish and dried in oven. The film was than immersed in an aqueous solution of sodium hydroxide (1 mol/L) for 12 h, and then washed with distilled water.

2.2.2. Adsorption of Pb(II) ion

10 mg Pec-CMC-PEGDE adsorbent was contacted with 10 mL of 50 ppm Pb (II) solution at pH 2,3,4,5, and 6, then stirred with shaker for 3 hours. Then Filtering the mixture using Whatmann filter paper and measure the absorbance of the filtrate by using Atomic Absorption Spectrophotometry (AAS).

3. Result And Discussion

3.1. *FTIR Characterization*

In order to prove that the -COOH group is the dominant active group in pectin, the $-NH_2$ group is the dominant active group on the chitosan while the dominant active group in the CMC are the $-NH_2$ and -COOH groups. Furthermore, the Polyelectrolyte membrane Pectin-CMC tested its functional group using infrared spectrophotometer. FTIR spectra results are presented in Figure 1.

From the FTIR spectra of pectin there is a wide absorption band at 3387 cm^{-1} which shows the stretch vibration of the -OH group. The band at 1065 cm^{-1} is the vibration of the -C-O-. The uptake at 1628 cm^{-1} is the absorption of the -COOH group and at the 1751 cm^{-1} is wave number of C=O ester.
The IR spectra of Pec-CMC-PEGDE adsorbent have the same adsorption as the the constituent compounds Pectin, CMC, and PEGDE. The constituent compounds of pectin, CMC, and PEGDE show the characteristic peaks. The peak of 1628 cm^{-1} which indicates for the -COOH group, whereas the –OH group at 3449 cm^{-1} and 1034 cm^{-1} indicate the –C–O group. The peak on wave number 1597 cm^{-1} indicates COOH group of CMC, and at wave number 3464 show the peak of –OH group. Increasing of peak intensity at wave number 1095 cm^{-1} which is –C–O– group of adsorbent Pec-CMC-PEGDE indicates increasing of ether on the Pec-CMC adsorbent. The increasing peak on 2908 and 1411 cm^{-1} indicates an increasing of CH_2 groups after cross-linking.

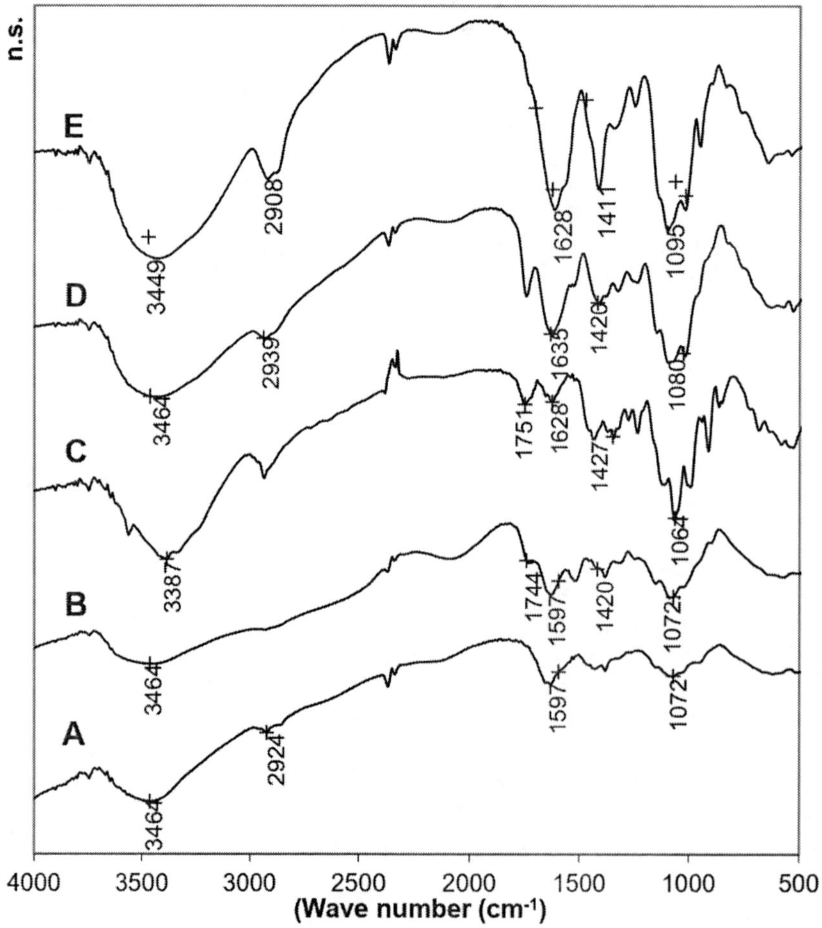

Figure 1 FTIR Spectra of: (A) Chitosan; (B) CMC; (C) Pectin; (D) Pec-CMC; and (E) Pec-CMC-PEGDE

3.2. Analysis of Pec-CMC-PEGDE adsorbent by XRD

Analysis with XRD spectrometry is intended to determine the crystallinity of the material after modification. Diffractogram Pectin, Chitosan, CMC, Pec-CMC and Pec-CMC-PEGDE shown in Figure 2. Initial chitosan showed peak at 2Θ = 10° and 20° degree showed chitosan was crystalline. Furthermore, in the CMC a wider peak is indicated at 2Θ = 10 ° and 20 °(Zang et al, 2011). This is due to a decrease in crystallinity degree due to the transplant process. The structure of chitosan crystals is strongly influenced by hydrogen bonds that occur in one chain (intra) and between chains (inter) chitosan. The grafting process to form the side group by changing hydrogen atom with acetic acid so make the hydrogen bond weaken. The difractogram pattern pectin shows only one sharp peak at an angle of 2Θ = 25 ° degrees while peak of Pec-CMC- PEGDE adsorbent doesn't found in XRD difractogram. This may indicate that after chemical modification the ability to form hydrogen bonds has decreased and the Pec-CMC- PEGDE adsorbent becomes amorphous. It shows Pec-CMC-PEGDE adsorbent is more stable.

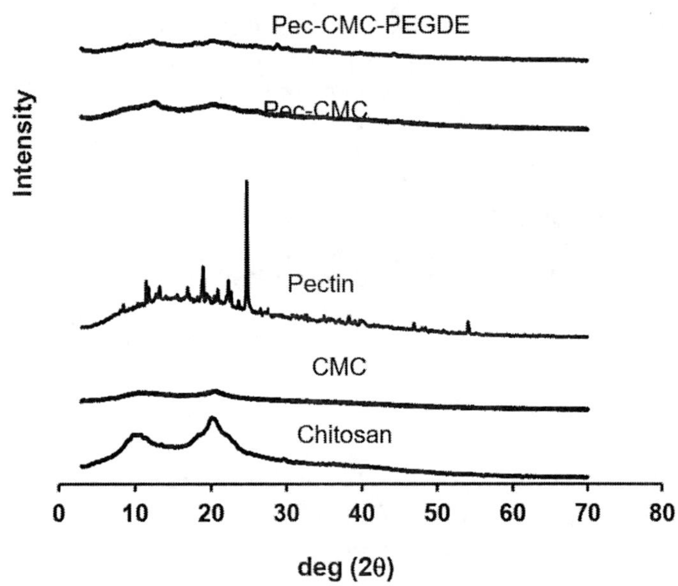

Figure 2. The diffractogram of Pectin, Chitosan, CMC, Pec-CMC and Pec-CMC-PEGDE

3.3 Thermal Analysis

DTA-TGA analysis aims to identify of the thermal caracteristic of chitosan, CMC, pectin and Pec-CMC-PEGDE adsorbent. The DTA-TGA termogram of CMC-Pec-PEGDE were shown in Figure 3.

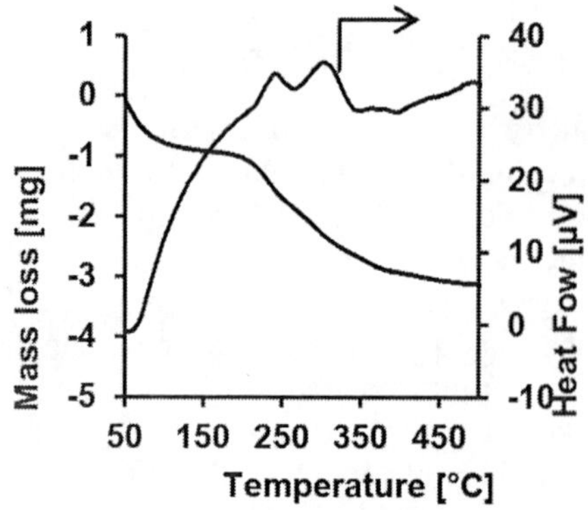

Figure 3. DTA termogram of: (a) chitosan; (b) CMC; (c) Pectin: and (d) CMC-Pec-PEGDE

DTA-TGA Thermogram of Pec-CMC-PEGDE adsorbent shows that two process stages are involved as temperature increase. The first stage at temperature range of 60° to 110° C is a weight loss of 8% which is loss of absorbed water due to the breaking of hydrogen bonds between water and adsorbent. The second stage is an initial degradation of the adsorbent with weight loss of 44% which is an exothermic reaction at temperature range of 150° to 370°C. This stage is an initial degradation, especially the side group of CMC and pectin. The third stage at temperature range of 370-500°C is an exothermic reaction from degradation of main chain of pec-CMC and continued degradation with weight loss of 5%. At this stage the degradation product of water, CO_2 and other small molecules are produced.

3.4 The aplication of Pec-CMC-PEGDE as adsorbent

Pec-CMC-PEGDE adsorbent was contacted with Pb(II) by varying of pH solution that show in figures 4. It show that the optimum condition of Pb(II) ion adsorption on carboxymethyl chitosan (CMC) occurs at pH 5, where the percentage absorbent in adsorpt Pb(II) was 91.12% with an adsorption capacity was 42.01 mg/g. The optimum condition of Pec-CMC-PEGDE in adsorpt of Pb(II) occurred on pH 5 with percentage absorbent was 99.99% with adsorption capacity was 46.11 mg/g.

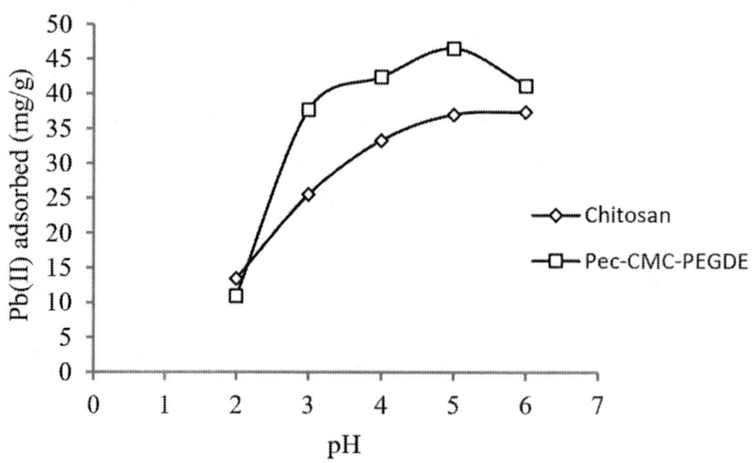

Figure 4. Effect of solution pH on adsorption capacity of Pb(II): Chitosan (upper) and Pec-CMC-EGDE (lower)

4. Conclusion

Pec-CMC-PEGDE were prepared by combine both polysaccharides Pectin and Chitosan with PEGDE as a crosslinker agent. Chitosan was grafted with acetate to form a carboxymetyl chitosan (CMC). Base on the analysis of XRD, DTA-TGA and photo SEM, the modified adsorbent Pec-CMC-PEGDE show that it stable. Based on FTIR spectroscopy, the adsorbent have active group comprise of carboxylate and amine groups. The result of characterization using XRD, shows that the adsorbent is semi-crystalline. Base on termo stability, the film adsorbent Pec-CMC-PEGDE stable up to 600°C. The modified adsorbent can be applied as an adsobent of Pb (II) ion remediation.

References

[1] Ardelean, E., Nicu, R., Asandei, D., Bobu, E., (2009), *European Journal of Science and Theology, 5(4), 67-75*
[2] Herman, D.Z. 2006. Jurnal Geologi Indonesia, Vol. 1 No. 1 Maret 2006: 31-36
[3] Siswati, D.N, Indrawati, T, Rhamah, M. (2006). *Jurnal Ilmiah Teknik Lingkungan*, 1(2), 67-72
[4] Blais, J.F., Dufresne, B., Mercier, G. (2000). *Rev, Sci, Eau* 12 (4), 687-711
[5] Kurniasari L., Riwayati I., Suwardiyono. (2012). *Jurnal Teknik Kimia*, 8(1), April 2012: 1- 5
[6] Herwanto, B dan Santoso, Eko. (2006*Akta Kimia Indonesia*, 2(1), 9-24
[7] Farda. E, Maharani .D. K. (2013). *Journal of Chemistry*, 2(1), 19-23
[8] Davies, D.H., C.M.Elso., E.R.Hayes. 1989. *Elsevier applied scince*. P. 474.

The Use of Electrochemical Impedance Spectroscopy Method for Corrosion Resistance Evaluation of 2205 and 410s Stainless Steels in Aqueous Sodium Chloride Solution

Andi Rustandi, Geraldo Sirait, Andy Saputra and Yunita Sadeli
Metallurgical and Material Engineering Departement, Faculty of Engineering, Universitas Indonesia, Depok-Indonesia

E-mail: siraitgeraldo@gmail.com[2]

Abstract. Corrosion behaviour of SS 2205 and SS 410s were investigated in various concentrations of aqueous sodium chloride solutions. Experimental work was carried out by using electrochemical impedance spectroscopy method at room temperature to evaluate the impedance of the system. Sodium chloride solutions were prepared with various concentration i.e. 1, 2, 3.5, 4 and 5 w.t%. The testing results were represented by Nyquist plot for both alloys. It was observed that the lowest corrosion resistance of both alloys was at 3,5% w.t NaCl which was similar to typical seawater solution. It was also observed that SS 2205 had better corrosion resistance along with greater impedance than the SS 410s in every NaCl concentration.

Index Terms. SS 2205, SS 410s, Corrosion Behaviour, NaCl aqueous solution, Electrochemical Impedance Spectroscopy.

1. Introduction

Stainless steel is widely used in industry and utilization in various environment because its ability to resist material degradation caused by reactive environment. Stainless steel performs better, compared with another kind of steel in resisting material degradation. Stainless steel has four class of specification, such as: ferritic, martensitic, duplex, and austenitic.

Martensitic stainless steel is composed by higher carbon and lower chromium content, compared with another kind of stainless steel. Martensitic stainless steel is used as cutting tool in industry, because of its hardness. This characteristic develops since the steel is processed through quenching, which forms the martensite structure.

Duplex stainless steel is composed with higher chromium and low nickel content. This composition made duplex stainless steel has two phase, which are austenite and ferrite, therefore duplex inherits both characteristics of the phase. Duplex stainless steel is mainly applied as offshore oil and gas equipment, in chemistry industry and petrochemicals, and application in maritime sector of industry.

Pitting corrosion is the most common corrosion mechanism that attacks stainless steel, and also the most severe. The mechanism of this kind of corrosion is conducted by destroying passive layer of stainless steel, then penetrate under steel the surface and make a pit-like hole under its surface [1].

Stainless steel is a kind of alloy which forms passive layer, therefore it is susceptible to be attacked by pitting corrosion. Pitting corrosion could occur because of decayed passive layer as an outcome of reactive halide ion attack such as Cl-,Br-, or I , which commonly contained in seawater [2].

Aside from halide ion containing environment, another condition which could destroy stainless steel's passive layer is oxygen content in the environment [3]. This occurs since oxygen is an electron receiver element of metals, which cause corrosion to occur [4].

Corrosion is a natural phenomenon, so it could not be stopped. However, corrosion phenomenon could be prevented by slowing the reaction rate of metals with the environment. Therefore, this research was aimed to observe corrosion behaviour of martensitic and duplex stainless steel in chloride ion with the variable of chloride concentration 1%, 2% 3.5%, 4% and 5% w/w of NaCl solution.

IOP Conf. Series: Materials Science and Engineering **299** (2018) 012053 doi:10.1088/1757-899X/299/1/012053

2. Details Experimental

2.1. Materials Specification

SS 2205 and SS 410s was used in this study. Both materials were machined and grinded well as preparation for EIS testing. Chemical compositions of both alloys shown in table I and table II and were tested using Optical Emission Spectrometer (OES).

Table 1. Composition of SS 2205

Chemical Composition (%)	C	Si	Mn	P
	0.028	0.483	1.41	0.038
	Ni	**Mo**	**Ti**	**Nb**
	5.32	2.81	0.005	<0.004
	S	**Cr**	**Cu**	**Fe**
	0.002	22.43	0.327	Bal.

Table 2. Composition of SS 410s

Chemical Composition (%)	C	Si	Mn	P
	0.015	0.740	0.504	0.017
	Ni	**Mo**	**Ti**	**Nb**
	0.382	0.007	0.0116	<0.004
	S	**Cr**	**Cu**	**Fe**
	<0.0005	11.82	0.064	Bal.

2.2. NaCl Solutions

Sodium chloride solutions were prepared at several concentrations, which is 1%, 2%, 3.5%, 4%, and 5% w/w, by dissolving 10, 20, 35, 40, and 50 grams of sodium chloride powder in 1 litre of water respectively. The solution was stirred using magnetic stirrer at room temperature (27^{0}C).

2.3. Samples Preparation

Both materials were cut into 5 pieces of 2cm x 2cm x 0,5cm plates from their original plates form. Samples then were grinded 60, 100, 180, 400, 500, 600, and 800 grades abrasive papers followed by polishing, cleaning and drying.

2.4. Polarization Cell

Electrochemical measurement by using EIS method was conducted in a cell chamber using carbon as counter electrode, stainless steel sample as working electrode and saturated Ag/AgCl electrode as reference electrode. Current range used was from 100 nA to 1A.

2.5. Electrochemical Impedance Spectroscopy Experiment

Electrochemical Impedance Spectroscopy experiment was carried out by Autolab potentiostat and NOVA 1.11 software.

3. Results and Discussion

3.1. Electrochemical Impedance Spectroscopy Testing

Electrochemical Impedance Spectroscopy methods was conducted to do qualitative comparison between impedances of SS 2205 and SS 410s. The result obtained from the methods were Nyquist plots of materials in different concentrations of chloride solution, which showed the impedance measured as comparison to observe corrosion resistance in different condition [5]. Nyquist plot was formed from the impedance of alternating current flowing back and forth on both samples surface.

Greater impedance correlates with less corrosion rate, as electron flow decreases with increasing resistance in the system. The data showed that straighter lines correlates with better corrosion resistance, as the lines have greater impedance [6].

IOP Conf. Series: Materials Science and Engineering **299** (2018) 012053 doi:10.1088/1757-899X/299/1/012053

Figure 1 and **Figure 2** showed that the corrosion resistance of the alloys ranked from the highest resistance to the lowest resistance was 1%,2%,5%,4% and 3,5% w/w sodium chloride concentration.

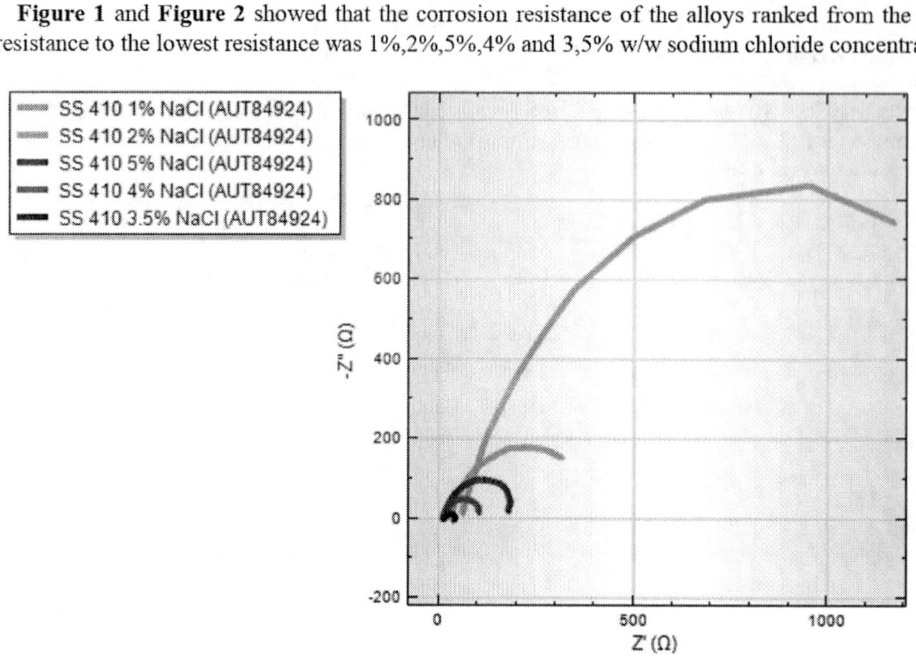

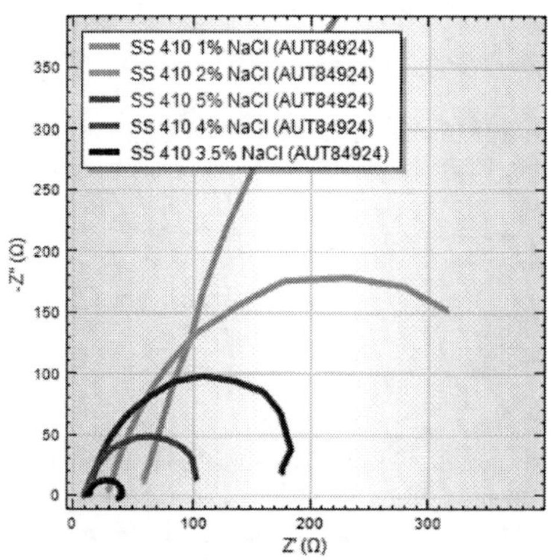

Figure 1. Electrochemical impedance spectroscopy curves of martensitic SS 410s in various sodium chloride concentration, zoomed **(bottom)**, and original dimension **(top).**

IOP Conf. Series: Materials Science and Engineering **299** (2018) 012053 doi:10.1088/1757-899X/299/1/012053

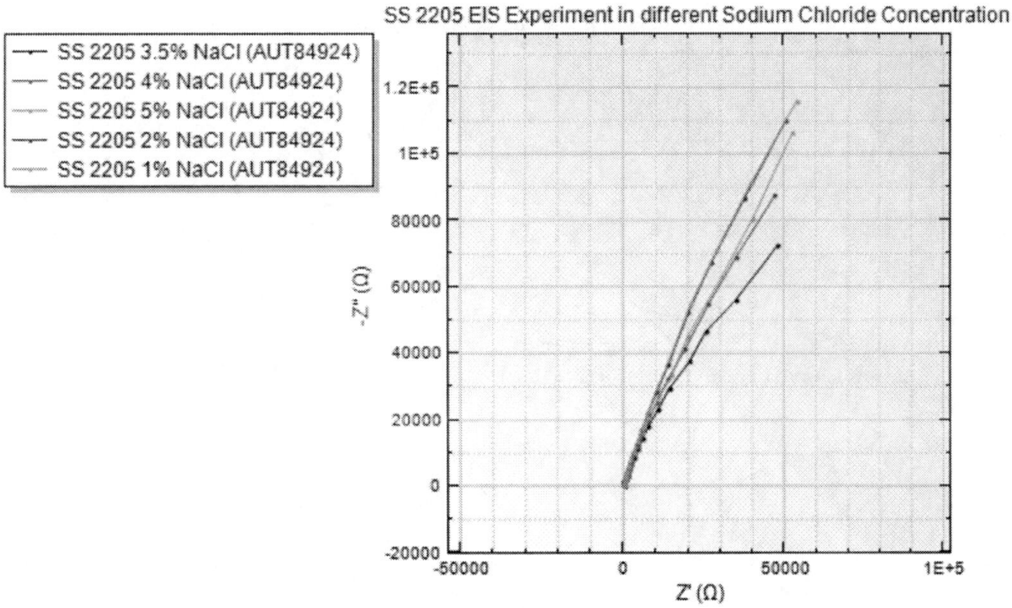

Figure 2. Electrochemical impedance spectroscopy curves of SS 2205 in various sodium chloride concentration.

3.2. Fit and Simulation

Besides the qualitative method conducted in this experiment, quantitative data was also provided by doing fit and simulation on provided nyquist graphs. Fit and simulation is a process that shows electrochemical equivalent circuit presenting the surface condition of the samples. Some of the elements that could be presented are the solution resistance (Rs), polarization resistance (Rp), double layer capacitance (CPE/Cdl), and constant phase element (Q). The difference between Rp and Rs is that the value of Rs is influenced by the condition of environment and material positioning, thus, the value of Rs does not represent the corrosion resistance of the sample. On the other hand, Rp is only influenced by the material itself. Thus, Rp should be considered as the corrosion resistance of the sample. The corrosion resistance of the sample has linear correlation with the value of Rp.

Table 3. Fit and simulation result of SS 410s

NaCl Concentration (%w/w)	Rs (Ω)	Rp (Ω)	CPE (μF)	N
1	56	448	29.7	0.968
2	28.9	185	34.2	0.979
3,5	10.2	19.9	69.1	0.876
4	8.64	33.9	54.6	0.996
5	13	51.6	17.7	1.05

Table 4. Fit and simulation result of SS 2205

NaCl Concentration (%w/w)	Rs (Ω)	Rp (kΩ)	CPE (μF)	N
1	6.50	466	75.6	0.853
2	8.14	457	78.4	0.85
3,5	26.1	197	93.6	0.84
4	14.3	265	88.4	0.854
5	9.35	332	83.0	0.852

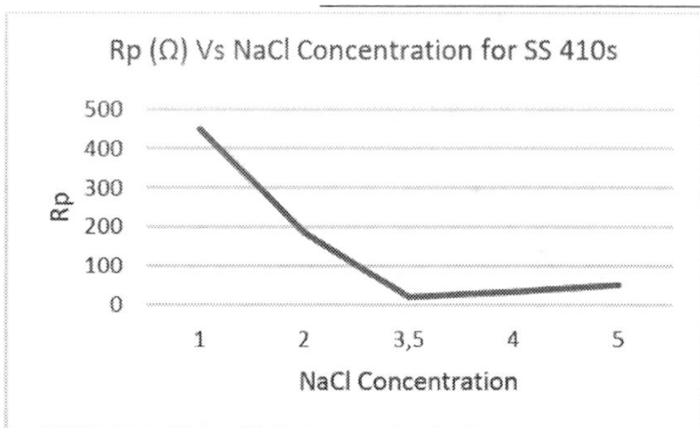

Figure 3. Rp of SS 410s in various sodium chloride concentration

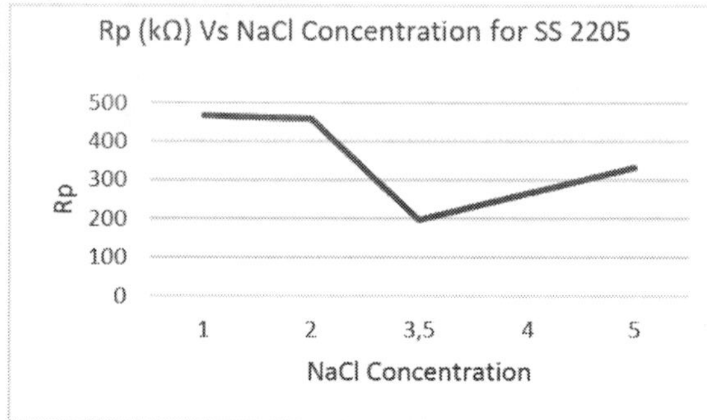

Figure 4. Rp of SS 2205 in various sodium chloride concentration

As shown in figure 3 and figure 4, according to conducted experiments both materials have less corrosion resistivity at 3.5% sodium chloride concentration. It is also shown that both materials at 1% and 2% sodium chloride concentration have higher resistance than at 4% and 5% sodium chloride concentration because the lack of sodium chloride as the electrolyte in the system. In addition, 4% and 5% sodium chloride concentration have better corrosion resistance than 3.5% concentration was occurred because of the lack of dissolved oxygen in the system [7].

3.3. Comparison of Nyquist Plots of SS 2205 and SS 410s

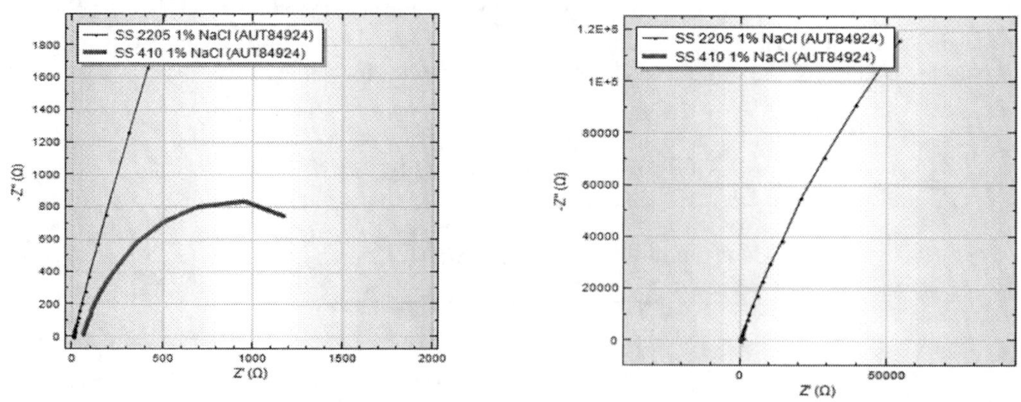

Figure 5. Nyquist plots of SS 2205 and SS 410s at 1% wt NaCl, original dimensions (Right) and zoomed (Left)

IOP Conf. Series: Materials Science and Engineering **299** (2018) 012053 doi:10.1088/1757-899X/299/1/012053

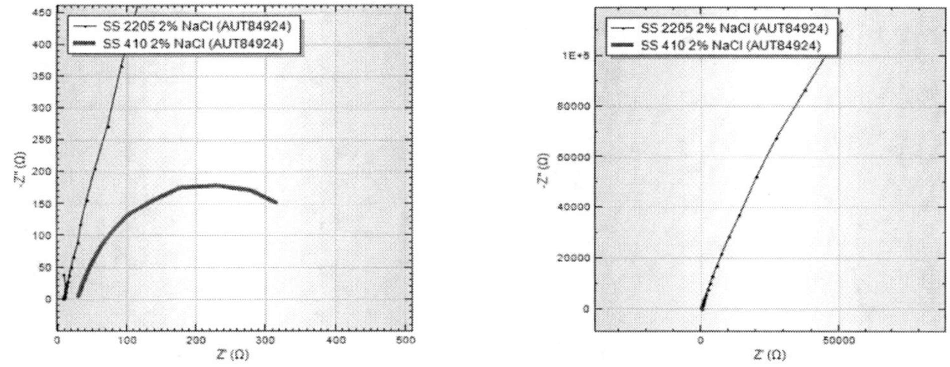

Figure 6. Nyquist plots of SS 2205 and SS 410s at 2% wt NaCl, original dimensions (Right) and zoomed (Left)

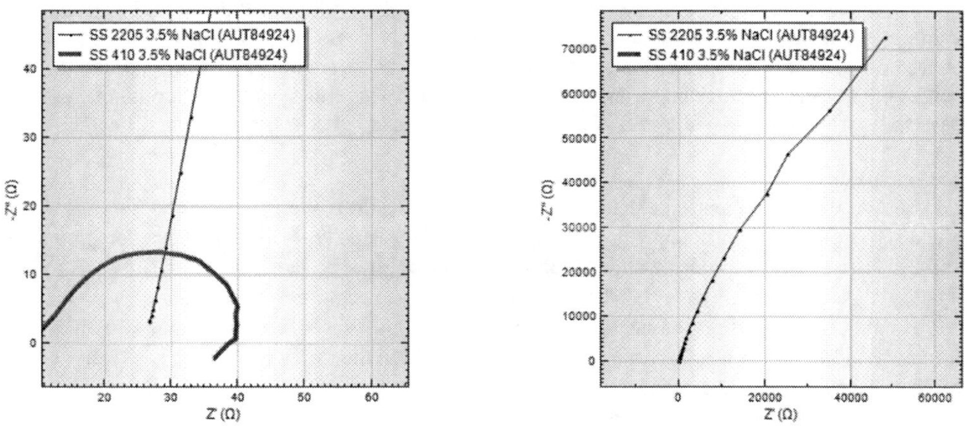

Figure 7. Nyquist plots of SS 2205 and SS 410s at 3.5% wt NaCl, original dimensions (Right) and zoomed (Left)

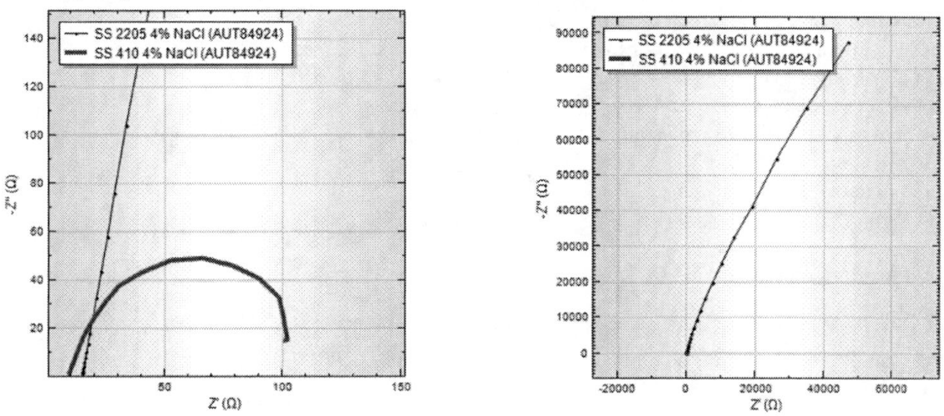

IOP Conf. Series: Materials Science and Engineering **299** (2018) 012053 doi:10.1088/1757-899X/299/1/012053

Figure 8. Nyquist plots of SS 2205 and SS 410s at 4% wt NaCl, original dimensions (Right) and zoomed (L

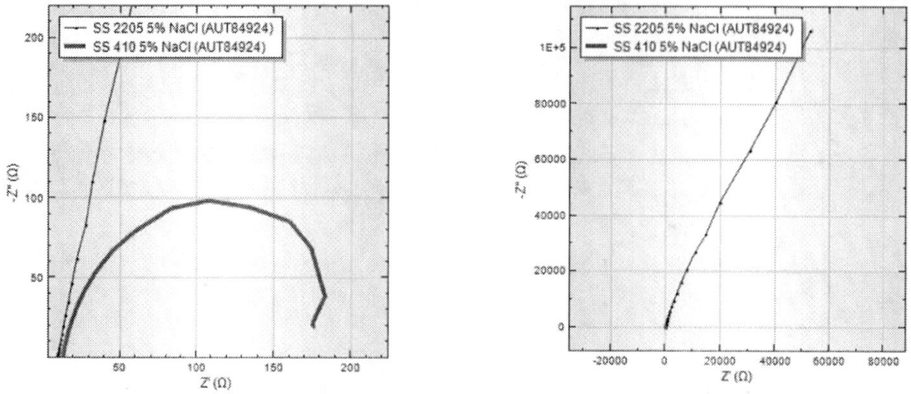

Figure 9. Nyquist plots of SS 2205 and SS 410s at 5% wt NaCl, original dimensions (Right) and zoomed (Left)

As shown in figure 5-9 above, the corrosion behaviour of both alloys could be considered by qualitatively comparing the graphics shown. As shown in figure 5-9 SS 2205 had far higher impedance value as well as better corrosion resistance compared to SS 410 at various sodium chloride concentration. The difference of the result and corrosion behaviour of the alloys was because the higher composition of chromium contained in the alloy of SS2205. This composition of chromium creates more passive layer to protect the alloy SS 2205. Thus, SS 2205 had better resistance than SS 410s.

4. Conclusions

Based on the experiment conducted in this study, we could make conclusion as follows using the Nyquist plot provided, we could conclude that the impedance and corrosion resistance for alloy SS 2205 and SS 410s at various chloride concentration from the highest to the lowest were 1%, 2%, 5%, 4%, 3,5% w.t NaCl consecutively. 3.5% sodium chloride concentration environment was the lowest corrosion resistance and impedance for both alloys, since the presence of maximum dissolved oxygen solubility and optimum electrolyte concentration occurred in the system. SS 2205 has higher corrosion resistance than SS 410s at any aqueous sodium chloride solution due to the effect of chromium addition which forms better passive layer in the alloy SS 2205.

REFERENCES

[1]. J R Davis 2000 *Corros. Underst Basics* **163–164** pp 62–66
[2]. A Ismail and N H Adan 2014 *Am. J. Eng. Res* **3** pp 64–67
[3]. Zaki Ahmad 2006 *Priciples of Corrosion Engineering and Corrosion Control*
[4]. J a Beavers, N G Thompson 2006 *Corros. Environ. Ind* **13** pp 1015–1026
[5]. R Chaves, I Costa, H G de Melo, S Wolynec 2006 *Electrochimica Acta* **51** pp 1842-1846
[6]. Metrohm Autolab *Electrochemical Impedance Spectroscopy (EIS) Part 1 - Basic Principles* Autolab Aplication Note EIS01.
[7]. Donghai D 2016 *Corrosion Science* **110** pp 134-142

5. Acknowledgements

The authors thank PT Voestalpine Bohler Welding Asia Pacific for technical support and Direktorat Riset dan Pengabdian Masyarakat Universitas Indonesia (DRPM UI) for financial support in this experiment.

IOP Conf. Series: Materials Science and Engineering **299** (2018) 012054 doi:10.1088/1757-899X/299/1/012054

Characterization Studies of Cyclotron CS-30 Carbon Puller Material Using Powder X-Ray Diffraction and SEM, EDX Cross Section Method

S Febriana[1], E S Riyanto[1], A Dimyati[2] and A Handayani[2]

[1]Center for Radioisotopes and Radiopharmaceuticals Technology, National Nuclear Energy Agency (BATAN), Puspiptek Area Serpong, Tangerang 15314, Indonesia
[2]Science and Technology Center for Advanced Materials National Nuclear Energy Agency (BATAN), Puspiptek Area, Serpong, South Tangerang 15314, Indonesia

Email: siska-febriana@batan.go.id

Abstract. The carbon puller material of Cyclotron CS-30 has been characterized to investigate the structure and composition of the puller. Two samples of pullers have been prepared in the form of powder from locally made carbon and the original cyclotron puller. X-ray diffraction (XRD) analysis of both samples powder gives identification of the structure and phase type of carbon, meanwhile the morphology and the elemental composition of both samples can be shown by Scanning Electron Microscopy (SEM) and Energy Dispersive X-Ray Analysis (EDX) technique. XRD samples are prepared by grinding the puller surface meanwhile the samples for SEM analysis are prepared by cross section method. XRD result shows amorphous carbon structure and a typical broad XRD profile's peaks and the diffused nature of the XRD profile indicate the presence of disorder in the samples. The first peak of the profile corresponds to the (002) peak of the hexagonal graphite structure, while the second peak corresponds to the (100) and (101) for both samples. Results of SEM-EDX showed that the carbon material from the original default puller of the CS-30 cyclotron contains mass percentage of lead significantly by 9.15%.

1. Introduction

Many of the important attributes for an effective process in cyclotron are very early in the acceleration process-in the central region. The performance of the cyclotron central region is essentially determined by first few turns of the accelerated ion beams[1]. For this reason, the central region has accordingly been the subject of a great deal for intensive study over a long period at our cyclotron CS-30 facility in national nuclear energy agency of Indonesia (Batan). Unfortunately, an in depth study and information correlating the structural characterictics of the material using in the central region is unavailable. Understanding the relationship between material properties, chemistry and atomic scale structural arrangement is critical for discovery and optimization of new functional materials[2]. In the present study, we report the characteristic material for one of the component in the central region, using one from the original manufacturer and one that have been locally made as a sample.

The sample component being used was puller, made of carbon. Since more than a quarter century, progresses in the development of carbon-based materials have been reported due to the importance of carbon in modern science, in high technology and also in a large domain of human activities. It is chemically inert and also resistant to corrosion and shock[3]. Carbon as a material possesses an excellent properties e.g. it can withstand very high temperature up to 3000°C, it exhibits low friction, low thermal expansion, good electrical conductivity and high mechanical strength[4].

Irradiation induced displacement of carbon atoms cause excess energy associated with vacancy/interstitial pairs to be stored up in the graphite crystallites[5]. The stored energy is released when the carbon atoms and the vacancies recombine and the release curves are characterized by a peak that becomes broader with increasing radiation dose whereas the maximum release rate of energy is reduced. So, the necessary and binding conditions are that the material carbon puller should be highly dense, amorphous and isotropic to achieve dimensional stability under irradiation.

The principal techniques that widely used for characterization of the material such as analysis of the microstructure and composition of a material are powder x-ray diffraction (XRD), scanning electron microscopy (SEM), and energy dispersive x-ray (EDX) spectroscopy. X-ray diffraction has been in use in two main areas, for the fingerprint characterization and the determination of the structure. Each solid material has its unique characteristic X-ray powder pattern which may be used as a "fingerprint" for its identification. Once the material has been identified, X-ray crystallography may be used to determine its structure, i.e. how the atoms pack together and what the interatomic distance and angle are, etc. X-ray diffraction is one of the most important characterization tools used in solid state chemistry and materials science[6]. Meanwhile, the EDX technique has been widely used to identify metals and/or contaminants while SEM is used to investigate the morphology and using the SEM technique, it is possible to observe the surface defects of the material[7,8]. Cross sectional method for powder sample in SEM measurement is rarely used by scientist, but in this paper we describe how to prepared powder sample with this method. The method contain metallurgical technique and can obtain to view the grains and particles clearly. The research findings would provide an important test basis for further studying of puller material under irradiation.

2. Experimental method

2.1. Chemicals and instrumentation

The carbon material used for the analysis is derived from the component called puller from the central region part of National Nuclear Energy Agency of Indonesia/Batan's cyclotron CS-30. Two sample are locally made carbon puller and original puller from the manufacturer with specific surface area 1.124cm^2/g and 1.060cm^2/g respectively. Local carbon puller made previously by the researcher in the cyclotron department using local graphite rod turn into the same shape of the original puller.

The structure of the sample was analyzed with X-ray powder diffraction (XRD Philips PW 1710 16) using Cu K_α radiation at room temperature. *Scanning Electron Microscopy (SEM); Electron Dispersive X-ray (EDX) measurement.* Experiments and analyses involving electron microscopy were performed in the Science and Technology Center for Advanced Materials of National Nuclear Energy Agency (BATAN), Indonesia. Secondary electron and backscattering electron images with qualitative and semi-quantitative chemical analyses in the EDX mode were obtained using a JEOL Type JSM - 6510 LA equipment.

2.2. Producedure reaction

Material for powder x-ray diffraction is finely ground to less than ~10 μm (or 200-mesh) in size. The resulted pattern was then matched using MATCH program to search the pattern structure approach.

SEM-EDX sample was prepared by cross sectional method to see through the cross sectional in each grain in the sample. The detailed sample preparation process of carbon for SEM;EDX measurement is as following :

(1) Mount the powder by suspending the powder in resin+hardener media to consolidate the sectioning.
(2) To expose a fresh surface the next step is grinding with the abrasive paper of 400, 600, 800, 1200 and 1500 in sequence.
(3) Polishing by alumina paste will removes the damages imparted of grinding operation and

followed by adding the sample into the water in ultrasonic cleaner for half an hour.

(4) Finally, sputtering technique (GL-1110X SPC-12-Compact Plasma Sputtering Coater) was performed to coat the sample with Au.

3. Result and discussion

Figure 1 shows the pattern of the xrd measurement for carbon puller which was locally made and the original carbon puller from manufacturer. As expected, the carbon material pattern shows some sharp diffraction peaks and also a broad features characteristics of an amorphous material. The diffused halos appearing at 2θ values i.e., 25 to 55 degrees are indicative of the amorphous nature of the puller material having some degree of arrangement (crystallinity) in the molecular chain of both samples. Both pattern fit well each other despite the fact that the local carbon has higher intensity and broader peak. The triangular peaks observed in the diffraction patterns result point to the presence of crystallinity.

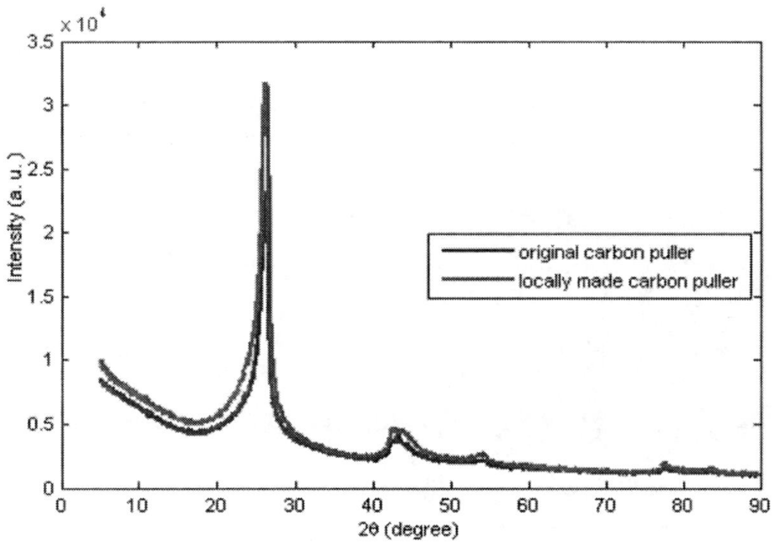

Figure 1. XRD Pattern of the carbon puller material

Figure 2 and 3 Show the identification of the peak correspond to the hexagonal structure of graphite in (002), (100) and (110) reflection using Match program. The diffraction profiles show the presence of a asymmetric (002) band around 26°, which suggests some graphite-like structure (crystalline carbon and a weak band at 43°. It is suggested that carbon has an intermediate turbostratic or randon layer lattice structure between graphite and amorphous[9]. This means that the carbon puller, amorphous carbon, contains significant amount of highly disordered material, which is responsible for the back ground intensity of the diffraction. Position of this peak is found to shift to higher 2θ value with increase in elemental content of carbon, this can be caused by the mechanical grinding of the sample without any heat treatment.

Refinement on the pattern data was not performed because of the lack of crystallinity and therefore next calculation on the percentage of crystallinity and amorphous part needs to be done. One of the basic assumptions of meaningful XRD is the homogeneous spatial and angular, distribution of the crystallite orientations in the different directions. Any deviation from homogenity makes sampling of intensities difficult (the reproducibility of the measurement).

In general, the comprehensive characterization of amorphous phases necessitates the use of multiple analytical techniques. Pair Distribution Function (PDF) analysis of total elastic scattering (Bragg and difuse) perhaps will be the most powerful method to access the Medium-Range Order (MRO) as seen in amorphous material. PDF can be used for exploring the effects of processing or strorage on amorphous materials and may be employed as a tool for comparing different processing methods and also to discern small structural differences[10].

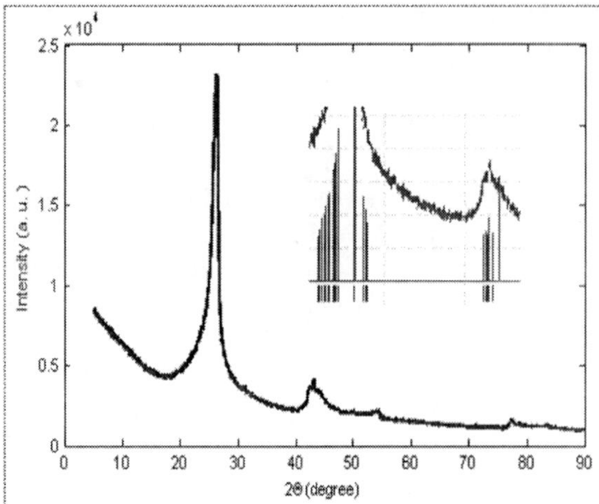

Figure 2. XRD Pattern of locally made carbon puller matched using Match program

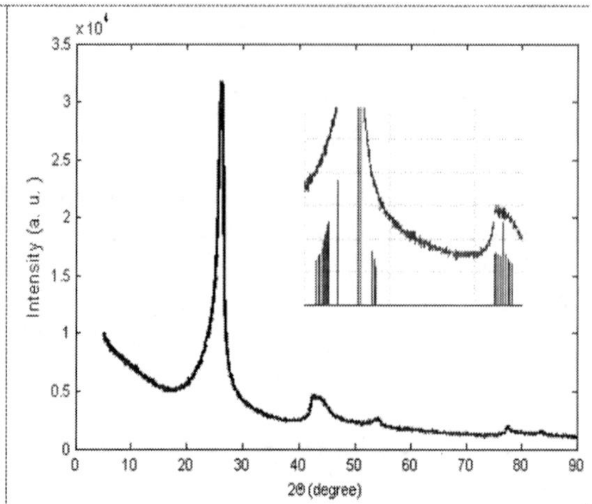

Figure 3. XRD Pattern of original carbon puller matched using Match program

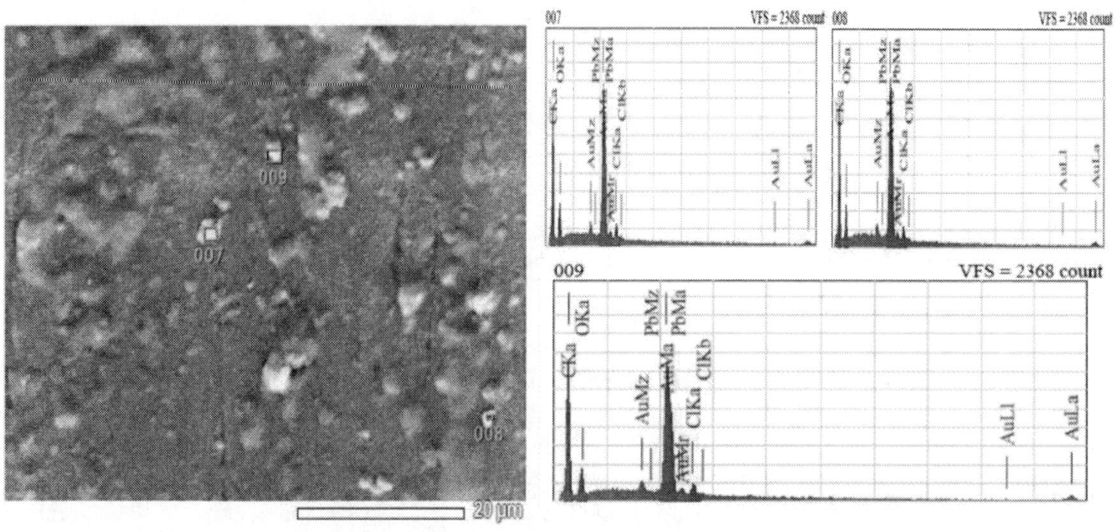

Figure 4. SEM image of the original manufacturer carbon puller with EDX measurement respectively

The morfology and the EDX measurement are shown respective in the Figure 4 from the original manufacturer carbon puller. As a comparison in Figure 5. it also shown the microscopy characteristics of the locally made carbon puller. Three point have been made in each image of SEM for best contrast that shows the grain cross section. The light and the darker color in contrast show the difference in resin media, grain and cross sectional inside the grain. Visually the grain of the local carbon is bigger than the original and in Figure 6 for higher magnification of 2500x, the agglomeration of the particle occurs. Adding a very small amount of powder into the resin media help reduce the aglomeration.

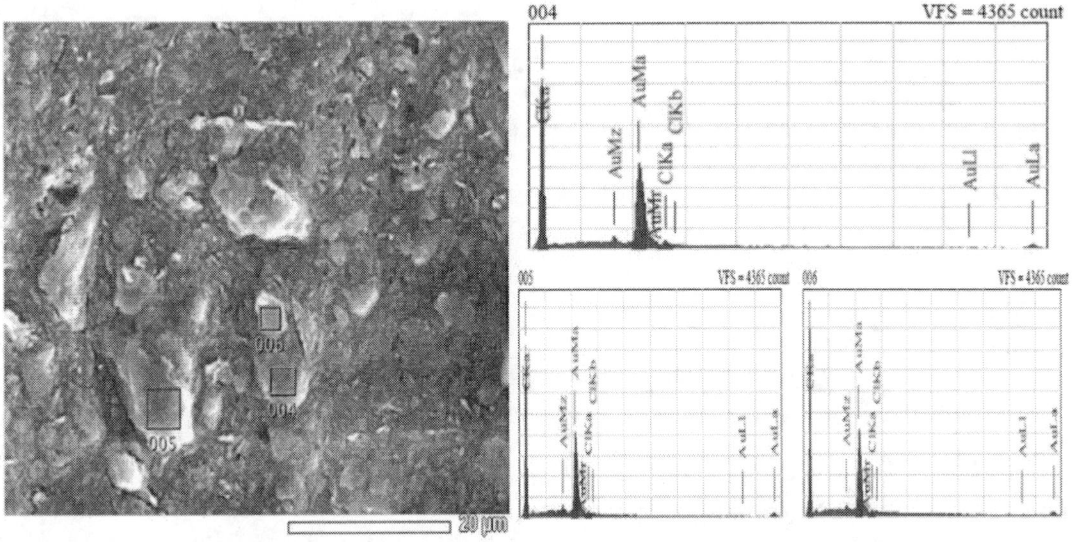

Figure. 5. SEM image of the locally made carbon puller with EDX measurement respectively

Figure. 6. Morfology of the locally made carbon with 2500x magnification

The spectrum of energy dispersive x-ray from the original puller show some amount of lead/Pb element which is not found in the locally made carbon puller. The detail of the composition is in Table.1. This element is suggested came from the irradiation in the cyclotron's central region during the process or it can naturally exist in puller. The Irradiation changes the crystal properties and indirectly the structure factors that can to be related the macroscopic to the microscopic properties. This can cause the displacement of the atoms and increase the disorder (amorphous) in the sample which correlated to the diffraction pattern result.

Table 1. Elemental composition (%) correspond to the EDX measurement

% Mass	Point	C	O	Cl	Au	Pb	Total
Original	007	36.33	10.78	2.35	41.97	8.58	100.00
Manufacturer	008	36.00	1.97	1.97	43.00	9.24	100.00
Carbon Puller	009	39.85	1.87	1.87	41.02	8.46	100.00
Locally Made	004	63.84	-	1.02	35.14	-	100.00
Carbon Puller	005	63.55	-	0.81	35.63	-	100.00
	006	64.76	-	0.36	34.88	-	100.00

4. Conclusions

The result of the structural investigation of the two sample of carbon puller showed that the diffraction pattern of such a material consists of two types of reflection. Crystalline type reflection which is graphite-like structure with diffraction peaks and two amorphous carbon which indicate the presence of disorder in the samples. SEM-EDX measurement of the original carbon material from Cyclotron CS-30 puller show 9.24% mass of lead.

Acknowledgment

The authors would like to express their gratitude to the staff members in the cyclotron division, Center for Radioisotopes and Radiopharmaceuticals Tecnology and in the Science and Technology Center for Advanced Materials, National Nuclear Energy Agency (BATAN) Indonesia for the help and guidance along this project.

References

[1] Bhandari R K and Dey M K 2011 Applications of accelerator technology and its relevance to nuclear technology *Energy Procedia* **7** 577–588.

[2] Shyam B Stone K H Bassiri R Fejer M M and Toney M F 2016 Measurement and modeling of short and medium range order in amorphous Ta 2 O 5 thin films *Nat. Publ. Gr.* **6:32170** 1–7.

[3] Krolow M Z Hartwig C A Link G C Raubach C W Pereira J S F Picoloto R S and Gonçalves M R F 2013 Synthesis and characterisation of carbon nanocomposites *NanoCarbon 2011* **3** 33–48

[4] Karpov A P and Mostovoy G E 2015 High-temperature mechanical properties of carbon and composite carbon-carbon materials *Inorg. Mater. Res.* **6** no. 5 454-460

[5] Sarkar A Dasgupta K Barat P Mukherjee P and Sathiyamoorthy D 2008 Studies on neon irradiated amorphous carbon using x-ray diffraction technique *Int. J. Mod. Phys. B* **22** no. 7 865–875

[6] Frost R L López A Scholz R Theiss F L and Wilson A 2015 Spectrochimica Acta Part A : Molecular and Biomolecular Spectroscopy SEM , EDX , Infrared and Raman spectroscopic characterization of the silicate mineral yuksporite *Spectrochim. ACTA PART A Mol. Biomol. Spectrosc.* **137** 607–611

[7] Han B Zhang L Zhang C Wang Y Yu X and Ou J 2016 Reinforcement effect and mechanism of carbon fibers to mechanical and electrically conductive properties of cement-based materials *Constr. Build. Mater.* **125** 479–489

[8] López A Scholz R Frost R L and Belotti F M 2015 SEM, EDX and vibrational spectroscopic study of the phosphate mineral ushkovite MgFe 2 3+(PO 4) 2 (OH) 2{\textperiodcentered} 8H 2 O--Implications of the molecular structure *J. Mol. Struct.* **1081** 329–333

[9] Manoj B and Kunjomana A G 2012 Study of stacking structure of amorphous carbon by x-ray diffraction technique *Int. J. Electrochem. Sci.* **7** 3127–34

[10] Thakral S Terbal M Thakral N and Suryanarayanan R 2016 Recent advances in the characterization of amorphous pharmaceuticals by x-ray diffractometry *Adv. Drug Delivery Reviews* **100** 183-193

Preparation and Characterization of Dabco (1,4-Diazabicyclo [2.2.2]octane) modified bentonite: Application for Congo red removal

Tarmizi Taher[1], Dedi Rohendi[2], Risfidian Mohadi[2], Aldes Lesbani[1,2]

[1] Environmental Science Study, Program Pascasarjana Universitas Sriwijaya. Jl. Padang Selasa, No. 524, Bukit Besar, Palembang, Indonesia
[2] Department of Chemistry, Faculty of Mathematics and Natural Science, Universitas Sriwijaya. Jl. Raya Palembang-Prabumulih, Km. 32, Indralaya, Ogan Ilir, Indonesia

Email: aldeslesbani@pps.unsri.ac.id

Abstract. Natural bentonite provided from Sarolangun deposit was modified with 1,4-Diazabicyclo[2.2.2]octane (Dabco) to form a new class of porous material. Prior further modification, the natural bentonite was cleaned up and activated by NaCl to remove the impurities and increase the bentonite nature. Dabco modified bentonite (Dabco-bent) was prepared by exchanging the inorganic cation placed in the interlayer space of the montmorillonite mineral structure with the 0.01 M $Dabco^{1+}$ at pH 6. The modified bentonite products were characterized using X-Ray powder diffraction and FT-IR to monitor the change of the bentonite crystallinity and function group due to the modification process. The XRD result confirmed that during the modification process, the *d*(001) of smectite peak at 2θ around 6° was shifted. After the modification, the *d*(001) reflection of the montmorillonite interlayer was shifted 0.36° to the left indicating that the interlayer space of the montmorillonite has been expanded during the modification process. The FTIR spectra of Dabco modified bentonite exhibit no significantly different with the host bentonite. However, the presence of the new band at the wavenumber around 3000 and 2800 cm^{-1} indicates that the Dabco molecule has been successfully inserted to the bentonite molecule. The Congo red adsorption experiment was performed onto Dabco-bent product by batch technique. The experiment data described that kinetic model for Congo red adsorption onto Dabco-bent was adequately followed the second-order kinetic model and well described by Freundlich adsorption isotherm model.

Keywords: *Natural bentonite, Dabco modified bentonite, Adsorption, Congo red*

1. Introduction

Bentonite, as one of the abundant natural clays, has broad application in industry sector for more than 40 years. The technology development has forced the bentonite utilization from the traditional toward higher value product like a catalyst, adsorbent, nanomaterials, medicine, and fine chemicals [1]. Nowadays, in many countries bentonite also used as raw material in cosmetic production due to its moisturizing, detoxification, decontamination, and itching relieving properties. The high demand of the bentonite clay in the various particular area such as science and technology lies in their high adsorption capacity, fire retardant, and naturally abundant and low-cost material [2].

Based on the previous literature and many published papers, the term of bentonite was used to describe the clay material that dominantly composed of smectite mineral wherever its origin [3]. The common bentonite used in industry was dominated by montmorillonite mineral, including calcium bentonite, and sodium bentonite. Montmorillonite mineral that composes the bentonite clay is a 2:1 layers class of smectite. It was constructed by silica tetrahedral sheets (T), and one central alumina octahedral sheet (O) [4]. The surface of bentonite has a permanent negative charge due to the isomorphous substitution of Al^{3+} with Fe^{2+} and Mg^{2+} in the octahedral sheet and Si^{4+} with Al^{3+} in the tetrahedral sheet. Naturally, the net negative charge on the bentonite surface was balanced by exchangeable inorganic cations located in the interlayers site [5].

Naturally, bentonite formed by the alteration of the volcanic ash. It consists dominantly by smectite minerals like montmorillonite and some others like saponite, nontronite, and hectorite [6]. Bentonite has hydrophilic properties on its surface. This property causing some restriction in its applications such as reduce the dispersion properties in the organic matrix and make it the as weak adsorbent for removal of organic compounds [6]. Although Bentonite is known as a high utilization natural material with comparatively low-cost than other materials, its restriction in the organic field will reduce the value of the bentonite material in the market. Till now, many methods have been developed in the way of bentonite modification to break the limitation of bentonite utilization.

The common modifications of bentonite material were physical activation, chemical activation, cation exchange, intercalation, plasma surface modification, and pillarization [7–10]. Based on these modification methods, pillarization was one of the most used method due to the high effectivity in changing the surface and interface physicochemical properties of the bentonite. Pillared clay has 20-fold higher surface area than the parent natural bentonite material [11]. In the pillaring process, the inorganic cation lied in the interlayer space of the bentonite was changed by another cation which has the better characteristic. Many kinds of cation have been studied to be used as a pillaring agent for the bentonite including organic cation, inorganic cation, organometallic cation, metallic cluster cation, polyoxide cation, and mix oxide cation [12].

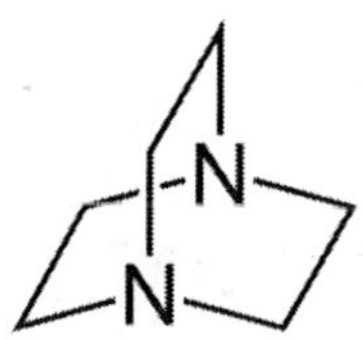

Figure 1. Chemical structure of 1,4-diazabicylo[2.2.2]octane (Dabco) compound used as bentonite modifier

In this study, a locally grounded bentonite from Sarolangun was modified by pillaring method using an organic based pillaring agent, i.e., 1,4-Diazabicyclo[2.2.2]octane (Dabco) (Figure 1). In order to investigate and to control the degree of modification process, X-Ray diffraction (XRD) and IR spectroscopy (FTIR) were used. The Dabco modified bentonite product then applied as an adsorbent to remove the Congo red dye from aqueous solution. Congo red dye is a well know colorant used in many textile industries and reported as one of the organic molecules that causing an environmental problem when contacted with the water body [13]. The result of this work is hoped to be a reference for the development of local Indonesian bentonite that has no significant economic cost in the market.

2. Materials and Methods

2.1. Materials

Bentonite as the host modified material used in this work was purchased from local bentonite mining located in Sarolangun district, Jambi Province, Indonesia. The natural bentonite pretreatment was conducted to reduce and remove the organic and inorganic impurities. 1,4-Diazabicyclo[2.2.2]octane (Dabco) was obtained from Kanto Chemical Company, Japan. It was used without further purification. Sodium hydroxide (NaOH) and hydrochloric acid (HCl) were purchased from Merck Millipore without further purification. Congo red dye used was in analysis grade provided by Sigma Aldrich and used as received.

2.2. Instrumentation

Chemical analysis of the natural, Na-exchanged bentonite, and modified bentonite were conducted using X-Ray diffraction on Rigaku Miniflex 600 instrument equipped with Ni-filtered CuKα radiation. The

XRD pattern was recorded at 40 kV and 40 mA with fix slit in 2θ range about 5 – 50° under scanning speed of 5°/min. The d_{001} value of the samples was measured using the Bragg's Law as follow:

$$d = \frac{n\lambda}{2\sin\theta} \quad (1)$$

Where n is the diffraction order, λ is the wavelength of the wavelength of the X-ray, θ is the diffraction angel, and d is the basal spacing [14]. The functional group of natural bentonite, Na-exchanged bentonite, and modified bentonite samples were characterized using FT-IR Shimadzu Prestige-21 instrument using KBr disc in the wavenumber 400-4000 cm^{-1}. The samples were contacted and mixed with the KBr powder and grinded finely in a porcelain mortar. The mixture then mounted to the FT-IR instrument, and the sample measurement started in the percent transmittance mode (%T). The Congo red concentration after contacted with Dabco-bent was measured using spectrophotometer UV-Vis EMCLab.

2.3. Natural Bentonite Preparation

Prior used, the raw bentonite was purified and cleaned by the following procedure. Approximately 500 g of raw bentonite was milled, and the fine-grained raw bentonite powder was collected and sieved through 100 mesh ASTM standard sieve. The fine-grained of raw bentonite then washed using distilled water for several times and dried in hot air oven for overnight. The clean bentonite sample then stores in an isolated flask for further use. The natural bentonite characterization was conducted using XRD and FTIR to study its chemical structure.

To optimize the modification process, the bentonite sample which is calcium based bentonite (Ca-Bentonite) was converted to the Na-bentonite by treating with NaCl. Approximately 100 g of clean bentonite sample was contacted with 500 mL of 1 M NaCl in 1000 mL beaker glass. The mixture then stirred for 24 hours with 200 rpm rotation speed at 60 °C. After stirring, the suspension washed using distilled water in a Buchner funnel equipped with a vacuum pump. The bentonite residue then rinsed several times using distilled water to remove the excess of NaCl and Cl^- ions. Light yellow bentonite was obtained than dried in the laboratory oven and grained with porcelain mortar. The change of bentonite structure during the Na-exchange process was monitored with XRD and FT-IR instrument.

2.4. Dabco modified bentonite preparation

Dabco modified bentonite was prepared based on the three steps procedure including preparation $Dabco^{1+}$ solution, Dabco insertion, and drying. $Dabco^{1+}$ solution (0.01 M) was prepared by diluting appropriate amounts of Dabco powder with distilled water at pH 6 ($K_1 = 10^{8.19}$) [15]. The insertion process was conducted in 500 mL beaker flask in a closed condition. Five grams of Na-bentonite and 250 mL of $Dabco^{1+}$ solution was mixed under vigorous stirring for 24 hours at room temperature condition. After stirring, the mixture then separated using filtration paper. The treated bentonite residue then washed with distilled water for several times and dried at room temperature. The dry intercalated bentonite then calcinated in a muffle furnace at 150 °C [6]. The Dabco modified bentonite product, assigned as Dabco-bent, then characterized using XDR and FT-IR to study the change of the crystallinity and the structure of the bentonite during the pillarization process.

2.5. Congo red adsorption study

The capability of the Dabco modified bentonite on Congo red removal was investigated in batch adsorption technique. The Congo red stock solution was prepared by dissolving 1 g of Congo red powder into 1000 mL of distilled water. The Congo red standard and working solution were made by diluting the Congo red stock solution into desired concentration with distilled water. As much as 0.1 g of Dabco-bent powder was contacted with 50 mL of Congo red solution under vigorous stirring with predetermined contact time in room temperature condition. After finished, the mixture then separated

by centrifugation then the remaining concentration of the Congo red solution was measured using UV-Vis spectrophotometer at wavelength 500 nm.
The Congo red removal percentage and the amount of equilibrium adsorption were calculated based on the following equations:

$$\%\text{Removal} = \frac{(C_i - C_t)}{C_i} x100 \quad (2)$$

$$q_t = \frac{(C_i - C_t)}{m} V \quad (3)$$

Where C_i is the initial concentration of the Congo red solution (mg L^{-1}), C_t is the remaining concentration of the Congo red after the adsorption process at any time (mg L^{-1}), q_t is the amount of the Congo red adsorbed on the adsorbent (mg g^{-1}), m is the mass of the adsorbent used (g), and V is the volume of the Congo red solution used (L).

3. Results and Discussion

3.1. Modified Bentonite Characterization

3.1.1. XRD Analysis

The primary technique that commonly used to study the mineralogical composition of the clay minerals is X-ray diffraction (XRD) powder. In this study, the structure of natural bentonite grounded from Sarolangun District was studied with XRD. The XRD powder pattern of the natural bentonite sample was depicted in Figure 2. The XRD pattern confirms that the typical characteristic peak of the montmorillonite (001) observed at the 2θ value around 6° and 27° [16]. As reported by Djukic et al. [17], the other peak observed at 2θ around 19, 20, and 26° were the associated mineral of bentonite, i.e., montmorillonite, calcite, dolomite, and quartz.

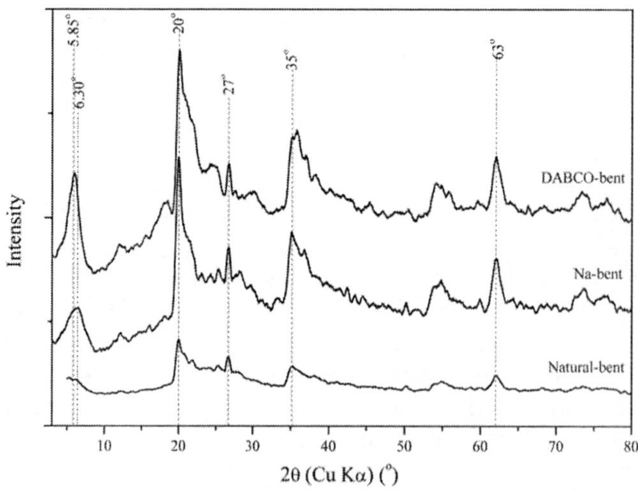

Figure 2. XRD powder pattern of Na-exchanged bentonite

The change of bentonite structure during the Na-exchange process was studied using XRD powder. The XRD powder pattern of the Na-bentonite described in Figure 2. The XRD pattern of the Na-bentonite exhibit the characteristic peak d(001) of the montmorillonite at 2θ value of 6.30°. Based on the value of the natural bentonite (Figure 2), the 2θ value of the Na-bentonite was shifted. This shifting also observed on the work done by Hao et al. [18] and this phenomenon is due to the exchanged of the interlayer cations (Ca^+) of the natural bentonite with Na^+ that has smaller cation size (Na^+ = 7.16Å < Ca^{2+} = 8.24 Å). This exchange process inflicts the constriction of the interlayer space [19].

As illustrated in Figure 2, the X-ray diffraction patterns of Na-bentonite and Dabco modified bentonite has the same typical diffraction peaks of bentonite at 2θ about 6° [18]. The peak of this pattern was shifted significantly by the Dabco intercalation. The Na-bentonite as the host material has 2θ value of 6.3° and shifted to 5.85° after modified by Dabco with concentration 0.01 M. The reduce of the 2θ value of the typical montmorillonite peak represent that the basal spacing of bentonite interlayer space increased. This result indicated that the Dabco molecule has intercalated into the interlayer space of the bentonite by replacing the Na^+ cation that constructed the host bentonite [20].

3.1.2. FTIR Analysis

One of the most powerful chemical analysis is Fourier transform infrared spectroscopy (FTIR). It is the most useful technique which is used to gain an infrared absorption spectrum of a material (liquid, solid, or gas). FTIR work by bombarding the IR radiation through a sample and some radiation passes through or transmitted, and some others are absorbed. Since it founded, the utilization of IR spectroscopy has been raised due to each chemical molecule produce different spectral fingerprints. In this study, the IR spectroscopy was used to investigate the spectral fingerprint of a molecule that constructed the bentonite sample.

The FTIR spectra of the natural bentonite from Sarolangun District was shown Figure 3. The natural bentonite spectrum exhibit an adsorption bands at 3448 and 1635 cm^{-1}. These bands were assigned to the stretching and bending respectively of the –OH group from the water molecule that contained in the interlayer space of the bentonite. A small bit band at 3626 cm^{-1} interpreted the stretching vibration of the –OH group that linked to the Al^{3+} in octahedral sheets. The strongest band was recorded at the wavenumber of 1033 cm^{-1}. This band could be easily being assigned as the Si-O stretching plane band that commonly recognized in the infrared spectrum around 1000 cm^{-1}. The relatively weak band at 532 and 470 cm^{-1} corresponded to the Si-O-Al in the octahedral sheet and Si-O-Si bending vibration, respectively [21].

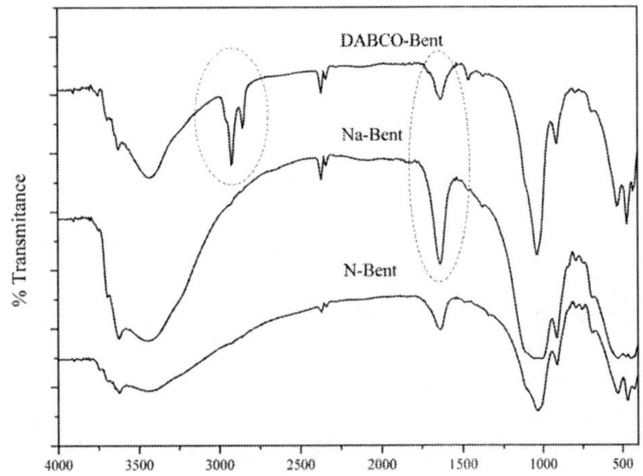

Figure 3. FT-IR spectra of natural bentonite (N-Bent), Na-exchanged bentonite (Na-Bent) and DABCO modified bentonite (DABCO-Bent)

The FTIR spectra of NaCl-treated natural bentonite shown in Figure 3. The figure showed that the characteristic band of the natural bentonite was not shifted. However, the intensity of the characteristic band at the wavenumber 3448 and 1635 and 1033 cm^{-1} as the characteristic peak of the water molecule and silica was significantly increased. This phenomenon described that during treated with NaCl, the water molecule content in the bentonite structure was increased. Moreover, at the wavenumber around 524 cm^{-1}, the band intensity of the Na-bent was reduced. This phenomenon indicated that the Si-O-Al group was reduced by the presence of the Na^{+} cation in the interlayer space of the bentonite structure [22].

After the modification process, as illustrated in Figure 3, the FTIR spectra of the Dabco-bent was slightly different with the Na-bent as the host. The main characteristic band of the Na-bent was not shifted significantly, but a new band at wavenumber around 3000 and 2800 cm^{-1} was recorded at the spectra of Dabco-bent. These band was the characteristic band of the C-H bending vibration. This finding describes that the Dabco molecule has successfully intercalated into the bentonite interlayer structure. Moreover, compared with the Na-bent spectra, Dabco-bent spectra has less intensity almost in all the recorded band. Particularly in the characteristic of the OH molecule. This finding indicated the removal of the water molecule in the calcination treatment during the modification process.

3.2. Adsorption Kinetics Study

The dependency of Congo red adsorption toward the contact time was investigated by conducting the adsorption experiment in various predetermined time, i.e., 0 to 60 minutes. In addition, this investigation is useful to calculate the kinetics of adsorption and the maximum adsorption capacity of the Dabco-bent for Congo red removal. The adsorption kinetics was investigated by calculating the constant of the adsorption rate and the adsorption capacity at equilibrium state. The experiment data was obtained by contacting 0.1 g of Dabco-bent with 50 mL of Congo red (50 mg L^{-1}) at room temperature with different contact time ranging from 0 – 60 minutes.

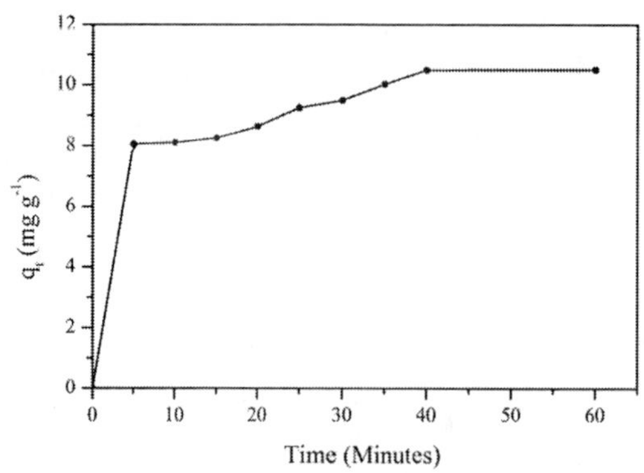

Figure 4. Effect of contact time on the adsorption capacity Dabco-bent

The result was presented as a graph of the correlation between the contact time to the adsorption capacity as depicted in Figure 4. The figure showed that the adsorption of Congo red on Dabco-bent was rapidly occurred in the beginning (around the first 15 minutes), then getting slowed down until reached the equilibrium after 60 minutes with maximum adsorption capacity 10.5 mg g^{-1}. As reported by Wang et al. [23], this phenomenon occurred due to in the begging of the adsorption process, the gradient concentration of the Congo red solution is larger. As the time went, the concentration gradient was reduced and the adsorption rate was gradually decreased until establishing the sorption equilibrium at 60 minutes. Therefore, the 60 minutes contact time was used for subsequent experiments.

Further, the obtained data of the contact time effect was evaluated to determine the kinetic adsorption parameters based on the adsorption kinetic models that have been established, i.e., the pseudo-first-order kinetic model, second-order kinetic model, and intraparticle diffusion model. The pseudo-first-order kinetic model is the first model that used to describe the adsorption kinetic model in a liquid/solid system. This model is mostly used in the aqueous solution adsorption system with weak concentration. The linear form of the pseudo-first-order kinetic model proposed by Lagergren [24] is presented as follows:

$$\log(q_e - q_t) = \log q_e \quad K_1 \frac{t}{2.303} \tag{4}$$

Where q_e is the amount of Congo red adsorbed onto Dabco-bent (mg g^{-1}) at equilibrium, q_t is the amount of the Congo red adsorbed (mg g^{-1}) at the respective time, and K_1 is the rate constant of the pseudo-first-order kinetic model. The value of the K_1 and predicted q_e based on the pseudo-first-order kinetic model could be determined based on the slope and intercept value from the plotted log (q_e-q_t) versus t.

The pseudo-second-order kinetic model is an adsorption kinetic model that refers to the capacity of the adsorption equilibrium. This model is express as follow:

$$\frac{t}{q_t} = \frac{1}{K_2 q_e^2} + \frac{1}{q_e} t \tag{5}$$

Where K_2 is the constant of the adsorption rate based on the pseudo-second-order model (g/mg min^{-1}). By plotting *t/qt* against t, the adsorption rate constant (K_1), and the predicted q_e can be calculated based on the intercept and slope obtained.

The mathematical equation for intra-particle diffusion is developed by Weber and Morris [25] as follow:

$$q_t = K_{id} t^{1/2} + C \tag{6}$$

Where K_{id} and C are the intraparticle diffusion rate constant (mg/g $min^{1/2}$) and a constant refers to the boundary thickness of the layer. If the plot of q_t against $t^{1/2}$ is a linear plot and passes through the origin, it can be described that the adsorption mechanism is only controlled by the intraparticle diffusion.

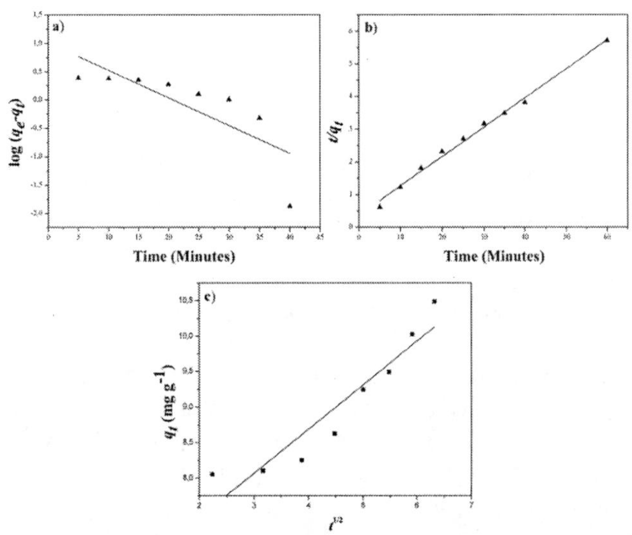

Figure 5. Fit of the experimental data to the pseudo-first-order (a), pseudo-second-order (b), and intraparticle diffusion (c) model

The plot and kinetic adsorption parameters based on the three kinetic models used are presented in Figure 5 and Table 1, respectively. The obtained data showed that the kinetic model for Congo red adsorption on Dabco-bent is best fitted by the pseudo-second-order model rather than the pseudo-first-order model. In this case, the R^2 value for the pseudo-second-order model is the closest to unity (0.9871), greater than the R^2 value for the pseudo-first-order model (0.8554). For the intraparticle diffusion model, the result showed that the C value is higher enough to declare the proportion to the boundary layer thickness. Moreover, the linear plot of the intraparticle-diffusion model (Figure 5c) was not pass through the origin. It indicates that the adsorption process was not completely controlled by the diffusion mechanism [26].

Table 1. Kinetic parameters for Congo red removal on Dabco-bent

Pseudo-first order		Pseudo-second order		Intraparticle diffusion	
q_e (mg g^{-1})	10.3121	q_e (mg g^{-1})	11.0358	K_{id}	0.5562
K_1 (min^{-1})	0.1126	K_2 (g mg^{-1} min^{-1})	0.0230	C	6.4659
R^2	0.8554	R^2	0.9871		

3.3. Adsorption isotherm study

The term adsorption isotherm is a way to explore the adsorption process through a graph that gives a correlation between the amount of the adsorbate and pressure when the system was in the equilibrium state with constant temperature. The investigations were done by conducting the adsorption experiment in various initial dye concentration with predetermined the other variable, i.e., contact time, temperature, and amount of the adsorbent. The result of the experiment was presented in Figure 6. As shown in the Figure, by increasing the initial dye concentration, the adsorption capacity was gradually increased. As reported in our previous study , this result was probably affected by the increase of the driving force due to the increasing the number of dye molecules in the solution. In the present work, the adsorption isotherm based on the experiment result was investigated by Langmuir and Freundlich model.

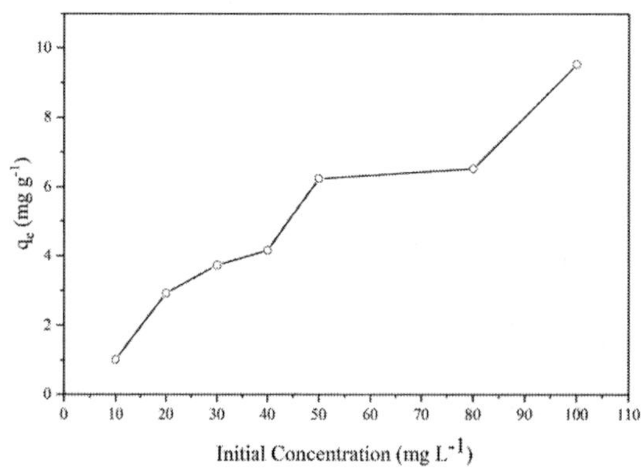

Figure 6. Effect of initial dye concentration

The Langmuir adsorption isotherm model is a model that describes the adsorption process as a monolayer sorption process. It means that the adsorbate adsorbed on the surface of adsorbent takes places in one active site of the adsorbent. The adsorption stops if the active site of the adsorbent has been covered by one adsorbate molecule or ion. The linear form of the Langmuir adsorption isotherm model is presented as follows:

$$\frac{C_e}{q_e} = \frac{1}{K_L q_m} + \frac{C_e}{q_m} \tag{7}$$

Where C_e is the concentration of the dye solution at the equilibrium state (mg L^{-1}), C_e is the concentration of the unadsorbed dye in the solution (mg L^{-1}), and the Langmuir constant (L mg^{-1}). The values of the Langmuir constant and the predicted maximum adsorption capacity could be determined from the intercept and the slope values of the plot C_e/q_e versus C_e.

The Freundlich adsorption isotherm model is the model that used to describe the sorption process on the heterogeneous surface with a multilayer sorption. The adsorption process is assume occurred in the heterogeneous surface, and the adsorbate molecule or ion is adsorbed on the surface of the adsorbent in the multilayer form [27]. The linear form of the Freundlich model is depicted in the following equation:

$$\log q_e = \log K_F + \tfrac{1}{n}\log C_e \quad (8)$$

Where KF (mg g^{-1}) is the Freundlich constant, and n is the constant refers to adsorption intensity. The $1/n$ values indicate the type on the adsorption based on the favorability, $1/n$ equals to zero means that the adsorption process is irreversible, adsorption process is favorable if the $1/n$ value more than zero and less than one, and unfavorable if the $1/n$ value more than one.

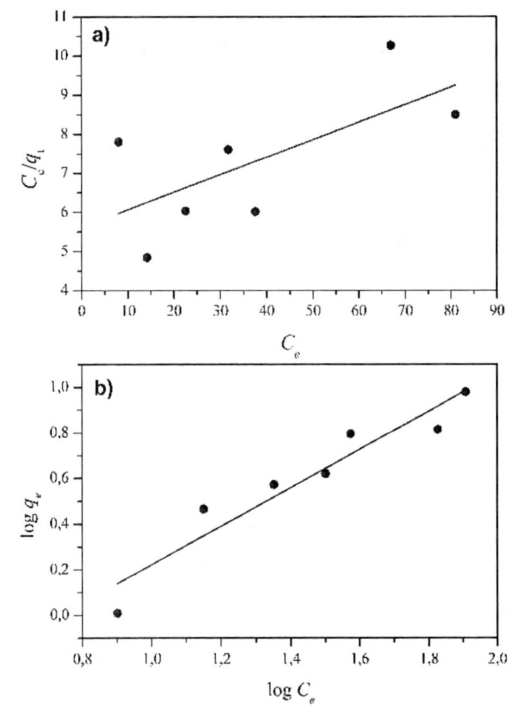

Figure 7. Fit of the experimental data to Langmuir a), and Freundlich b) isotherm model

The adsorption isotherm parameters based on the linear fitting of Langmuir and Freundlich model were summarized in Table 2. It can be seen that the best-fit adsorption isotherm model to describe the adsorption of Congo red on the Dabco-bent was Freundlich model with higher R^2 value than the Langmuir model. This finding is also in agreement with other results on dye adsorption on organo-modified clay [5,28]. This finding means that the adsorption process of Congo red on Dabco-bent was a nonideal adsorption mechanism occurred in the heterogeneous surface with a multilayer sorption form. Based on the linear fit graph of the adsorption isotherm (Figure 7) and the n value of the Freundlich model (Table 2) is higher than one, it can be concluded that the adsorption mechanism of the Congo red on Dabco-bent was dominantly a partition mechanism [5].

Table 2. Adsorption isotherm parameters for Congo red adsorption

Langmuir Isotherm model		Freundlich isotherm model	
q_m (mg g^{-1})	22.2816	K_F	0.2404
K_L (L mg^{-1})	0.0079	n	1.1893
R^2	0.4491	R^2	0.9118

4. Conclusion

In summary, the characterization of the natural bentonite grounded from Sarolangun District, Na-exchanged bentonite, and Dabco modified bentonite has been conducted using powder X-ray diffraction and FT-IR. The natural bentonite composition based on the XDR pattern result was montmorillonite with some associated mineral such as calcite, dolomite, and quartz. The characterization Dabco modified bentonite using powder XRD show that the peak of montmorillonite was shifted indicate that the Dabco molecule was successfully inserted into the montmorillonite interlayer space. The FTIR analysis of the Dabco modified bentonite represented that the specific band of the natural bentonite was not affected by the pillaring process and during the pillaring process the structure of bentonite host was not changed significantly. The application of Dabco-bent as an adsorbent for Congo red dye removal from wastewater shows that the Dabco-bent has a good adsorption capacity. The further study gives the result that the adsorption kinetic of Congo red on Dabco-bent fit well with the pseudo-second-order kinetic model. The result for adsorption mechanism study based on the adsorption isotherm model revealed that adsorption of Congo red on Dabco-bent follow the Freundlich isotherm model which is mean that the adsorption process was a non-ideal process.

Acknowledgements

Authors acknowledge Ministry of Research, Technology and Higher Education of Republic Indonesia (Kemenristek Dikti) for supporting this research through "*Program Magister Menuju Doktor Untuk Sarjana Unggul (PMDSU)*" Scholarship Grant with contract number 468/UN9.3/LT/2017 and 326/SP2H/LT/DRPM/IX/2016. .

References

[1] Gao X, Zhong H, Yao G, Guo W and Jin F 2016 *Catal. Today* **274** 49–54
[2] Gong Z, Liao L, Lv G and Wang X 2016 *Appl. Clay Sci.* **119** 294–300
[3] Vicente M A, Gil A and Bergaya F 2013 *Pillared Clays and Clay Minerals* vol 5(Elsevier Ltd.)
[4] Jeenpadiphat S and Tungasmita D N 2014 *Appl. Clay Sci.* **87** 272–7
[5] Yan L, Qin L, Yu H, Li S, Shan R and Du B 2015 *J. Mol. Liq.* **211** 1074–81
[6] Bilgiç C, Topaloğlu Yazici D, Karakehya N, Çetinkaya H, Singh A and Chehimi M M 2014 *Int. J. Adhes. Adhes.* **50** 204–10
[7] Taher T, Mohadi R, Rohendi D and Lesbani A 2017 *AIP Conf. Proc.* **1823** 20028
[8] Kaya M, Dilekog M F and Saka C 2016 1417–30
[9] Carriazo J, Guélou E, Barrault J, Tatibouët J M, Molina R and Moreno S 2005 *Catal. Today* **107–108** 126–32
[10] Vitanesa L, Purba R N, Lesbani A and Said M 2017 *Sci. Technol. Indones. Vol 2 No 1 January*
[11] Jeenpadiphat S and Tungasmita D N 2013 *Powder Technol.* **237** 634–40
[12] Cool, P. Vansant E F 1998 *Mol. Sieves* **1** 265–86
[13] Mohadi R, Hanafiah Z, Hermansyah H and Zulkifli H 2017 *Sci. Technol. Indones. Vol 2 No 4 Oct.*
[14] Ismadji S, Tong D S, Soetaredjo F E, Ayucitra A, Yu W H and Zhou C H 2016 *Appl. Clay Sci.* **119** 146–54
[15] Leemput L Van, Stul M S, Maes A, Uytterhoeven J B and Cremers A 1983 *Clays Clay Miner.* **31** 261–8

[16] Vhahangwele M and Mugera G W 2015 *J. Environ. Chem. Eng.* **3** 2416–25
[17] Djukić A, Jovanović U, Tuvić T, Andrić V, Grbović Novaković J, Ivanović N and Matović L 2013 *Ceram. Int.* **39** 7173–8
[18] Hao Y F, Yan L G, Yu H Q, Yang K, Yu S J, Shan R R and Du B 2014 *J. Mol. Liq.* **199** 202–7
[19] Jović-Jovičić N, Milutinović-Nikolić A, Banković P, Mojović Z, Žunić M, Gržetić I and Jovanović D 2010 *Organo-inorganic bentonite for simultaneous adsorption of Acid Orange 10 and lead ions* vol 47
[20] Jing P, Hou M, Zhao P, Tang X and Wan H 2013 *J. Environ. Sci.* **25** 1139–44
[21] Pawar R R, Lalhmunsiama, Bajaj H C and Lee S M 2016 *J. Ind. Eng. Chem.* **34** 213–23
[22] Wang L and Wang A 2008 *J. Hazard. Mater.* **160** 173–80
[23] Wang Y-Q, Zhang Z, Li Q and Liu Y-H 2012 *J. Radioanal. Nucl. Chem.* **293** 231–9
[24] Lagergren S 1898 *Handlinger* **24** 1–39
[25] Weber W J and Morris J C 1963 *J. Sanit. Eng. Div.* **89** 31–60
[26] Doulia D, Leodopoulos C, Gimouhopoulos K and Rigas F 2009 *J. Colloid Interface Sci.* **340** 131–41
[27] Bhatti H N, Jabeen A, Iqbal M, Noreen S and Naseem Z 2017 *J. Mol. Liq.* **237** 322–33
[28] Özcan A S, Erdem B and Özcan A 2004 *J. Colloid Interface Sci.* **280** 44–54

IOP Conf. Series: Materials Science and Engineering **299** (2018) 012056 doi:10.1088/1757-899X/299/1/012056

Study on the curie transition of P(VDF-TrFE) copolymer

Devi Eka Septiyani Arifin[1] dan J J Ruan[2,*]

[1]Department of Metallurgical Engineering, Faculty of Engineering, Jenderal Achmad Yani University, Bandung, Indonesia

[2] Department of Materials Science and Engineering, College of Engineering, National Cheng Kung University, Tainan, Taiwan

*Email: jrjeng@mail.ncku.edu.tw

Abstract. A systematic study was carried out to decipher the mechanism of Curie transition of piezoelectric crystals of poly(vinylidene fluoride trifluoroethylene) P(VDF-TrFE). The unique polarity of P(VDF-TrFE) crystalline phase below curie transition temperature is attributed to the lattice packing of all-trans molecular chains, which allocates all the substituted fluorine atoms on one side of molecular chains and hydrogen atoms on the other side. Therefore, a net dipole moment is created across the lateral packing of molecular chains. Nevertheless, due to the mutual repulsion among fluorene atoms, this all-trans conformation is not stable, and ready to change above Curie temperature, where thermal kinetic energy is sufficient to cause segmental rotation. As being illustrated by in-situ recorded X-ray diffraction and thermal analysis, the concerned curie transition is deciphered as a one-step process which is involved two process and this is different from conventional one-step solid-solid transitions. Accompanied with this one-step process during heating, the occurrence of lamellar bending is inferred for elucidating the decline of stacking regularity of crystalline lamellae, which reversibly recover during subsequent cooling. However, as the crystalline lamellae of P(VDF-TrFE) are confined in between the stacking of crystalline lamellae of PVDF, lamellar bending is restricted accordingly. As a result, a certain fraction of the piezoelectric crystalline lamellae was found to survive through the Curie transition. Thus, in addition to the suggestion of a one-step process as a new concept for understanding the Curie transition, the relationship between the lamellar stacking and transition of molecular packing is unveiled as well in this research.

Keywords-: Curie transition, lamellar bending, confinement effect

1. Introduction

Ferroelectric polymers have attracted lots of attention for decades due to the characteristics of their phase transitions. Among these polymers, poly(vinylidene fluoride) (PVDF) and its copolymer poly(vinylidene fluoride-trifluoroethylene) (P(VDF-TrFE)) are of particular interest. The unique ferroelectric feature of the PVDF and P(VDF-TrFE) crystalline phase is derived from the packing of molecular chains in an all-trans conformation, which arranges all the substituted fluorine atoms on one side of the molecular chain and the hydrogen atoms on the other side, thus producing a net dipole moment. The *trans* and *gauche* conformations are feasible to switch over within the ferroelectic phase transition, which is known as the Curie transition. Due to mutual repulsion among fluorine atoms, this all-trans molecular conformation is not stable. In order to lessen encountered steric hindrance, the *gauche* conformation is ready to develop above the Curie temperature, where thermal energy is sufficient to cause segmental rotation.

The ferroelectric polymer P(VDF-TrFE) has two crystalline forms, the low-temperature (LT) and the cooled (CL) ones. The LT form consist of an all-trans conformation with parallel arrangement of CF_2, as in the case of PVDF form I. In the CL form, the long trans segments consists of irregular trans-gauche conformation. This form can transform into the high-temperature (HT) form above the Curie temperature. The HT-form consists of TT, TG, and T$\bar{G}$ conformation, where the molecular chains are able to rotate through this trans-gauche conformation, as noted above.

As an attempt to widely enhance the use and stability of low-temperature piezoelectric crystalline form of P(VDF-TrFE) copolymer, this study focuses on further deciphering the mechanism of Curie transition of P(VDF-TrFE) copolymer. This research approach aims to establish the fundamental background of curie transition in order to learn essential knowledge needed for optimizing the use of precious piezoelectric feature of polymeric crystals.

2. Methodology

Blended samples of P(VDF-TrFE)/HMB and PVDF/ P(VDF-TrFE) were dissolved in N,N-dimethylformamide (DMF) (bp = 153 °C) for 1 day. The resulting, solution was then cast on a glass at 70 °C and the solvent evaporated immediately. The obtained film (1wt% P(VDF-TrFE), P(VDF-TrFE)/HMB 30/70 and PVDF/ P(VDF-TrFE) 30/70) was annealed at 80 °C (below Tc) for 1 hours and 120 °C (above Tc) for 45 minutes, then observed with DSC, 2D XRD, and SAXS. For the TEM sample, 2wt% P(VDF-TrFE)/HMB with a blend ratio of 2/98 was casted on glass at 70 °C and melted together. The HMB substrate was able to dissolved P(VDF-TrFE) at 180 °C. After melting, the sample was quenched by liquid nitrogen in order to grow P(VDF-TrFE) crystal first and then HMB crystal, and then annealed at 80 °C (below Tc) for 1 hours and annealed at 120 °C (above Tc) for 45 minutes.

3. Results and Discussion

3.1 Mechanism of curie transition of P(VDF-TrFE)

The DSC thermograms of VDF 70% - TrFE sample shows that two endothermic peaks are involved during Curie transition, including an endotherm shoulder and a sharp peak. Conventionally the curie transition is recognized as a kind of one-step transition of crystalline packing without involving the melting of original crystals. Thus the involvement of two endothermic processes during curie transition is curious, and further examinations were carried out via in-situ X-ray experiments, which will be discussed later. As the heating rate is increased, the occurrence of Curie transition shifted to higher temperatures. At the end of heating process, there is a very sharp endothermic melting peak. As the heating rate is increased, this melting peak shifts to higher temperatures. As the heating rate at 30 °C/min, two endothermic melting peaks were found. It appears that there is no sufficient time for the reorganization process, and thus crystalline domains of disparate stability are present.

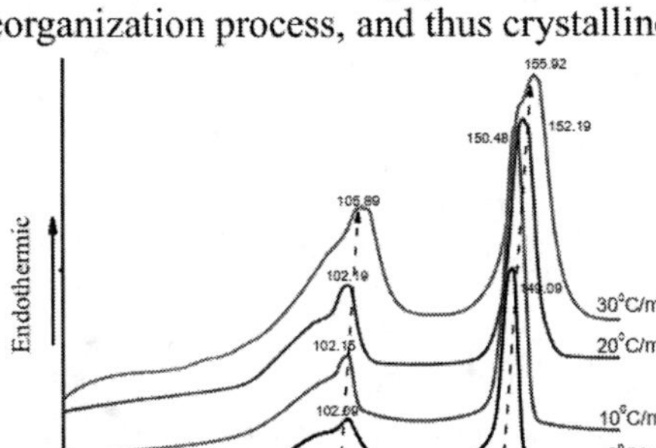

Figure 1. Thermal diagram of different heating rates for pure P(VDF-TrFE). The melting point and the Curie transition temperature of P(VDF-TrFE) occurs at higher temperature when increasing the heating rate.

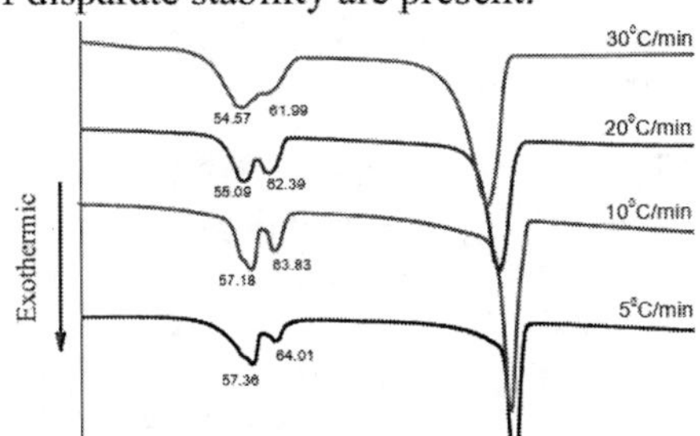

Figure 2. Thermal diagram of different cooling rates for pure P(VDF-TrFE). The crystallization point occurs at a higher temperature during the cooling process, and there are two exothermic peaks for the Curie

The cooling process of P(VDF-TrFE)-sample shows similar phenomena in Figure 2. Two exothermic peaks are involved during curie transition. After the occurrence of crystallization during cooling, further cooling results in a smaller exothermic peak first, which is followed by the presence of

IOP Conf. Series: Materials Science and Engineering **299** (2018) 012056 doi:10.1088/1757-899X/299/1/012056

a sharp exothermic peak. As the cooling rate is decreased, the Curie transition and crystallization temperature shifts to higher temperatures. The crystallization peak gradually become thinner as the cooling rate decreases. The sample has more time for crystallization at a lower cooling rate. During faster cooling processes, a wider temperature range is needed for the crystallization to finish, which is found as a common crystallization feature of polymeric materials.

The phase identification of P(VDF-TrFE) can be further studied by in-situ X-ray diffraction experiments. Figure 3 shows the in-situ X-ray diffraction profiles of pure P(VDF-TrFE) sample obtained during heating at 10 °C/min. This sample was prepared by solution casting and annealed at 80 °C for 1 hour afterwards. According to this result and based on the peak area, the curie transition can be classified as a one-step process during heating and cooling. The cooling process of pure P(VDF-TrFE)-showed similar phenomena. The (110)/(200) diffraction of the LT-form appears at 70 °C and the Curie transition during cooling begins with the gradual disappearance of the (001) diffraction of the HT-form, which finishes until reaching 50 °C.

This one-step process involved two process, the first process is the appearance of (001) diffraction of HT form and followed by the increase of lattice dimension along the *c*-axis. The second process is the rotational motion of molecular chains increases with the increase of temperature, which cause the disappearence of (110)/(200) diffraction of LT-form.

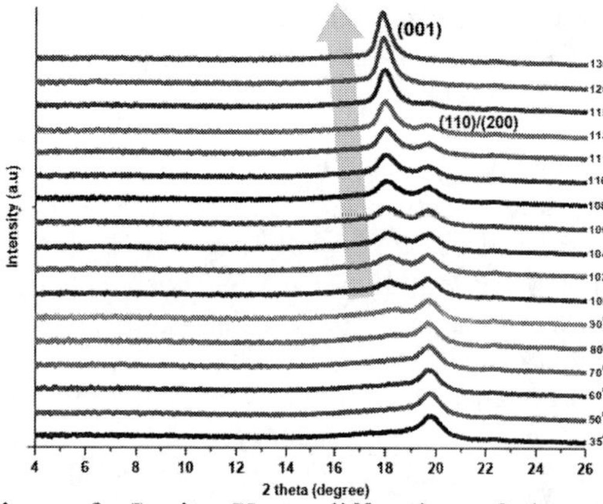

Figure 3. In-situ X-ray diffraction of the pure P(VDF-TrFE) profiles during heating at 10 °C/min. The arrow indicates the shift of the (001) diffraction peak of HT form during heating.

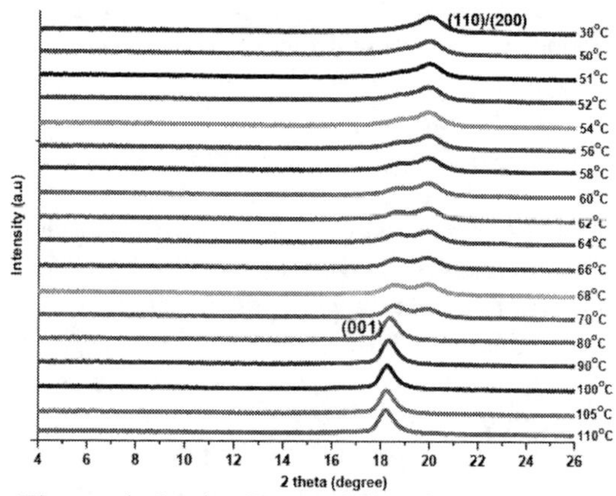

Figure 4. In-situ X-ray diffraction profiles of pure P(VDF-TrFE) sample during cooling at 10 °C/min.

3.1.1 The influence of curie transition on the evolution of lamellar stacking

For the VDF 70%-TrFE-sample without annealed, annealed at 80 °C and 120 °C for 1 hour and 45 minutes, the small angle x-ray scattering (SAXS) experiment was conducted to examine the influence of the Curie transition on the evolution of lamellar stacking. For the P(VDF-TrFE) sample without annealing process, as shown in Figure 5, a sharp peak was observed, indicating the established regularity of lamellae stacking upon annealing. As the temperature increased and approached around 115 °C, which is about the Curie transition temperature, the scattering peak shift to lower angle accompanied with the decrease of scattering intensity. The decrease of scattering intensity reflects the decline of stacking regularity. Hence the lamellar stacking was recognized to become less regular upon the occurrence of Curie transition.

Upon subsequent cooling process, the intensity of scattering peak increases again as the temperature approached the Curie transition temperature around 70 °C. Thus a reversible restoration of regular lamellar stacking is recognized. According to this result, irregular lamellar bending is proposed to occur during the Curie transition, which account for the decline of stacking regularity of crystalline lamellae, and thus the decrease of scattering intensity. The transformation of lattice packing is thus now unveiled capable of influencing the stacking of crystalline lamellae.

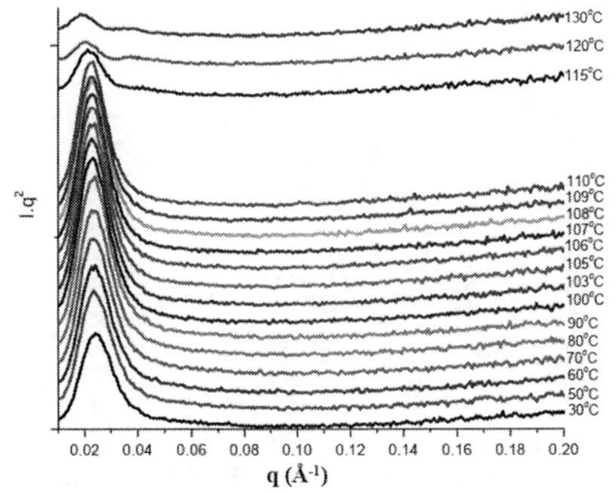

Figure 5 In-situ SAXS of pure P(VDF-TrFE) without annealing process during heating at 10 °C/min.

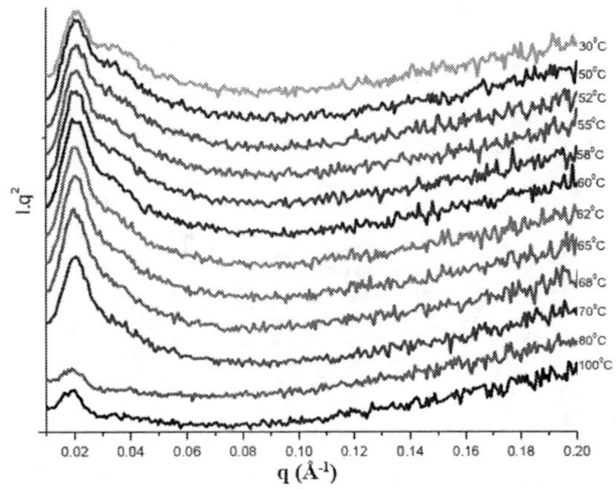

Figure 6 In-situ SAXS of pure P(VDF-TrFE) without annealing process during cooling at 10 °C/min.

Figures 7 and 8 show the in-situ SAXS experiments of the P(VDF-TrFE) annealed at 80 °C, which is below the Tc. A sharp peak was observed below curie transition and a new peak, shoulder peak, was observed above curie transition. This results is similar with without annealing process, but the shoulder peak was more higher when the sample annealed at 80 °C.

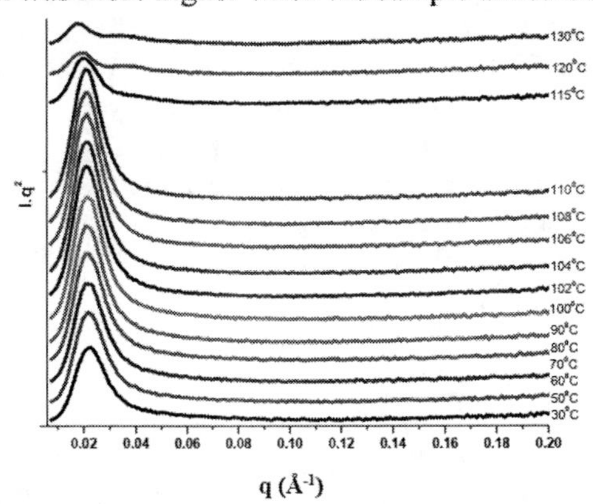

Figure 7 In-situ SAXS of pure P(VDF-TrFE) during heating at 10 °C/min. The sample was annealed at 80 °C for 1 hour.

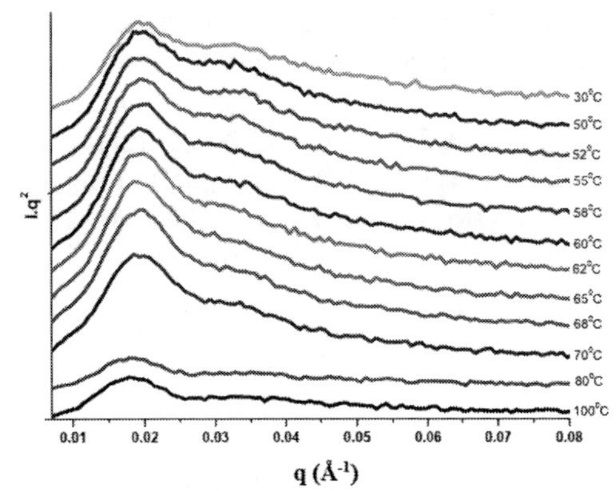

Figure 8 In-situ SAXS of pure P(VDF-TrFE) during cooling at 10 °C/min. The sample was annealed at 80 °C for 1 hour.

Figures 9 and 10 show the in-situ SAXS experiments of the P(VDF-TrFE) annealed at 120 °C, which is above the Tc. A sharp scattering peak is observed to slightly overlap with another scattering shoulder at higher angle. During the heating process at 10 °C/min, the intensity of scattering peak was found to increase first with the progressive intensity decay of scattering shoulder. The increase of SAXS intensity indicates the lamellar stacking becomes more regular, which appears accompanied with the melting of thinner lamellae based on the disappearance of scattering shoulder.

After reaching 110 °C during heating, the SAXS intensity decreases with the restoration of scattering shoulder at 115 °C. This result means the stacking of crystalline lamellae become less regular due to the occurrence of curie transition. As the temperature was decreased upon cooling process, the intensity increased again due to the restoration of regular lamellar stacking. The cooling process appears to recover the stacking of thinner crystalline lamellae also, as being indicated by the reversible presence of scattering shoulder.

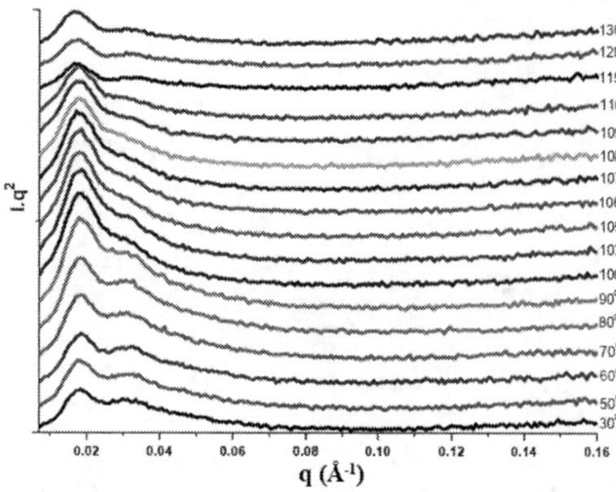

Figure 9 During heating at the rate of 10 °C/min, in-situ SAXS scattering profiles of P(VDF-TrFE) sample annealed at 120 °C for 1 hours are illustrated

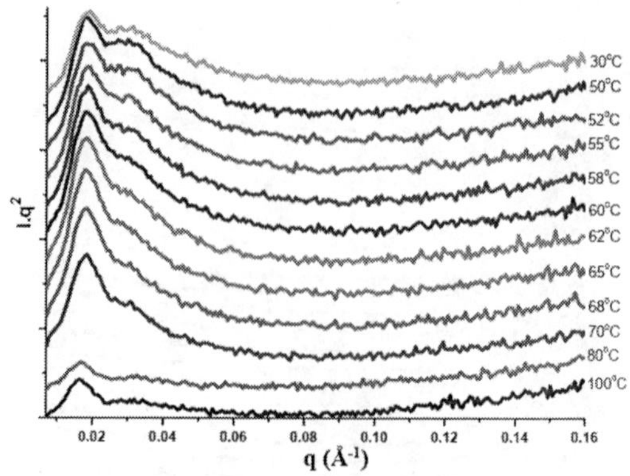

Figure 10 In-situ SAXS of Pure P(VDF-TrFE) during cooling at 10 °C/min. The sample was annealed at 120 °C for 1 hours.

3.2 Confinement effect on the melting of P(VDF-TrFE) crystals in binary mixtures

In order to explore the impact of the mixing ratio on the phase stability, both PVDF and P(VDF-TrFE) molecules were mixed together in different ratios. As shown in figure 11, multiple endothermic peaks were found. For the mixture containing 70 wt% of P(VDF-TrFE) component, the Curie transition of P(VDF-TrFE) results in one endothermic peak and one shoulder around 90 and 104 °C upon heating. The endothermic shoulder begins at 70 °C and overlaps with the second peak of the Curie transition. The PVDF component does not have a Curie temperature. The melting temperature of P(VDF-TrFE) crystals shifts to higher temperatures upon the increase of PVDF content, but the melting point of PVDF remains at 168 °C. Regarding the increase of melting temperature of P(VDF-TrFE) crystals upon the increase of PVDF content, the confinement effect yielded by surrounding PVDF crystals is proposed. As P(VDF-TrFE) crystalline lamellae are sandwiched in between PVDF crystals, the melting of P(VDF-TrFE) crystals is likely to be delayed and occur at higher temperatures.

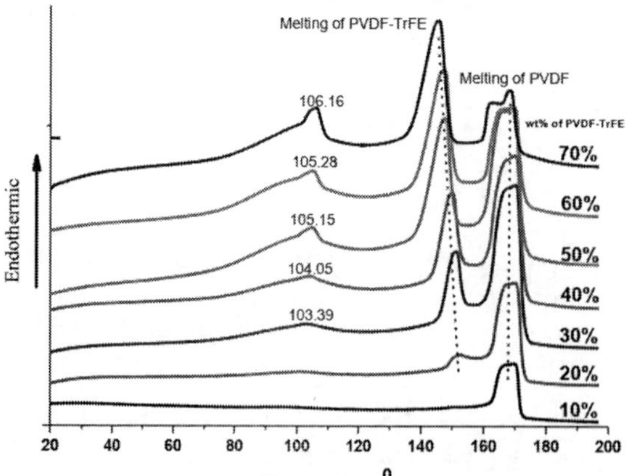

Figure 11 Thermal diagram of polymer blends with various compositions obtained via heating scans at 10 °C/min.

3.2.1 The influence of curie transition on the evolution of lamellat stacking PVDF/P(VDF-TrFE)

For the binary mixture PVDF/P(VDF-TrFE) samples annealed at 80 °C for 1 hour. As shown in Figure 12, in addition to the sharp scattering peak located at a lower angle, one broader scatter peak at higher angle was observed also. Figure 12 shows the increase of SAXS intensity of the sharper peak first upon heating, indicating the enhancement of stacking regularity of crystalline lamellae. As the temperature approached to 115 °C, the sharper scattering peak shifts to lower angle with the decrease of intensity, indicating the lamellar stacking become less regular upon Curie transition. Oppositely, the broader scattering peak becomes stronger above Curie transition. During the cooling process, the SAXS intensity of scattering peak at lower angle increased again after passing 70 °C, indicating the restoration of regular lamellar stacking. In contrast, the SAXS intensity of second scattering peak decreased as the sample is cooled below 70 °C.

The presence of the broader scattering peak is likely attributed to the less regular stacking of thinner crystalline lamellae of the minor PVDF component. Considering the crystallization history of mixture, thinner PVDF crystalline lamellae are derived to stack randomly in between thicker crystalline lamellae of copolymer. These thinner PVDF crystalline lamellae are all of nonpolar phase, which do not have curie transition. With the occurrence of curie transition during heating, the stacking of thicker copolymer crystalline lamellae becomes less regular and therefore occupies a larger amount of volume, which causes these thinner PVDF lamellae to stack in a closer manner. Thus the stacking of PVDF thinner lamellae becomes more regular above the Curie transition temperature, yielding a stronger scattering peak. As the sample is cooled below the Curie transition temperature, a certain fraction of free volume is restored via the regular stacking of copolymer lamellae. Thus the stacking of thinner PVDF crystalline lamellae become less regular again.

Thus these results shown on the Figure 12 and 13 are deciphered as a result of random stacking of thinner PVDF crystalline lamellae in between crystalline lamellae of copolymer, and also the impact of curie transition.

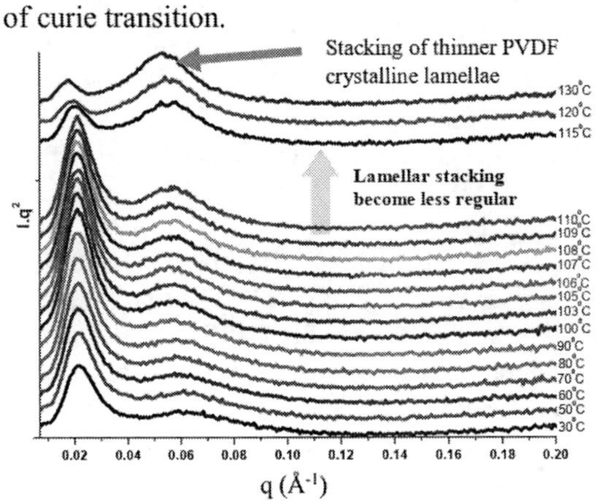

Figure 12 In-situ SAXS of PVDF/P(VDF-TrFE) mixture of the mixing ratio of 30/70 during heating at 10 °C/min.

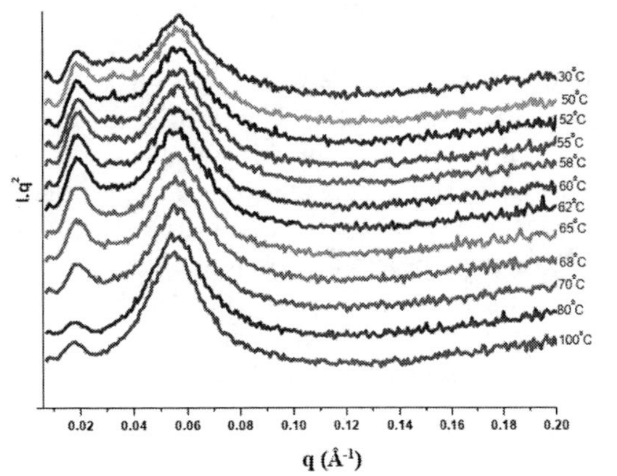

Figure 13 In-situ SAXS of PVDF/P(VDF-TrFE) 30/70 during cooling at 10 °C/min.

3.3 Curie transition of binary mixture P(VDF-TrFE)/HMB

The in-situ X-ray profile of the 30/70 mixture of P(VDF-TrFE) /HMB after annealing at 80 °C for 1 hour are shown in Figures 14 and 16 during heating and cooling. Consistent with in-situ x-ray of pure P(VDF-TrFE)-result, there are one-step process also for the mixture of P(VDF-TrFE)/HMB 30/70 which is involved two process. The first process is (001) diffraction of HT form appears at 90 °C, where this peak is 10 °C more earlier than pure P(VDF-TrFE). Nevertheless, the second process is the (110)/(200) diffraction of LT form is suddenly disappears at 100 °C, 15 °C different with (110)/(200) diffraction of LT form pure P(VDF-TrFE) and the (001) diffraction of HT have more sharp peak than pure P(VDF-TrFE). This result is consistent with DSC data.

During cooling process, Figure 15 shows the curie transition is happened from HT form to LT form. The (001) diffraction of HT form gradually disappear and the intensity decrease during cooling process. Indicate the disorder molecular structure of P(VDF-TrFE)-due to the HT form was randomly rotate along chain axis. The (110)/(200) diffraction of LT form gradually appears at 70 °C during cooling process. Indicate the order of all-trans conformation of LT form. In-situ X-ray experiment shows that reversible phase transition was happened in this process.

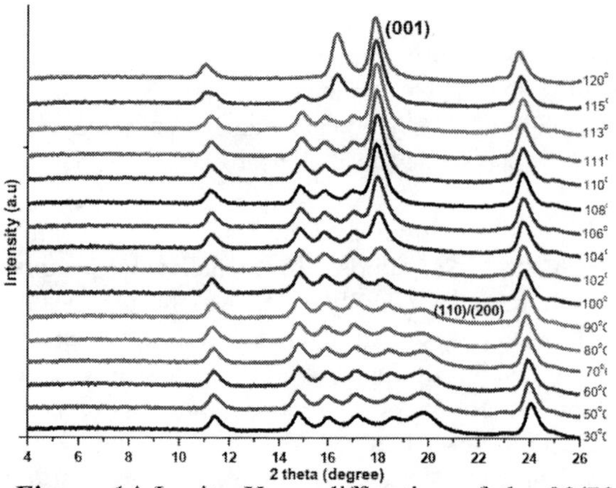

Figure 14 In-situ X-ray diffraction of the 30/70 mixture of P(VDF-TrFE)/HMB profiles during heating at 10 °C/min. The two different phases of P(VDF-TrFE)-appear in different ways as the temperature increases.

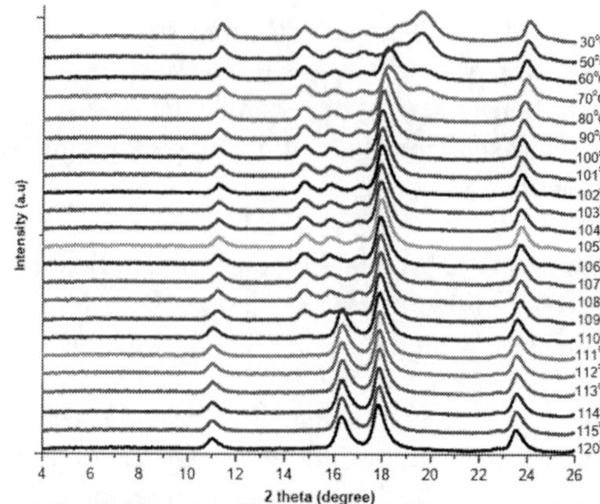

Figure 15 In-situ X-ray diffraction of the 30/70 mixtures of P(VDF-TrFE)/HMB profiles during cooling at 10 °C/min. The two different phases of P(VDF-TrFE)-appear in different ways as the temperature increases.

3.3.1 Epitaxial organization of P(VDF-TrFE) on HMB

A transmission electron microscope (TEM) was used to explore the epitaxial relationship between P(VDF-TrFE) and HMB. The HMB melt was able to dissolve semicrystalline P(VDF-TrFE). The P(VDF-TrFE) and HMB were dissolved in the dimethylformamide (DMF) and melted together when kept at 180 °C for 20 seconds and then directly quenched. The sample was annealed at different temperatures and then stored in a vacuum to evaporate the HMB crystals. While the HMB is included as the major component in the prepared eutectic mixture, the HMB crystalline platelets occurs first from molten state below 165 °C, which serve as the substrate for later epitaxial organization of P(VDF-TrFE). The quenching process at room temperature is able to preserve the epitaxial organization of P(VDF-TrFE). Epitaxy of P(VDF-TrFE) on HMB during quenching can be seen in Figure 17(a), P(VDF-TrFE) crystallized on HMB without annealing. Upon the development of P(VDF-TrFE) packing, arrays of oriented stripes were observed by TEM which is the crystals growing in two directions. The orientation of these stripes are marked by yellow arrow. The HMB crystalline platelets indicated by the white arrow, which is related to the fast growth direction. Figures 17 (b – d) show the epitaxy of PVDF-TrFE on HMB during annealing process. When the samples annealed at 60 °C, curving lamellae has been developed. Nevertheless, lamellae become more regular and very straight when the sample annealed at 80 °C. At 120 °C, lamellae continuously grow on HMB and thicker lamellae has been observed during this annealing process. These results are consistent with what have been observed from SAXS data, lamellar bending was observed above curie temperature, indicating a-very wavy lamellae at 120 °C.

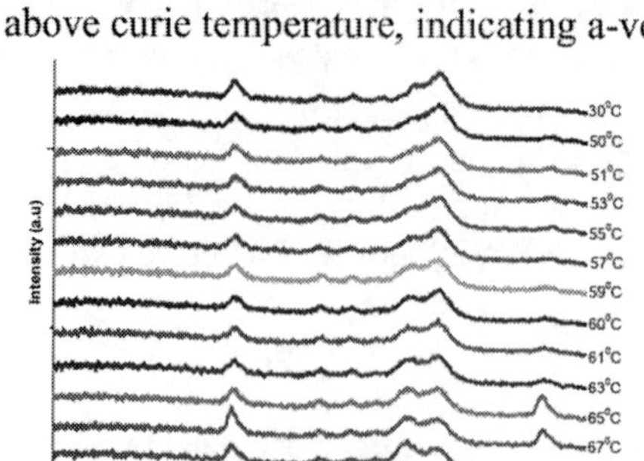

Figure 16 In-situ X-ray diffraction of the 30/70 mixtures of P(VDF-TrFE)/HMB profiles during cooling at 70 °C-30 °C.

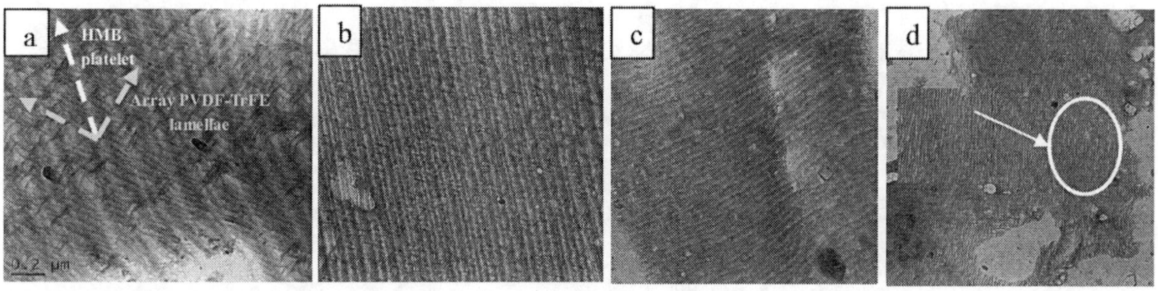

Figure 17 Electron micrograph of epitaxy P(VDF-TrFE) on HMB during (a) quenching (b) annealing at 60 °C (c) annealing at 80 °C and (d) annealing at 120 °C.

4. Conclusion

The curie transition of P(VDF-TrFE) copolymer has been investigated with a wide range of heating-cooling temperature. The Curie transition is deciphered as a one-step process which is involved two process and this is different from conventional one-step solid-solid transition. The first-process of curie transition: The increase of lattice dimension and intramolecular order and the second-process of curie transition: The decrease of intermolecular order within larger unit cells. The transformation of lattice packing upon Curie transition is accompanied with the adjustment of lamellar stacking. The presence of PVDF crystalline lamellae is able to stabilize the LT-form of P(VDF-TrFE). The stacking of thinner PVDF crystalline lamellae is influenced by the curie transition of surrounding crystalline lamellae of copolymer. Finally, epitaxy effect causes the LT-form of P(VDF-TrFE) suddenly disappears at Curie transition temperature.

5. References

[1] G. Teyssedre, A. Bernes, and C. Lacabanne 1995 Cooperative movements associated with the Curie transition in P (VDF-TrFE) copolymers, *Journal of Polymer Science Part B: Polymer Physics,* **vol. 33**, pp. 879-890
[2] Mayergoyz and G. Bertott 2005 Hysteresis in Piezoelectric and Ferroelectric Materials, *The Science of Hysteresis,* **vol. 3**, pp. 338-465
[3] P. Dobis, J. Bruestlova, and M. Bartlova 2010 Curie Temperature in Ferromagnetic Materials and Visualized Magnetic Domains, *3rd International Symposium for Engineering Education,* University College Cork, Ireland.
[4] R. Tanaka, K. Tashiro, and M. Kobayashi 1999 Annealing effect on the ferroelectric phase transition behavior and domain structure of vinylidene fluoride (VDF)–trifluoroethylene copolymers : a comparison between uniaxially oriented VDF 73 and 65% copolymers, *polymer,* **vol. 40,** pp. 3855–3865
[5] K. Tashiro and H. Hama 2004 Structural changes in isothermal crystallization processes of synthetic polymers studied by time-resolved measurements of synchrotron-sourced X-ray scatterings and vibrational spectra, *Macromolecular research,* **vol. 12,** pp. 1-10
[6] R. Gregorio and M. M. Botta 1998 Effect of crystallization temperature on the phase transitions of P (VDF/TrFE) copolymers, *Journal of Polymer Science-B-Polymer Physics Edition,* **vol. 36,** pp. 403-414
[7] A. Prabu, J. S. Lee, K. J. Kim, and H. S. Lee 2006 Infrared spectroscopic studies on crystallization and Curie transition behavior of ultrathin films of P(VDF/TrFE) (72/28) *Vibrational Spectroscopy,* **vol. 41**, pp. 1-13

IOP Conf. Series: Materials Science and Engineering **299** (2018) 012057 doi:10.1088/1757-899X/299/1/012057

Synthesis carbon foams prepared from gelatin (CFG) for cadmium ion adsorption

M. Ulfa[1] And D.K. Ulfa[1]
[1]Department of Chemistry Education, Facultyof Teacher Training and Education Sebelas Maret University
Ir. Sutami Street 36A Surakarta, Center of Java Indonesia 57126
E-mail: ulfa.maria2015@gmail.com

Abstract : In this paper, carbon foam from gelatin (CFG) was synthesized by acid-catalyzed carbonization of gelatin solution on mild condition by the simple method. Gelatin (Ge) were used as sacrificial template and source of carbon. Sulphuric acid was used as acid catalyst. Carbon foam CFG sample were characterized by scanning electron microscope (SEM), nitrogen adsorption desorption and FTIR for knowing textural and structural properties of the sample. Carbon foam CFG sample demonstrated macro pipes-channel like with pore size that varies between 30-40 μ and surface area in 60-100 m^2g-1. The carbon foams CFG sample were tested by using adsorption process for obtained their performance for decreasing Cd(II) ions from aqueous solutions. The adsorption capacities for cadmium was 46.7 mg/g obtained by using adsorbent dose 50 mg, initial concentration 50 ppm, contact time, 3 h; room temperature, stirring rate 150 rpm) which reached equilibrium at 55 min. Adsorption process fits using using Lagergren and Ho and McKay equation and measuring data
Keywords: carbon foam CFG, texture, structure, adsorption, cadmium ion, capacities

1. Introduction

Carbon foam as new generation of material applied in wider field such as adsorbent, catalyst, electrical material, radiator, fire resistant composite and electrodes [1]. Carbon foam have unique physical character such as high thermal stability, high thermal conductivity, large pore, large volume and high surface area. The high surface area of carbon foam playing urgently role as supporting part in catalysis and adsorption ter. The traditional technique for praparation of carbon foam using glassy and graphite step. The other method for carbon foam preparation explored by hard templating, carbonization, compression and soft templating. Some of source of carbon has been used on the previous paper such as coal, pitch carbon from petroleum aldehyde, sucrose, phenol, resin and polyimide, furfuril alcohol [2-3]. The using of gelatin as source of carbon foam very rarely investigated. Gelatin is alternative source of carbon foam due to the polymer contain with high of amine chain.

The carbonization of carbon foam source produce the unique character as graphitisable material [2]. The carbon with glassy part, graphitisable and have insulation character obtain after carbonization step. The technology to produce carbon foam develope rapidly in last decade trough hard templating methode [2-4]. However, this technique unfavourable due to the high cost and less effisient. The other side, the using soft templating method to replace hard templating method started to research in few years[3].

The present work aimed at producing carbonaceous porous matrices for hosting phase-change materials for seasonal heat storage, using a "green" precursor. Carbon foam from gelatin (CFG) was synthesized by acid-catalyzed carbonization. Characterization of Carbon foam CFG using scanning electron microscope (SEM), nitrogen adsorption desorption and FTIR Then were tested by using adsorption process which is fits using using Lagergren and Ho and McKay equation and measuring data. The adsorbate concentration is not influence the adsorption the desorption [9] but Ho and Mckay told that the adsorption rate is linearly related to the square of the number of adsorption vacant sites [10]. The equation of Lagergren and Ho and McKay describe in Table 4.

2. Experimental section

2.1 Materials

All chemicals including sulfuric acid, deionized water, gelatin and cadmium ion solution were purchased from Sigma Aldrich used without any purification. Gelatin technical grade (medium molecular weight (200 kDa) was purchased from Gel-Pro Australian supplier.

2.2 Instrumentation

The investigation of carbon foam samples using Philips X'Pert-MPD diffractometer by radiation of Cu Kα (k = 1.5418 Å). The operation of Cu anode at 40 kV and 30 mA. The isotherms adsorption desorption nitrogen of carbon foam measured by AUTOSORB-1 instrument (Quantachrome Co.) at -196 °C. The surface area and pore size of carbon foam sample measured by Brunauer–Emmet–Teller (BET) and Brunauer–Joyner–Halenda (BJH) method. The surface area were derived from nitrogen uptake from BET model at relative pressure 0-0.99. The pore size distributions were derived from the adsorption branches of the isotherms by using the Barett–Joyner–Halenda (BJH) model. The total pore volumes (Vt) were obtained from the amount of nitrogen uptake at P/P0 of 0.99

The images of carbon foam sample recorded by scanning electron microscope (SEM, Hitachi H800) combine with Electron Dispersive Spectrometer (EDX) to verify the presence of elemental inside the carbon foam samples. The analysis of functional group conducted at range 400–4000 cm^{-1} by Fourier transform infrared spectroscopy (FTIR) (Shimadzu spectrometer 2800).

2.3 Procedure: Preparation of Carbon foam from Gelatin (CFG)

Carbon foam from gelatin, designated as CFG, was prepared using gelatin as a structure-directing agent and carbon precursor. Gelatin CFG was synthesized by modified Pechini method [8]. At first, 0.01 mol sulfuric acid and 100 mL distillation water were mixed carefully. The solution of gelatin prepared by adding 1.5 g gelatin into sulfuric acid solution followed by stir process 60 min. The first mixture prepared by adding, 1.0 g gelatin into the solution of gelatin. The first mixture homogeninized at 150 rpm and heated at around 90 °C for 45 min. The foam like gel obtained by heateding process at 150 °C in oven for 15 h. Then, the next process followed by calcination at 900 °C with argon gas flow for 3 h. The synthesized CFG obtained by filtration with deionized water, and drying over night in oven at 100 °C.

2.4 Adsorption of cadmium ion into Carbon Foam of Gelatin (CFG)

The adsorption evaluation of cadmium ion as pollutant model using carbon foam from gelatin (CFG) as adsorbent was investigated by using the fixed-bed reactor. The test prepared by continuous flow term using column with a diameter and a length of 2.5 cm and 15 cm, respectively. The weight of adsorbent was 30 mg of carbon foam from gelatin (CFG) completed with paper filter in bottom of column(Whatman, USA) The cadmium ion as pollutant model was prepared by mixing the initial solution of cadmium ion and the effluent. The cadmium ion as pollutant model was adsorbed through the column at room temperature. The sampling effluent was obtained at range 5 to 800 minute (1 mL/5 minute) followed byn filtration step. After filtration, cadmium concentrations analyze using Atomic Absorption Spectrophotometry (AAS, Shimadzu AA-6300). The relationship between the dissolved K concentration in cadmium solution ($C_{solution}$, mg K L^{-1}) and the cadmium sorbed K concen- tration

IOP Conf. Series: Materials Science and Engineering **299** (2018) 012057 doi:10.1088/1757-899X/299/1/012057

(C_{Cd}, mg K kg^{-1}) can be described by either using Lagergren and Ho and McKay equation and measuring data

3. Result and Discussion

Figure 1 shows the typical picture of a single broken carbon foam from gelatin (CFG) that clearly has a foam-like structure by the simple camera and SEM. The SEM graph showed that carbon foam from gelatin (CFG) have foaming part after calcination step as first prediction. These phenomenon represented by sulfuric acid role that catalysis foam forming in gelatin chain during high temperature treatment. The resulting SEM image of carbon foam from gelatin (CFG) containing hard texture, soft line surface and disordered puzzle-like shape. The three dimension images of foam shape and texture exhibited by photograph are shown in Fig. 1. The photograph result show that the diameter size of CFG an average of ~100 μm, are match with sulfuric acid as catalyst factor. For a sum, the morphology of CFG is a mixture of puzzle-like pipe, disordered soft surface, and paralelical shapes. The texture of CFG surfaces not only exhibited by photograph but also SEM micrograph as shown in Fig 2. Water on gelatin vaporated by heat treatment and foaming process generate crack pore with diameter up to 10 μm within whole surface.

Gelatin structure as carbon precursor is very important to control characteristic of the carbon foam pore system. The Large pore of carbon foam obtained by reduction process of gelatin molecule which is hydrocarbons convert to elemental via losing of hydrogen, oxygen, and nitrogen atoms. The loss of this element from gelatin molecule describes the mass loss of gaseous species. The carbonization process cause mass loss of these gasses such as hydrogen, water, ammonia. Gelatin as organic material at high temperature decompose to be small molecul and element that generate microporous part with hard texture and easy broke in large dimension. However, the large molecules such water and amonia will be quickly decomposed by carbonization then can create macroporous part in CFG. The fact about microporous and macroporous part geenrated after gelatin decomposition shows that CFG could be a new carbon matrix woth unique molecular size and shape. The size and shape of porous carbon CFG predict appeared by the weight loss of water and ammonia as huge polar molecules in gelatin which is showed at table 1.

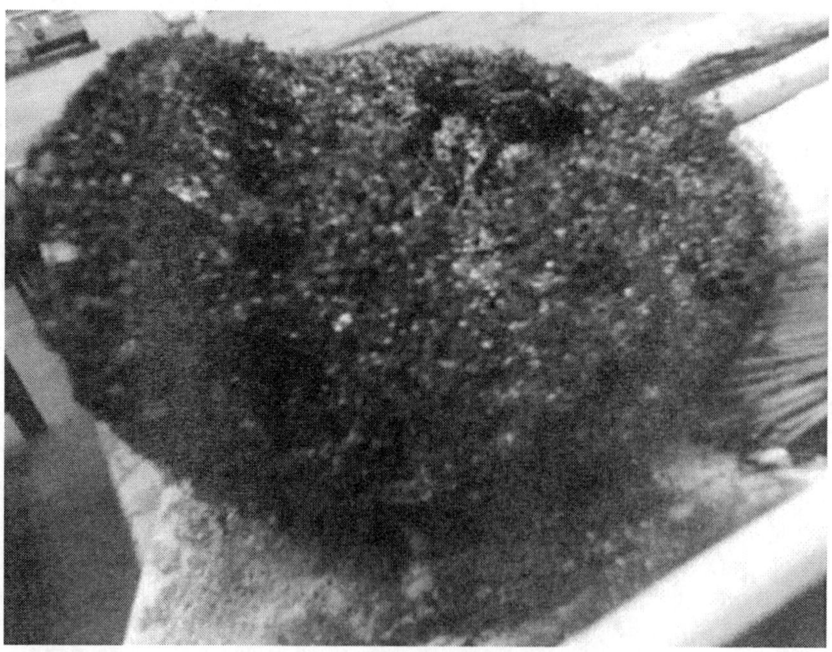

a.

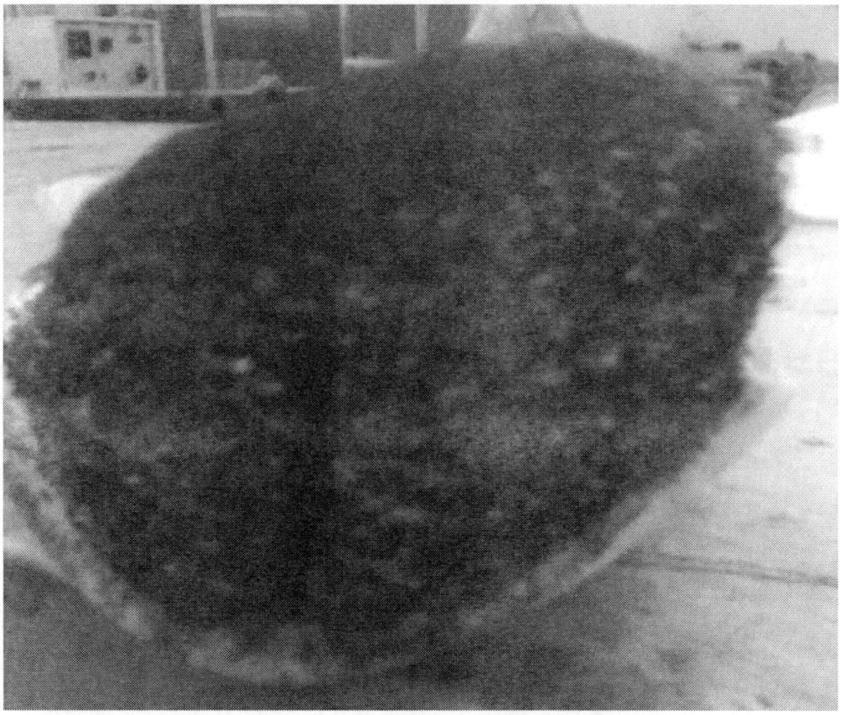

b.

Figure 1 Carbon Foam by Photograph a. Perpendicular view and b. paralel view

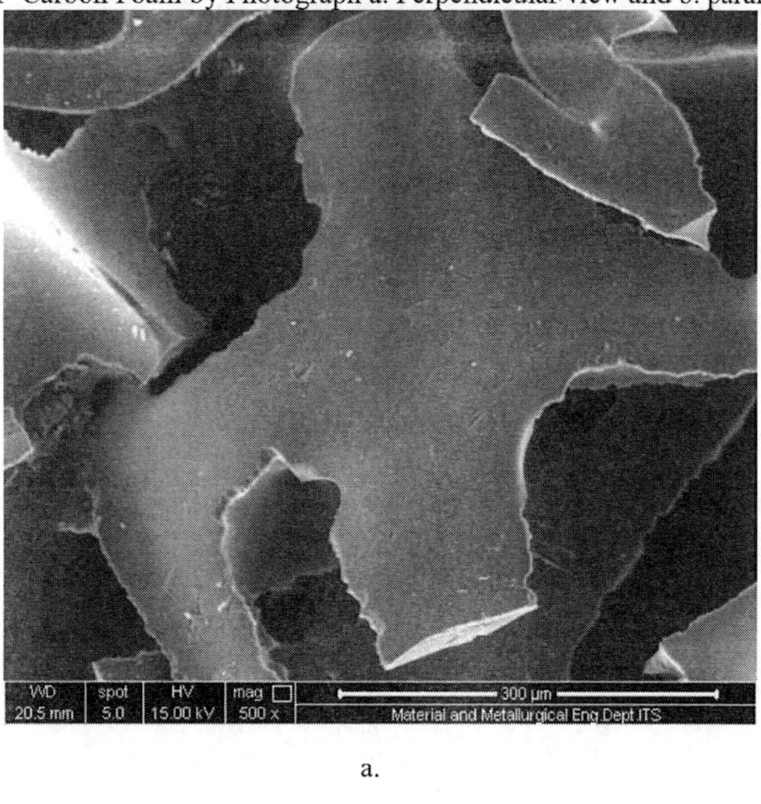

a.

b.

Figure 2 – SEM images of Carbon Foam of gelatin nanoparticles with magnification (a) 300μm and (b) 100μ

Table 1. The Trend of Weigth loss obtained by DTG Intrument on carbon foam sample

Temperature (°C)	% Wt	% Total Wt loss	Molecule assignment
0	100	-	-
0-100	88	12%	Water
100-150	75	25%	Water, amine
200-500	65	35%	Water, amine, oxygen, hydrogen
500-1300	5	95%	Water, amine, oxygen, hydrogen, carbon

Different thermogravimetry (DTG) measurements Table 1 and energy dispersive instrument exhibit that carbon and oxygen are the major elements of CFG. In the other side, gold element generate from Au grains that increase conductive and eletric character after coating step. In the major peak of C(79.9%) and O (18.9%) on carbon foam sample, indicating that the resulting carbon foam composed of carbon and oxygen phase as same as the previous research by Ge [8]. The mass loss of nirtogen from CFG shows decomposition process from gelatin during carbonization step. In addition, natrium peak in EDX spectrum show that preparation of gelatin using NaOH can create small presence of natrium still keep in gelatin chain during carbonization. The result of the elemental analysis shows that nothing sulfuric residue on CFGsuggests the presence of interaction between gelatin and sulfuric acid only on the preparation step but nothing on carbonization. The morphology of CFG describes of large numbers of hydroxyl groups presenting in gelatin inevitably interact with the negative charge of sulfuric acid by hydrogen bonding. That is the gelatin molecules should dominantly locate at the interface between sulfuric acid and gelatin chain (shown in Fig. 1). Some gelatin molecules may interact with sulfuric acid then sulfuric acid penetrates through the side chain of amino acid on whole gelatin molecule.

The CFG were examined by thermal gravimetric analysis (TGA) under nitrogen atmosphere. The TGA Table 1 exhibits a peak of large endothermic part at 110 °C that represent of water mass loss. These endothermic part was not only represent loss of twenty percent of water but also the deformation of helical structure and other bonding between gelatin molecules and catalyst. The melting temperature exhibit at 230 °C indicate the endothermic part appears [13]. The tranformation of tertiary structure of gelatin occured at range of oxidation and melting step. The splitting of peptide linkages of proteins starts at above 290 °C, and is indicated by several the exothermic peaks at 290 °C represent the peptide linkaged split within whole protein [7]. DSC information (not shown) shows the total weight loss at range 20 wt% up to 200 °C due to the water mass loss. In the other part, the weight loss of about 95 wt% from 200 °C to 1300 °C was associated the mass loss of majority of molecule and element high thermal treatment that generate carbon with high stability. As a sum, Table 1 explain about the huge change structure of during carbinization step that create some unique texture as SEM information due to the mass loss of water, amine, hydrogen and plenty of carbon as predicted before.

The FTIR reflectance spectrum of the CFG is showed in Table 2. The C–H vibrations showed by bands at 698, and 935 cm^{-1} with small intensity. The carboxylic groups, CH2 and CH3 wave number observe at 1154 and 1454 cm^{-1}The bands at 1524, 1541 and 1590 cm^{-1} represented to C=C and C=O vibrations in carbonyl groups respectively. The vibrations of –CH2 or –CH3 groups observed at at 1154 and 1454 cm^{-1}. The C–O stretching vibrations are represented by band at wave number 100–1300 cm^{-1}[14-16]. The difficulties of assign the bands in the range of 100–1300 cm^{-1} due to the overlapping of functional groups. The C–O–C stretching vibrations observed at at 1172 and 1268 cm^{-1} can indicated presence some functional group of alcohol, phenol and carboxyl. The band at 1103 cm^{-1} may be due to O–H. For a sum, the oxygen presented by spectra infra red explain about the mass loss of water which is match with TGA information

Table 2. Peak assignment of FTIR spectra of carbon foam gelatin

Peak assignment	Peak	Intensity	Corr. Intensity	Base (H)	Base (L)	Area
C–H vibrations	316.33	29.004	11.929	324.04	293.18	11.629
C–H vibrations	347.19	11.632	62.832	362.62	339.47	12.123
C–H vibrations	478.35	25.331	2.041	555.5	424.34	75.774
C–H vibrations	648.08	28.457	0.244	848.68	640.37	106.355
C–H vibrations	925.83	32.588	0.98	972.12	856.39	55.734
carboxylic	1080.14	32.486	1.333	1249.87	979.84	129.438
C–O stretching	1350.17	32.944	0.162	1357.89	1257.59	47.545
C=C vibrations	1627.92	30.131	0.777	1643.35	1558.48	43.074
C=O vibrations	1851.66	31.547	0.286	1859.38	1820.8	19.172
CH2	2337.72	24.701	0.995	2353.16	1936.53	223.817
CH3	2862.36	22.666	0.082	2870.08	2399.45	283.104
C–O–C stretching	2931.8	21.742	0.356	2947.23	2870.08	50.386
vibrations of alcoholic	3425.58	14.668	0.125	3433.29	2985.81	327.453
vibrations of phenolic	3749.62	16.601	1.323	3788.19	3726.47	46.927

Carboxylic	3873.06	16.923	0.134	3880.78	3834.49	35.448

Table 3. Cadmium ion adsorption performance of carbon foam gelatin

Contact time (min)	C_0 (ppm)	The end Concentration (ppm)	Concentration adsorbed %
2	100	75 .1	24,9
10	100	69 .5	30.5
30	100	57 .5	42,5
60	100	21 .7	78,3
120	100	20.5	79.5
240	100	19.9	80.1
600	100	19 .9	80.1
840	100	19 .9	80.1

The surface area of CFG sample observed by nitrgen adsorption-desorption is 162.9 m^2/g dominated by 65% macroporous carbon, 20% microporous and 15% mesoporous part.. The pore size distribution of CFG sample is the combination of mesoporous (20-50 nm) and macroporous range (up to 60 nm) which is obtained as good correlation data with SEM data (Fig 2). The small pore of CFG due to the loss of water and ammonia during high-temperature treatment as good correlation with the theory [21]. This CFG sample shows a relatively multiform with maxima around 20-30 nm and up to 60 nm. The broken pipe sizes of pore systems (as observed from the SEM image) cannot be distinguished by the BJH analysis. The high adsorption at high pressures explain that CFG have a lot of macropores. The CFG exhibits considerably low surface area due to its macropore system and more rigid void spaces, which makes it be the efficient matrix as the hard template in the nanomaterial synthesis.

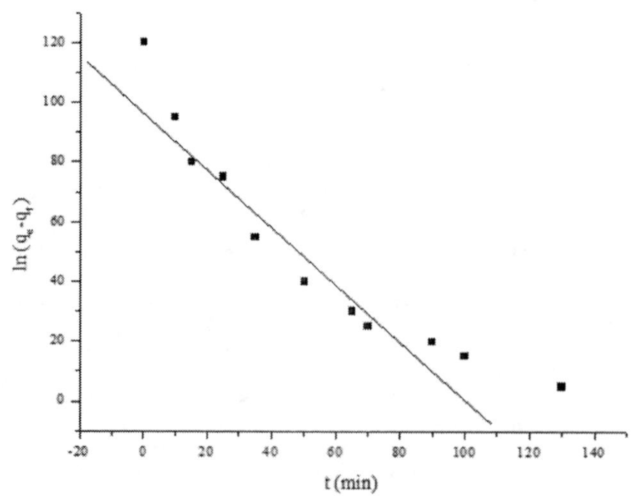

a.

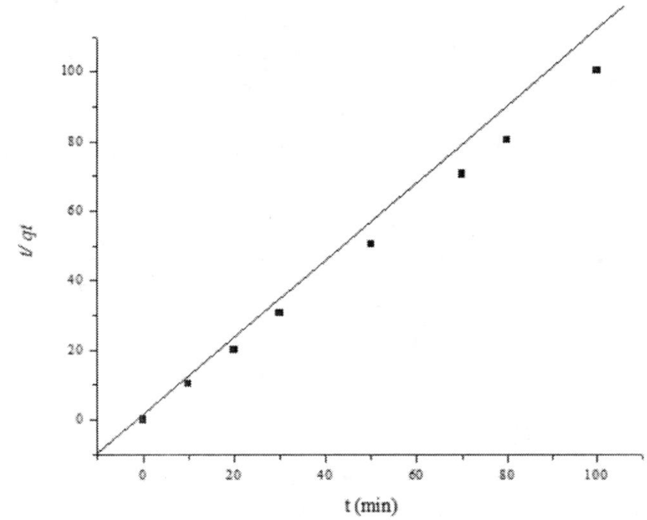

b.

Figure 3 The plot uses a. the Lagergren equation and b Ho Mckay equation in cadmium ion adsorption on the carbon foam of gelatin (CFG)

Table 3 showed the shaking time effect during cadmium adsorption into carbon foam of gelatin (CFG). The shaking time effect cadmiun ion adsorption onto carbon foam gelatin (CFG) was observed by mixing 50 mg of carbon foam of gelatin (CFG) into 10 mL of 100 ppm of cadmium solution the stirr for 0-840 min at room temperatures.. The results indicated that the equilibrium adsorption amount (qe) of cadmiun ion onto carbon foam gelatin (CFG) slightly decreases with shaking for 20-60 minute indicating that the adsorption process is exothermic in nature. However, the value of qe slightly stable from by increasing sking time from 60 to 840 minute, showing that saturation of adsorbent for the adsorption of cadmiun ion onto carbon foam gelatin (CFG) in this range. The capacities of adsorption for cadmium was 46.7 mg/g obtained by using adsorbent dose 50 mg, initial concentration 50 ppm, contact time, 3 h; room temperature, stirring rate 150 rpm) which reached equilibrium at 55 min. The homogenous adsorption of cadmium ion into carbon foam of gelatin (CFG) using Lagergren and Ho and McKay equation and measuring data described in table below

Table 4. Lagergren and Ho and McKay equation

Model	Equation
Lagergren.	$ln\,(q_e - q_t) = \ln q_e - k_1 t$
Ho and McKay	$\frac{t}{q_t} = \frac{t}{q_e} + \frac{1}{{q_e}^2 k_1}$

Table 3 shows the comparasion about Lagergren and Ho and Mckay equation. In the Lagergren kinetic model, the adsorbate concentration (A) is assumed to have no effect on the adsorption rate and the desorption rate is negligible. If q_t is an active site covered by adsorbat (A-S) at a given time and q_e is the adsorption capacity at equilibrium, the rate depends only on $(q_e - q_t)$ [9]. In the other side, Ho and Mckay approach appears based on the assumption that the adsorption rate is linearly related to the square of the number of adsorption vacant sites [10].

Broadly speaking, the whole coefficient of adsorption correlation of various models shows almost the same R value of 0.999 except the pseudo-order one model by Lagergren. The value of K Langmuir is closest to the value of K in the second order model of pseudo-Ho and McKay (1999), 1.82 and 1.72 g /mg respectively with a maximum value of 46.7 and 43.1 mg/g. This indicates that the adsorption of cadmium ion on the carbon foam of gelatin (CFG) approaches a pseudo second order model in which the adsorption rate is linearly related to the square of the number of vacant adsorption sites. From these indications it can be concluded that the carbon foam of gelatin (CFG) adsorption capacity of 46.7 mg / g represents the value representing the material's ability to adsorb cadmium ion. Carbon foam of gelatin (CFG) performance in this study did not differ much with previously research.The result of adsorption process fits using Lagergren, and Ho and McKay adsorption model conclude that carbon foam gelatin (CFG) is new adsorbent with the high potential for maintain environmental problem.

4. Conclusion

Carbon foam CFG prepared by carbonization of gelatin as carbon precursor and sulfuric acid as catalyst. The pores of CFG materials generated by the release of water and ammonia molecule from gelatin. The pore type of CFG material was the macroporous system with average size is 20-30 nm and contain the huge size-disordered pipe up to 10 μm. The CFG material contains oxygen and carbon as the major element. The thermal stability of CFG was up to 1000 °C indicating that CFG could be the new hard template candidate in the future. The adsorption performance summary describe that CFG material have high potential ability to adsorb cadmium pollutant in enviromental management.

Acknowledgements

The author acknowledge financial publication support from PNBP on Fundamental Research scheme Sebelas Maret University (FKIP-UNS) 2017.

References

[1] Qu J, Han F, Gao, and Qiu J., 2016, *New Carbon Mater*, **32**, 1, 86
[2] Ulfa M, Trisunaryanti W, Falah I, and Kartini I, 2016, *Indones. J. Chem*, **16**, 3, 239
[3] Tzvetkov G, Tsyntsarski B, and Balashev K, 2016, *Micron*, **89**, 34
[4] Jana P, Fierro V, and Celzard A, 2016, *Ind. Crops Prod,* **89**, 498
[5] Narasimman R, Vijayan S, and Prabhakaran K, 2014, *Mater. Sci. Eng. B*, **189**, 82
[6] Jiménez V, Sánchez P, Valverde J. L, and Romero A, 2010, *Mater. Chem. Phys*, **124**, 233
[7] Ge J., Shi J, and Chen L, 2009, *Carbon N. Y,* **47**, 4, 1192
[8] Jahanbakhshi M., 2017, *Mater. Sci. Eng. C*, **70**, 544
[9] Shan Ho Y, 2004, *Scietometrics*, **59**, 1, 171
[10] Ho Y. S, and Mckay G, 1999, *Procedia Technol*, **34**, 451

Theoretical Study of Electron Transfer Properties of Squaraine Dyes for Dye Sensitized Solar Cell

Ratna Juwita[1] **and Hui-Hsu Gavin Tsai**[1*]

[1]*Department of Chemistry, National Central University, Jhong-Li City, Tao-Yuan County 32001, Taiwan*

E-mail: raju_c4c4@yahoo.co.id and hhtsai@cc.ncu.edu.tw

Abstract. The environmental issues and high cost of Ru create many scientists to explore cheaper and safer sensitizer as alternative for dye sensitized solar cells (DSCs). Dyes play an important role in solar energy conversion efficiency. The squaraine (SQ) dyes has good spectral match with the solar spectra, therefore, SQ dyes have great potential for the applications in DSCs. SQ01_CA is an unsymmetrical SQ dye, reported by Grätzel and colleagues in 2007, featuring a D–π–spacer–A framework and has a carboxylic acid anchoring group. The electron donating ability of indolium in SQ01_CA and SQ01_CAA dyes is relatively weak, better performance may be achieved by introducing an additional donor moiety into indolium [1]. In this study, we investigate six unsymmetrical SQ dyes adsorbed on a $(TiO_2)_{38}$ cluster [2] using density functional theory (DFT) and time-dependent DFT to study electron transfer properties of squaraine dyes on their photophysical. SQ01_CA, WH-SQ01_CA, and WH-SQ02_CA use a carboxylic acid group as its electron acceptor. Furthermore, SQ01_CAA, WH-SQ01_CAA, and WH-SQ02_CAA use a cyanoacrylic acid group as its electron acceptor. WH-SQ01_CA and WH-SQ01_CAA have an alkyl, while WH-SQ02_CA and WH-SQ02_CAA have alkoxyl substituted diarylamines to the indolium donor of sensitizer SQ01_CA. Our calculations show with additional diarylamines in donor tail of WH-SQ02_CAA, the SQ dyes have red-shifted absorption and have slightly larger probability of electron density transferred to TiO_2 moiety. Furthermore, an additional –CN group as electron a withdrawing group in the acceptor exhibits red-shifted absorption and enhances the electron density transferred to TiO_2 and anchoring moiety after photo-excitation. The tendency of calculated probabilities of electron density being delocalized into TiO_2 and driving force for excited-state electron injection of these studied SQ dyes is compatible with their experimentally observed.

Key word: *Squaraine Dyes, Electron Injection Mechanism, Driving Force, Density Functional Theory, Electron Withdrawing Group*

IOP Conf. Series: Materials Science and Engineering **299** (2018) 012058 doi:10.1088/1757-899X/299/1/012058

1. Introduction

Dye sensitized solar cells (DSCs) is renewable energy source. It has promising technology because low production cost, large area, flexible, colorful, transparency and light weight devices [3],[4]. DSCs component consist of photosensitizer, a counter electrode, and a redox couple. Photosensitizer is one of the most important component because responsible for light harvesting and electron injection [5]. Grätzel and cowokers introducing ruthenium (Ru) complex as a photosensitizer in a DSCs such as the N3/N719 [6] dyes and the black dye N749 [7] with conversion efficiency of 10.8% [8]. They are promising sensitizers for DSCs, exhibit high performance and good stability [9]. However, Ru is very expensive, low natural abundance [10], and the difficulty for purification [9]. Recently, using porphyrin sensitizers dye SM315, which features structure of a donor-π bridge-acceptor with the cobalt (II/III) redox resulted Q-band absorption to 668 nm significantly red shifted. It is new result short-circuit current density (*Jsc*) of 18.1 mA cm^{-2}, and a power conversion efficiency (η) of 13% [11]. DSCs organic dyes (metal-free) is potential use as a photosensitizer. Ru-free organic dyes (e.g. triphenylamines, coumarins, and indolines) can absorb like as Ru dyes in 450-600 nm [12]. Replacing ruthenium based sensitizer with organic dyes is big chance for friendly environment and low cost. Organic sensitizer such as porphyrins, phthalocyanines, cyanines, and squaraine are the most interesting because they absorb strongly in the red and NIR regions [3]. Moreover, they are still growing for increased efficiency. The squaraine (SQ) dyes are interesting attention due to their unique photophysical and photochemical properties. These characterized by an aromatic four-membered ring system derived from squaric acid [13]. The SQ dye exhibit higher molar extinction coefficients [14], excellent performance in light harvesting [5], strong absorption because valency π-π* excitation in the long wavelength region and known their intense absorption in the red/near-IR region [13].

The SQ01_CA has planar molecule structure, its cause intermolecular close π-π aggregation. The additional of nonplanar triphenylamine groups to SQ01_CA can prevent the aggregation of the dyes and increase the distance between oxidizing hole and TiO_2 surface. The indolium donor has weak ability for injection on TiO2, so it better introducing an additional donor into indolium [15]. New unsymmetrical squaraine dyes SQ01_CA, SQ01_CAA, WH-SQ01_CA, WH-SQ01_CAA, WH-SQ02_CA, and WH-SQ02_CAA as potential alternative by intense absorption in the far red region. In most cases of photosensitizer employ electron withdrawing -COOH and CN-COOH as anchoring groups, due to a more favorable electronic stabilization [16]. WH-SQ01_CA and WH-SQ02_CA have modified donor moiety with additional diarylamines varied alkyl or alkoxyl chains, such as methyl and methoxyl groups [15] while anchor by additional -CN groups for WH-SQ01_CAA and WH-SQ02_CAA. Introducing -CN groups as a strongly electron withdrawing has effect on the squaraine absorption. In this research, we want to investigate the influence of indoline which containing diarylamines as donor and acceptor (anchor) with additional -CN group.

2. Computational Methods

The ground state geometries of all structure dyes have been optimized in deprotonated forms by density functional theory (DFT) theory using B3LYP functional couple with 6-31G(d,p) basis set then excitation energies were calculated by time-dependent-DFT (TD-DFT) in solvent (EtOH). The solvation effect is simulated using the conducting polarizable continuum model (C-PCM). Based on optimized geometries in neutral form, the optical properties were obtained by calculating singlet-singlet excited state in ethanol solution using TD-DFT with the long-range corrected functional CAM-B3LYP and 6-31G(d,p) basis set [17]. All calculation is performed with GAUSSIAN09 program package. The optimized geometry of dye on TiO_2 were performed in the gas phase with DFT calculation by using B3LYP 6-31G (d,p) basis set, while UV-Vis spectra by TD-DFT calculation use CAM-B3LYP 6-31G(d,p) basis set. The electronic density difference map (EDDM) to visualization electron density before and after transition connected using GaussSum. The DSCs performance consider the $(TiO_2)_{38}$ clusters as acceptor materials. The dyes adsorbed on the anastase (101) surface of the $(TiO_2)_{38}$ cluster [10].

3. Result and Discussion

3.1. Molecular Geometries

In this study, we investigate six unsymmetrical SQ-derived dyes. These dyes have similar π-conjugated moiety, but have different donor moieties and acceptor groups. The optimized structure parameters of studied SQ dyes computed at the B3LYP/6-31G (d, p) basis set with C-PCM ethanol solvent. Figure 1 shows structural parameter of the dyes, primary bond lengths and dihedral angles of two aromatic rings which give effect in absorption spectra. Moreover, the length of the bonds between the aromatic rings and electron donor groups are in the range 1.38 and 1.39 Å shorter than normal bond length C-C (1.54 Å), showing especially bond character which favors intra molecular charge transfer. Six unsymmetrical SQ-derived dyes have dihedral angles of backbone and anchor close to 0°. These dyes performed coplanar conformations. This coplanar molecule structure can be smoothly injected electron from the donor moiety to anchoring groups (carboxylic acid or cyanoacrylic acid) through the squaraic ring unit.

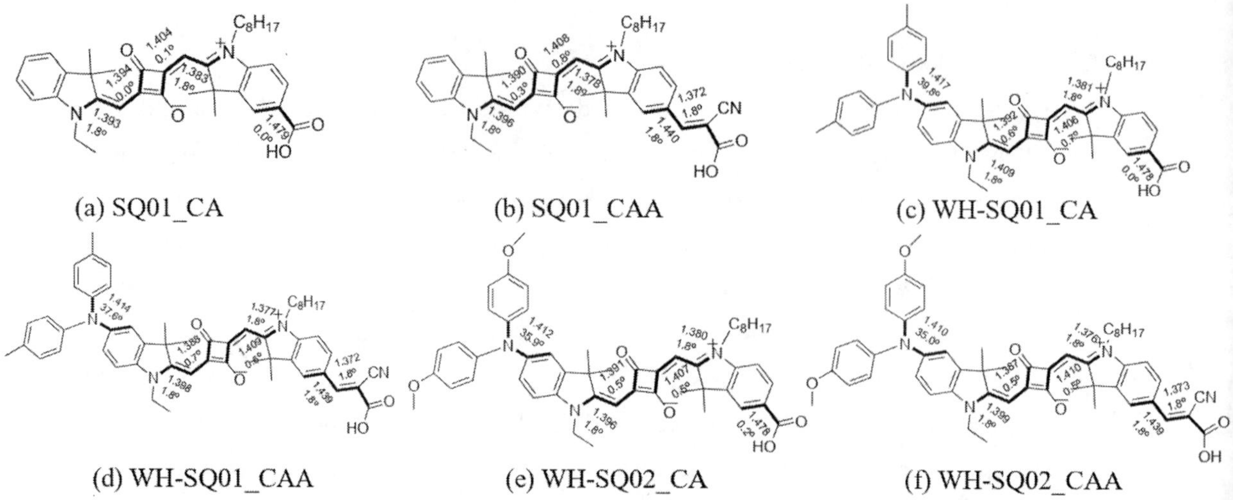

(a) SQ01_CA (b) SQ01_CAA (c) WH-SQ01_CA

(d) WH-SQ01_CAA (e) WH-SQ02_CA (f) WH-SQ02_CAA

Figure 1. Selected molecular geometry of SQ-derived dyes sensitizer. (a) SQ01_CA, (b) SQ01_CAA, (c) WH-SQ01_CA, (d) WH-SQ01_CAA, (e) WH-SQ02_CA, (f) WH-SQ02_CAA. The bond lengths are labeled Å. The dihedral angle is corresponding central bonds from four consecutive bonds.

3.2. Absorption Spectra of SQ Dyes

The most important of dyes for DSCs is an absorption spectrum matching the solar irradiation spectrum [18]. Moreover, we calculated by TD-DFT the UV-Vis absorption spectra in ethanol solvent for free molecule dyes, while water solvent for TiO_2. The optical data of the squaraine dyes SQ01_CA, SQ01_CAA, WH-SQ01_CA, WH-SQ01_CAA, WH-SQ02_CA and WH-SQ02_CAA in ethanol solution and adsorbed on TiO_2 are provided in Table 1. The computations electronic structure of dyes is performed using TD-DFT calculation CAM-B3LYP functional with 6-31G (d, p) basis set. Both experiment and calculation shows absorption spectra of addition an alkyl and alkoxyl substituted diarylamines to the indolium donor, for WH-SQ01_CA, WH-SQ01_CAA, WH-SQ02_CA and WH-SQ02_CAA.

The calculation result of studies free molecule dyes are an agreement with the Han and coworkers experimental report. The calculation with protonated state is more blue shift than experimental. The maximum absorption is 580 nm for WH-SQ02_CA with highest deviation from experimental -78. While, smallest one is -57 for SQ01_CA with λ_{max} 563 nm. Moreover, WH-SQ02_CAA has high absorption spectra in 599 nm which more red shift than WH-SQ02_CA. The dyes adsorbed on TiO_2 cluster calculated in water. These trends of UV-Vis spectra has similar trend with experimental spectra. The highest absorption spectra dye on TiO_2 in 583 nm for WH-SQ02_CA, and 604 nm for WH-SQ02_CAA. The additional strong electron withdrawing -CN group in anchor has effect to absorption band in the visible region which longer wavelength to the red shifted band. The six dyes have features a single strong absorption band in experiment and computational absorption spectra.

Table 1 Experimental and computational methods of studied SQ dyes in EtOH at their protonated state and adsorbed on TiO_2.

Dyes	In EtOH					Adsorbed on TiO_2		
	EXP.		Calculation (Protonated State)			EXP.	Calculation	
	ε_{max} (M^{-1} cm^{-1})	λ_{max} (nm)	f^a	λ_{max} (nm)	Deviation from EXP.	λ_{max} (nm)	λ_{max} (nm)	Deviation from EXP.
SQ01_CA	158,500[a]	620[a]	1.81	563	-57	622	564	-58
SQ01_CAA	—	—	2.13	582	—	—	585	—
WH-SQ01_CA	86,000	652	2.03	578	-74	626	581	-45
WH-SQ01_CAA	—	—	2.37	598	—	—	602	—
WH-SQ02_CA	122,000	658	2.05	580	-78	630	583	-47
WH-SQ02_CAA	—	—	2.39	599	—	—	604	—

[a] Oscillator strength

The value of energy HOMO and energy LUMO to indicate driving force for efficient electron injection into TiO_2 conduction band for photo excitation [5]'[19]. The driving force of SQ dye is adsorbed on TiO_2 are SQ01_CA (0.23), WH-SQ01_CA (0.31), and WH-SQ02_CA (0.36) (see Table 2). These driving force increase with additional diarylamines varied alkyl or alkoxyl chains in donor tail. The additional -CN group in acceptor will exhibit a relative low driving force in charge injection.

Table 2. Calculated driving force of SQ-dyes Absorbed on $(TiO_2)_{38}$ Cluster in water solution

Dyes	HOMO	E.abs	Aligned	Driving force
SQ01_CA	-5.91	2.2	-3.71	0.23
SQ01_CAA	-5.94	2.12	-3.82	0.12
WH-SQ01_CA	-5.76	2.13	-3.63	0.31
WH-SQ01_CAA	-5.79	2.06	-3.73	0.21
WH-SQ02_CA	-5.71	2.13	-3.58	0.36
WH-SQ02_CAA	-5.74	2.05	-3.69	0.25

3.3. Electron Transfer of Dyes upon Photo-Excitation

To analyze the propensity for electron transfer of the six dyes adsorbed on TiO_2 for open photo-excitation, we determined the contribution to electron density of each dye molecule. We use $(TiO_2)_{38}$ cluster with anatase 101 surface to calculate UV-Vis spectra. Table 3 shows the contribution of EDDM before and after photo-excitation on TiO_2. The SQ01_CA, WH-SQ01_CA, and WH-SQ02_CA in the excited state have the electron density higher than in the ground state. The absorption SQ01_CA is 564 nm with electronic transition located on TiO_2 9%. Moreover, SQ01_CAA is 13% with λ_{max} 585 nm. WH-SQ02_CAA as acceptor has higher absorption 583 nm and 17% electron density, while 604 nm and 15% electron density for WH-SQ02_CA after photo excitation, it show more red shift transition than others. These mainly transition of the dyes located in HOMO and anchor and some located on TiO_2. The presence of the donor tail chains (i. e methyl and methoxyl) and the CN groups encourages the electron transfer. The alignment energy levels of all dyes and electron density difference ground state and excited state are reported in Table 2 and Figure 2 (a-f) of SQ01_CA, SQ01_CAA, WH-SQ01_CA, WH-SQ01_CAA, WH-SQ02_CA, and WH-SQ02_CAA. Whereas, the excited state energy level is -3.71, -3.82, -3.63, -3.73, -3.58, and -3.69 eV, respectively. Furthermore, the energy level of experimental in conduction band TiO_2 semiconductor is -3.94 eV.

Table 3. Characters of Electron Density of SQ-dyes Absorbed on $(TiO_2)_{38}$ Cluster

Dyes	λ_{abs}/ f^a	Transitions[b]	Change	Percent Contribution (%)				
				Donor[c]	SQ	Indoline	Anchor	TiO_2
SQ01_CA	564 nm f=2.08	H → L (94%)	before	33	29	37	1	0
			after	28	35	24	3	9
			net	-5	6	-13	2	9
SQ01_CAA	585 nm f=2.37	H → L (88%)	before	31	28	38	3	0
			after	20	24	24	19	13
			net	-11	-4	-14	16	13
WH-SQ01_CA	581 nm f=2.34	H → L (86%) H-1 → L (5%)	before	51	20	27	1	0
			after	29	33	23	3	12
			net	-22	13	-4	2	12
WH-SQ01_CAA	602 nm f=2.64	H → L (80%) H-1 → L (6%)	before	50	21	27	3	0
			after	20	23	23	20	15
			net	-30	3	-4	17	15
WH-SQ02_CA	583 nm f=2.35	H → L (81%) H-1 → L (8%)	before	57	18	24	1	0
			after	27	32	21	3	17
			net	-30	14	-3	2	17
WH-SQ02_CAA	604 nm	H → L (76%)	before	56	18	24	2	0

f=2.67	H-1 → L (9%)	after	20	22	23	20	15
		net	-36	4	-1	18	15

(a) Oscillator strength. (b) The transitions with contribution larger than 5% are shown. (c) Indolium for SQ01_CA and SQ01_CAA; methyl substituded diarylamines to indoline donor for WH-SQ01_CA and WH-SQ01_CAA; methoxyl substituded diarylamines to indoline donor for WH-SQ02_CA and WH-SQ02_CAA.

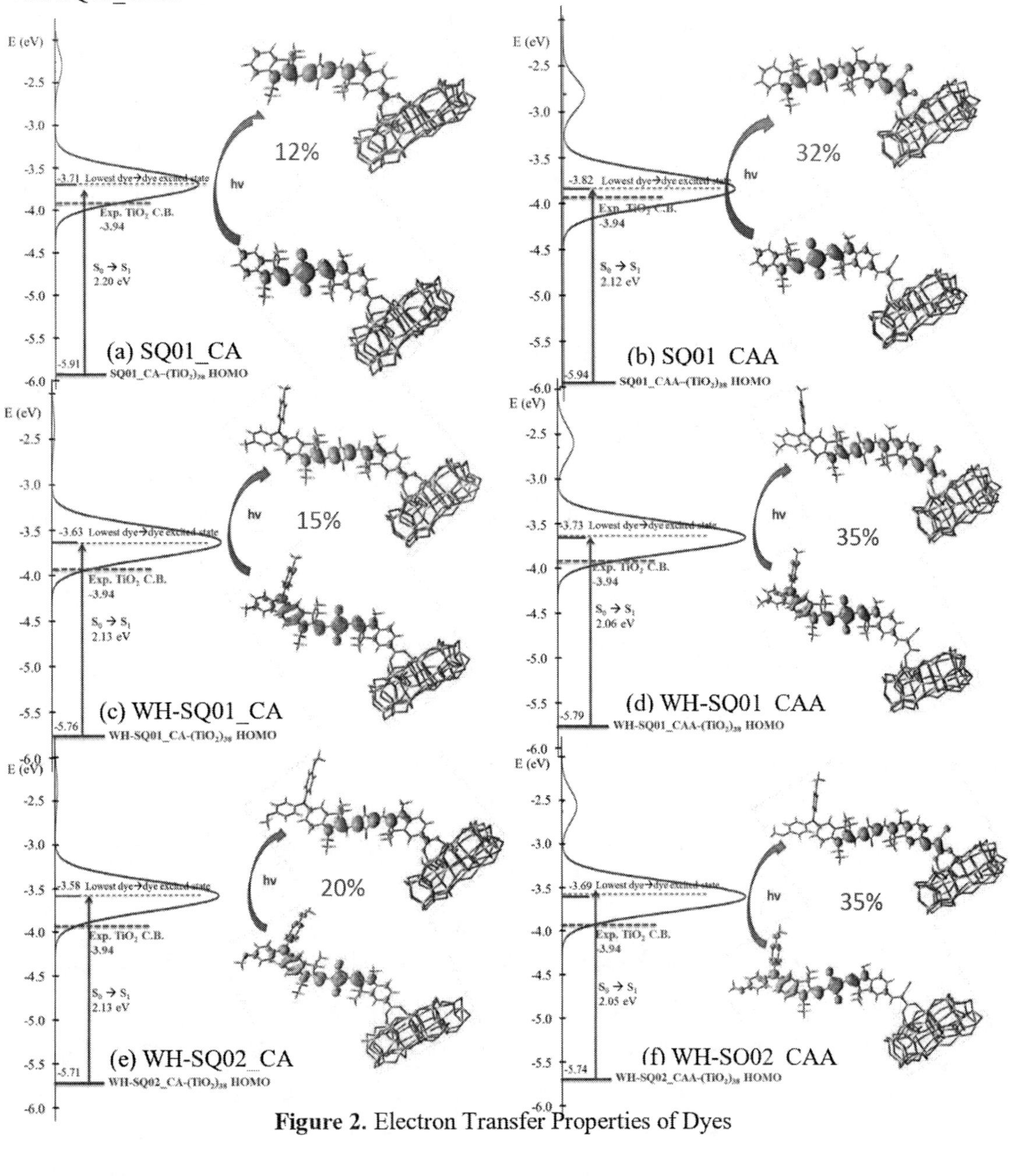

Figure 2. Electron Transfer Properties of Dyes

4. Conclusion

Our calculations show additional diarylamines in donor tail, the SQ dyes have red-shifted absorption and have slightly larger probability of electron density transferred to TiO_2 moiety. Moreover, the additional donor does not effectively push electron from donor to acceptor of TiO_2. Moreover, an additional –CN group as electron withdrawing group in the acceptor exhibits more red-shifted absorption even though –CN group attract electrons to itself and some electron density transfer to TiO_2.

5. Acknowledgment

We thank the Ministry of Science and Technology of Taiwan (Grant No. NSC 102-2113-M-008-008-MY2) for financial support and the National Center for High-Performance Computing and the V'ger computer cluster at the National Central University of Taiwan for allowing access to computer time.

6. References

[1] Linfeng Liu, J.C., Zhiliang Ku, Xiong Li, Hongwei Han, *Unsymmetrical squaraine sensitizers containing auxiliary arylamine donor for NIR-harvesting on dye-sensitized solar cell.* Dyes and Pigments, 2014. **106 (2014) 128e135**.

[2] Tsai, H.H.G., C.J. Tan, and W.H. Tseng, *Electron Transfer of Squaraine-Derived Dyes Adsorbed on TiO2 Clusters in Dye-Sensitized Solar Cells: A Density Functional Theory Investigation.* The Journal of Physical Chemistry C, 2015. **119**(9): p. 4431-4443.

[3] Qin, C.J., et al., *Novel Near-Infrared Squaraine Sensitizers for Stable and Efficient Dye-Sensitized Solar Cells.* Advanced Functional Materials, 2014. **24**(20): p. 3059-3066.

[4] Nazeeruddin, M.K., E. Baranoff, and M. Grätzel, *Dye-sensitized solar cells: A brief overview.* Solar Energy, 2011. **85**(6): p. 1172-1178.

[5] Li-Na Yang, Z.-Z.S., Quan-Song Li, Shi-Lu Chen a, Ze-Sheng Li, Thomas A. Niehaus, *Unsymmetrical squaraine dye containing dithieno[3,2-b:20,30-d] pyrrole as a p-spacer: A potential photosensitizer for dye-sensitized solar cells.* Journal of Power Sources, 2014. **137e145**.

[6] Fantacci, S. and F. De Angelis, *A computational approach to the electronic and optical properties of Ru(II) and Ir(III) polypyridyl complexes: Applications to DSC, OLED and NLO.* Coordination Chemistry Reviews, 2011. **255**(21–22): p. 2704-2726.

[7] Liu, S.-H., et al., *Theoretical Study of N749 Dyes Anchoring on the (TiO2)28 Surface in DSSCs and Their Electronic Absorption Properties.* The Journal of Physical Chemistry C, 2012. **116**(31): p. 16338-16345.

[8] Ozawa, H., et al., *Ruthenium sensitizers with a hexylthiophene-modified terpyridine ligand for*

dye-sensitized solar cells: synthesis, photo- and electrochemical properties, and adsorption behavior to the TiO2 surface. ACS Appl Mater Interfaces, 2015. **7**(5): p. 3152-61.

[9] Horiuchi, T., H. Miura, and S. Uchida, *Highly efficient metal-free organic dyes for dye-sensitized solar cells.* Journal of Photochemistry and Photobiology A: Chemistry, 2004. **164**(1–3): p. 29-32.

[10] Yakhanthip, T., et al., *Theoretical investigation of novel carbazole-fluorene based D-π-A conjugated organic dyes as dye-sensitizer in dye-sensitized solar cells (DSCs).* Journal of Computational Chemistry, 2011. **32**(8): p. 1568-1576.

[11] Mathew, S., et al., *Dye-sensitized solar cells with 13% efficiency achieved through the molecular engineering of porphyrin sensitizers.* Nat Chem, 2014. **6**(3): p. 242-247.

[12] Connell, A., et al., *A study of dye anchoring points in half-squarylium dyes for dye-sensitized solar cells.* Journal of Materials Chemistry A, 2014. **2**(11): p. 4055-4066.

[13] Bin Fan, Y.M., Marina Simeunovic, Simon Kuster, Thomas Geiger, and a.F.N.u. Roland Hany, *Squaraine Planar-Heterojunction Solar Cells.* Hindawi Publishing Corporation, 2009. **2009**: p. 7.

[14] Jradi, F.M., et al., *Near-Infrared Asymmetrical Squaraine Sensitizers for Highly Efficient Dye Sensitized Solar Cells: The Effect of π-Bridges and Anchoring Groups on Solar Cell Performance.* Chemistry of Materials, 2015. **27**(7): p. 2480-2487.

[15] Liu, L.F., et al., *Unsymmetrical squaraine sensitizers containing auxiliary arylamine donor for NIR-harvesting on dye-sensitized solar cell.* Dyes and Pigments, 2014. **106**: p. 128-135.

[16] Mª JOSÉ ALCALDE1, L.M., et al., *ELECTROCHEMICAL AND SPECTROSCOPIC PROPERTIES OF INDOLIZINO[1,2-B] QUINOLE DERIVATES.* J. Chil. Chem. Soc., 2013. **58**.

[17] Yanai, T., D.P. Tew, and N.C. Handy, *A new hybrid exchange–correlation functional using the Coulomb-attenuating method (CAM-B3LYP).* Chemical Physics Letters, 2004. **393**(1–3): p. 51-57.

[18] Oprea, C., et al., *Density Functional Theory (DFT) Study of Coumarin-based Dyes Adsorbed on TiO2 Nanoclusters—Applications to Dye-Sensitized Solar Cells.* Materials, 2013. **6**(6): p. 2372-2392.

[19] Garcia-Iglesias, M., et al., *Effect of bulky groups in ruthenium heteroleptic sensitizers on dye sensitized solar cell performance.* Chemical Science, 2012. **3**(4): p. 1177-1184.

IOP Conf. Series: Materials Science and Engineering **299** (2018) 012059 doi:10.1088/1757-899X/299/1/012059

Generating Two-Dimensional Repertoire of siRNA Linc-ROR and siRNA mRNA ARF6 from the lincRNA-RoR/miR-145/ARF6 expression Pathway that involved in the progression of Triple Negative Breast Cancer

Arli Aditya Parikesit[1*], Rizki Nurdiansyah[1]
[1]Department of Bioinformatics, School of Life Sciences, Indonesia International Institute for Life Sciences, Jl. Pulomas Barat Kav.88 Jakarta Timur 13210 Indonesia

*Corresponding author: arli.parikesit@i3l.ac.id

Abstract. The research for finding the cure for breast cancer is currently entering the interesting phase of the transcriptomics based method. With the application of Next Generation Sequencing (NGS), molecular information on breast cancer could be gathered. Thus, both in silico and wet lab research has determined that the role of lincRNA-RoR/miR-145/ARF6 expression Pathway could not be ignored as one of the cardinal starting points for Triple-Negative Breast Cancer (TNBC). As the most hazardous type of breast cancer, TNBC should be treated with the most advanced approach that available in the scientific community. Bioinformatics approach has found the possible siRNA-based drug candidates for TNBC. It was found that siRNA that interfere with lincRNA-ROR and mRNA ARF6 could be a feasible opportunity as the drug candidate for TNBC. However, this claim should be validated with more thorough thermodynamics and kinetics computational approach as the comprehensive way to comprehend their molecular repertoire. In this respect, the claim was validated using various tools such as the RNAfold server to determine the 2D structure, Barriers server to comprehend the RNA folding kinetics, RNAeval server to validate the siRNA-target interaction. It was found that the thermodynamics and kinetics repertoire of the siRNA are indeed rational and feasible. In this end, our computation approach has proven that our designed siRNA could interact with lincRNA-RoR/miR-145/ARF6 expression Pathway.

1. Introduction

As autocatalytic biomolecules, RNA has deemed as the possible origin of 'last universal common ancestor' (LUCA) [1]. Thus, slowly but sure, RNA-based flow of genetic information has supplemented the accepted protein-based dogma central theory [2,3]. However, this working pipeline still upheld by protein biomolecules such as DICER and DGCR [4]. Still, up to now, the role of RNA-based information flow is getting more important as the mechanism of non-coding (nc)RNA become unfolded [2,5]. Recent studies have shown that RNA or transcriptomics based diagnostics and drug candidates have been developed extensively for cancer [6–8]. Specifically, the role of silencing (si)RNA as short interference transcriptome, has been considered heavily as drug candidates [9,10]. The relative thermodynamics and kinetics versatility of siRNA are important contributing factors for the formulation integrity of the biomolecules [11]. Ground-breaking computational methods have been devised for designing the siRNA molecules, and it is greatly assisted with RNA Vienna Package [12,13]. This versatile package has been applied for transcriptomics annotations of virus, human, plants, and animals [14–16]. Those studies encourage us to apply the package into the breast cancer research.

As one of the most dangerous malady for woman, breast cancer is currently only curable via conventional medication approaches, namely radiotherapy and chemotherapy [17]. In Indonesia, breast cancer is also considered one of the deadliest diseases for woman [18]. However, current progress in breast cancer medication also points to the biomolecular approach, especially towards the transcriptomics methods [19–21]. Thus, in this respects, interest on how to obtained fine-grained repertoire on the exact biomolecular mechanism of the RNA molecules also grows [11,22]. Based on its biomolecular mechanism, breast cancer has four types, namely: luminal A, luminal B, Her-2, and triple negative breast cancer (TNBC). Concerning TNBC, which is considered the most dangerous type, it is regulated with lincRNA-RoR/miR-145/ARF6 expression pathway [23,24]. Computational measures to design siRNA based on lincRNA-RoR/miR-145/ARF6 pathway has been devised, hence without good resolution for its biomolecular trajectory [25]. Hence, Vienna RNA package already provided tools to measure the trajectory in efficient manner such as RNA fold for structure elucidation, Barriers server to comprehend the RNA folding kinetics, RNAeval server to validate the siRNA-target interaction [10,26]. In this end, the objective of this research is to annotate the trajectory of siRNA drug candidate for TNBC.

2. Material and Methods

The data was taken from existing publication, in the form of siRNA design [25]. It is elucidated as FASTA sequences and 'dot bracket' structural elucidation. Thus the online version of Vienna RNA Package in http://rna.tbi.univie.ac.at was accessed to utilized biomolecular repertoire annotation tools [27]. There are two siRNA designs that currently annotated, namely siRNA for lincRNA-RoR and ARF6. The sequences and structural elucidations were forwarded to Barriers server, RNA fold and RNAeval server which are parts of online Vienna RNA Package [11,14,28]. The tools would be utilized with their respective default parameters. The final step would be analyzing the thermodynamics and kinetics data in order to determine the exact trajectories of the biomolecules.

3. Results and Discussion

The repertoire data was shown with RNAfold and Barrier server program as part of Vienna RNA Package. RNAfold successfully provided 2D visualization of siRNA molecules, as shown in figure 1. The base pair probability in the both siRNA molecules of Figure 1 means the homology between query siRNA with the template on the database. The color scale from 0 to 1 refers to the probability of structural formation. As seen in Figure 1, due to the very high probability of structural formation, the round-shape structure is possibly occurred. The probability of structural formation indicator is comparable with the entropy of the biomolecules themselves.

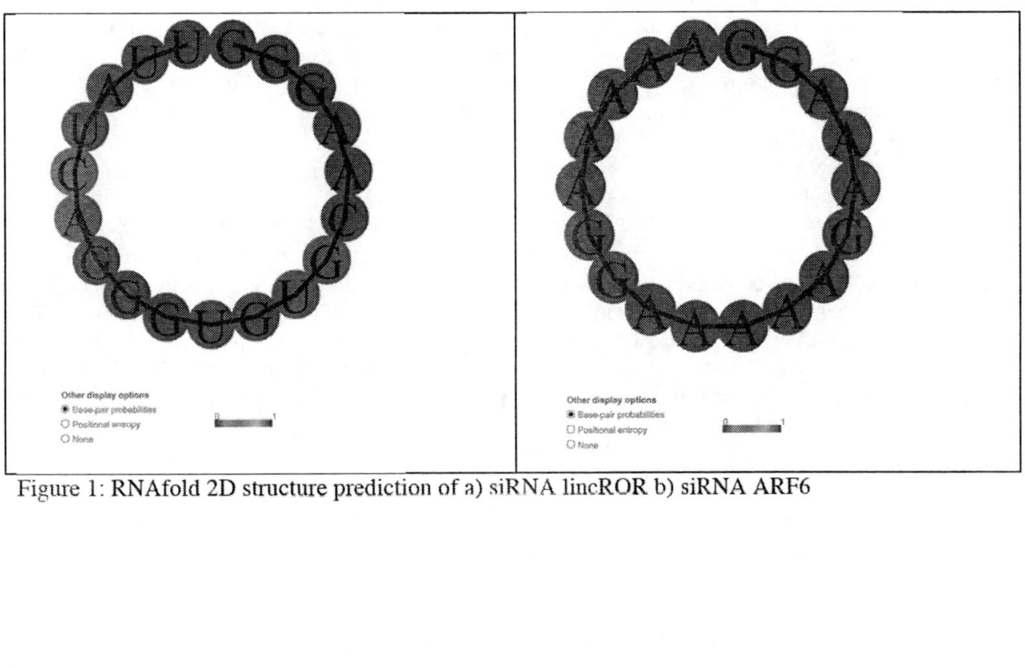

Figure 1: RNAfold 2D structure prediction of a) siRNA lincROR b) siRNA ARF6

IOP Conf. Series: Materials Science and Engineering **299** (2018) 012059 doi:10.1088/1757-899X/299/1/012059

Thus, the siRNA lincROR free energy of thermodynamics ensemble is -0.06 kcal/mol, while for siRNA ARF6 is 0.00 kcal/mol. Although the reaction is near to inert thermodynamics threshold, it is indeed spontaneous as shown in the folding repertoire. Moreover, the animation of the structural transition could be annotated and displayed as well. It is shown on Figure 2. In the left side of the figure, the base pairing occurred within respecting complementary bases. The pairing process would be projected to the structural formation on the right side. The top most figure is showing the thermodynamics repertoire of the structural transition. The structure that pointed in the peak shows the most feasible structure.

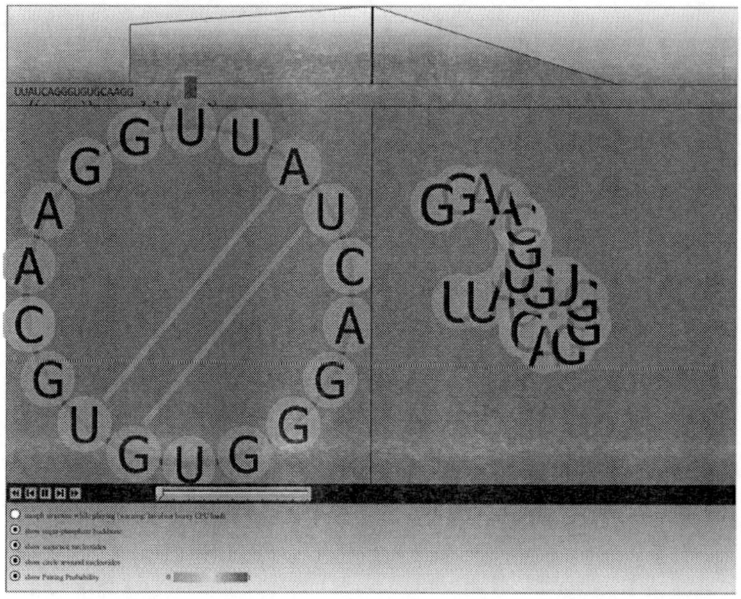

Figure 2: Barrier server trajectory of siRNA lincROR. The panel is animated in the computer.

For siRNA lincROR, RNAsubopt predicted 29 structures in an energy range of 40 kcal/mol above the minimum free energy. Then, for siRNA ARF6, RNAsubopt predicted 7 structures in an energy range of 8 kcal/mol above the minimum free energy. Whole predicted structures were spontaneous. The siRNA ARF6 was able to bind with its respective target RNA as shown from the total amount of total free energy binding of -19.80 kcal/mol. Meanwhile siRNA lincROR was able to bind with its respective target RNA as shown from the total amount of total free energy binding of -35.59 kcal/mol. In this end, the bindings were shown as spontaneous.

Although siRNA-based lead compound is very promising for drug development, there is certain concern that should be addressed. As RNA based compound, siRNA is prone to degradation by RNAse enzyme in our cells [29]. This concern should be addressed with more sophistication in drug delivery methods[30]. Some possible approaches are cyclization, using sterically stable liposome formation, and utilization of polymeric nanoparticles [31–33]. Although those methods were mainly developed for peptide and synthetic lead compounds, it is also feasible to incorporate them in siRNA lead compounds development. However, although in vitro and in vivo approach for siRNA drug delivery has provided useful results, the clinical trials are still need many improvements [9,34,35]. For further enhancements, scientist is trying to integrate nanotechnology with current drug design methods [36]. In this end, it would take some years from now until siRNA-based drugs become common in the market.

4. Conclusion

The siRNA structure elucidation for silencers of Linc-ROR and ARF 6 have proven to be feasible in thermodynamics and kinetics terms. Vienna RNA package has proven to be able to annotate the 2D structure of siRNA. The future venue of this research should be devised on the drug delivery methods of the lead compound, in order to reach the target cells safely.

Acknowledgement

The authors would like to thanks to The i3L Head of Research and Community Engagement (Ketua LPPM) Dr. Surjawan and Hibah Penelitian Berbasis Kompetensi DIKTI 2018 No. 0045/E3/LL/2018 for their heartfelt support towards this research. Thanks also goes to Faried Irmansyah from the IT Department of i3L for providing solid infrastructure for this computational research.

References

[1] Forterre P. The universal tree of life: an update. Front Microbiol 2015;6:717. doi:10.3389/fmicb.2015.00717.
[2] Mattick JS, Makunin I V. Non-coding RNA. Hum Mol Genet 2006;15 Spec No. doi:10.1093/hmg/ddl046.
[3] Mattick JS. RNA regulation: a new genetics? Nat Rev Genet 2004;5:316–23. doi:10.1038/nrg1321.
[4] Okamura K, Chung WJ, Ruby JG, Guo H, Bartel DP, Lai EC. The Drosophila hairpin RNA pathway generates endogenous short interfering RNAs. Nature 2008;453:803–6. doi:10.1038/nature07015.
[5] Bhat SA, Ahmad SM, Mumtaz PT, Malik AA, Dar MA, Urwat U, et al. Long non-coding RNAs: Mechanism of action and functional utility. Non-Coding RNA Res 2016;1:43–50. doi:10.1016/j.ncrna.2016.11.002.
[6] Gibb EA, Brown CJ, Lam WL. The functional role of long non-coding RNA in human carcinomas. Mol Cancer 2011;10:38. doi:10.1186/1476-4598-10-38.
[7] Mendell JT. Targeting a Long Noncoding RNA in Breast Cancer. N Engl J Med 2016;374:2287–9. doi:10.1056/NEJMcibr1603785.
[8] Eades G, Wolfson B, Zhang Y, Li Q, Yao Y, Zhou Q. lincRNA-RoR and miR-145 regulate invasion in triple-negative breast cancer via targeting ARF6. Mol Cancer Res 2015;13:330–8. doi:10.1158/1541-7786.MCR-14-0251.
[9] Oh Y-K, Park TG. siRNA delivery systems for cancer treatment. Adv Drug Deliv Rev 2009;61:850–62. doi:10.1016/j.addr.2009.04.018.
[10] Tafer H, Ameres SL, Obernosterer G, Gebeshuber CA, Schroeder R, Martinez J, et al. The impact of target site accessibility on the design of effective siRNAs. Nat Biotechnol 2008;26:578–83. doi:10.1038/nbt1404.
[11] Muckstein U, Tafer H, Hackermuller J, Bernhart SH, Stadler PF, Hofacker IL. Thermodynamics of RNA-RNA binding. Bioinformatics 2006;22:1177–82. doi:10.1093/bioinformatics/btl024.
[12] Gruber AR, Lorenz R, Bernhart SH, Neuböck R, Hofacker IL. The Vienna RNA websuite. Nucleic Acids Res 2008;36. doi:10.1093/nar/gkn188.
[13] Lorenz R, Bernhart SH, zu Siederdissen C, Tafer H, Flamm C, Stadler PF, et al. ViennaRNA Package 2.0. Algorithms Mol Biol 2011;6:26. doi:10.1186/1748-7188-6-26.
[14] Hofacker IL, Fekete M, Flamm C, Huynen MA, Rauscher S, Stolorz PE, et al. Automatic detection of conserved RNA structure elements in complete RNA virus genomes. Nucleic Acids Res 1998;26:3825–36.
[15] Marz M, Beerenwinkel N, Drosten C, Fricke M, Frishman D, Hofacker IL, et al. Challenges in

RNA Virus Bioinformatics. Bioinformatics 2014:btu105-. doi:10.1093/bioinformatics/btu105.
[16] Rose D. The long and the short of computational ncRNA prediction 2010.
[17] WHO. WHO Breast Cancer Prevention and Control. WHO Fact Sheet 2016. http://www.who.int/cancer/detection/breastcancer/en/.
[18] Kemenkes-RI. Profil Kesehatan Indonesia. Jakarta: 2013.
[19] Takahashi RU, Miyazaki H, Ochiya T. The roles of microRNAs in breast cancer. Cancers (Basel) 2015;7:598–616. doi:10.3390/cancers7020598.
[20] Kaboli PJ, Rahmat A, Ismail P, Ling K-H. MicroRNA-based therapy and breast cancer: A comprehensive review of novel therapeutic strategies from diagnosis to treatment. Pharmacol Res 2015;97:104–21. doi:10.1016/j.phrs.2015.04.015.
[21] Fan M, Krutilina R, Sun J, Sethuraman A, Yang CH, Wu Z-H, et al. Comprehensive analysis of microRNA (miRNA) targets in breast cancer cells. J Biol Chem 2013;288:27480–93. doi:10.1074/jbc.M113.491803.
[22] Boehr DD, Nussinov R, Wright PE. The role of dynamic conformational ensembles in biomolecular recognition. Nat Chem Biol 2009;5:789–96. doi:10.1038/nchembio.232.
[23] Hou P, Zhao Y, Li Z, Yao R, Ma M, Gao Y, et al. LincRNA-ROR induces epithelial-to-mesenchymal transition and contributes to breast cancer tumorigenesis and metastasis. Cell Death Dis 2014;5:e1287. doi:10.1038/cddis.2014.249.
[24] Huang J, Zhang A, Ho T-T, Zhang Z, Zhou N, Ding X, et al. Linc-RoR promotes c-Myc expression through hnRNP I and AUF1. Nucleic Acids Res 2015;44:3059–69. doi:10.1093/nar/gkv1353.
[25] Parikesit AA, Anurogo D. Prediksi Struktur 2-Dimensi Non-Coding RNA dari Biomarker Kanker Payudara Triple-Negative dengan Vienna RNA Package. Chim Nat Acta 2016;4:27. doi:10.24198/cna.v4.n1.10445.
[26] Hofacker IL, Stadler PF. Memory efficient folding algorithms for circular RNA secondary structures. Bioinformatics 2006;22:1172–6. doi:10.1093/bioinformatics/btl023.
[27] Parikesit AA, Anurogo D, Parikesit AA. Bioinformatics Approach towards Transcriptomics of Filaggrin. J Agromedicine Med Sci 2016;2:8–16.
[28] Hofacker IL, Fekete M, Stadler PF. Secondary structure prediction for aligned RNA sequences. J Mol Biol 2002;319:1059–66. doi:10.1016/S0022-2836(02)00308-X.
[29] Woodhams MD, Stadler PF, Penny D, Collins LJ. RNase MRP and the RNA processing cascade in the eukaryotic ancestor. BMC Evol Biol 2007;7 Suppl 1:S13. doi:10.1186/1471-2148-7-S1-S13.
[30] Bonetta L. RNA-based therapeutics: ready for delivery? Cell 2009;136:581–4. doi:10.1016/j.cell.2009.02.010.
[31] Tambunan USF, Parikesit AA, Dephinto Y, Sipahutar FRP. Computational design of drug candidates for influenza A virus subtype H1N1 by inhibiting the viral neuraminidase-1 enzyme. Acta Pharm 2014;64:157–72. doi:10.2478/acph-2014-0015.
[32] Du S-L, Pan H, Lu W-Y, Wang J, Wu J, Wang J-Y. Cyclic Arg-Gly-Asp peptide-labeled liposomes for targeting drug therapy of hepatic fibrosis in rats. J Pharmacol Exp Ther 2007;322:560–8. doi:10.1124/jpet.107.122481.
[33] Yuan X, Li L, Rathinavelu A, Hao J, Narasimhan M, He M, et al. siRNA Drug Delivery by Biodegradable Polymeric Nanoparticles. J Nanosci Nanotechnol 2006;6:2821–8. doi:10.1166/jnn.2006.436.
[34] Burnett JC, Rossi JJ. RNA-based therapeutics: current progress and future prospects. Chem Biol 2012;19:60–71. doi:10.1016/j.chembiol.2011.12.008.
[35] Castanotto D, Rossi JJ. The promises and pitfalls of RNA-interference-based therapeutics. Nature 2009;457:426–33. doi:10.1038/nature07758.
[36] Anurogo D, Parikesit AA, Ikrar T. Bionanomedicine: A "Panacea" In Medicine? Makara J Heal Res 2017;21:42–8. doi:10.7454/msk.v21i2.6524.

Ab-initio study on the absorption spectrum of color change sapphire based on first-principles calculations with considering lattice relaxation-effect

Mega Novita[1], Hikari Nagoshi[2], Akiho Sudo[2], Kazuyoshi Ogasawara[2]

[1]Faculty of Engineering and Informatics, University of PGRI Semarang, Jl. Sidodadi-Timur No.24 Semarang, Central Java 50232, Indonesia

[2]School of Science and Technology, Kwansei Gakuin University, 2-1 Gakuen Sanda, Hyogo 669-1337 Japan

novita@upgris.ac.id

Abstract. In this study, we performed an investigation on α-Al_2O_3: V^{3+} material, or the so-called color change sapphire, based on first-principles calculations without referring to any experimental parameter. The molecular orbital (MO) structure was estimated by the one-electron MO calculations using the discrete variational-Xα (DV-Xα) method. Next, the absorption spectra were estimated by the many-electron calculations using the discrete variational multi-electron (DVME) method. The effect of lattice relaxation on the crystal structures was estimated based on the first-principles band structure calculations. We performed geometry optimizations on the pure α-Al_2O_3 and with the impurity V^{3+} ion using Cambridge Serial Total Energy Package (CASTEP) code. The effect of energy corrections such as configuration dependence correction and correlation correction was also investigated in detail. The results revealed that the structural change on the α-Al_2O_3: V^{3+} resulted from the geometry optimization improved the calculated absorption spectra. By a combination of both the lattice relaxation-effect and the energy correction-effect improve the agreement to the experiment fact.

1. Introduction

Various crystals containing transition metal ions as impurities are used as optical materials. Corundum, or α-Al_2O_3 is an important rock-forming mineral, which is transparent, if it does not contain any impurities. If, however, it contains a small amount of chromium (α-Al_2O_3: Cr^{3+}), the color is changed to red and the material is called ruby. If, on the contrary, there is vanadium impurity instead of chromium (α-Al_2O_3: V^{3+}), it shows unique ability to change color when viewed under different light, thus the material is called color change sapphire. However, their applications are not limited to jewelry only, those materials can be utilized in the laser technology. Their performances mostly depend on the optical properties such as the absorption spectrum and the multiplet energies.

The prediction on the optical properties of the transition metal ions in various crystals has been feasible using the discrete variational multi-electron (DVME) method [1]. This method has been proved to be useful for the theoretical design of new optical spectra and new artificial jewelry. In the early development of DVME method, α-Al_2O_3: V^{3+} material has been investigated [1]. Ogasawara's group used $VAl_{14}O_{48}^{51-}$ model cluster which consists of 63 atoms and the energy correction namely configuration dependent correction and correlation correction (CDC-CC) was considered in the calculations [1]. Although the calculated absorption spectra have been successfully reproduced, the peak energies were overestimated. In our previous studies on the $3d^3$ ions such as V^{2+}, Cr^{3+} and Mn^{4+} doped in crystals based on DVME method [2-6], we figured that the consideration of lattice relaxation-effect in the calculations is important to improve the agreement with the experiment. Therefore, here we performed a similar calculation with some improvement in the α-Al_2O_3: V^{3+} structure by considering lattice relaxation effect.

In this work, we investigated the absorption spectra of V^{3+}-doped α-Al_2O_3 crystal based on the first-principles DVME method. The effect of lattice relaxation on the crystal structures was considered by performing geometry optimization using Cambridge Serial Total Energy Package (CASTEP) code [7-9]. Here, the CDC-CC energy correction was also considered to improve the accuracy.

2. Computational procedure

2.1. Model cluster

Two types of model clusters, non-optimized and optimized clusters consisting of 7 and 63 atoms were used in the calculations. The 7-atom model cluster consists of one trivalent vanadium atom surrounded by six oxygen atoms, VO_6^{9-}. The 63-atom model cluster consists of one trivalent vanadium atom surrounded by 14 Aluminum atoms and 48 oxygen atoms, $VAl_{14}O_{48}^{51-}$. The non-optimized cluster was simply constructed from the experimental crystal structure of α-Al_2O_3 which has a rhombohedral A_2X_3 structure (space group $R\bar{3}c$) reported by H. Sawada, et al., [10]. On the other hand, the optimized cluster including the lattice relaxation effect was constructed by the geometry optimization using CAmbridge Serial Total Energy Package (CASTEP) code [7-9] (CASTEP module in Materials Studio 6.0). This code is a first-principles band structure calculations program which is based on density functional theory (DFT) [11] combined with plane-wave basis sets and pseudopotentials [12]. The Vanderbilt [13] ultrasoft pseudopotentials were used. The orbitals explicitly treated as the valence state were 3s,3p, calculated with the $3s^2 3p^1$ configuration and a core radius of 2.00 a.u. for Al; 2s,2p calculated with the $2s^2 2p^4$ configuration and a core radius of 1.30 a.u. for O; and 3s,3p,3d,4s calculated with the $3s^2 3p^6 3d^{3.75} 4s^2$ configuration and a core radius of 1.80 a.u. for V. The exchange-correlation potential was considered within the Generalized Gradient Approximation (GGA) proposed by Perdew, Burke, and Ernzerhof (PBE) [14].

In order to construct the optimized cluster, we begin with performing a structural optimization of pure α-Al_2O_3 crystal. The cutoff energy of the plane wave was 500 eV and the Brillouin zone was sampled on the rhombohedral 3×3×3 Monkhorst-pack grids. Second, the external pressure from 0 to 110 GPa was applied and the Al-O bond lengths in the α-Al_2O_3 host crystal were estimated. Next, the optimized structure surrounding dopant- V^{3+} ion in α-Al_2O_3 crystal was estimated using the same method. The 2×2×1 supercell consisting of 120 atoms was constructed based on the optimized hexagonal structure. Cr^{3+} ion was then substituted into one Al^{3+} site. The cutoff energy of the plane wave was 500 eV and the Brillouin zone was sampled on 3×3×2 Monkhorst-pack grids. The numerical error was estimated to be less than 5×10^{-6} eV/atom by cutoff and *k*-point convergence tests. In this case, the effect of spin polarization was also taken into account. Under the calculations including the impurity ion, the local symmetry of Al site in pure α-Al_2O_3 crystal (C_3 symmetry) was kept to minimize the computational error.

2.2. One-electron calculations

In this work, the molecular orbital (MO) structure was estimated by the one-electron MO calculations using the discrete variational-Xα (DV-Xα) method which was developed by Adachi and co-worker [15]. In this case, the numerical atomic orbitals (AO) such as 1s-4p for Cr, 1s-3d for Al and 1s-2p for O were used as basis functions in the MO calculations. The Madelung potential was considered by locating 13,000-16,000 point charges at external atomic sites, i.e., +3 for Al and V sites and -2 for O site. In order to preserve the charge neutrality, these point charges were selected in such a way that the total charge of the cluster and the point charges equals zero. For the numerical integration, 160,000 and 216,000 sample points were used for 7- and 63-atom model clusters, respectively.

2.3. Many-electron calculations

After one electron approximation calculations, the many-electron calculations using the configuration interaction (CI) methods were performed. In this case, the electron-electron interactions are directly calculated. ${}_{10}C_2$ = 45 Slater determinants were constructed with MOs mainly consisting of V 3d orbitals were used as the basis functions. The many-electron wave functions and the electronic states were obtained by diagonalizing the many-electron Hamiltonian. The matrix element of the effective many-electron Hamiltonian H between two Slater determinants, Φ_p and Φ_q, can be generally expressed as

$$H_{pq} = \langle \Phi_p | H | \Phi_q \rangle = \sum_{i=1}^{L} \sum_{j=1}^{L} A_{ij}^{pq} \langle i | h | j \rangle + \sum_{i=1}^{L} \sum_{j=1}^{L} \sum_{k=1}^{L} \sum_{l=1}^{L} B_{ijkl}^{pq} \langle ij | g | kl \rangle, \tag{2.1}$$

where L is the number of impurity-state orbitals and A_{ij}^{pq} and B_{ijkl}^{pq} are coefficients obtained by the expansion of the Slater determinants.

Since multiplet energies according to CI calculations are generally overestimated [16], several corrections, known as configuration-dependent correction (CDC) and correlation correction (CC) in this case, were used. These corrections were introduced based on the consistency between the spin-restricted and the spin-unrestricted calculations, respectively. In this approach, the Hamiltonian including CDC and CC is expressed as

$$H_{pq}^{\text{CDC-CC}} = \sum_{i=1}^{L} \sum_{j=1}^{L} A_{ij}^{pq} \langle i | h | j \rangle + \sum_{i=1}^{L} \sum_{j=1}^{L} \sum_{k=1}^{L} \sum_{l=1}^{L} c \times B_{ijkl}^{pq} \langle ij | g | kl \rangle + D_{\text{CDC}}(m,n) \delta_{pq} \,. \tag{2.2}$$

where c is the CC factor estimated based on the consistency of the transition energy of the spin-flip transition from $(t_{2g}\uparrow)^3$ to $(t_{2g}\uparrow)^2(t_{2g}\downarrow)^1$ between the many-electron CI calculations and the spin-polarized one-electron MO calculations. On the other hand, $D_{\text{CDC}}(m,n)$ is the correction of the barycenter of the $(t_{2g})^3$, $(t_{2g})^2(e_g)^1$, $(t_{2g})^1(e_g)^2$, and $(e_g)^3$ configurations based on the one-electron MO calculation.

Using the eigenvector obtained by the diagonalization of the many-electron Hamiltonian, the many-electron wave functions corresponding to each multiplet state can be obtained explicitly as a linear combination of the Slater determinants. Therefore, the oscillator strength for the electric-dipole transition (transition probability) between multiplets can be calculated directly using [17]

$$I_{if} = 2(E_f - E_i) \left| \langle \Psi_f | \textstyle\sum_{k=1}^{n} \mathbf{r}_k | \Psi_i \rangle \right|^2 . \tag{2.3}$$

where Ψ_i and Ψ_f are the many-electron wave functions of the initial and final states, while E_i and E_f are the energy eigenvalues of these states. **e** denotes the unit vector parallel to the direction of the electric field of the incident light. The detailed procedure of DVME method is available in Ref. 1.

IOP Conf. Series: Materials Science and Engineering **299** (2018) 012060 doi:10.1088/1757-899X/299/1/012060

3. Results and Discussions

3.1. Geometry optimization

In the structural optimization, the lattice parameters of the unit cell of pure α-Al_2O_3 host crystal with 0 pressure was optimized and compared to the experimental values. The theoretical lattice constants on *a*, *b* and *c* axes were found to be 4.806, 4.806, and 13.119 Å while the experimental values [10] are 4.759, 4.759, 12.992 Å for *a*, *b*, and *c* axes, respectively. Since the difference between the experimental and theoretical values is about 1%, the optimized structures of the host crystals are considered to be in good agreement with the experiment. In the geometry optimization, we obtained the estimated V-O bond lengths of α-Al_2O_3: V^{3+} are d_1 = 1.956 and d_2 = 2.079 Å. The V-O bond lengths were elongated for about ca. 0.09 Å. Since the experimental data of Al-O bond lengths in the pure α-Al_2O_3 crystal are d_1 = 1.855 and d_2 = 1.971 Å, therefore the relaxation ratio is about 105.48%.

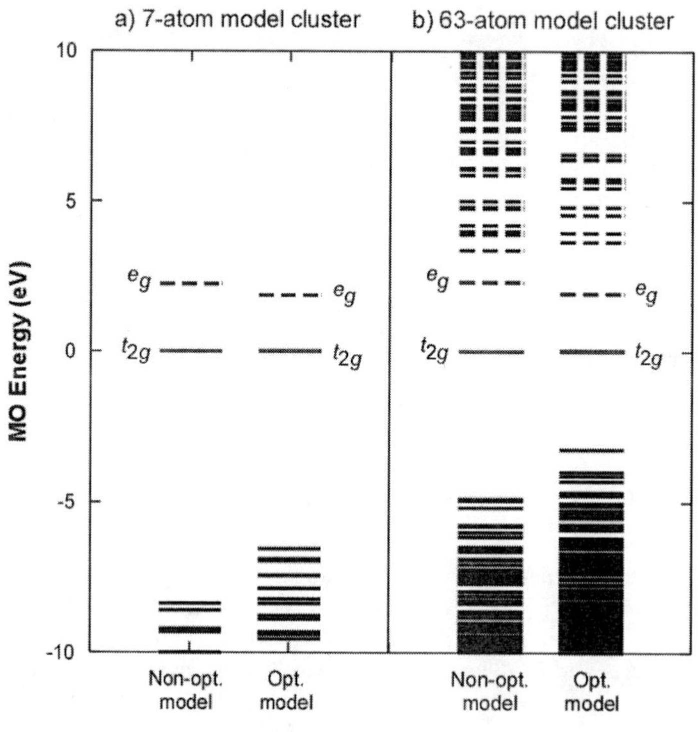

Figure 1. The molecular orbital (MO) energy of α-Al_2O_3: V^{3+} calculated based on (a) 7-atom model cluster (VO_6^{9-}) and (b) 63-atom model clusters ($VAl_{14}O_{48}^{51-}$) model clusters. Models without and with considering lattice relaxation-effect are indicated by non-optimized and optimized models.

3.2. Molecular orbital (MO) structure

The MO energy levels of α-Al_2O_3: V^{3+} calculated based on 7- and 63-atom model clusters using non-optimized and optimized clusters are shown in Fig. 1. The solid black lines indicate the valence band which are mainly consisting of O 2p orbital. The dashed black lines indicate the conduction band which are mainly consisting of Al 3s, 3p, and 3d orbitals. Moreover, the impurity levels split into t_{2g} and e_g set are indicated by solid blue lines and dashed red lines, respectively. The energy difference between them is denoted as $10Dq$, crystal field splitting. Although the t_{2g} level splits further into *a* and *e* levels due to the trigonal crystal field, however these splittings are negligibly small. Therefore, the notations of O_h symmetry were borrowed for simplicity. The calculated $10Dq$ obtained from the calculations based on 7-atom model clusters are 2.26 and 1.88 eV, for the non-optimized and the optimized models, respectively. On the other hand, the calculated $10Dq$ obtained from the calculations based on 63-atom model clusters are 2.32 and 1.93 eV, for the non-optimized and the optimized models, respectively. As we can see, when we use the larger size of model cluster, the energy difference between t_{2g} and e_g state becomes larger. Nevertheless, the consideration of lattice-relaxation effect decreases the crystal field splitting.

3.3. Absorption Spectra

Figure 2 shows the theoretical absorption spectra of α-Al_2O_3: V^{3+} calculated using the first-principles DVME method without referring to any experimental data. The theoretical absorption spectra estimated by several computational conditions involving lattice-relaxation effect and energy-correction effect are compared. The spectra obtained from the calculations based on 7-atom model clusters are shown on the left panel, Fig. 2a. On the other hand, the spectra obtained from the calculations based on 63-atom model clusters are shown on the right panel, Fig. 2b. Here, we used the Gaussian function with 0.2 eV FWHM (full width at half maximum) for broadening. The observed absorption spectra at 77 K obtained from the work reported by D.S. McClure [18] are shown at the bottom of each panel. The dashed red line indicates the σ spectrum, $E \perp c$. On the other hand, the solid blue line indicates the π spectrum, $E \parallel c$.

In the observed spectra, the first peak was assigned to the transition energy from the $^3T_{1a}$ ground state to the $^3T_2$ state, while the second peak was assigned to the transition energy from the $^3T_{1a}$ ground state to the $^3T_{1b}$ state. It was shown that the intensity of the observed $^3T_{1b}$ state is much stronger in the π spectrum compared to the σ spectrum. This tendency has been successfully reproduced in this work. The calculated spectra obtained based on either 7-atom or 63-atom model clusters. On the other hand, in the case of $^3T_2$ state, the observed intensity of σ spectrum is slightly stronger than the π spectrum. Unfortunately, this tendency could only be reproduced by the calculations based on 7-atom model cluster. Next, in the case of observed π spectrum, it was also shown that the relative intensity of $^3T_{1b}$ state is stronger than in those of $^3T_2$ state. Oppositely, in the case of σ spectrum, the relative intensity of $^3T_{1b}$ state is weaker than in those of $^3T_2$ state. In this work, we have successfully reproduced the tendency of π spectrum by various computational conditions. On the other hand, the calculated σ spectra of $^3T_2$ and $^3T_{1b}$ states are comparable.

Detailed investigation shows that although the lattice relaxation ratio was estimated to be about ca. 105.48% or elongated from the original structure of pure α-Al_2O_3, the calculated absorption spectra shift towards the lower energy region. On the other hand, the effect of energy correction, CDC-CC, plays an important role to also shift the peak position toward the lower energy side. Therefore, when the lattice relaxation-effect and the CDC-CC effect were combined, the agreement between the calculated and the observed absorption spectra was improved excellently.

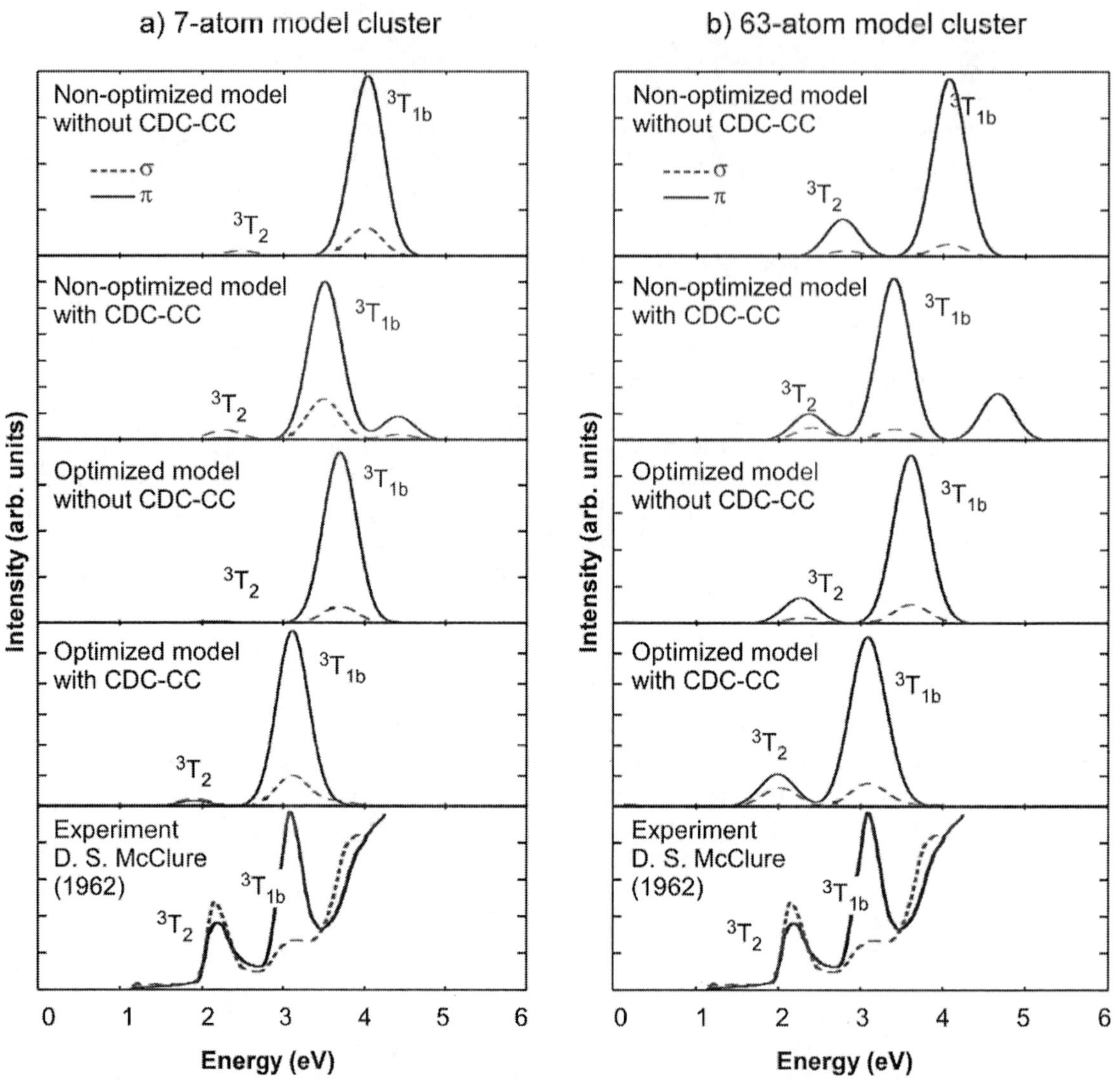

Figure 2. Theoretical absorption spectra of α-Al_2O_3: V^{3+} calculated based on (a) 7-atom model cluster (VO_6^{9-}) and (b) 63-atom model clusters ($VAl_{14}O_{48}^{51-}$) model clusters with several computational conditions involving lattice relaxation-effect and CDC-CC energy correction. The dashed red line indicates the σ spectrum, $E \perp c$. On the other hand, the solid blue line indicates the π spectrum, $E \parallel c$. The experimental absorption spectra reported by D.S. McClure [18] are shown at the bottom side.

3.4. Multiplet Energy Levels

Since V^{3+} belongs to $3d^2$ configuration, the ground state $^3T_{1a}$ has triply degenerate spatial functions. It splits into the $^3T_{1a}(A)$ and $^3T_{1a}(E)$ states due to the trigonal field, the energy of the $^3T_{1a}(A)$ state is set at zero. The transition from the $^3T_{1a}$ ground state to the triplets such as $^3T_2$, $^3T_{1b}$ are observed as strong broad bands, while the transition to the singlets such as $^1T_2$, ^1E and $^1A_1$ are observed as weak and sharp lines. The calculated multiplet energies both singlets and triplets are presented in Table 1 together with the experimental data obtained from Refs. 18-20. The calculations were based on 7-atom model cluster (VO_6^{9-}) and 63-atom model clusters ($VAl_{14}O_{48}^{51-}$) with various computational conditions such as lattice relaxation effect and CDC-CC energy correction effect. Table 1 shows that the consideration of CDC-CC energy correction reduces the multiplet energies of both singlets and triplets. Similarly, the consideration of lattice relaxation reduces the multiplet energies especially for the triplets. When we used the combination of both the lattice relaxation and the CDC-CC energy correction, the calculated $^3T_2$ states were underestimated and the other states were slightly overestimated. However, those energies reasonably agree with the experimental data. Compared to the calculations based on the 7-atom model cluster, the calculations based on the 63-atom model cluster reproduce the experimental data more suitably.

Table 1. The multiplet energy levels (eV) estimated based on 7-atom model cluster (VO_6^{9-}) and 63-atom model clusters ($VAl_{14}O_{48}^{51-}$) with various computational conditions involving lattice relaxation-effect and CDC-CC energy correction. The estimated values were averaged within each state in the O_h notation. The experimental multiplet energies obtained from several reports are listed on the top.

Multiplet state			$^1T_2$	^1E	$^1A_1$	$^3T_2$	$^3T_{1b}$
Experiment			1.09 [19]	1.21 [20]	2.61 [19]	2.17 (σ) [18] 2.16 (π) [18]	3.09 (σ) [18] 3.14 (π) [18]
7-atom model cluster	Non-optimized	Without CDC-CC	1.80	1.87	3.83	2.46	4.01
		With CDC-CC	1.45	1.51	3.10	2.27	3.51
	Optimized	Without CDC-CC	1.78	1.85	3.73	2.16	3.70
		With CDC-CC	1.43	1.51	3.01	1.91	3.13
63-atom model cluster	Non-optimized	Without CDC-CC	1.56	1.63	3.41	2.79	4.07
		With CDC-CC	1.26	1.36	2.75	2.40	3.41
	Optimized	Without CDC-CC	1.57	1.68	3.37	2.27	3.62
		With CDC-CC	1.31	1.43	2.80	1.98	3.11

4. Conclusion

The improved calculations for the absorption spectra of α-Al_2O_3: V^{3+} have been performed by DVME method without referring to any experimental parameter. Our results show that the lattice relaxation ratio was estimated to be ca. 105.48%. Due to the fact that the crystal radius of V^{3+} ion with 6 coordination number (0.78 Å) is larger than those of Al^{3+} ion (0.675 Å), the V-O bond lengths of α-Al_2O_3: V^{3+} were elongated about ca. 5.48% of Al-O bond lengths of pure α-Al_2O_3 crystal. Improvement on the α-Al_2O_3: V^{3+} structure through the geometry optimization evidently enhanced the accuracy of the calculated absorption spectra. Overall calculations show that the combination of the lattice relaxation and CDC CC energy correction is especially important to correct the overestimation in the absorption spectra of α-Al_2O_3: V^{3+}.

References

[1] K. Ogasawara, T. Ishii, I. Tanaka, and H. Adachi, 2000. "Calculation of multiplet structures of Cr^{3+} and V^{3+} in α-Al_2O_3 based on a hybrid method of density-functional theory and the configuration interaction", *Phys. Rev. B*, Vol. **61**, pp. 143-161.

[2] M. Novita, T. Honma, B. Hong, A. Ohishi, K. Ogasawara, 2016. "Study of Multiplet Structures of Mn^{4+} Activated in Fluoride Crystals", *J. Lumin.*, Vol. **169**, pp. 594-600.

[3] M. Novita and K. Ogasawara, 2014. "Study on Multiplet Energies of V^{2+}, Cr^{3+}, and Mn^{4+} in MgO Host Crystal Based on First-Principles Calculation with Consideration of Lattice Relaxation", *J. Phys. Soc. Jpn.*, Vol. **83**, pp. 124707-1 - 124707-6.

[4] M. Novita, H. Yoshida and K. Ogasawara, 2013, "Investigation of Ion Dependence of Electronic Structure for $3d^3$ Ions in $MgTiO_4$ Based on First-Principles Configuration-Interaction Calculations", *ECS Transaction*, Vol. **50**, pp. 9-17.

[5] M. Novita and K. Ogasawara, 2012. "Comparative Study of Multiplet Structures of Mn^{4+} in K_2SiF_6, K_2GeF_6, and K_2TiF_6 Based on First-Principles Configuration-Interaction Calculations", *Jpn. J. Appl. Phys.*, Vol. **51**, pp. 022604-1 - 022604-7.

[6] M. Novita and K. Ogasawara, 2012. "Comparative Study of Absorption Spectra of V^{2+}, Cr^{3+} and Mn^{4+} in α-Al_2O_3 Based on First-Principles Configuration-Interaction Calculations", *J. Phys. Soc. Jpn.*, Vol. **81**, pp. 104709-1 - 104709-9.

[7] M. C. Payne, M. P. Teter, D. C. Allan, T. A. Arias, and J. D. Joannopoulos, 1992. "Iterative minimization techniques for ab initio total-energy calculations: molecular dynamics and conjugate gradients", *Rev. Mod. Phys.*, Vol. **64**, pp. 1045-1097.

[8] M. D. Segall, P. J. D. Lindan, M. I. J. Probert, C. J. Pickard, P. J. Hasnip, S. J. Clark, M. C. Payne, 2002. "First-principles simulation: ideas, illustrations and the CASTEP code", *J. Phys.: Condens. Matter*, Vol. **14**, pp. 2717-2744.

[9] S. J. Clark, M. D. Segall, C. J. Pickard, P. J. Hasnip, M. I. J. Probert, K. Refson, and M. C. Z. Payne, 2005. "First principles methods using CASTEP", *Kristallogr.* Vol. **220**, pp. 567-570.

[10] H. Sawada, 1994. "Residual electron density study of D-aluminum oxide through refinement of experimental atomic scattering factors", *Mater. Res. Bull.*, Vol. **29**, pp. 127-133.

[11] P. Hohenberg and W. Kohn, 1964. "Inhomogeneous Electron Gas", *Phys. Rev. B*, Vol. **136**, pp. 864-871.

[12] R. Car and M. Parrinello, 1985. "Unified approach for molecular dynamics and density-functional theory", *Phys. Rev. Lett.*, Vol. **55**, pp. 2471-2474.

[13] D. Vanderbilt, 1990. "Soft self-consistent pseudopotentials in a generalized eigenvalue formalism", *Phys. Rev. B*, Vol. **41**, pp. 7892-7895.

[14] J. P. Perdew, K. Burke, and M. Ernzerhof, 1996. "Generalized Gradient Approximation Made Simple", *Phys. Rev. Lett.*, Vol. **77**, pp. 3865- 3868.

[15] H. Adachi, M. Tsukada, and C. Satoko, 1978. "Discrete Variational Xα Cluster Calculations. I. Application to Metal Clusters.", *J. Phys. Soc. Jpn.*, Vol. **45**, pp. 875-883.

[16] R. D. Cowan, The Theory of Atomic Structure and Spectra; University of California press: London, 1981; p 461.

[17] S. Sugano, Y. Tanabe, and H. Kamimura, Multiplets of Transition-Metal ions in Crystals (Academic, anew York, 1970)

[18] D. S. McClure, 1962. "Optical Spectra of Transition- Metal Ions in Corundum", *J. Chem. Phys.*, Vol. **36**, pp. 2757-2779.

[19] M. H. L. Pryce and W. A. Runciman, 1958. "The absorption spectrum of vanadium corundum", *Discuss faraday Soc.*, Vol. **26**, pp. 34-42.

[20] G. D. Jones and W. A. Runciman. 1960. "The infra-red absorption spectra of potassium manganicyanide and vanadium corundum", *Proc. Phys. Soc.* Vol. **76**, pp. 996-998.

IOP Conf. Series: Materials Science and Engineering **299** (2018) 012061 doi:10.1088/1757-899X/299/1/012061

Effects of alkanolamide and epoxidation in natural rubber and epoxidized natural rubbers compounds

I Surya*[1,2], S Maulina[1,2], H Ismail[3]
[1]Department of Chemical Engineering, Universitas Sumatera Utara, Medan, Indonesia
[2]Sustainable Energy and Biomaterial Center of Excellence, Faculty of Engineering, Universitas Sumatera Utara, Medan, Indonesia
[3]School of Materials and Mineral Resources Engineering, Universiti Sains Malaysia, Penang, Malaysia

*E-mail: indradanas@yahoo.com

Abstract. The effects of alkanolamide (ALK) loading and epoxidation on cure characteristics, crosslink density and mechanical properties of unfilled natural rubber (NR) and epoxidized natural rubbers (ENRs) compounds have been investigated. The NR, ENR-25 and ENR-50 having 0, 25 and 50 mol % of epoxidation were utilised. It was found that the cure and scorch times decreased with increasing ALK loading. At similar ALK loading, ENR-50 possessed the lowest scorch and cure times followed by ENR-25 and NR compounds. Compared to the control compound (without ALK), the mechanical properties of the three rubbers have been enhanced by the ALK addition. However at a similar ALK loading, the mechanical properties viz. tensile modulus, tensile strength and hardness of ENR-50 vulcanisates were the highest followed by ENR-25 and NR vulcanisates. The superiority in mechanical properties of ENR-50 was due to its highest degree of crosslink density.

1. Introduction

Natural rubber (NR) is produced from the latex of the *Hevea brasiliensis* tree. Before coagulation, the latex was stabilized with preservatives such as ammonia, formaldehyde, sodium sulfite and hydroxylamine which are added to produce technically-specified, constant viscosity grades of NR. NR contains 93 to 95% of cis-1, 4 polyisoprene as shown in Figure 1. NR is widely used in general purposes, industrial and engineering applications which include hoses, seals, sealant, O-rings, safety boots, tires, bushing and mountings materials.

Epoxidised natural rubber (ENR) has been introduced to the rubber world as a chemically modified form of NR, in which some of the unsaturations are converted into epoxide groups which are randomly distributed along the NR chains as shown in Figure 2. Whenever NR is epoxidised, both its physical and chemical properties change according to the epoxidation degree. Commercially, ENR is available in 25 mol% epoxidation (ENR-25) and 50 mol% epoxidation (ENR-50). Many investigations on the preparation of ENR from NR latex and its characteristics have been reported and, practically up to 90 mol% epoxidation is possible [1]. Increasing the degree of epoxidation increases the glass transition temperature (Tg) which results in decrease resilience (more damping) and air permeability, increase hysteresis and better wet traction. The epoxidation increases the polarity of the NR which improves its resistance towards hydrocarbon oils. ENR-25 and ENR-50 are strain crystallize [2], which reflected by their gum vulcanisates properties.

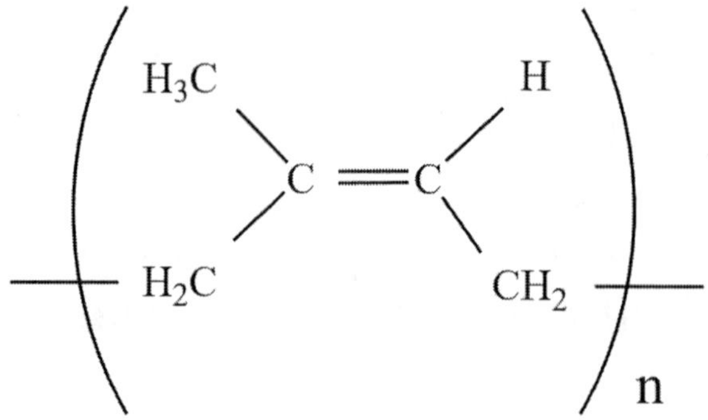

Figure 1. Molecular structure of natural rubber.

ENR, like NR, are cured by any sulphur accelerated vulcanisation system, but semi efficient (Semi-EV) system or efficien (EV) system type formulation is preferred [2-3]. In general, Semi-EV or EV vulcanisation of ENRs showed properties comparable to those of their NR counterparts. While Conventional (CV) system is not recommended for ENRs because of their poor ageing characteristics [4].

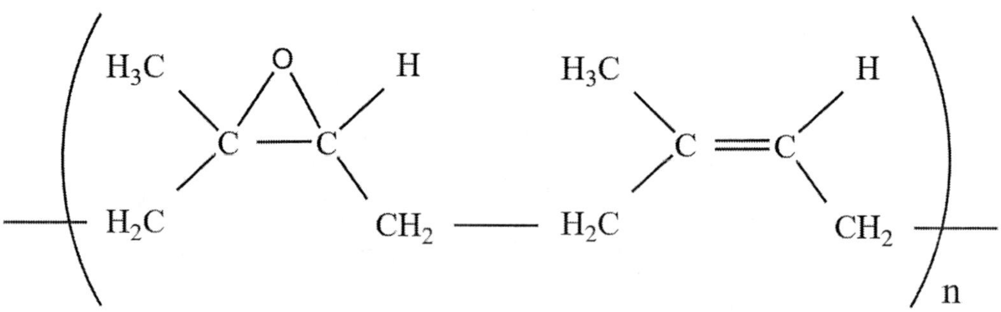

Figure 2. Molecular structure of epoxidised natural rubber.

This study reports the utilising of alkanolamide (ALK) as a rubber additive for NR and ENRs. The effects of ALK addition and degree of epoxidation on cure characteristics, crosslink density and mechanical properties of unfilled NR and ENRs compounds were investigated.

2. Materials and Methods

2.1. Materials

NR grade SMR-L was obtained from Guthrie (M) Sdn. Bhd., Seremban, Malaysia. ENR 25 and ENR 50 were supplied by the Rubber Research Institute Malaysia (RRIM). Other compounding ingredients such as sulphur (S), zinc oxide (ZnO), stearic acid, N-isopropyl-N'-phenyl-p-phenylenediamine (IPPD), Mercapto benzothiazolyl disulphide (MBTS) were supplied by Bayer Co. (M) Sdn. Bhd., Petaling Jaya, Selangor, Malaysia. All materials were used as supplied. The ALK was synthesized in our laboratory using Refined Bleached Deodorized Palm Stearin (RBDPS) and diethanolamine. The reaction procedure and molecular characterization of the ALK was given in the previous report [5] and the chemical formula of ALK is $CH_3(CH_2)_{14}CON(CH_2CH_2OH)_2$.

2.2. Compounding and cure characteristics

A semi-efficient vulcanisation system was applied for the compounding. The compounding procedure was performed on a two-roll mill (Model XK-160). Table 1 displays the compound formulation of unfilled NR and ENRs compounds with and without ALK. The compounding ingredients were mixed using a laboratory two-roll mill, Model XK-160. The cure characteristics of the unfilled rubbers compounds were determined at 150 °C using a Monsanto Moving Die Rheometer (MDR 2000). The respective scorch time (ts_2), cure time (t_{90}) and torque difference (M_H—M_L) were obtained from the rheograph according to ISO 3417. The compounds were subsequently compression moulded using a stainless steel mould at 150 °C with a pressure of 10 MPa using a laboratory hot-press based on the respective curing times.

Table 1. Composition of the rubber compounds

Ingredients	Content (phr)[a]
Rubber[b]	100.0
Zinc oxide	5.0
Stearic acid	2.0
IPPD	2.0
MBTS	1.5
Sulphur	1.5
ALK	0.0; 0.2; 0.4; 0.6; 0.8 and 1.0

[a] parts per hundred parts of rubber
[b] NR, ENR-25 or ENR-50

2.3. *Tensile and hardness properties*

Dumbbell-shaped samples were cut from the moulded sheets. Tensile tests were performed at a cross-head speed of 500 mm/min using an Instron 3366 universal tensile machine according to ISO 37. The tensile strength (TS) and stress at 100% elongation (M100), stress at 300% elongation (M300) and elongation at break (EB) were investigated. The hardness measurements of the samples were obtained according to ISO 7691-I using a Shore A type manual Durometer.

2.4. *Measurement of crosslink density*

Swelling tests on the unfilled NR and ENRs vulcanisates were performed in toluene in accordance with ASTM D471-12a. The cured test pieces (30 mm × 5 mm × 2 mm) were weighed using an electric balance and swollen in toluene until equilibrium, which took 72 hours at room temperature. The samples were taken out from the liquid, the toluene was removed from the sample surfaces and the weight was determined. The samples were then dried in the oven at 70°C until constant weights were obtained. The swelling results were used to calculate the molecular weight between two crosslinks (M_c) by applying the Flory-Rehner equation [6-7].

$$M_c = \frac{-\rho_r V_s V_r^{1/3}}{\ln(1-V_r)+V_r+\chi V_r^2} \quad (1)$$

$$V_r = \frac{1}{1+Q_m} \quad (2)$$

where ρ is the rubber density which are 0.92 g/cm^3 and 0.95 g/cm^3 respectively, V_s is the molar volume of the toluene (V_s = 106.4 cm^3/mol), V_r is the volume fraction of the rubber in the swollen specimen, Q_m is the weight increase of the vulcanisate in toluene and χ is the interaction parameter of the rubber network–solvent (χ of NR/ENRs = 0.393). The crosslink density is given by;

$$V_c = \frac{1}{2M_c} \quad (3)$$

3. Results and Discussion

3.1. *The effects of ALK addition on cure characteristics*

The effects of ALK addition on cure characteristics of the unfilled NR and ENRs compounds are shown in Table 2. As can be seen, the additions of each 0.2 phr of ALK decreased the scorch and cure times of those control compounds. Cure enhancement phenomena were observed. It was due to the amine content of ALK [8-11], because amine is an accelerator for rubber compounds [8].
The higher the ALK loading, the lower were the scorch and cure times. It meant that increases the ALK loading caused increases in amount of amine and hence, the faster were the curing processes.

Table 2. The effects of ALK on cure characteristics and crosslink density of NR and ENRs compounds

NR / ENRs Compounds	ALK Loadings					
	0.0 (Control)	0.2	0.4	0.6	0.8	1.0
SMR-L/NR						
ts_2, min.	4.59	4.52	4.24	3.96	3.74	3.73
t_{90}, min.	7.13	7.05	6.65	6.45	6.00	5.83
M_H—M_L, dN.m	4.77	4.85	4.97	5.21	5.12	4.89
Crosslink dens.*	33.77	33.78	34.16	36.18	35.82	33.93
ENR-25						
ts_2, min.	3.56	3.24	2.96	2.74	2.54	2.43
t_{90}, min.	6.03	5.59	5.55	5.51	5.45	5.30
M_H—M_L, dN.m	5.12	5.19	5.25	5.79	5.44	5.24
Crosslink dens.*	34.05	34.17	35.62	38.15	37.42	34.34
ENR-50						
ts_2, min.	2.53	2.47	2.44	2.38	2.33	2.19
t_{90}, min.	5.46	5.38	5.34	5.26	5.20	4.99
M_H—M_L, dN.m	6.31	6.53	6.83	6.85	6.89	5.96
Crosslink dens.*	39.84	40.28	41.96	44.08	46.06	43.65

* Crosslink density (x10^5 mol/cm^3)

At a similar ALK loading, the scorch and cure times of both ENRs were shorter than those of NR. It was due to the presence of epoxide groups in the ENRs molecules. The epoxide groups activated the adjacent double bonds and caused them a more susceptible and ready for curing [12]. Compared to ENR-25, ENR-50 has more epoxide groups. Therefore, the scorch and cure times of ENR-50 were shorter than those of ENR-25.

Table 2 also shows that the additions of each 0.2 phr of ALK into the control compounds increased the torque differences. NR and ENR-25 have similar trends of torque differences which were increased with the addition of ALK up to 0.6 phr, and decreased beyond the loading. In contrast to both rubbers, the torque differences of ENR-50 increased with the ALK addition at a relatively higher loading, 0.8 phr, and then decreased with further increase in the ALK loading.

Theoretically, torque difference represents shear dynamic modulus which indirectly related to the crosslink density of a rubber compound [9-12]. The increases of the torque difference up to ALK optimum loadings (0.6 phr for NR and ENR-25; 0.8 phr for ENR-50) were due to the function of ALK as a (secondary) accelerator during the curing of the rubbers compounds. Although an accelerator constitutes a very small part of a rubber compound, it has a profound influence on the nature of curing [8].

The decreases of the torque differences, beyond the optimum ALK loadings, were due to the excessive amount of ALK, which decreased the total crosslink density.

At a similar ALK loading, the torque differences of both ENRs were higher than that of NR. It was due to the degree of epoxidation which formed additional amine-epoxy crosslinks [12] in curing processes of the rubbers compounds. Compared to ENR-25, ENR-50 with more epoxide groups might form a higher degree of additional amine-epoxy crosslinks.

3.2. The effects of ALK addition on crosslink density

The crosslink density of the unfilled NR and ENRs compounds with and without ALK was determined by using Equations (1)-(3) [6] and the results are shown in Table 2. The results of crosslink density were in line with those of torque difference. The crosslink density increased up to the optimum ALK loadings, and then decreased beyond them.

The increases in crosslink density up to the optimum ALK loadings were due to the function of ALK as a secondary accelerator and the nature of the molecule. The nitrogen atom or amine constituent of ALK activated not only the rubbers but also the elemental suphur during curing process. Together with the others curatives, the amine formed intermediate complexes which attached the available elemental suphur to rubbers chains more efficient [13] causing higher in degree of crosslinking. The nitrogen atom also acted as a hydrogen acceptor and amine accelerators activated the elemental suphur and/or the rubber for the crosslinking reaction [13].

The decreases in crosslink density beyond the optimum ALK loadings were most probably due to the softening or lubricating effect of the excessive ALK which caused in decreases in crosslink density. This can be due to the phenomenon of dissolving a part of the elemental sulphur into the excessive ALK, and as a consequent less sulphur was attached to the rubbers chains.

3.3. The effects of ALK addition on mechanical properties

The effects of ALK loading on mechanical properties of the unfilled NR and ENRs vulcanisates are shown in Table 3. As can be seen, the additions of each 0.2 phr ALK into the control compounds increased the elongation at break (EB). Increases the ALK loading further increased the EB or the extensibility of the rubbers vulcanisates. It might be due to the function of ALK as an internal plasticiser [5]. A plasticiser provides a monolayer in the rubbers compounds which causes in free volume and hence, allowing a more mobility/flexibility for the chains of the rubbers [13-14]. Increases the ALK loading have the same effect as increases in free volume that enhanced the flexibility or the extensibility of the unfilled rubbers vulcanisates. Therefore, as shown in Table 3; the higher ALK loading the higher were the EB.

Table 3. The effects of ALK on mechanical properties of NR and ENRs compounds

NR / ENRs Vulcanisates	ALK Loadings 0.0 (Control)	0.2	0.4	0.6	0.8	1.0
SMR-L/NR						
M100, MPa	0.538	0.544	0.561	0.586	0.545	0.516
M300, MPa	1.235	1.236	1.311	1.364	1.267	1.231
EB, %	1070.0	1095.7	1116.5	1141.5	1154.0	1172.0
TS, MPa	17.7	19.3	19.5	20.1	18.4	16.5
Hardness, Shore A	36	37	39	40	38	37
ENR-25						
M100, MPa	0.549	0.556	0.569	0.595	0.554	0.542
M300, MPa	1.334	1.391	1.403	1.454	1.366	1.350
EB, %	803.4	863.3	877.2	886.6	893.4	913.4
TS, MPa	18.0	19.7	20.2	20.9	19.7	19.4
Hardness, Shore A	37	38	40	41	39	38

ENR-50						
M100, MPa	0.637	0.721	0.732	0.753	0.760	0.692
M300, MPa	1.610	1.771	1.781	1.793	1.891	1.689
EB, %	772.5	825.0	837.5	842.9	865.0	869.8
TS, MPa	19.5	20.4	21.0	21.4	21.8	20.0
Hardness, Shore A	39	40	41	42	43	40

At a similar ALK loading, The EBs of both ENRs were lower than that of NR. It was due to decreases in degree of crystallinity as the mole % epoxidation increased from NR to ENR 50 [12]

As shown in Table 3, the ALK additions increased the M100, M300, TS and hardness slightly up to some maximum levels (at 0.6 phr for NR and ENR-25; and 0.8 phr for ENR-50). Beyond the loadings caused in decreases those mechanical properties. Tensile strength and hardness exhibited similar trends. Both M100 and M300 are typical measures stiffness or hardness of a rubber vulcanisate [15] Modulus or stiffness/hardness and also tensile properties are dependent mainly on degree of crosslink [16-18]. The increases in mechanical properties up to the ALK optimum loadings were due to a higher degree of crosslink density. The decreases in the properties beyond the ALK optimum loadings were due to a lower degree of crosslink density and a more pronounced softening effect of ALK.

At a similar ALK loading, the mechanical properties of both ENRs were higher than those of NR. Presumably, it was due to the specific chemical crosslinking taking place between the epoxy groups and ALK during cure process and hence, contributed to a higher degree of crosslink density.

4. Conclusion

Alkanolamide can be utilised as a new curative additive in unfilled natural and epoxidized natural rubbers compounds. As an accelerator, it increased the cure rate. It was also increased the crosslink density, hardness and tensile properties up to 0.6 phr (for natural and epoxidized natural rubber 25% epoxidation); and up to 0.8 phr (for epoxidized natural rubber 50% epoxidation). The optimum loadings were 0.6 phr of alkanolamide for natural and epoxidized natural rubber 25% epoxidation; and 0.8 phr for epoxidized natural rubber 50% epoxidation. The alkanolamide also can be utilised as an internal plasticiser which increased the flexibility or elongation at break of the unfilled natural and epoxidized natural rubbers compounds.

Acknowledgement

The authors would like to thank Universiti Sains Malaysia for providing the research facilities for carrying out the experiment and for making this research work possible.

References

[1] Akiba M and Hashim A 1997 *Prog Polym Sci* **22**(3) 475-521.
[2] Gelling I *et al* 1988 *Natural rubber science and technology* Eirich FR Ed 367.
[3] Baker *et al* 1985 *Rubber Chem Technol* **58**(1) 67-85.
[4] Gelling et al 1985 *Rubber Chem Technol* **58**(2) 243-57.
[5] Surya I *et al* 2013 *Polym Test* 2013 **32**(8) 1313-21.
[6] Flory PJ and Rehner JJr 1943 *J Ch Ph* **11** 521-26.
[7] Flory PJ 1950 *J Ch Ph* **18**(1) 108-11.
[8] Long H 1985 *Basic Compounding and Processing of Rubber, Rubber Division* American Chemical Society Inc USA.
[9] Surya I and Ismail H 2016 *Polym Test* **50** 276-82.
[10] Surya I *et al* 2014 *Polym Test* **40** 24-32.
[11] Surya I *et al* 2015 *Polym Test* **42** 208-14.
[12] Ismail H and Chia H 1998 *Polym Test* **17**(3) 199-210.
[13] Rodgers, B. (2004). *Rubber Compounding: Chemistry and Applications*: CRC.
[14] Ismail H and Chia H 1998 *Polym Test* **17**(3) 199-210.

[15] Teh P *et al* 2004 *Eur Polym J* **40**(11) 2513-21.
[16] I Surya *et al* 2017 *IOP Conf Ser Mater Sci Eng* **223** 012006.
[17] Surya I *et al* 2017 *IOP Conf Ser Mater Sci Eng* **223** 012012.
[18] Surya I and Ismail H 2016 *Plast Rubber and Compos* **45**(7) 287-93.

Synthesis and Characterization of Fe_3O_4 Nanoparticles using Polyvinyl Alcohol (PVA) as Capping Agent and Glutaraldehyde (GA) as Crosslinker

Lale Budi Hutami Rahayu[1], Ika Oktavia Wulandari[1], Djoko Herry Santjojo[2], Akhmad Sabarudin[1*]

[1]Department of Chemistry, Faculty of Science, Brawijaya University, Jl Veteran 65145 Malang, Indonesia
[2]Department of Physics, Faculty of Science, Brawijaya University, Jl Veteran 65145 Malang, Indonesia

Email: sabarjpn@ub.ac.id

Abstract. The use of polyvinyl alcohol (PVA) as a capping agent and glutaraldehyde (GA) as a crosslinker for a synthesis of magnetite (Fe_3O_4) nanoparticles is able to reduce agglomeration of produced Fe_3O_4. Additionally, oxidation of Fe_3O_4 by air could be avoided. The synthesis is carried out in two steps: first step, magnetite (Fe_3O_4) nanoparticles were prepared by dissolving the $FeCl_3.6H_2O$ and $FeCl_2.4H_2O$ in alkaline media ($NH_3.H_2O$). The second step, magnetite nanoparticles were coated with polyvinyl alcohol (PVA) and glutaraldehyde (GA) to obtain Fe_3O_4-PVA-GA. The latter material was then characterized by FTIR to determine the typical functional groups of magnetite coated with PVA-GA. X-ray Diffraction analysis was used to determine structure and size of crystal as well as the percentage of magnetite produced. It was found that the produced nanoparticles have crystal sizes around 4-9 nm with the cubic crystal structure. The percentage of magnetite phase increases when the amount of glutaraldehyde increased. SEM-EDX was employed to assess the surface morphology and elemental composition of the resulted nanoparticles. The magnetic character of the magnetite and Fe_3O_4-PVA-GA were studied using Electron Spin Resonance.

1. Introduction

Nanoparticles are nano-sized materials that can be applied in several fields such as industrial and biomedical fields. Applications of nanoparticles in the biomedical field to enhance magnetic resonance imaging (MRI) magnification image and as drug delivery system for cancer therapy [1]. The use of the drug delivery system is capable of delivering the drug to the target cell [2]. The presence of magnetic properties in magnetic materials can be used to transmit targeted drugs using an external magnetic field. Iron oxide is a magnetic material that has low toxicity, cheap, biocompatible, biodegradable and unique magnetic properties [3], [4]. The magnetic field in the iron oxide occurs because of the four unpaired electrons in the d orbitals [5]. Iron oxide can be utilized as a target drug delivery system if it has a size below 20 nm. Some researchers have reported their works on iron oxide nanoparticles by varying the factors that affect the properties of nanoparticles [2], [6], [7].

The synthesis of magnetite nanoparticles is largely chemical made to obtain nanoparticles in small size more easily. The synthesis method with co-precipitation at room temperature is known to be the least expensive, easy method and obtains a high percentage of magnetite (Fe_3O_4) [8]. Magnetite nanoparticles have a low stability that tends to agglomerate and easy to oxidate. The agglomeration properties of magnetite nanoparticles influenced the presence of Van der Waals forces and the pull of magnetic dipole [9], [10]. To prevent aggregation of magnetite nanoparticles, a surface modification or a capping agent capable of conjugating with therapeutic (drug) molecules [1] is required. Usually, nanoparticle coatings may use polymers or surfactants, since the presence of crosslinked polymers can prevent coagulation and produce monodisperse particles [11].

Polyvinyl alcohol (PVA) can be used to stabilize nanoparticles due to their biodegradable and hydrophilic properties [12]. However, the hydrophilic nature of PVA can reduce its effectiveness as a capping agent because drug release can occur before it reaches the target cell associated with its interaction with the fluid in the body. Thus a crosslinker is required to inhibit the degradation of the nanoparticle-polymer so that degradation can occur after the drug reaches the target cell. Glutaraldehyde (GA) can be used as a crosslinker because it is stable in acidic conditions so that it can be obtained a nanoparticle that has a high stability in acidic conditions, and can be degraded in body fluids after reaching the target cell. The synthesis of nanoparticles using PVA as a capping agent and glutaraldehyde as a crosslinker has been done by Kim [13] and developed by Wahid [14]. The presence of PVA and glutaraldehyde will produce better Fe_3O_4 nanoparticles.

2. Materials and Methods

2.1. Chemicals and Instrumentation

The material used was Iron (III) chloride hexahydrate, Iron (II) chloride tetrahydrate, glutaraldehyde (Sigma Aldrich), from polyvinyl alcohol (PVA, Mw 3500) technical grade (Merck Indonesia), and aqueous ammonium hydroxide.

FTIR *(Fourier Transform Infrared)* spectrophotometer (Shimadzu FTIR QP89500, KBr pellet in wavenumber range 4000-400 cm^{-1}). The FTIR analysis used to determine the functional groups of nanoparticle samples. Freeze Dry used to remove water molecules on nanoparticle samples (35 XL Virtis SP Scientific) at temperature condition under -40°C and pressure up to ≤ 20 mT. ESR *(electron spin resonance)* used to determine the magnetic character of the nanoparticle materials (Leybold Didactic GMBH). SEM-EDX (*Scanning electron microscopy*) analysis used to determination of morphology and elemental composition in nanoparticles (Hitachi SU 3500 with magnification around 1000 - 20.000 kX). XRD analysis used to measure the crystal size and lattice parameters of Fe_3O_4 nanoparticles and the existence of Fe_3O_4 from nanoparticle samples were carried out using Diffractometer system XPERT-PRO with Cu-Kα λ= 1.5406 Å radiation was used for X-ray powder diffraction measurements. XRD data was recorded across a 2θ range of 5° to 60° using a step size of 0.001°.

2.2. Synthesis Procedure

The synthesis of Fe_3O_4 nanoparticles by co-precipitation method follows a procedure performed by Kim et al., (2012) with a briefly described modification wherein 0.125 g of $FeCl_2.4H_2O$ and 0.35 g $FeCl_3.6H_2O$ with molar ratio (1: 2) dissolved by 100 mL aquadem in a beaker glass and stirred at room temperature at 180 rpm under vacuum, added 5 mL of 3M NH_4OH solution dropwise by syringe pump at a rate of 10 mL/h while stirring using a magnetic stirrer, resulting in a Fe_3O_4 precipitate. Further 5 mL of PVA solution with concentration (5 g of PVA dissolved in 70 mL aquadem) dropwise at 10 mL/h and glutaraldehyde solution (50, and 150 μL) while stirring. The resulting precipitate is then separated by pulling it using a bar magnet placed beneath the beaker glass. The precipitate is then washed using aquadem until neutral pH or ± 7. The sample is then dried in a freeze dryer and the sample is stored in the desiccator before further analysis. The resulting samples are labeled with the name Fe_3O_4-PVA_5-GA_{50} and Fe_3O_4-PVA_5-GA_{150}. Fe_3O_4 nanoparticles were produced using the same method without the addition of PVA and glutaraldehyde.

3. Results and Discussions

The aim of this study was to synthesize Fe_3O_4 nanoparticles using Iron (II) chloride tetrahydrate hexahydrate and Iron (III) chloride hexahydrate with ratio mole (1:2) in alkali medium. The nanoparticles were coated with PVA to avoid oxidizing easily, and glutaraldehyde to stabilize the nanoparticles. To determine the effect of added glutaraldehyde, mole variation of glutaraldehyde were added to the sample of the nanoparticles were added. Interactions occurring on the surface of nanoparticles in the presence of PVA and glutaraldehyde can be seen in scheme 1. The presence of hydroxyl groups on the Fe_3O_4 surface serves to stabilize the surface of Fe_3O_4. Then, the presence of glutaraldehyde will bind to one PVA hydroxyl group with another PVA hydroxyl group forming a crosslinker. Glutaraldehyde may act as a crosslinker because in glutaraldehyde there are two carbonyl groups located at the ends of the glutaraldehyde group. In addition, the presence of glutaraldehyde will also produce an active compound glutaraldehyde which will directly bind to the surface of the nanoparticles, so it is important to organize the amount of glutaraldehyde mole [15].

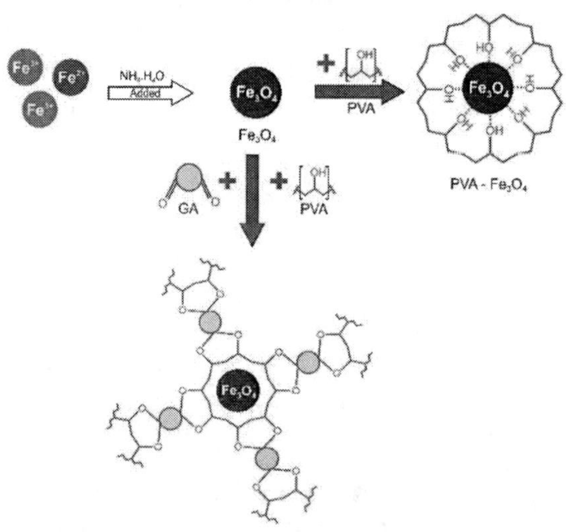

Scheme 1. Interaction of crosslinker formation between Fe_3O_4 nanoparticles with presence of PVA and glutaraldehyde

All synthesized nanoparticle samples indicate the presence of magnetic properties in aqueous water when the sample is closer to the external magnetic field as can be seen in Figure 1. After the sample is dried, the bare Fe_3O_4 and coated with PVA and GA shows different colors. The Fe_3O_4 nanoparticles are black in shape as sand, while the coated nanoparticles are blackish brown with finer shapes, as can be seen in Figure 2.

(a) (b)

Figure 1. Fe_3O_4 nanoparticles (a) before and (b) exposed closer to an external magnetic field

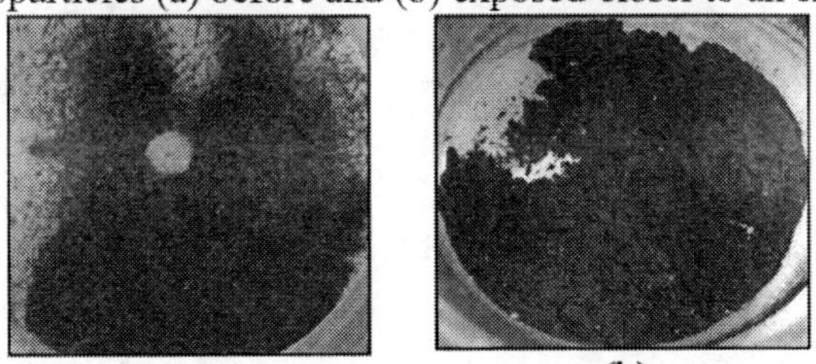

(a) (b)

Figure 2. The forms of Fe_3O_4 nanoparticles (a) bare (a) after coated PVA and glutaraldehyde

SEM analysis shows that the surface morphology of bare Fe_3O_4 and coated are shaped cluster (see Figure 3). This is due to the very small size of the crystals of Fe_3O_4 which is below 20 nm, so the surface energy is still very high. From Figure 3 it is known that the presence of Fe_3O_4 can be seen in the presence of small particles of light/white, due to the presence of high Fe atomic number.

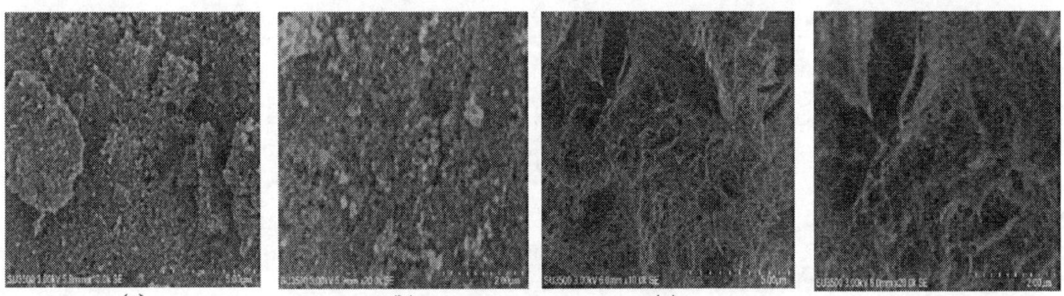

(a) (b) (c) (d)

Figure 3. SEM micrographs for Fe_3O_4 (a) bare in 10.000x magnification (b) bare in 20.000x magnification (c) after coated in 10.000x magnification and (d) after coated in 20.000x magnification

EDX analysis found that elements contained in the sample are Fe, O, and C. The presence of C in Fe_3O_4 bare is due to the sample being analyze coated with carbon. The EDX spectral peaks are in the region binding energy of Fe in the range of 0.8, 6.3, and 6.8 keV [16], with the percentage of atomic mass for bare F_3O_4, F_3O_4-PVA_5-GA_{50} and F_3O_4-PVA_5-$GA1_{50}$ presented in Table 1.

IOP Conf. Series: Materials Science and Engineering **299** (2018) 012062 doi:10.1088/1757-899X/299/1/012062

Table 1. The percentage of atomic mass for nanoparticles samples with the amount of mole of glutaraldehyde added

Samples	%Fe	%O	%C
Bare Fe_3O_4	61,45	32	6,55
Fe_3O_4-PVA_5-GA_{50}	34,73	46,20	19,06
Fe_3O_4-PVA_5-GA_{150}	41,00	48,09	10,91

Diffractogram for nanoparticle samples of XRD characterization can be seen in Figure 4 and the crystal size was determined by Williamson-Hall calculations [17]. Figure. 4 exhibits all samples where the diffraction peaks in the hkl plane are (220), (311), (222), (400), (422), (333) respectively corresponding to the COD standard (Crystallography Open Database) number # 96-900-2320 for Fe_3O_4. The Fe_3O_4 crystal is composed of Fe^{3+} composed of the tetrahedral side and half of Fe^{2+} from the caved octahedral side. The crystal size for bare Fe_3O_4, Fe_3O_4-PVA_5-GA_{50}, and Fe_3O_4-PVA_5-GA_{150} were respectively 4.66; 5.96; and 7.85 nm. So it can be seen that nanoparticles that are coated PVA and glutaraldehyde will increase the crystal size of Fe_3O_4.

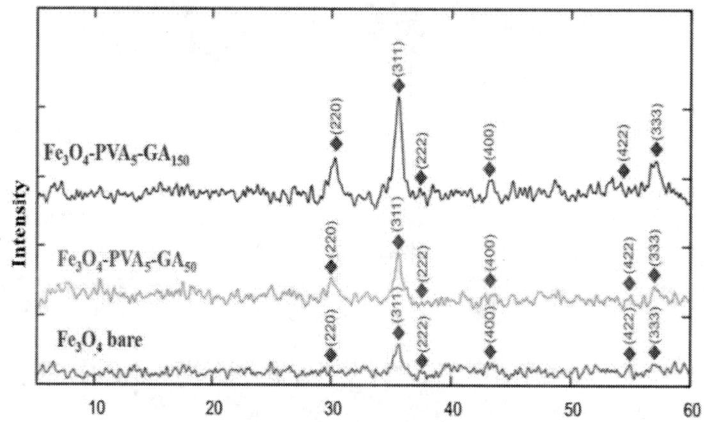

Figure 4. Diffractogram for bare Fe_3O_4, Fe_3O_4-PVA_5-GA_{50} and Fe_3O_4-PVA_5-GA_{150} nanoparticles

The percentage obtained in all samples, the highest percentage of magnetite, followed by the percentage of maghemite and hematite. The amount of mole of glutaraldehyde added, the percentage of magnetite increases, which can be seen in Figure 5.

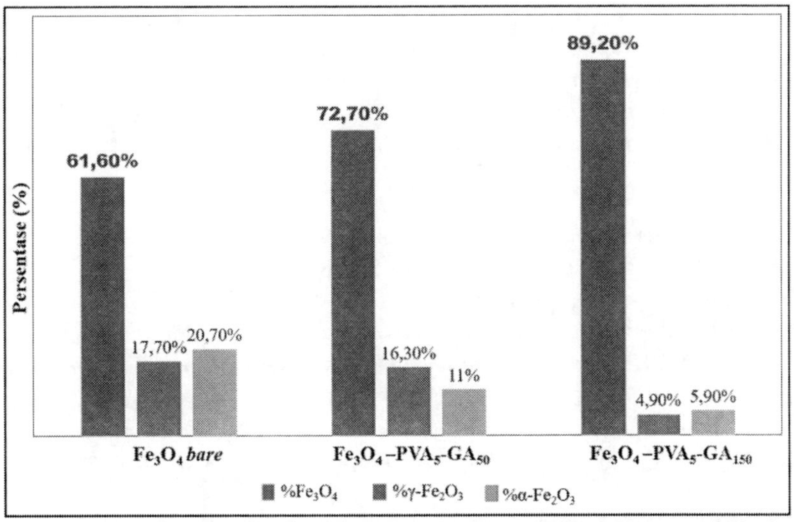

Figure 5. The percentage of magnetite from bare Fe_3O_4, Fe_3O_4-PVA$_5$-GA$_{50}$, and Fe_3O_4-PVA$_5$-GA$_{150}$ nanoparticles

Analysis of magnetic characters from nanoparticle samples by calculating the g-lande factor values obtained from ESR measurements. The g-lande factor values for bare Fe_3O_4, Fe_3O_4-PVA$_5$-GA$_{50}$ and Fe_3O_4-PVA$_5$-GA$_{150}$ nanoparticles were 1.94541; 1.90611; and 1.91903. From the g-lande value obtained, it is known that this value is close to the g factor value for Fe tend to ferromagnetic.

Figure 6 shows FTIR spectra from samples of Fe_3O_4 bare nanoparticles as well as coated PVA and glutaraldehyde. The fingerprint region for Fe-O appears in sharp areas of 450 cm^{-1} and ~ 585 cm^{-1} indicating the presence of magnetite compounds [18] and in the sample there is also a slight peak in the absorbent band of ~ 623 cm^{-1}, indicating that maghemite is also formed [19]. The presence of maghemite compounds that are also detected due to Fe_3O_4 that is not coated will easily oxidize in the presence of oxygen. It is known that in the presence of PVA and glutaraldehyde, the absorption bands in the area of ~ 623 cm^{-1} are decreasing as the amount of mole of glutaraldehyde increased.

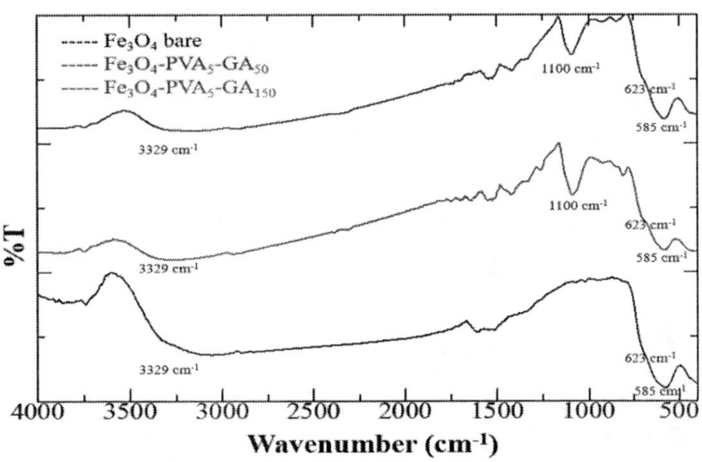

Figure 6. The Spectra of FTIR of bare Fe_3O_4 and coated with PVA and glutaraldehyde.

4. Conclusion

This study described the effect of the added number of glutaraldehyde moles to Fe_3O_4 nanoparticles in the presence of PVA as a capping agent. In this study, the crystal size ranges from 4 to 9 nm with cubic crystal structure with the percentage of magnetite obtained on the Fe_3O_4-PVA_5-GA_{150} sample of 89.20% which is the maximum percentage of the three samples. So it can be concluded that the addition of glutaraldehyde in Fe_3O_4 will increase the percentage of magnetite obtained.

References

[1] Santos M C et al 2016 *Applied Surface Science*, **367** 26–35
[2] Gupta A K and Gupta M 2005 *Biomaterials* **26** 18 3995–4021
[3] Oka C et al 2015 *Journal of Magnetism and Magnetic Materials* **381** 278–284
[4] Hyeonjeong J et al 2016 *Journal of Controlled Release* Accepted manuscript
[5] Teja, Amyn S and Koh P 2009 *Progress in Crystal Growth and Characterization of Materials* **55** (1-2) 22–45
[6] Qiu X et al 2000 *Chinese Journal of Polymer Science* **18** (6) 535–539
[7] Niu J M and Zheng Z G 2014 Effect of Temperature on Fe_3O_4 *Advanced Materials Research* **900** 172-176
[8] Wu W, He Q and Jiang C 2008 *Nanoscale Res Lett* **3** 397-415
[9] Masoudi A et al 2012 *International Journal of Pharmaceutics* **433** (1-2) 129–141
[10] Theerdhala S et al 2010 *Ultrasonics Sonochemistry* **17** (4) 730–737
[11] Kumar Ramesh, Inbaraj B S and Chen B H 2010 *Materials Research Bulletin* **45** 1603–1607
[12] Nadeem M et al *PLoS ONE* **11** (6) 1–12
[13] Kim SY, Ramaraj B and Yoon K R 2012 *Surface and Interface Analysis* **44** (9) 1238–1242
[14] Sabarudin A, Wahid R., Nalle F C, Shobirin R A and Santjojo D J D *RASAYAN J.Chem* **10** (4) 1261-1270
[15] Figueiredo Katia, Alves Tito and Borges Cristiano 2009 *Journal of Applied Polymer Science* **111** 3074–3080
[16] Mahdavi M, Ahmad Mansor B, Haron Md Jelas, Namvar Farideh, Nadi Behzad, Rahman Mohamad Zaki Ab, and Amin Jamileh 2013 *Molecules* **18** 7533-7548
[17] Prabu Y T, Rao K V, Kumar V S and Kumari B S 2014 *World Journal of Nano Science and Engineering* **4** 21-28
[18] Saxena N and Singh M 2017 *Journal of Magnetism and Magnetic Materials* Accepted manuscript
[19] Cornel R M and Schwertmann U 2003 *The Iron Oxides: Structure, Properties, Reactions, Occurrences and Uses* Second Edition. (Weinheim: WILEY-VCH Verlag GmbH and Co. KGaA) Chapter 7 p 146

The Role of Modification SBA-15 Mesoporous Silica with CPTMS in Cd Adsorptions

Donanta Dhaneswara, Jaka Fajar Fatriansyah, Aloysius Brahmarsi Mahagnyana, Farhan Delayori, Dewanto Adhy Putranto and Siti Utami Agustina Adriyani Anwar*

Metallurgical and Materials Engineering, Faculty of Engineering, Universitas Indonesia, Kampus UI Depok, Jawa Barat, 16424

Email: sitiutamilehe@yahoo.co.id

Abstract. Mesoporous silica SBA-15 was synthesized using Pluronic 123 triblock copolymers as surfactant and functionalized with CPTMS to modify its surface. The synthesis of SBA-15 was prepared at optimized concentration of Pluronic 123 surfactant at 60 mM and showed surface area of 831.996 m^2/g determined by BET N_2 adsorption isotherm test at 77 K. The functionalized product, CPTMS-SBA-15 on the other hand has lower surface area at 711.061 m^2/g. This very high surface of CPTMS-SBA-15 and pure SBA-15 were used as adsorbent of Cd from industrial waste water using laboratory water samples. CPTMS-SBA-15 showed lower surface and pore size in comparison with pure SBA-15 but the effectiveness in adsorption of Cd was found higher than pure SBA-15. In this study, we compared the percentage of Cd removal using CPTMS-SBA-15 and pure SBA-15 systems. The materials were characterized by XRD, the presence of organic group was demonstrated by FTIR, and the ions concentration in solution after adsorption process was determined by AAS.

1. Introduction

According to World Bank, the population growth in Indonesia has reached 1.2% per year by 2015, this resulted in rapid development of economic in Indonesia [1]. Therefore, the industrialization process is growing to meet the needs of the community. As the development in industrialization increase, the outcome shows negative effect to environment due to water pollution contains of heavy metals from industrial waste. Unlike any other metals, heavy metals such as Lead, Cadmium, Copper, Zinc, and Chromium has high density than water. These hazardous materials are very dangerous to living things surrounded by them. Especially Cadmium, low exposure to this metal can lead to various symptoms such as vertigo, abdominal pain, vomit, etc. [2]. Thus, remediation of heavy metal ions from industrial water is necessary to prevent further issues.

Nanotechnology has produced a wide range of advanced materials such as mesoporous materials. Mesoporous materials have many wide applications for example as catalysts [3], enzyme carriers [4], and adsorbent material [5]. According to IUPAC, mesoporous materials has pore diameter 2 nm to 50 nm, highly ordered mesoporous structures, and often has large pore volume up to 70% and large surface area (>700 m^2/g) [6], accordingly making it suitable to use in aqueous media as an adsorbent for metals ion adsorption.

Zhao [7[has developed and synthesized a mesoporous material called SBA-15 (stands for *Santa Barbara Amorphous 15)* in University of California, Santa Barbara. SBA-15 is made of silica and has well-ordered hexagonal mesoporous structure. The average diameter of the pores is 46 to 300 angstroms and the thickness of the silica walls is about 31 to 64 angstroms, therefore it has high mechanical and thermal stability making it a promising material to use in wide applications [7]. SBA-15 was synthesized using triblock copolymer as template because it has gained many interest regarding its simplicity to use in synthesis. In this work, a triblock copolymer surfactant Pluronic P123® ($EO_{20}PO_{70}EO_{20}$, EO = ethylene oxide, PO = propylene oxide) was used as a template because of its ability to form long cylindrical micelles and Tetraethyorthosilicate/TEOS as a silica precursor through sol-gel process. Pure SBA-15 is produced through this process and has surface area up to 746.70 m^2/g [8]. Moreover, large surface area of SBA-15 is preferred in adsorption application.

To increasing its capability as adsorbent materials so that it can interact with metallic cations then SBA-15 was modified or functionalized using functional groups such as $-NH_2$, -SH, -S- etc. Previously, [9] has used multi-amine grafted functional groups contains of amino, di-amino, and tri-amino to modify surface area of pure SBA-15 and obtained a remarkable adsorption capacity of 726 mg Hg^{2+}/g [9]. The SBA-15 capability used as a heavy metal absorber has been demonstrated by [10] on some heavy metals such as zinc (Zn), copper (Cu) and cobalt (Co) using 3-aminopropyl-triethoxysilane (APTES) and salicylaldehyde (SA) as functional group. The functionalized SA-SBA-15 showed a remarkable adsorption capacity for Cu^{2+} compared to other metals [10]. In this present study, we had two goals. Firstly, use (3-Chloropropyltrimethoxysilane) or CPTMS as salinization agent to modify the surface area of pure SBA-15 on Cd adsorption capacity. Other than functional groups mention earlier, CPTMS has Chlor atom to accommodate single electron pair to metallic cation so that it interacts strongly. The process of SBA-15 functionalization with CPTMS has been done by Asgari et. al [5] which is used as a lead metal ion absorbent material. Secondly, we studied the adsorption capacity of functionalized SBA-15 with CPTMS in Cd adsorption application.

2. Experimental

2.1 Materials

Materials were prepared to synthesize silica mesoporous SBA-15 through sol-gel process. Tetraethyorthosilicate/TEOS was purchased from Merck and used as silica precursor, as template Pluronic P123® was purchased from Sigma Aldrich, 2 M HCl, ethanol, and distilled water were used solvent. (3-Chloropropyltrimethoxysilane)/CPTMS and toluene were purchased from Sigma Aldrich and used for functionalization of SBA-15. All of materials were used without further purification.

2.2 Synthesis SBA-15 and functionalization method

SBA-15 material was synthesized using sol-gel process. The sol-gel process was conducted in very acidic condition to allow self-assembly of silica occurs in wet template. Firstly, three solutions were prepared. 5 ml ethanol was mixed with 31.25 gram of TEOS then stirred for 30 min in room temperature. To create acidic condition, second solution was made of 5 ml of ethanol and 10 ml of HCl (2 M). Then, another 10 ml of ethanol was mixed with 50 ml of distilled water and stirred with the second solution for 30 min in room temperature making it third solution. Secondly, the first solution was added to the third solution and stirred for another 30 minutes in room temperature. The mixed solution was refluxed for 2 h at 50-60°C. The TEOS solution was ready to be use in further step.

The next step was synthesizing Pluronic P123® surfactant template. Pluronic P123® with optimum concentration 60 mM based on Dhaneswara et. al [8] was dissolved in 10 ml HCl (2 M) and 25 ml ethanol. This solution was added dropwise into the previous TEOS solution and then the mixture was stirred until it turned to gel. The resulting gel was dried in furnace for 1 h at 100°C and then calcined for 5 hours. Mesoporous material SBA-15 was ready for adsorption measurement.

The surface modification of SBA-15 was obtained by following this procedure. First, 0.8 gram of synthesized SBA-15 was suspended in 25 ml of toluene. Second, 5.6 ml of CPTMS was added dropwise to the suspended SBA-15 solution. Then, the mixture was kept under reflux for 24 h at 110°C in a round bottom flask. The obtained precipitate was filtered, washed with toluene and ethanol, and then dried for another 24 h at 100°C. The final product will be referred as CPTMS-SBA-15.

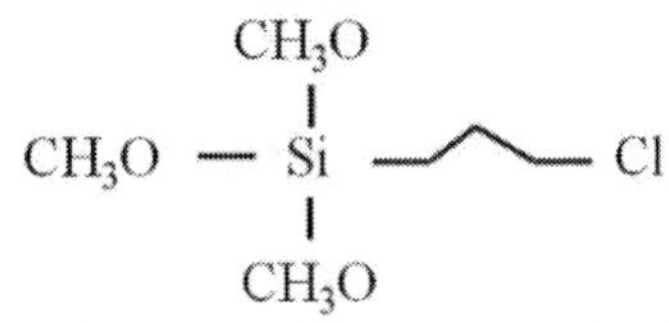

Figure 1. Schematic illustration of chemical structure of SBA-15.

2.3 Cd ion adsorption tests

To determine optimum amount of adsorbent for maximum removal of Cd^{2+} ion therefore initial Cd^{2+} ion concentration of 50 mg/L were prepared with various amount of adsorbent which is 40 mg/L, 80 mg/L, and 120 mg/L were added to 25 ml ion solutions. Hence, adsorption of Cd^{2+} ion with various concentrations were carried out using the determined optimum amount of adsorbent.

99,999% cadmium ions were dissolved in distilled water therefore the cadmium ion solutions with initial concentrations were prepared for adsorption tests. Initial concentration of cadmium ion solutions was 50 mg/L, 100 mg/L, 250 mg/L, 400 mg/L, and 500 mg/L. Optimum amount of adsorbent and 25 ml of cadmium ions solutions were placed in vials and stirred for 15 min The adsorption tests were carried out in room temperature at the rate around 100 rpm. Furthermore, the adsorbents were filtered from the resulting mixtures.

2.4 Characterization

To obtain characteristic of the samples, various characterizations should be done. Crystal structure and orientation of SBA-15 and CPTMS-SBA-15 were measured using small-angle x-ray diffraction. The XRD patterns were carried out within 2θ range from 0° to 10° with step size 0.02° and time 76.8 per step using Cu Kα radiation. Quantachrome adsorption-desorption equipment were used to obtain information about N_2 adsorption-desorption isotherm, mesoporous volume, and pore size of the products and performed at 77 K and specific surface area of the products were calculated using Brunauer-Emmer-Teller (Quantachrome) method. Fourier transform infrared (PerkinElmer) spectra were prepared to determine the functional groups. Transmission electron microscopy (Tecnai G2 spirit Win) were performed to obtain information about the mesostructure of the products. The initial and the remaining metal concentration in solutions were measured via atomic absorption spectrophotometry (Shimadzu).

3. Result and Discussion

3.1 SBA-15 and CPTMS-SBA-15 mesoporous materials comparison

In this paper, mesoporous material SBA-15 has been functionalized successfully with TEOS and Pluronic P123® at concentration 60 mM. As shown in figure 1, Cl lies within the end of the CPTMS functional group. Surface characteristic such as pore diameter, pore volume, and specific surface area of SBA-15 must be analyzed, to get information about the effect of Cl on the surface of SBA-15. The surface observation was conducted using BET. The results for SBA-15 and CPTMS-SBA-15 are shown in table 1.

Table 1. The surface characteristic of SBA-15 and CPTMS-SBA-15 using BET.

	Pore Diameter (Å)	Pore Volume (cc/g)	Specific Surface Area (m^2/g)
SBA-15	29.203	265.161	831.996
CPTMS-SBA-15	28.521	199.649	711.061

3.1.1 Structure characteristic of SBA-15 and CPTMS-SBA-15

Table 1 shows in comparison with surface characteristics of non-functionalized SBA-15, the surface area of CPTMS-SBA-15 is much narrow than pure SBA-15 so do its pore volume and pore diameter. The result shows that pore volume and surface area of CPTMS-SBA-15 were decreased by 25% and 15% compared to non-functionalized SBA-15. It indicated that CPTMS-SBA-15 gas uptake/adsorption capacity is lower than non-functionalized SBA-15. This outcome was obtained by the adsorption-desorption curve exhibits in figure 2.

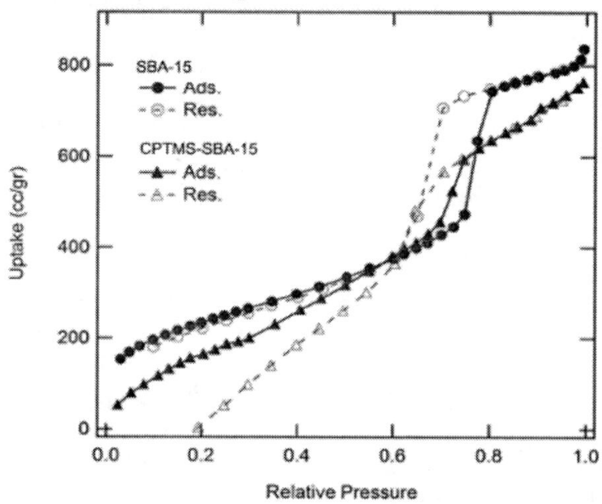

Figure 2. The N_2 adsorption-desorption isotherm curve of non-functionalized SBA-15 and CPTMS-SBA-15. Both curves exhibit type IV with H1 type of hysteresis adsorption isotherm.

Figure 2 reveals that all materials exhibit irreversible type IV adsorption-desorption isotherm with H1 type of hysteresis in relative pressure range from 0.6 to 0.8 for non-functionalized SBA-15 and from 0.6 to 0.73 for CPTMS-SBA-15 because of the high concentration of Pluronic P123®. According to IUPAC classification, type H1 is associated with porous materials consisting of well-ordered cylindrical-like pore channels [11]. As previously reported by (Dhaneswara et. Al, 2016), large pore size indicates more homogenous pore structure. Accordingly, to small surface area and pore volume of CPTMS-SBA-15, instead of having more homogenous pore, CPTMS-SBA-15 has more heterogeneity pore structure than non-functionalized SBA-15. It can be seen from both curves that before and after the functionalization, the curve shows stable shape indicating there was no pore structural changes occurred and the uniform mesoporous structure existing was still in its original state.

According to BET method, the following results of CPTMS-SBA-15 which is pore diameter, pore volume, and specific surface area has smaller value than non-functionalized ones. However, it indicates that the adsorbent surface was under modification. The reduction of this factors occurred because of the functionalization agent of CPTMS was found inside the pores too. Pore size distribution curves for CPTMS-SBA-15 and SBA-15 are shown in Figure 3 shows that SBA-15 has narrow pore size distribution with maximum pore diameter around 29.203 Å. On the other hand, the pore size of CPTMS-SBA-15 is decreased even though the distribution was nearly the same. Furthermore, the peak of CPTMS-SBA-15 in the curves is lower in comparison with non-functionalized SBA-15. It is indicating that CPTMS-SBA-15 has less order structure. This result has previously reported by Chong et. Al [12].

Figure 4 shows FTIR result for SBA-15 and CPTMS-SBA-15. The FTIR spectrum of SBA-15 and CPTMS-SBA-15 shows typical bands at 2950-2850 cm^{-1}, this belong to symmetric-asymmetric stretching of the C-H bonds. A band at 722 cm^{-1} was corresponded to vibration of Si-C bonds. A band at about 500 cm^{-1} was also identified as C-Cl stretching vibration. These results confirm the successful functionalization of SBA-15. Therefore, we can imply the CPTMS-SBA-15 structure form these result (shown in figure 7).

In figure 5, SAXS patterns of both materials SBA-15 and CPTMS-SBA-15 were shown. The XRD patterns clearly display broad peak of 2 θ at around 0.3683 degree and another peak at around 0.6138 for both materials. The SAXS patterns clearly display three peaks, first peak (100) is corresponding to surface walls. The other two peaks (110) and (200) show the hexagonal structures from SBA-15 materials. After functionalization, the height of the peak was reduced due to the presents of Cl and the possibility of the structure alteration.

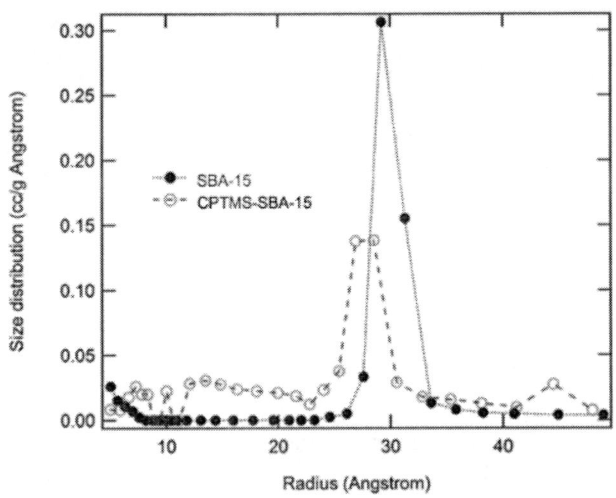

Figure 3. BET method for pore size distribution of SBA-15 and CPTMS-SBA-15. The pore size of CPTMS-SBA-15 is narrower compared to SBA-15 even though the distribution was nearly the same.

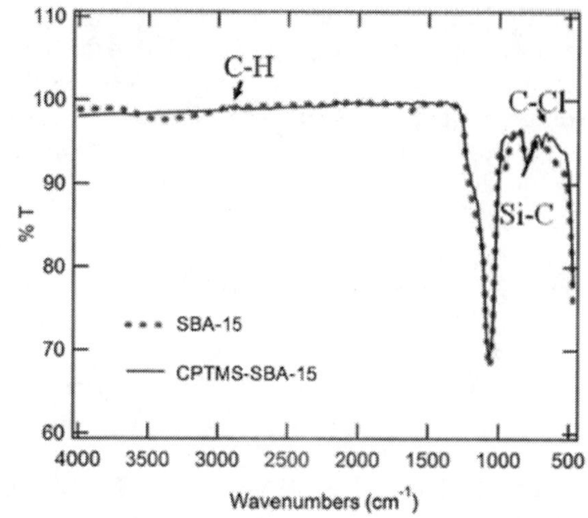

Figure 4. FTIR spectra for SBA-15 and CPTMS-SBA-15

3.1.2 Morphology of SBA-15 and CPTMS-SBA-15

In figure 6, figure 6(a) shows a TEM image of SBA-15 while figure 6(b) shows a TEM image of CPTMS-SBA-15. It seen in the figure that CPTMS-SBA-15 has vivid surface compared to SBA-15 due to modification process with Chlor and methyl. From both figures, we can see that both materials have nano-sized pore.

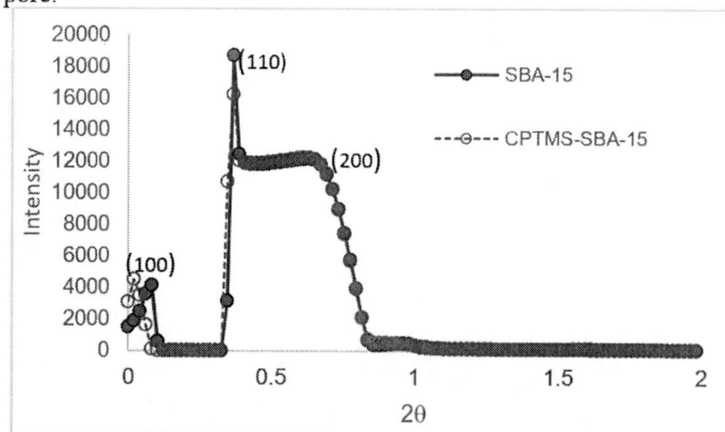

Figure 5. SAXS patterns of SBA-15 and CPTMS-SBA-15.

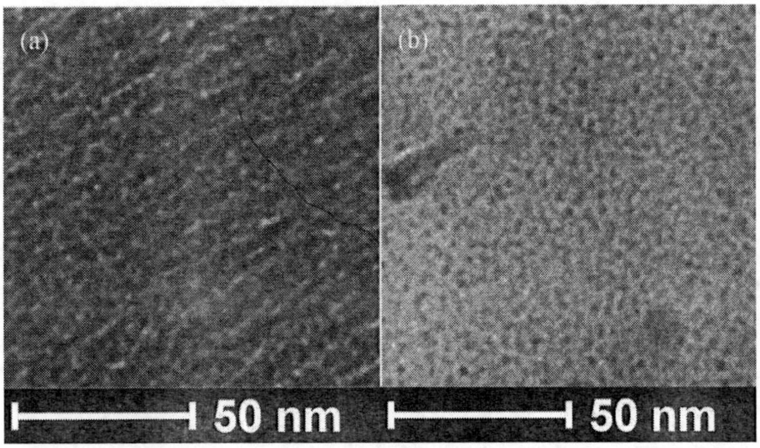

Figure 6. The TEM image of (a) SBA-15 and (b) CPTMS-SBA-15

3.2 Cd ion adsorption studies

Table 2 shows the percentage of removal for both adsorbent materials and the results show that Cd^{2+} ion removal for CPTMS-SBA-15 is higher in comparison with non-functionalized SBA-15. Result shown in Figure 8, it is observed that increasing the adsorbent amount will increase the percentage of removal and finally the maximum extraction of cadmium ions was achieved using 120 mg/L of CPTMS-SBA-15 with percentage of removal reaches 63.98%. Hence, 120 mg/L of CPTMS-SBA-15 were used in order to determine removal efficiency of Cd^{2+} ion.

In figure 8, as can be seen that CPTMS-SBA-15 has higher removal percentage in adsorbing heavy metal which is cadmium regarding this work compared to non-functionalized SBA-15, even though the surface area of CPTMS-SBA-15 is lower than non-functionalized SBA-15. As expected, functionalization in SBA-15 surface affects reduction in pore size and pore volume as obtained using BET method because of BET surface and volume are standardized towards pure silica weights as shown by Mureseanu et al. [10] . This is confirming the role played by organic functional groups from CPTMS were located both in the surface and inside the mesoporous pores. As result in reducing the BET surface area during modification. However, it does not affect the effectiveness in heavy metal adsorption as investigated in this work. The modification of SBA-15 generates O-Si-Cl bridge as shown in figure 7. Furthermore, this bridge may take Cd better than the surface with only Si-O.

The Cd^{2+} ion adsorption tests were carried out to determine removal efficiency of the adsorbent toward Cd^{2+} ion. The quantity of Cd removal was investigated by measuring the initial and final concentration of Cd^{2+} ion. In figure 9, the effect of initial concentration of Cd^{2+} ion shows that 120 mg/L of adsorbent has highest Cd^{2+} ion removal percentage with 50 mg/L Cd^{2} ion concentration. This result is in a good agreement with Asgari et al. [5] which is the adsorption process will increase as the increasing of metal ion concentrations if there are unoccupied sites. However, if all the sites are occupied, the amount of metal ion on adsorbent does not increase although the ion concentration in solutions increases.

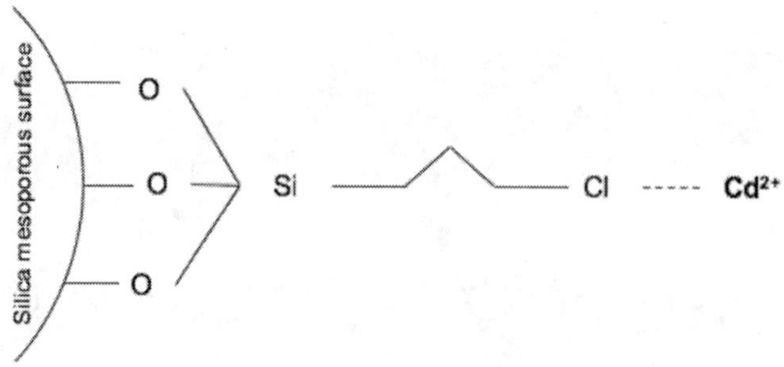

Figure 7. Schematic structure of CPTMS-SBA-15.

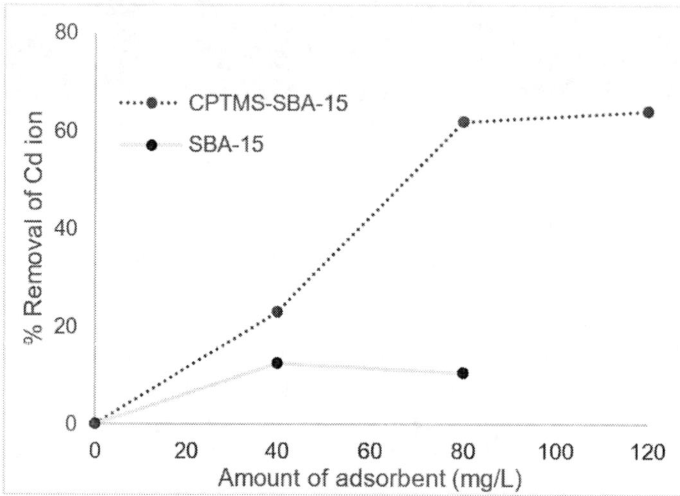

Figure 8. Effect of amount of adsorbent on the removal of Cd ions for SBA-15 and CPTMS-SBA-15.

Table 2. Cd removal of SBA-15 and CPTMS-SBA-15

Adsorbent	Initial concentration (mg/L)	Cd removal: Final concentration (mg/L)	Cd removal: Percentage (%)
SBA-15	40	12.39	9.8
	80	10.46	23.8
CPTMS-SBA-15	40	10.59	23
	80	5.24	61.8
	120	4.96	63.9

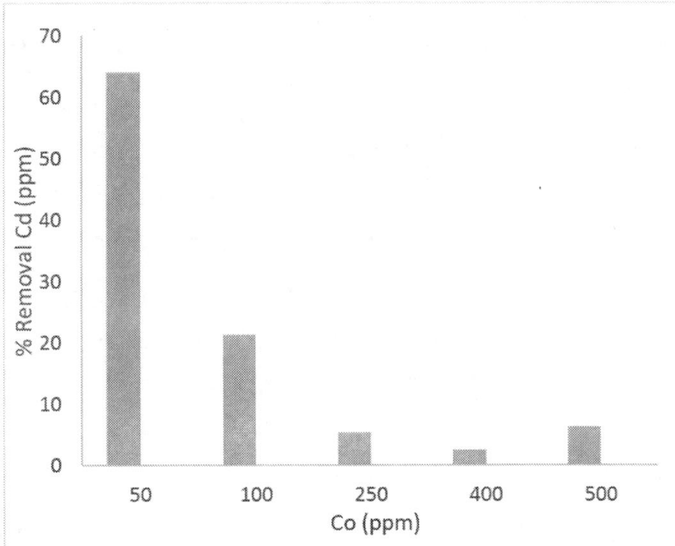

Figure 9. Effect of initial concentration on the removal of Cd^{2+} ions by CPTMS-SBA-15.

4. Conclusion

In this study, SBA-15 was successfully synthesized and functionalized with CPTMS. The functional groups grafted on the SBA-15 surface and inside the pore resulted in smaller surface area, pore size, and pore volume of CPTMS-SBA-15. However, it does not reduce its remarkable adsorption efficiency for heavy metal adsorption. The application of remediation of heavy metals from real samples were demonstrated using laboratory wastewater samples. The functionalized SBA-15 shows higher effectivity for adsorbing Cd^{2+} ion in comparison with non-functionalized SBA-15 using 120 mg/L of adsorbent. Regarding removal effectivity in the effect of initial concentrations of Cd^{2+} ion, 120 mg/L of CPTMS-SBA-15 has highest removal percentage for 50 mg/L initial concentrations.

Acknowledgments

The author thanks DRPM (Universitas Indonesia Research Directorate) for supporting this work under Hibah PITTA 2017 Tahun Anggaran 2017, No: 766/UN2.R31/HKP.05.00/2017 grant.

References

[1] Statistics Indonesia (BPS) 2008 Indonesia Population Projection 2010-2035.
[2] Baselt, R. C., & Cravey, R. H 1995 Disposition of Toxic Drugs and Chemicals in Man. In *Year Book Medical Publishers* (4th ed., Vol. 4th, pp. 105–107). Chicago: Year Book Medical Publishers.
[3] Hung, B.-Y., Kuthati, Y., Kankala, R., Kankala, S., Deng, J.-P., Liu, C.-L., & Lee, C.-H. 2015 *Nanomaterials* ***5***(4) 2169–2191.
[4] Thielemann, J. P., Girgsdies, F., Schlögl, R., & Hess, C. 2011 *Beilstein Journal of Nanotechnology* ***2*** 110–118.
[5] Asgari, M. S., Zonouzi, A., Rahimi, R., & Rabbani, M. 2015 *Oriental Journal of Chemistry*, ***31***(3) 1537–1544.
[6] Cao, G., & Wang, Y. 2004 Nanostructures and nanomaterials: synthesis, properties, and applications
[7] Zhao, D. 1998 *Science* ***279***(5350) 548–552.
[8] Dhaneswara, D., & Sofyan, N. 2016 *International Journal of Technology*, *7*(6), 1009–1015

[9] Zhang, L., Yu, C., Zhao, W., Hua, Z., Chen, H., Li, L., & Shi, J. 2007 *Journal of Non-Crystalline Solids* ***353***(44–46) 4055–4061.

[10] Mureseanu, M., Reiss, A., Stefanescu, I., David, E., Parvulescu, V., Renard, G., & Hulea, V. 2008 *Chemosphere* ***73***(9) 1499–1504.

[11] Thommes, M. 2004 *Nanoporous Materials - Science and Engineering* (p. 336). London: Imperial College Press.

[12] Chong, M., S., A., & Zhao, X. S. 2003 *The Journal of Physical Chemistry B* ***107*** 12650–12657.

IOP Conf. Series: Materials Science and Engineering **299** (2018) 012064 doi:10.1088/1757-899X/299/1/012064

Preparation and Characterization of Chitosan-coated Fe_3O_4 Nanoparticles using Ex-Situ Co-Precipitation Method and Tripolyphosphate/Sulphate as Dual Crosslinkers

Ika O Wulandari[1], Vita T Mardila[1], D J Djoko H Santjojo[2,3], Akhmad Sabarudin[1,3]*

[1]Department of Chemistry, Faculty of Science, Brawijaya University, Malang, 65145, Indonesia
[2]Department of Physics, Faculty of Science, Brawijaya University, Malang, 65145, Indonesia
[3]Research Center for Advanced System and Material Technology (ASMAT), Brawijaya University, Malang 65145, Indonesia

*Corresponding author: sabarjpn@ub.ac.id

Abstract. The unique properties of nanomaterial provide great opportunities to develop in several fields. Several types of nanoparticles have been proven beneficial for biomedical and therapeutic agent development. Particularly for clinical use, nanoparticles must be biocompatible and non-toxic. Iron oxide nanoparticles consist of either magnetite (Fe_3O_4) or maghemite (γ-Fe_2O_3) was eligible to use for in vivo application including targeting drug delivery. Due to their distinct properties, these nanoparticles could be directed to the specific target under external magnetic field. However, nanoparticles have a tendency to form agglomeration. Therefore, surface modification was required to reduce the agglomeration. In this study, nanoparticles of Fe_3O_4 were produced and coated by biomaterial (chitosan) using ex-situ co-precipitation method. Nanoparticles of Fe_3O_4 were synthesized by adding ammonia water into iron ferric and ferrous solution. Synthesis process of Fe_3O_4 was conducted prior to adding chitosan. Chitosan was then cross-linked by a combination of tripolyphosphate/sulphate. The different composition ratio and crosslinking time provide the different physical and magnetic characteristics of nanoparticles. Particle and crystallite size was determined by using Scanning Electron Microscopy (SEM) and X-Ray Diffraction (XRD) respectively, whereas magnetic characteristic was determined by Electron Spin Resonance (ESR). The results showed that the ratio enhancement between chitosan: Fe_3O_4 increase the particle size, while decreased the crystallite size. Morphology and particle size were influenced by the ratio of crosslinkers. It was found that the higher tripolyphosphate content was contributed to the small size and more spherical morphology. In addition, the influence of crosslinking time toward crystallite size was determined by altering stirring time. The longer duration of crosslinking time, provide the larger crystallite size of chitosan-Fe_3O_4. There was an interesting correlation between particle size and amount of Fe_3O_4 phase with ESR curve.

1.Introduction

Nanoparticles become an interesting topic to further investigate due to their unique characteristics. During the last decade, several studies show that nanoparticles are capable to be employed at cellular and molecular level. Nanoparticles are developed on numerous application fields. Nanoparticles can be synthesized by various methods which include microemulsion, thermal decomposition, hydrothermal synthesis, sonochemical, and coprecipitation. Coprecipitation method is mostly used due to low cost, high yield product, not requiring hazardous solvent and high temperature. Nanoparticles can be synthesized from either organic (polymer) or inorganic (metal) matter [1-3].

Magnetic nanoparticles are excellent for targeted drug delivery system. This material could be pointed to the specific target under control of the external magnetic field. However, for the biomedical purpose, nanoparticles must be biocompatible, biodegradable, and less toxic. Among several magnetic nanoparticles, only Fe_3O_4 and Fe_2O_3 possessing characteristics in vivo application [3].

Magnetic nanoparticles have less stability and tend to form agglomeration due to its small particle size and a large surface to volume ratio. It was also easily oxidized in air. Therefore, surface modification is needed to stabilize magnetic nanoparticle and avoid oxidation processes. Moreover, the surface modifier might offer additional functional group capable to attach another molecule such as protein and drug [2].

Chitosan is a product of deacetylated chitin which has biocompatible and biodegradable characteristics. Therefore, it can be used along with magnetic (Fe_3O_4) nanoparticle in drug delivery system. Chitosan has good stability in the neutral state, however less stable under pH 6.5. Chitosan nanoparticle can be manufactured through additional of crosslinker [4,5]. Crosslinker has also contributed to sustaining chitosan stability in acid condition. The pH of a tumor cell is about 5.5, without crosslinker agent chitosan will degraded. Therefore, drug release will rapidly occur. The existence of crosslinking agent that attached on chitosan-Fe_3O_4 nanoparticle structure utilized to inhibit chitosan degradation as well as controlling drug release from [6,7].

Tripolyphosphate (TPP) is a type of harmless anionic crosslinker. TPP can interact by inter- and intramolecular with chitosan to provide better spherical morphology. According to the previous report by Shu and Zhu (2002), the combination of dual crosslinker consisting of TPP/sulphate and TPP/citrate can give smaller particle size with spherical morphology [8]. Thereby as mentioned above, in this study nanoparticle chitosan-Fe_3O_4 will be conducted by using a combination of both crosslinker agents, TPP, and sulphate, for drug delivery system candidate.

Physical properties and magnetic characteristic of nanoparticles could be affected by various factors. Those factors include the composition of materials, crosslinker agents, pH, the existence of oxygen, temperature, and duration of reaction [8,9]. In this research, some parameters affecting particle size, crystallite size, and magnetic characteristic of nanoparticles were observed. Among the different parameters studied are a chitosan-Fe_3O_4 ratio,TPP: sulphate ratio, and crosslinking time. Nanoparticles are then characterized by using X-Ray Diffraction (XRD), Scanning Electron Microscopy-Energy Dispersive X-Ray Spectroscopy (SEM-EDX), and Electron Spin Resonance (ESR).

2.Material and Method

2.1 Chemicals

Chemicals consisting of ferric chloride ($FeCl_3{\cdot}6H_2O$, 99%) , ammonia (NH_3, 25% (v/v) in H_2O), sodium sulphate (Na_2SO_4, ≥99%), and acetic acid glacial (CH_3COOH, 100% were purchased from Merck, Indonesia); ferrous chloride ($FeCl_2{\cdot}4H_2O$, 98%), chitosan (LMW, 75-85 % deacetylated) and sodium tripolyphosphate ($Na_5P_3O_{10}$, 85%) were purchased from Sigma-Aldrich, Singapore.

2.2 Synthesis of magnetic iron oxide nanoparticles

Magnetite iron oxide (Fe_3O_4) nanoparticles without chitosan coating (bare Fe_3O_4) were synthesized by the co-precipitation of Fe(II) and Fe(III) salts, 1.72 g and 4.70 g respectively in 50 ml deionized water. Ammonia solution (3M) was added to the iron salts solution through syringe pump (10 mL h^{-1}) under mechanical stirring and the pH of the solution reached 10. The black precipitant was formed from this process. The process ends by washing with de-ionized water to remove any alkali metals until neutral state (pH±7). Samples then dried by freeze drying process before being characterized.

2.3 Synthesis of chitosan-coated magnetic nanoparticles

Magnetic nanoparticles of Fe_3O_4 coated by chitosan were synthesized using ex-situ co-precipitation method. In this term, Fe_3O_4 were synthesized prior to surface coating by chitosan layer. The first step is a synthesis of the Fe_3O_4 nanoparticle by mixing of $FeCl_2.4H_2O$ (1.72 g) and $FeCl_3.6H_2O$ (4.70 g) salts in 50 mL deionized water and followed by adding a drop of Ammonia solution (3 M) through syringe pump (10 mL h^{-1}). The pH of the solution was adjusted to 10. During this process, the mixture solution was stirred by a magnetic stirrer. The black precipitant obtained in this process was indicated as Fe_3O_4 nanoparticles. Before coated by chitosan, Fe_3O_4 nanoparticles were washed several times with de-ionized water to remove any alkali metals, and separated by magnetic decantation. Surface modification of Fe_3O_4 nanoparticles was carried out by adding the different amount of 1% chitosan solution (10 mL, 20 mL, and 30 mL) into 10 mL of water suspended Fe_3O_4 nanoparticles to form ferrogel solution. The mixture was stirred by a magnetic stirrer for 30 minutes. Afterwards, 10 mL of dual crosslinkers consisting of TPP and sulphate (with volume ratio TPP: sulphate 3:1 was added to the solution of chitosan-Fe_3O_4 through a syringe pump (10 mL h^{-1}) under mechanical stirring for 1 h. The yielded solution was then stirred for additional 1 h. The colloidal chitosan-coated magnetic Fe_3O_4 nanoparticles were extensively washed with de-ionized water and separated by magnetic decantation for several times. Samples were then dried by freeze drying process.

The Nanoparticles were prepared according to the procedure above. Nevertheless, instead of adding crosslinkers, 10 mL of dual crosslinker TPP: sulphate was added with different ratio 1:1, 3:1 and 5:1 into ferrogel solution with constant ratio composition of chitosan: Fe_3O_4 1:1. This different composition ratio of crosslinker was intended to understand the effect of crosslinker toward physical properties and magnetic characteristic of nanoparticles of chitosan-Fe_3O_4. The same procedure for preparing nanoparticles was also followed. But, instead of stirring after the addition of cross-linked, nanoparticles with a constant composition ratio of chitosan: Fe_3O_4 1:1 and ratio composition of TPP: sulphate 3:1 was stirred further for 1 h. Another batch was prepared and stirred for 2 and 3 hours to determine the effect of crosslinking time toward final properties of nanoparticles.

2.4 Characterization

Morphological characterization and particle size determination of dried nanoparticles chitosan-coated Fe_3O_4 and bare Fe_3O_4 were performed by Scanning Electron Microscope-Energy Dispersive X-Ray Spectroscopy (Hitachi High-Tech TM3000; Hitachi High-Technologies Corporation, Tokyo, Japan). Crystallite size of nanoparticles was determined by X-Ray Diffractometer (Philips type X'pert MPD; from PANalytical B.V., Eindhoven, Netherlands) at room temperature using Cu Kα radiation (λ =1.5406 Å) with the 2θ range of 15 – 65^0. Crystallite size of nanoparticles was achieved through calculation using modified Debye-Scherrer Equation as given in Equation 1 [10],

$$\ln\beta = \ln\frac{K\lambda}{L.\cos\theta} = \ln\frac{K\lambda}{L} + \ln\frac{1}{\cos\theta} \qquad (1)$$

Where β is full-width at half-maximum diffraction peak (FWHM), K is Scherrer constant related to crystallite shape, λ is x-ray wavelength (nm), cos θ is Bragg angle (rad), and L is crystallite size (nm). Crystallite size was obtained by plotting ln β against ln (1/cosθ).

Furthermore, the iron phase abundance was calculated by material analysis using diffraction (MAUD) software. Magnetic properties were observed by Electron Spin Resonance (Leybold Heraeus; Leybold, Huerth, Germany) at room temperature with electric current 0.3 A and frequency were 33.7-76.5 MHz, respectively. Magnetic properties were confirmed by calculating the slope of ESR curve. The higher slope of the curve indicates higher magnetism of the nanoparticles.

3.Results and Discussion

3.1 Formation steps of Fe_3O_4 nanoparticles

In this research, nanoparticles based on biomaterial of chitosan-Fe_3O_4 were synthesized as a candidate of drug delivery system by ex-situ co-precipitation method. Unsoy, *et.al*, (2014) described that coprecipitation method is a very advantageous method to synthesize nanoparticles for biomedical application due to the simple and mild synthesis conditions [6]. In addition, nanoparticles that are produced by coprecipitation have higher biocompatibility and biodegradability. Physical properties and magnetic characteristics of nanoparticles were influenced by several factors such as composition ratio and crosslinking time.

During synthesis of Fe_3O_4, the color of the iron solution changed slowly from light brown, dark brown, and finally to black by adding base solution continuously. The formation of black precipitant was indicating the formation of the Fe_3O_4 nanoparticle. The color appearance changing because of the change of phase compound that formed during synthesis processes [11]. The possible reaction for the formation of the Fe_3O_4 nanoparticle can be written as follows [3]:

$$Fe^{3+} + 3OH^- \rightarrow Fe(OH)_3 \quad (2)$$
$$Fe(OH)_3 \rightarrow FeOOH + H_2O \quad (3)$$
$$Fe^{2+} + 2OH^- \rightarrow Fe(OH)_2 \quad (4)$$
$$2FeOOH + Fe(OH)_2 \rightarrow Fe_3O_4 + 2H_2O \quad (5)$$

3.2 Crystallite size, iron phase abundance, morphology, and magnetic properties of nanoparticles
Dried powder of bare Fe_3O_4 and chitosan-Fe_3O_4 were then characterized by using X-Ray Diffractometer. **Figure 1**. shown XRD patterns of bare Fe_3O_4 and chitosan-Fe_3O_4 with various composition ratio and crosslinking time. Characteristic peaks were observed in the XRD pattern at 2θ of 9.6°, 30.1°, 35.5°, 43.1°, 54.5°, 57.6° and 63.6° corresponding to the diffractions of 220°, 311°, 400°, 422°, 511° and 440° crystal faces of Fe_3O_4 spinel structure (ICSD #26410). Peak broadening was indicating small size of nanoparticles. The crystallite size of nanoparticles was evaluated from the XRD data using modified Debye-Scherrer equation. Crystallite size of bare Fe_3O_4 and chitosan-Fe_3O_4 nanoparticles in different composition ratio and crosslinking time were presented in **Table 1**.

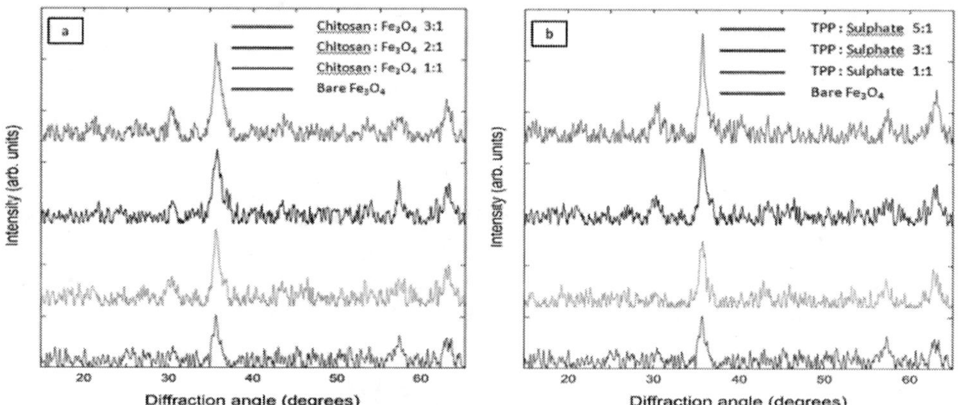

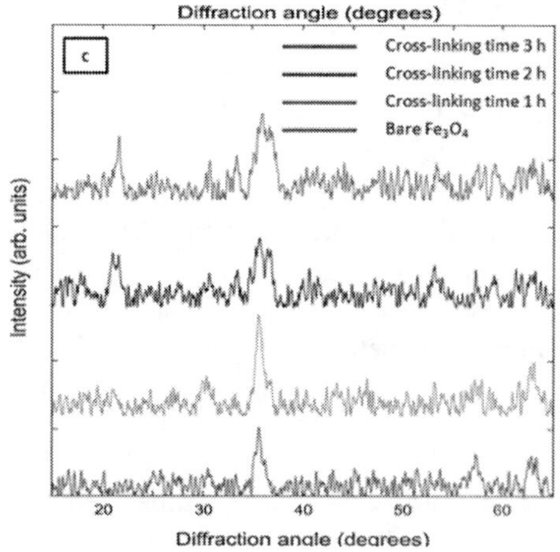

d

Fe_3O_4 (ICSD #26410)		γ-Fe_2O_3 (ICSD #172905)	
hkl	2θ	hkl	2θ
220	30.1012	220	30.06014
311	35.4492	311	35.40035
400	43.0746	400	43.01440
422	53.4321	422	53.35522
511	57.1035	511	57.02057
440	62.5445	440	62.45177

Figure 1. X-ray diffraction patterns of both bare Fe_3O_4 and chitosan-Fe_3O_4 nanoparticles with various composition ratio (a, b) and cross-linking time (c), miller indices and 2θ position of magnetite (Fe_3O_4) and maghemite (γ-Fe_2O_3) according to ICSD data.

The crystallite size of nanoparticle decreases as composition ratio consists of chitosan: Fe_3O_4 and TPP: sulphate ratio increases. However, crystallite size increases as crosslinking time increases. XRD patterns of chitosan-Fe_3O_4 in various crosslinking time (2 and 3 h) show a different pattern to the others since revealing a peak at 2θ 21.56^0. It was identified as the peak of chitosan with orthorhombic structure [12].

Table 1. Determination of crystallite size and particle size of nanoparticles.

Parameter	**L [a] (nm)**	**Particle Size[b] (nm)**
Bare Fe_3O_4	11.90	72.3-162
Ratio of Chitosan: Fe_3O_4		
1:1	6.20	253-482
2:1	4.27	456-793
3:1	3.78	583-882
Ratio of TPP: Sulphate		
1:1	7.79	368-593
2:1	6.20	253-482
3:1	5.34	202-273
Crosslinking Time (h)		
1	6.20	253-482
2	8.75	283-320
3	9.44	142-226

[a] Calculated by Modified Debye-Scherrer equation (1)
[b] Measured by SEM

The XRD data were also used for determining the iron content of each sample. The percentage of iron phase abundance of the chitosan-Fe_3O_4 nanoparticles were exhibited in **Table 2**. As shown in **Figure 2**, the samples contained not only a Fe_3O_4 phase but also γ-Fe_2O_3. The percentage of Fe_3O_4 on the bare Fe_3O_4 sample is less than chitosan-Fe_3O_4. Crystal of γ-Fe_2O_3 is arranged with a Fe^{3+} ion in octahedral and tetrahedral sites. Crystal of Fe_3O_4 arranged by Fe^{3+} ion and Fe^{2+} ion in octahedral sites as well as a Fe^{3+} ion in tetrahedral sites. Whereas, the Crystal structure of γ-Fe_2O_3 arranged by a Fe^{3+} ion in octahedral and tetrahedral sites. Crystal structure of Fe_3O_4 and γ-Fe_2O_3 were depicted in **Figure 2**.

Table 2. Percentage of iron phase abundance on various samples.

Reaction Parameter	**Magnetite (%)**	**Maghemite (%)**
Bare Fe_3O_4	87.98	12.02
Chitosan:Fe_3O_4 (1:1)	93.29	6.71
Chitosan:Fe_3O_4 (2:1)	92.73	7.27
Chitosan:Fe_3O_4 (3:1)	89.35	10.65
TPP:Sulphate (1:1)	89.88	10.12
TPP:Sulphate (3:1)	93.29	6.71
TPP:Sulphate (5:1)	96.3	3.7
Crosslinking Time 1 h	93.29	6.71
Crosslinking Time 2 h	94.52	5.48
Crosslinking Time 3 h	97.12	2.88

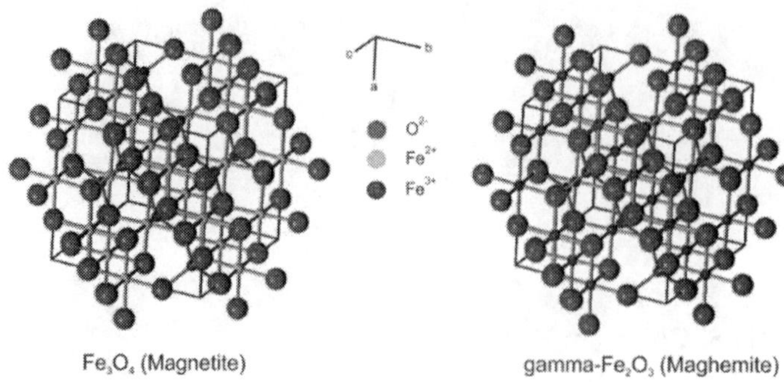

Figure 2. Model of crystal structure of magnetite (Fe_3O_4) and maghemite (γ-Fe_2O_3).

The composition ratio of crosslinker plays an important role during synthesis because it can be affecting physical properties of synthesis products. In this work, chitosan molecules were crosslinked by different volume of tripolyphosphate and sulphate as crosslinking agents. Particle size and morphology of synthesis products of bare Fe_3O_4 and chitosan-Fe_3O_4 nanoparticles in different crosslinker volume ratio have been observed by SEM-EDX and presented in **Figure 3**. According to SEM images, it was known that the particles are forming clusters. Clustering could occur during purification of Fe_3O_4 before chitosan coated. Surface morphology of chitosan-coated Fe_3O_4 is almost spherical that compared with bare Fe_3O_4 (without chitosan). Particles size determined by SEM-EDX were listed in **Table 1**. The particle size of chitosan-Fe_3O_4 decreased with the increasing of TPP: sulphate ratio.

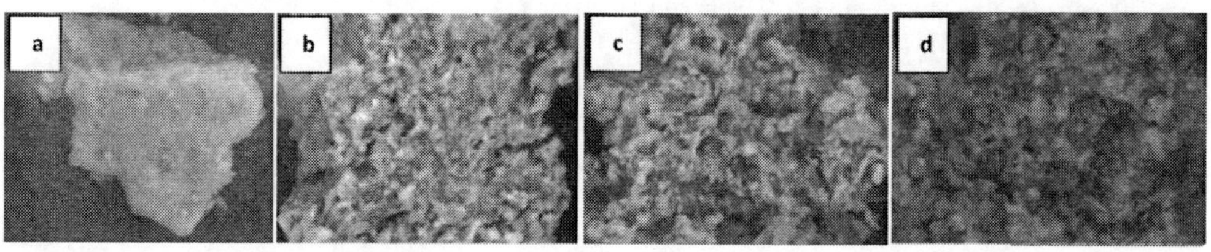

Figure 3. SEM images of nanoparticles (a) bare Fe_3O_4, and chitosan-Fe_3O_4 with various crosslinker ratio (b) 1:1, (c) 3:1, (d) 5:1 with 3000 times magnification.

Comparing between EDX spectra of bare Fe_3O_4 and chitosan coated Fe_3O_4 samples are presented in **Figure 4**. The peaks around 0.8, 6.3, and 6.8 keV are related to the binding energies of Fe [2]. The spectrum of bare Fe_3O_4 contained two peaks, which were assigned to Fe, and O. However, chitosan-Fe_3O_4 contained six peaks that assigned not only for Fe and O but also C, N, P, and S. The peak of C and N shows the existence chitosan layer on the surface of the iron oxide. Furthermore, the peak of P and S exhibits the existence of crosslinker agents.

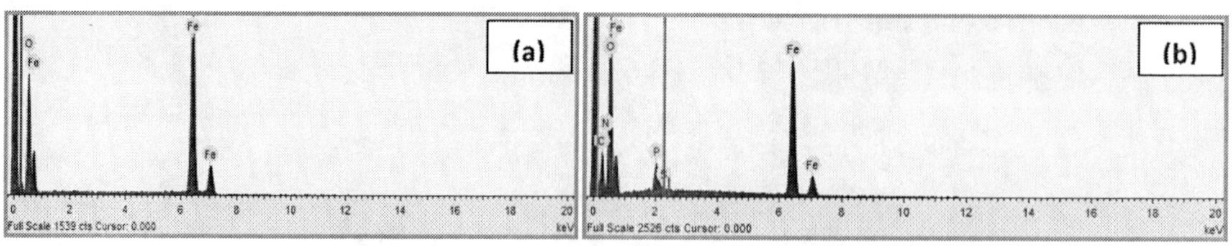

Figure 4. EDX spectra of (a) bare Fe_3O_4, (b) chitosan-Fe_3O_4 with constant crosslinker ratio 3:1, Chitosan:Fe_3O_4 1:1, and 1 h crosslinking time.

The composition ratio of chitosan and $Fe_3O_4$was influenced the magnetic characteristic of nanoparticles. Magnetic characteristic was determined by using electron spin resonance. The resonance frequency of each sample with different chitosan: Fe_3O_4 ratio was evaluated in the same frequency region (33.7-76.5 MHz). In contrast, ESR signal of bare Fe_3O_4 was obtained at 30.6 MHz. The differentiation of ESR curve profile can be observed by calculating the slope of each curve. The results show that the slope of ESR curve decreases as chitosan: Fe_3O_4 ratio increases. The slope of each ESR curve on various samples with different chitosan: Fe_3O_4 ratio was presented in **Figure 5(a)**, while the nanoparticle responses under the influence of external magnetic field were shown in **Figure 5(b)**.

The slope of ESR curve was increasing with the increase of Fe_3O_4 content on various samples. Samples have active response toward electromagnetic radiation waves. Considering the slope of ESR curves, asymmetrical Zeeman effects may be occurred caused by the changed of the energy system. Whole energy system will increase or decrease depends on the amount of Fe_3O_4 on various samples. The sharper slope of ESR curve was affected by the higher Fe_3O_4 content on various the samples. It was indicating the increasing energy of the overall system.

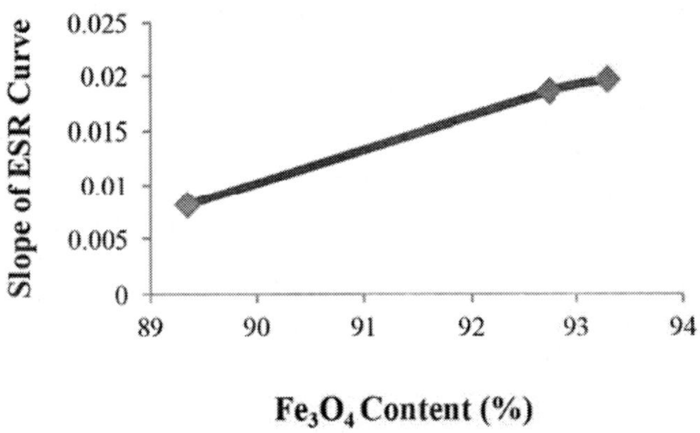

(a)

(b)

Figure 5. (a) Correlation curve between the slope of ESR curve toward the percentage of Fe_3O_4, in chitosan-Fe_3O_4 samples with various chitosan: Fe_3O_4 ratio, (b) magnetic properties test of the chitosan-Fe_3O_4 nanoparticle.

According to the result that given by SEM, XRD and ESR, there was an interesting correlation between particle size, the percentage of Fe_3O_4, and slope of ESR curve for nanoparticles produced by different composition ratio and crosslinking time. The smaller particle size would provide the higher percentage of Fe_3O_4 and the higher slope of ESR curve.

4. Conclusion

In this work, magnetic nanoparticles (Fe_3O_4) have been prepared by ex-situ co-precipitation method at room temperature. Composition ratio and crosslinking time were affecting physical properties and magnetic characteristic of nanoparticles product. The higher ratio of chitosan and Fe_3O_4 were produced materials with small crystallite size and low Fe_3O_4 content. Furthermore, the volume ratio of crosslinker was also affecting the particle size and morphology of the material. The smallest size of nanoparticles with spherical shape would be established in a higher ratio of TPP: sulphate. There was an interesting correlation between particle size, the composition of Fe_3O_4, and slope of ESR curve. The smaller particle size will lead to the increase of Fe_3O_4 content as well as increase the value of ESR slope.

References

[1] Hariani, P.L., Faizal, M., Ridwan, Marsi and Setiabudidaya D2013 Synthesis and Properties of Fe_3O_4 Nanoparticles by Co-precipitation Method to Removal Procion Dye *International Journal of Environmental Science and Development***4** 336-340

[2] Mahdavi, M., Ahmad, M., Haron, M.J., Namvar, F., Nadi, B., AbRahman MZ and Amin J2013 Synthesis, Surface Modification and Characterisation of Biocompatible Magnetic Iron Oxide Nanoparticles for Biomedical Applications*Molecules***18** 7533-7548

[3] Wu, W., He Qand Jiang C2008 Magnetic iron oxide nanoparticles: synthesis and surface functionalization strategies *Nanoscale ResLett*3397–415

[4] Kumar, N, Patel, A.K., Kumari N and Kumar A 2014 A Review on Chitosan Nanoparticles for Cancer Treatment*International Journal of Nanomaterials and Biostructures***4** 63-65

[5] Al-Remawi M M A 2012 Properties of Chitosan Nanoparticles Formed Using Sulfate Anions as Crosslinking Bridges*American Journal of Applied Sciences***9** 1091-1100

[6] Unsoy, G., Khodadust, R., Yalcin, S., Mutlu P and Gunduz U 2014Synthesis of Doxorubicin Loaded Magnetic Chitosan Nanoparticles for pH Responsive Targeted Drug Delivery, *European Journal of Pharmaceutical Science***62** 243-250

[7] Ibezim, E.C., Andrade, C.T., Marcia, C., Baretto, B., Odimegwu DC and De Lima FF2011 Ionically Cross-linked Chitosan/TripolyphosphateMicroparticles for the Controlled Delivery of Pyrimethamine*Ibnosina Journal of Medicine and Biomedical Sciences***3** 77-88

[8] Shu X Z and Zhu K J 2002 Controlled Drug Release Properties of Ionically Cross-Linked Chitosan Beads: The Influence of Anion Structure *International Journal of Pharmaceutics***233** 217–225

[9] Tresilwised, N., Pithayanukul P and PlankC 2005 Factors Affecting Sizes of Magnetic Particles formed by Chemical Co-precipitation *Mahidol University Journal of Pharmaceutical Sciences***32** 71-76

[10] Monshi, A., Foroughi MR and Monshi MR 2012Modified Scherrer Equation to Estimate More Accurately Nano-Crystallite Size Using XRD*World Journal of Nano Science and Engineering***2** 154-160

[11] Ahn, T., Kim, J.H., Yang, H.M., Lee JW and Kim JD2012 Formation Pathways of Magnetite Nanoparticles by Coprecipitation Method*The Journal of Physical Chemistry C***116** 6069−6076

[12] Gong, J., Hu, X., Wong, K.W, Zheng, Z., Yang, L., Lau WM and Du R 2008Chitosan Nanostructures with Controllable Morphology Produced by A Nonaqueous Electrochemical Approach*Advanced Materials***20** 2111-2115

IOP Conf. Series: Materials Science and Engineering **299** (2018) 012065 doi:10.1088/1757-899X/299/1/012065

Ex-Situ Synthesis of Polyvinyl alcohol(PVA)-coated Fe_3O_4 Nanoparticles by Coprecipitation-Ultrasonication Method

Imam Riva'i[1], Ika Oktavia Wulandari[1], Hermin Sulistyarti[1], Akhmad Sabarudin[1,2*]
[1]Department of Chemistry, Faculty of Science, Brawijaya University, Jl Veteran 65145 Malang, Indonesia
[2]Research Center for Advanced System and Material Technology (ASMAT), Brawijaya University, Malang 65145, Indonesia

*Corresponding author: sabarjpn@ub.ac.id

Abstract. In this study, the synthesis of Fe_3O_4 nanoparticles was done with surface modification using PVA with coprecipitation-ultrasonication method. Time variations and PVA concentrations were added to determine the effect on crystallite size and lattice parameters on the synthesis of Fe_3O_4-PVA nanoparticles. Fe_3O_4 characterization was done using X-Ray Diffraction (XRD) and Fourier Transform Infrared Spectroscopy (FTIR) instruments. FTIR was employed to determine PVA coating on the surface of Fe_3O_4 nanoparticles. The crystallite size and lattice parameters were analyzed using XRD. From the FTIR data, it is known that the interaction between PVA and Fe_3O_4 nanoparticles is characterized by Fe-O-C group at 1100 cm^{-1} region which is characteristic of Fe_3O_4-PVA nanoparticles, C-H groups of PVA in the range of 2950 cm^{-1} wave number, C-C of PVA regions of wave number 1405 cm^{-1}, Fe_3O_4 and Fe_3O_4-PVA samples are in the range of 565 cm^{-1}. In addition, the variation of ultrasonication time and the addition of PVA concentration have an effect on the crystallite size change and the lattice parameter observed from the XRD data. The use of ultrasonication time will affect the size of the crystallite become smaller and the grating lattice parameters obtained are wider. The effect of addition of PVA showed that higher concentration of PVA resulted in smaller crystallite size and larger lattice parameters. These results indicated that ultrasonication time and addition of PVA concentration greatly affect the characteristics of nanoparticles.

1. Introduction

Nanoparticle has become rapidly growing field of study recently. Nanoparticle has an important role in various fields such as industry, healthcare, technology, and biology. It is divided into two types, nanocrystal and nanocarrier that have a wide variety of types such as quantum dot, liposomes, solid nano lipid particles, and metal nanoparticles [1]. Size of nanoparticles is about 1-100 nm [2]. It has unique magnetic characteristics such as super-paramagnetic, high coersivity, low temperature, and high magnetism. There are several methods for nanoparticle synthesis for instance thermal decomposition, hydrothermal synthesis, sonochemical, and co-precipitation [3]. Co-precipitation is a bottom-up synthesis step used to form a particle into certain size like nanometer-sized [4]. The principle of the co-precipitation method is to release a continuous bond metal in liquid form has without regard to the specific mechanism that occurs. From the method, a solid-shaped material from its liquid-shaped precipitates is obtained [5]. In general, this method can be applied to the synthesis process of zinc, titanium and iron [6]. Fe_3O_4 is a combined oxide metal of Fe_2O_3 and FeO. The oxide metal has amphoteric properties, and is easily oxidized by the environment, that is FeO becomes Fe_2O_3 after oxydation. Therefore, further treatment is needed to prevent Fe_3O_4 from oxidation. One method frequently used to prevent oxidation is a coating method with a polyvinyl alcohol (PVA) polymer [6,7,8].

PVA is a biocompatible polymer that can absorb moisture content for hydrogel formation [9]. Therefore, coating nanoparticle surfaces with modification using polyvinyl alcohol can prevent agglomeration resulting in monodisperse particles, coagulation, and stabilizing nanoparticles from oxidation [10]. PVA has a good film functioning as a shaper, emulsion, and adhesive properties. The surface layer of PVA particles can prevent agglomeration causing monodisperse particles. In the process, nanoparticles may be coated by PVA to prevent the coagulation of particles during bioapplication process [16].

Fe_3O_4 nanoparticle is commonly obtained by several chemical synthesis methods, for example coprecipitation, reverse micelle method, plasma microwave synthesis, sol-gel technique, freeze drying, ultrasound irradiaton, hydrothermal methods, as well as laser pyrolysis techniques [11], [12]. Coprecipitation is the simplest, easiest and least expensive method, but it can produce Fe_3O_4 particles in the nanometer order. Surface modification of Fe_3O_4 nanoparticle can be done by ex-situ and in-situ. In surface modification, nanoparticles should have biocompatible properties and can be degraded by the body. Materials that can be used for surface modification are organic and inorganic materials. Some examples of the organic materials are poly (ethylene-co-vinyl acetate), polyvinylpyroolidone (PVP), poly-lactic-co-glycolic acid (PLGA), polyethylene glycol (PEG) or polyvinyl alcohol (PVA) which also functions as material for coating [11, 13].

Ultrasonication time affects crystal alteration size, lattice parameter and the morphology of a particle. Ultrasonication time also results in adsorption and desorption which indicate that extracting time is really fast [14]. The longer the sonication time is, particle size tends to be more homogeneous and shrink eventually leading to a stable nanoparticle size and less agglomeration. It occurs because shock wave in sonication method can separate agglomeration and there is a perfect dispersion with the addition of surfactant as stabilizer [15].

Based on all information aforementioned, in this work, Fe_3O_4-PVA nanoparticles were prepared by ex-situ synthesis using coprecipitation coupled with ultrasonication method. Some parameters influenced a crystal size as well as a lattice parameter were investigated through the optimization of ultrasonication time and PVA concentration.

2. Materials and Method

2.1 *Chemicals and instrumentation*

The materials used were (Iron (III) chloride hexahydrate, Iron (II) chloride tetrahydrate, PVA, Mw 3500 (Merck, Indonesia) and aquaoeus ammonium hidroxyde.

The instruments were FTIR *(Fourier Transform Infrared)* spectrophotometer (Shimadzu FTIR QP89500, KBr pellet of which wavenumber ranges between 4000 and 400 cm^{-1}), Freeze Dry (35 XL Virtis SP Scientific with the temperature under -40°C and pressure up to ≤ 20 mT). XRD analysis was carried out using Diffractometer system XPERT-PRO with Cu-Kα λ= 1.5406 Å. The XRD data was recorded across a 2θ range of 5° to 60° using a step size of 0.001°.

2.2 Procedure reaction

The procedure for preparation of Fe_3O_4 nanoparticles was adopted from Ghanbari [17] with some modifications. The Fe_3O_4 nanoparticles were prepared first by mixing $FeCl_2.4H_2O$ and $FeCl_3.6H_2O$ with mol ratio of 1: 2 into 100 milliters of distilled water. In the following step, polyvinyl alcohol (5, 7, and 9 gram) was dissolved into100 mL of distilled water and stirred using a magnetic stirrer. Then, the Fe_3O_4 solution was placed on a sonicator, and an ammonia solution (2M) was added drop wise using a syringe pump at flow rate of 20 mL/ h. These procedures resulted in the formation of black precipitates of Fe_3O_4-PVA5, Fe_3O_4-PVA7, and Fe_3O_4-PVA9 nanoparticles. The black precipitate was washed using distilled

IOP Conf. Series: Materials Science and Engineering **299** (2018) 012065 doi:10.1088/1757-899X/299/1/012065

water until it reached the neutral pH. The sample was dried using the freeze dryer and stored before being characterized.

3. Result and discution

During initial preparation of Fe_3O_4 synthesis, along with the addition of distilled water, $FeCl_3.6H_2O$ and $FeCl_2.4H_2O$ salt solution turned into yellowish color. The addition of ammonia solution (2M) gradually turned yellowish solution into brownish yellow-brown, then brown-blackish and at the end of the process it became solid black. The formation of black color and the black precipitate indicated Fe_3O_4 nanoparticle formation. During the synthesis, the color of the solution changed several times due to the influence of the compound phase formed [17].

Ammonia solution, which was added into Fe(III) and Fe(II) solutions, caused OH to bind to Fe(III) and Fe(II) solution to form Fe_3O_4 nanoparticles. Furthermore, adding PVA solution to Fe_3O_4 nanoparticles resulted in coating Fe_3O_4 with PVA. It happened since the PVA compound has an O-H group, the H atom would be released and the C-H chain interacted with the Fe_3O_4 surface. **Figure 1** described the Fe_3O_4-PVA interaction.

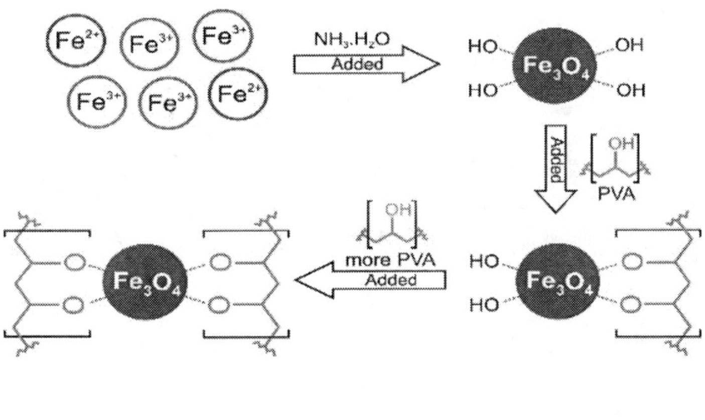

Figure 1. Fe_3O_4-PVA Interaction Mechanisms

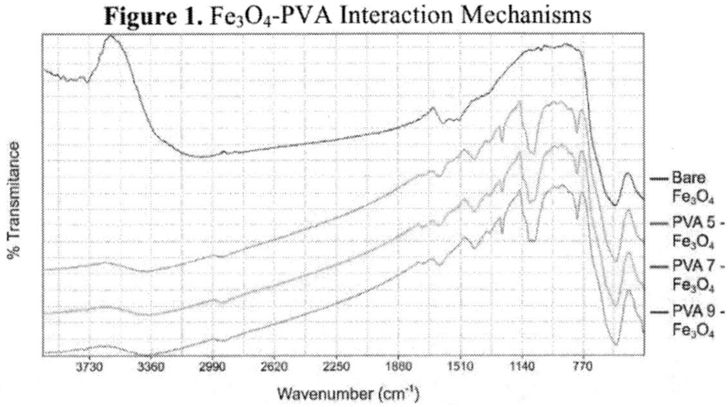

Figure 2. IR Spectra of Fe_3O_4-PVA (5,7,9) Nanoparticles and Bare Fe_3O_4 in 60-minute Ultrasonification Time

Figure 2 describes the Fe_3O_4-PVA and Fe_3O_4 nanoparticles spectra without coating. The spectra were obtained from Fe_3O_4-PVA samples with 60 minutes of ultrasonication time and the influence of (5, 7, and 9)%PVA concentrations. The absorption that appeared in the 3,600 cm^{-1} range showed the presence of an O-H group and it was identified as OH on the Fe_3O_4 surface. The 2980 cm^{-1} wave indicated the absorption of the C-H group of PVA and the absorption area of 1420 cm^{-1} was the C-C group of PVA. However, the FTIR spectra did not appear on the bare Fe_3O_4. The Fe-O-C bond interaction may be identified from uptake in a range of about 1100 cm^{-1} which indicated that Fe_3O_4 had been coated with PVA. The fingerprint range of 568cm^{-1}[18] showed sharp uptake, indicating the presence of Fe-O bond in bare Fe_3O_4 and Fe_3O_4-PVA samples.

Table 1. Influence of Fe_3O_4-PVA Ultrasonification Time towards Crystallite Size and Lattice Parameter

Samples	Sonication time (min)	Crystallite size	Lattice parameter
Bare Fe_3O_4	30	14.6043	8.3260
	60	7.8547	8.3490
	120	7.3354	8.3900
Fe_3O_4-PVA5	30	5.3941	8.3900
	60	8.9134	8.3610
	120	6.3616	8.3660
Fe_3O_4-PVA7	30	9.8878	8.3710
	60	19.5592	8.3550
	120	7.2919	8.3740
Fe_3O_4-PVA9	30	6.4016	8.3870
	60	12.5494	8.3470
	120	6.7617	8.3840

Table1 shows the changes in lattice parameters and crystallite size. Each variation had different trend or pattern. Generally the resulting pattern had similarities between each other making it easier to observe. It was also explained that lattice parameters with crystallite size were inversely proportional to the longer duration of ultrasonic usage time for the Fe_3O_4-PVA nanoparticlessynthesis.

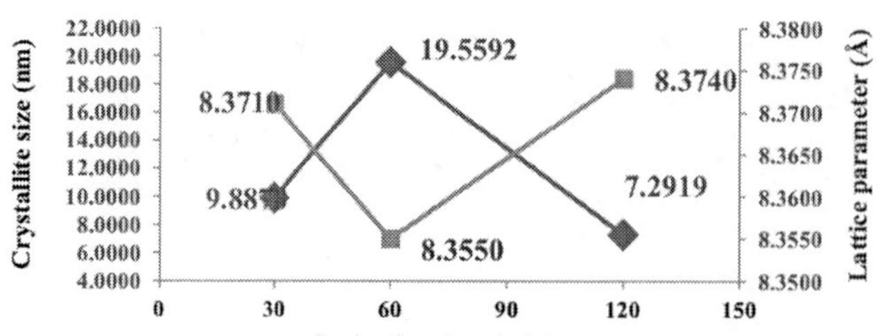

Figure 3. The effect of sonication time with crystallite size and Lattice parameter at 7% PVA of Fe_3O_4-PVA nanoparticles.

Figure 3 showed the smallest crystallite size is obtained at 7% PVA concentration and a 120-minute ultrasonication time. A 60-minute ultrasonication time resulted in crystallite with the largest size. Change in crystallite size obtained with 7% PVA concentration is similar to that when the PVA concentration was

5%. This may be due to the 30-minute ultrasonication time; during nucleation, the PVA polymer is immediately added to the nanoparticles. The PVA polymer could prevent formation of large sized Fe_3O_4 crystals. During the 60-minute ultrasonication, the nucleation took place and followed by growth of the nucleus after it was coated by PVA polymer. As an addition, the possibility of clustering nanoparticles also contributed larger-sized crystals. During the 120-minute ultrasonication, the size of the nanoparticle crystals decreased again since the Fe_3O_4 crystal clusters tended to interact with the PVA polymer and form the Fe_3O_4-PVA cluster. Yang (2005) stated that the longer the ultrasonication time would cause the isolation point (IEP) of Fe_3O_4 to lead to a more acidic condition (lower pH) [19]. In the pH conditions, the positive charge on the surface of Fe_3O_4 would increase. Consequently, the probability of a negatively charged PVA to interact with Fe_3O_4 positively charged would be greater. It would cause the crystallite size of Fe_3O_4-PVA with a 120-minute ultrasonication time became smaller. The lattice parameters of the 7% Fe_3O_4 nanoparticles with time the 30, 60 and 120 minute ultrasonication time were 8.3710 Å, 8.3550 Å and 8.3740 Å respectively. In the first 30 to 60 minutes, formation of Fe_3O_4 took place and after 60 minutes, deformation started taking place. Once the lattice narrowed down and the grid widened, deformation occurred and the bonds loosened.

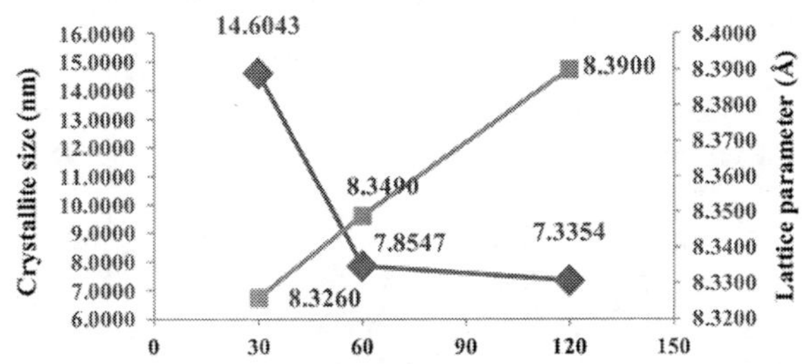

Figure 4. The effect of sonication time toward crystallite size and lattice parameter of bare Fe_3O_4 nanoparticles.

Figure 4 stated that the lattice parameters of the bare Fe_3O_4 nanoparticles when ultrasonication times were 30, 60 and 120 minutes were 8.3260 Å, 8.3490 Å dan 8.3900 Å respectively. The smallest crystallite size of bare nanoparticles was obtained at 120-minute ultrasonication time.It indicated that Fe_3O_4 sonication process has taken place from the start. However, deformation took place causing larger and longer lattice parameter as well as smaller crystallite size.

Table 2. Influence of PVA Concentration towards Crystallite Size and Lattice Parameter

Sonication Time (minutes)	PVA Mass (g)	Lattice Parameter	Crystallite Size
30	5	8.3900	5.3941
60	7	8.3610	8.9134
120	9	8.3660	6.3616
30	5	8.3710	9.8878
60	7	8.3550	19.5592
120	9	8.3740	7.2919
30	5	8.3870	6.4016
60	7	8.3470	12.5894
120	9	8.3840	6.7671

Based on **Table2** about the influence of PVA concentration on the crystallite size and lattice parameters, higher PVA concentration would result in smaller crystallite size and the larger lattice parameters.

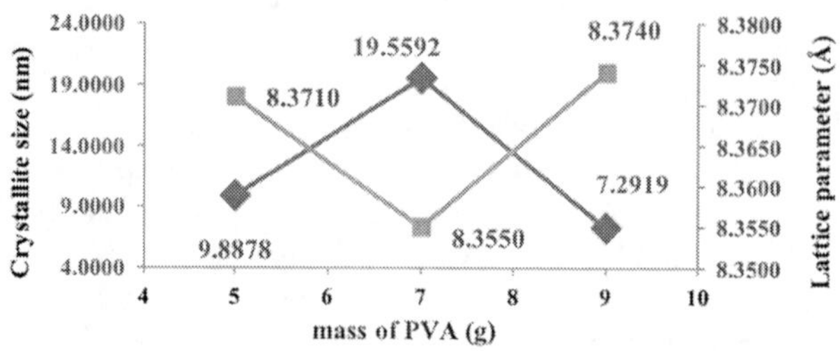

Figure 5. The effect of PVA Concentrations towards Lattice parameter and Crystallite Size

Figure 5 described the analysis on the influence of (5, 7 and 9)% PVA concentration and 60-minute ultrasonication time towards the crystallite size and lattice parameter in Fe_3O_4-PVA nanoparticle formation. The (5, 7 and 9)% PVA concentration resulted in crystals of whose sizes were 9.8878, 19.5592 and 7.2919 nm while the lattice parameter was 8.3710 Å, 8.3550 Å and 8.3740 Å. When the lattice parameter was 8.396 Å (JCPDS File19-629), a crystal was identified as magnetite but when the lattice parameter was 8.346 Å(JCPDS File 39-1346), it was categorized as maghemite [18]. On the other hand, the XRD analysis is between magnetite and maghemite.

The addition of PVA polymers on the surface of Fe_3O_4 nanoparticles greatly affected the lattice parameters of the nanoparticles. Larger lattice parameters indicated wider gap between Fe and O atoms. Based on the findings, it may be concluded that there was correlation between the crystallite size and lattice parameters of the Fe_3O_4-PVA nanoparticles. In this case, the nanoparticle crystallite size was inversely proportional to the lattice parameter. The smaller the crystallite size resulted in the bigger lattice parameter. The 7% PVA concentration resulted in larger sized crystal compared to the 5% PVA concentration. However, the higher the PVA concentration (9%) was, the smaller the crystallite size would be. In 9% PVA concentration, smaller crystallite size would have larger lattice parameter.

The diffraction pattern generated from the characterization process indicated the formation of Fe_3O_4. Based on **Figure 6** on one of the bare Fe_3O_4 variations, the diffraction peak at $2\theta = 30°, 35.6°, 43.2°$ and $57.2°$. The peak indicated that Fe_3O_4 nanoparticles had a cubic spinel structure based on the reference (JCPDS file, PDF No. 65-3107) [20].

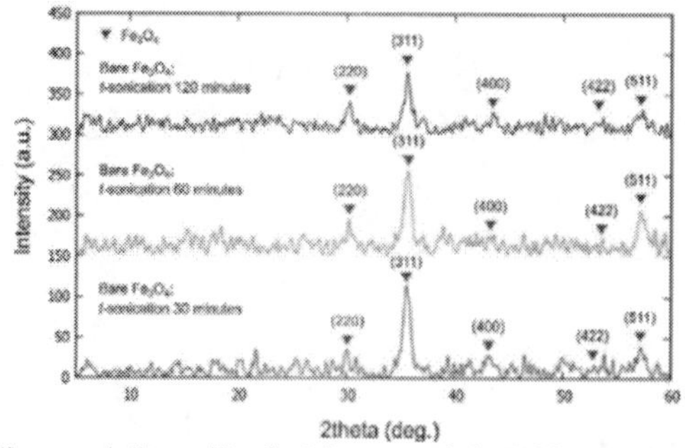

Figure 6. Bare Fe_3O_4 Nanoparticle Difractogram

Based on **Figure 7** on one of the 7% Fe_3O_4-PVAvariations, the diffraction peak at 2θ = 30.24°, 35.63°, 43.18°and 57.25°. The peak indicated that Fe_3O_4 nanoparticles had a cubic spinel structure based on the reference (JCPDS file, PDF No. 65-3107) [20].

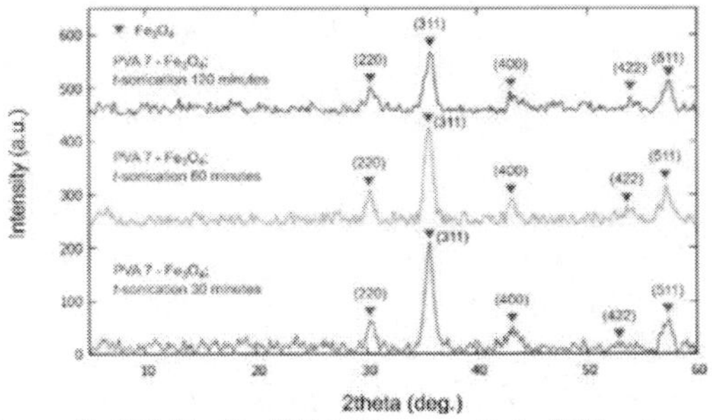

Figure 7. 7% Fe_3O_4-PVA Nanoparticle Difractogram

4. Conclusion

Based on the findings of the research on the ex-situ Fe_3O_4 nanoparticles synthesis and surface modification using PVA with the coprecipitation-ultrasonication method with various time variations and the addition of variation of PVA concentration towards the influence of lattice parameter and crystallite size using the X-ray Diffraction (XRD) instrument, Fourier Transformation Infra-Red (FT-IR), it can be inferred the characterization of Fe_3O_4-PVA nanoparticles using FTIR can be seen from the peak of Fe-O-C group, that becomes characteristic of Fe_3O_4-PVA nanoparticles. The time of ultrasonication of Fe_3O_4-PVA nanoparticles greatly influences change of crystallite size and lattice parameter. Longer ultrasonication time results in smaller crystallite size and larger lattice parameter. Based on the XRD data, the optimum ultrasonication time was 60 minutes and the most suitable level of PVA is 7%.

References

[1]Rawat M, Singh D and Saraf S 2006 Nanocarriers: Promising Vehicle for Bioactive Drugs *Biol. pharm. Bull* **29** 9 1790–1798

[2]Wu W, He Q and Jiang C 2008 Magnetic Iron Oxide Nanoparticles: Synthesis and Surface Functionalization Strategies *Nanoscale Res Lett* **3** 397-415

[3]Mahdavi M, Ahmad M, Haron M J, Namvar F, Nadi B, AbRahman M Z and Amin J 2013 Synthesis, Surface Modification and Characterisation of Biocompatible Magnetic Iron Oxide Nanoparticles for Biomedical Applications *Molecules* **18** 7533-7548

[4]Sau T K and Rogach A L 2012 Complex-shaped Metal Nanoparticle Wiley-VCH:Germany.

[5]Zhu K, Zhang M, Hong J and Yin Z 2005 Size Effect on Phase Transition Sequence of TiO_2 Nanocrystal *Mater. Sci. Eng. A* **403** 87–93

[6]Castro A L, Nunes M R, Carvalho A P, Costa F M and Florencio M H 2008 Synthesis of Anatase TiO_2 Nanoparticles with High Temperature Stability and Photocatalytic Activity *Solid State Sci* **10** 602–606

[7]Iriani D F 2012 Pengaruh Variasi Persentase Polyvinyl Alcohol (PVA) pada Ferrogel Berbasis Nanopartikel Magnetit (Fe_3O_4) *thesis* Physics Department Universitas Islam Negeri Sunan Kalijaga, Yogyakarta

[8]Maurizi, L, Claveau A and Hofmann H 2015 Polymer Adsorption on Iron Oxide Nanoparticles for One-step Amino-functionalized Silica Encapsulation *Nanomaterials* **2015** 1-6

[9]Kim SY, Ramaraj B and Yoon K R 2012 Preparation and Characterization of Polyvinyl alcohol-grafted

Fe_3O_4 Magnetic Nanoparticles through Glutaraldehyde *Surface and Interface Analysis* **44** 91238–1242

[10] Mahmoudi M et al 2008 Optimal Design and Characterization of Superparamagnetic Iron Oxide Nanoparticles Coated with Polyvinyl Alcohol for Targeted Delivery and Imaging *J. Phys, Chem* **112** 14470–1448

[11]Sholihah L K 2010 Sintesis dan Karakteristik Partikel Nano Fe_3O_4 yang Berasal dari Pasir Besi dan Fe_3O_4 Bahan Komersial (ALDRICH) Institut Teknologi Sepuluh November

[12]Zhao X and Harris J M 1998 Novel Degradable Poly (ethylene glycol) Hydrogels for Controlled Release of Protein *J. Pharm. Sci.* **87** 1450-1458

[13]Yan A, Liu X, Qiu G and Wu H 2008 Solvothermal Synthesis and Characterization of Size-controlled Fe_3O_4 Nanoparticles *Alloy. Compd* **458** 487–491

[14]Jiang Hong, Yan Zhong et al 2012 Zincon-Immobilized Silica-Coated Magnetic Fe_3O_4 Nanoparticles for Solid-Phase Extraction and Determination of Trace Lead in Natural and Drinking waters by Graphite Furnace Atomic Absorption Spectrometry *Journal Talanta* **92** 251-256

[15]Mason T J 2014 Introduction to this Special Edition of Ultrasonics Sonochemistry *Ultrasonics Sonochemistry* **25** 1-3

[16]Telefoncu A 2007 Improving the Stability of Cellulase by Immobilization on Modified Polyvinyl Alcohol Coated Chitosan Beads *Mol. Catal* **45** 10–14

[17]Ghanbari D, Salavati-niasari M and Ghasemi-kooch M 2014 A Sonochemical Method for Synthesis of Fe_3O_4 Nanoparticles and Thermal Stable PVA-Based Magnetic Nanocomposite *J. Ind. Eng. Chem* **20** 3970-3974

[18]Saxena N and Singh M 2017 Efficient Synthesis of Superparamagnetic Magnetite Nanoparticles Under Air for Biomedical Aplication *Journal of Magnetism and Magnetic Materials* Accepted manuscript

[19]Yang C, Wang G, Lu Z, Sun J, Zhuang J and Yang W 2005 Effect of Ultrasonic Treatment on Dispersibility of Fe_3O_4 Nanoparticles and Synthesis of multi-core Fe_3O_4/SiO_2 core/shell Nanoparticles *Journal of Materials Chemistry* **15** 4252-4257

[20] Shen W, Shi MM, Wang M and Chen H Z 2010A Simple Synthesis of Fe_3O_4 Nanoclusters and Their Electromagnetic Nanocomposites with Polyaniline *Mater ChemPhys* **122** 588–594

IOP Conf. Series: Materials Science and Engineering **299** (2018) 012066 doi:10.1088/1757-899X/299/1/012066

One-pot Synthesis and Surface Modification of Fe_3O_4 Nanoparticles Using Polyvinyl Alcohol by Coprecipitation and Ultrasonication Methods

Aditya D Nugraha[1], Ika O Wulandari[1], L B Hutami Rahayu, Imam Rivai, D J Djoko H Santojo[2,3], Akhmad Sabarudin[1,3,*]
[1]Department of Chemistry, Faculty of Science, Brawijaya University, Malang, 65145, Indonesia
[2]Department of Physics, Faculty of Science, Brawijaya University, Malang, 65145, Indonesia
[3]Research Center for Advanced System and Material Technology (ASMAT), Brawijaya University, Malang 65145, Indonesia

*Corresponding Author: sabarjpn@ub.ac.id

Abstract. Among the various substances developed through nanoparticles, iron oxide (Fe_3O_4) nanoparticle is one of the substances that have been widely used in various fields such as industry, agriculture, biotechnology and biomedicine. The synthesis of Fe_3O_4 nanoparticle can be carried out by two methods, consist of chemical and mechanical synthesis methods. Coprecipitation is one of the most commonly used methods for chemical synthesis. Fe_3O_4 compounds are easily oxidized because they are amphoteric. To avoid the continuous oxidation process, chemical modification process should be carried out with the addition of a solution of polyvinyl alcohol (PVA). In this study, PVA-coated Fe_3O_4 nanoparticles were synthesized by in-situ coprecipitation and ultrasonication methods through direct mixing (one-pot synthesis) of the iron (II) chloride tetrahydrate ($FeCl_2.4H_2O$), iron (III) chloride hexahydrate ($FeCl_3.6H_2O$), and PVA under alkaline condition. The effects of addition amount of NH_3solution (by adjusting its flow rate using automated syringe pump) and PVA concentration were gently studied. Interaction of PVA with Fe_3O_4 nanoparticle was identified by infrared spectroscopy whereas lattice parameters and crystallite sizes of the synthesized Fe_3O_4 nanoparticles and PVA-coated Fe_3O_4 nanoparticles were assessed by X-ray diffraction (XRD).

1. Introduction

Nanoparticle is very small, invisible and microscopic with diameter ranging from 1 to 100 nm composed of both organic and inorganic chemical substances [1]. Among the various nanoparticles developed, iron oxide (Fe_3O_4) compound is one of the nanoparticle substances that have been widely used in various fields such as agricultural industry, biotechnology, and biomedicine [2].

In general, the synthesis of Fe_3O_4nanoparticle can be done through two methods, namely chemical and mechanical syntheses methods, with coprecipitation method is one chemical synthesis method that is commonly used because the process is simple with lower cost [3]. One application of nanoparticles in the biomedical field is used as a drug delivery intermediate using Fe_3O_4 compounds. Fe_3O_4 compounds have several characteristics that are biocompatible, biodegradable, non-toxic, and containing magnetic properties [1].

On the other hand, Fe_3O_4 compounds also have some drawbacks that are easily oxidized and tend to agglomerate [4], so it is necessary to modify the surface chemically, of which is by the addition of polyvinyl alcohol (PVA) solution. Polyvinyl alcohol (PVA) is a polymer solution that has the properties to readily form a gel as it is reacted with water, so that the oxidation process in Fe_3O_4 compounds can be minimized [5].

In this study, the methods used in PVA-Fe_3O_4syntheses are the combination of coprecipitation and sonochemical (ultrasonication) with the rationale is that the combination of these two methods tends to be easy to conduct, requiring relatively lower cost, and the final amount obtained is likely increase. In addition, the combination of the two methods is expected to produce nanoparticles with the small size of distribution. Moreover, the ultrasonication synthesis method is believed to provide control over the more spherical forms/morphologically [6].

In this study, the surface of the modified Fe_3O_4 nanoparticles is PVA-based polymer which would be synthesized by a combination of coprecipitation-ultrasonication (co-precipitation assisted ultrasonication) method. The surface modification was carried out by addition of polyvinyl alcohol (PVA) solution at three concentration variations. In addition, the effect of time difference of sonication on the nanoparticle crystallite size was also studied in more detail. The effect of the addition of concentration of polyvinyl alcohol (PVA) solution and the duration of sonication on the crystallite size of Fe_3O_4 nanoparticles was analysed by using X-ray diffraction (XRD) and Fourier Transform Infrared (FTIR).

2. Material and Method

2.1 Chemicals

The materials used in this study include ammonia 25% v/v in H_2O, and polyvinyl alcohol (PVA) purchased from Merck (Indonesia), $FeCl_2.4H_2O$ 99% (w/w), $FeCl_3.6H_2O$ 98% (w/w) solids purchased from Sigma-Aldrich (Singapore).

2.2 Synthesis of Fe_3O_4 nanoparticles and Fe_3O_4-PVA Nanoparticles

In the process of synthesis of Fe_3O_4 bare nanoparticles, at first $FeCl_2.4H_2O$ and $FeCl_3.6H_2O$ solids were consecutively weighed as much as 0.2 g and 0.5406 g (1:2 molar ratio), and then both solids were dissolved in a beaker glass with 20 mL deionized water. The Fe (II) and Fe (III) solutions were mixed and then placed on a sonicator device. Meanwhile, 2M of $NH_3.H_2O$ solution was prepared at the same time. Furthermore, 10 mL of $NH_3.H_2O$ solution was flowed to Fe(II) and Fe(III) mixture with flow rate setting of 30 mL/h. During the process of addition of $NH_3.H_2O$ into the mixture of Fe salt, a black precipitate was formed. The black precipitate indicates Fe_3O_4 has been formed. Next it was washed with distilled water to neutral pH. Before that, the characterised Fe_3O_4 nanoparticles were dried with freeze drying method.

The 2% of PVA solution was prepared by weighing 2 g of polyvinyl alcohol solids and dissolved into 100 mL of distilled water. The solution was stirred with magnetic stirrer on a hotplate at temperature of ± 82 ^{0}C. The heating and stirring process was carried out for ± 15 minutes. Before further use, the PVA polymer solution was let cool to the room temperature.

The preparation of the Fe mixed solution between $FeCl_2.4H_2O$ and $FeCl_3.6H_2O$ was carried out by weighing the two solids at 0.2 g and 0.5406 g respectively (Fe(II): Fe(III) molar ratio was 1:2). After Fe(II) and Fe(III) solids were weighed and then put into 100 mL beaker glass, and 10 mL PVA polymer solution was added. The mixture was stirred until both Fe solids were mixed with PVA polymer, and then the surface of the beaker glass was covered using an airtight plastic.

The next step was the beaker glass containing the mixture of Fe(II), Fe(III) and PVA solution was placed in a sonicator batch. A 10m L of $NH_3.H_2O$ solution was added to the mixture of Fe and PVA solution through a syringe pump. The addition of ammonia water is carried out by adjusting the flow rate at 30 mL/h.

During the addition process of $NH_3.H_2O$ solution to the mixture of $FeCl_2.4H_2O$ and $FeCl_3.6H_2O$, a black precipitate was formed. After the sonication process, the precipitate obtained was washed using distilled water until it reached a neutral pH (pH=7). Then the sample was dried with a freezing dryer before the characterisation process.

The same procedures as above were done for syntheses of Fe_3O_4-PVA nanoparticles, but with different PVA concentrations of 3% and 4%. These three variations of PVA concentrations would be used to see how their effect on the size of the resulting crystallite nanoparticles.

2.3 Characterization of nanoparticles by FTIR and XRD

The characterization of Fe_3O_4 nanoparticles was employed by Fourier Transform Infrared (FTIR) (Shimadzu FTIR QP89500, KBr pellet in wavenumber range 4000-400 cm^{-1}) and was performed to determine the presence of functional groups found in Fe_3O_4 bare nanoparticles and Fe_3O_4-PVA nanoparticles. While characterization of Fe_3O_4 nanoparticles using XRD was performed to determine the size of nanoparticle crystals as well as lattice parameters of the Fe_3O_4 bare nanoparticles and Fe_3O_4-PVA nanoparticles. Measurements with X-ray diffraction (XRD-XPERT-PRO) were performed with Cu Kα radiation (30 mA, 40 KV) with a range of 2θ 5-600. The X-ray diffraction pattern resulted from the characterization of each sample variant was then compared with Inorganic Crystal Structure Database (ICSD) #26410 data to confirm the existence of $Fe_3O_4$of the sample variant of Fe_3O_4-PVA nanoparticles. The size of the Fe_3O_4-PVA crystals is determined by the modified Debye Scherer method with Equation as follows [1]:

$$\ln \beta = \ln \frac{K\lambda}{L.\cos\theta} = \ln \frac{K\lambda}{L} + \ln \frac{1}{\cos\theta}$$

Where :
L :average size of crystallite (nm)
λ :wavelength of X-Ray (nm)
β :FWHM
K :crystallite shape constant

3. Result and Discussion

Prior to the addition of $NH_3.H_2O$ solution, a mixture of $FeCl_2.4H_2O$ solids and $FeCl_3.6H_2O$ solids and polyvinyl alcohol (PVA) solution was bright yellow. After being added with $NH_3.H_2O$ solution, the colour of the mixture would turn into solid black which indicates the formation of the Fe_3O_4 compound.

Precipitation occurs in Fe^{3+} ions because of the reaction to OH group of $NH_3.H_2O$ solution to form Fe(III) hydroxide in the pH range of 3-4. In addition Fe^{2+} ions will form Fe(II) hydroxide in the higher pH range (8-9) [7]. Fe(III) hydroxide decomposition process becomes FeOOH compound, and then FeOOH compound will react with $Fe(OH)_2$ compound to form Fe_3O_4compound. The use of 1:2 ratio in $FeCl_2.4H_2O$ solids and $FeCl_3.6H_2O$ solids is intended to obtain higher purity magnetite compounds.

The samples of Fe_3O_4 nanoparticles were characterized using FT-IR. The result of the characterization using FT-IR was in the form of FT-IR spectra from the samples of Fe_3O_4 bare and Fe_3O_4-PVA nanoparticles presented in Figure 1. The presence of Fe_3O_4 can be confirmed from the peak that appears in the wave number area between 3514-3334 cm^{-1}. This peak indicates OH uptake on Fe_3O_4 surface.

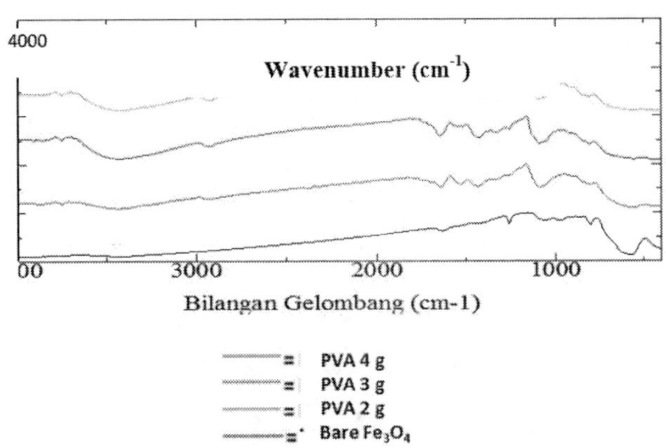

Figure 1. FT-IR spectra of Fe_3O_4-PVA nanoparticles with the mass variations of PVA in ultrasonication time of 60 minutes

Based on the results of characterization using FTIR on the samples of Fe_3O_4-PVA nanoparticles, 5 absorption peaks were obtained showing the functional groups indicating Fe_3O_4 and PVA. Adsorption peak at the 3400 cm^{-1} wave region is identified as OH group of Fe_3O_4, while the C-H and C-C groups of PVA are identified respectively at the wavelength area of 2950 cm^{-1} and 1400 cm^{-1}. The interaction between PVA and Fe_3O_4 is known from the presence of absorption in the 1100 cm^{-1} wave area which is identical to the Fe-O-C group interactions. In the fingerprint area, there is absorption at the 560 cm^{-1} wave number which is confirmed as the interaction between Fe-O from Fe_3O_4.

In the spectrum of Fe_3O_4 bare nanoparticles, it is observed that it does not contain C-C and C-O groups because there is no PVA content in this compound, while on the other three spectra, it is observed that the presence of C-C and C-O groups is because Fe_3O_4 compounds have been coated with PVA surface. As the PVA concentration increases, the intensity of the spectra of the Fe-O group formed becomes more evident, while the peaks obtained for the C-C and C-O groups even sharpen.

FTIR spectra showing the presence of Fe-O functional groups will be more evident with the increasing concentrations of PVA. Such condition will cause the PVA layer which coats Fe_3O_4 compounds to be more numerous, so the distance between Fe and O atoms produced is smaller. In the use of 2 to 3 g of PVA, there is a growing crystallite size due to the growing crystal of the Fe_3O_4 nanoparticles, while the use of mass of 3 to 4 grams of PV will cause the decrease of crystal size due to the ultrasonication process that will cause bonds between atoms increasingly tenuous. Ultrasonication process will provide energy that causes the breaking of the bonds so that the crystallite size resulted will be smaller.

The results of characterization with XRD are presented in Figure 2. Based on characterization using XRD, two identical peaks of Fe_3O_4 were obtained respectively at hkl (311) and (511). From Figure 2, we can get information that the decreasing diffractogram peaks due to the use of 2 to 3 grams of PVA, then sharp peaks reappear in the use of 3 to 4 grams of PVA at hkl position (311), while the peak of the diffractogram at hkl (511), the intensity decreases as the increasing mass usage of PVA.

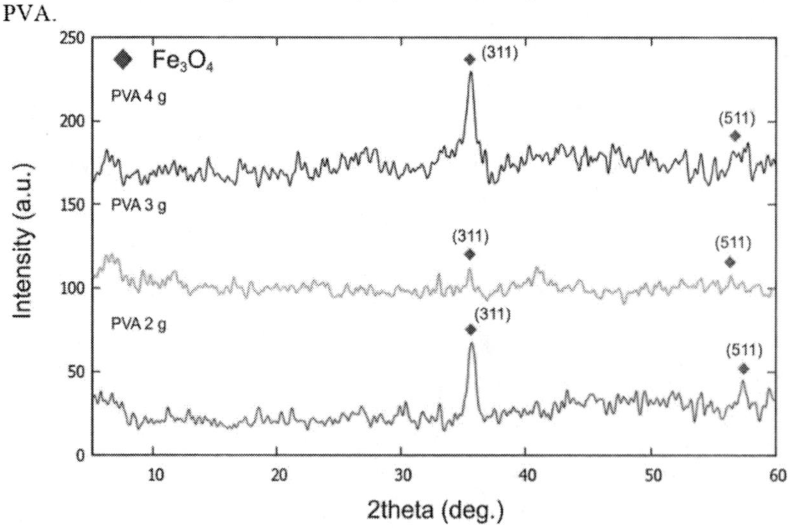

Figure 2. XRD pattern of PVA-Fe_3O_4 (2, 3 and 4 g PVA with ultrasonification time of 60 minutes)

The original large crystals will split and form smaller sizes. There is a comparable correlation between the size of the crystallite and the lattice parameter (Figure 3). Based on JCPDS file 19-629, the lattice parameter for Fe_3O_4 is 8.396. The lattice parameter calculations show the values close to the lattice parameters for standard Fe_3O_4.

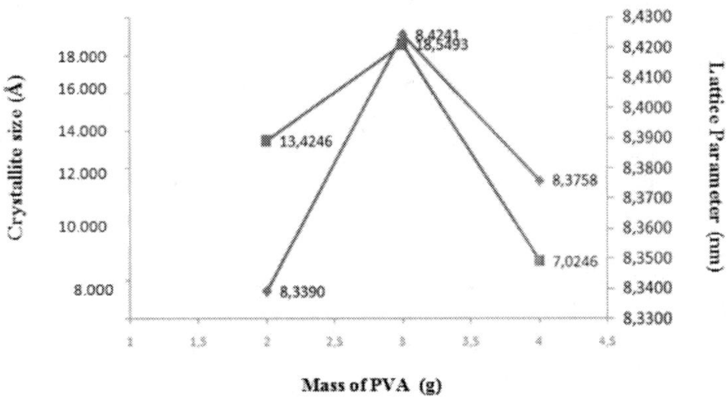

Figure 3. Graph of comparison between lattice parameter and the size of crystallite at PVA-Fe_3O_4 (2, 3 and 4 g in ultrasonication time of 60 minutes)

4. Conclusion

The Fe_3O_4-PVA nanoparticles have been successfully synthesized by coprecipitation assisted ultrasonication methods. The effect of variation of concentration of PVA on Fe_3O_4 surface was studied from FTIR spectra. The decrease in the intensity of the absorption peaks from the OH group in the wavelength area 3514-3334 cm^{-1} correlates with an increase in PVA concentration. In addition, the concentration of PVA added on Fe_3O_4surface affects the lattice parameter and the size of crystallite. The largest crystallite and lattice parameters were obtained in the addition of PVA with a concentration of 3%.

References

[1]Wu, W., He Q and Jiang C 2008 Magnetic Iron Oxide Nanoparticles: Synthesis and Surface Functionalization Strategies *Nanoscale Research Letters* **3** 397–415

[2]Liu, X.D., Chen, H., Liu, S.S., Ye LQ and Li Y P2015 Hydrothermal Synthesis of Superparamagnetic Fe_3O_4Nanoparticles with Ionic Liquids as Stabilize *Materials Research Bulletin* **62** 217-221

[3]Hariani, P.L., Faizal, M., Ridwan, Marsi dan Setiabudidaya D 2013 Synthesis and Properties of Fe_3O_4 Nanoparticles by Co-precipitation Method to Removal Procion Dye *International Journal of Environmental Science and Development* **4** 336-340

[4]Horak, D., Babic, M. Mackova Hand Benes M J 2007 Preparation and Properties Of Magnetic Nano And Microsized Particles For Biological And Environmental Separations *J. Sep. Sci.* **30**1751-1772.

[5]Ayyapan, S., Panneerselvam, G., Antony M P and Philip J 2011 High Temperature Stability of Surfactant Capped $CoFe_2O_4$ Nanoparticles *Materials Chemistry and Physics* **130** 1300-1306.

[6]Monshi, A., Foroughi MR and Monshi MR 2012 Modified Scherrer Equation to Estimate More Accurately Nano-Crystallite Size Using XRD *World Journal of Nano Science and Engineering***2** 154-160
[7]Liang, X., Shi, H., Jia, X., Yang Y and Lin X 2011 Dispersibility, Shape and Magnetic Properties of Nano-Fe_3O_4 Particles*Materials Sciences and Application* **2** 1644-1653

IOP Conf. Series: Materials Science and Engineering **299** (2018) 012067 doi:10.1088/1757-899X/299/1/012067

Uncaria gambir Roxb. mediated green synthesis of silver nanoparticles using diethanolamine as capping agent

A Labanni[1], Zulhadjri[1], D Handayani[2] and S Arief[1]

[1] Material Laboratory, Chemistry Department, Universitas Andalas, Limau Manis 25163, Padang, Indonesia.
[2] Laboratory of Sumatran Biota, Faculty of Pharmacy, Universitas Andalas, Limau Manis 25163, Padang, Indonesia.

Email: Syukriarief@gmail.com

Abstract. Studies of silver nanoparticles preparation has been developed increasingly due to the wide application in various areas and field, such as medicine, energy, catalysis, and electronic. An environmental-friendly method is needed to fabricate biocompatible silver nanoparticles without producing hazardous materials to the environment. In this study, we synthesized silver nanoparticles by green synthesis method, using leaf extract of gambir (*Uncaria gambir* Roxb.) as bioreducing agent and aqueous diethanolamine (DEA) solution as capping agents. The $AgNO_3$/DEA molar ratio was varied to investigate the effect of DEA concentration to the properties of silver nanoparticles. The formation of silver nanoparticles was indicated by colour changes to yellowish brown and confirmed by result of UV-Vis spectrophotometer analysis which shown absorption band at 400 to 410 nm. The absorbance was increased to the reaction time of 24 hours, and was decrease by the increasing of DEA concentration in reaction. TEM analysis showed that prepared silver nanoparticles were spherical in shape with diameter of 3,5 - 45,5 nm. The diameter of DEA capped silver nanoparticles was 13 nm, smaller than uncapped silver nanoparticles which was 26 nm It exhibited good stability to time reaction of one month which was potential to be developed in some fields.

1. Introduction

A nanoparticle is a microscopic particle with size of less than 100 nm [1]. In recent years, the study of nanoparticles has drawn the attention of scientists due to the excellent properties which bridge the gap between bulk and atomic or molecular structures. The unique properties include chemical stability, conductivity, and catalytic activity have made nanoparticles developed in some wide areas and application such as energy, environment, medical, sensor, catalysis [2]. Among all metal nanoparticles, silver has been increasingly developed due to the potential as antibacterial agents to be developed in medical utilizing. The preparation of silver nanoparticles have been carried out using various method, such as chemical reduction [3], electrochemical [4], sonochemical [5], microwave method [6], etc. Among all the methods, the chemical reduction method is the easiest and the most rapid way to conduct.

Unfortunately, the chemical reduction method employs chemical reagent to reduce Ag^+ ions to Ag nanoparticles which can endanger the environment and threaten the living organisms. In addition, the using of hazardous material will result in a low biocompatibility in the application. Therefore, an environmental method is developed by the scientists, called green synthesis using nonhazardous materials such as plants, bacteria, algae, etc. This method is considered as an environmentally friendly, economical, easy, low energy, and rapid method [7].

Uncaria gambir Roxb. is one of the commodities in West Sumatera used as materials for plywood. It is also used as traditional medicine for diarrhea, toothache, and fever due to the content of catechin which can act as a natural antioxidant [8]. In the synthesis of nanoparticles, catechin is expected to reduce Ag^+ ions to Ag^0 nanoparticles. Beside reducing agent, the reaction also involves capping agent to control the growth and size of the nanoparticles. In the previous study, we prepared silver nanoparticles using hydrothermal method with *Uncaria gambir* Roxb. as bioreducing agent, resulting colloidal silver nanoparticles with a diameter of 25 nm [9]. However, we need a simple route synthesis to investigate the optimum condition of capping agent and the effect on the characteristic of the prepared nanoparticles. In the present study, we synthesized silver nanoparticles using gambir leaf extract as bioreducing agent and diethanolamine in aqueous solution with various mole ratio of Ag^+/DEA in a facile and simple route.

2. Experimental

2.1. Materials

Fresh leaves of *Uncaria gambir* Roxb. were collected from Experimental Garden, Andalas University, Padang, Indonesia. Silver nitrate ($AgNO_3$) as precursor and diethanolamine (DEA) as capping agent were purchased from Merck, and distilled water.

2.2 Gambir leaf extract preparation.

Gambir leaf extract was prepared using modified method of Arief et al (2015) [9]. Firstly, the fresh and healthy leaves of *Uncaria gambir* Roxb. were washed thoroughly using fresh water and distilled water to remove contamination and unnecessary materials. The clean leaves then dried in shadow condition at room temperature for several days. The dried leaves then were ground to obtain gambir powder. The aqueous extract of gambir was made by dissolving the fine powder into deionized water with ratio of 1:10 g/ml. The mixture then was stirred for 2 hours at temperature of 65 ^{0}C. After cooling, the mixture then was filtered to get the extract and was stored at temperature of 4 ^{0}C for further use.

2.3 Silver nanoparticles preparation

Gambir leaf extract was added to a 100 ml Erlenmeyer flask containing diethanolamine while was stirring in a constant rate at room temperature. The silver nitrate then was added to the mixture with variation of $AgNO_3$: MEA molar ratio was 1:10, 1:20, and 1:30 which were coded as AgNps-DEA1, AgNps-DEA2, and AgNps-DEA3. The reaction was also carried out without capping agent as a comparator, coded as AgNps0. The mixture solution was continuously monitored using UV-Vis spectrophotometer in wavelength range of 200-700 nm. The obtained colloidal silver nanoparticles then stored in an airtight bottle at room temperature for further measurements.

2.4 Characterization

The optical properties of colloidal silver nanoparticles were obtained using Thermo Scientific Evolution 201 UV-Vis spectrophotometer at wavelength ranges of 200-700 nm. The morphological properties were analysed using JEOL-4000 Transmission Electron Microscopy (TEM) operated at 120 kV.

3. Results and Discussion

3.1 UV-Vis spectroscopy study

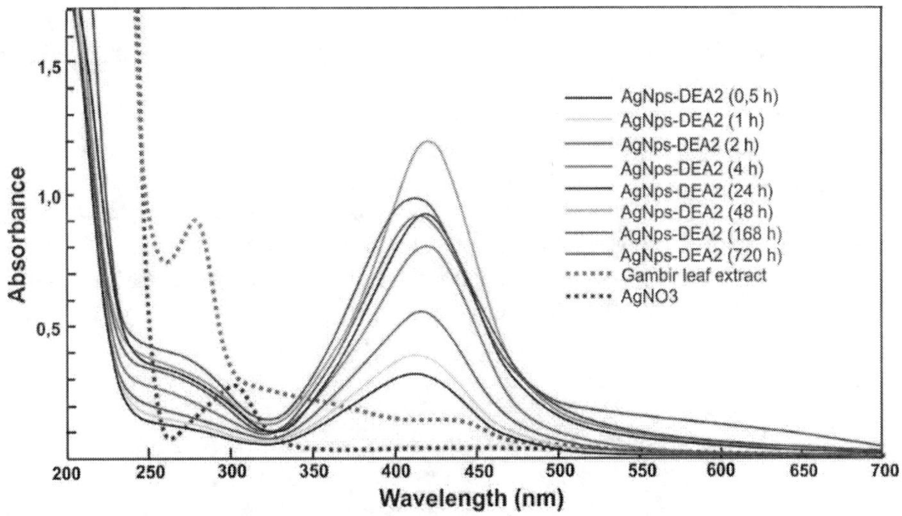

Figure 1. UV-Vis analysis of DEA stabilized gambir leaf extract mediated colloidal AgNps

The reaction was conducted in various molar ratio of AgNO3/DEA of 1:10, 1:20, and 1:30 based on the preliminary study. The formation of silver nanoparticles was visually recognized by colour changing from pale yellow to yellowish brown with the increasing of intensity as the increasing of reaction time. This was supported by UV-Vis spectrophotometer analysis. Gambir leaf extract showed a peak at 278 nm which referred to the presence of catechin [10], while $AgNO_3$ exhibited a peak at 301 nm (figure 1). These two peaks were shifted to the wavelength at 398-438 nm (Table 1) as colour changing, which strongly indicated a formation of silver nanoparticles. It was due to the reduction of Ag^+ to Ag^0 by catechin in gambir leaf extract. It provided a specific band of nanoparticles material, called Surface Plasmon Resonance (SPR) phenomenon which is specifically shown by a peak at wavelength of 395-420 nm for silver nanoparticles [11]. Previous reports about green synthesis of silver nanoparticles using *Catharanthus roseus* Linn. G. Donn. [12] and *Cleistanthus collinus* [13] showed suitable results where SPR bands peak were observed at 420 nm. In addition, the green synthesis studies have been also carried out using *Diospyros paniculata* [14] and *Alternanthera sessilis* Linn. [15], providing SPR band peaks at 418-430 and 435 nm, respectively.

Table 1. Absorbance and wavelength of DEA stabilized silver nanoparticles

Reaction time	Wavelength (nm)				Absorbance			
	AgNps-DEA1	AgNps-DEA2	AgNps-DEA3	AgNps0	AgNps-DEA1	AgNps-DEA2	AgNps-DEA3	AgNps0
0 hr	-	-	-	429	-	-	-	0,011
0,5 hr	409,63	413,14	429,0	434	0,648	0,317	0,213	0,017
1 hr	406,76	415,86	430,0	437	0,850	0,380	0,250	0,022
2 hr	405,00	415,98	436,0	435	1,237	0,546	0,290	0,023
4 hr	404,07	419,72	438,0	441	1,319	0,799	0,364	0,020
24 hr	399,68	419,91	422,0	434	1,286	0,942	0,798	0,036
48 hr	403,99	420,16	422,0	443	1,214	1,190	0,827	0,065
1 week	402,91	414,30	421,0	437	1,213	0,929	0,749	0,036
1 month	398,00	412,00	420,7	441	0,767	0,980	0,741	0,022

The UV-Vis spectrophotometer analysis also showed that the absorbance were gradually increased as the increase of time reaction to 24 hours (Table 1). This was strongly revealed that silver nanoparticles were continuously formed during the reaction. After 24 hours, the absorbance was decreased which allegedly because the particles began to settle at the bottom. This absorbance value was decreased as the increasing of DEA concentration. In addition, it was observed that the absorbance value of uncapped silver nanoparticles (AgNps0) was greater than DEA-capped silver nanoparticles. It revealed that DEA in aqueous solution acted as capping agent as well as reducing agent. Jia et al (16) reported a study on silver nanoparticles preparation using triethanolamine, another kind of alkanolamine compound as reducing agent with PEG and PVP as capping agents, resulting nanoparticles with mean diameter of 40 nm.

Different with absorbance, the wavelength value was increased as the increasing of the DEA concentration. Based on Mie Theory, the size of spherical nanoparticles can be estimated from the wavelength values in UV-Vis spectroscopy analysis. The larger the wavelength value, the larger the particle size [17]. These results showed that the diameter of silver nanoparticles was increased as the blueshift silver nanoparticles with bigger DEA concentration.

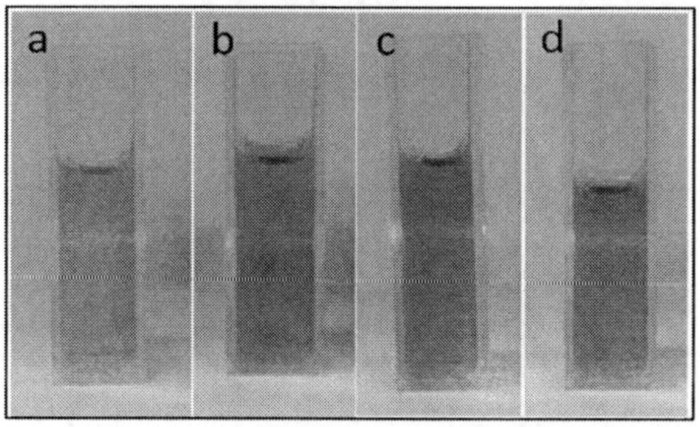

Figure 2. Laser beam radiation of silver nanoparticles in time reaction of a) 30 minute, b) 1 hour, c) 2 hours, and d) 24 hours

The laser beam radiation of colloidal silver nanoparticles was conducted to investigate the formation of silver nanoparticles colloid in the reaction based on the Tyndall Effect (figure 2). A homogenous solution is always transparent, light passes through with no scattering from solute particles which are molecule in size. While in colloid, a light scattering is occurs by the particles, resulting in a light line. Figure 2a showed a light line of silver nanoparticles in time reaction of 30 minutes which shown a light line indicating a formation of colloidal silver nanoparticles. It was observed that the intensity of the light was increased as the increase of reaction time to 24 hours (figure 2d), which confirmed an increase in the number of nanoparticles in the reaction. This sighting confirmed the result of UV-Vis spectroscopy analysis that silver nanoparticles were formed continuously during the reaction.

3.2 TEM morphology study

Figure 3 showed TEM analysis images of DEA stabilized silver nanoparticles with a molar ratio of $AgNO_3$/DEA was 1:10 and 1:30, and uncapped silver nanoparticles. All the particles were spherical in shape with diameter range of 4 - 52 nm for DEA capped and 9 - 39 nm for uncapped silver nanoparticles. This data support the result of UV-Vis spectrophotometer analysis and obviously showed a strong relation between wavelength and particle diameter based on TEM analysis result.

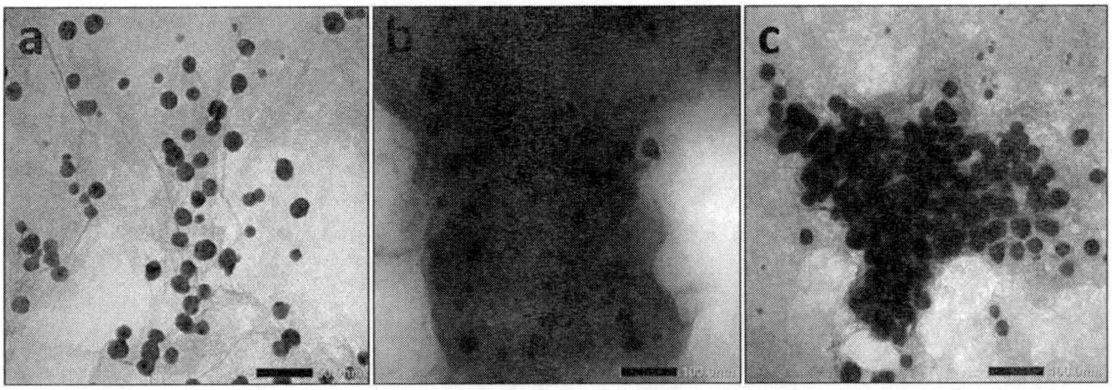

Figure 3. TEM images of a) AgNps-DEA1, b) AgNps-DEA3, and c) Uncapped AgNps

In addition, AgNps-DEA1 provided smaller mean diameter (13 nm) than AgNps-DEA3 (26 nm). The resulted silver nanoparticles were smaller than the silver nanoparticles prepared by Jia et al (16). Likewise, the size range of AgNps-DEA1 is less wide (4 - 20 nm) than AgNps-DEA3 (9 - 52 nm) (figure 4). This was due to the excessive using of capping agent. It was shown in figure 3b that there is a big clump between the particles which allegedly inhibit the interaction between silver ion with extract and decreasing the possible reduction reaction. Deficiency of nucleation reaction finally caused the growing reaction dominated and as a result, bigger particles were formed. In addition, AgNps0 exhibited bigger particle size than AgNps-DEA1, showed that the using of DEA as capping agent can reduce the particle size when used in an appropriate concentration. The most possible mechanism is that there was an interaction occurred between N and Ag surface, which then forms a protective monolayer on the surface of Ag nanoparticle. This layer can prevent an electrostatic interactions may occur among the particles, thereby control the growth and the particle size [18] These small-sized nanoparticles can provide better performance such as excellent conductivity, chemical stability, and catalytic activity, which can be developed in some areas of fields such as electronic, energy, catalysis, sensor, etc.

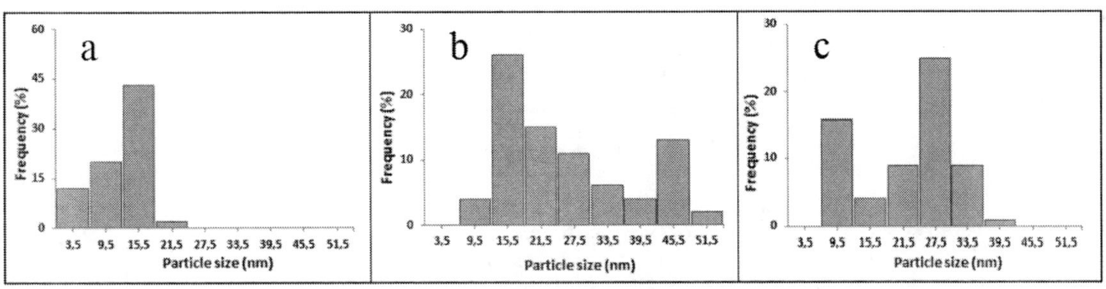

Figure 4. Size distribution of a) AgNps-DEA1, b) AgNps-DEA3, and c) Uncapped AgNps

4. Conclusions

Green synthesis of *Uncaria gambir* Roxb. leaf extract mediated silver nanoparticles had been successfully carried out using diethanolamine as capping agent. The formation of silver nanoparticles was confirmed by UV-Vis spectrophotometer analysis by a peak showed at wavelength range of 400 to 410 nm. The TEM analysis showed a formation of spherical silver nanoparticles with diameter of 3,5 - 45,5 nm. DEA capped silver nanoparticles provided smaller size than uncapped silver nanoparticles. This result showed an ability of DEA to control the growth and particle size, hence vindicate the stability of the silver nanoparticles. The increasing of DEA concentration caused the increasing of diameter. The prepared silver nanoparticles showed a good stability in aqueous solution more than of 1 month.

Acknowledment
The authors would like to thank the Directorate General of Higher Education, Ministry of National Education of Indonesia for funding this research.

References
[1] Iravani S 2011 *Green Chem.* **13** 2638
[2] Rao C, Muller A and Cheetam A 2004 *Further Titles of Interest Nanoparticles Molecular Clusters of the Main Group Elements Electrochemistry of Nanomaterials* (Weinheim: Wiley-VCH Verlag GmbH & Co. KGaA p 32
[3] Dipankar C and Murugan S 2012 *Coll. Surf. B Biointerfaces* **98** 112
[4] Nasretdinova G, Fazleeva R, Mukhitova R, Nizameev I, Kadirov M and Ziganshina A 2015 *Electrochem. commun.* **50** 69
[5] Kumar B, Smita K, Cumbal L, Debut A and Pathak R 2014 *Bioinorg. Chem. Appl.* **2014**
[6] Kahrilas G, Wally L, Fredrick S, Hiskey M, Prieto A and Owens J 2014 *ACS Sustain. Chem. Eng.* 2013 2 376
[7] Srikar S, Giri D, Pal D, Mishra P and Upadhyay SN. 2016 *Green Sustain. Chem.* **6** 34
[8] Rauf A and Zuliyanti A 2015 *Procedia. Chem.* **14** 3
[9] Arief S, Gustia V, Vanda D and Ban T 2015 *J. Chem. Pharm. Res.* **7** 189
[10] Uchida K, Ogawa K and Yanase E 2016 *Molecules* **21** 1
[11] Creighton J and Eadont D 1991 *J. Chem. Soc. Faraday Trans.* **87** 3881
[12] Subbaiah V, Rao Y, Aparna S, Prasad T, Reddy A and Gopal D 2013 *J. Coll. Surf. B Biointer.* **105** 194
[13] Kanipandian N, Kannan Sand Ramesh R 2014 *Mater. Res. Bull.* **49** 494
[14] Rao N, Lakshmidevi N, Pammi, Kollu P, Ganapaty S and Lakshmi P. 2016 *Mater. Sci. Eng. C* **62** 553
[15] Niraimathi K, Sudha V, Lavanya R and Brindha P. 2013 *Colloids and Surfaces B* **102** 288
[16] Jia Z, Sun H and Gu Q. 2012 *Colloids and Surfaces A* **419** 174
[17] Baset S, Akbari H, Zeynali H and Shafie M. 2011 *Dig. J. Nanomater. Biostructures.* **6** 709
[18] Yamamoto M, Kashiwagi Y and Nakamoto M 2006 *Langmuir* **22** 8581

Characterization of oil based nanofluid for quench medium

E P Mahiswara*, S Harjanto, W N Putra, G Ramahdita, S S Yahya and Kresnodrianto

Material and Metallurgy Departement, Faculty of Engineering, Universitas Indonesia, Depok, Indonesia

E-mail: evan.prasidya@ui.ac.id

Abstract- The choice of quench medium depends on the hardenability of the metal alloy, the thickness of the component, and the geometry of the component. Some of these will determine the cooling rate required to obtain the desired microstructure and material properties. Improper quench media will cause the material to become brittle, suffers from geometric distortion, or having a high undesirable residual stresses in the components. In heat treatment industries, oil and water are frequently used as the quench media. Recently, nanofluid as a quench medium has also been studied using several different fluids as the solvent. Examples of frequently used solvents include polymers, vegetable oils, and mineral oil. In this research, laboratory-grade carbon powder were used as nanoparticle. Oil was used as the fluid base in this research as the main observation focus. To obtain nanoscale carbon particles, planetary ball mill was used to ground laboratory grade carbon powder to decrease the particle size. This method was used to lower the cost for nanoparticle synthesis. Milling speed and duration were set at 500 rpm and 15 hours. Field Emission Scanning Electron Microscope (FE-SEM), and Energy Dispersive X-Ray (EDX) measurement were carried out to determine the particle size, material identification, particle morphology, and surface change of samples. The carbon nanoparticle content in nanofluid quench mediums for this research were varied at 0.1%, 0.2%, 0.3%, 0.4, and 0.5 % volume. Furthermore, these mediums were used to quench JIS S45C or AISI 1045 carbon steel samples which annealed at 1000°C. Hardness testing and metallography observation were then conducted to further examine the effect of different quench medium in steel samples.

1. Introduction

One of the process of steel hardening is quenching. Quenching refers to the fast cooling of a steel. In quenching, an austenized steel is immersed in a liquid, called quench medium or quenchant [1]. Commonly used quench medium are water, brine, polymers, oils, salts, and gases which have their own cooling characteristics. Brittle materials, geometric distortion, and high undesirable residual stresses on the components that lead to cracks could occur from improper use of quench media. The result of quenching, is highly effected by the cooling rates of quench media. In quenching, to get a good result, a cooling rate needs to be equal or higher than the critical cooling rate for martensite formation [2]. On the other hand, the geometry of the part is another factor that inluence the result of quenching. For components that has variation on their thicknes, quench medium with low cooling rates was used to avoid excessive residual stress, since a rapid cooling has a high possibility of the occurance of cracks, especially on the critical parts of the components. The example for selecting a quench medium is using oil as quench medium instead of water to get more evenly distributed cooling and to avoid cracking. However, the slow cooling rates of oil will result in lower hardness than the hardness achieved by water as quench medium.

Nanofluid refers to a fluid containing a small amount dispersed nano-sized (< 100 nm) solid nanoparticles. It has been proved that the nanoparticles will enhance thermal conductivity and improve thermal performance of heat transfer of the base fluid[3]. Studies on the performance of nanofluids with various types of particles has been done by many researchers[1, 4-15].

The present study aims to observe the quenching of carbon steel S45C using oil-based nanofluid with laboratory-grade carbon powder as nanoparticles. The samples were heated at 1000 °C for an hour and then quenched. The effect of nanofluid concentration was investigated to illustrate an optimal ratio of nanoparticle used in achieving the desired material properties.

2. Materials and methods

S45C carbon steel bar was used in this study. The chemical composition of the carbon steel sample was determined as given in Tables 1[16].

Table 1. Chemical composition of S45C carbon steel (%)

Fe	C	Si	Mn	P	S	Cr	Mo
98.3	0.47	0.287	0.718	0.0261	0.005	0.028	0.005
Ni	**Al**	**Co**	**Cu**	**Nb**	**Ti**	**V**	**W**
0.005	0.02	0.003	0.018	<0.002	0.035	<0.002	<0.01
Pb	**Sn**	**B**	**Ca**	**Zr**	**As**	**Bi**	
0.025	<0.002	0.0013	0.0005	<0.002	0.008	<0.03	

This study started by preparation of specimen of medium carbon steel about 30 samples. The samples were cut in the dimension of 15mm x 10mm x 10mm. Among this samples, non-heat treated were prepared as baseline comparison. Steel samples were pre-heated at 600 °C and then annealed at 1000 °C and holding it for an hour. Those samples were then quenched in oil-based nanofluid. A vickers hardness testing is conducted to measure the hardness of the samples. In the hardness testing, a force of 300 gf is applied to press the diamond indentor into the samples, and the force is maintained for 10 s. Finally, the hardness of the materials can be calculated by the size of the indentation. To obtain the average hardness, five different locations on each sample are selected to measure the hardness.

For nanofluids preparation, laboratory-grade carbon powder was used to synthesized oil-based nanofluids by two-step method. The nanofluids of 0.1%, 0.2%, 0.3%, 0.4% and 0.5% volume fraction of particles were prepared. The nanoparticles were first produced by high energy milling. The second step involves dispersing this nanopowder into the base fluid with the help of intensive ultrasonic agitation. The two-steps method was more economical than the one-step method to produce nanofluids commercially [15]. This method was suitable for wide range of particles such as oxide particles and carbon nanotubes and it is attractive to industry because it is simple for nanofluid preparation[17].

Nanoparticles synthesis was carried out at 500 rpm for 15 hours in the high energy ball mill. The nanoparticles were weighted according to the weight of each quenching medium based on % volume. After this step, oil was slowly added to 100 ml beaker glass which is filled wirh nanoparticles powder. SAE 5W-40 oil was used for this particular study. In order to increase the stability of solutions, each nanofluid suspension was sonicated in an ultrasonic apparatus for 280 seconds before the quenching experiments.

Microstructure examination were carried out for both treated and untreated samples. Each sample was prepared using standard metallography procedure. The phase of the specimens were made visible by etching using nital containing 2% Nitric acids and 98% methylated spirit on the polished surfaces.

Then the specimen was immersed into etching media for 3 seconds. Microscopic examination of the etched surface of various specimens was taken using a metallurgical microscope.

3. Results and discussion

Field Emission Scanning Electron Microscope (FE-SEM) imaging was conducted to examine the everage particle size of the laboratory-grade carbon powder before milling (Figure 1). From this observation, the average size of the powder used in this study was 15.04 μm.

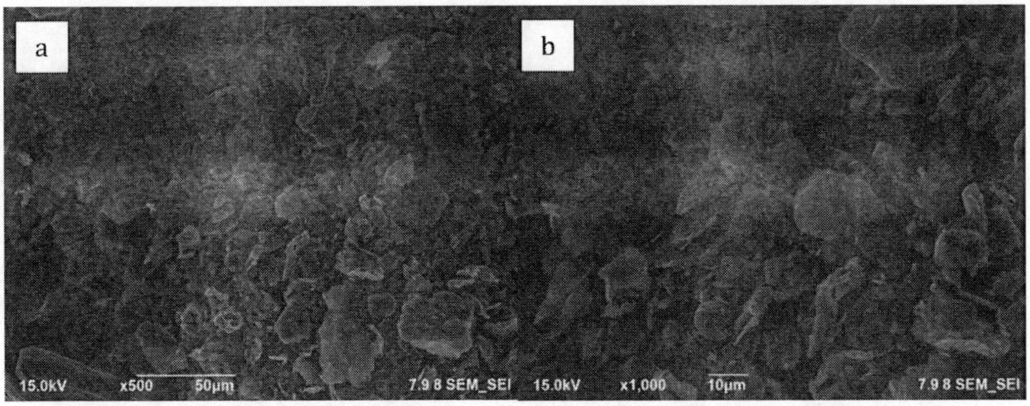

Figure 1. FE-SEM image of laboratory-grade carbon powder with (a) 500x magnification and (b) 1000x magnification.

The laboratory-grade carbon powder consist of 99.9 wt% carbon and some impurities of 0.1% Cu. This result confirms that the powders was mainly consist of carbon. The microstructure of the non-heat treated sample was shown in Figure 2[16]. Figure 2 showed that the non-heat treated sample mainly consist of ferrite and pearlite. The ferrite structure is the white area, while the darker area is pearlite.

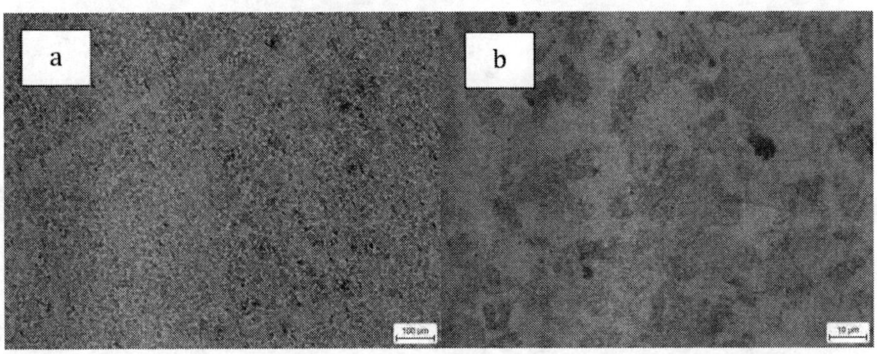

Figure 2. Microstructure of non-heat treated sample (a) 100x magnification (b) 1000x magnification

Figure 3 showed the microstructure of S45C after quenching in oil based laboratory grade nanofluid. Martensite structure was observed in the samples which were quenched in 0.1% vol. and 0.2% vol. nanodluid. As for the rest of the samples, the martensite structure was few. In sample 0.3% and 0.5 %, many pearlite structures was observed. This may because of particle agglomeration occurred in the nanofluid. Sample that quenched in 0.4% vol. nanofluid shows many martensite structure. This anomaly happens because the agitation by ultrasonic is poorly done, so the nanoparticle was not dispersed fully which affect the nanofluid performance.

The hardness of the sample quenched in oil is 363 HV. The maximum hardness for oil-based nanofluid was obtained at 0,2% vol. nanofluid with 570,98 HV (Figure 4). This value was linear to its microstruture which showed the most martensite structure. From the result, it could be found that carbon nanoparticles inside the fluid can enhance the quench severity with appropriate concentration. This happen due to transformation of martensite with rapid cooling. Transformation of martensite is driven by nanoparticles in the fluid which is having higher thermal conductivity than in normal fluid. The heat transfer enhancement phenomenon can be illustrated as the formation of nanoparticle embedded surface due to scattered deposition of nanoparticles. As a result, it increases the effective surface area of heat transfer[18]. Samples quenched in water have a hardness of 675 HV. From this, we can conclude that the oil based nanofuild has a severity of quenching between that of water and oil.

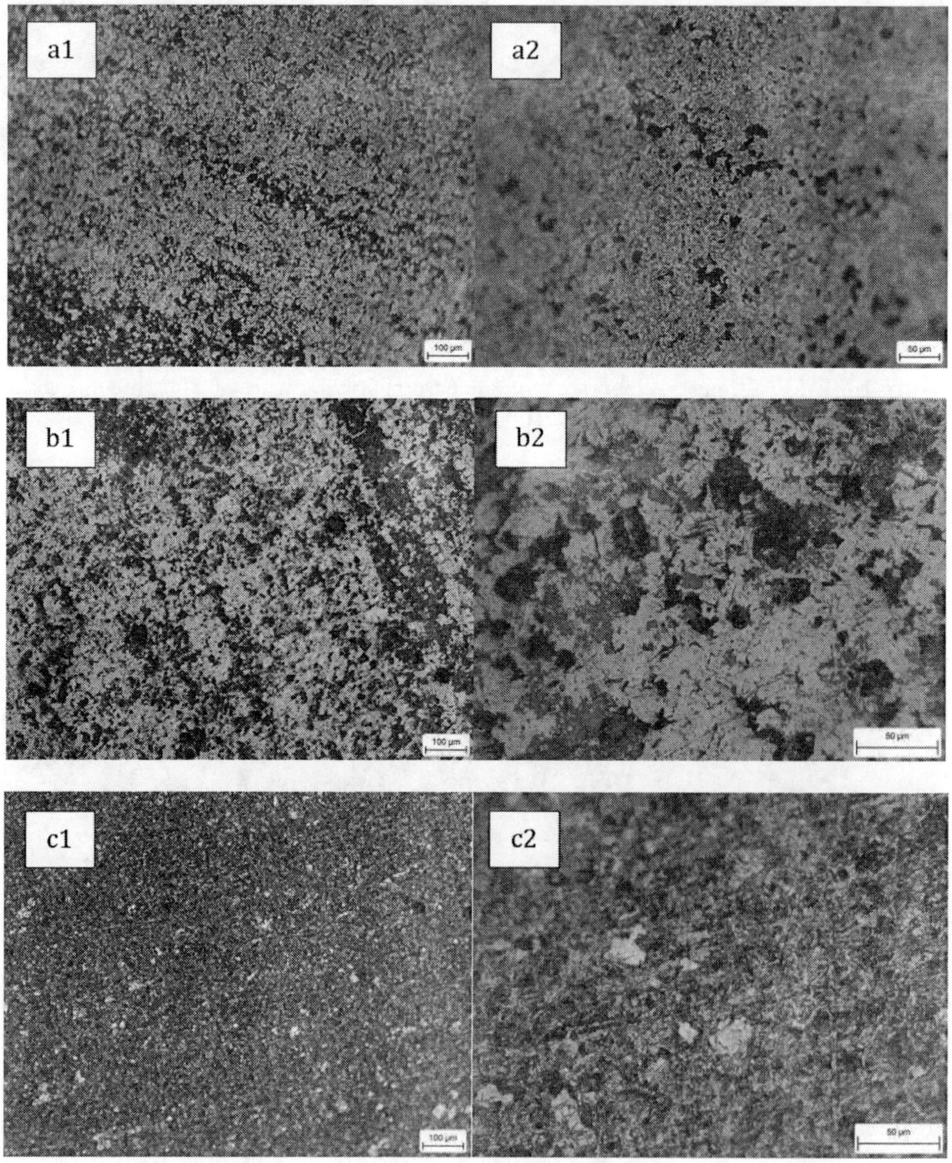

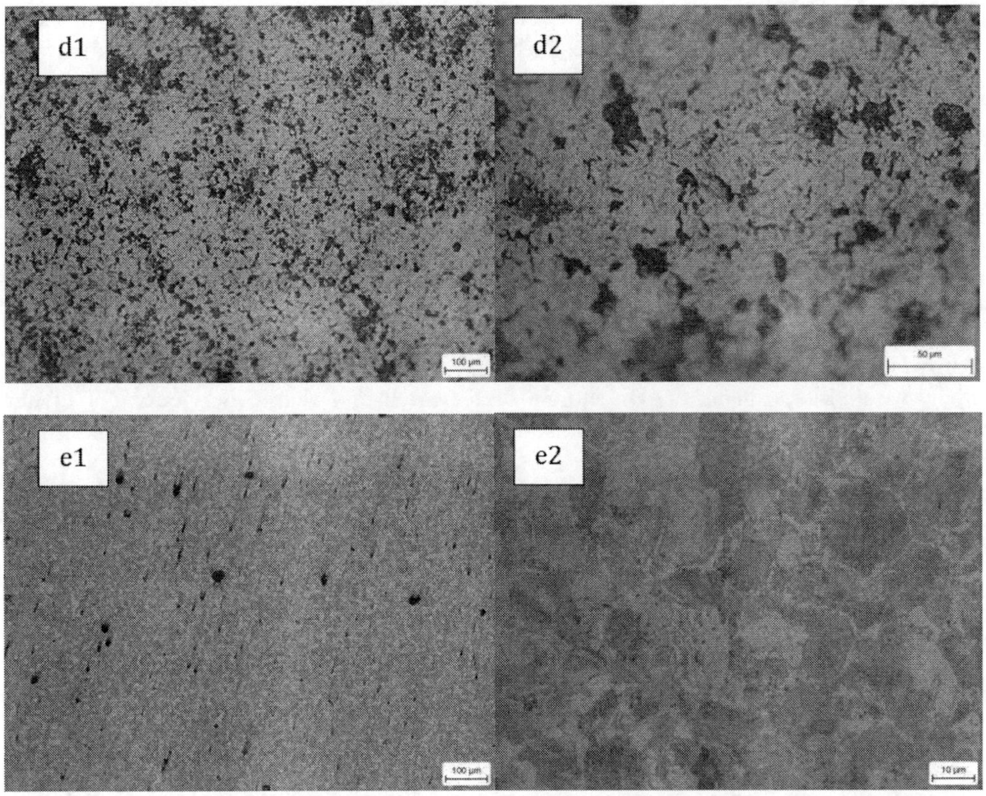

Figure 3. Microstructure of S45C after quenching in oil based laboratory grade nanofluid (a) 0.1% vol. (a1 for 100x magnification and a2 for 200x magnification), (b) 0.2% vol. (b1 for 100x magnification and b2 for 500x magnification), (c) 0.3% vol. (c1 for 100x magnification and c2 for 500x magnification), (d) 0.4% vol. (d1 for 100x magnification and d2 for 500x magnification) and (e) 0.5% vol. (e1 for 100x magnification and e2 for 1000x magnification).

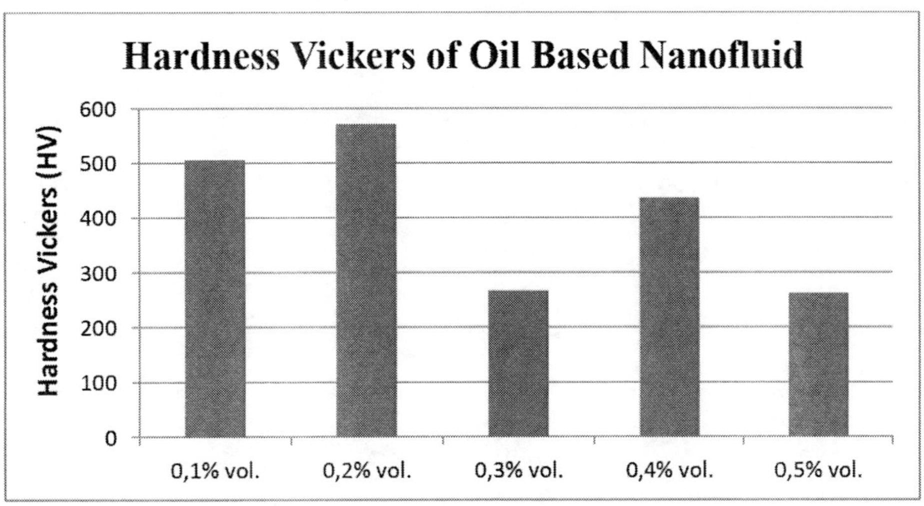

Figure 4. Result of Hardness Vickers Testing

4. Conclusion
The effect of laboratory-grade carbon in oil as quench medium was experimentally investigated using medium carbon steel, JIS S45C as specimen. For this purpose, various concentration of nanofluids were prepared with two step method. The variation of concentration is 0.1%, 0.2%, 0.3%, 0.4% and 0.5% volume. The microstucture and hardness value shows that volume fraction of nanoparticles used in nanofluids considerably affected the quenching process. Quenching with oil-based nanofluid will produce martensite structure. However, beyond 0,2% vol. the material phase mainly consist of pearlite structure. The maximum hardness is observed in 0,2% vol. with 570,98 HV. Higher particle concentration showed lower hardness which probably due to particle agglomeration. Overall, the hardness value was still higher than cooling without addition of nanoparticles. This concluded that nanoparticles inside the fluid can enhance the quench severity with appropriate concentration.

Acknowlegment
Authors would like to thank the Directorate of Research and Public Services, Universitas Indonesia for financial support and project administration respectively to conduct this research under PITTA Project on 2017 fiscal year.

Reference

[1] Nayak V U and Prabhu N K 2016 *J. Mater. Eng. Perform.* **25** 1474
[2] Grum J, Bozic S and Zupancic M *J. Mater. Process. Technol* **114** 57
[3] Dhinesh K D and Valan A A 2016 *Renew. Sust. Energ. Rev.* **60** 21
[4] Xie H, Wang J, Xi T, Liu Y, Ai F and Wu Q 2002 *J. Appl. Phys* **91** 4568
[5] Lee S, Choi S U S, Li S and Eastman J A 1999 *J. Heat Transf.* **121** 280
[6] Wang X, Xu X and Choi S U S 1999 *J. Thermophys. Heat Tr.* **13** 474
[7] Eastman J A, Choi S U S, Li S, Yu W and Thompson L J 2001 *Appl. Phys. Lett.* **78** 718
[8] Hong T K, Yang H S and Choi C J 2005 *J. Appl. Phys.* **97** 1
[9] Ramesh, Gopalan and Prabhu N K 2011 *Nanoscale Res. Lett.* **6** 334
[10] Xuan Y and Li Q 2000 *Int. J. Heat Fluid Fl.* **21** 58
[11] Babu K and Kumar T P 2011 *Int. J. Heat Mass Transf.* **54** 106
[12] Kim H, DeWitt G, McKrell T, Buongiorno J and Hu L W 2009 *Int J. Multiphas. Flow* **35** 427
[13] Khoshmehr H, Saboonchi A and Shafii M B 2014 *Int. J. Therm. Sci.* **75** 95
[14] Paul G, Das P K and Manna I 2015 *Int. J. Heat Mass Transf.* **137** 121009
[15] Jama M, Singh T, Gamaleldin S M, Koc M, Samara A, Isaifan R J and Atieh M A 2006 *J. Nanomater.* **2016** 6717624
[16] Yahya S S, Harjanto S, Putra P N, Ramahdita G, Kresnodrianto and Mahiswara E P. (*in press*) Characterization and observation of water-based nanofluids quench medium with carbon particle content variation *AIP Conf. Proc.*
[17] Ghadimi A, Saidur R and Metselaar H S C 2011 *Int. J. Heat Mass Tran.* **54** 4051
[18] Patra N, Gupta V, Singh R, Singh R S, Ghosh P and Nayak A 2017 *Resource-Efficient Technologies* **3** 378

One-Step Condensation and Hydrogenation of Furfural-Acetone Using Mixed and Single Catalyst Based on Ni/M-Oxide [M=Al; Mg]

S M Ulfa*, I N Pramesti and H Mustafidah

Chemistry Department, Faculty of Science, Brawijaya University, Jl. Veteran Malang, East Java, 65145

*E-mail: ulfa.ms@ub.ac.id

Abstract. Modification of furfural by condensation and hydrogenation reaction is a promising approach to produce higher alkane derivatives (C8-C13) as diesel fraction. This research investigated the catalytic activity of Ni/MgO as bifunctional catalyst compared with MgO-Ni/Al_2O_3 mixed catalyst for condensation-hydrogenation reaction. The Ni/MgO and Ni/Al_2O_3 with 20% Ni loading were prepared by wet impregnation methods using $Ni(NO_3)_2.6H_2O$ salt, calcined and reduced at 500°C. The catalyst performance was tested for one-step condensation-hydrogenation reaction using autoclave oil batch reactor. The reaction was conducted by reacting furfural and acetone in 1:1 ratio using water as solvent. Condensation reaction was performed at 100°C for 8 hours, followed by hydrogenation at 120°C during 7 hours. Analysis by gas chromatography showed that C=C double bond of furfurylidene acetone were successfully hydrogenated. Using Ni/MgO catalyst at 120°C, the products were identified as 1,5-bis-(2-furanyl)-1,4-penta-1-ene-3-one (2.68%) and 1,5-bis-(2-furanyl)-1,4-pentan-3-one (trace amount). On the other hand, reaction using mixed catalyst, MgO-Ni/Al_2O_3 showed better activity over bifunctional Ni/MgO at the same reaction temperature. The products were identified as 4-(2-furanyl)-3-butan-2-one (27.30%); 1,5-bis-(2-furanyl)-1,4-penta-1-ene-3-one (3.82%) and 1,5-bis-(2-furanyl)-1,4-pentan-3-one (1.11%). The impregnation of Ni on MgO decrease the physical properties of catalyst, confirmed by surface area analysis (SAA).

1. Introduction

Recently, alternative energy resources have been extensively explored due to the depletion of fossil fuels. Utilization of biomass as renewable material for bioenergy source has attracted considerable attention among other energy sources. The most potential candidate is furfural which derived from dehydration and hydrolysis of lignocellulosic biomass. Modification of furfural by condensation and hydrogenation reaction is a promising approach to produce higher alkane derivatives (C8-C13) as diesel fraction [1]. The aldol condensation of furfural with ketone/aldehyde would prolong the carbon chain and produces monomer/dimer of furanyl derivatives with carbon chain varied from C8-C15. The condensation adducts which contain ketonic C=O, aliphatic C=C, and furanic C=C bond can be converted into valuable liquid hydrocarbons through hydrogenation and followed by further deoxygenation reaction [2,3]. Combining the first-two reaction in one-step process is a big challenge.

The aldol condensation reaction is generally carried out in a homogeneous system using a strong base catalyst [4]. Homogeneous base catalysts such as NaOH and KOH have high reactivity and low cost production. However, there are several drawbacks such as the difficulty on the product separation, non-reusable and catalyzing side reactions such as oxidation and Cannizaro reactions [5]. Therefore, the usage of heterogeneous catalyst *e.g.*, MgO, CaO, Al_2O_3, and ZrO_2 give more advantages since it is reusable and easy to separate. MgO was reported as a good catalyst for aldol condensation reaction due to the high basic property [6].

On the other hand, the hydrogenation reaction use metal catalysts such as, copper (Cu), nickel (Ni), platinum (Pt), chromium (Cr) and palladium (Pd). Among different metal loading, the noble metal such as Pt and Pd have been revealed as suitable metals for hydrogenation reaction [7–9]. However, these nobel metal are expensive and high cost for industrial scale. As a consequence, the development of non-precious metal catalyst seems more rational as economical alternatives. It was reported by Gandia (1993) [10], that Ni metal has high activity for the hydrogenation of aromatics, alkenes and carbonyl compounds. The usage of support also important to increase the activity of catalyst.

Presently, bifunctional catalyst has been widely developed. The bifunctional catalyst is a double function catalyst which make the reaction is possible to run more efficiently in one-step reaction. This catalyst has two active sides which can be used for two different reactions [3]. Combining Ni as a metal catalyst and MgO as the support material is expected effectively catalyze condensation-hydrogenation reaction. Herein, we investigated the catalytic activity of Ni/MgO as bifunctional catalyst compared with MgO-Ni/Al_2O_3 mixed catalyst for condensation and hydrogenation reaction in one-pot reaction. The usage of MgO-Ni/Al_2O_3 mixed catalyst is expected to be specific for each reaction. MgO catalysts in condensation reactions is capable for deprotonating acetone to form enolate ions, whereas Ni/Al_2O_3 catalysts are expected to be specific for hydrogenation reactions capable of weakening bonds between hydrogen atoms to reduce the C=C double bond alkene.

2. Experimental

2.1. Materials

Furfural (FUR) was obtained from Sigma-Aldrich Singapore with >98% purity and used without further purification. Analytical grade of $Ni(NO_3)_2.6H_2O$, Al_2O_3 (120-190 m^2/g), methanol, and dichloromethane were bought from Merck Germany. Acetone was purchased from Smart Lab Tbk. Technical grade of MgO was obtained from the local vendor and calcined at 650°C for 2 hours under N_2 flowing before used. N_2 gas in >99.99% purity for calcination and condensation reaction, Ultra-High-Purity (UHP) H_2 gas for hydrogenation reaction from local vendor was used as received.

2.2. Methods

2.2.1. Preparation of Ni/Al_2O_3 and Ni/MgO

Ni/Al_2O_3 catalyst was prepared by wet impregnation method with 20 wt.% of Ni loading. Amount of $Ni(NO_3)_2.6H_2O$ (9.9 g) was dissolved in 100 mL methanol added with 8 g of Al_2O_3 and stirred for 24 h at room temperature. After that, the excess of solvent was removed by rotary evaporator and the solid dried in the oven for a night. Ni/Al_2O_3 catalyst was obtained after calcination under N_2 gas and reduction using H_2 at 500°C for 4 h. The Ni/MgO catalyst was prepared with similar procedure by dissolved 20 wt.% of Ni equal to 12.38 g of $Ni(NO_3)_2.6H_2O$ in 50 mL methanol added with 8.3 g of MgO and stirred for 24 h at room temperature. Calcination and reduction of Ni/MgO were conducted in similar condition.

2.2.2. Characterization

The catalysts obtained from previous methods were analyzed by X-ray diffraction (XRD) Pan Analytical Expert MPD. The measurements using Cu-Kα radiation at λ=1.5406 Å with the voltage and current held at 40 kV and 30 mA (2θ = 20-80°) at a scan rate of 1°/min. Scanning Electron Microscopy (SEM) was performed by Hitachi TM3000 connected with EDX Swift ED3000 to determine the surface morphology of the samples and qualitative calculation of metal loading in the catalyst. The surface area and pore volume of the catalyst are determined from BET adsorption isotherms using nitrogen at 77 K were measured by Quantachrome NOVA1200 analyzer after degassed at 300°C. Conversion and selectivity of the products were analyzed using Gas

Chromatography Mass Spectrometer (GCMS) injected on GCMS QP2010S Shimadzu with FID detector and AGILENT HP 5MS column (30 m with ID = 0.25 mm)

2.2.3. Catalytic activity test

The condensation of furfural and acetone was performed by autoclave oil batch reactor. Prior to reaction, 0.2 gram catalysts (MgO-Ni/Al_2O_3 (1:1) as mixed catalysts; Ni/MgO as a single catalyst) were initially activated *in situ* in the autoclave at 150°C under N_2 atmosphere for 1 h. The ratio of the furfural and acetone is 1:1. Furthermore, the furfural of 1.16 g (0.012 mol; 1 mL) and acetone of 0.70 g (0.012 mol; 0.882 mL) were introduced into the autoclave under N_2 gas atmosphere. The condensation reaction was carried out for 8 h (Ni/MgO catalyst) and 6 h (MgO-Ni/ Al_2O_3 catalyst) at 100°C under stirring condition [7]. After cooling down, N_2 gas was removed through the side valve and H_2 gas was introduced into autoclave. The hydrogenation reaction was carried out for 7 hours at temperature range of 120, 150 and 180°C under stirring condition. Then the autoclave was cooled to the room temperature, the liquid product was removed and filtered to separate the catalyst. The organic phase was separated by extraction using dichloromethane.

3. Results and discussion

3.1. Characterization of Ni/Al_2O_3 and Ni/MgO

Fig. 1 showed the morphology of the catalysts measured by Scanning Electron Microscopic (SEM). The particle size of the support material (MgO and Al_2O_3) is larger than after impregnation of Ni. It means that impregnation of nickel broke the support and resulted an irregular shape with smaller particle size. Agglomeration with irregular shapes is easily formed when impregnation method is used for the preparation of catalysts [11].

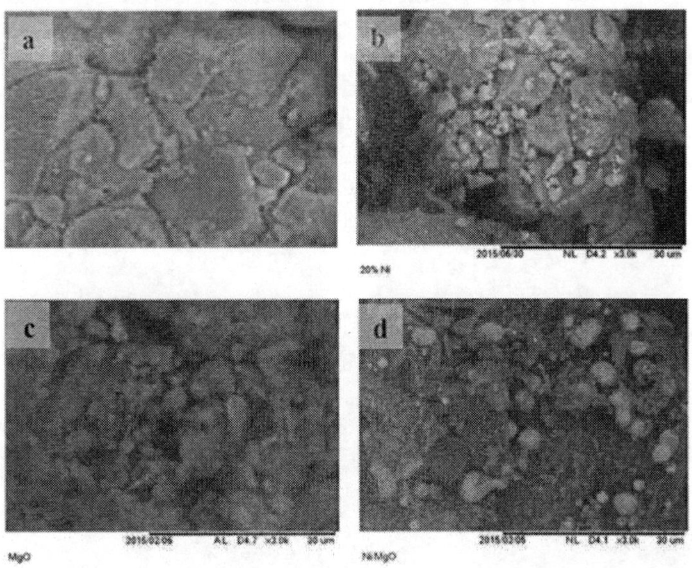

Figure 1. SEM micrograph of (a) Al_2O_3; (b) Ni/Al_2O_3; (c) MgO; (d) Ni/MgO (3000x optical zoom)

The crystal structure of Ni/Al_2O_3 and Ni/MgO were analyzed by X-ray diffraction (XRD) and the patterns were compared with the support material. The XRD spectra of the catalyst after reduced at 500 °C are shown in **Fig. 2**. The crystalline structure of Al_2O_3 and MgO were detected. This suggests that the impregnation of Ni does not change the crystalline structure of the support to an amorphous. The appearance of new peaks at $2\theta = 44.27°$; $51.59°$ and $75.89°$ are assigned for Ni metal formed from the reduction of Ni^{2+} ions at 500°C under H_2 gas flow. The crystallinity index of Ni is identified for (111), (200) and (202).

Fig. 2(a) showed another peak at 2θ = 37.4°; 44.9° and 66.7° were subjected for $NiAl_2O_4$ phase from the defect spinel of Ni and Al_2O_3 according to the preparation methods of catalyst [12]. The NiO peaks at 2θ = 43° and 63° were not detected because it is easily reduced to produced Ni°. In **Fig. 2(b)**, the 42.8° peak is detected in high intensity representing the specific peaks of MgO and NiO [13]. In addition, it can be seen that the specific peak of MgO undergoes a shift to a larger number after impregnation of Ni metal. This shift is assumed due to a reaction between MgO and Ni formed a $Ni_xMg_{1-x}O$ solid as reported by Yamazaki et al. [14].

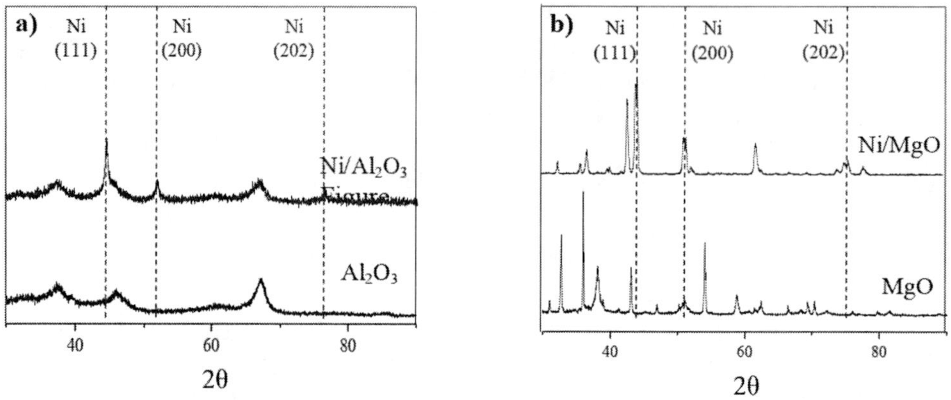

Figure 2. XRD profile of (a) Al_2O_3 and Ni/Al_2O_3; (b) MgO and Ni/MgO

Table 1. Physical properties of catalysts

Catalyst	$S_{BET}{}^{a}$ (m^2/g)	$V_{meso}{}^{b}$ (cc/g)	$D_{meso}{}^{b}$ (nm)	$D_{micro}{}^{c}$ (nm)[a]
MgO	96.106	0.184	3.818	0.432
Ni/MgO	1.984	0.014	3.838	0.821
Al_2O_3	132.697	0.284	4.903	0.432
Ni/Al_2O_3	74.185	0.216	5.649	0.432

[a]S_{BET} (total surface are) by BET method; [b]V_{meso} (mesoporous volume), D_{meso} (mesoporous diameter) by BJH method; [c]D_{micro} (microporous diameter) by HK method

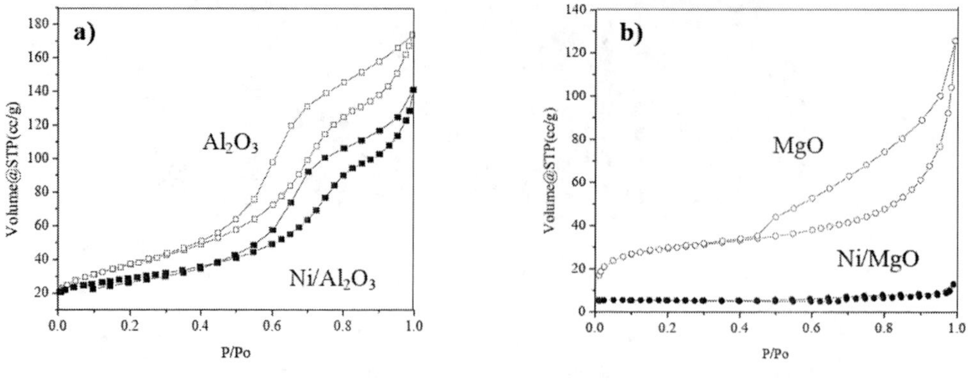

Figure 3. Isotherms adsorption/desorption of (a) Ni/Al_2O_3 and (b) Ni/MgO

Fig. 3 showed the isotherms of Ni/Al_2O_3 and Ni/MgO with each support materials. Compared with Al_2O_3 and Ni/Al_2O_3, the impregnation of 20% Ni into MgO reduced the surface area (S_{BET}) significantly, from 96.106 m^2/g to 1.984 m^2/g. Likewise, the pore volume of MgO was drastically decreased from 0.184 cc/g to 0.014 cc/g after the impregnation (**Table 1**). On the other hand, the surface area of Al_2O_3 was not significantly reduced. The decreasing of the surface area of MgO might be due to the impregnation method which decreased the crystallinity of the support material and the pores were covered by the active metal [15]. Mesoporous diameter (D_{meso}) of Ni/MgO was increased by 0.02 pm compared with MgO, and microporous diameter (D_{micro}) increased after the impregnation process. It indicated that there is a strong interaction between Ni and MgO metals. Therefore, most of the MgO pores are blocked by Ni metal and produce microporous material [16].

3.2. Catalytic activity for condensation-hydrogenation reaction

Catalytic activity test of mixed (MgO-Ni/Al_2O_3) and single catalyst (Ni/MgO) for one-step condensation-hydrogenation reaction was carried out in the presence of water using autoclave oil batch reactor is depicted in **Scheme 1**. The condensation products were identified as monomer (I) and dimer (II). Since the obtained product from condensation is a liquid in room temperature making it possible for hydrogenation reaction without additional treatment [17]. The mixture of condensation product was directly hydrogenated using H_2 gas. GC-MS analysis showed that C=C double bond of monomer (I) and dimer (II) were successfully hydrogenated. The formation of hydrogenation product (III) can be elucidated from reduction of C=C exocyclic of monomer (I). Whereas, bis-furanyl derivatives (IV) and (V) should be hydrogenation product of dimer (II).

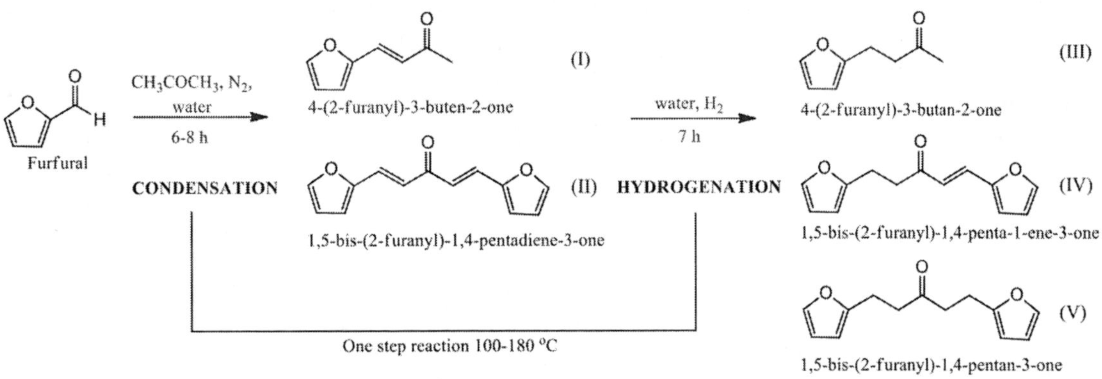

Scheme 1. One-step condensation-hydrogenation reaction of furfural-acetone

Fig. 4 showed the comparison yield of one-step condensation-hydrogenation reaction over MgO-Ni/Al_2O_3 mixed catalyst and Ni/MgO single catalyst. Using Ni/MgO catalyst at 120°C, the products were identified as 1,5-bis-(2-furanyl)-1,4-penta-1-ene-3-one (IV) (2.68%) and 1,5-bis-(2-furanyl)-1,4-pentan-3-one (V) (trace amount). On the other hand, reaction using MgO-Ni/$\gamma-Al_2O_3$ mixed catalyst, showed better activity over bifunctional Ni/MgO catalyst at same reaction temperature. The products were identified as 4-(2-furanyl)-3-butan-2-one (III) (27.30%); 1,5-bis-(2-furanyl)-1,4-penta-1-ene-3-one (IV) (3.82%) and 1,5-bis-(2-furanyl)-1,4-pentan-3-one (V) (1.11%).

Based on N_2 physisorption analysis, the impregnation of Ni on MgO decrease the physical properties of catalyst. The decreasing of the surface area and pore volume of MgO caused the condensation reaction does not proceed effectively. Yingxin and co-workers (2007) reported that there is a strong interaction between Ni metals and MgO making Ni^{2+} ions are totally sticked on the surface and cover MgO lattice [18]. Therefore, the conversion of furfural as starting material was lower on Ni/MgO catalyst than MgO-Ni/Al_2O_3. Mixed catalyst offer more advantages for one pot reaction, since it has two specific active sites for each reaction. Further, reaction using Ni/MgO 10-15% Ni loading is under consideration.

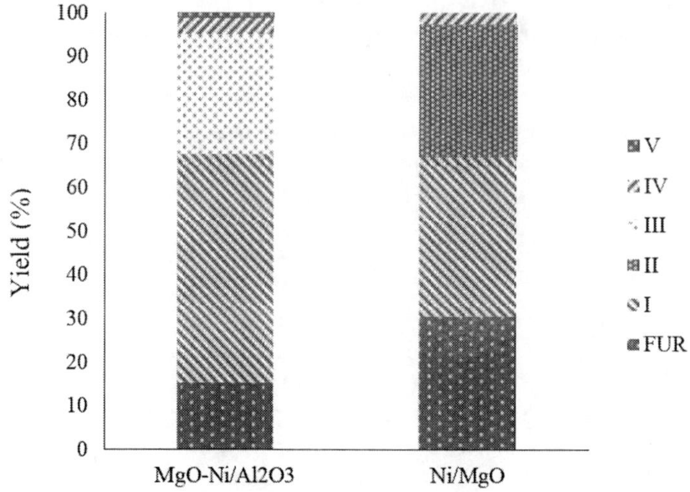

Figure 4. Condensation-hydrogenation reaction of furfural-acetone at 120°C

3.3. Effect of reaction temperature on hydrogenation

$MgO-Ni/Al_2O_3$ showed better activity for one-step condensation-hydrogenation of furfural-acetone at 120°C. Therefore, it was needed to investigate effect of temperature on hydrogenation reaction using $MgO-Ni/Al_2O_3$ mixed catalyst. The conversion and selectivity of condensation-hydrogenation products are summarized in **Table 2**. Increasing temperature should only affect the hydrogenation products. The reaction at 150°C (entry 2) only gave hydrogenation product IV (2.07%). Increasing reaction temperature until 180°C (entry 3) produced the highest yield of hydrogenated monomer III (33.78%). Based on these results, hydrogenation reaction using $MgO-Ni/Al_2O_3$ mixed catalyst were not significantly influenced by the reaction temperature. However, it was reported by Sharma (2014) that higher hydrogenation reaction temperature would initiate the cracking of C-C bond [19].

Table 2. The conversion and selectivity of condensation-hydrogenation reaction [a]

Entry	T (°C)	Conversion (%)	Selectivity (%)					
			FUR	**I**	**II**	**III**	**IV**	**V**
1	120	84.79	15.21	52.55	-	27.3	3.82	1.11
2	150	89.60	10.40	26.16	61.37	-	2.07	-
3	180	56.09	43.91	22.31	-	33.78	-	-

[a] Reaction condition: catalyst: $MgO-Ni/Al_2O_3$; condensation reaction during 6 h and hydrogenation reaction during 7 h, 40 psi H_2, water solvent

4. Conclusion

One-step condensation-hydrogenation reaction of furfural-acetone was successfully conducted using $MgO-Ni/Al_2O_3$ mixed catalyst and Ni/MgO single catalyst in autoclave batch reactor. $MgO-Ni/Al_2O_3$ mixed catalyst showed better activity over Ni/MgO catalyst at 120°C. The impregnation of Ni on MgO decrease the physical properties of catalyst, confirmed by SAA.

Acknowledgments
This work was supported financially by the PUPT research grant from Directorate General of Research, Technology and Higher Education number: 063/ADD/SP2H/LT/DPRM/VIII/2017.

References

[1] Huber GW, Iborra S, Corma A. 2006 *Chem Rev* **106** 4044–98.
[2] Barrett CJ, Chheda JN, Huber GW, Dumesic JA. 2006 *Appl Catal B Environ* **66** 111–8.
[3] Dedsuksophon W, Champreda V, Laosiripojana N. 2010 *Eng J* 2010 **14** 1–10.
[4] Tsuji H, Yagi F, Hattori H, Kita H. 1994 *J Catal* **148** 759–70.
[5] Vashishtha M, Mishra M, Shah DO. 2013 *Appl Catal Gen* **466** 38–44.
[6] Choudary BM, Kantam ML, Ranganath KVS, Mahendar K, Sreedhar B. 2004 *J Am Chem Soc* **126** 3396–7.
[7] Mahfud A, Utomo EP, Ulfa SM. 2016 Hydogenation/deoxygenation (H/D) reaction of furfural aceton condensation product using Ni/Al_2O_3-ZrO_2 catalyst **5** 108–15.
[8] Faba L, Díaz E, Ordóñez S. 2014 *Appl Catal B Environ* **160–161** 436–44.
[9] Faba L, Díaz E, Vega A, Ordóñez S. 2016 *Catal Today* **269** 132–9.
[10]Gandía LM, Montes M. 1993 *Appl Catal Gen* **101** L1–6.
[11]Rahemi N, Haghighi M, Babaluo AA, Jafari MF, Estifaee P. 2013 *J Ind Eng Chem* **19** 1566–76.
[12]Kim P, Kim Y, Kim H, Song IK, Yi J. 2004 *Appl Catal Gen* **272** 157–166.
[13]Zhang M, Ma H, Gao Z. 2015 *J Nanomater* **16** 59.
[14]Yamazaki O, Tomishige K, Fujimoto K. 1996 *Appl Catal Gen* **136** 49–56.
[15]Park Y, Kang T, Lee J, Kim P, Kim H, Yi J. 2004 *Catal Today* **97** 195–203.
[16]Mohan V, Pramod CV, Suresh M, Reddy KHP, Raju BD, Rao KR. 2012 *Catal Commun* **18** 89-92.
[17]Ulfa SM, Mahfud A, Nabilah S, Rahman MF. 2017 *IOP Conf. Ser. Mater. Sci. Eng* **vol. 172** IOP Publishing p. 012053.
[18]Yingxin L, Jixiang C, Jiyan Z. 2007 *Chin J Chem Eng* **15** 63–7.
[19]Sharma A. Production of xylitol by catalytic hydrogenation of xylose. Pharma Innov 2014;2.

IOP Conf. Series: Materials Science and Engineering **299** (2018) 012070 doi:10.1088/1757-899X/299/1/012070

Green Synthesis of Silver Nanoparticles using Extract of *Pinus merkusii* Jungh & De Vriese Cone Flower

N I Azkiya, M Masruri*, S M Ulfa
Chemistry Department, Faculty of Science, University of Brawijaya

E-mail: masruri@ub.ac.id

Abstract. The paper studies recent application of cone flower waste from Pinus merkusii Jungh & De Vriese for an environmentally unclear method for synthesis silver nanoparticle. Phytochemical characterization using iron trichloride solution showed the extract of Pinus merkusii cone flower contains of phenolic group of secondary metabolite. This group acts as both reducing and stabilizing agents. For the synthesis of silver nanoparticle, solution of silver nitrate is added to the extract at 60°C. The effect of extract concentration (5-20%) and time reaction (15-60 min) is investigated. The formation of silver nanoparticle is confirmed by the color change from yellowish to brown. Meanwhile, UV-Vis characterization of silver nanoparticle in extract 20% and 60 min reaction showed surface plasmon resonance (SPR) at 431 nm, and transmission electron microscope (TEM) revealed the particle size range in between 8 and 23 nm with a spherical in shape.

1. Introduction

Silver nanoparticles (AgNP) in recent years has aroused a growing interest due to their unique properties and potential application such as catalytic, antibacterial and anticancer drug delivery [1–3]. Although synthesis of silver nanoparticles (AgNP) have been applied by chemical methods, the reactants used in these reaction are potentially hazardous. Synthetic method based on environmentally benign is need for AgNP synthesis, do not only an alternative that means friendly of obtaining these nanoparticle, but also have a well-defined morphology and size.

Plant extract have been reported to bioreduce metal ion to metal nanoparticle currently. Some extract plants have achieved success in the synthesis of metal nanoparticle [4–6]. The previous reference exhibited plant wastes can be utilized for the synthesis of nanoparticle too. *Pinus merkusii* are one of Indonesian's original pine forest, and the cone flower are waste. According to the reference, the *Pinus merkusii* contains of α-pinene, β-pinene [7], phenolic and flavonoid compound like pinosylvin, pinosylvin monomethyl ether, dimethyl ether pinosylvin, and pinocembrin [8]. The phenolic compound is capable of functioning in reducing metal ions to form nanoparticle (reducing agent). It has also been reported that metal nanoparticle was successfully synthesized from *Pinus merkusii* cone flower extract [9]. So, this study is proposed to study of silver nanoparticle formation using cone flower extract of *Pinus merkusii*, which is a simple and environmentally benign method.

2. Experimental

2.1. Chemical and instrumentation

Cone flower of *Pinus merkusii* were collected from Coban Rais, Batu city, East Java. $AgNO_3$ was purchased from Sigma Aldrich. UV-visible spectral analysis was recorded on a double beam spectrophotometer (SHIMADZU 1601). IR measurements were performed with a SHIMADZU 8400.

The size and shape of the AgNP were identified by TEM JEOL JEM 1400 at an accelerating voltage of 100 kV.

2.2. Procedure

The extract was produced by 50 g of cone flower powder boiled with 250 mL distilled water at 70°C for 2 hours. After that, the extract was filtered and stored at 4°C for further use, and the content of phenolic compound was tested phytochemically [9]. The synthesis of silver nanoparticles using *Pinus merkusii* cone flower extracts follows the procedure of Mo et al.,[5]. The reaction mixture was prepared by adding 25 mL of 0.1 M silver nitrate solution to 50 mL of extract (5%, 10%, 15%, 20%) at 60°C for 15, 30, 45, and 60 minutes. The color of the solution changed to brown which confirmed the reduction of silver nitrate to silver nanoparticles. The synthesized AgNP were separated by centrifugation at 10,000 rpm for 15 minute. Products were further analyzed by FTIR, UV-Vis and TEM.

3. Results and discussion

Cone flower extract of *Pinus merkusii* was composed of phenolic compound and confirmed by phytochemical test using $FeCl_3$. The extract color change from light brown to blackish-green that indicates the presence of phenolic compounds. The cone flower extract were used to reduce silver ion to silver nanoparticle. The reduction was confirmed by the color change to brown (Fig.1).

Figure 1. Color change of AgNP

Different extract concentration were used for the synthesis of silver nanoparticles and studied using UV-Vis spectroscopy (Fig. 2). The extract concentration varied from 5%, 10%, 15% and 20% in 0.1 M silver nitrate solution. As a result, larger extract concentration lead to an increase in peak absorbance in UV-Vis spectrum. By increasing the extract concentrations, indicates nanoparticle production was also increased. The effect of reaction time on the synthesis of AgNP was showed at Figure 3. It can be seen that AgNP began to obtain within 30 min after the addition of silver nitrate to the *Pinus merkusii* cone flower extract solution. As the increased of reaction time, the intensity of peak absorbance increased gradually, indicating an increasing amount of AgNP formed. The maximum result of synthesis AgNP obtained in extract concentration 20% and reaction time 60 minutes, with absorbance maxima at 431 nm indicated the characteristic surface plasmon resonance (SPR) peak of AgNP.

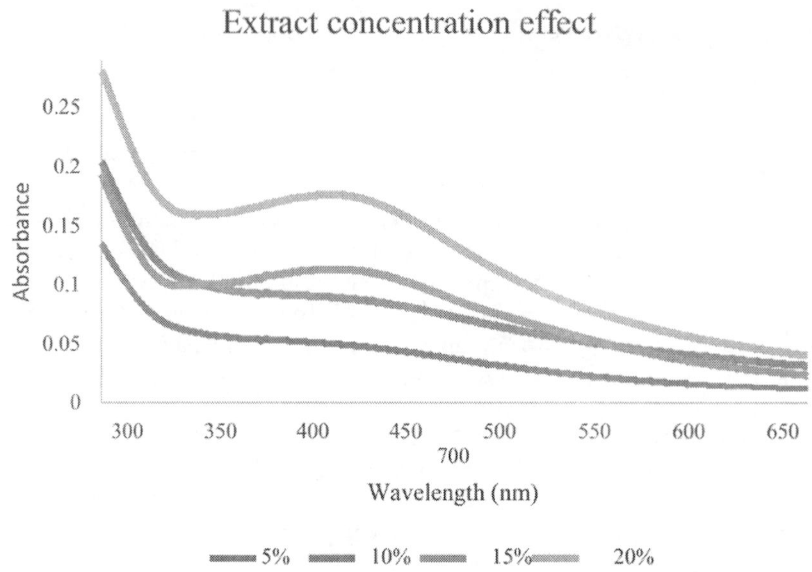

Figure 2. UV-Vis absorption of AgNP with different concentration of extract

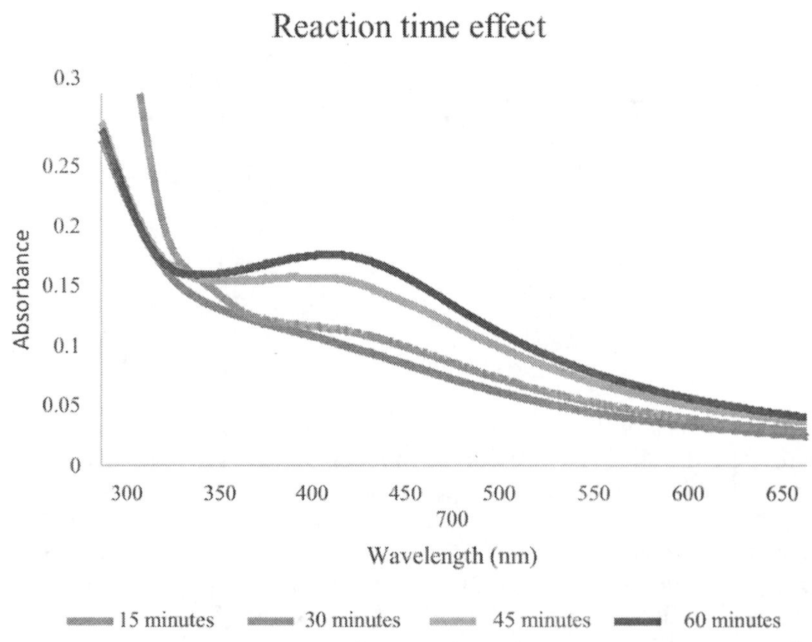

Figure 3. UV-Vis absorption spectra of AgNP as a function of the reaction time

The functional groups involved in the formation of AgNP and extract determined using FTIR spectroscopy were shown in Figure 4. The water extract of *Pinus merkusii* cone flower showed the peaks at 1640.14 cm^{-1} and 3451.18 cm^{-1}. The broad peak around 3451.18 cm^{-1} in the spectra indicates the existence of O-H group of phenolics compound, and the band at 1640.14 cm^{-1} is characteristic of C=C aromatic. After nanoparticles synthesis, the band at 1640.14 cm^{-1} and 3451.18 cm^{-1} shifted to 1638.21 cm^{-1} and 3431.89 cm^{-1}, and the band at 1383.63 cm^{-1} and 1032.61 cm^{-1} appear, which could be assumed that the phenolic hydroxyl groups are involved in the synthesis as reducing agent.

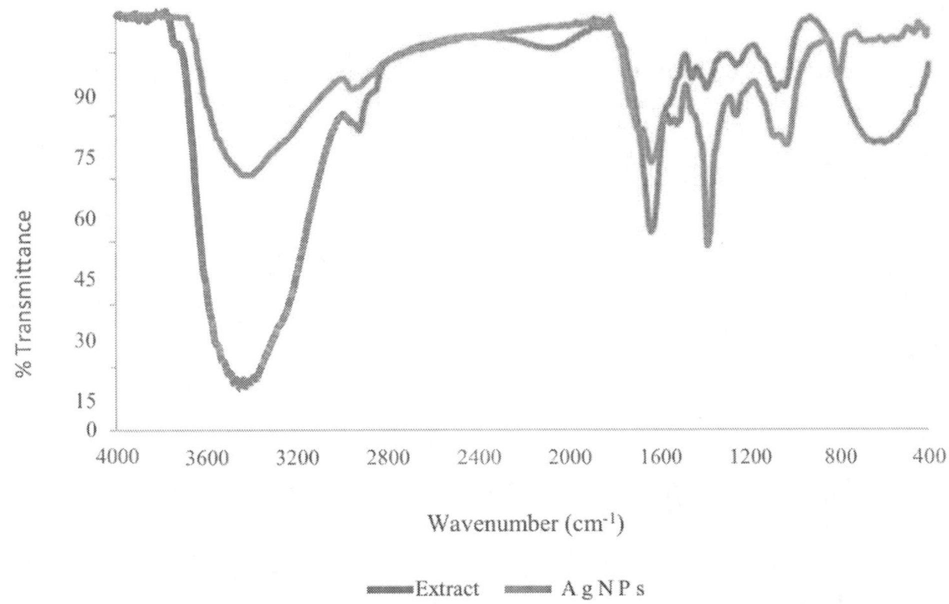

Figure 4. FTIR spectra of Extract (blue) and AgNP (orange)

The morphology of the synthesized AgNP using the water extract of *Pinus merkusii* cone flower were further confirmed by the TEM micrograph image. The images reveal that the AgNP are spherical in shape, with the particles size distribution between 9 to 23 nm (Fig. 5).

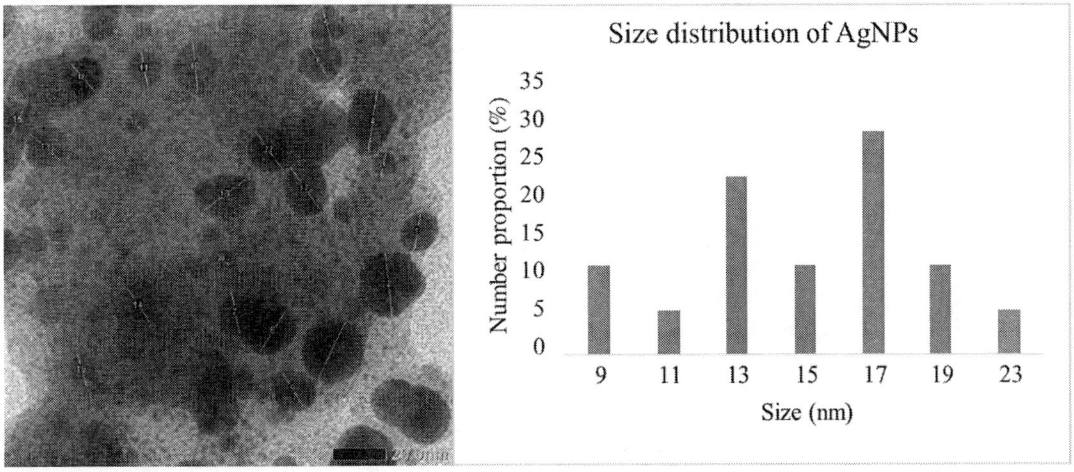

Figure 5. TEM image and size distribution of AgNP using cone flower extract of *Pinus merkusii*

4. Conclusion

Cone flower waste extract from *Pinus merkusii*, which are locally obtainable and environmentally benign were successfully used for the synthesis of silver nanoparticles. The green synthesis can obtain smaller particle size and avoided aggregation of AgNP. The characterization of AgNP was verified by FTIR, UV-Vis, and TEM. The AgNP were spherical in shape with particle size between 9 to 23 nm, respectively.

5. Acknowledgement
The authors acknowledge financial support provided by Direktorat Riset dan Pengabdian Masyarakat, Direktorat Jenderal Penguatan Riset dan Pengembangan Kementrian Riset, Teknologi dan Pendidikan Tinggi, according to the research contract number 063/SP2H/LT/DRPM/2017. The TEM image was recorded at Chemistry Department, Faculty of Science, UGM, Yogyakarta.

Reference
[1] N. Muniyappan and N. S. Nagarajan, 2014 *Process Biochem* **vol. 49** no. 6 pp. 1054–1061.
[2] C. S. Espenti, K. S. V. K. Rao, and K. M. Rao, 2016 *Mater. Lett.* **vol. 174** pp. 129–133.
[3] S. Patra, S. Mukherjee, A. K. Barui, A. Ganguly, B. Sreedhar, and C. R. Patra 2015 *Mater. Sci. Eng. C* **vol. 53** pp. 298–309.
[4] M. Masruri, M. A. Baihaqi, S. Riyanto, and A. Srihardyastutie 2017 Improving antibacterial activity of Spathodea campanulata Beauv's water extract with copper nanoparticle on Staphylococcus aureus p. 020015.
[5] Y. Mo, Y. Tang, S. Wang, J. Lin, H. Zhang, and D. Luo 2015 *Mater. Lett.* **vol. 144** pp. 165–167.
[6] V. Ravichandran, S. Vasanthi, S. Shalini, S. A. Ali Shah, and R. Harish 2016 *Mater. Lett.* **vol. 180**, pp. 264–267.
[7] Masruri, M. F. Rahman, and B. N. Ramadhan, 2016 *IOP Conf. Ser. Mater. Sci. Eng.* **vol. 107**, p. 012060.
[8] A. Wijayanto, S. Dumarçay, C. Gérardin-Charbonnier, R. K. Sari, W. Syafii, and P. Gérardin 2015 *Ind. Crops Prod.* **vol. 69** pp. 466–471.
[9] M. Masruri, D. N. Pangestin, S. M. Ulfa, and S. Riyanto 2017 A High Antibacterial Activity of A Mixture From Flower Extract of Pinus merkusii Jungh. & De Vriese And Copper Nanoparticle on Staphylococcus aureus presented at the International Conference on One Health.

Synthesis and Proapoptotic Activity on Cervical Cancer Cell of Ester Eugenol 1-(3-Methoxy-4-hydroxy)phenyl-2-propylmethanoate

Moh. Farid Rahman[1]*, Muhammad Nazhif Haykal[2)], Novi Andriani Siagian[2)], Priscilla Maiselina Sriepindonnta[2)], Norman Alexander Tampubolon[2)]

[1]Chemistry Department, Faculty of Sciences, Universitas Brawijaya
[2] Medical Department, Medical Faculty, Universitas Brawijaya

*E-mail: m_farid@ub.ac.id

Abstract. Proapoptotic activity of ester eugenol,1-(3-methoxy-4-hydroxy)phenyl-2-propylmethanoat, which synthesized from eugenol is reported. Eugenol as starting material in the synthesis of ester eugenol was obtained from fractional distillation of clove oil with the yield of 70.66%. Synthesis of ester eugenol was camed out by addition-esterification reaction through reaction between eugenol and formic acid with mol ratio of 1:27 and reaction time for11 h. GC-MS analysis showed ester eugenol was afforded purity of 92.42% and the yield in of 93.34%. UV spectra of ester eugenol was observed the formation of carbonyl group at λmax 290 nm and supported by FT-IR analysis at 1714.60 cm-1 (carbonyl group), 1193.65 cm-1 (C-O-C ester group) and the absence of vynil group in eugenol structure at region 914.20 and 995.20 cm-1. Mass spectra showed ion molecule at m/z 210 was accordance with molecular weight of ester eugenol. Afterward, HeLa cell culture media was prepared for cervical cancer antiproliferative test. The result which showed in histogram indicated that LC50 of ester eugenol was reached at concentration below 0.01% while eugenol was up to 0.01% that observed cervical cancer cell apoptotic activity. LC50 value of ester eugenol was obtained at concentration 48.73 ppm. This research reported that natural product modified its structure has potency to cure cervical cancer.

Keyword: Eugenol, cervical cancer, apoptotic, antiproliferative

1. Introduction

Cervical cancer is a desease which is generally suffered by woman and attributes the highest human death in Indonesia. Annually, about 500,000 new patients all over the world especially in developing country [1]. In 2013, 98,692 people were diagnosed with cervical cancer and becomes the highest prevalence level in Indonesia which increased 0.8% [2]. Cervical cancer is caused by human papilloma virus (HPV) which responsible for all of cervical cancer cases. Cervical cancer is vicious tumor that grows in women reproductive organs which located in upper genital tract between uterus and vagina [3]. Persistant infection of oncogenic HPV types explicitly assigned as causing agent of vicious process in the cervix epithel lesions. Women who is infected persistant HPV may indicated increased risk to the cervical cancer [4].

Therapy provision like operation, radiotherapy, and chemotherapy may lead to double disadvantages into patients, such as lower immune systems, high cost treatment due to import medicine. Therefore, it requires novel innovation to provide medicines from natural products which spread over in Indonesia and has antiproliferative agent toward cancer cell. A compound possesses antiproliferative activity and easy to find is eugenol (4-allyl-2-methoxyfenol) which comprised in clove oil (80-90%) with strong aroma [5]. Eugenol has been applied for antiseptic and anesthesia in medical science [6]. Eugenol also has pharmacology effect such as antihelminthic, antiinflammatory, antiviral, antibacteria, analgesic, antioxidant, anticancer, and antimutagenic [7].

Eugenol has been reported going to tumor growth by 24.35% with dose of 100 ppm [8]. Eugenol also reveals its significant effect toward hepar cancer cell growth. HCT 15 and HT 29 cell was suppressed by eugenol with IC_{50} value 300 and 500 μM, respectively [9]. Eugenol possesses good reducibility and deployment of antioxidant activity against DPPH and lipid peroxidation [10]. Moreover, DNA oxidation process which exposed by hydroxyl radical that afford fenton reaction can be protected by eugenol. Those findings support eugenol becomes one of candidate for disease prevention which induced by stress oxidative effect. Thus, eugenol has good prevention effect toward diseases such as cancer, coronary heart, inflammatory disorder, and neurologic degeneration [11]. This obviously revealed that natural antioxidant which contained phenolic structure, such as eugenol plays an important role in protecting tissues toward free radical damages [12].

Antiproliferative activity of eugenol toward cancer cell is corresponding to the aromatic ring in eugenol structure. Methylated eugenol has ability to inhibit cervix cancer [13]. This ability can be improved by modifying allylic group which bonded with aromatic ring of eugenol to be an ester group that useful for increasing radical scavenging ability [14] and reduce the flavor of eugenol. Allylic group of eugenol converted to carbonyl group by addition reaction gave ester eugenol, 1-(3-methoxy-4-hydroxy)phenyl-2-propylmethanoat, that expected to enhance antiproliferative effect toward cervix cancer cell. Ester group in the molecule such as caffeic acid phenylethyl ester (CAPE) which contained in propolis extract gives strong antioxidant activity [15].

Assymmetric alkene addition of eugenol which utilized asymmetric reactant principally occur regiospecifically which means reactant involving addition to the double bond in one direction. In this reaction, it prevailing the Markovnikov rules that addition of positive charges, such as H^+, reactant will bond double bond C atom which contain more hydrogen atom, thus it forms the stable carbocation. Therefore, according to observation only one product is abundant from two of obtained products [16].

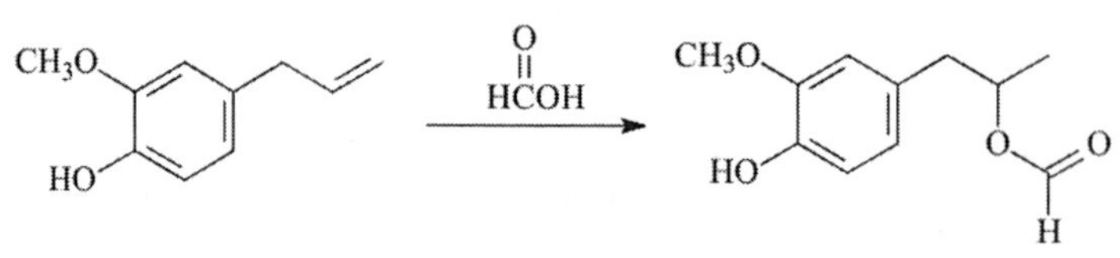

Figure 1. Synthetic route of ester eugenol through addition reaction

Cytotoxycity test toward cancer is a general basic test for anticancer drug and chemopreventive compounds. One of method that used generally for in vitro cytotoxycity test is MTT method [17]. This method is based on reduction reaction of MTT reagent (3-(4,5-dimethylthiazole-2-yl) 2,5-diphenyltetrazolium bromide) which catalyzed by dehydrogenase succinate enzyme in human cell. This test is initiated by transfer of 104 test cell into complete culture media which consist of FBS as a main cell nutrition, penicillin-streptomycin as a bacteria contaminant prevention, amphoterizin-B as a fungi contaminant prevention and RPMI 1640 as carrier media which is each volume of 100 μL) into 96-well plate and incubated in CO_2 incubator 5% overnight. Afterward, test samples were given using yield series and replicated 3 times, and then reincubated overnight. In the third test day of MTT reagent addition, it forms formazan crystal after 4 h in living cell. And then, it was added SDS stopper to quench the MTT reaction. The absorbance was recorded using ELISA reader at wavelength of 595 nm. Death level apoptotically can be analyzed using flow cytometry methods [18].

IOP Conf. Series: Materials Science and Engineering **299** (2018) 012071 doi:10.1088/1757-899X/299/1/012071

2. Material and Methods

2.1. Material
Sample materials used in this research is clove leaves oil collected from distillers. Chemical used in this research, was distilled water, sodium hydroxide (Merck), sodium sulfate anhydride (Merck), hydrochloric acid (Merck), dichloromethane (Merck), formic acid (Merck), diethyleter (Merck), sodium carbonate (Merck), eugenol (Merck). Media Rosewell Park Memorial Institute (RPMI 1640) (Gibco), Fetal Bovine Serum (FBS) (Sigma), Penicillin–streptomycin (Sigma), Amphotericin B (Sigma), dimethyl sulfoxide (DMSO), trypsin-EDTA (Sigma) (trypsin 0.25%), Trypan Blue (Sigma), Sodium Dodecyl Sulfate (SDS) and Phosphat Buffer Saline (PBS) (Invitrogen).

2.2. Methods

2.2.1. Eugenol isolation from clove oil
150 mL of clove oil was distillied in vaccumm at 100 mmHg. Distillate was collected at 80-82°C. The density was also measured and analyzed using Infra-Red spectrophotometer and GC-MS.

2.2.2. Synthesis of ester eugenol
Method was modified from synthesis of ester methyleugenol [19]. 248,40g of formic acid 98% (203,90 mL, 5,40 mol) was put into 2-neck flask 500 mL which set by condenser, stirrer, and dropper funnel which contained 32.80 g ofeugenol (30.83 mL, 0.20 mol). Eugenol was added into the flask stirred refluxed for 11 h at 130°C. Afterward, distillation was conducted under 100-105°C (1 atm) to obtain a residual reaction of formic acid. The residu washed using concentrated $NaHCO_3$ until pH reached netral and then was extracted using 30 mL of diethyl ether for 2 times. Organic phases was collected and dried by anhydrous Na_2SO_4. The organic phases was filtered and concentrated. The result was analyzed by UV-Vis spectrophotometer, Infra-Red spectrophometer, and GC-MS.

2.2.3. Apoptotic test toward HeLa cervix cancer cell
Anticervical cancer test was modified from MTT method [20]. 100 μL of cervical cancer cell suspension (HeLa) using 3x104 cell/100 μL media was distributed into 96-well plate and incubated for 24 h. After incubation, 100 μL of test solution in various concentrations was added into the plate. For positive control, 100 μL of culture media was added, and then 100 μL of cisplatin in various concentrations was added into the plate. For cell control, 100 μL of culture media was added into 100 μL of cell suspension and for solvent control, 100 μL of DMSO was added into culture media 100 μL and 100 μL of cell suspension with proper dilution which is equivalent with dilution of test solution concentration. And then, the solution was incubated for 24 h with CO_2 flow 5% and O_2 flow 95% at 37°C.

In the final incubation, flowcytometry procedure was conducted by washing cell for 2 times using cold cell staining buffer then cell re-suspension was conducted in annexin v binding buffer in concentration of 0.25-1.0 x 107 cell/mL. Afterward, 100 μL of cell suspension was transferred into 5 mL reaction tube and was added 5 μL of FITC annexin v and 10 μL of propidium iodide solution with concentration of 20 μg/mL. cell was rotated slowly and was incubated for 15 mins under room temperature (25°C) in dark container. Finally, 400 μL of annexin v binding buffer in each tube was analyzed by flowcytometer with proper machine setting. Data was analyzed by Cell Pro Quest software. Afterward, uncolored cell usng FITC annexin v and PI will not recorded as apoptotic, cell colored using FITC annexin v will be measured as early apoptotic, cell positively colored using FITC annexin v and PI will be recorded as late apoptotic or necrotic.

3. Result and Dicussion

3.1. Synthesis of ester eugenol

Synthesis of ester eugenol was carried out by reacting eugenol from fractional distillation and formic acid using mol ratio of eugenol:formic acid = 1 : 27 for 11 h. The result tabulated in Table 1.

Table 1. Physical properties of the product

No	Physical parameter	Eugenol	Ester Eugenol
1.	Color	colorless	brown
2.	Odor	spicy	scent
3.	Density	1.065 g/mL	1.26 g/mL

A specific chromatogram peak in t_R 19.31 mins which indicated the purity level of ester eugenol reached 92.42%. This result based on the mass spectra in Figure 4 which observed molecule ion at m/z 210 was accordance with molecular weight of ester eugenol. According to the purity grade, the yield percentage of ester eugenol was determined as high. This result indicated that reaction time treatment for 11 h showed high efficiency level compared to the earlier report for reaction of methyleugenol and formic acid with reaction time for 21 h [19].

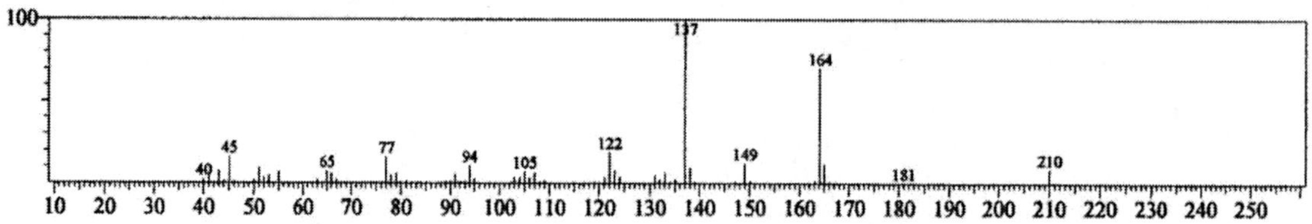

Figure 2. Mass spectra in t_R 19.31 mins

Ester eugenol was successfully synthesized by analyzing using UV-Vis which observed the carbonyl group (C=O) according to the absorbance at λ_{max} of 290.40 nm. This result also corresponds to the FT-IR spectra (Figure 3) which showed the carbonyl group with high intensity at 1714 cm^{-1} and C-O-C ester at 1193 cm^{-1} which indicated the formation of ester eugenol. It also supported the absence of absorbance at 914.20 which indicated the absorbance from out of plane of terminal alkene in the eugenol.

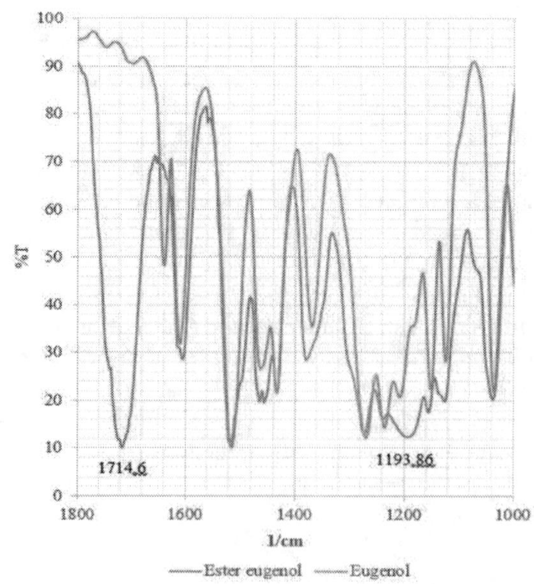

Figure 3. Infrared spectra overlay of ester eugenol and eugenol

3.2. Aptototic test of ester eugenol toward cervical cancer HeLa cell

Apoptotic test of ester eugenol toward cervix cancer cell was conducted by preparing the HeLa cell media culture in certain time and compared to treatment of 0.1 % ester eugenol. Figure 4 showed the observation result of HeLa cell growth after left for 2 days.

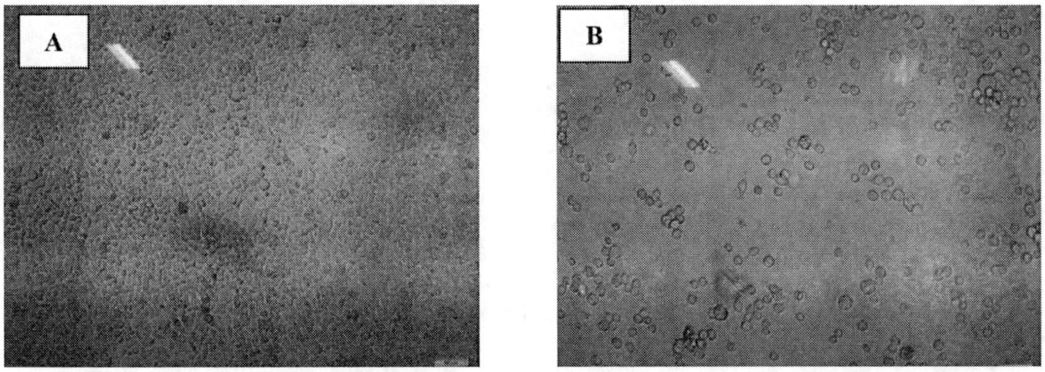

Figure 4. Cervical cancer HeLa cell growth: (A) without treatment (B) treatment with 0,1% ester eugenol.

Cell death was observed in ester eugenol treatment which indicated the decrease of cell rigid compared without treatment. It indicated that ester eugenol has toxic effect toward cervix cancer cell. Moreover, inhibition effectivity of ester eugenol can be analyzed from LC_{50} of each ester eugenol and eugenol toward cervix cancer cell as shown in Figure 5.

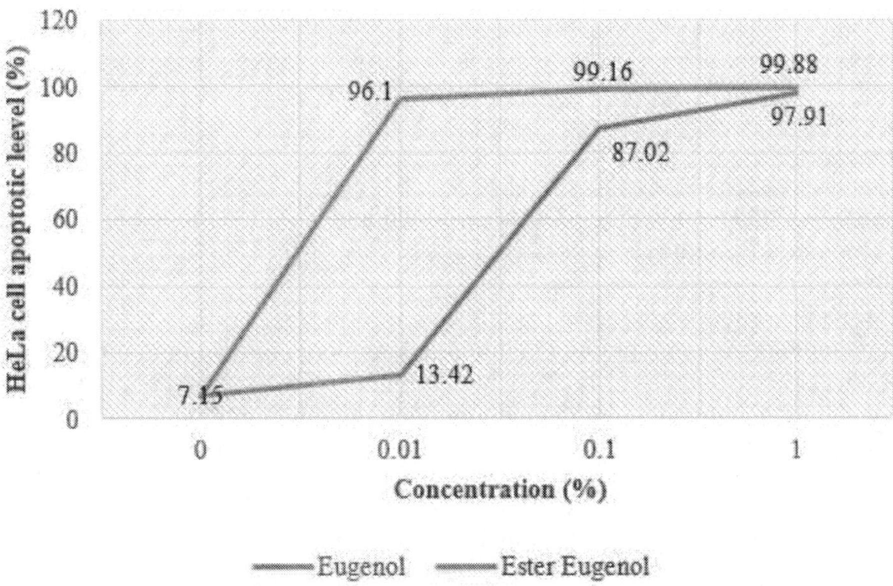

Figure 5. Comparison of eugenol and ester eugenol effect toward cervix cancer cell HeLa apoptotic

According to the Figure 5, it observed qualitatively that LC_{50} value of ester eugenol was below than 0.01% (→), while eugenol was up to 0.01% (→). It showed that ester eugenol ability in the cervix cancer HeLa cell was greater than eugenol. It showed that vinyl group (alkene) of eugenol which converted to ester contain carbonyl group that affected significantly to enhance apoptotic toward cervix cancer cell besides its ability as radical scavenger to inhibit cancer cell growth. Moreover, the test result of LC_{50} determination of ester eugenol toward cervix cancer cell below 0.01% (after compared to the control and solvent effect) can be seen in Figure 6.

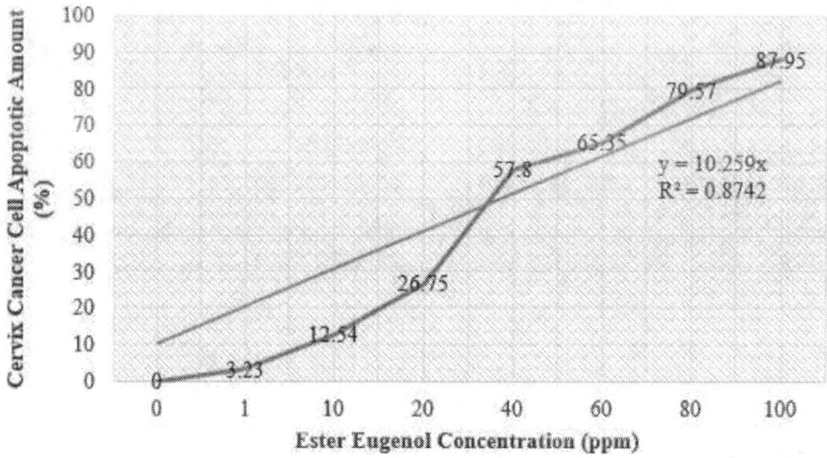

Figure 6. Correlation between ester eugenol concentration and cervix cancer HeLa cell death

Figure 6 showed linear correlation between ester eugenol concentration and cervix cancer cell apoptotic level. According to equation of $y = 10.259x$, LC_{50} of ester eugenol was determined at concentration 48.73 ppm. It indicated that ester eugenol ability toward cervix cancer HeLa cell apoptotic was very strong in concentration below 50 ppm that was able to induce cervix cancer cell apoptotic until 50%.

4. Conclusion

Ester eugenol was successfully synthesized from eugenol isolated from clove oil mol ratio of eugenol : formic acid = 1:27 for 11 h purity 92.42% and yield of 93.34%. Ester eugenol showed apoptotic antiproliferative level toward cervix cancer cell greater than eugenol with LC_{50} value at concentration 48.73 ppm.

Acknowledgement
Authors are grateful to Ministry of Research, Technology and Higher Education of the Republic of Indonesia for for their financial support to fund this research.

References

[1] Zarchi, M. K., Behtash, N., Chiti, Z., & Kargar, S. 2009. Cervical cancer and HPV vaccines in developing countries. *Asian Pacific Journal Cancer*. 10: 969-974.
[2] Kemenkes. 2015.*Situasi Penyakit Kanker*. Jakarta: Kementerian Kesehatan Republik Indonesia.
[3] Ibeanu, O. A. 2011. Molecular pathogenesis of cervical cancer. *Cancer biology & therapy*. 11(3): 295-306.
[4] Castellsagué, X. 2008. Natural history and epidemiology of HPV infection and cervical

cancer. *Gynecologic oncology*. 110(3): 4-7.

[5] Gülçin, İ. 2011. Antioxidant activity of eugenol: A structure–activity relationship study. *Journal of medicinal food*. 14(9): 975-985.

[6] Markowitz, K., Moynihan, M., Liu, M., & Kim, S. 1992. Biologic properties of eugenol and zinc oxide-eugenol: A clinically oriented review. *Oral surgery, oral medicine, oral pathology*. 73(6): 729-737.

[7] Kamatou, G. P., Vermaak, I., & Viljoen, A. M. 2012. Eugenol—from the remote Maluku Islands to the international market place: a review of a remarkable and versatile molecule. *Molecules*. 17(6): 6953-6981.

[8] Jaganathan, S. K., Mazumdar, A., Mondhe, D., & Mandal, M. 2011. Apoptotic effect of eugenol in human colon cancer cell lines. *Cell biology international*.35(6): 607-615.

[9] Jaganathan, S. K., Mondhe, D., Wani, Z. A., Pal, H. C., & Mandal, M. 2010. Effect of honey and eugenol on ehrlich ascites and solid carcinoma. *BioMed Research International*. 2010.

[10] Nam, H., and Kim, M. M. 2013. Eugenol with antioxidant activity inhibits MMP-9 related to metastasis in human fibrosarcoma cells. *Food and Chemical Toxicology*. 55: 106-112.

[11] Reynolds, J. J. 1996. Collagenases and tissue inhibitors of metalloproteinases: a functional balance in tissue degradation. *Oral diseases*. 2(1): 70-76.

[12] Fujisawa, S., Atsumi, T., Kadoma, Y., & Sakagami, H. 2002. Antioxidant and prooxidant action of eugenol-related compounds and their cytotoxicity. *Toxicology*. 177(1): 39-54.

[13] Yi, J. L., Shi, S., Shen, Y. L., Wang, L., Chen, H. Y., Zhu, J., & Ding, Y. 2015. Myricetin and methyl eugenol combination enhances the anticancer activity, cell cycle arrest and apoptosis induction of cis-platin against HeLa cervical cancer cell lines. *International journal of clinical and experimental pathology*. 8(2): 1116.

[14] Mesías, M., Navarro, M., Martínez-Saez, N., Ullate, M., del Castillo, M. D., & Morales, F. J. 2014. Antiglycative and carbonyl trapping properties of the water soluble fraction of coffee silverskin. *Food Research International*. 62: 1120-1126.

[15] LeBlanc, L. M., Paré, A. F., Jean-François, J., Hébert, M. J., Surette, M. E., & Touaibia, M. 2012. Synthesis and antiradical/antioxidant activities of caffeic acid phenethyl ester and its related propionic, acetic, and benzoic acid analoguesc. *Molecules*. 17(12): 14637-14650.

[16] March, J. 1992.*Advanced Organic Chemistry: Reaction, mechanisms, and structure*. 4th ed. John Wiley & Sons Inc. New York.USA. pp 970-971.

[17] Tian, Q. and Hong Zang, Y. 2015. Antiproliferative and apoptotic effects of ethanolic herbal extract of Achillea falcata in human cervical cancer cells are mediated via cell cycle arrest and mitochondrial membrane potential loss. *JBUON*. 20(6):1487-1496.

[18] Gomes, D.L., C.B.S. Telles, M.S.S.P. Costa, J.A. Lima, L.S. Costa, T.S.L. Keesen, H.A.O. Rocha. 2015. Methanolic extracts from brown seaweeds Dictyota cilliolata and Dictyota menstrualis induce apoptosis in human cervical adenocarcinoma HeLa cells. *Molecules*. 20(4): 6573-6591.

[19] Rahman, M.F. 2002. Sintesis *N*-[1-metil-2-(3,4-dimetoksifenil)etil] asetamida.*Natural*. 6(2): 3-7.

[20] Ola, A. R., Ikawati, Z., Meye, E. D., & Tawo, B. D. 2008. Molecular identification and anticancer activity of alkylphenol from cashew nut shell oil (Anacardium occidentale) grown In Timor Island. *Indonesian Journal of Pharmacy*. 137-144.

A Potent *Staphylococcus Aureus* Growth Inhibitor Of A Dried Flower Extract Of *Pinus Merkusii* Jungh & De Vriese And Copper Nanoparticle

Masruri Masruri[1*], Dinna Norani Pangestin[1], Siti Mariyah Ulfa[1] and Slamet Riyanto[2], Arie Srihardyastutie[1], Moh. Farid Rahman[1]

[1]Chemistry Department, Faculty of Mathematics and Natural Sciences, Brawijaya University. Jl. Veteran 65145 Malang, Indonesia
[2]Laboratory of Microbiology, Faculty of Medicine, Brawijaya University. Jl. Veteran 65145 Malang, Indonesia

Email: masruri@ub.ac.id

Abstract. The paper report antibacterial activity of flower extract from *Pinus merkusii* Jungh Et De Vriese and its mixture with copper nanoparticle on *Staphylococcus aureus*. This finding revealed the potency of pine forestry waste to overcome a bacterial-resistance problem on some commercially antibiotics. The extract was prepared by hot water extraction of a dried powder of pine flower. Copper nanoparticle was synthesized following "green synthesis technique" using phenolic-rich extract of pine's flower as a reduction and capping agent. In short, a mixture of pine's flower extract and copper nanoparticle importantly was able to inhibit the growth of *Staphylococcus aureus* four times higher than that using water extract.
Keywords: pinus merkusii, phenolic, antibacterial, nanoparticle, green synthesis, staphylococcus aureus

1. Introduction

Resistant of pathogenic bacterial such as *Staphylococcus aureus* in some commercially antibiotic have become a serious health problem in Indonesia [1–3], not only for human but also for animal [4,5]. This leads the way on finding new kind of antibacterial compounds and antibacterial herb from local natural sources. In the same time, the forests of *Pinus merkusii* Jungh et De Vriese was expanded recently. In total more than 800 kHa in East Java region, and this has paid development strategy to improve domestic income. Turpuntine oil and oleoresin has become the main product, but the forests also provide a huge amount of flower cone-waste. Previous finding indicated that antibacterial activity of essential oil isolated from pine's sap. Some terpenoid groups of alpha-pinene, beta-pinene, delta-carene, d/l-limonene consist of oil [6,7]. However, this mixtures still provide a low antibacterial activity. In addition, other compounds were also reported from the extract of spruce and knot pine tree. Some phenolic groups detected such as abietic acid, dehydroabietic acid, pimaric acid, isopimaric acid, notrachelogenin, liovil, pinosylvin, pinocembrin, methoxypinosylvin, dimethoxypinosylvin[8], hydroxymatairesinol, allo-hydroxymatairesinol [9]. However, these had low fungal activity and also no antibacterial activity reported. Recently, we have found that incorporating of compound as well as in the extract with metal and metal nanoparticle could improve antibacterial activity [7,10]. The paper reports a new finding incorporating of water extract of pine's flower with copper nanoparticle for inhibiting the growth of pathogenic *Staphylococcus aureus*.

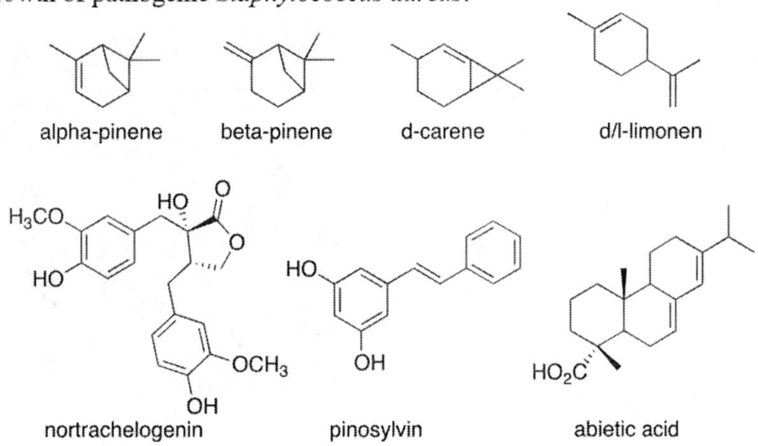

Figure 1. Some representative compounds isolated from pine tree

2. Experiment

2.1. Material and Chemicals
A dried-pine flower prepared from a pine's forestry waste of *Pinus merkusii* Jungh et De Vrise from local forestry in Malang, East Java. Plant taxonomy of the sample was determined by Dr. Djati Batoro (Laboratory of Plant Taxonomy, Brawijaya University). Chemicals used for research include copper sulfate hexahydrate (Merck), ethanol (Merck), acetone (Merck), and demineralized water (Bratachem)

2.2. Extraction Procedure
Water extract was prepared by extracting of a 50 g of dried powder of pine's flower with 250 mL water. This was boiled for 30 min and filtered of the aqueous extract of pine flower. Phenolic compounds in the extract was detected as precipitate or color changing of solution to violet or green-black [11], and its quantity was determined compared to gallic acid.

2.3. Nanoparticle Synthesis
Green synthetic of copper nanoparticle was undertaken by using phenolic-rich of pine's flower extract following reported procedure [12]. Nanoparticle analysis was performed using infrared spectrophotometry (Shimadzu FTIR 8400 series) and SEM-EDAX (FEI Inspect-S50).

2.4. Antibacterial Testing
Antibacterial activity was tested on *Staphylococcus aureus* local strain following Kirby-Bauer testing method [13]. Activity was determined by measuring (mm) clear zone surround the disc.

3. Result and Discussion

Extraction of a dried powder of pine flower cone produced a yellowish-brown of concentrated extract. Previous researchers reported some secondary metabolite composed the extracts from some part of tree, such as lignin [9], phenolic [8], terpenoide [6,14] groups, and oleoresin [15]. Moreover the result is found phenolic content 341.8 GAE/g which is able to be used for reduction and capping copper nanoparticle. Figure 2 (A) shows infrared spectra both of the extract and mixture of extract and copper nanoparticle. Incorporating nanoparticle was detected in wavenumber 550 and 760 cm^{-1} for bond stretching of Cu--O. These both peaks are not observed in the extract sample. Interaction of copper nanoparticle with functional group of extract was detected in 3400 cm^{-1} (Cu---OH) and 1750 cm^{-1} (C=O---Cu). In addition, SEM image (Fig. 1B) display a clear nanoparticle size between 20-35 nm.

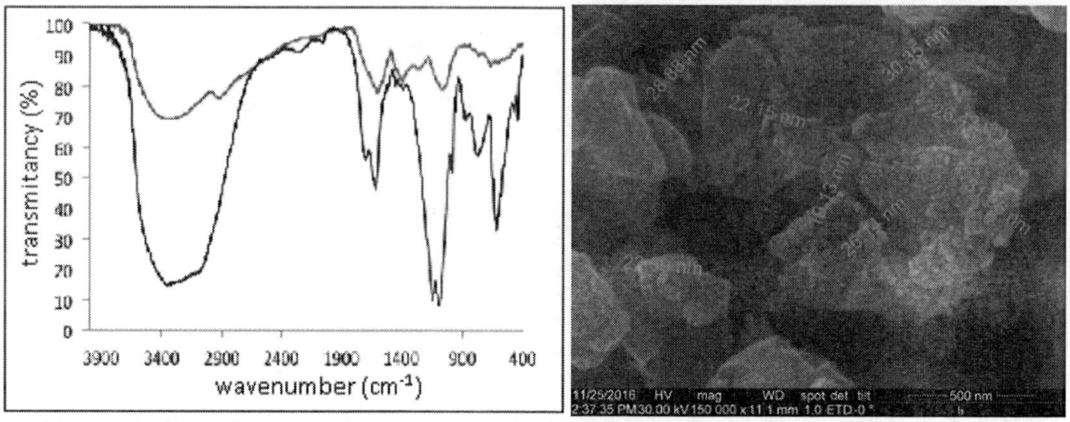

Figure 2. Infrared spectra of water extract pine's flower (red line) and its mixture with copper nanoparticle (black line) (A), and SEM image of copper nanoparticle (B)

The antibacterial activity was evaluated on *Staphylococcus aureus* and Kirby-Bauer Test [13] was applied to measure biological activity of the extract and its mixture with copper nanoparticle. The result is tabulated in Table 1 and compared to other finding previously [10,16]. The growth inhibition activities of the extract in various concentrations are recorded between 6.08 and 12.21 mm. The highest value using 100% of pine flower cone's extract, and the lowest using extract in 12.5%. This activity increases to 29.67 mm for 100% concentration of mixture the extract with copper nanoparticle. Similarly, results are provided for concentration 12.5% doubles to 12.21 mm. Visualization of the growth inhibition is displayed in Fig. 3. A clear zone surrounds the disc-loaded sample of extract contained-copper nanoparticle getting wider by increasing the concentration of sample from 12.5% to 100%. Conversely, a control-contained solvent did not have activity for *S. aureus*.

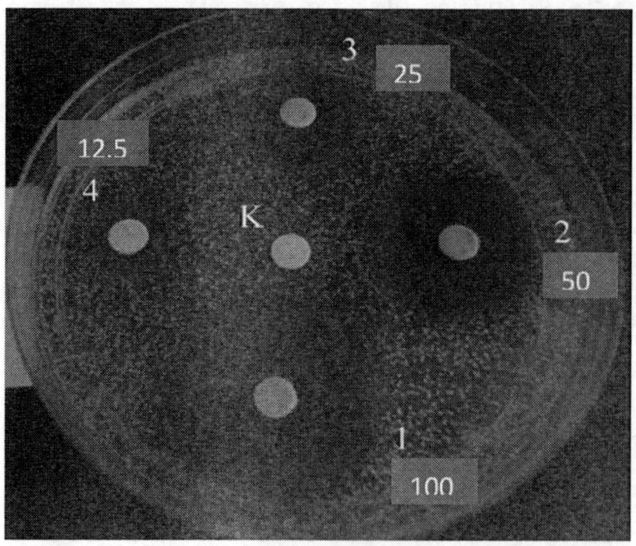

Figure 3. Antibacterial evaluation on *Staphylococcus aureus* of a mixture of pine flower cone's extract with copper nanoparticle in various concentrations, K is symbol for control.

Table 1. Growth inhibition activity of extract and its mixture with copper nanoparticle comparing to some references

Sample	Growth inhibition diameter (mm) at concentration			
	100%	**50%**	**25%**	**12.5%**
Extract	12.21±0.02	9.48±0.01	7.12±0.01	6.08±0.01
Mixture of extract-copper nanoparticle	29.67±0.01	25.44±0.01	22.46±0.00	12.21±0.02
Copper nanoparticle	6.15 [16]	-	-	-
Amoxillin 10 mg	7.50 [10]	-	-	-
Amoxillin-silver nanoparticle	14.0 [10]	-	-	-
Erytromycin 5 mg	10.0 [10]	-	-	-

4. Conclusion
The water extract of pine flower cone contained phenolic compound can be used for reduction and capping copper nanoparticle. Its mixture with copper nanoparticle importantly improved antibacterial activity by inhibiting the growth of *Staphylococcus aureus*. This finding opens the potency of pine flower waste as antibacterial agent.

5. Acknowledgment
Author acknowledges to some institution for fund the research; PUPT Universitas Brawijaya and Ministry of Research, Technology and Higher Education from Republic of Indonesia. Mr. Widji Sulistjo and Hadi Kurniawan from Instrumentation Laboratory, Chemistry Department of Brawijaya University.

References

[1] Hadi U, Duerink D, Lestari E S, Nagelkerke N, Keuter M, Huis in't Veld D, Suwandojo E, Rahardjo E, Van Den Broek P and Gyssens I 2008 a *Clin. Microbiol. Infect.* **14** 698–707
[2] Hadi U, Duerink D O, Lestari E S, Nagelkerke N J, Werter S, Keuter M, Suwandojo E, Rahardjo E, van den Broek P and Gyssens I C 2008 *Int. J. Infect. Dis.* **12** 622–9
[3] Hadi U, Qiptiyah M and Paraton H 2014 *Indones. J. Trop. Infect. Dis.* **4** 5–8
[4] Locatelli C, Cremonesi P, Caprioli A, Carfora V, Ianzano A, Barberio A, Morandi S, Casula A, Castiglioni B and Bronzo V 2017 *J. Dairy Sci.* **100** 608–19
[5] Ge B, Mukherjee S, Hsu C-H, Davis J A, Tran T T T, Yang Q, Abbott J W, Ayers S L, Young S R and Crarey E T 2017 *Food Microbiol.* **62** 289–97
[6] Masruri M, Rahman M F and Prasodjo T I 2007 *J. Ilmu-Ilmu Hayati* **19** 32–5
[7] Masruri M, Amini R W and Rahman M F 2016 *Indones. J. Chem.* **16** 59–64
[8] Wijayanto A, Dumarçay S, Gérardin-Charbonnier C, Sari R K, Syafii W and Gérardin P 2015 *Ind. Crops Prod.* **69** 466–71
[9] Holmbom B, Eckerman C, Eklund P, Hemming J, Nisula L, Reunanen M, Sjöholm R, Sundberg A, Sundberg K and Willför S 2003 *Phytochem. Rev.* **2** 331–40
[10] Shahverdi A R, Fakhimi A, Shahverdi H R and Minaian S 2007 *Nanomedicine Nanotechnol. Biol. Med.* **3** 168–71
[11] Masruri M, Lutfillah M, Sumaryanto A and Retnowati R 2014 *J. Trop. Life Sci.* **4** 161–5
[12] Iravani S 2011 Green synthesis of metal nanoparticles using plants *Green Chem.* **13** 2638–50
[13] Boyle V J, Fancher M E and Ross R W 1973 *Antimicrob. Agents Chemother.* **3** 418–24
[14] Amini R W, Masruri M and Rahman M F 2014 *J. Ilmu Kim. Univ. Brawijaya* **1** 140–6
[15] da Silva Rodrigues K C, Apel M A, Henriques A T and Fett-Neto A G 2011 *Biomass Bioenergy* **35** 4442–8

[16] Masruri M, Baihaqi M A, Riyanto S and Srihardyastutie A 2016 *Proceeding of International Conference on Chemistry, Chemical Process and Engineering* Proceeding of International Conference on Chemistry, Chemical Process and Engineering (Yogyakarta, Indonesia)

IOP Conf. Series: Materials Science and Engineering **299** (2018) 012073 doi:10.1088/1757-899X/299/1/012073

Performance of Natural Dye and Counter Electrode from Robusta Coffee Beans Peel Waste for Fabrication of Dye-Sensitized Solar Cell (DSSC)

T Setiawan, W Y Subekti, S S Nur'Adya, K Ilmiah, *S M Ulfa

Chemistry Department, Faculty of Science, Brawijaya University, Jl. Veteran Malang, East Java, 65145-Indonesia

*E-mail: ulfa.ms@ub.ac.id

Abstract. The DSSC prototype using activated carbon (AC) and natural dye from Robusta coffee bean peels have been investigated. The natural dye obtained from the extraction of Robusta coffee bean peels is identified as anthocyanin by UV-Vis spectrophotometer at maximum wavelength 219.5 nm and 720.0 nm in methanol. From the FT-IR analysis, the vibration of O–H observed at 3385 cm^{-1}, C=O at 1618 cm^{-1}, and C–O–C at 1065 cm^{-1}. The counter electrode prepared by calcined the peels at 300°C. Surface analyser of AC showed the larger surface area compared prior activation. The DSSC prototype was prepared using FTO glass (2x2 cm) coated with carbon paste in various thickness. The working electrode is coated with the TiO_2 paste. The optimum voltage measured was 395mV (300 μL of CA), 334 mV (200 μL AC), and 254 mV (100 μL AC). From this result, we understand that the thickness of counter electrode influent the voltage of the DSSC.

1. Introduction

Solar cell is a device which capable to convert solar light energy into electrical energy as photovoltaic system. Solar energy is one of the energy sources for solar cells. The study of solar cells was developed as the diminishing source of fossil energy. The type of solar cells is divided into inorganic-based and organic-based materials. The development of inorganic-based solar cells, such as by the used of silicon, GaAs, InP, CdTe, and other semiconductor materials were popular. Currently, conventional silicon-based solar cells have been able to achieve up to 30% efficiency. However, silicon solar cells have a very high fabrication cost cause the silicon must have in high purity. In addition, silicon is a very mine-stacked material [1,2].

Dye-Sensitized Solar Cell (DSSC) is one of the most promising strategy as an alternative energy since solar energy is unlimited. Although the efficiency is lower than conventional solar cells, DSSC have several advantages among others, such as, lower production costs and simple manufacturing process. The use of natural dye source is also the interesting factor since Indonesia has many natural products resources [3].

Anthocyanin and β-carotene were reported as prospective natural dye for DSSC. Anthocyanin found in extracts of strawberries, mangos teen skin, red spinach leaves, Rosella flowers, super red dragon fruit, purple sweet potatoes and black sticky rice [3,4]. Another potential biomass as a source of anthocyanin is red coffee skin waste. Coffee skin waste also contains natural antioxidants such as β-carotene, polyphenols and vitamin C [5]. Since the coffee production in Indonesia reached 675,882 tons and consist of 35% of coffee skin waste, it became a potential source for natural dye.

The most important factor for fabrication of DSSC is counter electrode which coated with platinum metal (Pt) as a catalyst. The comparative electrode is the electron capture electrode from the electrolyte so that the DSSC cell can produce an electric current. The use of Pt is considered less economical because the price is expensive, cannot be reuse and not environmentally friendly. Activated carbon is cheaper material that can be used to substitute Pt with almost equal effectiveness. Raw carbon from carbonization of biomass is reported to produce active carbon with a large surface area size and can be applied as a comparative electrode equivalent to Pt metal [6,7].

The mechanism of transfer electron on DSSC system using dye and TiO_2 is depicted in **Fig. 1**. Anthocyanin bind with TiO_2 to form complexes junction involved the carbonyl (C=O) and hydroxyl (O–H) groups causing excited electrons to be injected into TiO_2 band gap [3]. Here, we reported the used of Robusta coffee bean peels for the sensitizer source that is anthocyanin and activated carbon as counter electrode become DSSC based on natural resources. By varied the thickness of counter electrode from 100-300 μL/cm^2 gave 254-395 mV under the UV sunlight equivalent.

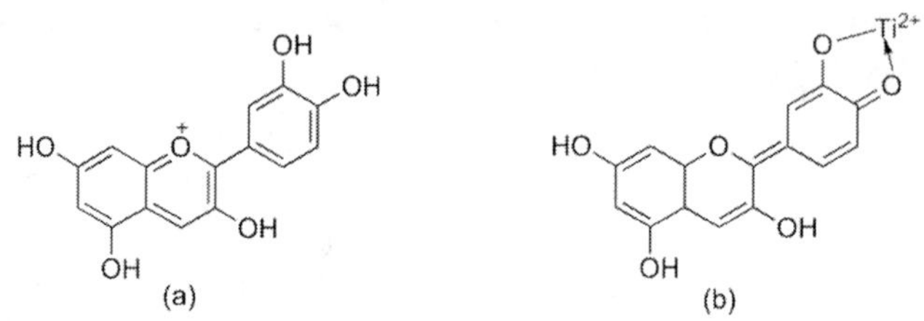

Figure 1. Structure of anthocyanin (a); complex binding anthocyanin–TiO_2 [3].

2. Experimental

2.1. Isolation of Anthocyanin
Coffee beans peel was dried under sunlight for four days and ground into a powder. Extraction of anthocyanin were carried out using methanol:citric acid:water in the ratio of 50:8:42 (v/v) for 24 hours. Residue and filtrate were separated using vacuum filtration. The excess of solvent was evaporated in vacuum. The crude anthocyanin extract was obtained in 12.23 mL/80 g.

2.2. Characterization of Anthocyanin Extract
Characterization of the extract was conducted using UV–Vis Spectrophotometry in methanol. Functional group of extract was identified using Fourier Transform Infrared Spectroscopy (FTIR) using NaCl window. The Liquid Chromatography Mass Spectroscopy (LCMS) was used for the identification of anthocyanin types based on mass analysis.

2.3. Fabrication of DSSC

2.3.1. Preparation of Carbon Active as Counter Electrode
After the extraction process, the solid waste from coffee beans peel is carbonated by drying and calcining at 300°C for two hours. The carbon was activated by stirring in HCl solution 0.1 M for three hours, followed by filtration. Drying the activated carbon in oven for three hours. Counter electrode was prepared from 0.3 g activated carbon and added with 5 mL ethanol. Activated carbon (AC) was characterized using Scanning Electron Microscope (SEM) to identify morphology and size distribution of compared with carbon. To prepare counter electrode, an activated carbon paste was deposited on the conductive substrate glass in various thickness (300 μL, 200 μL, 100 μL) and dried at 100°C for 15 min.

2.3.2. Preparation of conductive glass FTO coated with TiO_2

A fluorine-doped tin oxide (FTO) glass was used as conductive glass. FTO was first cleaned and coated with TiO_2 paste. The paste was prepared by diluted 2.8 g TiO_2 in 12 mL ethanol, stirred until homogeneous, then coated the paste on to FTO by slip casting method. Dried the layer at 350°C for 15 min. The conductive glass was then immersed into anthocyanin extract during 24 h. The working electrode and counter electrode were assembled into a cell. A drop of electrolyte solution containing triiodide injected in the cell and the hole of counter electrode was sealed using sealing spacer.

2.3.3. Characterization and measurement of voltage and current
The efficiency of DSSC was measured by multi-meter under halogen lamp (150 watt) as solar source model with radius 60 cm from the lamp. The test was conducted for various thickness counter electrode in FTO glass in 300 μL, 200 μL, 100 μL of the carbon solution.

3. Results and Discussion

3.1. Dye Characterization
The absorption spectrum of coffee beans peel extracts was measured using spectrometer UV–Vis 1601 Shimadzu on 200–800 nm. The result showed that the maximum wavelength at λ_{max} = 219.5 and 720 nm in the presence of methanol. This result showed the crude extract of coffee bean peels consist of anthocyanin properties [4]. From the analysis of UV-Vis spectrum showed that the extract can absorb not only ultraviolet but also visible region. It indicated that natural dye fulfil the characteristics to become potential dye sensitizers because it has absorption ability in the visible region under the illumination of the halogen lamp and also can convert light energy to electrical energy. The minimum energy needed by the electrons to the excitation states is around 720 nm then anthocyanin is suitable precursor for DSSC sensitizer.

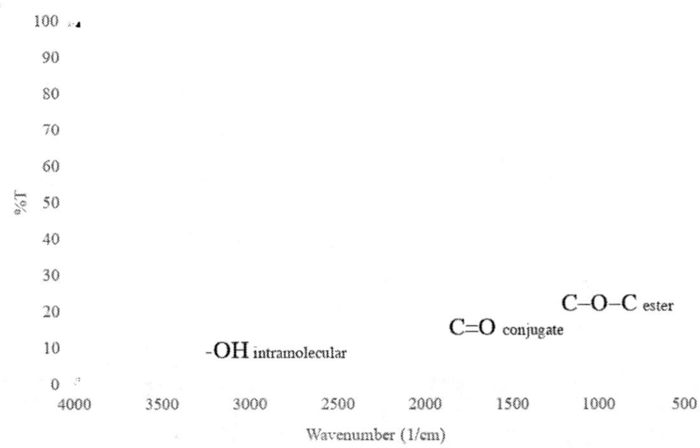

Figure 2. FTIR spectrum of natural dye

Functional groups of natural dye from coffee beans peel extracts were identified by Spectrophotometer FTIR Shimadzu 8400S. **Fig 2** showed that there are three specific functional groups were detected, that is, O–H group at 3380 cm^{-1}, C=O group at 1618 cm^{-1} and C–O–C group at 1065 cm^{-1}. Analysis from FTIR confirmed the structure of anthocyanin which consist of benzene skeleton, conjugated double bond, carbonyl group, C–H bond, and O–H bond [8].

Structure identification was also carried out by LC-MS to identify the anthocyanin type based on its aglycon moiety. **Fig. 3** showed that anthocyanin type is cyanidin with m/z 286.50. The aglycon type is identified as cyanidin-3-glucoside (m/z = 449), cyanidin-3-rutinoside (m/z=581), and cyanidin-3-sambubioside (m/z=595) as reported by Trikas, et al (2016) [9]. Cyanidin was more stable than another type of anthocyanin, especially cyanidin-3-glucoside.

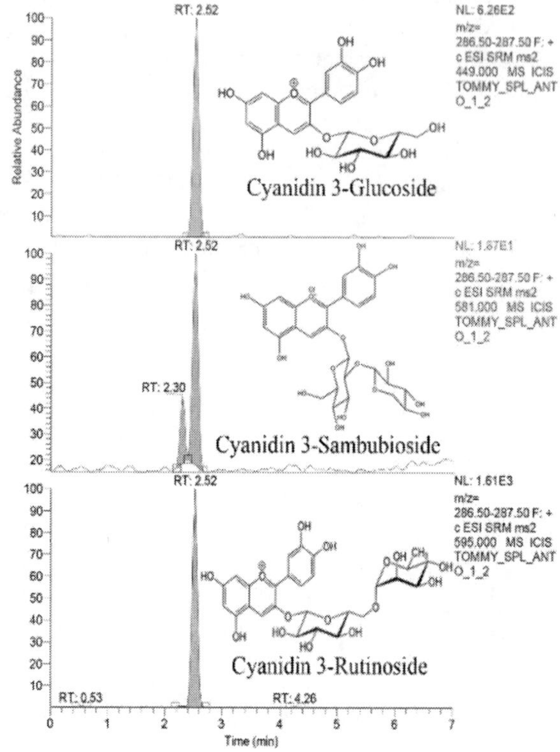

Figure 3. LC-MS spectrum of natural dye type

3.2. Activated Carbon Characterization

Activated carbon (AC) was characterized using Scanning Electron Microscope (SEM) to identify surface morphology of the carbon, before activated **(Fig. 4a)** and after activated **(Fig. 4b)**. The activated carbon has a uniform shape and small size distribution, while non-activated carbon has an irregular shape. The activated carbon has larger surface area and absorb sensitizer effectively to reduce triiodide electrolyte solution to iodide, then the electron transport cycle is faster [6].

Figure 4. SEM micrograph of (a) activated carbon and (b) non-activated carbon

3.3. Voltage and Current Measurement

The measurement of voltage and current of DSSC prototype was done using digital multimeter and halogen lamp 150 watt. Halogen lamp can absorb visible area so that it is said to be analog with solar light. **Table 1** showed the photovoltaic result from DSSC prototype using antocyanin and activated carbon in different thickness of counter electrode. Antocyanin gave high intensity and can absorb light due to cyanidin moiety is the most stable anthocyanin type [4].

On the other hand, the thickness of counter electrode influences the voltage of DSSC. The thicker the layer of the activated carbon provide the homogenous distribution of the pores. It would make triiodide as an electrolyte solution easily reduce to iodide, and then the cyclic of electron transport proceed faster than the thinner one in DSSC system. Anthocyanin as sensitized dye in DSSC system can also easily adsorb light from the solar energy. Therefore, the combination of anthocyanin and activated carbon from coffee beans peel is prospective material for DSSC system.

Table 1. Photovoltaic parameters for different samples

No	Thickness variation of counter electrode	Volt (mV)	Resistance (mΩ)	Current (mA)
1	300 μL	395	158	2.50
2	200 μL	334	199	1.67
3	100 μL	254	271	0.93

4. Conclusion

Coffee beans peel can be used for natural dye and counter electrode in DSSC system. The best result was obtained from 300 $\mu L/cm^2$ counter electrode with voltage 395 mV and 2.50 mA. The used of anthocyanin and activated carbon from coffee beans peel supposed to be the prospective material for improvement of DSSC system.

Acknowledgments

The authors wish to thank to Ministry of Research, Technology and Higher Education for the research funding through PKMP research project.

References

[1] Hau SK, Yip H-L, Jen AK-Y. A Review on the Development of the Inverted Polymer Solar Cell Architecture. Polym Rev 2010;50:474–510. doi:10.1080/15583724.2010.515764.
[2] Gong J, Sumathy K, Qiao Q, Zhou Z. Review on dye-sensitized solar cells (DSSCs): Advanced techniques and research trends. Renew Sustain Energy Rev 2017;68:234–46. doi:10.1016/j.rser.2016.09.097.
[3] Calogero G, Di Marco G, Caramori S, Cazzanti S, Argazzi R, Bignozzi CA. Natural dye senstizers for photoelectrochemical cells. Energy Environ Sci 2009;2:1162. doi:10.1039/b913248c.
[4] Chien C-Y, Hsu B-D. Optimization of the dye-sensitized solar cell with anthocyanin as photosensitizer. Sol Energy 2013;98:203–11. doi:10.1016/j.solener.2013.09.035.
[5] Prata ERBA, Oliveira LS. Fresh coffee husks as potential sources of anthocyanins. LWT - Food Sci Technol 2007;40:1555–60. doi:10.1016/j.lwt.2006.10.003.
[6] Maiaugree W, Lowpa S, Towannang M, Rutphonsan P, Tangtrakarn A, Pimanpang S, et al. A dye sensitized solar cell using natural counter electrode and natural dye derived from mangosteen peel waste. Sci Rep 2015;5. doi:10.1038/srep15230.
[7] Moolsarn K, Tangtrakarn A, Pimsawat A, Duangsa K, Mongkolkachit C, Maiaugree W, et al. A Dye-Sensitized Solar Cell Using a Composite of PEDOT:PSS and Carbon Derived from Human Hair for a Counter Electrode. Int J Photoenergy 2017;2017:1–11. doi:10.1155/2017/1064868.

[8] Wahyuningsih S, Wulandari L, Wartono MW, Munawaroh H, Ramelan AH. The Effect of pH and Color Stability of Anthocyanin on Food Colorant. IOP Conf Ser Mater Sci Eng 2017;193:012047. doi:10.1088/1757-899X/193/1/012047.
[9] Trikas E, Papi R, Kyriakidis D, Zachariadis G. A Sensitive LC-MS Method for Anthocyanins and Comparison of Byproducts and Equivalent Wine Content. Separations 2016;3:18. doi:10.3390/separations3020018.

β-Cryptoxanthin and Zeaxanthin Pigments Accumulation to Induce Orange Color on Citrus Fruits

Inanpi Hidayati Sumiasih[1,2], Roedhy Poerwanto[1], Darda Efendi[1,3], Andria Agusta[4], Sri Yuliani[5]

[1.] Agronomy and Horticulture Department, Faculty of Agriculture, Bogor Agricultural University. Jl. Meranti, IPB Darmaga Campus, Bogor 16680, Indonesia.

[2.] Agroecotechnology Program, Faculty of Bioindustry, and Trilogi University, Jakarta. Jl. TMP. Kalibata No. 1, Jakarta 12760 Indonesia

[3.] Center for Tropical Horticulture Studies (PKHT) Baranangsiang Campus of IPB, Jl. Raya Pajajaran, Bogor 16144

[4.] Research Center for Biology, Indonesia Institute of Sciences (LIPI), Jl. Raya Bogor Km. 46, Cibinong Science Center, Cibinong 16911, West Java, Indonesia

[5.] Center for Agricultural Postharvest Research and Development, Jl. Tentara Pelajar Cimanggu No.12, Ciwaringin, Bogor, West Java 16124

E-mail: inanpihs@trilogi.ac.id

Abstract. Degreening, a transformation process of green color on citrus peel to be orange color on tropical low-land citrus fruits often fails. Orange color of the citrus peel comes from the mixture carotenoid pigments, such as zeaxanthine and mainly β-cryptoxanthin and β-citraurin. The accumulation of β-citraurin occurs when the fruits are exposed to low temperature, and otherwise, it will fail to occur. Precooling treatment on lowland tropical citrus fruits is expected to stimulate the accumulation of β-citraurin. The results showed the most favorable color obtained from precooling and 24-hour ethylene exposure duration. This treatment could decrease total chlorophyll and β-carotene content as well as proven to increase 3 times the accumulation of β-cryptoxanthin in accelerating the appearance of bright orange color on citrus peel. Degreening gave no significant effect to internal quality of *Citrus reticulata*.

Keywords: *Citrus reticulate; degreening; β-cryptoxhantin; zeaxanthin; β-carotene; Citrus color Index*

1. Introduction

The increasing public awareness of the nutritional value effects on the increasing demand for citrus fruits. This increase led to importing large amounts of fresh citrus, which made Indonesia as the second largest importer in ASEAN after Malaysia. Increasing consumer demand for imported citrus fruits was due to consumer interest in the fruit appearance and the taste of orange citrus instead of non-uniform citrus (yellowish green). [15] and [7] reported that 85% of respondents in Bogor chose to buy imported citrus, while 15% of respondents preferred to buy local ones. This decision was determined by the attractive appearance of imported citrus fruits compared to local fruits. Another existing problems are very low availability of orange *Citrus reculata* and low yield which could not meet the meet the demand. Therefore, this research is needed to improve color quality to make it more consumer-attractive/orange.

Efforts to increase the quantity and quality of citrus fruits are needed to meet the market demand. One of the efforts to meet the consumer favorable taste is the formation of orange color on citrus fruits. Therefore, degreening treatment with ethylene should be done after harvesting. Degreening is the destruction process of green color on citrus peel with orange color. Degreening so far can not stimulate the formation of orange color because the fruits only managed to turn into yellow. Yellow-colored fruits are considered unripe even almost rotten so that it is less desired by consumers. Yellow-colored fruit peel formed by accumulation of cryptoxanthin [5, 2, 19], orange-colored fruit peel comes from the mixture of β-cryptoxanthin with β-citraurin on carotenoids. Pigments contributed in the formation of β-citraurin is β-cryptoxanthin and zeaxhantin [17, 3, 13].

The research regarding the transformation of pigments on citrus peel in Indonesia has never been done before. According to the literature mentioned above, the failure of degreening in tropical regions to form orange color due to the temperature which β-citraurin will be activated, which is at 18 – 24 °C. Citrus tree planted in different elevation will be exposed to different temperature level during growth. Precooling treatment right after harvest is expected to substitute low temperature not found during its growth time. It is expected to induce the β-citraurin and β-cryptoxanthin synthesis, because once harvested, fruit metabolisms persist.

Basically, the role of ethylene hormones to stimulate the rapid degreening process in color change is enormous [18]. Ethylene is a substance commonly used in degreening citrus fruits [13, 19, 9, 10]. [12, 11] reported that ethylene has been shown to play a role in reducing chlorophyll content in fruit peel, causing discoloration of citrus peel from green to orange

Therefore, a degreening technology should be developed to induce orange color formation, as well as the identification of β-cryptoxanthin and zeaxantin on carotenoids as the agent of orange color formation. Here, we reported the effect of precooling and precise ethylene exposure duration in the formation of orange color in citrus peels, and the changes of β-cryptoxanthin, zeaxanthin, and β-carotene pigmets content in citrus peel before and after degreening.

2. Research Methodology

2.1 The degreening treatment and fruit quality observation were done at PKHT IPB and Center for Biological Research-LIPI Cibinong. The research took place from January 2016 to July 2017.

The materials used were citrus fruits and ethylene gas, *Citrus reticulata* "tejakula". The tools used consists of refractometer, degreening containers, cool storage, syringe, colorreader, HPLC (High Performance Liquid Chromatography), spectrophotometry, and chromatography.

Randomized Factorial Block Design was used as the research design. The treatment consists of 2 factors, precooling treatment and exposure duration. There were 2 levels of precooling factor that is precooling and without precooling. There were 3 levels of exposure duration factor, without exposure, 24 hour-exposure, and 48-hour exposure. As a result, there were 6 combinations repeated 3 times and 18 units of experiment.

Citrus reticulata "tejakula" was obtained from smallholdings in Tuban, East Java. The estates located in lowland region at 40 m AMSL. The fruits were distributed to Laboratory of PKHT IPB Bogor. Fruits were placed into degreening containers then tightly sealed and kept airtight. Ethylene gas at 100 ppm concentration was injected into the degreening containers using 5 ml-syringe. The containers then put in cold storage at 15 °C. The fans were turned on during the degreening treatment, for control, 24 hours, and 48 hours. The fruits then were inspected and the color changes were observed.

2.2 Hue Angle identification

Changes in the color quality of citrus peel was measured with Minolta Color Reader type 310. This tool color notation system is hunter notation system (color system L, a, and b). Hunter color notation system is characterized by three parameters of color, the brightness symbolized by L, the chromatic colors symbolized by 'a' notation and color intensity with the notation 'b' (1). The measurement results were expressed in the hue angle. The results of the measurements of a and b values are converted into chromatic units of hue angle (°hue). The °hue angle describes the pure color which indicates the dominant color in a mixture of several colors. According to [8], °hue can be obtained using the following equation:

$$°hue = \tan^{-1}\left(\frac{b}{a}\right)$$

2.3 Material Preparation

Citrus peels are separated from the edible portion, then immediately cooled with nitrogen liquid in the dewar tube until next step.

Carotenoid Extraction

The identification of carotenoids was performed by the method described by Kato *et al* (2004). The pigment was extracted from the sample with hexane solution: aceton: ethanol (2:1:1, v/v) containing 0.1% (w/v) 2.6 ditert butyl 4 methylphenol and 10% (w/v) magnesium carbonate. After the organic solvent evaporates entirely, the carotenoid-containing extract, which is esterified into fatty acid, is disacorated with 20% (w/v) methanolic KOH. Water-soluble extracts are removed by saturated NaCl. The pigments were partitioned in the dhietyl ether phase and evaporated.

2.4 Analysis of β-cryptoxanthin, zeaxanthin and β-carotene

The residue is redissolved in methyl tert butyl ether solution: methanol (4:6, v/v). An aliquot (30 μL) is separated by an HPLC reversed phase (Shimadzu, Ascentis) equipped with type C-18, at a 1 mL min^{-1} flow rate. The standards of β-cryptoxanthin, zeaxanthin and β-carotene were prepared based on the method described with fresh weight micrograms per gram (preparing for calibration of the standard). Quantification of β-cryptoxanthin, zeaxanthin and β-carotene was performed in 3 replications. The data were obtained from average HPLC values and calculated using the calibration equation. With standard pigments (Figure 1).

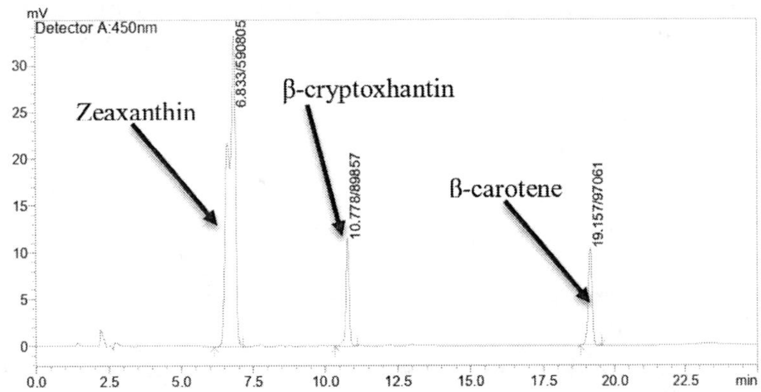

Figure 1. HPLC chromatogram for mixture of authentic sample of Zeaxhantin β-cryptoxanthin, and β-carotene.

3. Results and Discussions

There were two different experiments conducted in this research: to observe the effect of precooling and ethylene exposure duration on the change of citrus reticulata color. The effect of precooling and exposure duration to *Citrus reticulata* of Tuban origin showed the best color change was on the duration of 24 hours exposure with precooling and degreening treatment. Citrus peel color with precooling treatment on 24-hour ethylene exposure resulted in orange peel color, whereas without precooling showed a yellowish orange color (Figure 2).

Citrus reticulata "tejakula" with precooling and degreening treatment on 24-hour ethylene exposure duration showed uniform orange color, whereas the control fruits (without precooling and degreening) showed the fruits color stayed green (Figure 3).

Total soluble solids and total acid

The measurement for measuring citrus sweetness level is expressed in the content of total soluble solids. Based on the precooling treatment and ethylene exposure duration at the final observation, it was shown that the treatments gave no significant effects to total soluble solids, total acid and citrus firmness (the data is not shown).

Hue Value

The decrease in Hue value indicates that there is a change of color from green to yellow to orange. Orange color on citrus peel was shown with Hue ≤ 71.5. Hue (°Hue) is the conversion result of a and b values and becomes an important parameter in the citrus measurement method (6).

Citrus reticulata "tejakula" treated with precooling treatment with degreening on 24-hour exposure duration on day 9 after treatment showed rapid decrease in hue value compared with other treatments, with a value of 68.07. Hue values with treatment without precooling on 24-hour exposure duration on day 12 showed a decrease in hue value compared to control and 48-hour exposure duration, with a value of 68.11 (Figure 4).

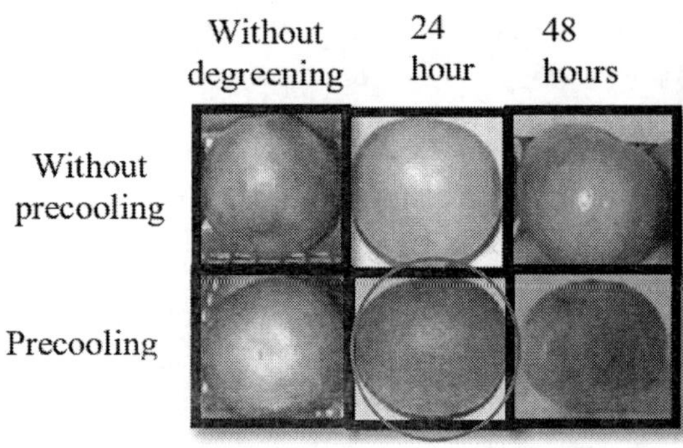

Figure 2. The change of *Citrus reticulata* "tejakula" peel color with degreening treatment on several ethylene exposure durations (9 days after treatment)

Figure 3. The color of *Citrus reticulata* "tejakula" peel, precooling without degreening (a), precooling with degreening (b) 24-hour ethylene exposure duration (9 days after treatment)

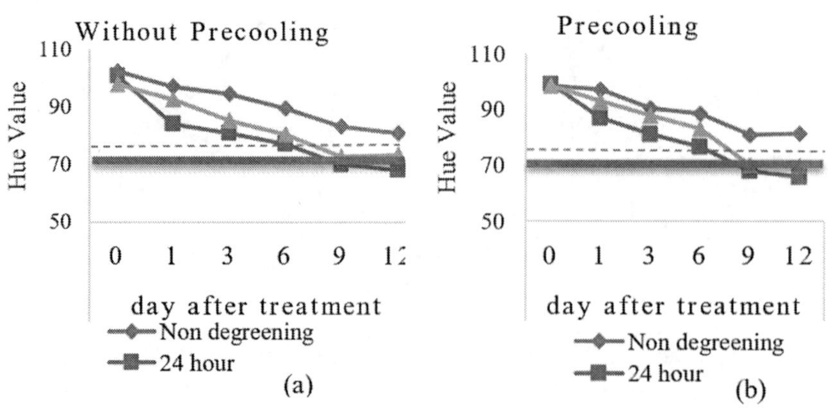

Figure 4. The trend of Hue Value changes of *Citrus reticulata* "tejakula" peel color on degreening without precooling treatment (a) and with precooling (b) at several exposure duration

Identification of β-Carotene, β-Criptoxanthin and Zeaxantin

Ccarotenoid biosynthesis contributes to the increase of citrus color from green to orange, the amount of β-cryptoxanthin and zeaxanthin content increase while chlorophyll and β-carotene content decrease during storage and these changes are accelerated by precooling and degreening applications with 4-hour exposure duration in *Citrus reticulata* "tejakula". The HPLC chromatograms for β-carotene, β-cryptoxanthin and zeaxanthin pigments are shown below (Figure 5 and 6).

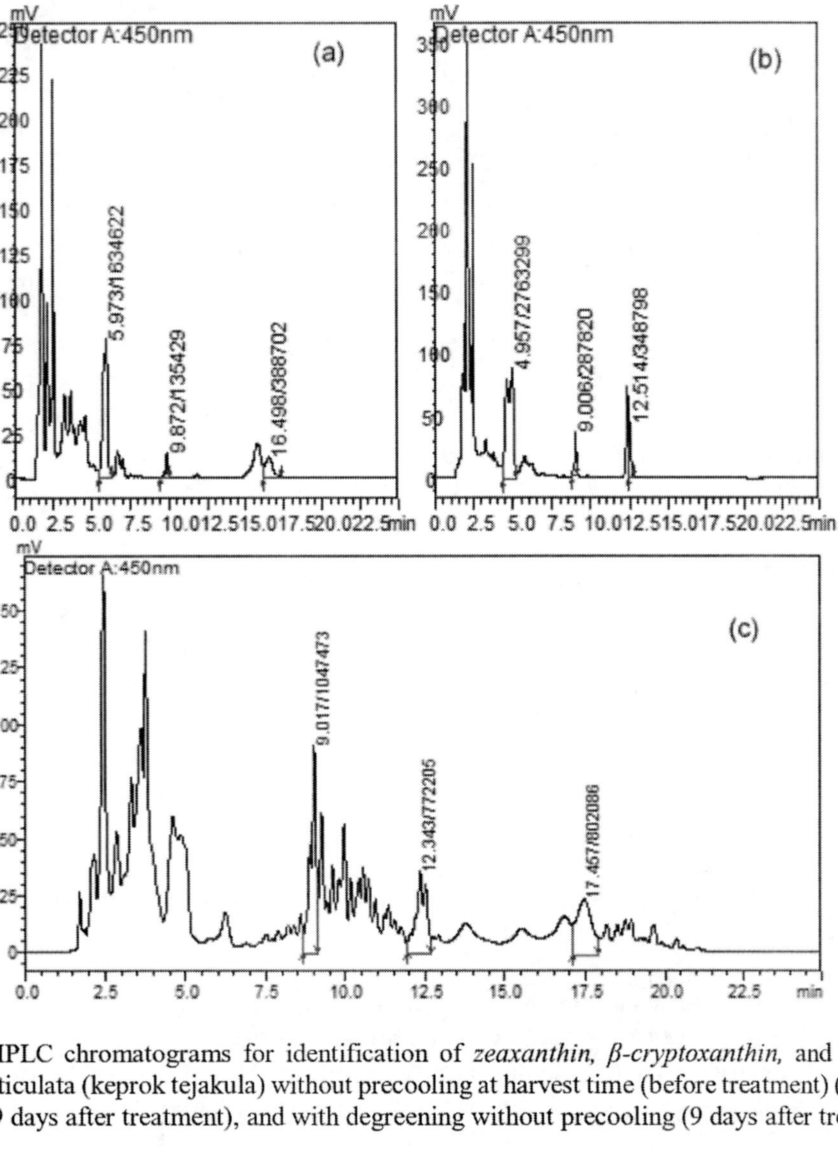

Figure 5. HPLC chromatograms for identification of *zeaxanthin, β-cryptoxanthin,* and *β-carotene* pigment of reticulata (keprok tejakula) without precooling at harvest time (before treatment) (a), without degreening (9 days after treatment), and with degreening without precooling (9 days after treatment).

IOP Conf. Series: Materials Science and Engineering **299** (2018) 012074 doi:10.1088/1757-899X/299/1/012074

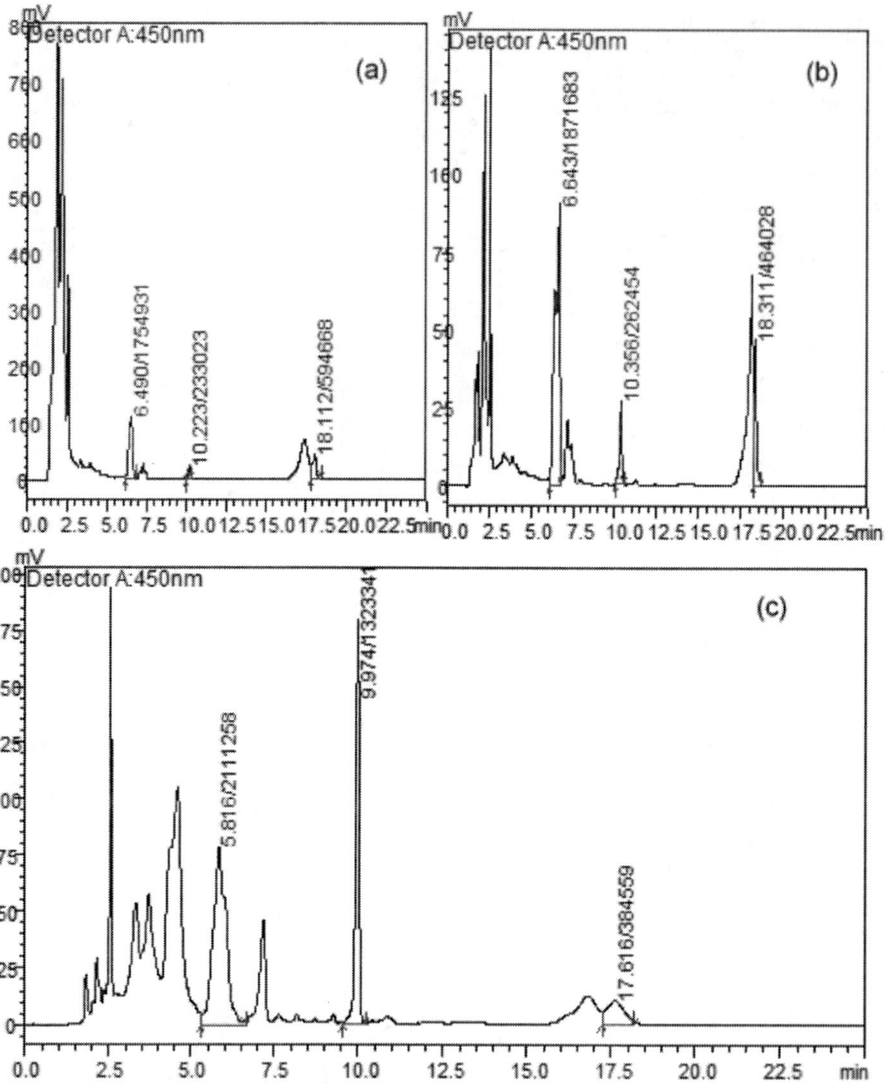

Figure 6. HPLC chromatograms for identification of *zeaxanthin, β-cryptoxanthin,* and *β-carotene* pigment of reticulata (keprok tejakula) with precooling at harvest time (before treatment) (a), without degreening (9 days after treatment), dan precooling with degreening (9 days after treatment).

The value of β-cryptoxanthin content in citrus peel with precooling without degreening treatment before treatment was 1.93 ug g FW^{-1}, while treatment of precooling and degreening could significantly increase the β-cryptoxanthin content to 37.16 ug g FW^{-1} on day-9 after treatment. The content of β-cryptoxanthin in citrus without precooling also increased but only 6.35 ug g FW^{-1} (Figure 7). [4, 13] also observed similar results that the orange peel color changed from green to orange due to the nonphotosintetic characteristic of carotenoid synthesis, which is β-citraurin which causes reddish orange color on Mandarin citrus peel. The accumulation of these compounds is determined by the availability of photosynthetic carotenoid precursor such as zeaxanthin, β-cryptoxanthin, nonphotosintetic carotenoids (β-citraurin) which occur only at low temperatures.

The results showed that the formation of orange color on citrus peel required precooling treatment prior to degreening treatment. Without precooling, degreening of *Citrus reticulata* "tejakula" would be slower. Precooling was needed because β-citraurin could be induced only at low temperature. The content of β-carotene on citrus peel after precooling and degreening treatment sowed rapid decrease compare to the content of β-carotene on citrus peel without precooling. Before degreening treatment, the content of β-carotene was 5.32 ug g FW-1, while 9 days after precooling treatment and 24-hour ethylene exposure duration caused the content to decrease to 44.82 ug g FW-1.

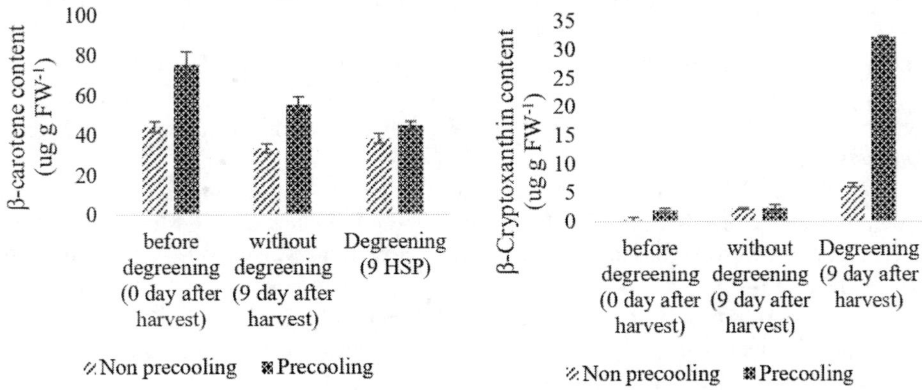

Figure 7. The change of β-carotene and β-Cryptoxanthine accumulation in citrus peel with degreening and 24-hour exposure duration

The value of zeaxhantin content on precooling and degreening treatment with 24-hour exposure duration gave result 37.65 ug g FW-1 on day-9 after treatment. The value of zeaxhantin treatment content without degreening showed a lower value, with a value of 12.22 ug g FW-1. While the zeaxhantin value at initial condition before degreening without precooling (26.96 ug g FW-1), with precooling (27.88 ug g FW-1) can be seen in Figure 8.

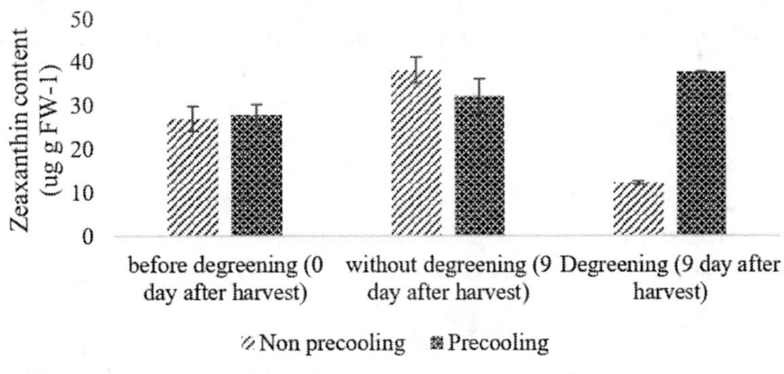

Figure 8. The change of Zeaxanthin accumulation on citrus peel with degreening 24-hour exposure duration

4. Conclusions

The most favorable result was obtained from precooling and 24-hour ethylene exposure duration. This treatment could decrease the β-carotene content and was proven to increase 3 times the content of β-cryptoxanthin pigment to accelerate the formation of citrus peel to turn orange. Degreening treatment gave no significant effect to internal quality of citrus peel.

5. Acknowledgments

The author would like to express gratitude to PKHT IPB and LIPI Cibinong which have facilitated the research. The author also would like to express gratitude to Ministry of Research, Technology, and Higher Education for the Doctoral Dissertation Research Grant with research contract number 0432/K3/KM/2017 the author would like to thank to Ministry of Agriculture for the partnership grant number 714/LB.620/1.1/2/2013.

6. Reference

[1] Andarwulan N, Kusnandar F, Herawati D 2011 *Analisis Pangan* Jakarta (ID): Dian Rakyat

[2] Fanciullino A L, Cercos M, Mayer CD, Froelicher Y, Talon M, Ollitrault P dan Morillon R 2008 Changes in carotenoid content and biosintesis gene expression in juice sacs of four orange varities (*Citrus sinensis*) differing fruit color *Journal Agricultural of Food Chemestry* (Washington DC) **54** 4397-4406.

[3] Gang Ma, Zhang L, Matsuta A, Matsutani K, Yamawaki K, Yahata M, Wahyudi A, Motohashi R, Kato M 2013 Enzymatic Formation of b-Citraurin from b-Cryptoxanthin and Zeaxanthin by Carotenoid Cleavage Dioxygenase4 in the Flavedo of Citrus Fruit *American Society of Plant Biologists* 682–695.

[4] Kato M, Ikoma Y, Matsumoto H, Sugiura M, Hyodo H, Yano M 2004 Accumulation of Carotenoids and Expression of Carotenoid Biosynthetic Genes during Maturation in Citrus Fruit. Accumulation of Carotenoids and Expression of Carotenoid Biosynthetic Genes during Maturation in Citrus Fruit **134**.

[5] Kato M, Matsumoto H, Ikoma Y, Okuda H, Yano M 2006 The role of carotenoid cleavage dioxygenase in the regulation of carotenoid profiles during maturation in citrus fruit. *Journal of Exprimental Botany* (JP: Japan) **57**. No.10. pp. 2153-2164.

[6] Lee H S 2000 Objective measurement of red grapefruit juice color *J. Agric. Food Chem* **48**; 1507–1511.

[7] Nafisah S N 2013 Sikap dan Persepsi Konsumen Terhadap Jeruk Lokal dan Jeruk Impor di Pasar Modern Kota Bogor *Skripsi* (Institut Pertanian Bogor).

[8] Manera J, Brotons JM, Conesa A, Porras I 2012 Relationship between air temperature and degreening of lemon (*Citrus lemon* L. Burm. F) peel color during maturation. *Australian Journal of Crop Science*. 6 (6):1051-1058.ISSN: 1835-2707.

[9] Mayuoni L, Tietel Z, Patil S, Porat R 2011 Does ethylene degreening affect internal quality of citrus fruit *J. Postharvest Biol and Technol* **62**; 50-58.

[10] Peng G, Xie XL, Jiang Q, Song S, Xu CJ 2013 Chlorophyll a/b binding protein plays a key role in natural and ethylene-induced degreening of ponkan (*Citrus reticulate* Blanco) *J Sci Hortic* **160**; 37-43.

[11] Purvis A C 1980 Sequenci of Chloroplast Degreening On Calamondin Fruit As Influenc By Ethylene And Agno3 *Plant pgysiol* **68**.

[12] Purvis A C Barmore C R 1981 Involvement of ethylene in chlorophyll degradation in peel of citrus fruit *Plant Physiol* **68**, 854–856.

[13] Rodrigo M J, Alquezar M, Alos E, Medina V, Carmona L, Bruno M 2013 A Novel Carotenoid Cleavatage in the Biosynthesis of Citrus fuit specific apocarotenoid pigments *Experimental Botany* **64**, No 4461-4478.

[14] Rodrigo M J and Zacarias L 2007 Effect of postharvest ethylene treatment on carotenoid accumulation and the expression of carotetenoid biosynthetic genes in the flavedo of orange (*Citrus cinensis* L. Osbeck) fruit *Scientia Hortikultura* **43**. pp.14-22

[15] Shanti S I 2007 Analisis Keputusan Konsumen Dalam Mengkonsumsi Jeruk Lokal Dan Jeruk Impor Di Ritel Modern (Kasus Konsumen Giant Botani Square Bogor) *Skripsi* (Institut Pertanian Bogor).

[16] Stewart I and TA Wheaton. 1971. Effent of Ethylene and Temperature On Carorenoid Pigmentation of Citrus Pell. Proc. Fla. State hort. Soc **84**:264-266.

[17] Wardowsky WM, Miller WF, Grierson D. 2006. Degreening. In: Wardosky WF, Miller WM, Hall DJ, Grierson D, eds. *Fresh citrus fruits*. Longboat Key: Florida Science Source, 277–298.

[18] Zhou YJ, Sun CD, Zhang LL, Dai X, Xu CJ, Chen KS. 2010. Preferential accumulation of orange-colored carotenoids in Ponkan (*Citrus reticulata*) fruit peel following postharvest application of ethylene or ethephon. J. Scientia Horticulturae **126**; 229-235

IOP Conf. Series: Materials Science and Engineering **299** (2018) 012075 doi:10.1088/1757-899X/299/1/012075

Methylation of Brazilein on Secang (*Caesalpinia sappan Linn*) Wood Extract for Maintain Color Stability to the Changes of pH

Zeni Ulma[a], Edia Rahayuningsih [a*], Tutik Dwi Wahyuningsih[b]

[a, a*] Department of Chemical Engineering, Faculty of Engineering,
Universitas Gadjah Mada, JL. Grafika 2, Kampus UGM, Yogyakarta 55281, Indonesia

[b]Department of Chemistry, Faculty of Mathematics and Natural Sciences,
Universitas Gadjah Mada, JL. Grafika 2, Kampus UGM, Yogyakarta 55281, Indonesia

*Corresponding Author : edia_rahayu@ugm.ac.id

Abstract. The stability of natural dyes to the changes of pH is really necessary when the natural dyes are applied either on fabric or food. This research aimed to increase the stability of brazilein, a compound contained within the secang wood extract, to the changes of pH. The methylation process was done by reacting Dimethyl Carbonate (DMC) with the brazilein on the secang wood extract. DMC acts as a substance that substitute hydroxyl group on brazilein. The methylation reaction of brazilein on secang wood extract was operated on a three-necked round-bottomed flask fitted with mercury-sealed stirrer and reflux condenser under 80°C temperature and 250 rpm stirring speed. There were two variables observed in this research; the DMC amount ratio to the amount of secang wood extract and the time of the methylation process. The research showed that at the 1:10 the DMC amount ratio to the amount of wood extract and 8 hours of the methylation process give the better stability of color of the secang wood extract than the variation of the other variables.

Keyword : Secang wood, *Caesapinia sappan L.*, *brazilein*, pH, methylation.

1. Introduction

Indonesia has thousands of types of natural dye-producing plants that have been used since a long time ago. All this time, the natural red dye can be obtained from root bark of *Morinda citrifolia.* Therefore, the natural red dye for textiles is still difficult to produce in large capacity, due to the limited amount of raw materials that can be obtained [5]. Mean while secang wood is very potential to be used as a source of natural red dye because the amount of the secang wood which is very abundant in Indonesia and has not been utilized optimally. The negative side of secang wood is its instability of the *brazilien,* a compound contained within the secang wood extract itself. Hence engineering enhancement of the quality of the *brazilein* as a source of red dye is required. From all of the plant species which can be used as natural dye producers, not all of them have been tested for their stability and discoloration yet. Some plants that already evaluate for its stability according to re Merbau wood *(Instia bijuga)*, Tegeran wood *(Cudraina javanensis),* Tingi bark *(Terminalia bellirica),* and Coconut coir (*Cocus nucifera*) [9].

Secang (*Caesalpina sappan L.*) wood is a member of the *Caesalpiniaceae* which mostly exist in Indonesia. Aside from its use as healthy drink ingredients, secang (*Caesalpina sappan L.*) wood can also be used as raw material of natural dyes for food or textile. The part of the tree that taken as a natural dyes is the stem. Secang wood extract produces a red traffic color at neutral pH (pH 6-7) and tend to be red purple by the increase of pH [8].

Brazilin is a dahlia yellow colored compound while *brazilein* is a reddish colored compound. The change-over of *brazilin* into *brazilein* is due to the delocalization of the electrons because of the presence of carbonyl groups as a result from the oxidation process of *brazilin*. Therefore the pigments from secang wood extract contain more *brazilein* [3]. The pigments or the dye molecules is a composite of three constituent components i.e unsaturated organic compounds, chromophore groups, and auxochrome groups. The color carrier chromophores group were unsaturated covalent bond compounds, which lead to the electron uptake. The substance attached to a chromophore group may be impaired or reinforced. It called as the auxochrome group, which is saturated, and it is a part of the dye molecular structure that binding with the fabric [11]. The *brazilein* pigment is a chemical structure with a hydroxyl group (OH), which is an auxochrome group.

One of the strategies for binding oxygen atoms due to the loss of hydrogen atoms on the brazilein hydroxyl groups is methylation method. The methylation reaction is a replacement reaction of an atom or molecule with a methyl group [13]. The methyl group (CH_3) binds to the unstable oxygen atoms in the *brazilein* auxochrome group, and then it forms a new auxochrome group; methyl brazilein which is more stable. Commonly used methyl electrophile source is dimethyl sulfate (DMS) reagent with 0,44 g/kg LD_{50}, dimethyl halide (CH_3-X) with 0.079 g/kg LD_{50}, and dimethyl carbonate (DMC) with 13,8 g/kg LD_{50}. Based on the LD_{50} value of the three methyl electrophile source, DMC is a green methylation agent that is environmentally friendly[12]. In this research used DMC as a methylation material for secang wood extract structure is applied. The methylation reaction between *brazilein* and dimethyl carbonate will produce a methyl group attached to an oxygen atom which was previously a hydroxyl group. The mechanism of this reaction is that the methyl group derived from dimethyl carbonate reagent will attack the hydroxyl group that binding with a benzene compound which has no electron resonance in its structure. The hydrogen atom will be easily substituted by the methyl group because of the hydroxyl group attached to the benzene compound that has no electron resonance in its structure tends to remove the hydrogen atom [1], so with the changes of the *brazilein* auxochrome group which has three hydroxyl groups before, as the effect of the methylation reaction, it will turn into two hydroxyl groups with one oxygen atom attached to a methyl group. The reduction of hydroxyl groups in the *brezilein* functional group will result on the increased of the color stability of *brazilein* to the changes of pH. The mechanism of the methylation reaction of *brazilein* as presented in Fig. 1.

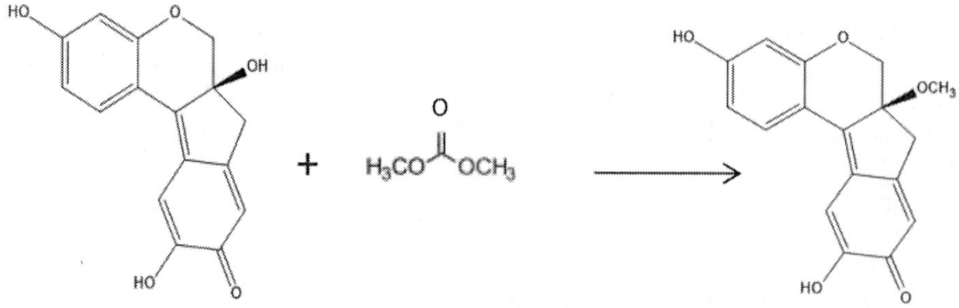

Figure 1. The reaction of DMC with brazilein [10;14]

2. Material And Method

2.1 Materials

The secang wood used in this research was obtained from the Beringharjo market Kodya Yogyakarta. High purity chemicals were used in this research, which included dimethyl carbonate 99 % (Merck, Germany), HCl with purity of 37% was taken from C.V. Chemix Pratama, Yogyakarta, Indonesia, NaOH in the form of crystalline was taken from C.V. Chemix ,Yogyakarta, Indonesia.

2.2 Method

The experiment was operated in laboratory scale batch system usiang a stirred reactor with mantle heater and condenser (Fig. 2).

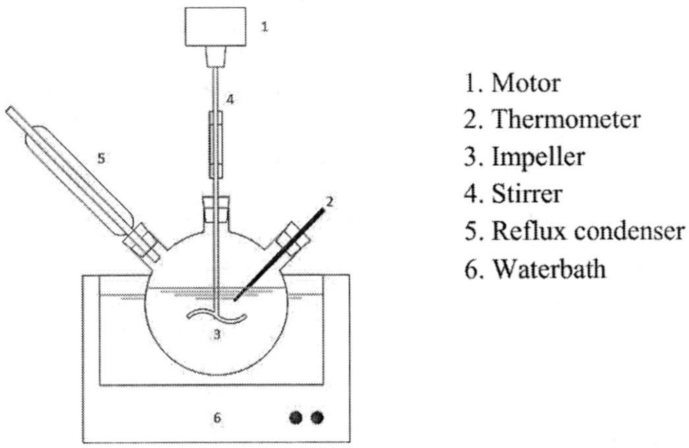

Figure 2. The experimental set-up used for *brazilein* extraction

To obtain the *brazilein* extract from secang wood, the extraction process was done under optimum condition based on the previous research by Wahyuningrum[8] that conducted using water as solvent, 100 °C temperature, and 70 minutes of the extraction time. However, the ratio used in this extraction was 30 g : 500 mL of the weight of the secang wood to the volume of water as solvent. It is intended to obtain the *brazilein* concentration in the extracts suitable for the methylation reaction. The methylation reaction of brazilein on secang wood extract was operated on a three-necked round-bottomed flask fitted with mercury-sealed stirrer and reflux condenser as shown in Fig. 2. The methylation reaction was done under 80°C temperature, 250 rpm stirring speed with two operation variables; the DMC volume ratio to the secang wood extract volume and the time of the reaction. The variations of the DMC volume ratio to the secang wood extract volume were 1:100; 1:20; 1:10 while the variations of the methylation reaction time were 1, 5, and 8 hours. Each one of the methylation reaction results was tested for its stability to pH; 10 mL of the methylation result was taken to tested in pH 2, 3, 5, 7, and 10. Furthermore 5 mL of the solution was taken, diluted to 100 mL, and then its absorbance was measured using a UV-Vis spectrophotometer at 541 nM wavelength.

3. Result And Discussion

3.1 The brazilein extract stability from secang wood

The *brazilein* extract gives a certain color for every pH condition. The color of the brazilein at vaious pH conditions stated in the classic *Reichs-Aussch für Lieferbedingungen* (RAL) system for color standardization qualitatively, as shown in Fig. 3.

Figure 3. Classic RAL System for color standardization[4]

The brazilein extract around pH 6 to 7 or neutral pH gives pearl ruby red and ruby red color , and changed into purple red until wine red by the increase of pH. The secang extract produce ruby red color. For the low pH condition (pH 2-5) secang extract produce of color dahlia yellow until traffic red [4]. The changes of color of the secang wood extract for various pH condition presented in Table 1 as well as the absorbance value which was measured using a UV-Vis spectrophotometer at 541 nm wavelength presented in Fig. 4. Based on the result, it can be declared that the secang wood extract was unstable by the changes of pH. This was in line with the statement of Mohan et al. [7] that the color stability of the secang wood extract was quite low because of its –OH functional group or conjugated cyclic bond so the attempt to obtain more stable extract color need to be done.

Table 1. The changes of the secang wood extract color for various pH

pH	**COLOR** (RAL SYSTEM)
2	Dahlia yellow
3	Dahlia yellow
5	Traffic red
7	Ruby red
10	Purple red

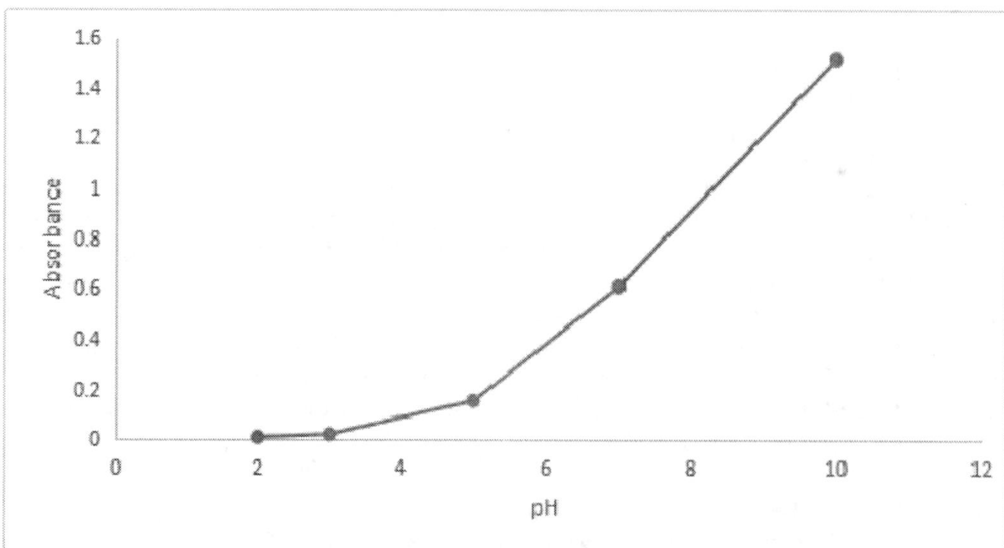

Figure 4. The absorbance values of secang wood extract for various pH condition which was measured using a UV-Vis spectrophotometer at 541 nm wavelength.

3.2 The variation of the DMC volume ratio to the extract volume for methylation reaction

The changes of methyl *brazilein* color; the methylation result of secang wood extract for various values of DMC ratio and the secang wood extract to the changes of pH presented in Table 2. As for the absorbance value which was measured using a UV-Vis spectrophotometer at 541 nm wavelength presented in Fig. 5. The successful of methylation reaction can identified by stabilize of absorbance value which was measured for various pH conditions (pH 2, 3, 5, 7, and 10).

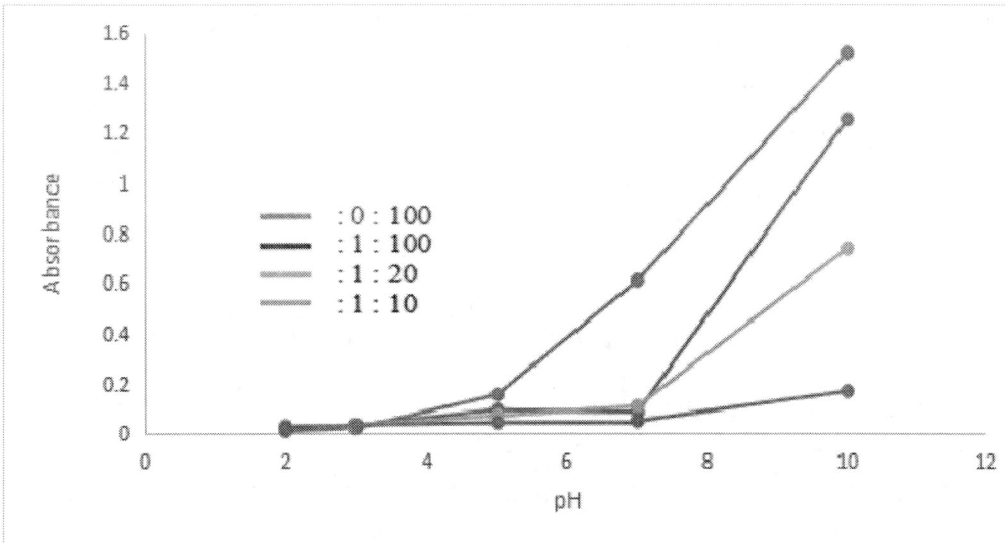

Figure 5. The absorbance values of the methylation result of secang wood extract in the variation of DMC ratio to the secang wood extract for various pH conditions.

Table 2. The methyl *brazilein* color; the methylation result of secang wood extract for various values of DMC ratio to the secang wood extract for various pH conditions

pH	The variation of DMC ratio to the secang wood extract			
	0	1 : 100	1:20	1:10
2	Dahlia yellow	Dahlia yellow	Dahlia yellow	Dahlia yellow
3	Dahlia yellow	Dahlia yellow	Dahlia yellow	Dahlia yellow
5	Traffic red	Traffic red	Traffic red	Traffic red
7	Ruby red	Ruby red	Traffic red	Traffic red
10	Purple red	Purple red	Ruby red	Traffic red

Figure 5 and Table 2 present the 1:10 DMC volume ratio to the secang wood extract gives the most stable methyl *brazilein* color and the absorbance value to the changes of pH. The research showed that the color stability was increased regardless of the various pH conditions due to the reduction of the hydroxyl group on the *brazilein* because of the methylation reaction. The hydroxyl group attached with chemical structure will reduce the color stability, while the methylation will increase the color stablity to the changes of pH [2].

3.3 The variation of the methylation time

The changes of methyl *brazilein* color and absorbance value; the methylation result of secang wood extract for various times presented in Fig. 6 and Table 3

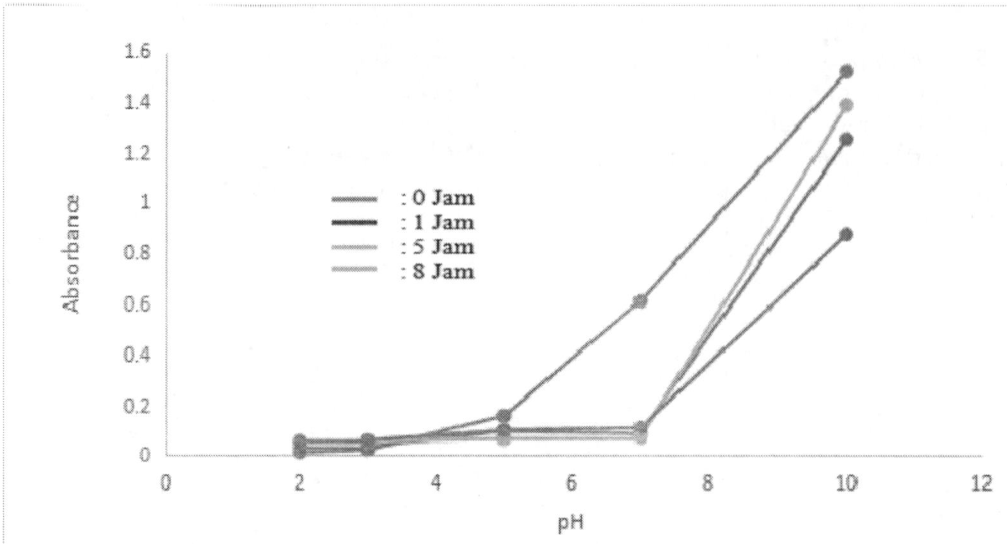

Figure 6. The absorbance values of the methylation result of secang wood extract in the variation of methylation time for various pH condition.

Table 3. The color changes of the secang wood extract for various methylation times

pH	The variation of time			
	0	1 hour	5 hour	8 hour
2	Dahlia yellow	Dahlia yellow	Dahlia yellow	Dahlia yellow
3	Dahlia yellow	Dahlia yellow	Dahlia yellow	Dahlia yellow
5	Traffic red	Traffic red	Traffic red	Traffic red
7	Ruby red	Ruby red	Ruby red	Traffic red
10	Purple red	Purple red	Ruby red	Ruby red

Figure 6 and Table 3 present the color stability to the changes of pH which was increased by the longer of the methylation time. The best color stability condition was performed under 8 hours of the methylation time, where the absorbance value did not significantly different on the 2, 3, 5, 7 pH range. As for the 10 pH value, the increase of the absorbance value is quite high, where this condition may be caused of the methyl bond on the *brazilein* did not strong enough, so the color stability under this pH value has not been reached the optimum condition yet. Therefore, the addition of the methylation time should be done on the next research. According to [6], DMC is a non-toxic methylation agent and it is a green chemical as it takes a long time in the methylation reaction.

4. Conclusion

The best condition achieved from this *brazilein* methylation on secang wood extract was on 1:10 ratio of the DMC to the secang wood extract, where the stability of color in different pH did not resulted absorbance value that significantly different. Meanwhile the best condition in the variation of the methylation time was the one under 8 hours.

Reference

[1] Brady, J. E.; Maun, S. Anas, K. Sally, S. T. , Kimia Universitas; Jilid I, **2010**, Binarupa Aksara, Jakarta Indonesia.
[2] Fennema., Food Chemistry, 3th Edition, **1996,** New York, Marcel Dekker, Inc.
[3] Ghobadian, B., and Chegeni, G. R., Spray Drayer Parameters for Fruits Juice Drying, *World Journal of Agriculture Science*, **2007**,3 (2), 230-236.

[4] H. Prasasti, "Peningkatan intensitas dan perluasan spektrum warna dari ekstrak kayu secang (Caesalpinia sappan L)," Yogyakarta, **2013**.
[5] Heyne, K., Tumbuhan Berguna Indonesia, Yayasan sarana Wana Jaya, **1988,** Departemen Kehutanan, Indonesia.
[6] Hou, C., Chen, Y., Chen, W., and Li, Wei., Microwave-Assisted Methylation Cassava Starch With Dimethyl Carbonate, *Journal Carbohydrate Research*, **2011**, 346, 1178-1181.
[7] Mohan, G., Anand, S.P., Doss, A., Efficacy of Aqueous and Methanol Extracts of Caesalpinia sappan. L and Mimosa pudica L. for their potential Antimicroba activity, *South As. J. Biol.Sci*, **2011**, 1 (2), 48-57.
[8] M. T. Wahyuningrum, "Penentuan Kondisi Operasi Optimum Ekstraksi Brazilin dari Kayu SecangNo Title," Yogyakarta, 2003
[9] Rahayuningsih, E. Wijayanto, A. Nurfitasari, P., Preservation on Natural Colorant Extract of Jelawe Fruit Peel *(Terminalia bellirica)* in Water-Solution. Indonesia, *J. Chem.* **2016,** 16 (3): 315-321.
[10] Sang, H. P., Ji, H. P., Chang, Tae-Sun., Rajni, Hatti-kaul., Dimethyl Carbonate as Green Chemical, *Current Opinion in Green and Sustainability Chemistry,* **2017**,5, 61-66.
[11] Suheryanto, Dwi., Pengaruh Konsentrasi Zat Warna Basa Terhadap Ketuaan dan Ketahanan Warna Pada Serat Sabut Kelapa, *Proceeding Seminar nasional penelitian, pendidikan, dan penerapan MIPA* Universitas Negeri Yogyakarta, **2009**, ISBN, 978-979
[12] Tundo P., New Developments in Dimethyl Carbonate Chemistry, *Pure & Appl. Chem*, **2001**, 73 (7), 1117-1124 .
[13] Yulinda, L. R., Wahyuningsih, T. D., Pranowo, H. D., Metilasi Asam Galat Menggunakan Agen Metilasi Dimetil Sulfat (DMS) dan Dimetil Carbonat (DMC), **2013,** *Bekala MIPA*, (2), 23.
[14] Wongsookin, K., Saowanee R., Malee T., Vichitr R., and John B. B., Study of Al(III) Complex With The Plant Dye Brazilein From Caesalpinia Sappan Linn, *Journal Science Technology*, **2008**, 1.

Microwave-assisted Synthesis of Benzimidazole Derivatives from Citronellal in Kaffir Lime (*Citrus hystrix* DC.) Oil

Dwi Sapri Ramadhan,[1,2] Warsito,[1,2] Elvina Dhiaul Iftitah[1,2]

[1] Department of Chemistry, Faculty of Science, Brawijaya University, Jl Veteran Malang, 65145, Indonesia
[2] Essential Oil's Institute, Brawijaya University, Jl Veteran Malang 65145, Indonesia

*Corresponding author: warsitoub88@yahoo.com

Abstract. Benzimidazoles are an important group of heterocyclic aromatic organic compound in the field of medicinal chemistry. It plays a very important role in pharmacological activity such as, analgesic, antiinflammatory, antibacterial, anti ungal, anti iral, antithelmenthic, anti convulsant, antihypertensive, and antiulcer.The synthesis of benzimidazole derivative under microwave irradiation of citronellal as raw material which is extracted from *Citrus hystrix* DC. (kaffir lime) leaves. The reaction using methanol and dichloromethane as solvent with variation of reaction time at 30, 40, 50, 60, and 70 minutes and mole ratio of citronellal to 1,2-phenylenediamine, 1:1, 1:1.5, and 1:2, respectively. Synthesis products was characterized by FT-IR and GC-MS. The optimum reaction time was obtained at 60 minutes in the present of dichloromethane and 1:2 mole ratio of citronellal to 1,2-phenylenediamine. The yield of the product is 42.19%. Structure of the synthesized compound were assigned on spectrum bands for C=N str at 1661 cm^{-1}, C-N str at 1271 cm^{-1}, C=C str of benzene ring at 1622 and 1456 cm^{-1} and C=C str of alkene at 1379 cm^{-1}. GC-MS analysis showed single peak with retention time of 15.339 minute and m/z 242.

1. Introduction

Benzimidazole is a heterocyclic organic compound, which important in medicinal chemistry. This compound is bicyclic compound which consist of benzene ring fused with imidazole, which possesses many pharmacological properties [1]. The most prominent benzimidazole compound in nature is N-ribosyl-dimethylbenzimidazole, which serve as an axial ligand for cobalt in vitamin [1, 2, 3]. Benzimidazoles were reported to have antimicrobial properties against bacteria or fungi [4]. Benzimidazole and its derivatives are reported to be physiologically and pharmacologically active and some applications are found in the treatment of several diseases including epilepsy, diabetes and infertility [5]. This important group of substances has found practical applications in a number of fields: analgesic [6,7,8], antiinflammatory [7,8,9,10], antibacterial [11], antifungal [12], antiviral [13,14], anti thelmenthic [15], anticonvulsant [16,17], anticancer [18,19], antiulcer [20], antihypertensive [21], drugs against parasites [22,23] while some derivatives have been synthesized and evaluated for inhibition of HIV-1 [24].

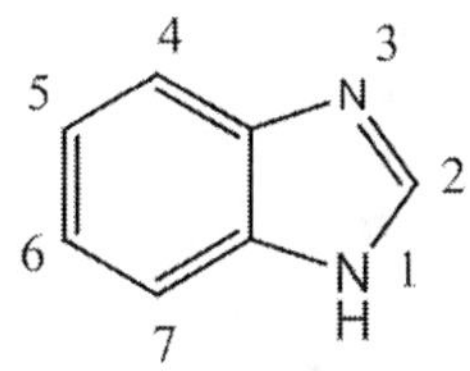

Figure 1. H-Benzimidazole

The efficient and economical methods to synthesis benzimidazole by condensation reaction between 1,2-phenylenediamine with various compounds in the presence of various reaction condition as reflux for 8h in ethanol, reflux for 1.5h in iodine/water, and using methanol proceeds at room temperature with only natural sources, molecular oxygen and visible light irradiation with blue LEDs [25]. Nowadays, microwave assisted organic synthesis is gaining widespread acceptance in drug discovery laboratories. Microwave technology, by accelerating chemical reactions from hours or days into minutes, provides quick results [26]. Saberi has reported synthesis of 2-benzimidazoles under microwave irradiation and solvent-free condition which is catalyzed by alumina, silica gel and zeolite HY the reaction mixture was then irradiated in a domestic microwave oven for 5-9 min at 160-560 W [27]. Nagula et all has reported Microwave-assisted synthesis caused a significant reduction in the reaction times and improvement in the yields of new benzimidazole bearing thiazolidinedione derivatives [28]. Current drug development is focused on compounds that are synthesized from natural materials. In this study, benzimidazole derivative from citronellal in kaffir lime oil were synthesized utilizing microwave assisted. The effect of reaction time and solvent was studied by comparing the yield of product.

2. Material and Methods

2.1. Materials

Kaffir lime oil was obtained from Essential Oil's Institute, Brawijaya University. The oil is extracted from a mixture of leaves and twigs with steam distillation. 1,2-phenylenediamine was provided by Sigma-Aldrich. Dichlorormethane, and methanol solvents were provided by Merck Chemical Company and were used without further purification.

2.2. Methods

The kaffir lime oils were obtained by steam distillation of 375 Kg (w/w) 1:1 mixture between the leaves and twigs (w/w) 1:1 distilled with steam pressure 2.5-4 bar for 4h. The essential oil was characterized by refractometer, picnometer, FT-IR and GC analysis was performed on Agilent 5890, equipped with HP-608 (30m x 320μm i.d.; film thickness 0.50μm) capillary column. The carrier gas was nitrogen (velocity: 1.5 mL/minutes). The column oven temperature was programmed as follows: initial temperature 100 °C, 5 minute; ramp to 250 °C at 5 °C/minutes. Quantitative analysis of GC were used to determines the percentage of citronellal.

The synthetic strategy is illustrated in Fig. 3. The benzimidazole derivatives was synthesized by mixing 0.0065 mole of 1,2-phenylenediamine and 0.013 mole of citronellal (1:2) in 5 mL dichlorometane and reacted under domestic microwave oven (LG 1200W 245 MHz) for 60 minutes. The mixture was cooled to room temperature for 15 minutes and put in the refrigerator (-5 °C) for 24 hours. The crystals formed are filtered and washed with dichloromethane, dried at room temperature and stored. Variations of reaction time have been carried out at 30, 40, 50, 60, and 70 minutes. The mole ratio of citronellal and 1,2-phenylenediamine was modify at 1:1, 1:1.5, and 1:2 ratio. The reaction is also carried out using a methanol solvent.

2.3 Characterization of synthesized products

The IR spectra of the compounds were recorded on Perkin Elmer FT-IR spectrometer with KBr pellets. GCMS analysis was performed on Agilent 7890B (GC) and 5977B (MS), equipped with HP-5 MS capillary column. The inlet temperature was 200 °C. The carrier gas was helium (flow: 54.087

mL/minutes). The column oven temperature was programmed as follows: initial temperature 150 °C, 1 minute; ramp to 300 °C at 10 °C/minutes. GC peak areas were used to determines the percentage of product, and MS spectra to determine m/z.

3. Result and Discussion

3.1 Identification of kaffir lime oil

The kaffir lime oil gives the yellow liquid and have a fresh, aromatic, explosive fragrance which is floral and citrus-like. The refractive index, specific gravity, yield of distillation, percentage of citronellal are 1.439, 0.85 g/ml, 0.5% and 60.44%, respectively. The kaffir lime oil was characterized by FT-IR spectroscopy compared with standard citronellal spectrum (Fig. 2). It was found that there was a peak 1726 cm^{-1} (C=O stretching), 2922, 2874 cm^{-1} (C-H stretching of aldehyde) and 1454, 1379 cm^{-1} (C=C stretching of aromatic), similar with citronellal spectrum.

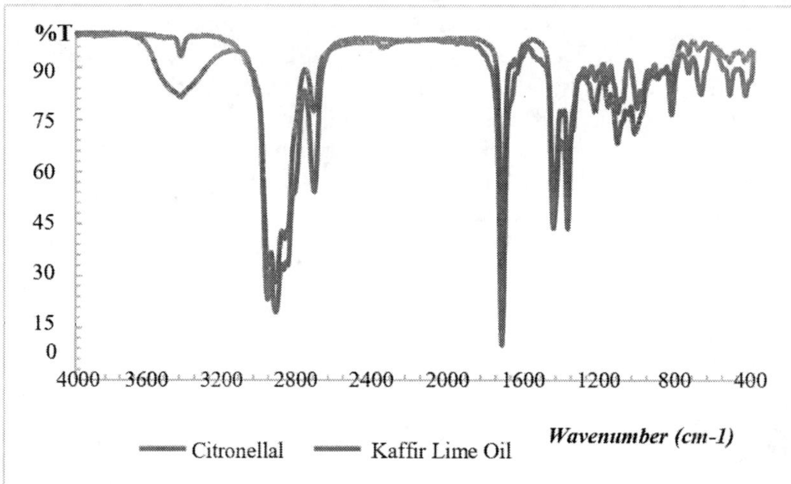

Figure. 2 Spectra of kaffir lime oil and citronellal

3.2 The synthesis of Benzimidazole Derivative

The benzimidazole derivative were synthesized by microwave assisted synthesis (Figure 3). When the reaction does not use a solvent, a crystal product is not formed. The products of benzimidazole derivative compounds synthesized using dichloromethane solvent appear purer in the form of white crystals, while using methanol obtained yellow crystals (Fig. 4). The condensation reaction of 1,2-phenylenediamine with citronellal requires the medium to interact, therefore the reaction carried out without the use of the solvent does not produce the product in solid form, its different when the reaction using a solvent. Methanol is a more polar solvent than dichloromethane, in which solvent polarity affects the product. in this case the solvent with high polarity is not suitable as a medium of interaction between molecules, so the resulting product is not pure (Fig 4.a). Otherwise when using a less polar solvent can produce products with high purity.

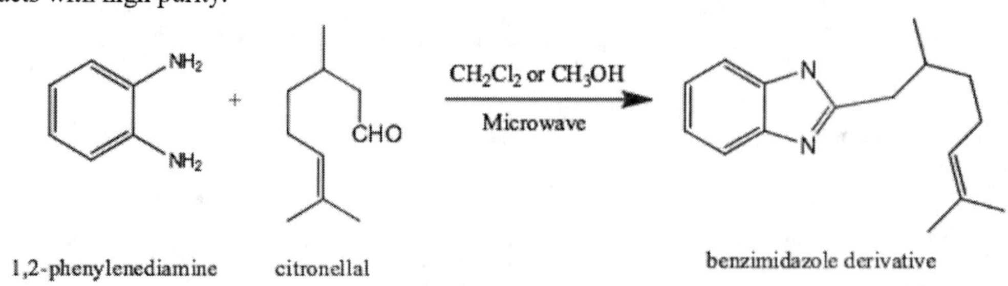

Figure 3 The synthesis of benzimidazole derivative from citronellal

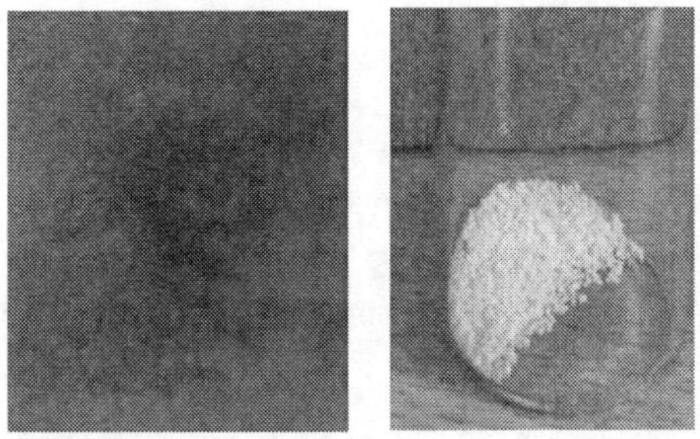

Figure 4. Product of benzimidazole derivative in : (a) methanol (b) dichloromethane

The structures of the synthesized compounds were confirmed using FTIR (Fig. 5), and GC-MS (Fig. 6). Characteristics of benzimidazole derivative from FTIR spectra (Fig. 5) contained a distinctive band at 1661 cm^{-1} and 1271 cm^{-1} as C = N str and C-N str of imidazole group, C = C str of benzene group at 1622 cm^{-1} and 1456 cm^{-1}, C = C str of alkene group at 1379 cm^{-1} and the character peak of aldehyde (C=O) was missed due with the reaction. GC-MS analysis showed single peak (tR 15.339 minute) and m/z 242.2 as molecular ion (M^+) and m/z 145 as base peak of spectra mass (Fig 6).

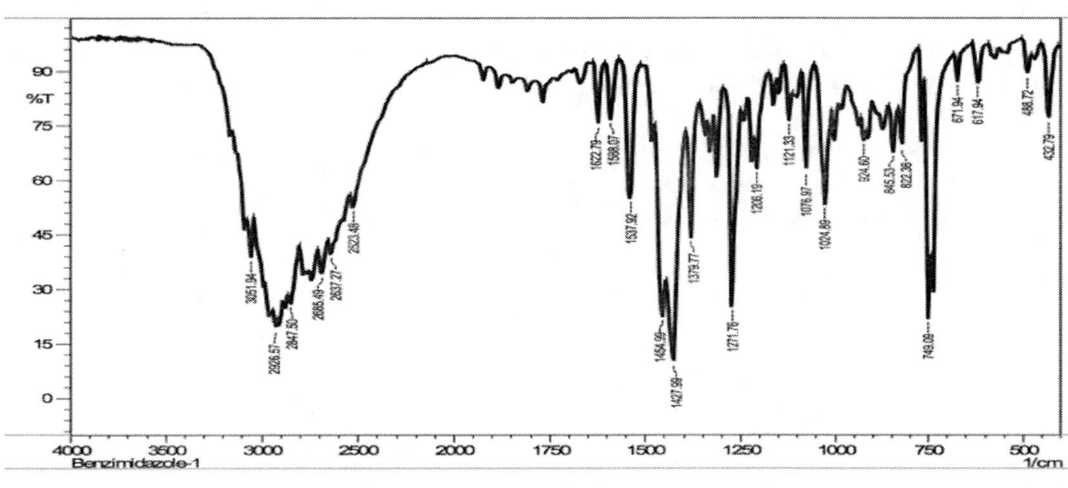

Figure 5 IR spectra of synthesized product

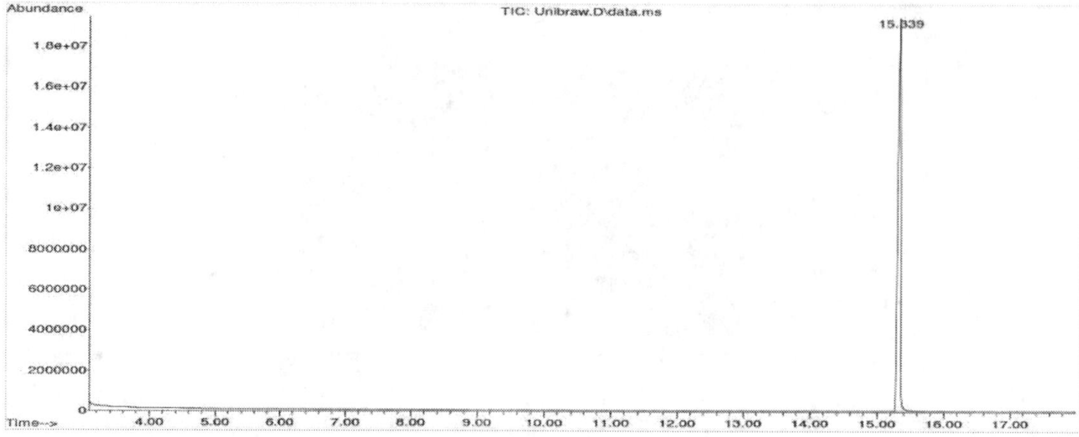

Figure 6 GCMS spectra of synthesized product

The comparative data of the synthesized compounds are provided graph in Fig. 7. The result shows the time required to react the whole compound using microwave irradiation is 30 to 70 minutes in the presence dichloromethane and methanol. The yield of benzimidazole derivative by dichloromethane solvent increased significantly and is greater than methanol solvent. The time optimum of microwave-assisted synthesis using dichloromethane solvent at 60 minute with yield 19.23%. When the reaction conducted in the presence of methanol at 50 min gave 18.59%.

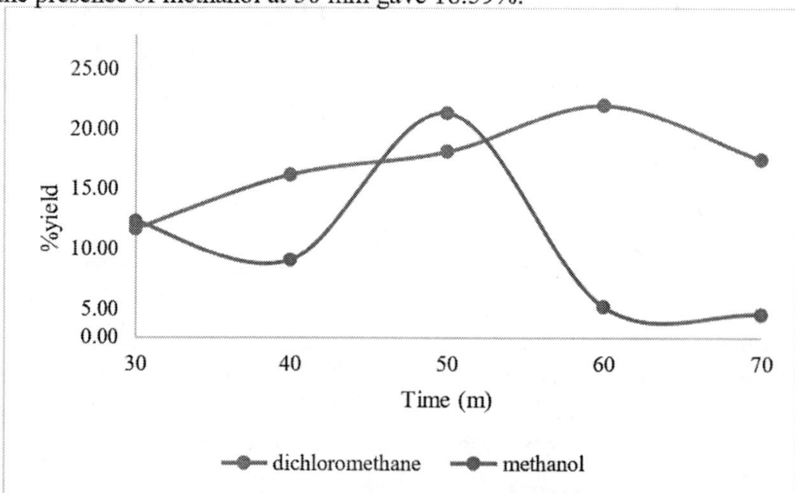

Figure 7 Variation of solvent in the synthesis of benzimidazole derivative

In addition, the time optimum of microwave-assisted synthesis using dichloromethane solvent at 60 minute is used at variations of mole ratio of citronellal to 1,2-phenylenediamine, respectively 1:1, 1:1.5, and 1:2 shown in Table 1. The optimum reaction time of microwave-assisted synthesis using dichloromethane solvent at mole ratio 1:2 with yield 42.19% shown in Table 1. The number of mole ratio starting material will affect the resulting product, the more moles of 1,2-phenylenediamine increasing the amount of product with a fixed mole of kaffir lime oil.

Table 1. Optimization of the synthesis of benzimidazole derivative from citronellal [a]

Entry	Mole ratio		CH_2Cl_2 (mL)	Time (minutes)	Yield (%)
	Kaffir lime Oil	1,2-phenylene diamine			
1	1	1	-	60	NR[b]
2	1	1	5.0	60	10.94
3	1	1.5	5.0	60	16.41
4	1	2	5.0	60	42.19

[a] Reaction conditions: using microwave assist on the reflux process (frequency 2450 MHz, wattage 1200W, cavity temperature 120 ^{0}C)
[b] No crystal was formed

4. Conclusion
The optimum yield of benzimidazole derivative which was synthesis by condensation between citronellal and 1,2-phenylenediamine under microwave irradiation is 42.19% at 60 minutes and 1:2 mole ratio using dichloromethane solvent. Further studies have to be carried out to test the antibacterial activity of benzimidazole derivative synthesized from citronellal.

Acknowledgments
The author gave a gratitude to Directorate General of Higher Education for financial support and Essential Oil's Institute, Brawijaya University for providing all necessary facilities to conduct the experiments.

References
[1] Goswami 2012 *IJSID* **2** pp 121-136
[2] Manna K., Aggarwal Y 2009 *Bioorg. Med. Chem. Lett.* **19** pp 2688-2692
[3] Ramanpreet 2011 *IJRPC* **1**
[4] David, W., Zhijan, K., & Christine, C 2004 *Molecules* **9** pp 158-163
[5] Pavel Kopel, Dorota Wawrzak , Vratislav Langer, Kristyna Cihalova, Dagmar Chudobova, Radek Vesely, Vojtech Adam and Rene Kizek 2015 *Molecules* **20** pp 10360-10376
[6] Shukla 2012 *IJPSR* **3** pp 922-927
[7] Tupe AP, Pawar PY, Mane BY, and Magar SD 2013 *RJPBCS* **4** pp 928
[8] Sanahanbi N. and T. Sivakumar 2013 *CODEN(USA) : AJBPAD* **4**
[9] Marriappan G 2011 *Indian Journal of Chemistry* **50B** pp.1216-1219.
[10] Sandeep Waghulde1, Ganesh Lonsane4, Varsha Borwandkar3, Sachin Laddha2, Synthesis and anti-inflammatory activity of novel 2-(Substituted alkyl or aryl pyridynyl) benzimidazole derivatives.
[11] K F Ansari and C Lal 2009 *J. Chem. Sci* **121** pp. 1017–1025
[12] Hamdan S. Al-Ebaisat 2011 *J. Appl. Sci. Environ. Manage* **15** pp 451 – 454
[13] Chimirri A et al *Antiviral Chemistry & Chemotherapy* **10** pp 211–217.
[14] M.S. Shingare, D.V. Mane, D.B. Shinde, S.N. Thore and S.B. Bhawsar 1996 *Asian Journel of Chemistry* **8** pp 225-228
[15] K Sreena 2009 *HYGEIA* **1**
[16] Ramanpreet 2011 *IJRPC* **1**
[17] B Cakir, E Yildirim, T Ercanli, K Erol, MF Sahin 1999 *Farmaco* **54** pp 842–845.
[18] Kalyankar TM, Pekamwar SS, Wadher SJ, Tiprale PS and Shinde GH 2012 *International Journal of Chemical and Pharmaceutical Sciences* **3**
[19] Kapuriya K., Ganure A., Davda S., Kitawala M., Topiya H 2013 *UJP* **2** pp 57-62

[20] Patil, A., Ganguly, S., Surana, S *Rasayan Journal of Chemistry*, 1 pp 447-460
[21] Reddy BA 2010 *JPRHS* **2** pp 103-113
[22] Ceballos, L.; Virkel, G.; Elissondo, C.; Canton, C.; Canevari, J.; Murno, G.; Denegri, G.; Lanusse, C.; Alvarez, L 2013 *Acta Trop* **127** pp 216–225.
[23] Perez-Villanueva, J.; Hernandez-Campos, A.; Yepez-Mulia, L.; Mendez-Cuesta, C.; Mendez-Lucio, O.; Hernandez-Luis, F.; Castillo, R 2013 *Bioorg. Med. Chem. Lett* **23** pp 4221–4224
[24] Roth T, Morningstar ML, Boyer PL, Hughes SH, Buckheit RW, Michejda CJ 1997 *J Med Chem* **40** pp 4199-4207
[25] Shatha Ibrahim Alaqeel *Journal of Saudi Chemical Society* **21** pp 229–237
[26] Shah, J.J and Krishnapriya Mohanraj 2014 *Indian J Pharm Sci* **76** pp 46-53.
[27] A. Saberi 2015 *Iran J. Sci. Technol.* **39A1** pp 7-10.
[28] Pankaj S., T. Srinivasa R., Nuggula P. K., Kishna R. S., Suresh K. B., Nagula S 2017 *Euro. J. Med. Chem.* **138** pp 234-245

IOP Conf. Series: Materials Science and Engineering **299** (2018) 012077 doi:10.1088/1757-899X/299/1/012077

A Green Synthesis of Chalcones As an Antioxidant and Anticancer

Elfi Susanti VH, Widiastuti Agustina Eko Setyowati

Department of Chemistry FKIP Universitas Sebelas Maret, Surakarta

Email: elfisusantivh@staff.uns.ac.id

Abstract. Three chalcones (4'-amino-4-methoxy chalcone, 4'-amino-3,4-dimethoxy chalcone and 4'-amino-3,4,5-trimethoxy chalcone) has been synthesized by a green chemistry approach using grinding technique. Antioxidant activity of the chalcones were assessed using 1,1-biphenyl-2-picrylhydrazyl (DPPH) radical scavenging method. Cytotoxicity of chalcones sythesized was evaluated using a tetrazolium (MTT) colorimetric assay against cervical cancer cell line, HeLa. The antioxidant activity test showed that 4'-amino-4-methoxy chalcone had a stronger activity than the 4'-amino-3,4-dimethoxy chalcone and 4'-amino-3,4,5-trimethoxy chalcone, respectively with IC_{50} 58.85, 64.79 and 210.3 μg/mL. These results indicate that there is a relationship between the structure of chalcone with antioxidant activity, the more methoxy groups in the ring B of the chalcone, antioxidant activity is getting smaller. The chalcone synthesized showed cytotoxicity against HeLa cell line with IC_{50} value of 31.75, 36.65, 49.04 μg/mL, respectively. It was observed that the highest cytotoxic activity was found at 4'-amino-4-methoxy chalcone (IC_{50} 31.75 μg/mL). Lower activity was showed by 4'-amino-3,4,5-trimethoxy chalcone with IC_{50} value of 49,04 μg/mL. There is a relationship cytotoxicity with chalcone structure, the more the number of methoxy groups in ring B chalcone, will decrease the activity of cytotoxicity.

1. Introduction

Chalcone (1,3-diphenyl propenon) is known to have a variety of pharmacological activities, such as anticancer [1], antiproliferative [2], antimicrobial [3] and [4], insecticides [5] and antioxidants [6. This compound is the main precursor in the synthesis of a variety of heterocyclic compounds, therefore it is very important to develop the latest strategies in the synthesis of chalcone.

Chalcone can be synthesized by Claisen-Schmidt condensation, the condensation reaction between aromatic aldehydes and ketones to form aromatic ketone α, β-unsaturated. Solvent in the synthesis reaction chalcone derivatives are hazardous to health and not environmentally friendly. One of the principles of green synthesis is the removal of solvent in the synthesis process, can be made by grinding technique. Modification of synthesis of chalcone with grinding techniques ever conducted for the synthesis of chalcone of 2-acetyl-1-naphtol and benzaldehyde [7]. Synthesis using this technique requires no solvent, the reaction time is short (4-8 minutes) and high yield (84-95%). Grinding technique has also been used for the chalcone synthesis of cyclohexanone and benzaldehyde, yield 96-98% [8]. Susanti *et al.*, (2014) have synthesized hydroxy chalcone of hydroxy acetophenone with grinding techniques, giving 70-84% yield [9][10]. However, the grinding technique has never been used for the synthesis of amino chalcone derivative from methoxy benzaldehyde and amino acetophenone.

Kumar et al., (2008) have synthesized several chalcone derivatives by several methods, i.e reflux, ultrasonic irradiation and grinding techniques [16]. Kumar explained that the synthesis with grinding technique gives better results than using ultrasonic irradiation and reflux, yield 83-92%, 5-80% and 24-89%. Rateb and Zohdi (2009) have synthesized several chalcone derivatives by grinding technique using NaOH catalyst, with good results (yield 71-95%) [14]. Zangade et al. (2011) also have synthesized derivatives of 2-hydroxy chalcone with grinding technique using KOH catalyst, yield 85-94% [7]. Susanti, et al. (2014) have also been synthesized chalcone derivatives of 4-hydroxy acetophenone and veratraldehyde using grinding techniques and provide greater yield (70-84%) compared to the conventional chalcone synthesis (45-75%) [9].

Chalcone activity as anticancer and antioxidant is a topic of interest in recent years. Antioxidants are substances that protect cells from damage caused by free radicals. Some chalcones reported to have antioxidant activity include 2'-hydroxy-2,4'-dimethoxy chalcone and 4'-hydroxy chalcone [11], 2'-hydroxy chalcone, 2'-hydroxy-4-methoxy chalcone, 2'-hydroxy-4-chloro chalcone, 2'-hydroxy-2-chloro chalcone, 2'-hydroxy-2-bromo chalcone dan 2'-hydroxy-3,4,5-trimethoxy chalcone [6].

It has been widely reported that flavonoids have potential as positive and negative regulators so that potential as anticancer. One class of flavonoids that have the potential as an anticancer is chalcone and its derivatives [12]. Some chalcones reported to have anticancer activity include 2'-methyl chalcone, 4'-bromo-4,6-dimethoxy chalcone, 4'-phenoxy-2-methoxy-6-flouro chalcone, and 4'-methoxy-3-bromo-6-methoxy chalcone (Upendra, et al, 2014).

Chalcone pharmacological activity generally depends on the number and position of the hydroxy group, methoxy and other groups on the rings A and B, so that the research will be synthesized 3 chalcone derivatives that have variations of methoxy grup in ring B of chalcone. This Chalcone will be studied its activities as antioxidant and anticancer. Synthesis of amino chalcone derivatives is very important, as the scientific basis for the development of antioxidant and anticancer, and study the relationship the chalcone structure with antioxidant and anticancer activities.

2. Experimental

2.1. *General procedure for synthesis of chalcone*

The chalcone were synthesized as mentioned in the reference [9]. A mixture of 4-aminoacetophenones (5 mmol), benzaldehyde (10 mmol) and solid pallete of NaOH (20 mmol) was ground with a mortar and pestle for 15 minutes at room temperature. The obtained solid mixture was diluted with cold water, acidified with cold HCl (10%), and extracted with ether. The ether layer was washed with water, followed by drying over anhydrous Na_2SO_4 and the solvent was evaporated. The reaction mixture was purified by column chromatography on a silica gel column with hexane:ethyl acetat (6:4) as eluent, followed by recrystallization. Chalcone synthesized were characterization used by GC-MS.

2.2. *In vitro antioxidant activity assay*

The chalcone synthesized were assessed for antioxidant activity using 1,1-biphenyl-2-picrylhydrazyl (DPPH) radical scavenging method. The DPPH assay was based on the reported methods [10]. Samples were dissolved in methanol and prepared in various concentrations (0, 10, 30, 50 and 70 ppm). Each solution is put into a test tube. Into each tube was added 500 µL solution of 1 mM DPPH in methanol, refined to 5.0 mL, and then incubated at 37°C for 30 minutes. The solution was measured at λ 515 nm. Positive controls use is vitamin C.

2.3. *In vitro anticancer activity assay*

Cell treatment was performed by adding each of the 100 ul-cell suspensions with a density of 2X104 / 20,000 cells / well (at 96 well plates). Then let stand for 1-2 hours. After that add 100 µl sample with various concentration dose. Incubation in CO_2 incubator for 24 hours (5% CO_2, 37°C, 98% moisture), after 24 hours seen under a microscope and photographed. Then discard the existing medium (in reverse way) on the tissue paper (for the attached cell). Add 100 ul MTT (5 mg MTT, 1 ml PFS, 9 ml medium RPMI complete /medium grower) at each well. Incubate for 4-6 hours. Add Stop Solution 100ul at each well, and incubation overnight. Read on ELISA Reader at 550 nm wavelength. The number of living cells is proportional to the absorbance. Calculate live cell percentage and IC_{50}.

3. Results And Discussion

Chalcone been synthesized through Claisen-Schmidt condensation reaction of acetophenone and benzaldehyde in the presence of NaOH. Three amino chalcone derivative was synthesized through Claisen-Schmidt condensation reaction between 4-amino acetophenone and benzaldehyde derivatives (Figure 1). Benzaldehyde derivatives used were 4-methoxybenzaldehyde, 3,4-dimethoxybenzaldehyde, and 3,4,5-trimethoxybenzaldehyde.

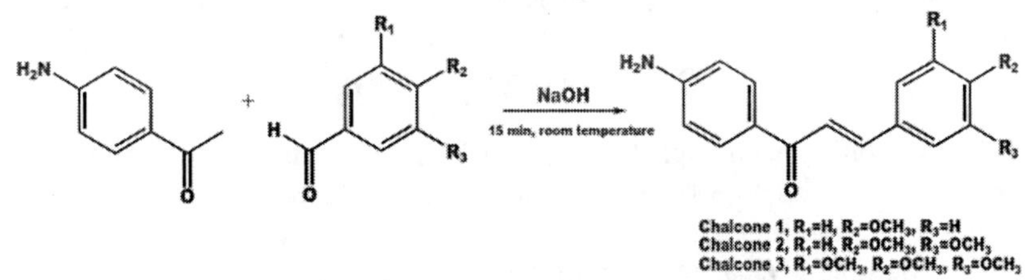

Figure 1. Synthesis Scheme of chalcone derivatives

The compounds synthesized in the form of giving a yellow orange color with Mg/HCl, showed positive flavonoids. The yield of the compounds 31.2 to 74.4% (Table 1). Structure of the synthesized compounds were confirmed by GC-MS.

Table 1. The yield of chalcone synthesized

Compounds	Yield (%)
Chalcone **1** (4'-amino-4-methoxy chalcone)	54.4
Chalcone **2** (4'-amino-3,4-dimethoxy chalcone)	74.4
Chalcone **3** (4'-amino-3,4,5-trimethoxy chalcone)	31.2

Identification of chalcone with GC-MS gave chromatogram and mass spectra. GC chromatogram for chalcone **1** showed 9 peaks, the most dominant being peak 8 at retention time (tR) 29.09 min with a relative purity of 71.4%. The peak was indicated as chalcone **1**, evidenced by the mass spectra (Figure 2) showed the presence of molecular ions (M+) as the peak base corresponding to the molecular weight of chalcone **1** (253). The fragmentation pattern of the Chalcone **1** shows molecular ions (M +) with loss CH_3 radicals gave m/z 238. The fragment m/z 161 with the loss of $C_6H_4NH_2$ radical, while the m/z 92 fragments release the radical $C_3H_9O_2$.

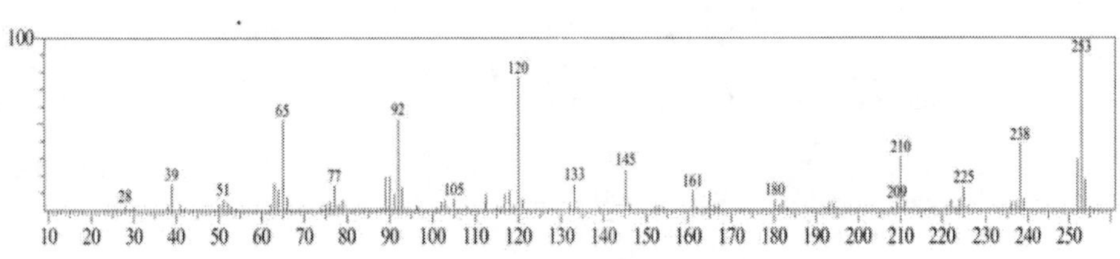

Figure 2. Mass spectra of chalcone **1**

IOP Conf. Series: Materials Science and Engineering **299** (2018) 012077 doi:10.1088/1757-899X/299/1/012077

GC chromatogram for chalcone **2** shows 5 peaks, with peak 5 as the dominant peak at retention time (tR) 30.6 min and relative purity 8.68%. The peak is indicated as chalcone **2**, shown from the mass spectra (Figure 3) of the presence of molecular ions (M +) as a peak base corresponding to the molecular weight of chalcone **2** (283). The fragmentation of the Chalcone **2** shows molecular ions (M +) with loss CH_3 radicals gave m/z 268. The fragment m/z 253 with the loss of OCH_3 radikal. The loss of $C_6H_4NH_2$ radical gave fragment m/z 191.

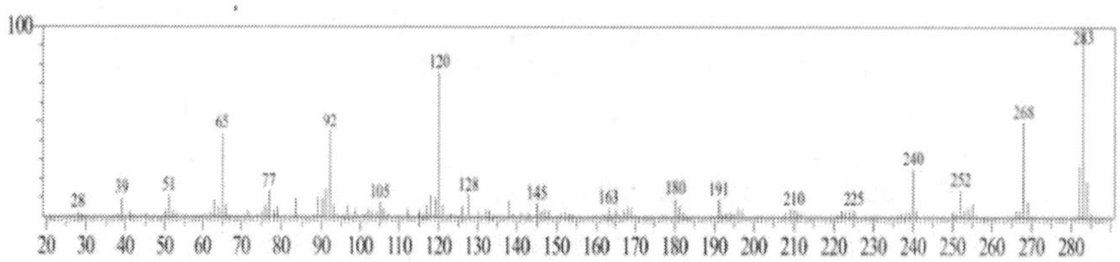

Figure 3. Mass spectra of chalcone **2**

Chalcone **3** gave a GC chromatogram (Figure 6), where peak 7 at retention time (tR) of 32.8 minutes and relative purity of 19.35% was suspected as chalcone 3, shown from the mass spectra (Figure 4) showing molecular ions (M +) At 313 as a peak base corresponding to the molecular weight of chalcone **3**. The dominant peak at retention time of 16.2 was the peak of the 3,4,5-trimethoxy benzaldehyde, showing the reactants have not been fully synthesized. The fragmentation of the Chalcone **3** shows molecular ions (M +) with loss CH_3 radicals gave m/z 298. The fragment m/z 282 with the loss of OCH_3 radikal.

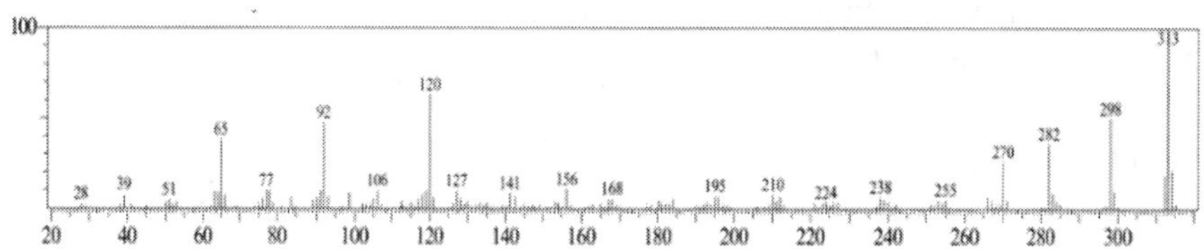

Figure 4. Mass spectra of chalcone **3**

The antioxidant activity of chalcone synthesized exhibited good antioxidant properties, with the strongest being observed in Chalcone **1 (Table 2)**. However, all the synthesized compounds were less potent than vitamin C as the reference. The potencies for the antioxidant activity of the test compounds to the reference drug are in the following order: Vitamin C > Chalcone **1** > Chalcone **2** > Chalcone **3.**

Table 2. The antioxidant activity of chalcone synthesized

No	Compounds	IC_{50} (µg/mL)	Categories
1	Chalcone **1** (4'-amino-4-methoxychalcone)	58.85	strong
2	Chalcone **2** (4'-amino-3,4-methoxychalcone)	64.79	strong
3	Chalcone **3** (4'-amino-3,4,5-trimethoxchalcone)	210.27	weak
4	Vitamin C	2.59	Very strong

The antioxidant activity of the three chalcone showed that chalcone **1** and **2** have strong antioxidant activity with IC_{50} values 58.85 and 64.79 mg/mL respectively, chalcone **3** has a weak antioxidant activity. These results also indicate that there is a relationship between the structure of chalcone with antioxidant activity. The more methoxy groups in the ring B of the chalcone, antioxidant activity is getting smaller. Chalcone **3** with three methoxy groups in ring B provides antioxidant activity decrease is very large.

The chalcone synthesized showed cytotoxicity against HeLa cell line with IC_{50} value of 31.75, 36.65, 49.04 μg/mL, respectively (Table 3). It was observed that the highest cytotoxic activity was found at chalcone **1** (IC_{50} 31,75 μg/mL). Lower activity was showed by Chalcone **3** with IC_{50} value of 49,04 μg/mL. There is a relationship cytotoxicity with chalcone structure, the more the number of methoxy groups in ring B chalcone, will decrease the activity of cytotoxicity.

Table 3. Cytotoxic Activity of Chalcone synthesized on HeLa Cell

No	Compounds	IC_{50} (μg/mL)
1	Chalcone **1** (4'-amino-4-methoxy chalcone)	31.75
2	Chalcone **2** (4'-amino-3,4-methoxy chalcone)	36.65
3	Chalcone **3** (4'-amino-3,4,5-trimethox chalcone)	49.04

4. Conclusion

In summary, a series of novel chalcones have been synthesized by grinding technique. Their in vitro antioxidant and cytotoxic activities against HeLa cell line were also reported. On this basis, the SAR analysis and potential antioxidant and anticancer antitumor activities of these chalcone are under investigation. 4'-amino-4-methoxychalcone with a substitution of methoxy at position 4 at ring B of chalcone were the most effective antioxidant and cytotoxic compounds. These results may be helpful for the design of future antioxidant and anticancer reagents, and offer potential application in the discovery of anticancer drugs.

References

[1] Xia Y, Yang Z Y, Xia P, Bastow K F, Nakanishi Y and Lee K H 2000 *Bioorg. and Med. Chem. Lett.* **10**, 699-701
[2] Patil C B, Mahajan S K and Katti S A 2009 *J. Pharm. Sci. & Res.* **1**(3), 11-22
[3] Mandge S, Singh H P, Gupta D and Moorthy H R 2007 *Trend Applied Sci. Res.* **2**, 52-56
[4] Prasad, Y. R, Lakshmana A R and Rambabu R, 2008 *E-J. Chem.* **5**(3), 461-466
[5] Nalwar Y S, Sayyed M A, Mokle S S, Zanwar P R and Vibhute Y B 2009 *World J. Chem* **4**(2), 123-126
[6] Belsare D P, Pal S C, Kazi A A, Kankate R S and Vanjari S S 2010 *Int. J. ChemTech. Res.* **2**(2), 1080-1089
[7] Zangade S, Mokle S, Vibhute A and Vibhute Y 2011 *Chem. Sci. J* **13**, 1-6
[8] Rahman M A F M, Ali R, Jahng Y and Kadi A A 2012 *Molecules* **17** 571-583
[9] Susanti E V H, Matsjeh S, Wahyuningsih, T D, Mustofa and Redjeki T 2014 *Indo. J. Chem* **4** (2) 174 – 178
[10] Susanti E V H, Matsjeh S, Wahyuningsih T D, Mustofa and Redjeki T 2014 *Indo. J. Pharm.* **25** 17–24
[11] Oyedapo O A, Adewunmi C O, Iwalewa E O and Makanju V O 2008 *J. Biol. Sci.* **8**(1), 131-136
[12] Nijveldt R J, Nood E, Hoorn D E C, Boelens P G, Norren K and Leeuwen P A M 2001 *Am. J. Clin. Nutr* **74** 418-425
[13] Rateb N M and Zohdi H F 2009 *Synthesic Communications* **39** 2789-2794

[14] Zangade, S., Mokle, S., Vibhute, A., and Vibhute, Y. 2011 *Chem. Sci. J* **13** 1-6.
[15] Kumar, S., Lamba, M.S., & Makrandi, J.K. Green Chem. Lett. and Rev 1**(2)**, 123-125 2008

IOP Conf. Series: Materials Science and Engineering **299** (2018) 012078 doi:10.1088/1757-899X/299/1/012078

Green Chemistry: Strategy in Essential Oils Sustainability by Development of Insecticide Using Docking Method

Warsito[1, 2,*], **EP Utomo**[1, 2], **SM Ulfa**[1], **BN Kholila**[1], **P Nindyasiwi**[1]
[1] Department of Chemistry, Faculty of Science, Brawijaya University, Jl Veteran Malang, 65145, Indonesia
[2] Essential Oil's Institute, Brawijaya University, Jl Veteran Malang 65145, Indonesia

E-mail warsitoub88@yahoo.com

Abstract. Sustainable agricultural applications in green chemistry was associated with the development of insecticide production based on secondary metabolites, such as essential oils. This research used In Silico modeling for insecticide formulation based on essential oils. The insecticidal formula was made on the basis of the Ki value of multiple docking results between the major components of essential oils as ligand with *Spodotera litura* receptor (2DJC) studied using Autodock Tools software. Insecticide formula activity test was done by contact method of toxic and leaf contact with essential oils concentration at level 0% - 1%. The results of the in silico study showed that the inhibition constants (Ki) of citronellal and anethol ligands combination were 1.6 mM however of citronellal and eugenol as ligands were 1.75 mM and formulated rasio (v/v), respectively 5 : 1 and 4 : 1. In addition, in vitro activity of insecticide formula with the ratio of 5: 1 possess LC_{50} value 0.10% (toxic contact) and 0.35% (leaf contact). While the formula with a ratio of 4: 1 possess LC50 value 0.05% (toxic contacts) and 0.31% (leaf contact).

1. Introduction

The green chemistry revolution is providing an number of challenges to those who practice chemistry in industry, education and research. With these challenges however, there are an equal number of opportunities to discover and apply new chemistry, to improve the economics of chemical manufacturing and to enhance the much-tarnished image of chemistry.

The challenge for chemists and others is to develop new products, processes and services that achieve the societal, economic and environmental benefits that are now required, as well as in agriculture to make natural insecticides. This requires a new approach which sets out to reduce the materials and energy intensity of chemical processes and products, minimise or eliminate the dispersion of harmful chemicals in the environment [1].

Many areas of science and engineering have progressed in terms of rapid data generation, data analysis, building mathematical models, and computer simulations. In the biomolecular field, in silico research is a form of computational modeling to study biological aspects as well as interactions at the molecular level [2]. Docking is a computational and modeling method used to determine the interaction between ligands with a receptor or protein [3]. *AutoDock* is a docking program that has a flexible docking technique and a grid-based quick method for energy evaluation [4]. This computational method uses the Iterated Local Search (ILS) algorithm [5]. The usefulness of the docking program has evolved in areas such as the pharmaceutical field in the development of new drug discovery and design, the industrial field in the manufacture of perfumes and agriculture for example in the manufacture of pesticides. In the biomolecular field by knowing the interaction on the desired compound we can make a natural pesticide formula with natural base material ie essential oil. Where essential oils also have the opportunity to be developed into other derivative products such as pesticides. The use of docking molecule modeling as an effort to make natural pesticide formulations can reduce the conventional pesticide manufacture conventionally, the preparation of a series of formulation with variations of the bioactive component, making it more time efficient and cost.

Essential oil obtained from citronella oil has the main components of citronellal (36,19%), geraniol (32,82%) and citronellol (11,37%). The main components of citronella have phytotoxic and antimicrobial properties that can be used as pest control [6]. The active components of oil basil are methyl chavicol (estragole), methyl eugenol, linalool, champer. Basilicum oil (*Oscimum basilicum L.*) contains an insecticidal bioactive element [7]. In Chiou Ling Chang's study, et al (2009) basilicum oil (tested on three species for insecticidal activity showed a steep dose-response relationship. Deadly periods for 90% mortality (LT_{90}) [8]. While the main component of clove oil is eugenol (80-90%) which is as antioxidant and antimicrobial. The antimicrobial mechanisms of eugenol compounds include disrupting cell membrane function, inactivating enzymes, inhibiting chitin synthesis, nucleic acids and proteins and inhibiting energy production [9]. Hummelbrunner and Murray (2001) studies reported that eugenol and citronellal compounds are toxic to grayak caterpillar *S. litura* with each LD_{50} being 2.5-157.6 mg / insect and 66.0-111.2 mg / insect [10]. Grayak caterpillars are pests that attack crops such as tomatoes, chilli and tobacco, causing the leaves to become perforated with 20-40% damage to the plants, so that with the manufacture of pesticides from essential oils can help reduce damage to plants.

2. Materials and Methods

2.1 Materials

The materials used in the study in silico are the 3D receptor structure of the grayak caterpillar growth protein (S. litura) downloaded from www.rcsb.org with the 2DJC pdb access code. The ligands used are ligands of the main constituent components of citronella oil (citronellal, geraniol, citronellol), basil oil (anethole) and clove oil (eugenol) with 3D structure downloaded from www.chemspider.com. While the ingredients used in research in vitro are citronella oil (CO), basil oil (BO), clove oil (CLO), acetone, tween 80 and aquades. Insect test used is grayak caterpillar (S. litura) instar 2.

2.2 Methods

2.2.1 Plant Refinery Basil.

The process of distillation begins by inserting water as much as 1- 2 liters into the kettle. Then put chestnut on water and basil leaves weighing 1 kg put on chestnut. Then close the kettle tightly so that no steam will be removed. After all the tools of the distillation process is done by heating. The distillation process is performed 4-5 hours.

2.2.2 Analysis of Essential Oil Components

Essential oil contains volatile compounds, so it can be analyzed by gas chromatography mass spectrometry (GC-MS) method. Gas Chromatography-Mass Spectrometry is a combination of two instruments, namely gas chromatography that serves as a component separator on the sample mixture and mass spectrometry that serves to detect compounds that have been separated on gas chromatography. CO, BO, CLO were characterized using GC-MS with GC-MS type QP2010S / shimadzu. This type of GC-MS has a column length of 30 meters with a column temperature of 60-300 ° C. Temperature ijector and temperature detector 280°C which has a gas flow rate of 90 mL/min with a helium gas.

2.2.3 Determination of Ki Value Using In Silico Method

Ligands with 3D structures used are ligands of major constituent components CO (geraniol, citronellal and citronellol), BO (anethole) and CLO (eugenol). Then the ligand is optimized on *Hyperchem* software. Ligands are stored in (.hin) format. Ligand that has been optimized, changed its extension from (.hin) to (.pdb) in *OpenBabel 2.3.2*. The macromolecule (receptor) used is *Spodoptera litura* with

2DJC access code with pdb extension. The protein macromolecules are separated from solvents and ligands using the *Discovery Studio Visualizer 3.5*. Then do molecular optimization with .pdb format. Macromolecules and ligands were docking using *AutoDock Tools* software to determine the molecular interactions between ligands and macromolecules to obtain Ki values and known formulas for the manufacture of insecticides. The docking process consists of 4 stages: ligand and protein receptor preparation (file format ".pdb" changed to ".pdbqt"), running Autogrid (change file format to ".gpf"), running AutoDock (file format changed to ".dpf ") and analysis of docking results. Docking process is done that is single and multiple ligand docking. Each ligand of the main constituent components CO (geraniol, citronellal and citronellol) and BO (anethole) were docking with 2DJC receptors as well as the main component ligands CO and CLO (eugenol) were docking with 2DJC receptors. Based on Ki value of docking result of each ligand of main component of CO, selected one ligand with lowest Ki value which then docking by multiple ligand docking with anetol ligand and eugenol interchangeably. Multiple ligand docking interaction modeling is done in two stages that produces Ki_1 and Ki_2 values. First, macromolecular docking with ligand 1 is generated Ki_1 value. Then the docking results are stored with the file format ".pdbqt" to redocking with ligand 2 so Ki_2 is generated.

2.2.4 Making Natural Insecticide Formulas

The preparation of the formulation solution is based on the results obtained from the in silico test. In silico test, Ki value is generated with unit mM (milli Molar) so that IC_{50} can be determined. CO and BO formulations as well as CO and CLO can be determined by multiplying the IC_{50} values by the molecular weight of each ligand with the following equation:

$$[M] + [L] \rightleftarrows [M-L] \quad (a.1)$$

$$Kd = \frac{[M][L]}{[ML]} \quad (a.2)$$

Jika [M] ⋙ dari [L], maka [ML] = [M]

Jadi $Kd = \frac{\cancel{[M]}[L]}{\cancel{[M]}} = [L]$

$Kd = [L]$ (Ki∽Kd)

$$Ki = \frac{IC_{50}}{1+\frac{LT}{Kd}} \quad (a.3)$$

$$Ki = \frac{IC_{50}}{1+\frac{Kd}{Kd}} \quad (a.4)$$

$$Ki = \frac{IC_{50}}{1+1} = \frac{IC_{50}}{2} \quad (a.5)$$

$$IC_{50} = 2\ Ki \quad (a.6)$$

According to equation a.6, then are citronellal concentration in CO $=IC_{50}.M_{citronellal,}$ anethole concentration in BO = $IC_{50}.M_{anethole}$ and eugenol concentration in CLO = $IC_{50}.M_{eugenol}$. If the value of IC50 and mass of molecule unrtuk each is included in the equation then obtained the ratio of oil in insectide formula, that is CO-BO (5: 1) and CO-CLO (4: 1). The volume composition of the formula is assumed in the research of Pinheiro *et al* (2013), with tween 80 (mL): oil (mL): acetone (mL) ratio is 1: 11600: 20000, so the comparison of natural insecticides is obtained. The natural insecticide formulation was prepared by dissolving CO in the BO (ratio 5: 1), acetone, tween 80. Then put into 100 mL poultry flask and added aquades up to 100 mL volume then shaken until homogeneous. The second formulation is made in the same way but uses CO and CLO (ratio 4: 1). In this study, the concentration interval based on insecticide formula above was used 0%, 0,0625%, 0,125%, 0,25%, 0,5% and 1% concentration.

2.2.5 Test of Natural Insecticide Activity

Activity test of natural pesticidal formula was carried out by contact method of toxic and leaf contact adopted from the research of Pinheiro, *et al* (2013) [10]. Leaf contact method test is method that done by way of grayak caterpillar instar 2 instilled into container of test (19 x 19 x 30 cm) which contain jatropha leaves ± 15 cm diameter previously sprayed natural insecticide as much as 1.5 mL per leaf side . While the toxic contact method is done by natural insecticides directly sprayed into the grayak caterpillar in a 250 mL glass of glass (diameter 8 cm, height 10 cm). Observation of grayak caterpillar death after 1 hour application for 24 hours.

3. Result and Discussion

3.1 Plant Refinery Basil

Basil oil obtained from the distillation of basil plant (*Ocimum basilicium L*) result of determination from Biology Department, Faculty of Science Brawijaya University. The distillation process is carried out by the method of steam-water distillation. Basil leaves are used as much as 1221 grams, distillation lasts until all the essential oil runs out of the marked by not dripping oil from the condenser. Essential oils obtained have a yield of 1.16% (w/w) then collected and stored in tightly sealed containers. The oil of basil plant distillation has a clear yellow color with a specific gravity of 0.952 g / mL.

3.2 Analysis of Essential Oils Composition

Based on the analysis using GC-MS it appears that in CO detected 30 peaks. The three main components of CO composition with the largest component of geraniol were 32.30%, citronellal 17.26% and citronellol 10.56% **(Table 1)**. Subsequently on the oil of basil was detected 8 peaks with the dominant peak of anethole compound 83,98% **(Table 2)**. The CLO showed that there were 8 peaks with 1 peak showing the highest content of clove oil, ie eugenol compound as much as 65.90% **(Table 3)**. The peaks described in the chromatogram can be determined by their constituent components based on the available mass spectra. The mass spectra was compared to patterns of fragmentation of similar component patterns in libraries based on WILLEY7.LIB library by ensuring SI values above 94. Simulation interaction of citronellal and anetol molecules with 2DJC receptors as well as citronellal and eugenol molecules with 2DJC receptors (*S. litura*) in silico.

Table 1. Major components of citronella oil

Peak	tR (min)	% Area	m/z	SI	Compound
8	12.589	17,26	154,140, 136, 121, 111, 95, 83, 69, 55, 41, 38	94	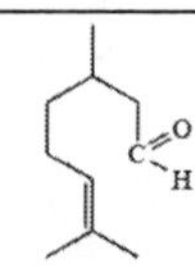 Citronellal
11	13.800	10,56	156, 138, 123,109, 95, 81, 69, 55, 41	96	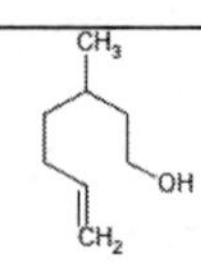 Citronellol
13	14.304	32,30	154, 136,123, 111, 93, 70,69, 55, 41	96	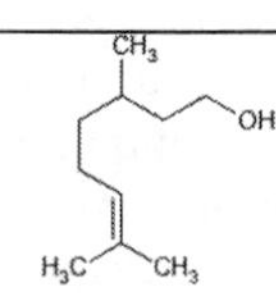 Geraniol

Table 2. Major component of basil oil

Peak	tR (min)	% Area	m/z	SI	Compound
8	13.351	83,98	148,147, 133,117, 105	96	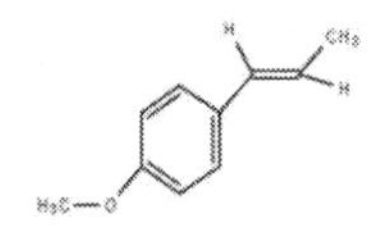 Anetol

Table 3. Major component of clove oil

Peak	tR (min)	% Area	m/z	SI	Compound
3	15,938	65,90	55, 77, 91, 103, 131, 149, 164	97	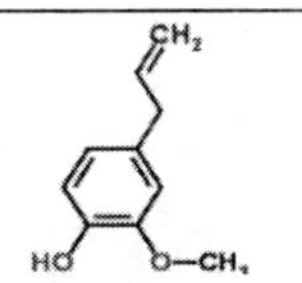 Eugenol

Geometric structure optimization using *HyperChem software* is done to obtain ligand conformation with the most stable geometry structure. The most stable conformation of the most stable structure of the ligand geometry is represented by several parameters, ie the lowest value of Ki indicating the ligand to interact with the active side of the strong receptor.

Table 4. Result optimized on *Hyperchem* and docking with *Autodock Tools* for each ligand

Essential Oil	Ligand	Log P	Surface Area (Å^2)	Hidration Energy (kcal/mol)	Ki Values (mM)
	Geraniol	2,46	444,73	-3,89	3,29
CLO	Citronellal	2,26	451,41	14,12	2,18
	Citronellol	2,75	434,70	-2,78	5,13
BO	Anethole	1,29	357,35	-2,87	1,18
CO	Eugenol	0,31	338,28	-7,58	1,14

Docking each ligand with the receptor protein that has been done will result in 10 models of ligant interaction conformation with macromolecules. From 10 kinds of conformation model of docking result obtained one of the most stable conformation model is the lowest price of Ki. By knowing the smallest value of Ki from each component of compilers CO, BO and CLO can be done multiple ligand docking is docking two ligands.

Multiple docking results between pairs of two ligand molecules from citronellal-anetol and citronellal-eugenol with receptors from the *S. litura* insect macromolecules are presented in Figure 1. Ki's value results from the interaction of these molecules are presented in Table 5.

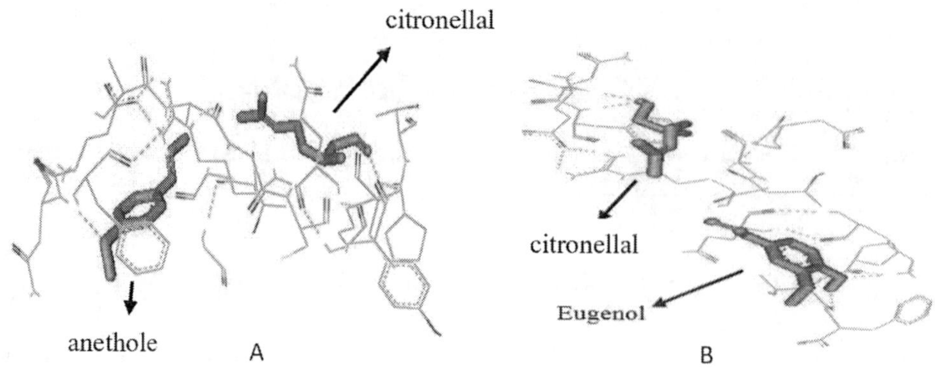

Figure 1. Multiple docking of two ligands
A. Citronellal-anethole B. Citronellal-eugenol

Table 5. Ki value of ligands multiple docking

Ligand	Ki_2 Value (mM)
Citronellal dan Anethole	1.6
Citronellal dan Eugenol	1.75

3.3 Insecticide Activity Test based on Essential Oils

The results of testing of natural insecticide activity showed that CO-BO and CO-CLO based insecticides caused mortality of *Spodoptera litura* larva. The larval mortality of *S. litura* is shown in **Figure 2** and **Figure 3.** Based on the image can be seen that at 0% concentration no mortality occurs in *S. litura*, this indicates that the solvent has no effect on mortality. So this 0% concentration can be used as a control for natural insecticides. In the graph of insecticidal concentration relationship (a) CO-BO (b) CO-CLO with present mortality *Spodotera litura* by toxic contact if we compare percent mortality of natural insecticide formula CO-CLO has a faster percent mortality. The same is true of leaf contact methods. From the mortality percentage of each test, it is known that LC_{50} values within 24 hours are based on probit analysis on toxic contact method and leaf contact. LC_{50} with leaf contact method on CO-BO natural insecticide formula that is 0.35% concentration and on natural insecticide formula of CO-CLO 0,308%. While LC_{50} with poison contact method on CO-BO natural insecticide formula that is 0,101% concentration and on natural insecticide formula CO-CLO 0,052%.

IOP Conf. Series: Materials Science and Engineering **299** (2018) 012078 doi:10.1088/1757-899X/299/1/012078

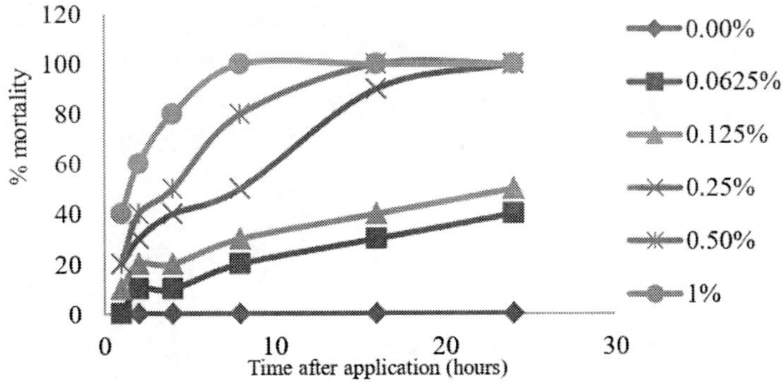

A

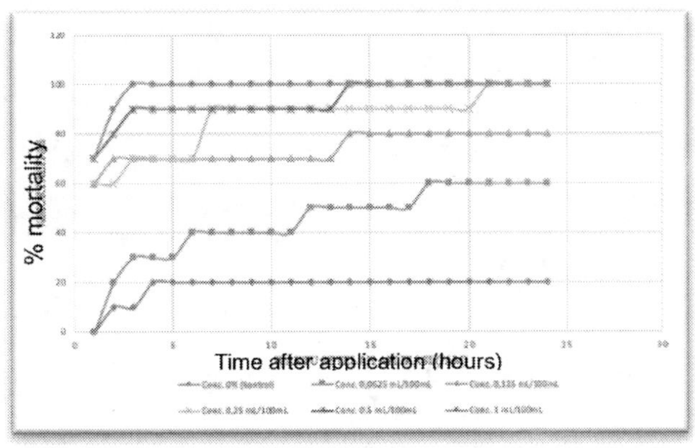

B

Figure 2. Time correlation after application with % mortality (Toxic Contact Method)
(a) CO-BO (b) CO-CLO

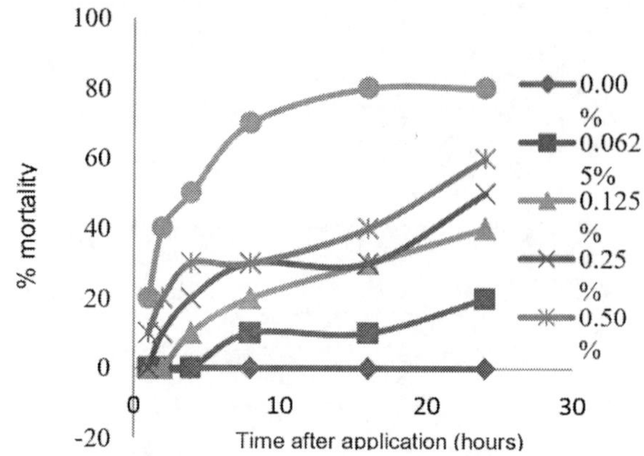

A

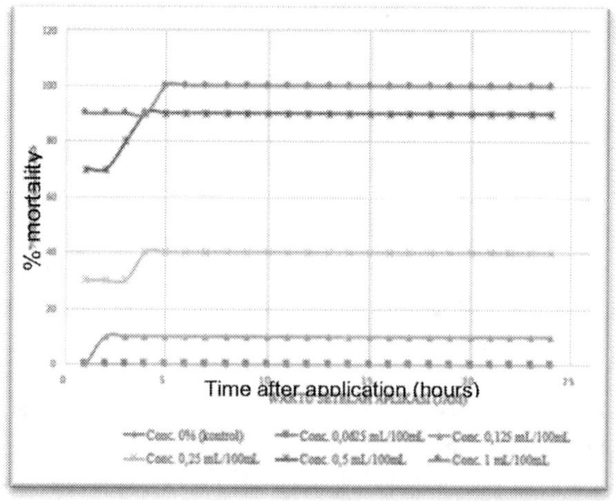

B

Figure 3. Time correlation after application with % mortality with% mortality (Leaf Contact Method) (a) CO-BO (b) CO-CLO

4.Results and Discussion

The results of the in silico study showed that the inhibition constants (Ki) of citronellal and anethol ligands combination were 1.6 mM however of citronellal and eugenol as ligands were 1.75 mM and formulated rasio (v/v), respectively 5 : 1 and 4 : 1. In addition, in vitro activity of insecticide formula with the ratio of 5: 1 possess LC_{50} value 0.10% (toxic contact) and 0.35% (leaf contact). While the formula with a ratio of 4: 1 possess LC50 value 0.05% (toxic contacts) and 0.31% (leaf contact).

5.Acknowledgments

On this occasion, thank you to : Department of Chemistry, Faculty of Science, Brawijaya University and Essential Oil's Institute, Brawijaya University for providing all necessary facilities to conduct the experiments.

References

[1] J. H. Clark 1999 *Green chemistry : challenges and opportunities* 1–8
[2] B. Palsson, The challenges of in silico biology Moving from a reductionist paradigm to one

that views cells as systems will necessitate 2000 **vol. 18** no. November, pp. 1147–1150.
[3] S. B. Leach ar, P. C. 2006 *J. Med Chem* **vol. 20.**
[4] W. Luo 2014 *Med. Chem. (Los. Angeles)* **vol. 4** no. 9.
[5] A. J. Morris, G.M., Goodsell, D. S., Halliday, R.S., Huey, R., Hart, W. E., Belew, R. K. and Olson 1998 *J. Comput. Chem.* **vol 19** pp 1639-1662.
[6] B. L. S. Leite 2011 *African J. Biotechnol* **vol. 10** no. 42 pp. 8312–8319.
[7] H. P. T. Deshpande, R. S. 1977 *Pestic.* **11** 11Ð12.
[8] C. L. Chang and I. L. K. Y. U. Cho 2009 Insecticidal Activity of Basil Oil , trans -Anethole , Estragole , and Linalool to Adult Fruit Flies of Ceratitis capitata , Bactrocera dorsalis , and Bactrocera cucurbitae.
[9] O. A. and A. Oyedemi, O. O., Adewusi, E. A., Aiyegoro 2011 *Asian Pac. J. Trop. Biomed.,* **vol. 1** no. , pp. 353–358.
[10] M. B. Hummelbrunner, L. A and Isman 2001 *J. Agric. Food Chem.* **vol. 49** no. 2 pp. 715–720.
[11] P. F. Pinheiro, V. T. De Queiroz, V. M. Rondelli, A. V. Costa, T. D. P. Marcelino, and D. Pratissoli 2013 Insecticidal Activity Of Citronella Grass Essential Oil On Frankliniella schultzei And Myzus persicae Atividade inseticida do óleo essencial de capim-citronela sobre Frankliniella schultzei e Myzus persicae pp. 138–144.

Study on Esterification Reaction of Starch Isolated from Cassava (*Manihot esculeta*) with Acetic Acid and Isopropyl Myrtistate Using Ultrasonicator

Helda Wika Amini* [1,2], **Masruri** [2], **Siti Mariyah Ulfa** [2]

[1] Department of Pharmacy, Sekolah Tinggi Ilmu Kesehatan Karya Putra Bangsa, Jl. Raya Tulungagung-Blitar KM 4, Tulungagung, Indonesia
[2] Depertment of Chemistry, Faculty of Mathematics and Natural Sciences, Brawijaya University, Jl.Veteran 65145, Malang, Indonesia

Email: heldawikaamini@gmail.com

Abstract. Cassava starch is a polysaccharide consists of amylose and amylopectin. This research was purposed to modify the starch isolated from local cassava (*Manihot esculenta*). Modification was undertaken to study the esterification reaction of cassava starch with acetic acid and with isopropyl myristate. Moreover, morphology observation was also conducted both for original starch and its modification yields. It was found that cassava's starch was isolated in 16.4% yield as a white powder. Esterification on the starch provided DS value 0.549 for ratio 1:2 of starch-acetic acid. It gave DS value 0.356 for ratio 1:3 of starch-isopropyl myristate. Treatment by ultrasonication from 0 to 60 minutes was significantly improved the DS value to 0.549 for starch-acetic acid. But it gave DS value to 0.413 for 30 minute ultrasonication of starch-isopropyl myristate. In addition, morphology of the starch observed by microscope gave different features with starch ester acetate and starch ester myristate. The original starch consists of granules, but starch ester acetate indicates a non-granules shape (amorf solid). Moreover for starch ester myristate shows a rather bigger size of granules, and all of the granules afforded were round and oval.

1. Introduction

Cassava plant is one of Indonesia's main agricultural crops [1,2]. In 2015, Indonesia produced 21,801,415 tons of cassava and production in East Java reached up to 3,161,573 tons [3]. Fresh cassava is consists of 35% starch [1] even dried cassava can contain up to 80% of starch [4]. Starch is a polysaccharide that has a high molecular weight and is composed of a polymer of glucose that binds to glycoside bonds [5].

Modification of starch is a process to change the structural of the amylose and amylopectin. Modification or functionalization is to produce better properties of starch. One of procedure for functionalization of starch is esterification [6]. Regiec and coworkers esterified potato starch with fatty acid of oleic acid called by starch ester oleate that could be a thickening agent owing to its hydrophilic and hydrophobic properties [7]. In 2012, Luo et al. [8] synthesized starch ester palmitate which improved the hydrophobicity performance exhibiting good applied prospect. In 2013 Putri et al. [9] reported the pre-gelatinization and phtalazation process of cassava starch resulting pre-gelatinized of cassava starch phthalate (PCSPh) that had a good characteristic as film-forming excipient for transdermal film for ketoprofen. Moreover, starch ester laurate was produced by esterification reaction between cassava starch and lauric acid. High fatty acid starch esters (chain length of fatty substituent ≥ C8), which have special hydrophobicity, thermo-plasticity, and biodegradability properties, are one kind of the most important starch derivatives [10]. Tang et al. [11] synthesized esterified cellulose nanocrystal (E-CNC) by ultrasonication method. Esterification reaction without ultrasonication had a low DS value (degree of substitution) of 0.22. Using ultrasonication method, the DS value of E-CNC increased to 0.46. This paper will be reported recent finding on the esterification reaction of the cassava starch using acetic acid and isopropyl myristate assisted by ultrasonication. Visual and microscopy morphology, infrared spectra profile, and their DS value will be reported as well.

2. Experimental

2.1. Chemicals and reagent.
Raw cassava was purchased from Dinoyo traditional market, Malang, East Java, Indonesia. Chemical reagents were used such as glacial acetic acid (Smart Lab), isopropyl mirystate (Bratachem), concentrated sulfonic acid (Merck), acetone (Smart Lab), and ethanol (LIPI).

2.2. Instrumentation
We used instrumentation such as infrared spectrophotometer (Shimadzu FTIR-8400S, sample was analyzed using KBr plate (5 mg sample + 95 mg KBr)), ultrasonicator (Branson 2210), microscope (Olympus BX51 with camera DP20), vacuum pump, a set of glasses, stirrer, a sieve with 100 mesh size, oven, grater, Buchner funnel, and analytical balance (Ohaus Pioneer PA214).

2.3. Isolation starch from cassava
The fresh cassava was cleaned and grated into a bowl of water. The grated cassava was squeezed out to get starchy water. It needs to stand starchy water for around 30 minutes. The starch which has white color will settle at the bottom of bowl. The water was separated with the starch. Then the starch was dried and sifted with 100 mesh size. Then we calculated the starch percentage (weight of starch is divided by weight of fresh cassava and multiplied by 100%).

2.4. Esterification reaction between starch-acetic acid and starch-isopropyl myristate
The esterification procedure was conducted based on the research of Tang et al. [11] and Xu et al. [12] with some modification which was published in 2014 [13]. Esterification reaction was firstly conducted by ratio variation between starch and acetic acid (g/mL): 1:1, 1:2, 1:3, and 1:4. Then the mixture of starch-acetic acid was stirred for 60 minutes and then stood for 15 hours at room temperature. The mixture was added 1 drop of concentrated sulfuric acid as catalyst and stirred at 80° C for 60 minutes. The resulting suspension was kept at 69° C and stirred for 60 minutes followed by treatment in an ultrasonicator. After the reaction, the mixture was washed with cold water (10 mL) for 3 times. Then the result of esterification starch dried in an oven at a 55° C. The esterification reaction between starch and isopropyl myristate was conducted following procedure above. The solvent of ethanol:aceton (1:1) of 15 mL was used to wash the esterification yield between starch and isopropyl myristate. The esterification yields are starch ester acetate and starch ester myristate.

2.5. Esterification reaction with time variation of ultrasonication
The time variation of ultrasonication was conducted for ratio 1:2 between starch and acetic acid which has highest DS for 0, 30, 60, and 360 minutes. Moreover, we used ratio 1:3 between starch and isopropyl myristate with varying the time of ultrasonication. The results of esterification reaction are analyzed by infrared spectrophotometry. Based on research Garcia and Vidal [14], the DS value was determined by FTIR spectra with the formula:

$$DS = \frac{0.76\,x\,A_{CO}}{A_{OH}}$$

Where DS is the degree of substitution, ACO is the absorbance of carbonyl groups, and AOH is the absorbance of hydroxyl groups in the FTIR. Determination of the morphology of starch and esterification yields was performed by a microscope with magnification of 100x, 200x, and 400x.

3. Result and Discussion

3.1. Isolation of Cassava's Starch

Table 1 listed the isolation of starch from cassava. The cassava has different starch percentage. The highest starch percentage is at entry 3 which has 25.5% of starch at 500 gr of fresh cassava. The starch percentage is affected by age, location, climate and cassava farming area [25]. The average percentage of cassava starch is 16.4%.

Table 1. Isolation of starch from cassava

Entry	The weight of fresh cassava (g)	The weight of starch (g)	Starch percentage (%)
1	500	54.49	10.9
2	500	65.59	13.1
3	500	126.5	25.3

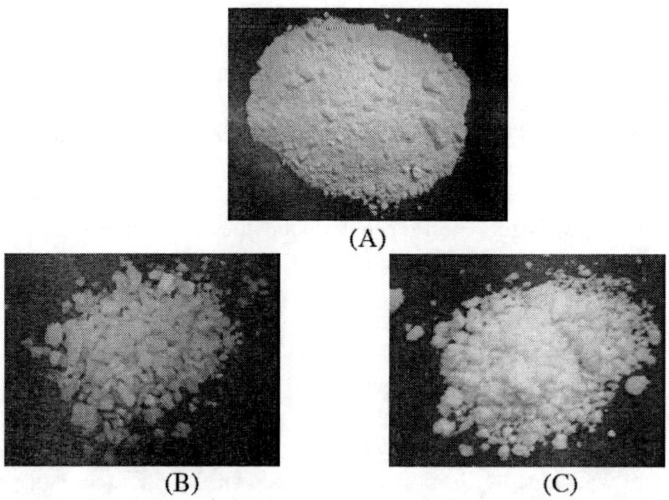

(A) (B) (C)

Figure 1 Starch (A) and esterification starch: starch ester acetate (B), starch ester myristate (C)

3.2. Esterification of Starch

Starch and esterification yields are presented in Figure 1. Figure 1.A shows the original starch forms white powder. Esterification reaction causes the characteristics of starch turns into solids (B) and white powder agglomerated (C). In addition, the success of starch esterification can be demonstrated by the changes of absorption area and absorption intensity between starch and starch esterification yields. Starch spectra (Figure 2 black line) at 3357, 2931, and 1155 cm^{-1} show the starch-containing O-H, C-H and C-O-C asymmetric groups, respectively. Esterification of starch produce new absorption at 1740 cm^{-1} which indicates the C=O ester group. In addition, it is observed that there is reduction of hydroxyl group peak at 3357 cm^{-1} region of esterification yields spectra in relative to that of its original starch. It is seen that the esterification of starch by acetic acid/isopropyl myristate was successful. The esterification reaction carried out in accordance Figure 3.

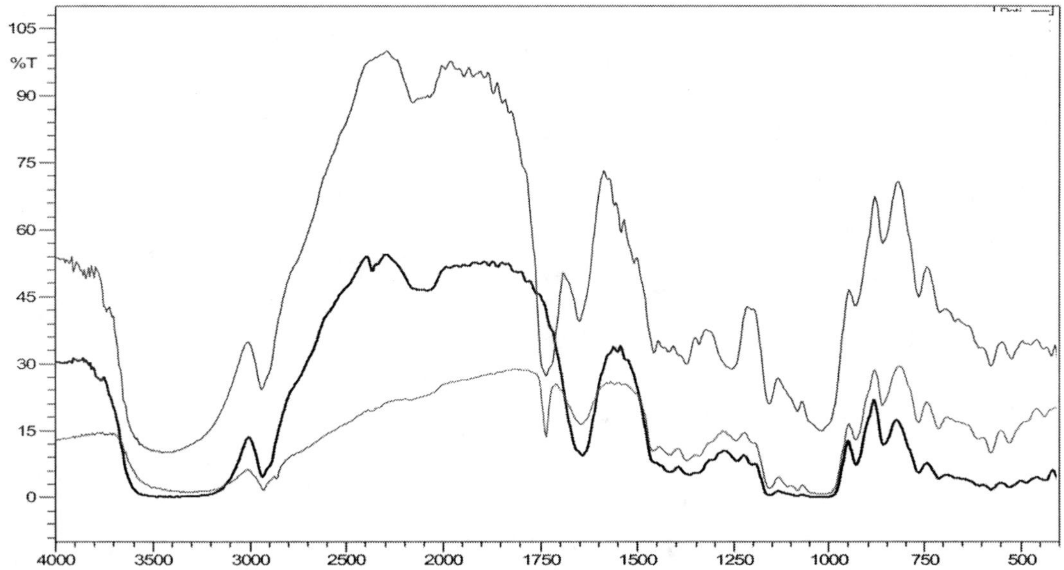

Figure 2 Infrared spectra of starch and esterification yields (black: native starch, blue: starch ester acetate, and red: starch ester myristate)

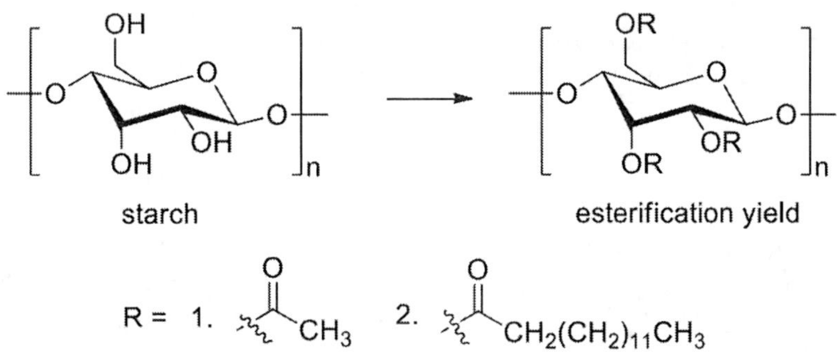

Figure 3 The esterification reaction of starch (esterification reaction between starch and acetic acid: (1) or isopropyl myristate (2))

3.3. Esterification reaction between starch-acetic acid and starch-isopropyl myristate

Figure 4 presented the DS value of esterification yields with ratio variation. The DS of starch ester acetate increase from ratio 1:1 to 1:2. However, the DS of starch ester acetate decrease from ratio 1:2 to 1:4. On the other hand, the DS of starch ester myristate slightly increase with increasing the ratio of isopropyl myristate acid from ratio 1:1 to 1:3 and turn to decrease for ratio 1:4. Theoretically, increasing the amount of reagent can increase the DS of starch because it will increase the rate of collision between reactant molecules and starch. Immobility or limited movement of the hydroxyl groups of starch also causes the esterification reaction depends on the availability of reagents in close to starch molecule [12]. However, starch acetate ester product declined in the ratio of 1:3 and 1:4, while the starch ester product myristate in a ratio of 1:4. It may be accused by equilibrium reactions that lead to backlash reaction.

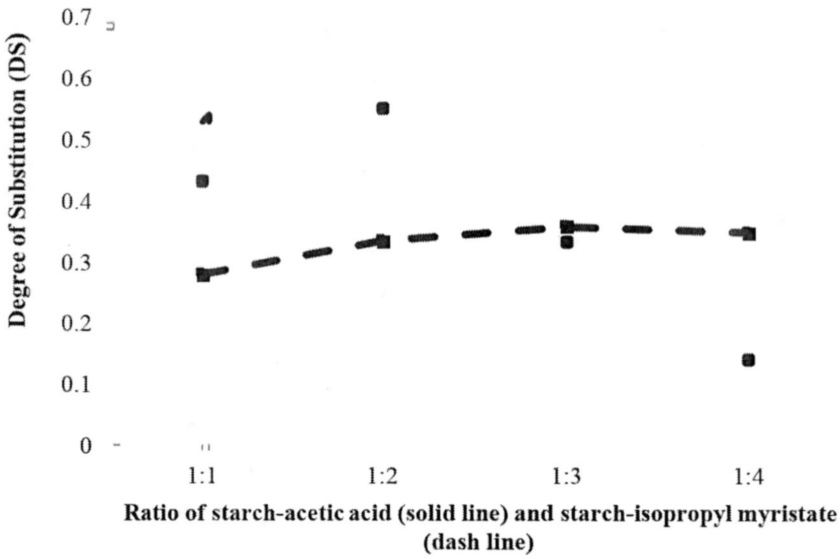

Figure 4 The DS of starch ester acetate and starch ester myristate with ratio variation between starch and acetic acid/isopropyl myristate

3.4. Esterification reaction with variation of ultrasonication time

Ratio 1: 2 for starch-acetic acid and 1: 3 for starch-isopropyl myristate was used to investigate the effect of variation of ultrasonication time. Figure 5 showed the correlation between ultrasonication time and DS value. The treatment without ultrasonication (0 min) yields lowest value of DS 0.462 for starch ester acetate and 0.278 for starch ester myristate. Ultrasonication treatment of esterification reaction generally increased the DS. It is seen that the highest DS value of starch ester acetate is at 60 minutes. Moreover, starch ester myristate has highest value of DS at 30 minutes. However, the increasing the ultrasonication time at 360 minutes for the reaction of the starch-acetic acid and after 30 minutes of reaction starch-isopropyl myristate give decreasing DS value. The longer ultrasonication time could damage to the interior of starch [11]. Thus, only few hydroxyl groups may be substituted by acetyl and myristyl groups.

3.5. Morphology of starch and esterification yields

Morphology of starch and esterification yields were observed by a microscope which were presented in Figure 6. Figure 6 shows the starch granules are composed of visible dots in black and white. This is because the granules have birefringence properties [15] that can reflect light. The starch granule (Figure 6.A) is round and oval. While starch esterification reaction products with acetic acid (starch ester acetate) composed of granulated and non-granular (amorphous solids) (Figure 6.B-C). Starch ester myristate granule is composed of round and oval shapes (Figure 6.D-F).

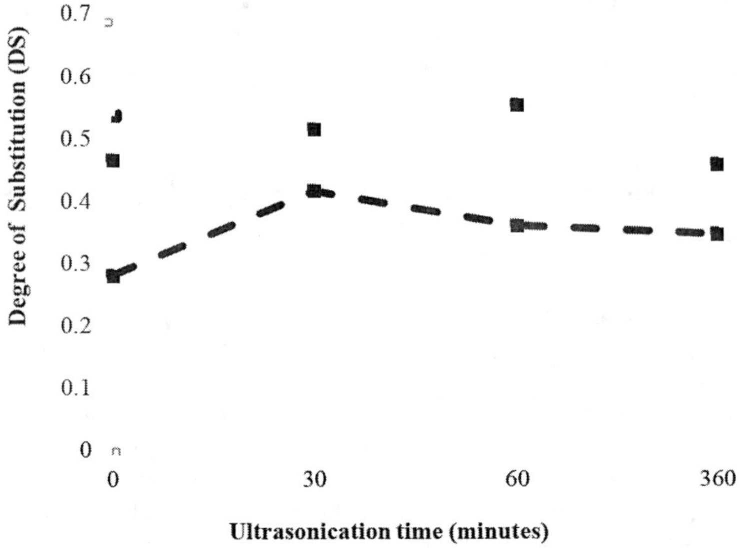

Figure 5 Correlation of ultrasonication time with the DS value (solid line: acetic acid modified starch, dash line: isopropyl myristate modified starch)

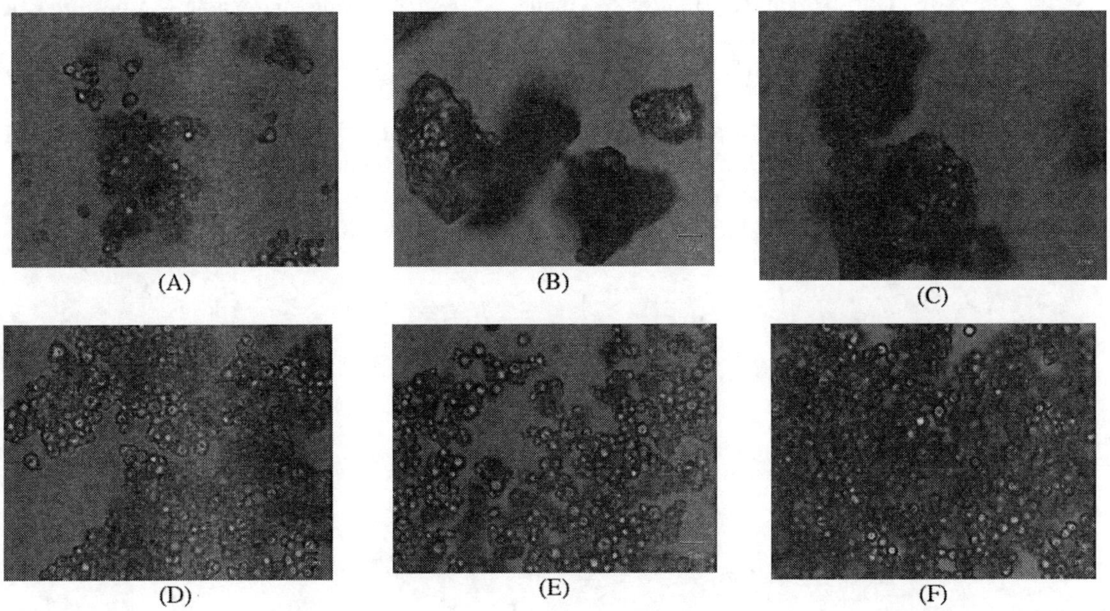

Figure 6 The image was captured with magnification of 400x on a sample of starch (A), starch ester acetate (B. ratio 1:2, C. without ultrasonication treatment), and starch ester myristate (D. ratio 1: 3, E. without ultrasonication treatment, F. 30 minutes of ultrasonication).

4. Conclusion

The esterification yields of starch are starch ester acetate and starch ester myristate. Ratio variation of reactants and ultrasonication time affect the DS value. The highest DS value for starch ester acetate is 0.549 at ratio 1:2 for 60 minutes of ultrasonication. On the other hand, the highest DS value of starch ester myristate is 0.413 at ratio 1:3 for 30 minutes of ultrasonication. Starch consists of granules with round and oval shapes. Starch ester acetate is mostly composed of nongranules (amorphous solid) while starch ester myristate composed of granules.

Acknowledgment

This work was supported by Organic Chemistry Laboratory from Chemistry Department, Faculty of Mathematic & Natural Sciences, Brawijaya University.

Reference

[1] Prabawati, S., N. Richana, Suismono Inovasi pengolahan singkong menigkatkan pendapatan dan diversifikasi pangan http://www.litbang.deptan.go.id/download/one/104/file/Manfaat-Singkong.pdf/ Accessed date 30 July 2016.
[2] Askar Daun singkong dan pemanfaatannya terutama sebagai pakan tambahan Balai Pertanian Ternak Bogor http://peternakan.litbang.deptan.go.id/fullteks/wartazoa/wazo51-5.pdf Accessed date 30 July 2016.
[3] Statistik (BPS) Produksi tanaman ubi kayu seluruh provinsi Badan Pusat Statistik Indonesia https://www.bps.go.id/linkTableDinamis/view/id/880/ Accessed date 30 July 2016.
[4] Onyenwoke and Simonyan Afr J Agric Res 2014 **9** 3853-3863.
[5] Yu, H., Cao, Y. Fang, Q., Liu, Z. BioResources 2015 **10** 3520-3530.
[6] Aberoumand, A. World J Dairy Food Sci 2011 **6** 115-124.
[7] Boruczkowska, H., T. Boruczkowski, W. Leszczyński, E. Tomaszewska-Ciosk, J. Miedzianka, W. Drożdż, P. Regiec, Zywn-Nauk Technol J 2012 **4** 98-107.
[8] Lu, X., Luo, Z., Yu, S., Fu, X., J Agric Food Chem 2012 **60** 9273-9279.
[9] Putri, K. S. S., Surini, S., Anwar, E., Asian J Pharm Clin Res 2013 **6** 62-66.
[10] Zhang, Y., Gan, T., Hu, H., Huang, Z., Huang, A., Zhu, Y., Feng, and Yang, M. Ind Eng Chem Res 2014 **53** 2114–2120.
[11] Tang, L., B. Huang, Q. Lu, S. Wang, W. Ou, W. Lin, X. Chen, 2013 Bioresour Technol 2013 **127** 100-105.
[12] Xu, Y., Miladinov, V., Hanna, M. A. Cereal Chem 2004 **81** 735-740.
[13] Amini, H. A., Masruri, Ulfa, S. M. Jurnal Ilmu Kimia Universitas Brawijaya 2014 **1** 140-146.
[14] R. B. Garcia and R. R. L. Vidal POLIMEROS 2000 **10** 155-161.
[15] S. C. Alcázar-Alay and M. A. A. Meireles Food Sci Technol 2015 **35** 215-236.

IOP Conf. Series: Materials Science and Engineering **299** (2018) 012080 doi:10.1088/1757-899X/299/1/012080

Synthesize of zinc nanoparticles using Indonesian velvet bean (*Mucuna pruriens*) extract and evaluate its potency in lowering catalepsy in mice

Ratnaningsih Eko Sardjono[1*], Fitri Khoerunnisa[1], Iqbal Musthopa[1], Dinar Khairunisa[1], Putri Astuti Suganda[1], Rahmi Rachmawati[1]

Department of Chemistry, Faculty of Mathematics and Natural Science Education, Universitas Pendidikan Indonesia, Jalan Setiabudi No. 229, Bandung 40154, Indonesia

Email: ratnaeko@upi.edu

Abstract. This study aims to synthesize zinc nanoparticles using Indonesian velvet bean (*Mucuna pruriens*) seed extract and evaluate its potency in lowering catalepsy in mice. The research conducted consist of extraction of *M. pruriens* seed powder, synthesis of zinc-*M. pruriens* seed extract nanoparticles (Zn-MPn), characterization of Zn-MPn, and catalepsy test of Zn-MPn. *M. pruriens* seed powder was extracted by maceration using ethanol-water (1:1) at *p*H 3 adjusted with citric acid. The Zn-MPn was synthesized by reacting zinc acetate dihydrate ($Zn(CH_3COO_2)_2.2H_2O$) solution with *M. pruriens* seed extract for 40 min, dispersibility of the reaction was controlled by using sonication and ultrasonic homogenizer. The Zn-MPn obtained was characterized by scanning electron microscopy with energy dispersive X-ray spectroscopy (SEM-EDX), transmission electron microscopy (TEM), and Fourier-transform infrared (FTIR). Catalepsy test of Zn-MPn was conducted at doses of 5, 10, 15, 20 and 25 mg/kg body weight. The results of SEM-EDX and TEM analysis showed that the Zn-MPn formed nanoparticles with a particle diameter of 55 nm. Based on FTIR analysis, the absorption band at 464.8 cm^{-1} was a typical absorption indicated the Zn-O interaction on Zn-MPn. Catalepsy test showed that Zn-MPn on the all five doses were able to lower the catalepsy in mice with the best dose was 10 mg/kg body weight.

Keywords: *Mucuna pruriens*, velvet bean, nanoparticle, zinc, antiparkinson, catalepsy

1. Introduction

Parkinson is a neurodegenerative disease and the second most common after Alzheimer. Parkinson disease is allegedly caused by the damage of nerve cells that causes the decrease of dopamine in the brain and leads a person to have less ability to regulate the movements, bodies, and emotions [1]. One of the symptoms in people with Parkinson is catalepsy, a disorder of the central nervous system, characterized by muscle rigidity, immobile posture and decreased sensitivity to pain [2], [3]. The efforts to lower catalepsy are still dominated by using synthetic drugs, such as carbidopa, levodopa, and benserazide. However, these drugs have serious side effects, especially in long-term use [4].

Traditional medicine using plant extracts has been chosen as an alternative due to the lower side effects. *M. pruriens* or velvet bean is considered as a potential plant to lower catalepsy [5], [6]. *M. pruriens* seed contains L-dopa, a precursor of dopamine that is able to cross the blood-brain barrier and help to improve the lack of dopamine [7]–[9]. *M. pruriens* seed from Indonesia has been reported to contain alkaloids, steroids, saponins, tannins and 7.56 to 13.9% of L-dopa [10]. Various studies have reported the activity of *M. pruriens* seed as a promising antiparkinson [11]–[13]. Antiparkinson pharmacological studies of *M. pruriens* seed extract on mice (*Mus musculus*) showed that the extract at doses of 200 mg/kg and 400 mg/kg body weight could lower the catalepsy significantly [6].

The changing of the extract into a nanoparticle is claimed to have advantages. Nanoparticles ameliorate the compatibility and bioavailability of the extract, hence it improve the performance of the extract as drugs in the treatment of several diseases, including Parkinson. The ability of the nanoparticle drugs to reach the target directly makes the drugs are not widely distributed in the body hence it impacts to the low the side effects of the drugs. Nanoparticle drugs are also easily suspended in a liquid and capable to cross the organs and tissues [14]. Since nanoparticles enhanced advantages over the larger particles due to their size, distribution, and morphology, synthesis of nanoparticles has been a great demand. Among the metallic nanoparticles, zinc nanoparticles are very interesting to explore because of their physical and chemical properties. Zinc nanoparticles have low-toxicity, high biocompatibility, and are biodegradable that make them widely applied in the biomedical field. Some of them have been used in a number of drug formulations due to their antibacterial, antifungal, and disinfecting properties [15], [16].

A method to synthesize nanoparticles involving biological components such as microorganisms and plant extracts as reducing agents is more environmentally friendly because it does not use the toxic chemicals, can be used in large-scale production, and is energy saving. Synthesis nanoparticles mediated by plant extract is faster, easier and cheaper compared to it mediated by microorganism [17]–[19]. *M. pruriens* seed extract has been reported to be used in the nanoparticle synthesis, and the nanoparticle obtained has been proved could improve motoric disorders in animals induced by 1-methyl-4-phenyl-1,2,3,6-tetrahydropyridine (MPTP) significantly compared to *M. pruriens* seed extract itself [20]. This result suggests that nanoparticle provides a better bioactivity.

The empirical evidence shows the potency of *M. pruriens* in the development of a promising antiparkinson medicine. Unfortunately, there is no scientific study on the biosynthesis of zinc nanoparticles-*M. pruriens* seed extract (Zn-MPn) from Indonesia. In this paper, we report the biosynthesis and characterization of Zn-MPn from Indonesia as a candidate of nanoparticle herbal medicine in lowering catalepsy as one of the symptoms of Parkinson disease.

2. Materials and Methods

2.1. Material

M. pruriens seeds were obtained from Bantul, Yogyakarta, Indonesia. The chemical used included: ethanol (96%), citric acid p.a, ($Zn(CH_3COO_2)_2.2H_2O$) p.a, distilled water, haloperidol, PGA (Pulvis Gummi Arabicum) 1% and animal feed (P551).

2.2. Extraction of M. pruriens seeds

M. pruriens seeds were sun-dried and ground into a powder. The powder (5.9 kg) was macerated with ethanol-water (1:1) with the addition of citric acid (up to *p*H 3) for 3×24 h. The 12 Liters of the extract was evaporated at 40 °C under low pressure in a rotary vacuum evaporator then dried using a freeze dryer to obtain 242.70 gram (4.11%) of the dry extract.

2.3. Synthesis of Zn-MPn

5 gram of *M. pruriens* seed extract was dissolved in 100 mL of distilled water then stirred for 15 min and filtered to give 50,000 ppm of extract solution. Zn-MPn was prepared by adding the extract

solution into 0.04 M ($Zn(CH_3COO_2)_2.2H_2O$) solution dropwise with a volume ratio of 1:1. The mixture was stirred for 20 min with sonication and 20 min with an ultrasonic homogenizer and kept at room temperature for 24 h. The suspension was centrifuged at 5,000 rpm and washed several times using ethanol, then dried in an oven at a temperature of 40 °C to produce a Zn-MPn powder.

2.4. Characterization of Zn-MPn

2.4.1. Scanning Electron Microscopy-Energy Dispersive X-ray (SEM-EDX). Scanning electron microscopy (SEM) and elemental analysis (energy dispersive X-ray analysis: EDX) were carried out using Hitachi SU3500 with coating ion sputter Hitachi MC1000. It was conducted to determine the morphology, size and elemental composition of Zn-MPn. Samples were crushed, coated with Au and placed on the sample container then analyzed.

2.4.2. Transmission Electron Microscopy (TEM). Characterization using TEM conducted to determine the shape and size of the Zn-MPn. Samples were mixed with a dispersing agent, placed on the sample grid then analyzed.

2.4.3. Fourier Transform Infrared (FT-IR). Characterization of functional groups contained in the extract and AgMPn was done by FT-IR spectroscopy Shimadzu 8400. Samples were prepared in KBr pellets. The measurement was conducted in a wave number range 4000-400 cm^{-1}.

2.5. Catalepsy Test

2.5.1. Preparation of Animals. 3-months-old healthy male mice with 18-35 body weight were employed in catalepsy test. Animals were housed in polypropylene cages and acclimated for a week prior to the experimentation period under standard conditions (± 22 °C). A standard diet (CP551) and water were provided to animals.

2.5.2. Experimental Design. The mice were randomly distributed to nine groups (Table 1) with three mice each. Observations were done 30 min after administration of haloperidol suspension. Haloperidol was given to mice orally 30 min after oral administration of the vehicle (PGA 1%) or L-dopa, or the extract, or Zn-MPn at the respective dose. The intensity of catalepsy was measured as the length of time the mice hang on at 50 cm in height with both forelegs holding the horizontal wire with 0.5 cm in diameter (Fig. 1). Mice were catalepsy if hang on the wire for more than 15 seconds [34, 35].

Table 1. Group distribution for catalepsy test

No	Group	Treatment
1	normal	water
2	negative control	haloperidol 5 mg/kg body weight
3	positive control	L-DOPA 5 mg/kg body weight
4	the extract	*M. pruriens* seed extract at 200 mg/kg body weight
5	dose I	Zn-MPn at 5 mg/kg body weight
6	dose II	Zn-MPn at 10 mg/kg body weight
7	dose III	Zn-MPn at 15 mg/kg body weight
8	dose IV	Zn-MPn at 20 mg/kg body weight
9	dose V	Zn-MPn at 25 mg/kg body weight

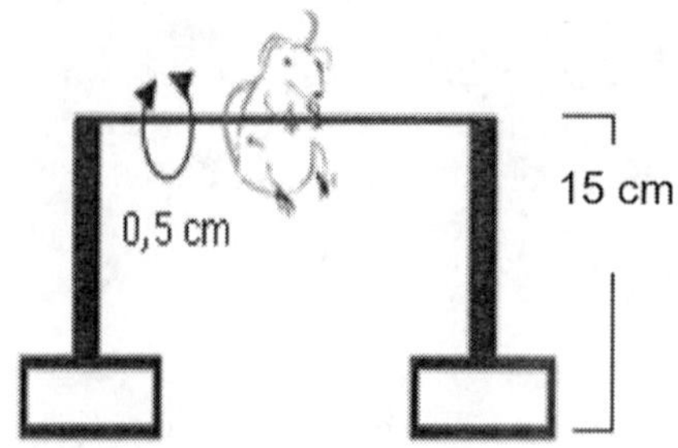

Figure 1. Catalepsy Test

2.5.3. Data analysis. The data of catalepsy test were analyzed statistically using one-way analysis of variance (ANOVA) followed by Dunnett's test to evaluate significant differences between the control and the treated groups. All statistical analysis was performed using SPSS 22.0 software. $P < 0.05$ was considered statistically significant.

3. Results and discussion

3.1. Synthesis and Characterization of Zn-MPn

The colour of zinc acetate dihydrate ($Zn(CH_3COO_2)_2.2H_2O$) solution changed from colourless into black when the extract solution was added. The colour change of the solution occurs during the sonication process. It indicated the formation of Zn-MPn qualitatively. Based on SEM and TEM images, Zn-MPn has a particle size of 55 nm with an oval in shape. The EDX spectrum showed the elemental composition contained in Zn-MPn, where the composition of zinc was 18.3% and the others were carbon (38.5%) and oxygen (26%). The composition of carbon and oxygen may be derived from biomolecules contained in *M. pruriens* seed extract that contributed to the formation of zinc nanoparticles. This result indicated the reduction of zinc ions into zinc metal.

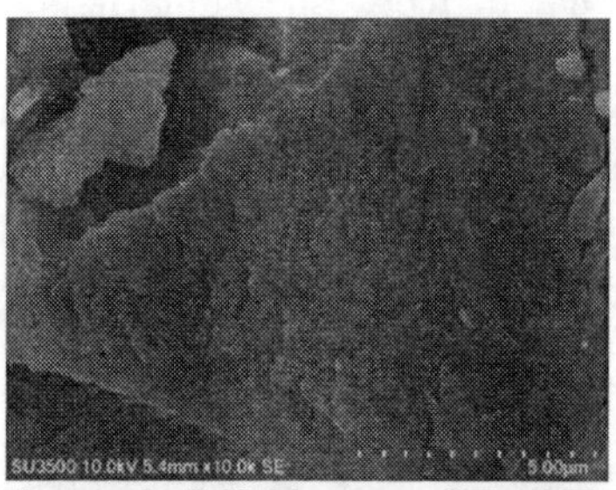

Figure 2. SEM image of Zn-MPn

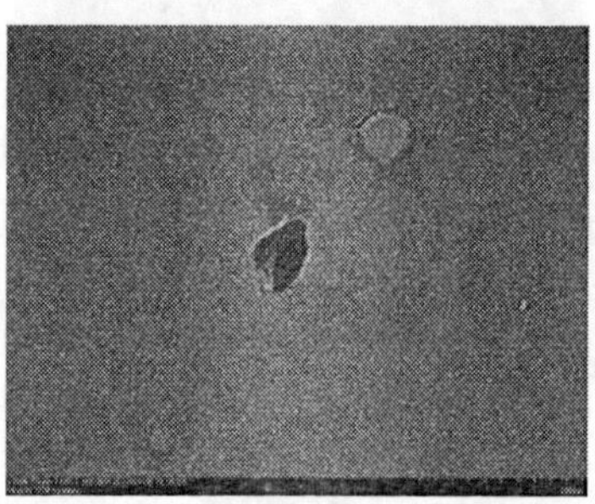

Figure 3. TEM image of Zn-MPn

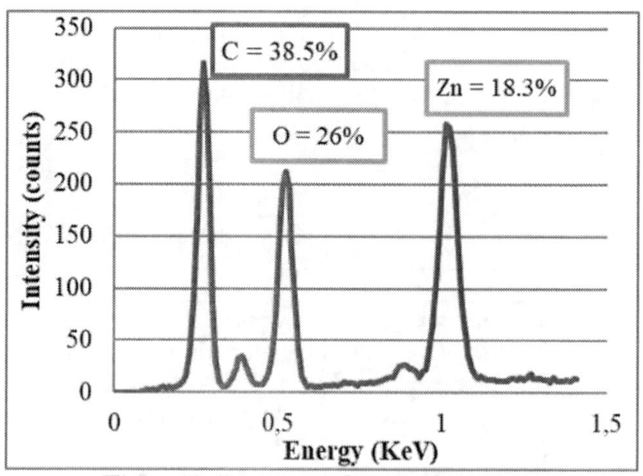

Figure 4. Spectrum of EDX of Zn-MPn

IOP Conf. Series: Materials Science and Engineering **299** (2018) 012080 doi:10.1088/1757-899X/299/1/012080

Characterization with FTIR was performed on both *M. pruriens* seed extract and Zn-MPn. It aims to determine the biomolecules contained in *M. pruriens* seed extract involved in the formation of Zn-MPn. Based on FT-IR spectra of the extract (Fig. 5), an intense and broad absorption band at 3384.8 cm^{-1} indicated the overlapping of O-H stretching vibration of flavonoids, alkaloids, polyphenols, alcohols or water with N-H stretching vibration of amine compounds due to hydrogen bonding. The absorption band at 1627.8 cm^{-1} referred to N-H bending vibration of amine which is possible to be derived from the L-DOPA, while the absorption band at 1529.4 cm^{-1} referred to C=C stretching vibration. The absorption band at 1400.2 cm^{-1} referred to C-H bending vibration of sp^2 carbon, which both were possible to be derived from the aromatic ring of amino acid. The absorption band in absorption area of 1288.4 cm^{-1} indicated the C-O stretching vibration of aromatic compounds and the absorption band at 1074.3-1118.6 referred to C-O stretching vibration of amino acid.

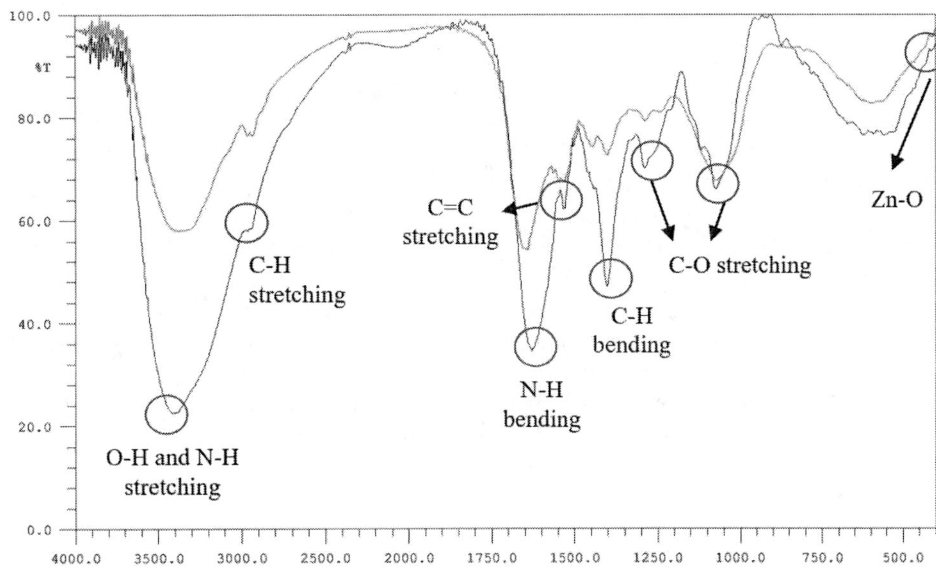

Figure 5. Spectrum of the extract (black line) and Zn-MPn (green line)

In the FTIR spectrum of Zn-MPn, there are several changes in the intensity of the absorbance and shifts in absorbance area. There was a new absorption band in absorption area of 464.8 cm^{-1} which is a typical band for the Zn-O interaction. This finding indicated the presence of interactions of the compounds in the extracts with the zinc metal to form Zn-MPn. The presence of N-H, O-H, C=C, and C-O functional groups referred to the presence of amino acid or L-dopa that possible involved in the synthesis of Zn-MPn as reducing agents. The hydroxyl groups contained in L-dopa could act as a reducing agent, which oxidized to form dopaquinone, indicated by the decrease in the intensity of the absorbance.

Table 2. The wavelength numbers in the FTIR spectra of *Mucuna pruriens* seed extract and Zn-MPn

Functional froup	Wavelenght (cm^{-1})	
	M. pruriens **seed extract**	**Zn-MPn**
O-H (stretching), N-H (stretching)	3384.8	3315.4
C-H (stretching)	2960.5-2931.6	2960.5-2931.6
N-H (bending) (NH_2)	1627.8	1639.4
C=C (stretching)	1529.4	1531.4-1542.9
C-H (bending)	1400.2	1440.7-1398.3
C-O aromatic	1288.4	1286.4-1240.1
C-O (stretching)	1074.3-1118.6	1072.3
Zn-O	-	464.8

3.2. Catalepsy Test

The catalepsy observation was performed 60 min after the administration of haloperidol and L-DOPA, or the extract, or Zn-MPn, orally. The intensity of the catalepsy was measured as the time of the mice hang on a 15 cm height rod with both front legs hold a wire with 0.5 cm in diameter. Mice were catalepsy if they hang on the wire for more than 15 sec. The data of the catalepsy test are shown in Fig. 6. Based on Fig. 6, the negative control group had a hanging time of more than 20 sec, means the mice were catalepsy. Group treated by L-dopa, the extract and Zn-MPn on the all five doses had an average of hanging time of fewer than 20 sec. Although the hanging time was longer than the normal control group which was 2 seconds, this indicated the intensity of catalepsy in mice of these groups decreased. Fig. 5 also shows that the hanging time of the groups treated by Zn-MPn on the all five doses was shorter than the group treated with the extract. This showed that Zn-MPn on the all five doses lowered the catalepsy better than the extract. The hanging time of group treated by Zn-MPn at a dose of 10 mg/kg body weight was shorter than its of group treated with L-dopa, hence Zn-MPn at a dose of 10 mg/kg body weight lowered catalepsy better than the positive control group.

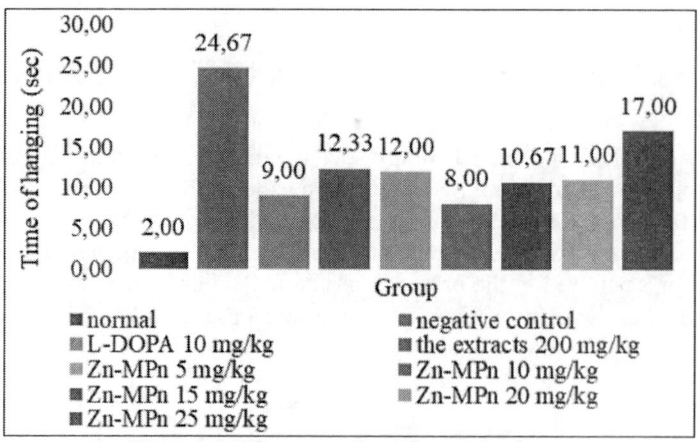

Figure 6. Diagram of catalepsy decrease

A statistical test using one-way ANOVA followed by Dunnett post hoc test was conducted to determine the significance of Zn-MPn in lowering catalepsy in mice. The limit of significance in this study was 0.05 with a 95% confidence level. If the *P* value less than 0.05, means that the data is significantly different. Zn-MPn has a significant effect in lowering catalepsy if it has *P* less than 0.05 compared to negative control. The results of the statistical test are shown in Table 3.

Table 3. The results of statistical test for Zn-MPn groups

Group	Zn-MPn 5 mg/kg	Zn-MPn 10 mg/kg	Zn-MPn 15 mg/kg	Zn-MPn 20 mg/kg	Zn-MPn 25 mg/kg
Normal	**0.047***	0.323	0.093	0.079	**0.002***
Negative control	0.126	**0.019***	0.068	0.079	0.949
Positive control	0.844	0.998	0.982	0.962	0.067
The extracts	1,000	0.613	0.983	0.994	0.336

*$P<0.05$ is considered as significantly different. Statistical test was conducted by one way ANOVA followed by Dunnett post hoc test.

Group treated with Zn-MPn at a dose of 10 mg/kg body weight showed the *P* value less than 0.05 compared to negative control, means the condition of the mice was significantly different with the condition of the mice in negative control group. It demonstrated that Zn-MPn at a dose of 10 mg/kg body weight lowered the catalepsy significantly. This group also showed the *P* value more than 0.05 compared to normal and positive control groups. This showed that the administration of Zn-MPn at a dose of 10 mg/kg body weight lowered catalepsy until the mice being normal (healthy) and the activity of Zn-MPn at a dose of 10 mg/kg body weight in lowering catalepsy was equal to its of L-DOPA. In the other hand, although the administration of Zn-MPn at a dose of 5, 10, 15, and 20 mg/kg body weight lowered the catalepsy better than the extract, statistically its activity was not significantly different.

4. Conclusion

Zn-MPn has successfully synthesized from zinc acetate dihydrate ($Zn(CH_3COO_2)_2.2H_2O$) solution with *Mucuna pruriens* seed extract for 40 min. The Zn-MPn obtained has a particle size of 55 nm with an oval in shape. Zn-MPn has a typical band of Zn-O interaction in the absorption area of 464.8 cm^{-1} on the FTIR spectrum. Zn-MPn at a dose of 10 mg/kg body weight demonstrated to lower the catalepsy in mice significantly.

5. References

[1] Lewitt PA 2012 Levodopa for the treatment of Parkinson's disease *N. Engl. J. Med.* **359** 1–9
[2] Dar PA *et al* 2012 Concept of catalepsy (jamood/shakhoos) in Greeko-Arab medicine: a review *Internationale Pharmaceutica Sciencia* **2** 26–31
[3] Banasikowski TJ and Beninger RJ 2012 Haloperidol conditioned catalepsy in rats: a possible role for D_1-like receptors *Int. J. Neuropsychopharmacol* **15** 1525–1534
[4] Brooks DJ 2008 Optimising levodopa therapy for Parkinson's disease with levodopa/carbidopa/entacapone: implications from a clinical and patient perspective *Neuropsychiatr. Dis. Treat.* **4** 39–47
[5] Champatisingh D *et al* 2011 Anticataleptic and antiepileptic activity of ethanolic extract of leaves of *Mucuna pruriens*: a study on role of dopaminergic system in epilepsy in albino rats *Indian J. Pharmacol.* **43** 197–199
[6] Sardjono RE *et al* 2016 Evaluation of antiparkinson's activity of Indonesian velvet bean (*Mucuna pruriens*) extract *ARPN J. Eng. Appl. Sci.* **11** 10856–10861
[7] Hefti F and Melamed E 1980 L-dopa's mechanism of action in Parkinson's disease *Trends.*

Neurosci. **3** 229–231

[8] Kumar P and Saha S 2013 An updated review on taxonomy, phytochemistry, pharmacology and toxicology of *Mucuna pruriens J. Pharmacogn. Phytochem.* **2** 306–314

[9] Misra L and Wagner H 2007 Extraction of bioactive principles from *Mucuna pruriens* seeds *Indian J. Biochem. Biophys.* **44** 56–60

[10] Sardjono RE *et al* 2012 The physicochemical composition of Indonesian velvet bean (*Mucuna pruriens* L) *Global J. Res. Med. Plants Indigenous* **1** 101–108

[11] Cassani E *et al* 2016 *Mucuna pruriens* for Parkinson's disease: low-cost preparation method, laboratory measures and pharmacokinetics profile *J. Neurol. Sci.* **365** 175–180

[12] Kavitha C and Thangamani C 2014 Amazing bean '*Mucuna pruriens*': a comprehensive review *J. Med. Plants Res.* **8** 138–143

[13] Kasture S *et al* 2009 Assessment of symptomatic and neuroprotective efficacy of *Mucuna pruriens* seed extract in a rodent model of Parkinson's disease *Neurotox. Res.* **15** 111–122

[14] Mittal AK *et al* 2013 Synthesis of metallic nanoparticles using plant extracts *Biotechnol. Adv.* **31** 346–356

[15] Shekhawat MS *et al* 2014 A biomimetic approach towards synthesis of zinc oxide nanoparticles using *Hybanthus enneaspermus* (L.) F. Muell *Tropical Plant Research* **1** 55–59

[16] Manokari M and Shekhawat MS 2016 Production of zinc oxide nanoparticles using extracts of *Passiflora edulis* Sims. f. flavicarpa Deg *World Scientific News* **47** 267–278

[17] Akhtar MS *et al* 2013. Biogenic synthesis of metallic nanoparticles by plant extracts *ACS Sustain. Chem. Eng.* **1** 591–602

[18] Njagi EC *et al* 2014 Biosynthesis of iron and silver nanoparticles at room temperature using aqueous sorghum bran extracts *Langmuir* **27** 264–271

[19] Herlekar M *et al* 2014 Plant-mediated green synthesis of iron nanoparticles *J. Nanoparticles* **2014** 1–9

[20] Arulkumar S and Sabesan M 2015 The behavioral performance tests of *Mucuna pruriens* gold nanoparticles in the 1-methyl 4- mouse model of Parkinsonism *Asian Pac. J. Trop. Dis.* **2015** 499–502

IOP Conf. Series: Materials Science and Engineering **299** (2018) 012081 doi:10.1088/1757-899X/299/1/012081

Treatment of wastewater batik by electrochemical coagulation using aluminium (Al) electrodes

Riyanto[1,3] and Eny Puspitasari1

[1]Department of Chemistry, Faculty of Mathematics and Natural Science, Universitas Islam Indonesia, Jalan Kaliurang KM 14,5 Sleman Yogyakarta Indonesia.

e-mail: riyanto@uii.ac.id

Abstract. Treatmentof wastewater batik by electrocoagulation method using aluminium (Al) electrodes has been done. Electrocoagulation method was chosen for treatment of wastewater batik because it is cheap, easy and efficient waste degradation. This research was conducted using aluminium (Al) electrodes and sodium chloride as an electrolyte solution. The purity of aluminium electrode was analysis using Scanning Electron Microscopy-Energy Dispersive X-ray (SEM-EDX). Electrochemical coagulation has been done using wastewater batik volume 50 mL with variation of time (10, 30, 50, 70, and 90 minutes), variation of voltage (5, 7, 9, 10, and 11 V), and variation of salt addition (0.5; 0.75; 1.00; and 1.25 g). Batik wastewater was analyzed before and after electrocoagulation by Spectrophotometer UV-Vis and the content of Pb was analyzed by Atomic Absorption Spectrophotometer (AAS). The research results show that optimum conditions electrolysis time, voltage and sodium chloride was 90 minutes, 10 V and 1.25 g, respectively. The results of this study showed the longer the electrolysis time, the higher the voltage, and the increasing number of salt added, then the batik waste decreased absorbance, alteration of color from black to clear yellow. The content of Pb in batik waste has decreased from 0.5844 mg/L to 0.1630 mg/L.

1. Introduction

The batik industry has produced organic wastes with large volumes, solid colors, strong odor and has a high temperature, pH, Biochemical Oxygen Demand (BOD), High Suspended Solid. This is due to the use of chemicals and dyes in the batik production process. The chemicals used include NaOH, Na_2CO_3, $NaHCO_3$, sulphuric acid, sulfite, nitrite, while the dyestuffs used include acid dyes, bases, direct dyes, reactive dyes, naphthol dyes, and vat dyes. Indigosol dye is often used because it produces bright colors and does not easily fade, but the water used by the laundry can cause disruption to the environment. Batik waste containing indigosol compounds is very dangerous because it can cause some health impacts. These dyes can cause skin diseases and are very harmful can result in skin cancer.

The industrial batik has produced liquid waste contains heavy metals which are hazardous wastes, which can cause damage to the environment. The existence of industrial waste can be known in the form of physical contamination, such as smelling sting, and contaminants will make the water becomes turbid. The emergence of these symptoms can absolutely be used as one of the signs of high levels of water pollution. Colored wastes arise from the use of dyestuffs that are still attached after use.

Batik industrial liquid waste directly discharged into the environment without being processed first then it can cause pollution, especially aquatic ecosystems. High temperatures will result in dissolved oxygen content or Dissolved Oxygen (DO) in declining water that will kill the organism thus disrupting the balance of the water ecosystem. In addition, organic waste will increase the nitrogen content into nitrate compounds that cause the bad smell.

Wastewater dyes processing can be done chemically, physics, and biology. Some researchers do research, processing textile waste is adsorption of methylene blue by using activated carbon [1], electrochemical oxidation [2,3,4,5]. Several studies have been done to treatment dyes waste. Chemically there are several methods of liquid waste treatment including photocatalytic reactions [6], and coagulation [7,8]. In addition, other methods are color degradation with oxidation reactions, anaerobic reactions, and photocatalysis reactions. Electrocoagulation is a simple and easy method for reducing various pollutants and pollutants, both organic and inorganic materials. In the research that has been done used aluminum and iron plate, and a combination of both. This electrocoagulation method is also able to decrease dissolved solids content, COD, and BOD.

The electrocoagulation method is an alternative method for reducing pollutants in wastewater. The multiple electrode application is done by assembling the Fe and Al electrode plates. Variations used in this study has the variation of plate number, size, voltage, and time. From the research it can be seen that electrocoagulation method of multiplexed electrocoagulation can to improve the quality of domestic batik wastewater, using the number of electrode plate effect on the improvement of the quality of domestic industrial wastewater. In addition, this electrocoagulation method can reduce the content of COD, BOD, TSS, pH and has met the water quality standard.

Liquid waste is classified as one of the main contaminants in batik industry activity. The liquid waste of this batik industry has a concentrated color and is carcinogenic, so it needs to get serious attention. This problem is caused by pollution by the textile dye is difficult to be overcome by biology (microorganism) and chemicals (coagulation and flocculant). Many researchers have been done treatment dyes wastewater using electrocoagulation [9,10,11,12,13]. Electrocoagulation method have done for textile wastewater [14,15,16,17]. The advantage of this electrocoagulation method is that it can replace the use of expensive (more economical) chemical reagents, resulting in a collection of suspended particles more durable and more easily filtered, producing cleaner water than with chemical coagulation or sedimentation, and can be used for processing waste that has high turbidity. In addition, the implementation of waste treatment by this method is quite simple and controlled rather than using chemical coagulants. Although it requires more complicated equipment preparation and greater investment, in the long run, this electrocoagulation method is very profitable.

In this paper has been the study of the treatmentof wastewater batik by electrocoagulation method using aluminium (Al) electrodes. Electrochemical coagulation has been done using wastewater batik with the variation of electrolysis time, variation of potential and variationof sodium chloride. The aluminium electrode was used as an anode and cathode has the analysis of purity using SEM-EDX. Batik wastewater was analyzed before and after electrocoagulationby Spectrophotometer UV-Vis and the content of Pb was analyzed by Atomic Absorption Spectrophotometer (AAS).

2. Experimental Section

2.1. Material and Instrumentation

The sample used is liquid waste from batik industry in Yogyakarta Indonesia. The instruments used in this study are one set of an electrolysis cell, magnetic stirrers, UV-Vis Spectrophotometers (Hitachi U-2010), SEM-EDX (Phenom TMProx/Pro/Pure) for characterization of aluminium electrode, and Atomic Absorption Spectrometers (Perkin Elmer Pin Aade 900 T). The materials used in this research are aluminium metal, batik waste, sodium chloride, HNO_3 analytical grade, the solution of Pb 1000 ppm (Merck), and distillate water.

2.2. Experimental Method

2.2.1. Electrocoagulation of batik waste with variation of electrolysis time and potential

A total of 50 mL of batik waste was put into the beaker glass and added 0.5 g of sodium chloride. Then electrolyzed with a voltage of 5 V, using electrolysis time variations 10, 30, 50, 70, and 90 minutes. After the electrolysis was completed, the sample solution was filtered using filter paper and the filtrate was measured for absorbance by using a UV-Vis spectrophotometer. A total of 50 mL of batik waste was put into the beaker glass and added 0.5 gram of sodium chloride. Then electrolyzed with voltage variations (5, 7, 9, 10, and 11 V) using electrolysis time is 50 minutes. After the electrolysis was completed, the sample solution was filtered using filter paper and the filtrate was measured for absorbance by using a UV-Vis spectrophotometer.

2.2.2. Electrocoagulation of batik waste using sodium chloride

A total of 50 mL of batik waste was inserted into a beaker and added a kitchen salt with variations (0.25, 0.5, 0.75, 1.0 and 1.25 grams). Then electrolyzed with a voltage of 5 V. The time used is 50 minutes. After the electrolysis was completed, the sample solution was filtered using filter paper and the filtrate was measured for absorbance by using a UV-Vis spectrophotometer.

2.2.3. Determination of Pb using Atomic Absorption Spectrophotometer (AAS)

The sample has destruction before analysis using AAS. A total of 25 mL of batik waste was put into the Kjeldahl flask. The sample of batik waste was destruction using the HNO_3 solution as much as 5 mL. The solution is heated to a digestor block at a temperature of 100 °C. The solution was destructed to about 0.5 mL remaining. The solution in the cooled Kjeldahl flask is then fed into a 25 mL measuring flask and diluted with distilled water to the limit marker. The solution is shaken out to homogeneous. The sample solution was filtered through Whatman filter paper 41. The standard solution obtained had concentrations of 0.5, 1.0, 2.0 and 3.0 ppm. The sample solution and standard solution of Pb series were measured using atomic absorption spectrometers.

3. Result and Discussion

3.1. Characterization of Aluminium using SEM-EDX

Prior to the electrolysis of batik waste, the first aluminum electrode used was analyzed using SEM-EDX to identify and determine the composition of the aluminum electrode element. The results of aluminum electrode analysis with SEM are presented in Figure 1.

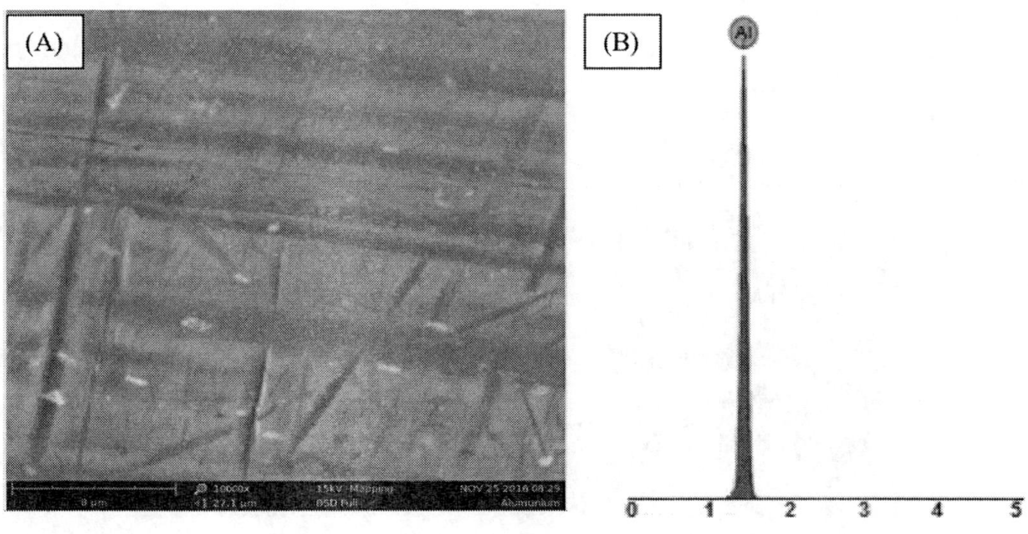

Figure1. SEM image of aluminium surface with magnification 10000x (A) and EDX spectra the aluminium surface (B)

Figure 1A shows that based on SEM analysis result with 10000x magnification it is seen that aluminum surface morphology is not uniform. The surface of aluminum is scratched so uneven lines. Figure 1B shows the results of aluminum electrode analysis using EDX. Figure 1B shows that the aluminium has a high purity that is with the atomic concentration of the Al is 100%. In Figure 1B it appears that the peak that appears there is only one is aluminum, this means that the element content contained in aluminummetal contains only Al and no other elements.

3.2. *Effect of electrolysis time to electrocoagulation of batik wastewater*

Figure 2 shows the results of electrolysis of batik waste using a potential of 5 V with the addition of 0.5 g of sodium chloride, and variations of electrolysis time are 10, 30, 50, 70, and 90 minutes. Spectra analysis of batik waste results before and after electrolysis is shown in Figure 2.

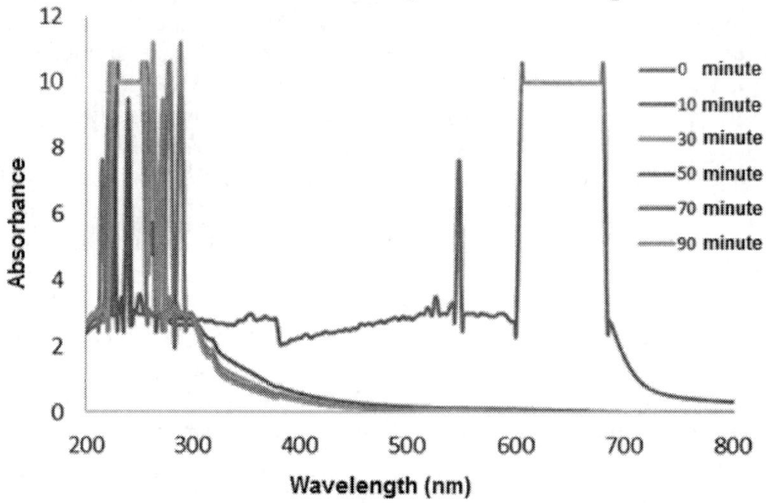

Figure 2. Spectra analysis batik waste before and after electrocoagulation with variation electrolysis time

Figure 2 shows that the initial batik waste (before electrolysis) has a peak in the visible region of 400-800 nm, this is because batik waste before electrolysis is still colored and the solution is concentrated. After 10 minutes of electrolysis the color of batik waste began to degraded, then after electrolysis for 90 minutes batik waste solution became the clearest, so peak generated at 400-800 nm wavelength began to decline. In these circumstances, it can be seen in accordance with the results of the analysis of the batik waste solution still contains compounds that have a specific wavelength in the UV region (200-400 nm). Figure 3 shows that the longer the electrolysis process, the compound content in batik waste decreases, this is indicated by the lower absorbance rate produced.

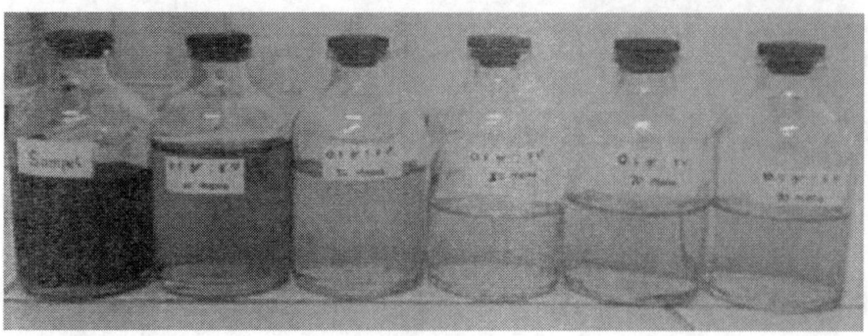

Figure 3. Image of degradation color of batik waste before and after electrocoagulation with variation electrolysis time 0-90 minutes

3.3. Effect of potential to electrocoagulation of batik waste

Electrolysis of batik waste has been done using electrolysis time 50 minutes, 0.5 grams of salt, and with variations of 5, 7, 9, 10, and 11 Volt voltage. The spectra of the analysis of batik waste that has been electrolysis are shown in Figure 4.

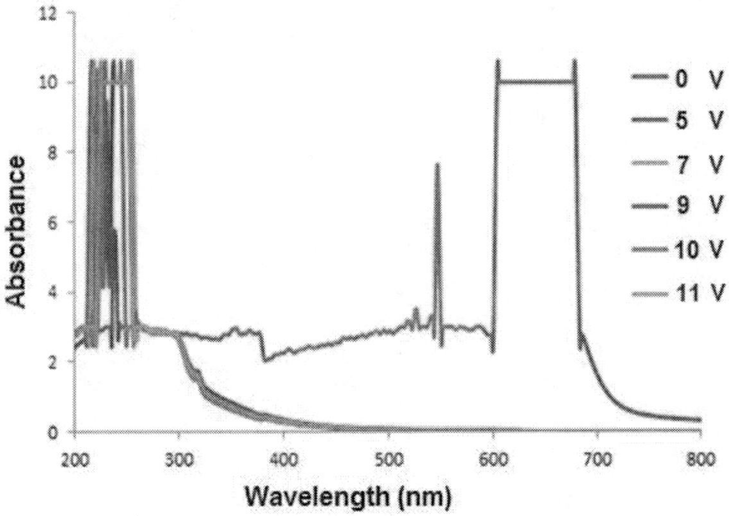

Figure 4. Spectra analysis batik waste before and after electrocoagulation with variation potential

Figure 4 shows that the initial batik waste (before electrolysis) has a peak in the visible region of 400-800 nm, this is because batik waste before electrolysis is still colored and the solution is concentrated. After electrolysis done for 50 minutes, the color of batik waste starts clear, so electrolysis done for 50 minutes. In this research, electrolysis process with the variation of 5, 7, 9, 10, and 11 Volt. The results of the analysis can be seen that the electrolysis process at a voltage of 5 V causes the solution of batik waste starts clear so that the resulting absorbance lower. It can be seen that the electrolysis process influences the content of compounds in batik waste. This process can degrade the dye present in batik waste. In Figure 5 shows the color change of electrolysis result of batik waste with potential variation with electrolysis with voltage 5-11 V.

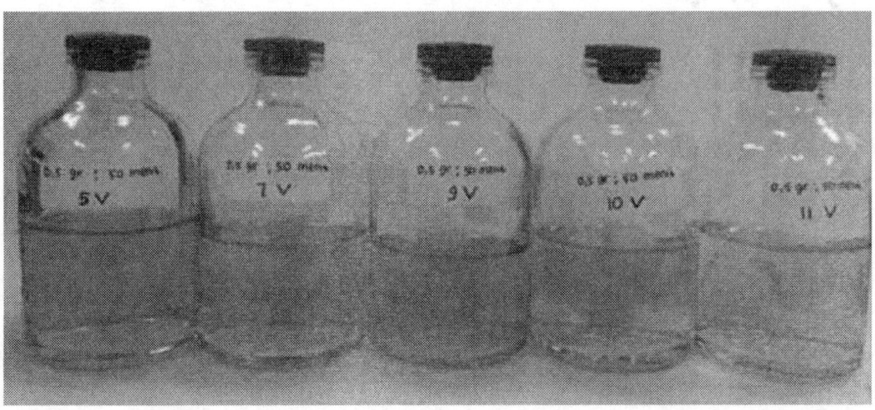

Figure 5. Image of degradation color of batik waste before and after electrocoagulation with variation potential

IOP Conf. Series: Materials Science and Engineering **299** (2018) 012081 doi:10.1088/1757-899X/299/1/012081

3.4. Effect of sodium chloride to electrocoagulation of batik waste

Electrolysis of batik waste using potential 5 V, electrolysis time 50 minutes, and variation of addition of NaCl are 0; 0.25; 0.50; 0.75; 1.00; and 1.25 g. The spectra of the analyzed batik waste are shown in Figure 6.

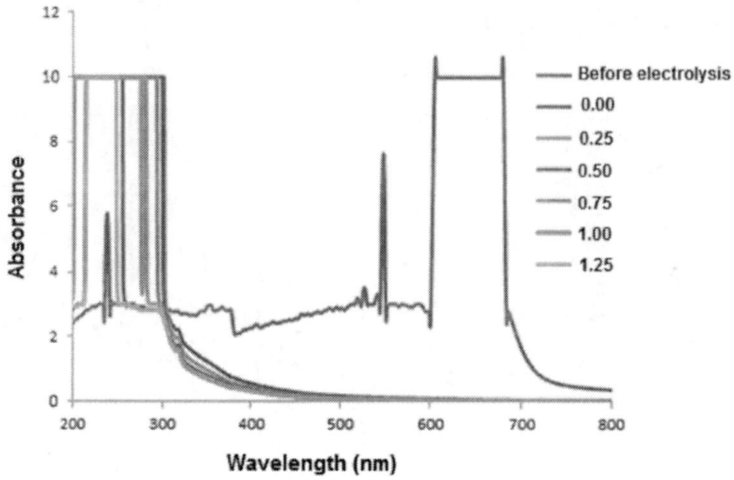

Figure 6. Spectra analysis batik waste before and after electrocoagulation with variation of NaCl

Figure 6 shows that the initial batik waste (before electrolysis) has a peak in the visible region of 400-800 nm, this is because batik waste before electrolysis is still colored and the solution is concentrated. After electrolysis done for 50 minutes, the color of batik waste starts clear, so electrolysis done for 50 minutes. In this research, electrolysis process is done with the variation of the salt addition of kitchen 0, 0.25, 0.50, 0.75, 1.00 and 1.25 g. Based on the results of the analysis, it can be seen that when the electrolysis process without using salt peak produced began to fall and the solution of batik waste also began to clear, resulting in lower absorbance. The sodium chloride (NaCl) is a powerful electrolyte that can dissociate completely in solution into Na^+ and Cl^- ions. The greater is the NaCl concentration the greater the current flowing in the solution, thus causing a greater percentage of dye degradation.

At the cathode, there will be reduction reactions to the cations, which are included in this cation are H^+ ions and metal ions. The H^+ ion of an acid will be reduced to a hydrogen gas that will be free as gas bubbles. Reaction: $2H^+ + 2e \rightarrow H_2$

If the solution contains alkali, alkaline earth metal ions, then these ions cannot be reduced from the reduced solution is the solvent (water) and hydrogen gas (H_2) formed at the cathode.

Reaction: $2H_2O + 2e \rightarrow 2OH^- + H_2$

The reaction on the anode used in the aluminum metal will oxidize, the reaction:

$$Al^{3+} + 3H_2O \rightarrow Al(OH)_3 + 3H^+$$

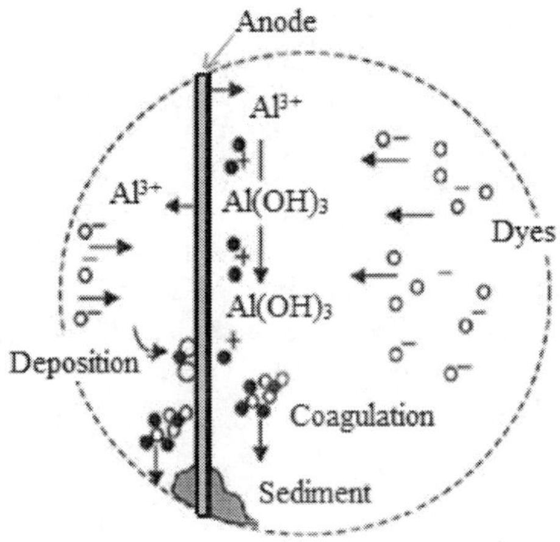

Figure 7. Electrocoagulation process on anode electrode using aluminium[18]

NaCl in the electrocoagulation process plays an important role in dye oxidation by reacting as follows [7]:

Cathode:
$NaCl \rightarrow Na^+ + Cl^-$
$Cl_2- + H_2O \rightarrow H^+ + Cl^- + HOCl$
$HOCl \rightarrow H^+ + OCl^-$
Anode:
$Dye + OCl^- \rightarrow CO_2 + H_2O + Cl^-$

The Cl^- ion will react with water to form chlorine hydroxy, and then oxidized to form chlorine oxide and react with the dye occurring in the anode and produce CO_2 gas, water, and Cl^- ions.

3.5. Determination of Pb in batik waste before and after electrolysis using AAS

Heavy metal contained in batik industrial waste is Pb. Therefore, research on the analysis of Pb content in batik waste before and after electrolysis. The absorbance of the standard solution of Pb has been obtained, then made calibration curve to be obtained by the equation. The equation is used to determine the concentration of Pb in batik waste.

The obtained line equation is then used to calculate the concentration of Pb in batik waste. The concentration of Pb in batik waste prior to electrolysis was 0.5844 mg/L. In the analysis of Pb content in batik waste after electrolysis, the samples used are the variation of sodium chloride, that is 0.5 gof sodium chloride with potential 5 V and electrolysis time is 50 minutes. The absorbance is then used to make the standard solution calibration curve of Pb. The calibration curve of the concentration of standard Pb solution with its uptake is shown in Figure 8.

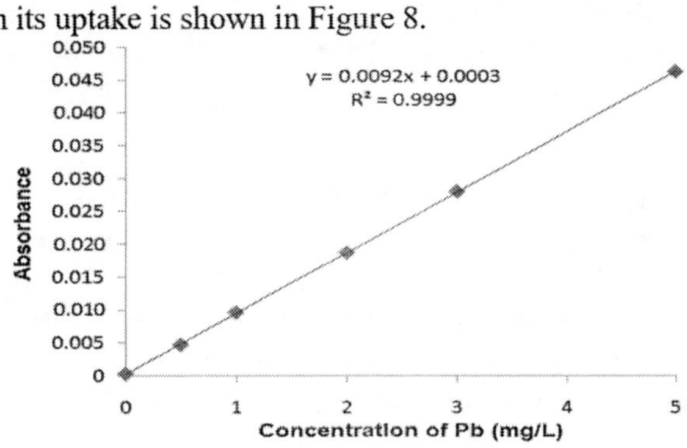

Figure 8. The calibration curve of the concentration of standard Pb solution

The obtained line equation is then used to calculate the concentration of Pb in batik waste. The concentration of Pb in batik waste after electrolysis was 0.1630 mg/L. The concentration of Pb in batik waste water before and after electrolysis process is 0.5844 mg/L and 0.1630 mg/L so that Pb content in batik waste fulfil the maximum requirement of Pb content in water is 1.0 mg/L.

4. Conclusion

The result of characterization of aluminum (Al) with SEM EDX shows surface is no uniform with high purity is 100%. The optimum result shows that batik wastewater treatment using electrolysis time, potential and sodium chloride is 90 minutes, 10 volts and 1.25 g respectively, seen from the color of batik waste which originally dark black become yellow clear. Aluminum electrode is more economical and is a good electrical conductor. The content of Pb in batik wastewater after electrolysis decreased the concentration of 0.5844 mg/L to 0.1630 mg/L.

References

[1] Yasin 2007 *The Malaysian Journal of Analytical Sciences.* **11** 400
[2] Nordin N,Amir S F M, Riyanto and Othman M R 2013 *Int. J. Electrochem. Sci.* **8** 11403
[3] Nordin N, Amir S F M and Yusop M R 2015 *Acta.Chim. Slov.* **62** 642
[4] Panizza M and Huitle M 2013 *Elsevier J.* **90** 1455
[5] Sala M and Gutierrez B M C 2012 *International Journal of Photoenergy.* **10** 1
[6] Rashed M N and El-Amin A A 2007 *Int. J.Phys.Sci.* **2 3** 73
[7] Chatzisymeon E, Xekoukoulotakis N P, Coz A, Kalogerakis N and Mantzavinos D 2006 *Journal of Hazardous Materials*, **B137** 998
[8] Kobaya M, Tan O C and Bayramoglu M 2003 *Journal of hazardous Materials.* **B100** 163
[9] Nandi B K and Patel S 2017 *Arabian Journal of Chemistry.* **10** 2961
[10] Fajardo A S, Martins R C, Silva D R, Martínez-Huitle C A and Ferreira R M Q 2017 *Journal of Electroanalytical Chemistry.* **801** 30
[11] Bassyouni D G, Hamad H A, El-Ashtoukhy E S Z, Amin N K and Abdel-Latif M M 2017 *Journal of Hazardous Materials.* **335** 178
[12] Segura S G, Eiband M M S G, Melo J V and Huitle C A M 2017 *Journal of Electroanalytical Chemistry.* **801** 267
[13] Eyvaz M, Kirlaroglu M, Aktas T S and Yuksel E 2009 *Chemical Engineering Journal.* **153 1-3**,16
[14] Aoudj S, Khelifa A, Drouiche N, Hecini M and Hamitouche H 2010 *Chemical Engineering and Processing: Process Intensification.* **49** 1176
[15] Kobya M, Demirbas E and Can O T 2006 *Journal of Hazardous Materials.* **132** 183
[16] Khandegar V and Saroha A K 2013 *Journal of Environmental Management.* **128** 949
[17] Phalakornkule C, Polgumhang S, Tongdaung W, Karakat B and Nuyut T 2010 *Journal of Environmental Management.* **91** 918
[18] Den W and Huang C 2005 *Colloids and Surfaces A: Physicochemical Engineering Aspects.* **254** 81

Acknowledgements

This research was supported by Ministry of Research, Technology and Higher Education of the Republic of Indonesia, Directorate General of Research and Development Reinforcement, Directorate of Research and Community Services through *Penelitian Terapan Unggulan Perguruan Tinggi* Research Grant 2017 for the financial support. We would like to thank the Integrated Instrumentation Laboratory and the Environmental Quality Laboratory at the Islamic University of Indonesia for the ease of use of the equipment to make this research successful.

Kinetics and Thermodynamics Interaction Between Bentonite Inserted Organometallic Compounds [$Cr_3O(OOCH)_6(H_2O)_3$](NO_3) With Methylene Blue Dye In Aqueous Medium

Risfidian Mohadi[1*], Radja Nardo Purba[1], Dedi Rohendi[1], and Aldes Lesbani[1, 2]

[1] Department of Chemistry, Faculty of Mathematics and Natural Sciences, Sriwijaya University. Jl. Raya Palembang Prabumulih Km 32 Ogan Ilir 30662, South Sumatera, Indonesia
[2] Integrated Research Laboratory, Graduate School, Sriwijaya University, Jl. Padang Selasa, Bukit Besar Palembang 30139, South Sumatera, Indonesia

[*]Corresponding email : risfidian.mohadi@unsri.ac.id

Abstract. Bentonite insertion was conducted with organometallic compound [$Cr_3O(OOCH)_6(H_2O)_3$](NO_3)) and then tested its stability at various pH, then its interaction between bentonite inserted by organometallic compound [$Cr_3O(OOCH)_6(H_2O)_3$] with methylene blue dye. The interaction between organometallic compound [$Cr_3O(OOCH)_6(H_2O)_3$] with methylene blue showed an adsorption rate of 15.49 min^{-1} at the concentration of methylene blue 25 mg/L. The adsorption capacity and adsorption energy (ΔE) were 2.74 mol/g and 5.58 kJ/mol, respectively, at temperature 70 °C where visible trends of enthalpy ($\Delta H°$) and entropy ($\Delta S°$) values decrease with increasing concentration of methylene blue.

Keyword: bentonite, adsorption, methylene blue, organometallic, [$Cr_3O(OOCH)_6(H_2O)_3$] (NO_3)

1. Introduction

Bentonite is a type of mineral smectite composed by an aluminosilicate, forming a layered structure and is a good cation exchanger. The main content of bentonite is montmorillonite. Montmorillonite has a cavity which causes a very large surface area of up to 700-800 m^2/g [1]. This specific surface area is exposed to water dispersion, due to its high expansion ability and causes montmorillonite to accept organic compounds and metal ions.

The ability of bentonite as an adsorbent or catalyst can be improved by widening the distance between layers through insertion process that will produce a pillared or inserted bentonite [2]. Many researchers have made modifications through bentonite insertion. Characterize modified bentonite with Al as a polycation, performs inserted bentonite process using Aluminum metal. The results showed an increase in the surface area of bentonite when the addition of micro cation solution of Al into the suspension of Ca-bentonite[3]. However, microcations that have been developed have not been effectively used as adsorbents or catalysts, since the obtained surface area is still small. The layer of bentonite increased when the ammonium salt is added to the bentonite simultaneously with microcation Al. However, the results obtained are still not effectively used as adsorbents, since they only have small surface area. Therefore in order to produce bentonite which has a large surface area requires a large positively charged compound (macrocation) [4]. This macrocation is derived from the synthesis results performed in the laboratory through a process of chemical synthesis.

In this study, bentonite macrocation is inserted with an organometallic compound $[Cr_3O(OOCH)_6(H_2O)_3]$ (NO_3) and subsequently used as adsorbents to adsorb methylene blue dye. The inserted bentonite is characterized using XRD analysis, and identified of their functional group by a spectrophotometer FTIR. The parameters tested in the dye adsorption process are: the influence of adsorption time, the effect of initial concentration of adsorbate and adsorption temperature.

2. Methodology

Materials used in this research were natural bentonite, chromium nitrate, formic acid, hydrochloric acid, sodium hydroxide, nitrogen gas, methylene blue and distilled water, while instrumentation applied for analysis were *X-Ray* diffractometer Shimadzu lab X-type 6000, spectrophotometer FTIR Shimadzu Prestige-21, and UV-Vis Spectrophotometer Thermo Scientific Genesis 20.

2.1 Preparation of Bentonite and Characterization

Natural bentonite samples were heated using a furnace at a temperature of 400 °C. Next, a 100 g bentonite product was immersed in 500 mL of H_2SO_4 0,5 M and stirred using magnetic stirrer for 2 h. then filtered and dried. Bentonite acidification that has been dried and was mashed with crushed and characterized using XRD diffraction, FTIR spectrophotometer analysis.

2.2 Synthesis of Organometallic Compound $[Cr_3O(OOCH)_6(H_2O)_3]$ (NO_3) and its characterization

Organometallic compounds $[Cr_3O(OOCH)_6(H_2O)_3]$ (NO_3) synthesized by mixing $Cr(NO_3)_3{\bullet}9H_2O$ (40 g) in 50 mL of distilled water and stirred using a *magnetic stirrer.* 100 ml of NaOH 3 M was added to the solution for 10 minutes to obtain a solution with green color. The obtained solution was filtered using filter paper. 12.5 mL of formic acid was added into the residue and refluxed for 2 hours, then allowed to stand at room temperature a few minutes and was filtered to obtain green color crystals [5]. The dried green crystals product was analyzed using a FT-IR Spectrophotometer.

2.3 *Insertion of Organometallic Compounds* $[Cr_3O(OOCH)_6(H_2O)_3]$ (NO_3) *into Bentonite and its Characterization*

Bentonite insertion process was made in several weight ratio of activated bentonite: organometallic compounds $[Cr_3O(OOCH)_6(H_2O)_3]$ (NO_3) at (2.5:1), (1:1), (1:2), (1:3), respectively. A 2.5 g of activated bentonite was dissolved in 200 mL of distilled water into 4 different flasks. 100 mL of NaOH 1 M was added into the flasks containing 1; 2.5; 5, and 7.5 g of organometallic compounds $[Cr_3O(OOCH)_6(H_2O)_3]$ (NO_3). Next, the mixtures were poured into the flask which already contains activated bentonite and stirred for 24 hours using a magnetic stirrer at room temperature under nitrogen air condition [6]. The results obtained were then filtered and dried at room temperature. Results of bentonite inserted by organometallic compounds$[Cr_3O(OOCH)_6(H_2O)_3]$ (NO_3) at every weight ratio subsequently characterized using XRD diffraction, FT-IR Spectrophotometer analysis.

2.4 Adsorption of methylene blue dye by Activated Bentonite and Bentonite inserted Organometallic Compounds $[Cr_3O(OOCH)_6(H_2O)_3]$ (NO_3)

2.4.1 Effect of Adsorption Time and Kinetics Parameters

The methylene blue adsorption kinetics was studied by performing the adsorption process at constant room temperature by varying the contact time to monitor the effect of the time change on the adsorption percent. Amount of 0.01g activated bentonite, and 0.05 g of bentonite inserted organometallic compounds $[Cr_3O(OOCH)_6(H_2O)_3]$ (NO_3) added into 50 mL of 25 mg/L methylene blue solution then homogenized with a horizontal shaker at a predetermined interval time. The adsorption time varied from 10; 20; 30; 40; 50; 60; 70; 80, and 90 min. After adsorption process the methylene blue concentration in solution was measured by using a UV-Vis Spectrophotometer at a wavelength of 665 nm.

2.4.2 Effect of Initial Concentration and Temperature and Thermodynamics Parameters

Thermodynamic parameters of adsorption of methylene blue by activated bentonite and bentonite inserted organometallic compounds $[Cr_3O(OOCH)_6(H_2O)_3](NO_3)$ through various initial concentrations of methylene blue and adsorption temperature. Amount of 0.01g activated bentonite and 0.05 g of bentonite inserted organometallic compounds $[Cr_3O(OOCH)_6(H_2O)_3](NO_3)$ each adsorbent interacted with 50 mL of methylene blue solution at various concentrations of 10; 15; 25; 30 and 40 mg/L and then homogenized using a horizontal shaker for 30 min at various temperatures 30; 40; 50; 60 and 70 °C. The mixture then measured using a UV-Vis Spectrophotometer at a maximum wavelength of 665 nm to determine the residual concentration of the methylene blue.

3. Results and Discussions

3.1 Synthesis and Characterization of Organometallic Compound $[Cr_3O(OOCH)_6(H_2O)_3](NO_3)$ *Using FT-IR Spectrophotometer*

Synthesis of organometallic compounds $[Cr_3O(OOCH)_6(H_2O)_3](NO_3)$ based on the procedures carried out by Lesbani [7]. At the end of the synthesis process obtained green colored crystals had formed indicating the organometallic compound $[Cr_3O(OOCH)_6(H_2O)_3](NO_3)$ were then characterized using FT-IR spectrophotometer to identify functional groups of the organometallic compound.

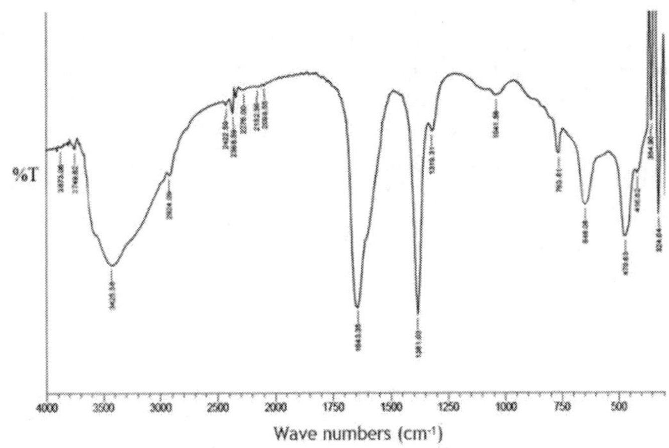

Figure 1. The FT-IR spectra of Organometallic Compound $[Cr_3O(OOCH)_6(H_2O)_3](NO_3)$

The FT-IR spectra in Figure 1. shows the organometallic compound $[Cr_3O(OOCH)_6(H_2O)_3](NO_3)$ have a distinctive vibration at wave number 648 cm^{-1} (Cr_3-O), and broadband at wave number 3425 cm^1 which is stretching vibration of -OH group of molecules of H_2O were detected in the solid organometallic compound $[Cr_3O(OOCH)_6(H_2O)_3](NO_3)$. The existence of a formic acid group that serves as a ligand contained in the wave number 2800-2950 cm^{-1} and the cluster of NO_3^- appears at wave number 1381 cm^{-1} which is a counterion [5]. Furthermore, the organometallic compound $[Cr_3O(OOCH)_6(H_2O)_3](NO_3)$ used in the process of insertion into the bentonite activation results that the product will be applied as an adsorbent in the adsorption process of methylene blue dye.

3.2 Characterization of Bentonite and Bentonite Inserted Organometallic Compound $[Cr_3O(OOCH)_6(H_2O)_3](NO_3)$ *using XRD Diffraction, FTIR Spectrophotometer Analysis*

Bentonite has a distinctive basal spacing in the range of diffraction angle 2θ between 2-6°[7]. It is characterized by the diffraction angle (2θ) in natural bentonite, bentonite heating to 400 °C, and bentonite results of acidification using sulfuric acid are diffracted at 4.1°; 3.3 °; and 4.1 ° with basal spacing by 21.2 Å; 26.2 Å; and 21.2 Å, respectively. Diffractogram of natural bentonite, bentonite heated at a temperature of 400 °C, and bentonite acidification using sulfuric acid results shown in Figure 2.

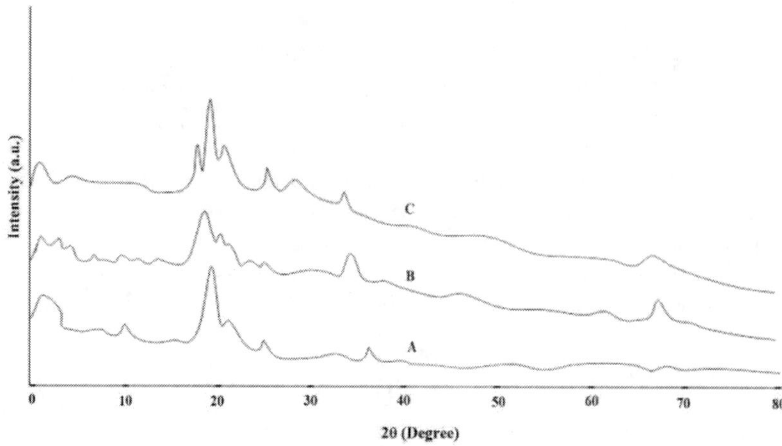

Figure 2. The XRD pattern in (a) Natural bentonite, (b) Bentonite heated at a temperature of 400 °C, and (c) Bentonite acidification results using sulfuric acid 0.5M

Based on data from the XRD measurement results of the analysis in Figure 2 shows the large diffraction angles appear on the area diffraction angle of 19°, 20°, and 34° which indicates the bentonite is a class of layered material that contains several minerals including mineral illite, montmorillonite, and quartz [8]. Diffraction typical of natural bentonite, bentonite heating to 400 °C and acid bentonite showed that the bentonite has a high crystallinity.
Bentonite inserted of the organometallic compound $[Cr_3O(OOCH)_6(H_2O)_3](NO_3)$ further characterized using XRD as shown in Figure 3.

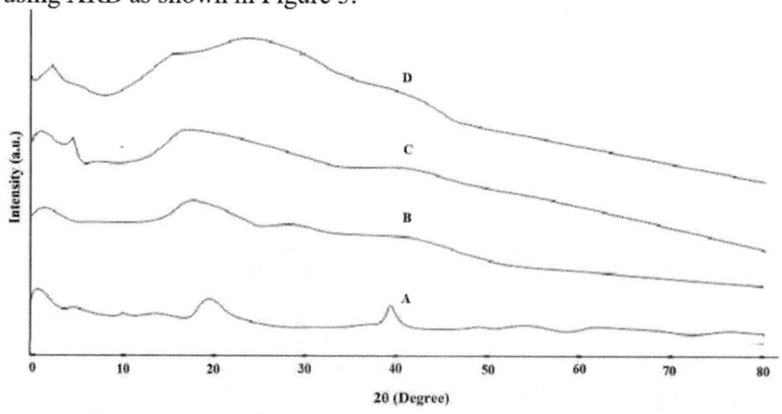

Figure 3. The XRD pattern of bentonite inserted organometallic compound $[Cr_3O(OOCH)_6(H_2O)_3](NO_3)$ in the ratio of (A) 2.5:1; (B) 1:1; (C) 1:2 and (D) 1:3

The results of XRD measurements on bentonite inserted of the organometallic compound $[Cr_3O(OOCH)_6(H_2O)_3](NO_3)$ change in the angle of diffraction 2θ agree. The bentonite and the organometallic compound $[Cr_3O(OOCH)_6(H_2O)_3](NO_3)$ ratio (2.5:1), (1:1), (1:2) and (1:3) shows a typical diffraction angle on bentonite which appeared in the area that is 3.3° 2θ with basal spacing of 26.4 Å to Figure 3.A, 26.5 Å in Figure 3.B, 26.2 Å in Figure 3.C, while in Figure 3.D. 26.2 Å, respectively. In the insertion bentonite, the diffraction tends to widen which decrease of crystallinity due to the presence of water molecule in the organometallic compound. The FT-IR spectrophotometer data shows the presence of hydroxyl groups caused by water molecules absorbed OH on the surface of the octahedral layer and Si-O-Si groups in the tetrahedral layer[9]. FT-IR spectra of natural bentonite, bentonite heating to 400 °C, and a sulfuric acid activated bentonite is shown in Figure 4.

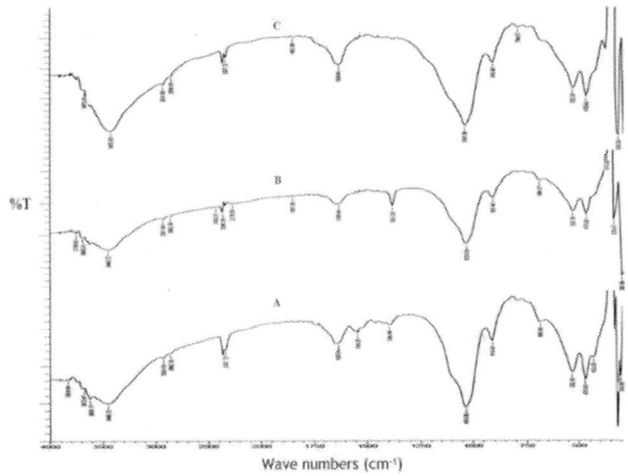

Figure 4. The FT-IR spectra of (a) Natural bentonite, (b) Bentonite heated at a temperature of 400 °C, and (c) Bentonite results in acidification using sulfuric acid 0.5M

The activation process by heating causes reduces the intensity at the area of wave numbers 3448, 7 cm^{-1} and 1635.6 cm^{-1} indicate the water content in bentonite was reduced. At the OH bending vibration of water molecules was observed at 1635.6 cm^{-1} region which is an OH vibration of water molecules trapped within the silicate framework (water of crystallization). The comparison data of wavenumber FT-IR spectra of natural bentonite, bentonite heating at a temperature of 400 °C, and bentonite acidification using sulfuric acid 0.5 M results outlined in Table 1.

Table 1. Data of wave number of natural bentonite, bentonite heated at a temperature of 400 °C, and bentonite results acidification by sulfuric acid 0.5 M

Functional group	Wavelength (cm^{-1})		
	Natural bentonite	Bentonite heated at 400 °C	Bentonite acidification
Stretching Si-O-Si	1033.8	1033.8	1041.5
Stretching HOH	3448.7	3448.7	3425.5
Buckling Al-O-Al	910.4	910.4	910.4
Buckling HOH	1635.6	1635.6	1635.6
Buckling Al-O-Si	532.3	532.3	532.3
Buckling Si-O-Si	470.6	470.6	470.6

Based on Table 1 shows that the shift of wavenumber of 1033.8 cm^{-1} to 1041.5 cm^{-1} that indicates the vibrational stretch from the group Si-O-Si was observed as a broad absorption peak with a sharp intensity indicating a high content of montmorillonite on bentonite[10]. The wave numbers 3626.1 cm^{-1} indicates stretching vibration of the OH group in the Al layer octahedral Al-OH or Mg-Al.

The product of the insertion process is then characterized using a FT-IR spectrophotometer. Figure 5 shows the absorption peaks in the insert of the organometallic compound bentonite $[Cr_3O(OOCH)_6(H_2O)_3](NO_3)$ with multiple comparisons. The Cr-O vibration of the organometallic compound that inserted into bentonite indicated by absorption band at 640-700 cm^{-1} region and can also be seen that there is a shift wavenumber 1635.6 cm^{-1} to 1597 cm^{-1} in buckling vibration of OH in bentonite resulting from insertion of organometallic compounds by comparison

(1:2) and (1:3). The shift wavenumbers also occur in the bending vibration of Si-O-Si on bentonite the insert of the organometallic compound [$Cr_3O(OOCH)_6$ $(H_2O)_3$] (NO_3) with a ratio (1:3) the wave number of 470.6 cm^{-1} to 478.6 cm^{-1}.

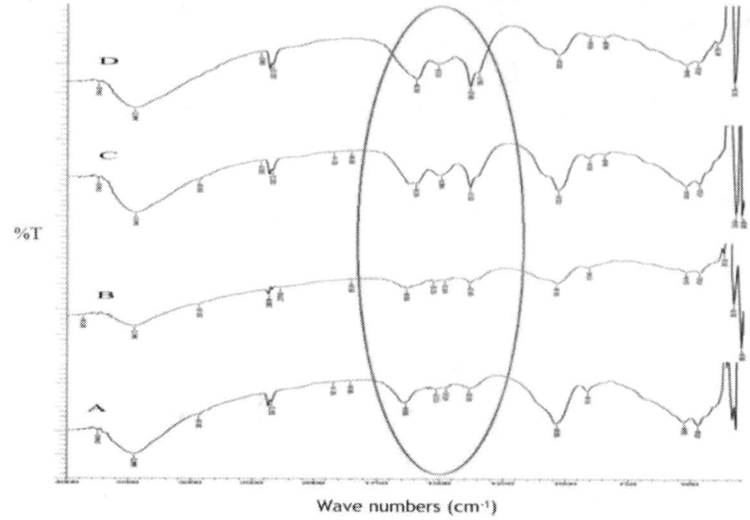

Figure 5. FT-IR spectra bentonite inserted organometallic compound [$Cr_3O(OOCH)_6$ $(H_2O)_3$] (NO_3) in the ratio of (A) 2.5: 1, (B) 1:1, (C) 1:2, and (D) 1:3

Based on the FT-IR spectra the insertion process is shown in the ratio (1:2) and (1:3) with sharp spectrum compared to a ratio (2.5:1) and (1:1). Therefore, the optimal insertion was in the ratio (1:2) and (1:3). The FT-IR wavenumber data of Figure 5. are summarized in Table 2.

Table 2. Data of wave number of bentonite inserted organometallic compound [$Cr_3O(OOCH)_6$ $(H_2O)_3$] (NO_3) in the ratio of (A) 2.5: 1, (B) 1:1, (C) 1:2, and (D) 1:3

Functional group	Ratio of Bentonite : [$Cr_3O(OOCH)_6$ $(H_2O)_3$] (NO_3)			
	2.5:1	1: 1	1: 2	1: 3
	Wave number (cm^{-1})			
Stretching HOH	3448.7	3448.7	3441	3448.7
Buckling HOH	1635.6	1635.6	1597	1597
Stretching Si-O-Si	1033.8	1041.5	1033.8	1033.8
Buckling Al-O-Al	910.4	910.4	910.4	910.4
Buckling Al-O-Si	524.6	524.6	524.6	524.6
Buckling Si-O-Si	470.6	470.6	470.6	478.6
Cr-O	670	700	673	673

3.3 Effect of Adsorption Time of Methylene Blue adsorption by bentonite inserted of the organometallic compound [$Cr_3O(OOCH)_6$ $(H_2O)_3$] (NO_3)

Effect of adsorption time of methylene blue adsorption was conducted to determine the optimum time interaction between methylene blue with an adsorbent activated bentonite and bentonite inserted of the organometallic compound [$Cr_3O(OOCH)_6$ $(H_2O)_3$] (NO_3). An equilibrium state characterizes the optimum interaction between the concentration of the methylene blue and the adsorbent. The duration of adsorption time affects the absorption process [11]. This is because the longer the adsorbent interaction with the adsorbate allows the number of collisions to occur. According to the theory of collision, the reaction rate takes place on the number of collisions of the unity of time, the more collisions that occur, the faster the reaction takes place. Chart the influence of time methylene blue adsorption to the adsorption capacity of the adsorbent activated bentonite and bentonite inserted of the organometallic compound [$Cr_3O(OOCH)_6$ $(H_2O)_3$] (NO_3) shown in Figure 6.

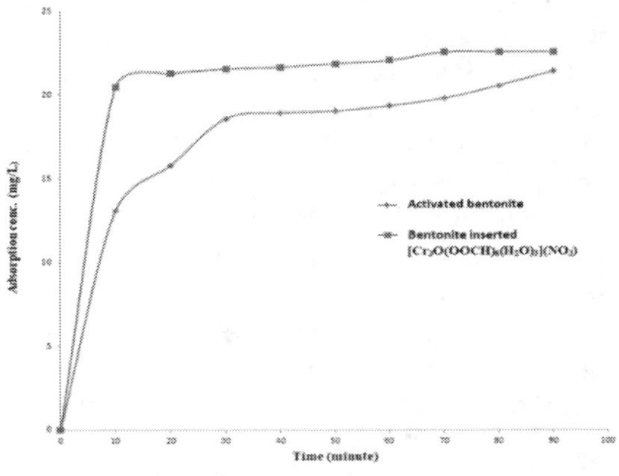

Figure 6. Effect of adsorption time of methylene blue by activated bentonite and bentonite inserted of the organometallic compound $[Cr_3O(OOCH)_6\,(H_2O)_3]\,(NO_3)$

Based on Figure 6. seen on activated bentonite (control) the amount of methylene blue that adsorbed at first 10 minutes relatively slow and increase gradually and then to stable absorption at an additional time. Adsorption of methylene blue on bentonite inserted of the organometallic compound $[Cr_3O(OOCH)_6\,(H_2O)_3]\,(NO_3)$ is high absorption rate and close to stable. According to Jankwoska (1991), if the adsorbate has filled all the pores of the adsorbent will occur resulting in saturation of adsorption decreased reactivity.
Adsorption Kinetics is used to determine the adsorption rate occurring in the adsorbate to the adsorbate and is affected by time[12]. The rate of adsorption kinetics parameter obtained from Langmuir equations

$$\frac{\ln\frac{C_0}{C}}{C} = k_1\frac{t}{C} + K \qquad (1)$$

where:

C_o = Initial concentration (mg/L)
C = Conc. after interaction time (mg/L)
t = Adsorption time (minute)
k_1 = Adsorption rate (menit^{-1})
K = Adsorption equilibrium contant.

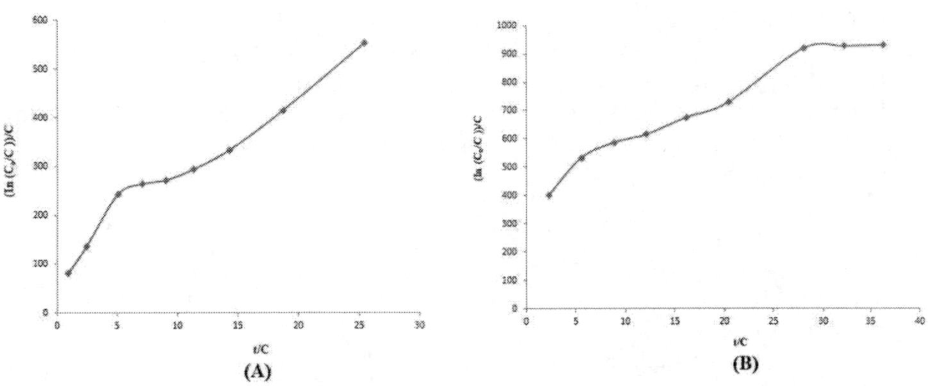

(A) (B)

Figure 7. Kinetic parameters adsorption of methylene blue by (A) activated bentonite, and (B) bentonite inserted of the organometallic compound $[Cr_3O(OOCH)_6\,(H_2O)_3]\,(NO_3)$

IOP Conf. Series: Materials Science and Engineering **299** (2018) 012082 doi:10.1088/1757-899X/299/1/012082

The curve of kinetic parameters for the activated bentonite and bentonite inserted of the organometallic compound $[Cr_3O(OOCH)_6(H_2O)_3](NO_3)$ are presented in Figure 7. The rate of adsorption constants can be calculated from Langmuir equation by plotting ln(Co/C)/C versus t/C. The slope as a adsorption rate (k_1) and correlation coefficient (R^2) of the plot ln(Co/C)/C versus t/C in adsorption of methylene blue by activated bentonite (A), and bentonite inserted of the organometallic compound $[Cr_3O(OOCH)_6(H_2O)_3](NO_3)$ (B). The adsorption rate (k_1)where 17.25 min^{-1} and 15.49 min^{-1}, respectively. The correlation coefficient for both adsorbents about (> 0.9) indicates that the process of adsorption kinetics of methylene blue meets to Langmuir kinetics equation [13].

3.4 Effect of Temperature and Initial Concentration of Methylene Blue and the Determination of Thermodynamic Parameters

The effect of temperature in the adsorption process of methylene blue was studied at various temperature interaction of 30, 40, 50, 60, and 70 °C, respectively. The effect of the initial concentration of the methylene blue eq.10, 15, 25, 30, and 40 mg/L, respectively to adsorbents (A) activated bentonite, and (B) bentonite inserted of the organometallic compound $[Cr_3O(OOCH)_6(H_2O)_3](NO_3)$ as shown in Figure 8.

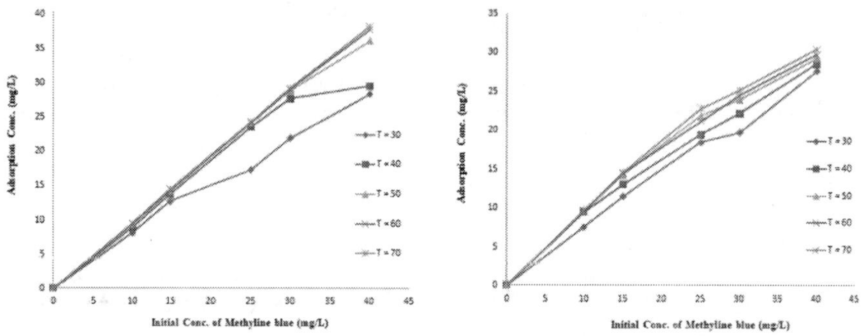

(A) (B)

Figure 8. Effect of adsorption temperature and initial concentration of methylene dye by (A) activated bentonite, and (B) bentonite inserted of the organometallic compound $[Cr_3O(OOCH)_6(H_2O)_3](NO_3)$

In Figure 8. shows the effect of temperature and initial concentration of methylene dye by (A) activated bentonite, and (B) bentonite inserted of the organometallic compound $[Cr_3O(OOCH)_6(H_2O)_3](NO_3)$. It is seen that the higher the temperature and the concentration, it will increase the amount of methylene blue adsorbed. Based on the adsorption of methylene blue at various temperature and initial concentration give thermodynamic parameters in the adsorption system. Parameters such adsorption capacity (b), the adsorption energy (E), enthalpy (ΔH) and entropy (ΔS) is presented in Table 3.

In Tabel 3. showed increased adsorption energy (E) as increased of temperature on activated bentonite, and bentonite inserted of the organometallic compound $[Cr_3O(OOCH)_6(H_2O)_3](NO_3)$. The activated bentonite has adsorption energies of 10.69 kJ/mol and for bentonite inserted of the organometallic compound $[Cr_3O(OOCH)_6(H_2O)_3](NO_3)$ has adsorption energy of 5.58 kJ/mol, respectively at a temperature of 70 °C.

Thermodynamic parameters on the adsorption process methylene blue, such as enthalpy (ΔH) and entropy (ΔS) can be calculated from the *slope* and *intercept* of 1/T as the x-axis against ln Kd as the y-axis obtained from Equation 2.

$$\ln K_d = \frac{\Delta S}{R} - \frac{\Delta H}{RT} \quad (2)$$

Where:

ΔH = enthalpy
ΔS = entropy
Kd = coefisien distribution of adsorbat (*Qe/Ce*)
R = constanta of gas (8.314 J/mol K)
T = temperature

Table 3. The adsorption capacity (b), the adsorption energy (E), enthalpy (ΔH) and entropy (ΔS) of the adsorption of methylene blue by (A) activated bentonite, and (B) bentonite inserted of the organometallic compound $[Cr_3O(OOCH)_6 (H_2O)_3] (NO_3)$

Adsorbent	T (°C)	b (mg/g)	E (kJ/mol)	C_o (mg/L)	R^2	ΔH (kJ)	ΔS (kJ/K)
Activated Bentonite	30	10.32	3.60	10	0.914	31.60	0.131
	40	16.23	3.40	15	0.916	32.39	0.136
	50	59.52	8.32	25	0.717	50.84	0.194
	60	158.73	10.66	30	0.833	51.02	0.194
	70	192.31	10.69	40	0.938	51.03	0.188
Bentonite Inserted $[Cr_3O(OOCH)_6 (H_2O)_3] (NO_3)$	30	1.57	5.15	10	0.660	42.49	0.155
	40	1.68	3.62	15	0.934	50.59	0.178
	50	2.09	4.97	25	0.859	26.04	0.094
	60	2.15	4.77	30	0.953	21.13	0.075
	70	2.74	5.58	40	0.987	7.29	0.031

Enthalpy value (ΔH) and entropy (ΔS) in the methylene blue adsorption by activated bentonite and bentonite inserted of the organometallic compound $[Cr_3O(OOCH)_6 (H_2O)_3] (NO_3)$ as presented in Table 3. increased with increasing concentrations of methylene blue. The entropy (ΔS) at a concentration of methylene blue 40 mg/L was decreased for the bentonite inserted of the organometallic compound $[Cr_3O(OOCH)_6 (H_2O)_3] (NO_3)$. The enthalpy (ΔH) of activated bentonite with methylene blue concentration of 40 mg/ is 51.03 kJ, and the bentonite inserted of the organometallic compound $[Cr_3O(OOCH)_6 (H_2O)_3] (NO_3)$ with a concentration of 15 mg/L is 50.59 kJ, whereas the entropy (ΔS) the greatest on bentonite activation results in the concentration of the methylene blue 30 mg/L is 0.194 kJ/K and the bentonite inserted with a concentration of 15 mg/L is 0.178 kJ/K with adsorption capacity (ΔE) were 2.74 mol/g at temperature of 70 °C [5].

4. Conclusions

Bentonite inserted of the organometallic compound $[Cr_3O(OOCH)_6 (H_2O)_3] (NO_3)$ has been successfully synthezised in various proportions. Characterization using XRD analysis did not show that the optimal insertion occurs at a various weight ratio. Further characterization using FT-IR spectrophotometer showed that the optimal insertion process occurs on a comparison of (1:2) and (1:3) showed a wider spectrum and sharper compared with a ratio (2.5:1) and (1:1). Bentonite inserted of the organometallic compound $[Cr_3O(OOCH)_6 (H_2O)_3] (NO_3)$ absorption of methylene blue at various contact time and temperatures gave the adsorption rate at a concentration of 25 mg/L of 15.49 min^{-1}. The adsorption capacity (b) and the adsorption energy (E), the largest at a temperature of 70 °C with respective values of 2.74 mg / g and 5.58 kJ/mol and at variable concentrations showed enthalpy (ΔH), and entropy (ΔS) decreased with increasing concentrations of methylene blue.

Acknowledgement

This research supported from RISTEKDIKTI, the Republic of Indonesia through Hibah Profesi of Sriwijaya University 2017. The authors wish to acknowledge gratefully to Prof. Aldes Lesbani, Ph.D and all research group members of Integrated Research Laboratory, Graduate School, Sriwijaya University.

References

[1] Bendou S and Amrani M 2014 *J. Miner. Mater. Charact. Eng.* **2** 404–13
[2] Bergaya F, Jaber M and Lambert J-F 2012 (Springer, London) pp 41–75

[3] Nagendrappa G 2011 *Appl. Clay Sci.* **53** 106–38
[4] Zhou C H 2011 *Appl. Clay Sci.* **53** 87–96
[5] Paluta Utami H and Marihot Siburian D 2016 *Sci. Technol. Indones.* **1** 20–4
[6] Vitanesa L, Nardo Purba R, Lesbani A and Said M 2017 *Sci. Technol. Indones.* **1** 9–16
[7] Lesbani A, Kawamoto R, Uchida S and Mizuno N 2008 *Inorg. Chem.* **47** 3349–57
[8] Kouraichi S, El-Hadi Samar M, Abbessi M, Boudouh H and Balaska A 2015 *Catal. Sci. Technol.* **5** 1052–64
[9] Uddin F 2008 *Metall. Mater. Trans. A* **39** 2804–14
[10] Palin E J, Dove M T, Redfern S A T, Saniz-Díaz C I and Lee W T *Phys. Chem. Miner.* **30** 293–304
[11] Mohadi R, Kurniawan C, Yuliasari N, Hidayati N and Lesbani A 2015 *Prog. Chem. Appl. Chitin its Deriv.* **20** 213–21
[12] Taher T, Mohadi R, Rohendi D and Lesbani A 2017 p 20-28
[13] Professor I, Moradi M, Dehpahlavan A, Rezaei Kalantary R, Ameri A, Farzadkia M and Izanloo H 2015 *Environ. Heal. Eng. Manag. J.* **2** 99–106

IOP Conf. Series: Materials Science and Engineering **299** (2018) 012083 doi:10.1088/1757-899X/299/1/012083

Synthesis of Zeolite-X from Bottom Ash for H_2 Adsorption

R Y Kurniawan[1], T Q Romadiansyah[1], A D Tsamarah[1], and N Widiastuti[1*]

[1]Chemistry Department, Faculty of Science, Institut Teknologi Sepuluh Nopember (ITS), Jl. Arief Rahman Hakim, 60111, Surabaya.

Email : nurul_widiastuti@chem.its.ac.id

Abstract. Zeolite-X was synthesized from bottom ash power plant waste using fusion method on air atmosphere. The fused product dissolved in demineralized water and aluminate solution was added to adjust the SiO_2/Al_2O_3 molar ratio gel prior hydrothermal process. The synthesis results were characterized using X-Ray Diffraction (XRD), Scanning Electron Microscope (SEM), and Fourier Transform Infrared (FTIR). The results showed that the zeolite-X has a high crystallinity with octahedral particle. The pure-form zeolite-X then was characterized and tested for H_2 gas adsorption by gravimetric method to determine the H_2 gas adsorption capacity of zeolite-X from bottom ash and it was compared to synthetic zeolite-X.

1. Introduction

Hydrogen is potentially ideal energy sources for fossil fuel replacement. The reasons for this is due to highly reactive, bumpy gas, relatively small size, thus resulting in easily stored in a particular material and promosing to be applied in fuel cell vehicle. It also occurs at high level of energy, which reaches up to 142 MJ/kg [1]. Besides, hydrogen is environmentally friendly energy source, clean and only produce water residue as a result of hydrogen combustion with O_2 gas in fuel cell. Therefore, hydrogen technology has been developed in various developed countries such as such as United State of America, Republic of South Korea and Japan [2-3]. Nevertheless, the development of hydrogen technology as an alternative energy source requires a variety of considerations. One of the challenges in the development of hydrogen technology is finding materials with large amounts of hydrogen storage up to 6-9 wt% in accordance with the targets proposed by the Department of Energy (DoE) in USA in 2010-2015 [4].

Porous materials has been the subject of research in developing hydrogen storage such as zeolites, carbon nanotubes, nano fibers, metal organic frameworks and other porous materials. Among the materials, zeolite-X, a porous alumina-silicate material is often highlighted to be a hydrogen storage material owing to having pore diameter size of 3-8 Å [7, 8]. With this pore size, hydrogen gas with the size of 2.890 Å can be well absorbed into the zeolite-X pores [9]. This material meets with the requirement of hydrogen storage material, that has high porosity, high surface area, diameter pore size equal or greater than the size of the gas molecule [6].

However, zeolite-X is generally synthesized from synthetic materials, which has disadvantages, such as a high cost for synthesis and waste production. Therefore, it is necessary to look for another way of synthesizing zeolites from the lowest cost material. Research on the synthesis of zeolites that has been developed recently is utilizing fly ash as the precursor. Various types of synthesized zeolites from fly ash are zeolite-X and zeolite-A [10-11]. Another study also reported that zeolite-X with high crystallinity can be synthesized from fly ash by fusion method followed by hydrothermal treatment [12].

Fly ash is a hazardous waste derived from power plants, which is a non-carbonaceous and has mild mass fraction. Instead of fly ash, other types of waste derived from power plant is bottom ash, which is also classified as a hazardous waste [13]. Unlike fly ash, bottom ash has a heavier mass fraction because it still contains carbon residues as well as impurities such as metal oxides. This condition makes the utilization of bottom ash lower compared to fly ash.

The chemical components contained in the bottom ash are silicon (Si) 24.10%, aluminum (Al) 6.80%, iron (Fe) 33.59%, calcium (Ca) 26.30% and carbon (C) is around 11.2% of the total weight [14]. The proportion of the chemical content of the bottom ash shows that the bottom ash has the same content as the fly ash. Hence, the bottom ash can be used as an alternative material for zeolites especially for zeolite-X.

The synthesis of zeolite-X from the bottom ash are also performed through a fusion process using alkali followed by a hydrothermal process like the synthesis of zeolite-X from fly ash. In the previous research, zeolite-X with high crystallinity can be synthesized at temperature of 100°C for 15 hours and its crystallinity decrease as the hydrothermal time increases [15]. Therefore, in the present study, the zeolite-X was synthesized from the bottom ash at the temperature of 100°C for 15 hours to produce a high crystallinity of zeolite for hydrogen adsorption.

2. Materials and methods

2.1. Equipments and Materials

The equipments used were hydrothermal reactor (stainless steel autoclave), gravimetry equipment. The materials used were bottom ash from power plant, solid $NaAlO_2$ (Sigma Aldrich), Na_2SiO_3 solution (Sigma Aldrich), NaOH pellet (99% p.a) (Sigma Aldrich), HCl 37% (SAP) and H_2 Ultra High Purity (UHP) gas. Characterization was performed using X-Ray Diffraction (XRD) Expert Pan Analytical, Scanning Electron Microscope (SEM) SU 3500 and Fourier Transform Infrared (FTIR) Thermo Scientific Nicolet iS10.

2.2. Procedures

2.2.1. Preparation of Bottom Ash. The bottom ash was first sieved with a size of 60 mesh sieve to obtain fine and uniform particles. This fine bottom ash was then heated at 110°C for 3 hours. Meanwhile, to reduce the excess of Fe and Ca contents in the bottom ash, Fe and Ca separations were performed using ferrite magnet and HCl 37%.

2.2.2. Extraction of Si and Al from Bottom Ash. In this study, zeolite-X was synthesized using a fusion method, followed by hydrothermal treatment. The bottom ash melting process was carried out by mixing the bottom ash with alkali (NaOH) with a proportional ratio of the bottom ash: NaOH =1:1, then heated in a blast tube at 750°C in the air atmosphere. The solids obtained from the melting process were then weighed and dissolved in 12 mL/g of demineralized water. The solution was stirred and aged in a polyethylene bottle for 2 hours at room temperature. The solution was filtered to separate the residue from the solution. The filtrate obtained from the filtration process was the source of Na, Si and Al as a precursor for synthesizing zeolite. *2.2.3. Synthesis of Zeolite-X.* Zeolite-X was synthesized from the extract solution obtained from the melting process by adjusting the molar composition to reach $NaAlO_2$: 4 SiO_2: 16 NaOH: 325 H_2O [16] by adding $NaAlO_2$-NaOH solution. The zeolite-X mixture or slurry was put into a sealed hydrothermal reactor for the crystallization process of zeolite-X. The hydrothermal process of zeolite-X was carried out at temperature of 100°C for 15 hours. The solid obtained from the hydrothermal crystallization was separated from the filtrate, then washed with distilled water until pH 9-10 and dried in an oven at 105°C for 24 hours.

2.2.4. Characterization. The solids obtained from the synthesis process were characterized by X-Ray Diffraction (XRD) to determine the crystallinity of the zeolite formed. The structural morphology of the zeolite-X is observed by Scanning Electron Microscopy (SEM) and the fungsional groups of the zeolite-X after adsorption was observed by Fourier Transform Infrared (FTIR).

2.2.5. H_2 gas Adsorption. Determination of H_2 gas adsorption capacity on the bottom ash based zeolite-X was performed using gravimetric method as presented in Fig.1. Firstly, the sample of zeolit-X was degassed at 350°C under vacuum condition. Gravimetric method was performed at 30°C. The hydrogen gas was fed into the cell sample through the zeolite-X sample and the gas flow was set constant at a rate of 20 ml/min. Data are taken every one minute until a constant weight was obtained. The weight was recorded as the final weight after the adsorption process (m_t)

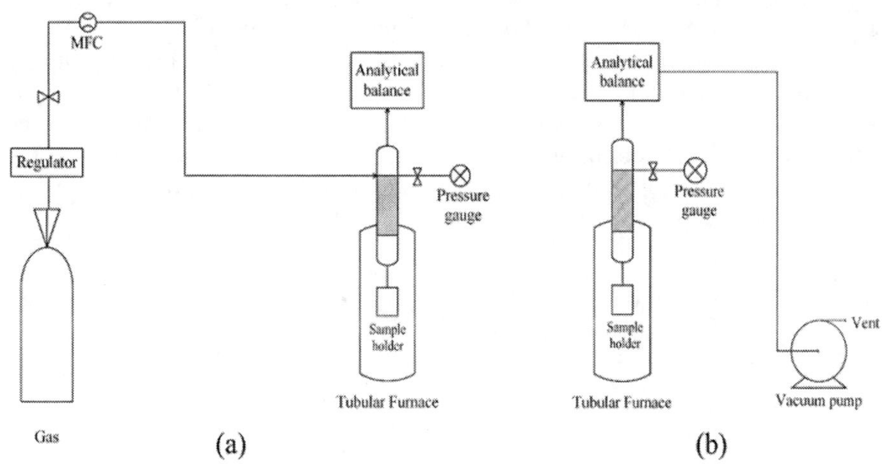

Fig.1 – Schematic diagrams of (a) adsorption and (b) desorption

3. Results and discussions

3.1. Preparation of Si and Al from Bottom Ash

This research was started by the determination of chemical composition of the bottom ash using X-Ray Flourescence Instrument (XRF). The determination of chemical composition contained in the bottom ash was very important to obtain the percentage of Si and Al weight content in the bottom ash as a precursor in the synthesis of zeolite. The result of bottom ash analysis with X-Ray Flourescence (XRF) can be seen in Table 1.

Table 1. Result of bottom ash chemical composition analysis.

Sample	Component (%)			
	Si	Al	Fe	Ca
Bottom Ash Before Separation	15.2	5.5	48.4	25.5
Bottom Ash After Separation	37.1	3.8	21.5	9.1

Table 1 shows that bottom ash contains a high percentage of Si and Al in their oxide forms, they are 15.2% and 5.5%. Due to the presence of Si and Al content, the bottom ash has the potential to be utilized as a precursor in the synthesis of zeolites. The value of Si and Al content was used in determining the Si/Al ratio in the synthesis of zeolite. The ratio value affected the type of zeolite formed. The low ratio of Si/Al is suitable for zeolite synthesis with low Si values such as Zeolite-X. Beside Si and Al, the bottom ash also contains a high percentage of Fe and Ca. Previous research also presented the same result that Fe and Ca were the main content of bottom ash beside Si and Al [17]. The high content of Ca and Fe can be separated using ferrite magnet and 37% HCl. The results of the separation were then re-analyzed using X-Ray Flourescence (XRF) to determine the value of Ca and Fe content after the separation process, the result of the bottom ash after separation is shown in Table 1.

The amount of components in the bottom ash changes after the separation process. Fe content decreased by 55.57%, while Ca content decreased by 64.31%. The reduction of the composition shows that Ca in calcite phase ($CaCO_3$) and calcium oxide (CaO) have reacted with HCl to form water-soluble $CaCl_2$. The decrement of Fe and Ca content shows that the separation process using ferrite magnet and concentrated HCl decreased the Fe and Ca content from the bottom ash effectively to reduce the oxide impurity of bottom ash.

In addition to the reduction amount of Fe and Ca content, Al content also decreased because Al also reacted with the concentrated HCl to form $AlCl_3$. On the other hand, the Si content has increased after the separation because of reducing the impurity. The results is in accordance with the result of bottom ash characterization using XRD after the separation as shown in Figure 2. The mineral phase of bottom ash after separation is dominated by quartz phase (SiO_2), while the impurity phases such as hematite in the form of Fe_2O_3 and calcite in the form of $CaCO_3$ and CaO have a fairly low intensity and distribution. This is in accordance with the results of XRF characterization, which showed that Fe and Ca content decreased in bottom ash after the separation process. In addition, there was also a mullite phase containing Al. Given the Si and Al content found in the bottom ash, it allows the zeolite to be synthesized from the bottom ash.

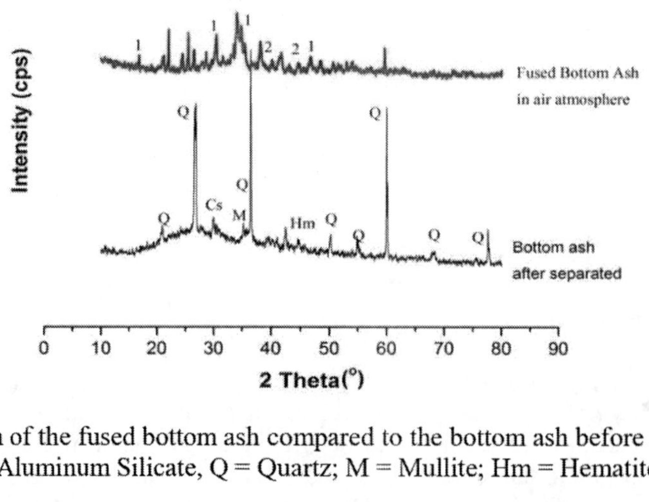

Figure 2. Difractogram of the fused bottom ash compared to the bottom ash before fused (1 = Sodium Silicate, 2 = Sodium Aluminum Silicate, Q = Quartz; M = Mullite; Hm = Hematite; Ks = Calcite).

However, quartz (SiO_2) and mullites are minerals that have small solubility values in water, thus, the bottom ash must be firstly fused in air atmosphere using alkali at high temperatures. The bottom ash-melting solids are then characterized by XRD to determine the mineral phase formed after the melting as shown in Figure 1. The difractogram shows the difference of the bottom ash before the melting process, that is the loss of peaks of hematite and calcite. In addition, the other crystalline phase peaks such as quartz and mullites are largely disappearing as well. This indicates that the quartz and mullite have reacted with the base (NaOH) and activated to produce a soluble salt in the form of sodium silicate (peak 1) (PDF 16-0818) and aluminum silicate salt (peak 2) (PDF 33-1203). This proves that the melting process at high temperatures was an effective method to extract the Si and Al content from the bottom ash into a soluble salt.

Previous researches have also reported that the fusion process of fly ash using alkali at high temperatures can change the fly ash mineral phases that are difficult to dissolve into a soluble sodium silicate and amorphous aluminum silicate [20-22]. Sodium silicate and sodium aluminum silicate were produced through the melting process, which were the major components in the formation of zeolites. After obtaining sodium silicate and sodium aluminum silicate from the fusion process, then extraction process of the solid (obtained from the fusion process) was conducted in order to obtain the dissolved Si and Al components. The extraction process in this study was conducted by mixing solids of demineralized fermentation. The obtained filtrate then tested by ICP-AES to determine the level of Si, Al and Na components contained in the filtrate, it is used for the determination of Si/Al ratio. The results of component analysis of Si, Al and Na using ICP-AES are shown in Table 2.

Table 2. The results of the fused bottom ash extract concentration analysis.

Sample	Component (ppm)		
	Si	Al	Na
Extract of fused bottom ash	3160.7	35.19	5833.3

In Table 2, it was found that the concentration of Al in the extract solution tended to be smaller than the Si and Na concentrations. This is because Al has a lower solubility value in an alkaline solution compared to Si, thus Al was more retained in the residual residue of the extract. With the low solubility value of Al in the extract, the Si / Al molar ratio value were high indicating not suitable for zeolite-X synthesis. Therefore, to adjust the Si/Al molar ratio composition according to the zeolite-X molar ratio, it was necessary to add $NaAlO_2$-NaOH solution as an external Al source. The addition of Al source aims to minimize the molar ratio of the aluminosilikat gel.

3.2. Synthesis of Zeolite-X from Bottom Ash

The bottom ash smelting extract with alkali was then used as the precursor in the synthesis of zeolite-X. The extracts are arranged in the composition of the molar ratio corresponding to the Na-Zeolite-X NaO: 4 SiO_2: 16 NaOH: 325 H_2O with the addition of $NaAlO_2$-NaOH solution to obtain mixed or slurry solution. The result of hydrothermal process was white solid and colorless filtrate. The solid formed has a pH of 14 indicating that the zeolite-X was formed because the zeolite-forming anion, Si $(OH)^{4-}$at pH> 12 [23].

3.3. Characterization of Zeolite-X from Bottom Ash

The XRD characterization of the zeolite-X has a typical zeolite-X peak at 2θ = 6.25 °; 10,35 °; 15.80 °; 20.43 °; 23.62 °; 23.73 °; 27.05 ° and 31.30 °as shown in Figure 3. In addition, the pattern of the synthesized zeolite diffractogram also has the same pattern as the peak standard (the X mark). This indicates that the bottom ash based zeolite-X has been successfully synthesized.

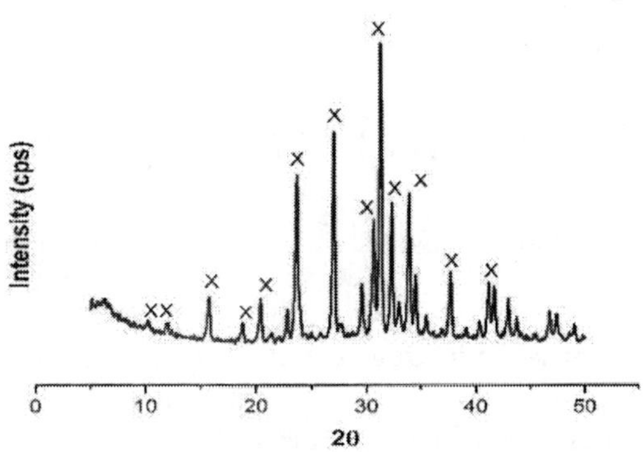

Figure 3. Difractogram of zeolite-X from bottom ash

IOP Conf. Series: Materials Science and Engineering **299** (2018) 012083 doi:10.1088/1757-899X/299/1/012083

The high level of crystallinity of the zeolite-X is also supported by the Zeolite-X morphology from the SEM characterization as shown in Figure 4. Figure 4 shows that the morphology of the zeolite-X exhibit octahedral particle with a uniform particle shape. This indicates that the hydrothermal process at 100°C for 15 hours produced a homogeneous crystalline zeolite-X with a high degree of crystallinity.

Figure 4. SEM Figure of Zeolite-X from bottom ash.

3.4. Hydrogen Adsorption

The hydrogen adsorption capacity of the zeolite-X was determined using the gravimetric method. The graph of the hydrogen adsorption capacity is shown in Figure5. Figure 5 shows that the hydrogen adsorption capacity of the zeolite-X increased continuously until 20 minutes, and then Levelled off and attained equibrium as the time was further increased. It might be due to the fact that initially there were abundant vacant adsorbent sites in the zeolite-X pore. With the progress of adsorption process, hydrogen gas adsorbed into the zeolite pore until completely reached equilibrium.

.

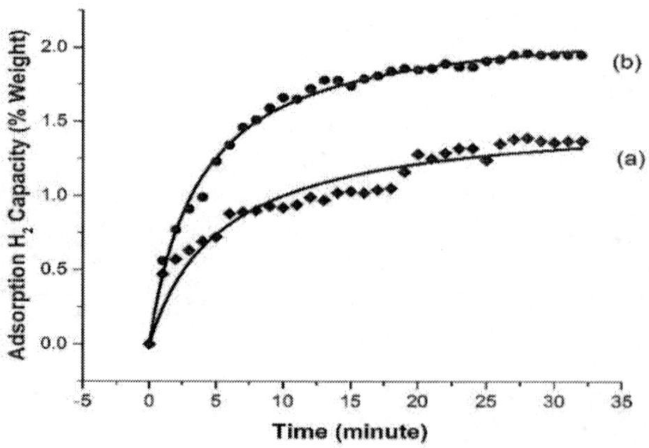

Figure 5. Graph of H_2 adsorption on (a) zeolite-X synthetic and (b) zeolite-X from bottom ash

Zeolite-X synthesized with pure chemicals has a lower value of adsorption capacity compared to the zeolite-X of the bottom ash which was about 1.03%wt of the capacity of the hydrogen gas. The zeolite-X of the bottom ash has an adsorption capacity of 1.60% wt. This value is greater 56% than zeolite-X from pure chemicals. The hydrogen storage capacity in the zeolite-X was 0.4% to 0.9 wt% under conditions of 30°C/100 bar [24-25]. This results indicate that the synthesized zeolite-X from bottom ash was more appropriate for hydrogen gas storage

Hydrogen gas adsorption process that occurs in zeolite-X belong to physisorption. Hydrogen gas physisorption process is highly dependent on the affinity of the gas on the surface of the adsorbent under certain conditions without the formation of chemical bonds. Adsorbents can absorb hydrogen gas easily to the surface of the material with weak interaction. In reciprocal interaction, asymmetric polarization of the electron charge is induced in the molecule that forms a temporary dipole moment and the hydrogen molecule becomes attracted by the electrostatic force. The interaction of the pull of the gas and the adsorbent comes from the force generated by changes in the charge distribution of the gas molecule and the surface of the adsorbent, thereby generating a pull between the induced and fluctuating dipole. However, at close distances the occurrence of overlap between the electron clouds of the gas molecules and the adsorbents results in repulsions that significantly increase rapidly [26].

FTIR instrument was used to study the interaction between hydrogen gas and the zeolite as shown in Figure 6.Figure 6 showed that the synthesized zeolite-X from the bottom ash have spesific character for zeolite-X. Spectra in 447 cm^{-1} region is a sign of bending vibration of TO_4. Double 6 ring structure connected with sodalite arrangement by absorbing band at 556 cm^{-1}. Vibration of symmetry from (Si-O-Al) at 666 cm^{-1}, while asymmetric strain vibration (Si-O-Al) at 949 cm^{-1}. In a previous study reported that the infrared spectra obtained on commercial zeolite-X occurred in regions 450 cm-1 to 1060 cm^{-1} with details of the region of 458 cm^{-1} is an indication of the bending vibration of TO, vibration symmetry in the area of 668-690 cm^{-1} and asymmetric vibrations occurred in the region 971-1060 cm^{-1}[27]. Other studies have also reported that zeolite-X has a characteristic functional group on the FTIR spectra at 456-1053 cm^{-1}. In addition to identify the typical character of zeolite-X, FTIR spectra results can also be used to determine the presence of solids such as zeolite-P and sodalite in the formed zeolite-X. Both impurities will block the double ring area on zeolite-X, thus the Infrared spectra for the double ring will not appear [28].

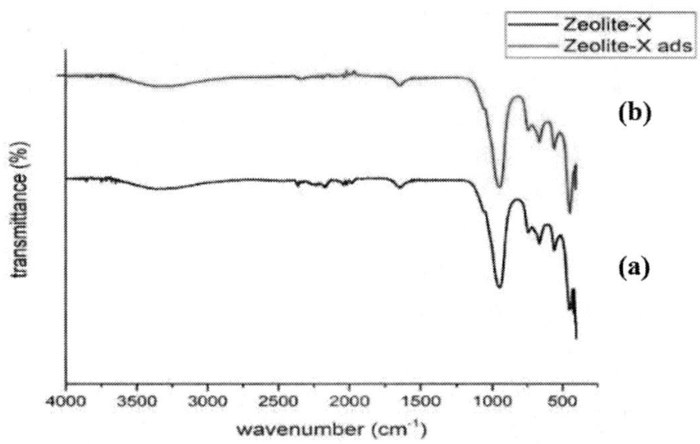

Figure 6. Spectra Zeolite-X before (a) adsorption, (b) after

The FTIR spectra of zeolite-X after the H_2 gas adsorption process in Fig. 6 (b) shows a difference with the zeolite-X initial spectra prior to adsorption. The difference is in the form of peak shift spectra and decrease in intensity. The peak in the 1640 cm^{-1} area decreased by 0.38%, while peak shift occurs for spectra at 3316 cm^{-1} to 3356 cm^{-1} along with low intensity decrease. The phenomenon of shifting and decreasing the intensity of the zeolite-X spectra results from the presence of a typical functional group interaction with the adsorbate used. FTIR spectra that experience a shift or decrease in intensity is the specific spectra for OH groups in H_2O molecules.

4. Conclusion

According to the discussion results, it can be concluded that high purity zeolite-X can be synthesized from the bottom ash through a fusion process with alkali at high temperatures, followed by hydrothermal processes for 15 hours at 100°C. The result showed that zeolite-X synthesized from the bottom ash has hydrogen gas adsorption capacity greater than zeolite-X from pure chemicals, which is up to 1.60%.

5. References

[1] Hirscher M 2009 *Handbook of Hydrogen Storage: New Materials for Future Energy Storage* (Germany: Wiley-VCH Verlag GmbH and Co. KgaA) p 12

[2] Schlapbach L and Zuttel A 2001 Hydrogen- storage materials for mobile applications *J. Nature.* **414** p 353

[3] Zuttel A, Borgschulte A and Schlapbach L 2008 *Hydrogen as a Future Energy Carrier* (Germany: Willey – VCH)

[4] Strobel R., Garche J, Moseley P T, Jorissen L and Wolf G 2006 Hydrogen storage by carbon materials *J. Journal of Power Sources* **159** p 781

[5] Morris R E and Wheatley P S 2008 *Gas Storage in Nanoporous Materials* vol 47 (Germany: Wiley-VCH Verlag GmbH Co.) p 4966

[6] Breck W D 1974 *Zeolit Molecular Sieves* (New York: John Wiley & Sons, Inc.)

[7] Li J R, Ma Y, McCarthy M C, Sculley J, Yu J, Jeong H K, Balbuena P B, and Zhou H C 2011 Carbon dioxide capture-related gas adsorption and separation in metal-organic frameworks *J. Coordination Chemistry Reviews* **255** p 1791

[8] Hardie S M L, Garnett M H, Fallick A E, Rowland A P and Ostle N J 2005 Carbon Dioxide Capture Using a Zeolite Molecular Sieve Sampling System for Isotopic Studies (^{13}C and ^{14}C) of Respiration *J. Radiocarbon* **47** p 441

[9] Mosher K 2011 *The Impact of Pore Size on Methane and CO_2 Adsorption in Carbon* (Stanford: Stanford University)

[10] Chang H L and Shih W H 2000 Synthesis of Zeolite A and X from fly ash and Their Ion Exchange Behaviour With Cobalt Ion *J. Industrial Engineering Chemical Research* **39** p 4185

[11] Molina A and Poole C 2004 A Comparative Study Using Two Methods To Produce Zeolit from Fly Ash *J. Mineral Engineering* **17** p 167

[12] Miyake M, Kimura Y, Ohashi T, and Matsuda M 2008 Preparation Of Activated Carbon-Zeolite Composite Materials From Coal Fly Ash *J. Microporous and Mesoporous Materials* **112** p 170

[13] Kementerian Lingkungan Hidup 2006 *Pengelolaan Bahan dan Limbah Berbahaya dan Beracun* www.Lingkunganhidup.com.

[14] Yanti Y 2009 *Sintesis Zeolit A dan Zeolit Karbon Aktif dari Abu Dasar PLTU Paiton dengan Metode Peleburan : Thesis Magister* (Surabaya: Institut Teknologi Sepuluh Nopember)

[15] Zhely N H M 2012 *Sintesis Zeolit-X dari Abu Dasar Batubara dan Karakterisasinya sebagai Material Penyimpan Hidrogen : Prosiding Tugas Akhir* (Surabaya: Institut Teknologi Sepuluh Nopember)

[16] Robson H and Lillerud K P 2001 Verified Syntheses of Zeolitic Materials, second Revised Edition *J. Elsevier Science B. V.* p 150

[17] Faridah A M, Widiastuti N and Prasetyoko D 2012 *Karakterisasi Abu Dasar PLTU PAITON: Pengaruh Perlakukan Magnet, HCl, dan Fusi dengan NaOH : Prosiding Seminar Nasional Kimia Unesa* (Surabaya: Universitas Surabaya) p C198

[18] Wolf K J, Smeda A, Muller M, and Hilpert K 2005 Investigation on the in Fuence of Additives for SO2 Reduction during High Alkaline Biomass Combustion *J. Energy Fuels* **19** p 820

[19] Brady J E 1998 *Kimia Universitas, Asas dan Strukur Jilid I (Terjemahan Sukmariah, M. Kamianti, A., dan Tilda, S)* (Jakarta: Binarupa Aksara)

[20] Chang H and Shih W 2000 Synthesis of Zeolite A and X from fly ash and Their Ion Exchange Behaviour With Cobalt Ion *J. Industrial Engineering Chemical Research* **39** p 4185

[21] Chunfeng W, Jiansheng L, Xia S, Lianjua W, and Xiuyun S 2009 Evaluation of zeolites synthesized from fly ash as potential adsorbents for wastewater containing heavy metals *J.Enviro. Sci* **21** p 127

[22] Yaping Y, Xiaoqiang Z, Weilan Q, and Mingwen W 2008 Synthesis of Pure Zeolites from Supersaturated Silicon and Aluminium Alkali Extracts from Fused Coal Fly Ash *J. Fuel* **87** p 1880

[23] Handayani R F 2012 *Sintesis ZSM-5 Menggunakan Prekursor Zeolit Nano Kluster : Pengaruh Waktu Hidrotermal : Proseding Skripsi* (Surabaya : Departemen Kimia, Institut Teknologi Sepuluh Nopember)

[24] Lee H J, Kim Y M, Kweon O S and Kim I I 2007 Structural and morphological transformation of NaX zeolite crystals at high temperature *J. Journal of the European Ceramic Society* **27** p 561

[25] Walton K S, Abney M B and LeVan M D 2006 CO_2 adsorption in Y and X zeolites modified by alkali metal cation exchange *J. Microporous Mesoporous Mater* **91** p 78

[26] Saha B B, Jribi A, Koyama S and El-Sharkawi I I 2010 Carbon Dioxide Adsorption Isotherm on Activated Carbons *J. Journal of Chem.Eng* **56** p 1974

[27] Ozdemir D O 2013 Zeolite-X Synthesis with Different Sources *J. Journal of Chemical, Enviromental & Biological Sciences* **1** p 229

[28] Balkus J K and Ly T K 1991 The Preparation and Characterization of an X-Type Zeolite *J. Journal of Chemical Education* **68** p 1

Metal Oxide Supported Vanadium Substituted Keggin Type Polyoxometalates as Catalyst For Oxidation of Dibenzothiophene

Aldes Lesbani*, Sarah Novri Meilyana, Nofi Karim, Nurlisa Hidayati, Muhammad Said, Risfidian Mohadi, Miksusanti

Department of Chemistry, Faculty of Mathematic and Natural Sciences, Sriwijaya University, Sumatera Selatan, Indonesia

*e-mail : aldeslesbani@pps.unsri.ac.id

Abstract. Supported polyoxometalatate $H_4[\gamma\text{-}H_2SiV_2W_{10}O_{40}]\cdot nH_2O$ with metal oxide *i.e.* silica, titanium, and tantalum was successfully synthesized via wet impregnation method to form $H_4[\gamma\text{-}H_2SiV_2W_{10}O_{40}]\cdot nH_2O$-Si, $H_4[\gamma\text{-}H_2SiV_2W_{10}O_{40}]\cdot nH_2O$-Ti, and $H_4[\gamma\text{-}H_2SiV_2W_{10}O_{40}]\cdot nH_2O$-Ta. Characterization was performed using FTIR spectroscopy, X-Ray analyses, and morphology analyses using SEM. All compounds were used as the catalyst for desulfurization of dibenzothiophene (DBT). Silica and titanium supported polyoxometalate $H_4[\gamma\text{-}H_2SiV_2W_{10}O_{40}]\cdot nH_2O$ better than tantalum due to retaining crystallinity after impregnation process. On the other hand, compound H $H_4[\gamma\text{-}H_2SiV_2W_{10}O_{40}]\cdot nH_2O$-Ta showed high catalytic activity than other supported metal oxides for desulfurization of DBT. Optimization desulfurization process resulted in 99% conversion of DBT under a mild condition at 70 °C, 0.1 g catalyst, and reaction for 3 hours. Regeneration studies showed catalyst $H_4[\gamma\text{-}H_2SiV_2W_{10}O_{40}]\cdot nH_2O$-Ti was remaining catalytic activity for desulfurization of DBT.
Keywords: vanadium, Keggin type polyoxometalate, metal oxide, dibenzothiophene

1. Introduction

Fossil fuel naturally contains organic and inorganic compounds [1]. Inorganic element such as sulfur in fossil fuel caused environmental problems [2]. Therefore, clean fossil fuel without sulfur is crucial to keep green environment. The conventional process to remove sulfur from fossil fuel using hydrodesulfurization (HDS) [3]. In this process, sulfides, disulfides, and thiols were successfully removed but organocompounds such as benzothiophene, dibenzothiophene and its derivatives with the high refractory index cannot be reduced [4]. Thus, other methods are needed for the removal of organosulfur.

Recent technology to remove organosulfur has been developed such as oxidative desulfurization (OD) under mild conditions [5-6]. In OD method, organosulfur with high steric hindrance is converted to the corresponding polar sulfone which can be easily removed by extraction or by adsorption. The successful key for OD is a catalyst which is important for activation of a green oxidant such as hydrogen peroxide [7]. Hydrogen peroxide is extensively used as oxidant due to produce water as a by-product after catalytic process [8].

Various catalysts have been tested for OD process such as ionic liquid [9], mesoporous silica [10], transition metal complex [11], tungsten supported on resin [12], tungsten and molybdenum based catalysts such as polyoxometalate compounds [13]. Polyoxometalates are early transition metal-oxygen cluster with various structures, oxidation states, and acidity level [14-15]. These compounds were intensively used as a catalyst for organic synthesis and transformation including OD process. Polyoxometalates [16], peroxometallate [17], supported polyoxometalate [18], and also polyoxometalate nanocomposites [19] were applied as a catalyst for OD process with moderate to excellent yield under mild conditions. Lesbani *et.al.* also used metal oxides supported polyoxometalate as a catalyst for OD of benzothiophene with high yield conversion under mild conditions using hydrogen peroxide as oxidant [20].

Herein, polyoxometalate $H_4[\gamma\text{-}H_2SiV_2W_{10}O_{40}]\cdot nH_2O$ was supported with various metal oxides such as silica, titanium, and tantalum to know the efficiency and catalytic activity of supported various metal oxides on polyoxometalate for OD of DBT. Several factors such as catalyst weight, temperature, reaction time, and also the regeneration of catalyst were investigated systematically to obtain high conversion of DBT into the corresponding sulfone as shown in Figure 1.

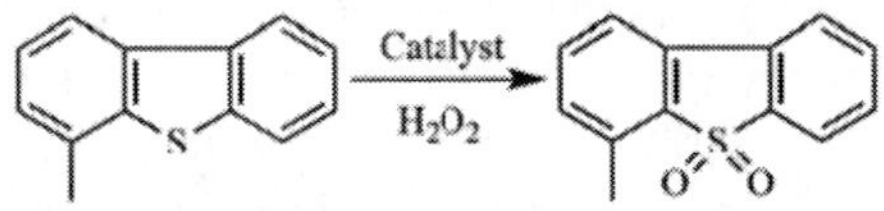

Catalyst : Metal oxides supported polyoxometalate

Figure 1. Oxidative desulfurization of DBT into corresponding sulfone.

2. Materials and Methodology

2.1. Chemical

Chemical reagents were purchased from Merck and Sigma-Aldrich such as dibenzothiophene (Sigma-Aldrich), sodium tungsten (Merck), titanium oxide (Merck), tantalum(V) chloride (Sigma-Aldrich), silicon dioxide (Merck), sodium metavanadate (Aldrich), hydrochloric acid (Merck), potassium chloride (Merck), diethyl ether (Merck), potassium carbonate (Merck), acetone (Sigma-Aldrich), ethanol (Merck), methanol (Merck), hydrogen peroxide (Merck), and ammonia (Merck).

2.2. Instrumentation

FTIR analysis was conducted using FTIR Shimadzu Prestige-21 spectrophotometer with KBr disc, and the sample was analyzed in the range of 300-4000 cm^{-1}. XRD powder analysis was performed using XRD Shimadzu LabX type-6000. Sample was scanning at $1^{o}.min^{-1}$. Morphology was analyzed using Jeol JED-2300 SEM analysis station. Catalytic reaction was monitored using Shimadzu GC 2010 Plus equipped with the RTX-1 capillary column.

2.3. Synthesis of $H_4[\gamma\text{-}H_2SiV_2W_{10}O_{40}]\cdot nH_2O$ and Characterization [21]

Compound $H_4[\gamma\text{-}H_2SiV_2W_{10}O_{40}]\cdot nH_2O$ was synthesis from $Rb_2K_2[\gamma\text{-}SiV_2W_{10}O_{40}]\cdot nH_2O$ [22] by extraction as follow: 5 g of $Rb_2K_2[\gamma\text{-}SiV_2W_{10}O_{40}]\cdot nH_2O$ was dissolved in 20 mL HCl 1 M, and the solution was transferred to a separatory funnel. Diethyl ether (10 mL) was added slowly into funnel followed by addition of a cold mixture of 15 mL diethyl ether and 15 mL concentrated HCl. The solution was mixed slowly. The heaviest layer was collected, and diethyl ether layer was removed under vacuum. The yellow solid was dissolved in minimum amount of water and solution was a vacuum to produce $H_4[\gamma\text{-}H_2SiV_2W_{10}O_{40}]\cdot nH_2O$. Characterization of $H_4[\gamma\text{-}H_2SiV_2W_{10}O_{40}]\cdot nH_2O$ was carried out using FTIR spectroscopy, XRD and SEM analyses.

2.4. Preparation of $H_4[\gamma\text{-}H_2SiV_2W_{10}O_{40}]\cdot nH_2O$-Metal Oxides and Characterization

a. $H_4[\gamma\text{-}H_2SiV_2W_{10}O_{40}]\cdot nH_2O$-Si [23]

Preparation of $H_4[\gamma\text{-}H_2SiV_2W_{10}O_{40}]\cdot nH_2O$ -Si was adopted from Newman *et.al* (2006) with slight modification was followed: 1 g of $H_4[\gamma\text{-}H_2SiV_2W_{10}O_{40}]\cdot nH_2O$ was dissolved with 50 mL methanol in 100 mL Beaker glass equipped with magnetic stirring. To the solution, 25 g of SiO_2 was added slowly, and the mixtures were stirred at medium speed for one h. The mixtures were kept overnight, and volume of solvent was reduced under vacuum to form $H_4[\gamma\text{-}H_2SiV_2W_{10}O_{40}]\cdot nH_2O$-Si, which was washed with acetone several times.

b. $H_4[\gamma\text{-}H_2SiV_2W_{10}O_{40}]\cdot nH_2O$-Ti [24]

Compound $H_4[\gamma\text{-}H_2SiV_2W_{10}O_{40}]\cdot nH_2O$-Ti was prepared as follow: 1 g of titanium oxide was mixed with 25 mL of ethanol. During stirring process, 1 g of $H_4[\gamma\text{-}H_2SiV_2W_{10}O_{40}]\cdot nH_2O$ was added. The reaction mixtures were heated at 60 °C for 30 min. The solid was washed with acetone and dried under vacuum to produce $H_4[\gamma\text{-}H_2SiV_2W_{10}O_{40}]\cdot nH_2O$-Ti.

c. $H_4[\gamma\text{-}H_2SiV_2W_{10}O_{40}]\cdot nH_2O$-Ta [25]

Compound $H_4[\gamma\text{-}H_2SiV_2W_{10}O_{40}]\cdot nH_2O$-Ta was prepared by a sol-gel method according to Leilei *et. al.* [25] as follow: 0.5 g of $H_4[\gamma\text{-}H_2SiV_2W_{10}O_{40}]\cdot nH_2O$ was dissolved with five mL of ethanol and two mL of water (solution A). Tantalum(V) chloride (0.6 g) was dissolve with 5 mL of ethanol (solution B). Solution A was mixed with solution B with stirring for one h until homogeneous sol formed. The homogeneous sol was kept overnight at 40 °C. Sol was heated at 150 °C for 3 h to form $H_4[\gamma\text{-}H_2SiV_2W_{10}O_{40}]\cdot nH_2O$-Ta. The solid was washed with ethanol and water until the filtrate was neutral. Characterization of $H_4[\gamma\text{-}H_2SiV_2W_{10}O_{40}]\cdot nH_2O$-metal oxides was conducted using FTIR, XRD, and SEM analyses.

2.5. Catalytic Oxidation of Dibenzothiophene [26]

Oxidation of dibenzothiophene was conducted using 250 mL Schlenk flask under oxygen atmospheric condition equipped with a magnetic bar and hot plate. In the flask, 0.1 g of catalyst, 0.1 g of dibenzothiophene, five mL of hexane, and 0.5 mL hydrogen peroxide were added slowly. The reaction mixtures were stirred for several hours and formation of sulfone was detected using GC at various times. Conversion of dibenzothiophene to sulfone was calculated by comparison chromatogram area of dibenzothiophene after reaction at the definite time and initial time. Oxidation of dibenzothiophene was studied through the selection of catalysts, reaction times, a weight of catalysts, and temperature reaction.

3. Result and Discussion

3.1. Preparation of Catalyst

The FTIR spectrum of the synthesized materials is shown in Figure 2. Vanadium substituted Keggin type polyoxometalate $H_4[\gamma\text{-}H_2SiV_2W_{10}O_{40}]\cdot nH_2O$ has typical IR bands at 972 cm^{-1} (band for W=O in the exterior of WO_6), 910 cm^{-1} (band for Si-O), 864 cm^{-1} and 779 cm^{-1} (bands for the W-Ob-W, and W-Oc-W, respectively) [27]. The peaks of polyoxometalate-metal oxides as shown in Figure 2 for $H_4[\gamma\text{-}H_2SiV_2W_{10}O_{40}]\cdot nH_2O$-Si (B) $H_4[\gamma\text{-}H_2SiV_2W_{10}O_{40}]\cdot nH_2O$-Ti (C), and $H_4[\gamma\text{-}H_2SiV_2W_{10}O_{40}]\cdot nH_2O$-Ta (D) has a unique vibration for each material due to metal oxides. The main peaks of $H_4[\gamma\text{-}H_2SiV_2W_{10}O_{40}]\cdot nH_2O$ after dispersion with metal oxides almost was shifted to the lower wavelength. $H_4[\gamma\text{-}H_2SiV_2W_{10}O_{40}]\cdot nH_2O$-Si has a vibration at 964 cm^{-1}, 910 cm^{-1}, and 779 cm^{-1} for Keggin type and 1088 cm^{-1} and 1381 cm^{-1} for additional peaks related with strong interaction between silica and polyoxometalate. The vibration of polyoxometalate $H_4[\gamma\text{-}H_2SiV_2W_{10}O_{40}]\cdot nH_2O$ in $H_4[\gamma\text{-}H_2SiV_2W_{10}O_{40}]\cdot nH_2O$-Si almost low and to be disappeared. Similar phenomena were found for vibration of $H_4[\gamma\text{-}H_2SiV_2W_{10}O_{40}]\cdot nH_2O$-Ti in Figure 2 C where metal oxide was dominant after the interaction. The use of tantalum as metal supported polyoxometalate gave board vibration of $H_4[\gamma$-

$H_2SiV_2W_{10}O_{40}]\cdot nH_2O$-Ta at 670-780 cm^{-1} as shown in Figure 2D attributed to the dispersion of tantalum into W-Ob-W and W-Oc-W in the both interior and exterior of polyoxometalate cage.

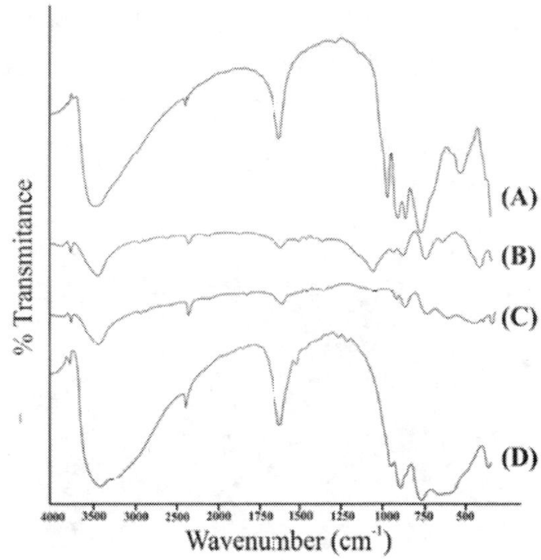

Figure 2. FTIR spectrum of $H_4[\gamma\text{-}H_2SiV_2W_{10}O_{40}]\cdot nH_2O$ (A) and $H_4[\gamma\text{-}H_2SiV_2W_{10}O_{40}]\cdot nH_2O$ -Si (B), $H_4[\gamma\text{-}H_2SiV_2W_{10}O_{40}]\cdot nH_2O$-Ti (C), and $H_4[\gamma\text{-}H_2SiV_2W_{10}O_{40}]\cdot nH_2O$-Ta (D)

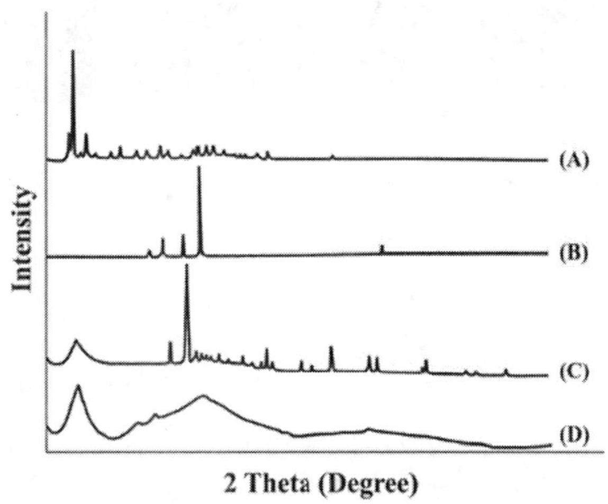

Figure 3. XRD Powder patterns of $H_4[\gamma\text{-}H_2SiV_2W_{10}O_{40}]\cdot nH_2O$ (A) and $H_4[\gamma\text{-}H_2SiV_2W_{10}O_{40}]\cdot nH_2O$-Si (B), $H_4[\gamma\text{-}H_2SiV_2W_{10}O_{40}]\cdot nH_2O$-Ti (C), and $H_4[\gamma\text{-}H_2SiV_2W_{10}O_{40}]\cdot nH_2O$-Ta (D).

Figure 3 shows XRD powder patterns of $H_4[\gamma\text{-}H_2SiV_2W_{10}O_{40}]\cdot nH_2O$ (A) and $H_4[\gamma\text{-}H_2SiV_2W_{10}O_{40}]\cdot nH_2O$-Si (B), $H_4[\gamma\text{-}H_2SiV_2W_{10}O_{40}]\cdot nH_2O$-Ti (C), and $H_4[\gamma\text{-}H_2SiV_2W_{10}O_{40}]\cdot nH_2O$-Ta (D). Crystalline polyoxometalate $H_4[\gamma\text{-}H_2SiV_2W_{10}O_{40}]\cdot nH_2O$ was observed in Figure 3a. On the other hand, although crystallinity of $H_4[\gamma\text{-}H_2SiV_2W_{10}O_{40}]\cdot nH_2O$-Si still retain, diffraction peaks of polyoxometalate at 2θ of 3-8 deg was lost. Probably the main Keggin structure was collapsed due to the strong effect of silica to exchange the heteroatom. The balancing effect between polyoxometalate and metal oxides was appeared on $H_4[\gamma\text{-}H_2SiV_2W_{10}O_{40}]\cdot nH_2O$-Ti and $H_4[\gamma\text{-}H_2SiV_2W_{10}O_{40}]\cdot nH_2O$-Ta as shown in Figure 3 C and D, respectively. In that case, diffraction of polyoxometalate at 2θ of 3-8 deg was broad, and vibration of metal oxide appeared at 2θ of 20-30 deg. Titanium and tantalum gave low exchange effect compare with silica to change heteroatom in Keggin structure.

The morphology of metal oxides supported vanadium substituted Keggin type polyoxometalate $H_4[\gamma\text{-}H_2SiV_2W_{10}O_{40}]\cdot nH_2O$ were shown in Figure 4A-D. SEM images indicated block irregular shape of $H_4[\gamma\text{-}H_2SiV_2W_{10}O_{40}]\cdot nH_2O$. SEM images of metal oxides supported $H_4[\gamma\text{-}H_2SiV_2W_{10}O_{40}]\cdot nH_2O$ were different with original polyoxometalate with the size of images in the range of 220-830 nm while polyoxometalate has 5.2 μm. As an application of materials for OD catalyst, small, regular shape and low size of materials is appropriate due to large contact interaction between substrate and catalyst.

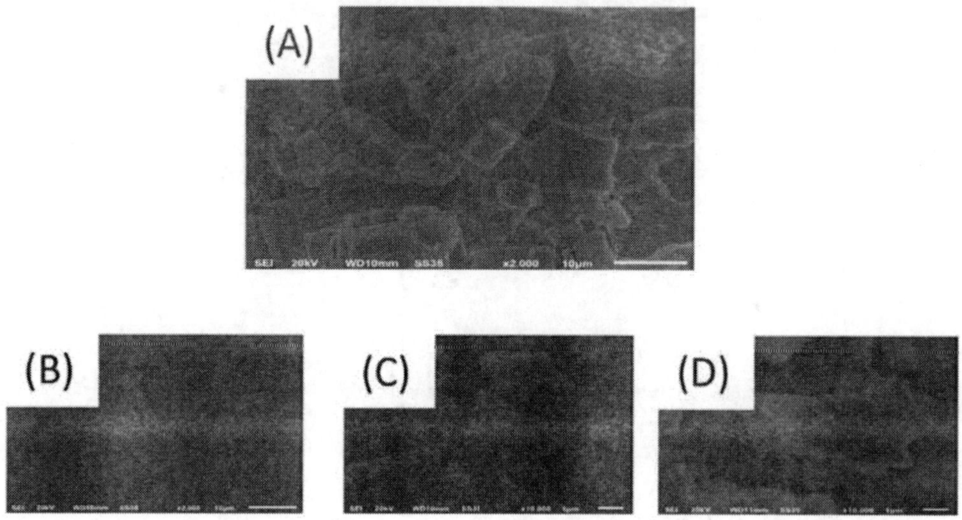

Figure 4. Morphology of $H_4[\gamma\text{-}H_2SiV_2W_{10}O_{40}]\cdot nH_2O$ (A) and $H_4[\gamma\text{-}H_2SiV_2W_{10}O_{40}]\cdot nH_2O$-Si (B), $H_4[\gamma\text{-}H_2SiV_2W_{10}O_{40}]\cdot nH_2O$-Ti (C), and $H_4[\gamma\text{-}H_2SiV_2W_{10}O_{40}]\cdot nH_2O$-Ta (D).

3.2. Catalytic Activity

3.2.1. Effect of Various Metal Oxide

The effect of metal oxides on the conversion of DBT was evaluated to know the role of metal supported polyoxometalate as a catalyst. The results are shown in Table 1. Tantalum supported Keggin type polyoxometalate $H_4[\gamma\text{-}H_2SiV_2W_{10}O_{40}]\cdot nH_2O$ has higher catalytic activity than other metal oxides supported polyoxometalate at reaction time 3 h. This result was similar with OD of dibenzothiophene using tantalum supported Dawson type polyoxometalate $(NH_4)_6[\beta\text{-}P_2W_{18}O_{62}]\cdot nH_2O$-Ta as a catalyst [28]. Original polyoxometalate $H_4[\gamma\text{-}H_2SiV_2W_{10}O_{40}]\cdot nH_2O$ shows lower catalytic activity than metal oxides supported polyoxometalate. Thus metal oxides supported polyoxometalate gave an effect into the addition of catalytic activity of desulfurization of DBT. The effect of catalytic activity depended on kinds of metal oxides as support materials.

Table 1. Oxidation of Dibenzothiophene Using Various Catalysts

No	Catalyst	% Conversion
1	$H_4[\gamma\text{-}H_2SiV_2W_{10}O_{40}]\cdot nH_2O$	31.50
2	$H_4[\gamma\text{-}H_2SiV_2W_{10}O_{40}]\cdot nH_2O$ -Si	41.40
3	$H_4[\gamma\text{-}H_2SiV_2W_{10}O_{40}]\cdot nH_2O$ -Ta	56.60
4	$H_4[\gamma\text{-}H_2SiV_2W_{10}O_{40}]\cdot nH_2O$ -Ti	38.10

3.2.2. Effect of Catalyst Weight

Desulfurization of dibenzothiophene was conducted systematically using catalyst $H_4[\gamma\text{-}H_2SiV_2W_{10}O_{40}]\cdot nH_2O$-Ta. In the first experiment, the effect of catalyst weight was investigated, and the results are shown in Figure 5.

When the catalyst weight was increased from 0.01 to 0.05 , the conversion of dibenzothiophene was increased from 19% to 32%. Further increase of the weight of the catalyst to 0.1 g led to 58.5% of dibenzothiophene conversion. On the other hand, the conversion of dibenzothiophene was sharply decreased to 36% when 0.15 g catalyst was used. Therefore, the catalyst weight of 0.1 g was preferable for oxidation of dibenzothiophene. Similar results was found for desulfurization of dibenzothiophene using $(NH_4)[\beta\text{-}P_2W_{18}O_{62}]$/Ta as catalyst [28].

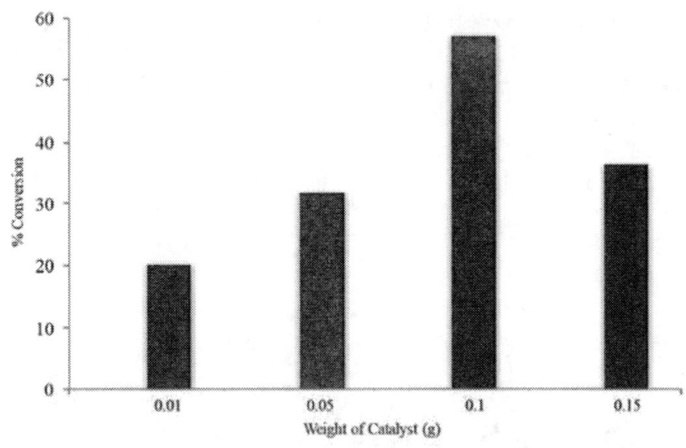

Figure 5. Effect of Catalyst Weight in the Oxidation of Dibenzothiophene Using $H_4[\gamma\text{-}H_2SiV_2W_{10}O_{40}]\cdot nH_2O$-Ta

3.2.3. Effect of Temperature

Temperature oxidation is an important parameter for oxidation of dibenzothiophene. In this research, the effect of temperature on oxidation of dibenzothiophene was conducted at mild conditions ranging from 30 – 75 °C as shown in Figure 6. Oxidation of dibenzothiophene was slowly increased from 30 – 50 °C and dramatically changed from 50 – 70 °C. The conversion of dibenzothiophene reached up to 60% at 70 °C and did not change with increasing temperature to 75 °C. Thus oxidation of dibenzothiophene was successfully conducted at mild conditions using catalyst $H_4[\gamma\text{-}H_2SiV_2W_{10}O_{40}]\cdot nH_2O$-Ta.

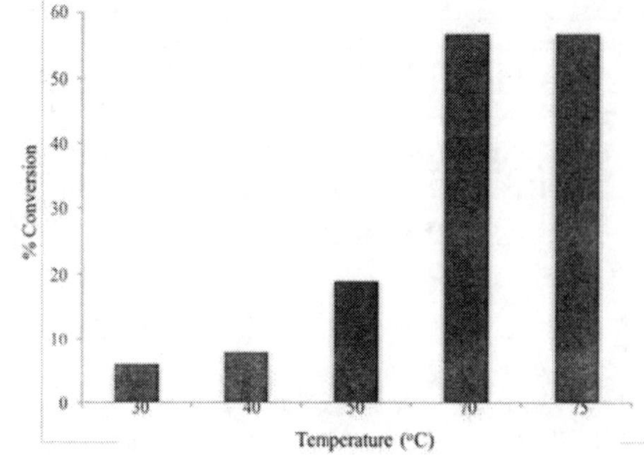

Figure 6. Effect of Temperature in the Oxidation of Dibenzothiophene Using $H_4[\gamma\text{-}H_2SiV_2W_{10}O_{40}]\cdot nH_2O$-Ta

3.2.4. Effect of Reaction Time

Desulfurization process of dibenzothiophene was continued using $H_4[\gamma\text{-}H_2SiV_2W_{10}O_{40}]\cdot nH_2O$-Ta as a catalyst in various reaction time. Figure 7 shows the effect of reaction time to desulfurization of 4-MDBT using $H_4[\gamma\text{-}H_2SiV_2W_{10}O_{40}]\cdot nH_2O$-Ta. Reaction time was studied from 1 – 4 h. At the reaction time increase, the conversion of dibenzothiophene was gradually increased until three h to give almost 60% conversion. It can be seen form Figure 7 that at 1 h reaction time, the conversion of dibenzothiophene was 39% and almost similar with reaction at 2 h. Reaction time at four h was slightly decreased from 60% at three h to 49% at four h. Therefore, the optimum oxidation of dibenzothiophene using $H_4[\gamma\text{-}H_2SiV_2W_{10}O_{40}]\cdot nH_2O$-Ta was achieved at 3 h reaction time.

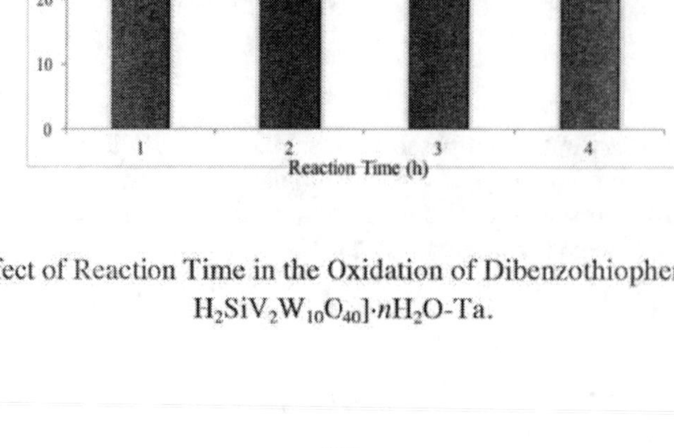

Figure 7. Effect of Reaction Time in the Oxidation of Dibenzothiophene Using $H_4[\gamma\text{-}H_2SiV_2W_{10}O_{40}]\cdot nH_2O$-Ta.

3.2.5. Reusability of Catalyst

The $H_4[\gamma\text{-}H_2SiV_2W_{10}O_{40}]\cdot nH_2O$-Ta catalyst can be removed from reaction mixtures due to heterogeneous properties of $H_4[\gamma\text{-}H_2SiV_2W_{10}O_{40}]\cdot nH_2O$-Ta. Catalyst $H_4[\gamma\text{-}H_2SiV_2W_{10}O_{40}]\cdot nH_2O$-Ta was collected from the reaction mixtures by conventional filtration. The catalyst was dried and washed using methanol and water, respectively. Furthermore, the catalyst was dried at 110 °C before use.

Desulfurization of dibenzothiophene using reuse catalyst was conducted using 0.1 g catalyst at 70 °C, for three hours reaction time. The results as shown in Table 2 showed that reuse catalyst was slightly decreased the catalytic activity of $H_4[\gamma\text{-}H_2SiV_2W_{10}O_{40}]\cdot nH_2O$-Ta about 24%. This condition may be attributed to inactivation of several active sites on $H_4[\gamma\text{-}H_2SiV_2W_{10}O_{40}]\cdot nH_2O$-Ta during recovery or oxidation processes.

Table 2. Recycling of Catalyst $H_4[\gamma\text{-}H_2SiV_2W_{10}O_{40}]\cdot nH_2O$-Ta

Recycling Catalyst	% DBT Conversion
Fresh catalyst	60.00
1st recycle	45.06

4. Conclusion

Supported polyoxometalate $H_4[\gamma\text{-}H_2SiV_2W_{10}O_{40}]\cdot nH_2O$ has high crystallinity after impregnation process with bulky structure. Supported polyoxometalate $H_4[\gamma\text{-}H_2SiV_2W_{10}O_{40}]\cdot nH_2O$ with metal oxides showed moderate to high catalytic activity with regeneration catalyst ability for first reuse process.

5. References

[1]. Dincer, I., Zamfirescu, C. 2014. Fossil Fuel and Alternatives: Advanced Power Generation Systems. Chapter 3, 95-141.

[2]. Zhao, Y., Ma, Q., Liu, Y., He, H. 2016. Influence of Sulfur in Fuel on the Properties of Difussion Flame Soot. *Atmospheric Environment*. 142, 383-392.

[3]. Xiao, J., Wu, L., Wu, Y., Liu, B., Dai, L., Li, Z., Xia, Q., Xi, H. 2014. Effect of Gasoline Composition on Oxidative Desulfurization Using a Phosphotungstic Acid/Activated Carbon Catalyst with Hydrogen Peroxide. *Applied Energy*, 113, 78-85.

[4]. Alvarez-Amparán, M.A., Cedeño-Caero, L. 2017. MoOx-VOx Based Catalysts For the Oxidative Desulfurization of Refractory Compounds: Influence of MoOx-VOx Interaction on the Catalytic Performance. *Catalysis Today*. 282(2), 133-139.

[5]. García-Gutiérrez, J.C., Laredo, G.C., García-Gutiérrez, P., Jimenez-Cruz, F. 2014. Oxidative Desulfurization of Diesel Using Promising Heterogeneous Tungsten Catalysts and Hydrogen Peroxide. *Fuel*. 138, 118-125.

[6]. Yan, X-M., Mei, P., Xiong, L., Gao, L., Yang, Q., Gong, L. 2013. Mesoporous Titania-Silica-Polyoxometalate Nanocomposite Materials For Catalytic Oxidation Desulfurization of Fuel Oil. *Catalysis Science & Technology*. 3, 1985-1992.

[7]. Jalilian, F., Yadollahi, B., Farsani, M.R., Tangestaninejad, S., Rudbari, H.A., Habibi, R. 2015. New Perspective to Keplerate Polyoxomolybdates: Green Oxidation of Sulfides with Hydrogen Peroxide in Water. *Catalysis Communications*. 66, 107-110.

[8]. Fraile, J.M., Gil, C., Mayoral, J.A., Muel, B., Roldán, L., Vispe, E., Calderón, S., Puente, F. 2016. Heterogeneous Titanium Catalysts For Oxidation of Dibenzothiophene in Hydrocarbon Solutions with Hydrogen peroxide: On the Road to Oxidative Desulfurization. *Appl. Catal. B: Environmental*. 180, 680-686.

[9]. Zhang, J., Wang, A., Li, X., Ma, X. 2011. Oxidative Desulfurization of Dibenzothiophene and Diesel Over $[Bmim]_3PMO_{12}O_{40}$. J. Catal. 279, 269-275.

[10]. Wang, D., Liu, N., Zhang, J., Zhao, X., Zhang, W., Zhang, M. 2014. Oxidative Desulfurization Using Ordered Mesoporous Silicas as Catalysts. *J. Mol. Catal. A: Chemical*. 393, 47-55.

[11]. Trehoux. A., Roux, Y., Guillot, R., Mahy, J-P., Avenier, F. 2015. Catalytic Oxidation of Dibenzothiophene and Thioanisole by a Diiron(III)Complex and Hydrogen Peroxide. *J. Mol. Catal A: Chemical*. 396, 40-46.

[12]. Long, Z., Yang, C., Zeng, G., Peng, L., Dai, C., He, H. 2014. Catalytic Oxidative Desulfurization of Diebnzothiophene Using Catalyst of Tungsten Supported on Resin D152. *Fuel*. 130, 19-24.

[13]. Wang, R., Zhang, G., Zhao, H. 2010. Polyoxometalate as Effective Catalyst for the Desulfurization of Diesel Oil. *Catalysis Todey*. 149, 117-121.

[14]. Hill, C.L. 2007. Progress and Challenges in Polyoxometalate-Based Catalysis and Catalytic Materials Chemistry. *J. Mol. Catal. A: Chemical*. 262, 2-6.

[15]. Sinaga. L., Lesbani. A. 2017. Thermal Sbatility Effect of $H_4[PVMo_{11}O_{40}]/SiO_2$. Science and Technology Indonesia. 2,1, 25-28.

[16]. Trakarnpruk, W., Rujiraworawut, K. 2009. Oxidative Desulfurization of Gas Oil by Polyoxometalates Catalysts. *Fuel Processing Technology*. 90, 411-414.

[17]. Torres-García, E., Galano, A., Rodriguez-Gattorno, G. 2011. Oxidative Desulfurization (ODS) of Organosulfur Compounds Catalyzed by Peroxo-metallate Complexes of WOx-ZrO_2 : Thermochemical, Structural, and Reactivity Indexes Analyses. *J. Catal*. 282, 201-208.

[18]. Qiu, J., Wang, G., Zhang, Y., Zeng, D., Chen, Y. 2015. Direct Synthesis of Mesoporous $H_3PMo_{12}O_{40}/SiO_2$ and Its Catalytic Performance in OxidativeDesulfurization of Fuel Oil. *Fuel*. 147, 195-202.

[19]. Rezvani, M.A., Shojaie, A.F., Loghmani, M.H. 2012. Synthesis and Characterization of Novel Nanocomposite, Anatase Sandwich Type Polyoxometalate, as a Reusable and Green Nano Catalyst in Oxidation Desulfurization of Simulated Gas Oil. *Catalysis Communications*. 25, 36-40.

[20]. Lesbani, A., Marpaung, A., Fithri, N.A., Mohadi, R. 2016. 12-Tungstophosphoric Acid/Silica catalyst for Oxidation of Benzothiophene. *Asian. J. Chem*. 28(3): 617-621.

[21]. Nakagawa, Y., Mizuno, N. 2007. Mechanism of $[\gamma\text{-}H_2SiV_2W_{10}O_{40}]^{4-}$ Catalyzed Epoxidation of Alkanes with Hydrogen Peroxide. *Inorg. Chem*. 46 (5): 1727-1736, doi: 10.1021/ic0623258.

[22]. Canny, J., Thouvenot, R., Tézé, A., Hervé, G., Leparulo-Loftus, M., Pope, M.T. 1991. Disubstituted Tungstosilicates. 2.γ- and β-Isomers of $[SiV_2W_{10}O_{40}]^{6-}$: Syntheses and Structure Determinations by ^{183}W, ^{51}V, and ^{29}Si NMR Spectroscopy. *Inorg. Chem*. 30(5): 976-981, doi: 10.1021/ic00005a020.

[23]. Newman, A.D., Brown, D.R., Siril, P., Lee, A.F., Wilson, K. 2006. Structural Studies of High Dispersion $H_3PW_{12}O_{40}/SiO_2$ Solid Acid Catalysts. *Phys. Chem. Chem. Phys*. 8(24): 2893-2902, doi: 10.1039/b603979k.

[24]. Poźniczek, J., Lubańska, A., Micek-Ilnicka, A., Mucha, D., Lalik, E., Bielański, A. 2006. TiO_2 and SiO_2 Supported Wells-Dawson Heteropolyacid $H_6P_2W_{18}O_{62}$ as the Catalyst for MTBE Formation. *Appl. Catal. A:General*. 298(1):217-224, doi: 10.1016/j.apcata.2005.10.013.

[25]. Leilei, X., Yihang, G., Huiru, M., Jianguo, G. 2013. Heterogeneous Acid Catalytic Esterification by Porous Polyoxometalate-Tantalum Pentoxide Nanocomposites. *Journal of Wuhan University of Technology-Mater.Sci.Ed*. 28(3):580-585, doi: 10.1007/s11595-013-0734-1.

[26]. Lesbani, A., Agnes, A., Saragih, R.O., Verawaty, M., Mohadi, R., Zulkifli, H. 2015. Facile Oxidative Desulfurisation of Benzothiophene Using Polyoxometalate $H_4[\alpha\text{-}SiW_{12}O_{40}]/Zr$ Catalyst. *Bull. Chem. React. Eng. Catal*. 10(2): 185-191, doi: 10.9767/bcrec.10.2.7734.185-191.

[27]. Lesbani. A., Gultom. E.O.J. 2017. Synthesis and Characterization of Metal Oxides Supported Keggin Type Polyoxometalate $Rb_2K_2[H_2SiV_2W_{10}O_{40}].nH_2O$. *Science and Technology Indonesia*. 2.2, 50-55.

[28]. Mohadi, R., Teresia, L., Fithri, N.A., Lesbani, A., Hidayati, N. 2016. Oxidative Desulfurization of Dibenzothiophene Using Dawson Type Heteropoly Compounds/Tantalum as Catalyst. *Indones. J. Chem*. 16(1): 105-110.

6. Acknowledgement

This research is a part of Hibah Profesi 2017. AL thank to Universitas Sriwijaya for "Hibah Profesi 2017" contract number 987/UN9.3.1/PP/2017. Authors thank Integrated Research Laboratory of Graduate School, Sriwijaya University for GC equipment for this experimental research.

Adsorption of Free Fatty Acid (FFA) in Low-Grade Cooking Oil Used Activated Natural Zeolite as Adsorbent

Monika Larasati Tres Ayu Putranti, Sang Kompiang Wirawan, I Made Bendiyasa

Department of Chemical Engineering, Universitas Gadjah Mada, Yogyakarta 55281 Indonesia

Corresponding author: skwirawan@ugm.ac.id

Abstract. Adsorption of free fatty acid (FFA) in low-grade cooking oil using active natural zeolite adsorbent was done as an effort to improve the quality of low-grade cooking oil so that it can fulfill the standard of fried oil which has been set on SNI 01-3741-2013. Adsorption was carried out with natural zeolite which activated with HCl and NaOH solution followed by the calcination process. The results showed that the NaOH activated zeolite decreased FFA content in low-grade cooking oil more than the HCl activated natural zeolite, with optimum NaOH concentration was 0.75 M. In the adsorption equilibrium analysis with temperature variation (25 °C, 40 °C, 80 °C), obtained that adsorption of FFA with NaOH activated natural zeolite follows Adsorption Isotherm Freundlich Model with equilibrium constant value was 20,5873; 0,9629 dan 0,8053.

1. Introduction

Cooking oil is one of the important foodstuffs used to process foodstuffs. Cooking oil serves as a heat conductor, provides calorific value, and provides a savory taste in food. The main content of cooking oil is triglycerides (triacyl glycerol) formed from glycerol and various fatty acids [1]. In addition to triglycerides, cooking oil also contains small amounts of complex lipids, sterols, free fatty acids (FFA), waxes, oil-soluble pigments, and hydrocarbons [2]. Based on the number of the refining process, there are two types of cooking oil on the market, they are packaged cooking oil and low-grade cooking oil. Low-grade cooking oil has low quality because it only has one filtering that causes the color is not clear and contains FFA higher than packaged cooking oil. Budiyanto, et al. [3], Tangkudung, et al. [4], and Lempang, et al. [5] have also proved that low-grade cooking oil has high FFA content, even exceeding the upper limit of FFA levels set at SNI 01-3741-2013 i.e. 0.3%.

The content of free fatty acids in oil is one measure of the quality of cooking oil, where FFA is expressed by the acid number. The high content of FFA can cause cooking oil to become rancid faster, where FFA can react further into aldehyde and ketone which is an indication of the occurrence of rancidity on cooking oil. In addition to causing rancidity in oil, the high consumption of cooking oil with a high free fatty acid content will increase levels of Low-Density Lipoprotein (LDL) or bad cholesterol in the blood.

The data from the government obtained in 2015 indicates that consumption of low-grade cooking oil is about 70%, and the rest already use packaged cooking oil [6]. The amount of low-grade oil consumption is not matched with the quality of low-grade cooking oil, it is necessary to improve the quality of low-grade cooking oil that meets the SNI standard with low operating cost. In general, the separation of FFA from the oil is carried out by reacting with an alkaline solution or another reactant to form a soap, or it may be called a neutralization process. The process of neutralizing cooking oil is usually done by using NaOH or Na_2CO_3. However, the use of NaOH with too high concentrations can result in saponification of small amounts of triglycerides, thereby reducing the yield of oil. While neutralization with Na_2CO_3, it produces soap that is difficult to separate, because the formation of CO_2 gas released from carbonate so as to cause foam in the oil [7]. In addition, when viewed in terms of green chemistry, the neutralization process is still using reagents that are toxic.

Based on that, the alternative to remove FFA from low-grade cooking oil is needed, where the decrease of FFA content in low-grade cooking oil using adsorbent is considered simple, cheap, and efficient. One interesting material to be developed as an adsorbent is a natural zeolite that has a large surface area and high selectivity. However, natural zeolites have some disadvantages, due to the irregular impurity and structure of the aluminosilicate framework. The presence of such impurities can reduce the activity of zeolites as adsorbents. To increase the adsorption capacity of natural zeolite, the activation will be done physically and chemically.

Widayat and Haryani [8] used natural zeolites that are only physically activated (crushed) in the study of acid number reduction, peroxide number and absorbance in the process of refining used cooking oil. Handoko et al [9] also improved the quality of cooking oil by using activated natural zeolite. In his research, adsorption was performed with two types of active zeolites, namely NZA and H5-NZA. NZA is a natural zeolite that is physically activated (crushed and heating) and chemically by using HF and HCl acid solutions. While H5-NZA is calcined NZA in Muffle Furnace (calcination without nitrogen gas) followed by a hydrothermal process, calcination with nitrogen gas and oxidation with oxygen gas. Astuti [10] decreased FFA of palm oil (CPO) using natural zeolite of Lampung which was activated using HCl solution.

The prediction of the interaction between FFA and surface of adsorbent delivered by some researchers [9] [11] states that the carboxyl group in FFA interact with the active side of the adsorbent so that FFA can be absorbed on the adsorbent. The active side of the zeolite that can interact with the FFA molecule is the hydroxyl group which is the Bronsted site and the cations contained in the silica-alumina framework of zeolite.

Acid-activated zeolites such as HCl or H_2SO_4 will undergo a dealumination process, i.e. the process of reducing the alumina atoms from the zeolite framework causing an increase in the Si/Al ratio and the acidity of the zeolite. While base activation causes bond dissolution due to Si extraction from the zeolite framework (desilication) resulting in a decrease of Si/Al ratio in zeolite. Zeolites with low Si/Al ratios tend to be more hydrophilic [12]. Treatment with acids on zeolites is also shown to cause the zeolite to become more hydrophobic so that the adsorption capacity of the polar compound will decrease [13]. The carboxyl group is a functional group having polar properties, so if the active side of the adsorbent is hydrophilic then FFA can be absorbed optimally. In this study comparing the adsorption power of FFA in natural zeolite without activation and that activated with HCl and NaOH, where the active natural zeolite having better adsorption will be determined the optimum condition and its equilibrium constant.

2. Methods

2.1. Activation of Natural Zeolite

The natural zeolite from Klaten was washed with distilled water periodically, then decanted and dried in an oven at 100 °C. Natural zeolite was then washed, crushed, and sieved with 100-mesh filtration. Natural zeolite powder was activated with 1 M HCl and NaOH solution with stirring for 24 hours at 25 °C. Active natural zeolite washed with distilled water to neutral and calcined at 400 °C for 3 hours.

2.2. FFA Adsorption Using Activated Natural Zeolite

Natural zeolite and activated natural zeolite were mixed with low-grade cooking oil in a vial bottle with 2:8 ratio and put in a shaker. IT was mixed in a shaker for 5 hours at 25 °C. Samples of adsorbed low-grade oil were then taken for analysis of their FFA amount.

The better zeolite in having FFA adsorption was analyzed equilibrium by adsorption on low-grade cooking oil with a variation of acid/base solution concentration and adsorption temperature (25 °C, 40 °C, 80 °C).

2.3. Analysis

2.3.1. Analysis of free fatty acid content (% FFA)
The free fatty acid amount was determined by titration of an alcoholic oil solution. A one-gram sample of low-grade cooking oil is weighed in Erlenmeyer and added with neutral ethanol and 3 drops of PP indicator, then heated to boiling on an electric heater. The sample solution is titrated with 0.01 M NaOH solution until the color change from clear to pink and it is not lost for 30 seconds. The free fatty acid content is calculated by the following calculations

$$\%FFA = \frac{M_{NaOH} x V_{NaOH} x Mr_{FFA}}{m_{sampel}} x100\% \quad (1)$$

where:

M_{NaOH} : NaOH concentration (M)
V_{NaOH} : Volume of NaOH (mL)
Mr_{FFA} : The molecular weight of FFA (oleic acid = 282 g/mol)
M_{sample} : Sample weight of low-grade cooking oil

2.3.2. Analysis of isothermal adsorption
The data obtained from the research were FFA concentrations at equilibrium (C_e), and the q_e value (fraction of FFA adsorbed on the pore surface of zeolite) can be calculated using the following equation

$$q_e = \frac{V(C_{A_0} - C_e)}{m} \quad (2)$$

Then make a plot between C_e and q_e to get the isothermal adsorption curve. The obtained graph was compared with the equilibrium equation used to obtain an isotherms adsorption model which according to the experiment. The determination of the isothermal adsorption model is performed by comparing the relative coefficient of each isothermal adsorption linear graphics.

3. Results

3.1. Characterization of Natural Zeolite
Character changes after treatment of both thermal and alkaline activation were proved by characterization methods using FT-IR and EDX instruments. The spectra of infrared spectroscopy result against natural zeolite without activation, thermal activation and acid-activated are shown in Figure 1.

Figure 1 Infrared Spectra of (a) No Activation, (b) Acid Activated and (c) Base Activated Natural Zeolite

It can be seen that there are some major absorptions that arise as a result of the vibration of the functional groups in samples of natural zeolite. In the spectra of natural zeolite, there is an absorption on the wavelength 462.92 cm^{-1} indicating the presence of Si-O-Al vibration, whereas in the spectra of acid and base activated natural zeolite has shifted to 470.63 and 455.20 cm^{-1}. The shift proves that there is a change of character of Si-O-Al in the framework of natural zeolite where the activation process is capable of causing desillation and dealumination reactions on the aluminosilicate structure.

Interpretation of the covalent bonds of the functional groups composing the natural zeolite framework is presented in Table 1.

Table 1 Interpretation of Infrared spectra of Natural Zeolite

Wave Number (cm^{-1})	Interpretation of Vibration Absorption Type from Functional Group
462.92	Si-O-Al
794.67	Si-O
1211.29	TO_4 internal asymmetric stretch vibration
1049.28	Si-O-Si
3448.72	stretch vibration of O-H from H_2O, Si-OH and/or free OH

The change of Si/Al ratio is determined by analysis using Electron Dispersion X-ray Spectroscopy (EDX). This technique of analysis was performed to detect the composition of the elements contained in the sample of zeolite. By characterization of Si and Al elements contained in natural zeolite structure using EDX, data of Si/Al ratio change on natural zeolite before and after activation shown in Table 2. It can be seen that in HCl activation there is an increase of Si/Al ratio, whereas at NaOH activation decreased Si/Al ratio.

Table 2 Si/Al Ratio of Natural Zeolite before and after activation

Natural Zeolite	Si/Al
No Activation	7,599
HCl activated	8,128
NaOH activated	6,788

3.2. Adsorption of FFA with Natural HCl and NaOH Activated Zeolite

In this study, free fatty acids on low-grade cooking oil were adsorbed using natural zeolites that had not been activated, and which were activated using HCl and NaOH. The results of free fatty acid content analysis before and after the adsorption process are shown in Table 3.

Table 3 The Result of FFA content analysis before and after adsorption process

Adsorbent	FFA content (%)	
	before adsorption	after adsorption
Natural Zeolite	0,4	0,33
ZAA HCl 1 M	0,4	0,2
ZAA NaOH 1 M	0,4	0,15

In Table 3 it can be seen that the greatest decrease in FFA content in low-grade cooking oil is by using activated NaOH zeolite as the adsorbent. Decreased amount of FFA with NaOH activated zeolite can reach up to 62.5%, whereas with natural zeolite without activation and HCl activated zeolite has the percent decrease only 17.5% and 50%. Natural zeolites are a porous natural mineral that has a large surface area and high selectivity. However, natural zeolites have disadvantages, due to the impurity and the irregular structure of alumina-silicate framework, the adsorption capacity of natural zeolite was very low. The existing impurities of organic, inorganic and water molecules in the natural zeolite cover the active site on the pore surface so that the adsorbate molecule cannot be absorbed. This is evidenced by the low FFA adsorbed on natural zeolite without activation. Activation zeolite by the acid solution can lead to dealumination where the Si/Al ratio will increase, and the zeolite becomes hydrophobic. While natural zeolite which is activated with a base can cause desilication where the ratio of Si / Al will decrease, and zeolite become more hydrophilic. Free fatty acids are carboxylic compounds which are having a polar side, that is in the carboxyl group (-COOH) which also is electropositive (acid) [14]. Based on this it can be explained that the FFA will be more interested in the surface of hydrophilic adsorbents, which in this study is a natural zeolite that is activated with NaOH base solution. The Prediction of interactions between FFA molecules with the polar zeolite side is shown in Figure 2.

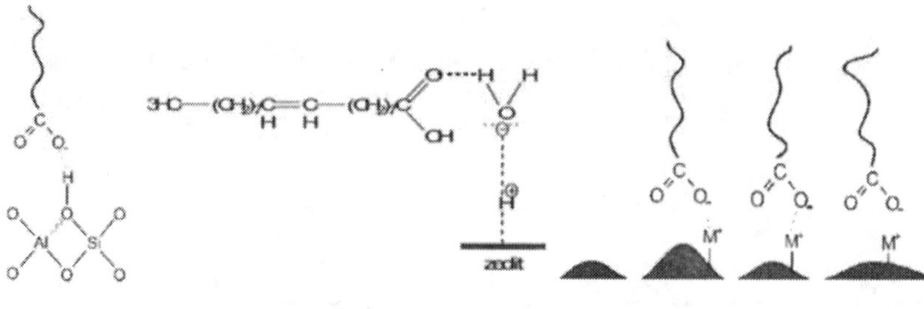

Figure 2 The Prediction of interactions between FFA molecules with natural zeolite

3.3. Adsorption of FFA with NaOH Concentration Variation for Natural Zeolite Activation

It is known that the NaOH activated natural zeolite has a better adsorption capacity against FFA compared to natural zeolite without activation and activated natural zeolite by HCl. A variation of NaOH concentration was performed to determine the optimum concentration for the activation of natural zeolite. The effect of NaOH concentration on the FFA adsorption process in low-grade cooking oil is shown in Figure 3.

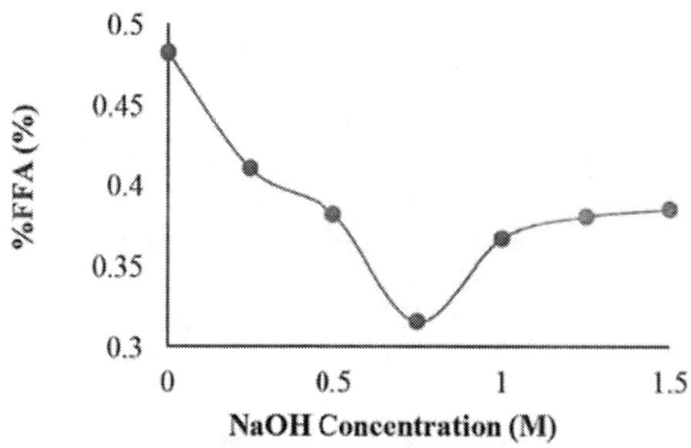

Figure 3 The effect of variation NaOH concentration on decrease of FFA amount

The results are shown in Figure 3 state that the optimum concentration of NaOH for the activation of natural zeolite is 0.75 M. It can be seen that the higher the NaOH concentration the more FFA molecules adsorbed on the zeolite pore, which is indicated by the less FFA content contained in the low-grade cooking oil. However, at concentrations of 1 M and 1.25 M NaOH levels of FFA in low-grade cooking oil after adsorption increased again. Concentrations of 1.0 and 1.25 M have made the viscosity of the NaOH solution to be very high or become more viscous so that the diffusivity of the solution into the pores of the zeolite will decrease. Decreasing the diffusivity of NaOH solution into the pores causes a lot of impurities in the pores so that the zeolite surface area will decrease and cause the adsorption power to be small [15]. In addition, the activation of natural zeolites with high NaOH concentrations may result in deformation of structural deformation structures or substantial loss, i.e. the presence of a small portion of Al^{3+} that is detached due to the high concentration of NaOH resulting in decreased micropore surface area.

3.4. Analysis of isothermal adsorption

The isothermal adsorption data is expressed as the amount of adsorbate absorbed by the adsorbent as a function of concentration and temperature so that in this research isothermal adsorption analysis is used initial concentration of FFA and temperature as an independent variable. The adsorption equilibrium depends on the interaction between the adsorbents (polar, non-polar, hydrophilic, hydrophobic, etc.) and operating conditions such as temperature, pressure and concentration [16], while the adsorption capacity of the adsorbent involves the interaction of 3 factors, i.e. the concentration of adsorbate in fluid, the concentration of adsorbate in solids and system temperature [17].

The isothermal adsorption is the relation of the number of molecules was absorbed in the adsorbent with the concentration absorbate on the fluid. Many isotherms adsorption models have been found to model a variety of adsorption equilibrium. Theoretical models are used to approach the experimental results from theoretical concepts. To evaluate the adsorption data, an appropriate isotherm equation is required for an adsorption system. From some existing isotherms equations, each has its limitations. In this research used some isotherm equations include:

Henry's Isothermal Adsorption Model

$$q_e = H.C_e \tag{3}$$

Langmuir's Isothermal Adsorption Model

$$q_e = \frac{q_m.K_L.C_e}{1+K_L.C_e} \tag{4}$$

Freundlich's Isothermal Adsorption Model

$$q_e = K_F.C_e^{\frac{1}{n}} \tag{5}$$

Based on FFA concentration data at equilibrium (C_e) it can be obtained q_e (Equation 2) and plot isothermal adsorption curve based on isothermal adsorption model Henry, Langmuir and Freundlich presented in Figure 4. The correlation coefficient obtained from the three isotherms model, on the adsorption of FFA using NaOH-activated zeolite shows that the experimental data is closest to the Freundlich model. The Freundlich approach indicates that the heterogeneous surface of the heterogeneous activated zeolite has an unequal FFA adsorption capability. This may occur because there are two possible interactions between the FFA molecules and the active site on the pore surface of the zeolite, i.e. with the hydroxyl groups at the Bronsted site and Na^+ ions. In addition to indicating the occurrence of heterogeneous adsorption, the Freundlich model approach also indicates the formation of a multilayer coating on the zeolite pore surface. Long carbon chains in FFA molecules are nonpolar, allowing to have an affinity with the nonpolar side of other FFA molecules which then form a multilayer coating.

The Freundlich adsorption isotherm model is the closest approximation of the FFA adsorption equilibrium data using the alkaline activated natural zeolite, where Freundlich's frequency linearization is performed to obtain the equilibrium constant value shown in Table 4.

Table 4 Parameter Value of Linear Equations Model of Freundlich Isothermal adsorption

Parameter	Temperatur Adsorpsi		
	25 °C	40 °C	80 °C
R^2	0,9492	0,9271	0,9628
K_F	0.5873	0.9629	0.8052
n	1.2931	1.2297	1.3706

The K_F value shows the active natural zeolite capacity, while n is an indicator of the efficiency of the adsorption process at various FFA concentrations. Increasingly larger values of n indicate that more and more zeolites absorb FFA with rising temperatures.

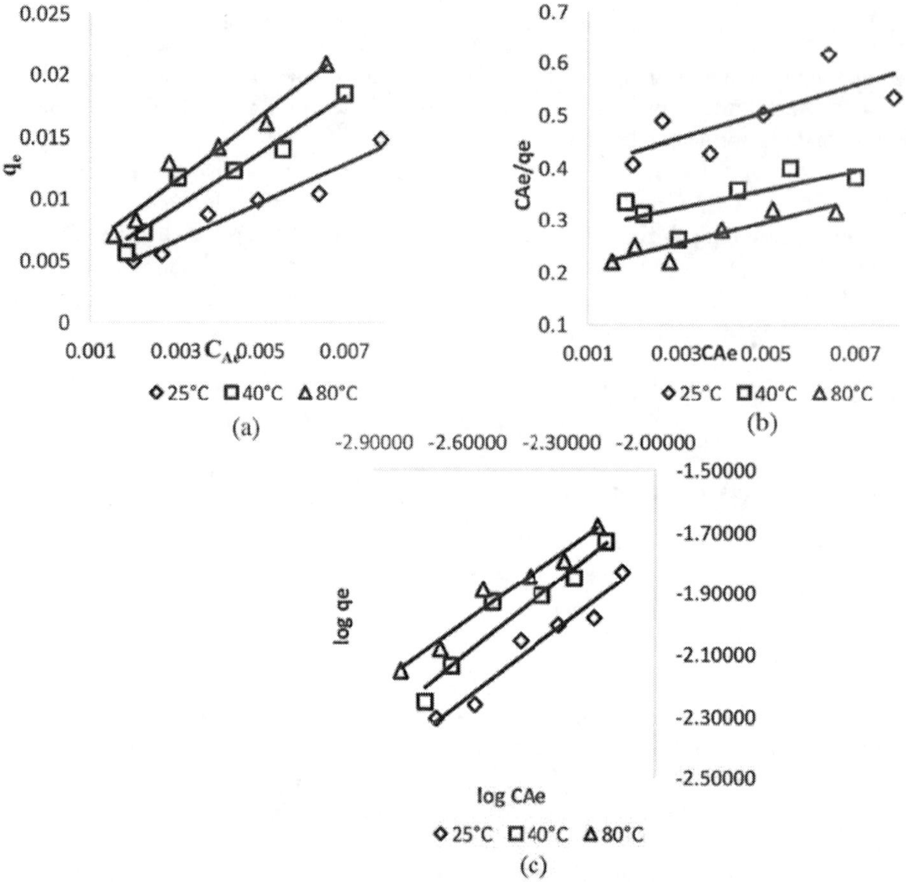

Figure 4 (a) Henry's (b) Langmuir's and (c) Freundlich's Isothermal adsorption Model Curve

4. Conclusions

The adsorption capacity of FFA in low-grade cooking oil using NaOH activated zeolite is better than natural zeolite without activation and HCl activated natural zeolite which can achieve decrease of FFA content up to 62.5%. The optimum concentration of NaOH in the activation of natural zeolite as FFA adsorbent in low-grade cooking oil is 0.75 M. The free fatty acid adsorption equilibrium model on the NaOH activated zeolite follows the Adsorption Isotherm Freundlich Model with equilibrium constant is 0.5873; 0.9629 dan 0.8053 and n value is 1.2931; 1.2297 dan 1.3706.

References

[1] Winarno, F.G., 1997, *Kimia Pangan dan Gizi*, PT. Gramedia Pustaka, Jakarta Another reference
[2] Buckle, K.A., dkk., 1987, *Ilmu Pangan*, UI Press, Jakarta
[3] Budiyanto, Zuki, M., dan Hutasoit, M.S., 2012, Ketahanan Minyak Goreng Kemasan dan Minyak Curah pada Penggorengan Kerupuk Jalin, *Jurnal Agro Industri*, **Vol 2** No. 1, ISSN 2088-5369, Bengkulu
[4] Tangkudung, M., Kadir, S., Pateda, S.M., 2013, Perubahan Kadar Asam Lemak Bebas pada Minyak Goreng Curah, Minyak Jagung dan Minyak Zaitun setelah Proses Penggorengan Berulang, *Skripsi*, Jurusan Kesehatan Masyarakat, Fakultas Ilmu Kesehatan dan Keolahragaan, Universitas Gorontalo
[5] Lempang, I.R., Fatimawali, Pelealu, N.C., 2016, Uji Kualitas Minyak Goreh Curah dan Minyak Goreng Kemasan di Manado, *Pharmacon Jurnal Ilmiah Farmasi*, **Vol. 5** No.4, 155-161
[6] Laksito, I., 201**Vol 3** No.1, 18-23, 5, Konsumsi Minyak Goreng Curah Masyarakat Masih Tinggi,

www.bisnis.news.viva.co.id/news/, diakses pada tanggal 15 Maret 2017

[7] Ketaren, S., 1986, *Pengantar Teknologi Minyak dan Lemak Pangan*, UI Press, Jakarta

[8] Widayat, S., dan Haryani, K., 2006, Optimasi Proses Adsorpsi Minyak Goreng Bekas dengan Adsorbent Zeolit Alam : Studi Pengurangan Bilangan Asam, *Jurnal Teknik Gelagar*, **Vol 17**, No.1, 77-82

[9] Handoko, D.S.P., Triyono, Narsito, Dwi, T., 2009, Peningkatan Kualitas MInyak Jelantah Menggunakan Adsorben H5-NZA dalam Reaktor Sistem Fluid Foxed Bed, *Jurnal Ilmu Dasar*, **Vol 10** No.2, 121-132

[10] Astuti, W., Junaedi, A., Suryani, E., Ismail, R., 2006, Penurunan Kadar Asam Lemak Bebas Minyak Kelapa Sawit (CPO) Menggunakan Zeolit Alam Lampung, *Prosiding Seminar Nasional Iptek Solusi Kemandirian Bangsa*, ISBN 979 368859 9, Yogyakarta

[11] Sari, V. I. P., 2007, *Perbandingan Aktivitas NZA dan NCA pada Peningkatan Beberapa Variabel Kualitas (Refreshing) Minyak Goreng Bekas dengan Reaktor Fluid Fixed-Bed*, Universitas Jember.

[12] Jozefaciuk, G., & Bowanko, G.,2002, Effect of acid and alkali treatments on surface areas and adsorption energies of selected minerals, *Clays and Clay Minerals*, **50** 771–783

[13] Sumin, L., Youguang, M., Shuhua, S., & Chunying, Z., 2009, The effect of hydrophobic modification of zeolites on Co2 absorption in different solvents, *Chinese Journal of Chemical Engineering*, **17** 36–41

[14] Rahayu, L. H., Purnavita, S., & Sriyana, H. Y., 2014, Potensi Sabut dan Tempurung Kelapa sebagai Adsorben untuk Meregenasi Minyak Jelantah. *Jurnal Momentum*, **10** 47–53

[15] Treybal, R. E., 1981, *Mass-Transfer Operations*, J. V. Brown & M. Eichberg, Eds., *McGraw-Hill, Inc*, Third Edition, **Vol. 5** Singapore.

[16] Thomas, W.J., dan Crittenden, B., 1998, *Adsorption Technology and Design*, Elsevier Science and Technology Books

[17] Richardson, J.F., Harker, J.H., Backhurst, J.R., 2002, *Particle Technology and Separation Process* in *Coulson and Richardson's Chemical Engineering*, Vol 2, Fifth edition, Butterworth-Heinemann

Insertion of bentonite with Organometallic $[Fe_3O(OOC_6H_5)_6(H_2O)_3(NO_3).nH_2O]$ as Adsorbent of Congo Red

Muhammad Said[1*], Hasja Paluta Utami[1], Ferlina Hayati[1]

[1]Dept. Of Chemistry, Faculty of Natural Science and mathematics, Sriwijaya University

*E-mail: saidusman2011@gmail.com

Abstract: The adsorption of Congo red using bentonite inserted organometallic has been investigated. The insertion bentonite was characterized using FT-IR Spectrophotometer, XRD and XRF analysis. The FT-IR characterization showed the higher intensity of peak wavenumber at 470.6 cm^{-1} for Fe_3O on the ratio 1:3. While the XRD characterization showed the shift of diffraction angle of 2θ was 5.2° and has a basal spacing of 16.8 Å. In the XRF characterization, the insertion process of organometallic occurred optimally with the percentage of metal oxide reached 71.75 %. The adsorption process of bentonite inserted organometallic compound $[Fe_3O(OOC_6H_5)_6(H_2O)_3(NO_3)\cdot nH_2O]$ showed the adsorption rate (k) is 0.050 min^{-1}, the largest adsorption capacity (b) at 70°C is 4.48 mol/g, the largest adsorption energy at temperature 30°C which is 6.4 kJ/mol Organometallic compounds. The value of the enthalpy (ΔH) and entropy (ΔS) decreased with increasing concentrations of the Congo red. Effect of pH on the adsorption on at pH 3 shows the biggest of number Congo red absorbed is 19.52 mg/L for insertion of bentonite.

Keywords: Bentonite, $[Fe_3O(OOC_6H_5)_5(H_2O)_3(NO_3)\cdot nH_2O]$, Congo red dye

1. Introduction

Organics pollutant such as coloured dye can be found industries effluent such as textiles, rubber, paper, plastic, cosmetics, etc. Discharging dyes into river although in a small amount can affect the organisme life. As dyes are designed resist with time and exposure to sunlight, they cannot be easily removed by conventional wastewater treatment. Thus, dye removal has been an important but challenging area of wastewater treatment. To remove dyes and other colored contaminants, the famous method used was adsoption [1].

Bentonite is a layered material or clay containing inorganic minerals found in nature [2]. Bentonite is used as an adsorbent [3]. Bentonite in its application as an adsorbent has deficiency among others containing impurities minerals so that its absorption is not optimal. Therefore, to optimize the absorption of bentonite activation needs to be done. The activation process aims to separate impurities naturally on bentonite. In the activation process, there are 2 ways done that are by physical way or chemical way.

IOP Conf. Series: Materials Science and Engineering **299** (2018) 012086 doi:10.1088/1757-899X/299/1/012086

In the physical process, combustion at high temperatures causes the removal of water molecules from the crystal sequence, so that two adjacent OH groups release one molecule of water [4], while in the chemical process using sulfuric acid aims to remove the metals of size small. This is due to the metals present in bentonite bound to the H^+ ions derived from the acid thus the layer acting as the insertion admission will be more open. Bentonite modification is expected to produce bentonite which has high adsorption ability and can perform insertion process that produces insulated atomic bentonite, molecule or complex compound.

Said and Palapa[5] found that the insertion process using Al_2O_3 metal oxide shows a slight increase in the distance between layers. This is indicated by the variation of contact time of adsorption of bentonite inserted metal oxide compound Al_2O_3 to ammonia at 4, 8, 16, 24 hours resulted that at 24 hours bentonite inserted Al_2O_3 able to adsorb ammonia with the largest percentage, while in an adsorption process, time 24 hours is so long that bentonite inserted Al_2O_3 metal oxide compounds can be said not yet optimally used as an adsorbent because of the size of the cation possessed by small metal Al_2O_3 oxide to become inserted bentonite.

Therefore, in this research, bentonite insertion process using organometallic compound as insertion will be done. The organometallic compounds $[Fe_3O(OOC_6H_5)_6(H_2O)_3(NO_3)\cdot nH_2O]$ have advantages such as high reactivity properties and larger cation size (macrocation). It is expected that bentonite insertion of organometallic compounds capable of inserting optimally and able to increase high adsorption capacity.

Bentonite inserted macrocation of organometallic compounds characterized using FT-IR, XRD, and XRF spectrophotometers to see the metal oxide composition. Further, bentonite inserted an organometallic compound to be used as an adsorbent to adsorb the Congo red dye. In this study, the selected adsorbate is Congo red dye, because this dye includes hazardous waste. The Congo red adsorption process that occurs can be affected by several conditions including the influence of adsorption time, the influence of concentration and temperature, and the pH of the solution. Determination of equilibrium concentration on the adsorption of activated bentonite and bentonite insertion of the organometallic compound was analyzed using UV-Vis spectrophotometer.

2. Materials and Method

2.1. Material

The materials used in this study are natural bentonite, Congo red dye, sulfuric acid, aquadest, iron (III) nitrate nonahydrate $[Fe (NO_3)_3.9H_2O]$, benzoic acid, Sodium hydroxide 0.1 M, and hydroxide acid 0.1 M. The characterization of bentonite was done with X-Ray spectroscopy Diffractometer (Shimadzu lab X-type 6000), FT-IR spectrophotometer (Shimadzu-Prestige-21), X-Ray fluorescence, PAN Analytical Type Minipod 4) and Uv-Vis spectrophotometer (Thermo Scientific Geneysis 20).

2.2. Bentonite inserted with Organometallic Compounds $[Fe_3O(OOC_6H_5)_6(H_2O)_3(NO_3)\cdot nH_2O]$

A total of 5 g of bentonite was dissolved in 200 mL of aquadest. A 15 g of the organometallic compound acting as the insertion compound is prepared by adding 100 mL of 1 M NaOH, then mixed in reflux and stirred vigorously for 24 hours at room temperature for 1 day while being fed by nitrogen gas. The resultant insertion of organometallic compounds is then washed with aquades and dried in open-air for 24 hours. The results of bentonite insertion of organometallic compounds were characterized using XRD, FTIR, and XRF spectroscopy.

2.3. Study of Adsorption of Congo Red

2.3.1. Effect of adsorption time

A total of 0.03 grams of bentonite adsorbent inserted organometallic compound and activated bentonite (control) were added to 50 mL Congo red with a concentration of 150 mg / L then stirred with a horizontal shaker at 30 min intervals. Variations in absorption time varied from 10, 20, 30, 40, 50, 60, 70, 80, and 90 min. Congo red which has gone through the adsorption process is silenced then measured its concentration by using Uv-Vis spectrophotometer to find out

2.3.2. Effect of Concentration and Temperature
The thermodynamic studies of dye adsorption on bentonite inserted macrocation adsorbents were carried out through a series of experiments by varying the concentration of dyestuff and adsorption temperature. A total of 0.03 g of adsorbent was mixed with 50 mL of 25, 50, 75, 100 and 150 (mg/L) dye solutions used as controls and for bentonite insertion of its organometallic compound using dye solutions of 10, 20, 30, 40 and 50 (mg/L), then stirred using a horizontal shaker for 1 hour at varying temperatures (30, 40, 50, 60, and 70°C). The solution was silenced and measured using a UV-Vis spectrophotometer to determine the residual concentration and dyestuff adsorbed concentration after adsorption.

2.3.3. Effect of pH
The effect of pH on dye adsorption on bentonite inserted organometallic compounds was studied by varying the initial pH of the dye solution. Adsorbent bentonite inserted as much as 0.03 grams was added to 50 mL dye solution at 25 mg / L concentration and then stirred with a horizontal shaker at room temperature for one hour. The initial pH of the dye solution is set at 3, 4, 5, 6, 8, 9, 10, 11, and 12 for inserted bentonite while for the activated bentonite (control) the initial pH of the dye solution is set to 1, 2, 3, 4, 5, 6, 8, 9, 10, 11, and 12 were added with 0.1 M NaOH addition and 0.1 M HCl of solution was allowed to stand for 10 min and measured using UV-Vis spectrophotometer to determine residual concentration and adsorbed concentration of the substance Color after the adsorption process.

2.3.4. Congo red Interaction Studies
The interaction between the adsorbent and the adsorbate is studied spectroscopically by looking at changes occurring in the adsorbent before and after the adsorption process. As much as 0.03 g of adsorbent bentonite insertion of the organometallic compound was mixed with 50 mL of dye with concentration 50 mg/L. The mixture is stirred using a horizontal shaker for 1 hour, then the mixture is separated through the filtration process. The adsorbents are then dried with an oven at 80°C. Adsorbents that have been used for the adsorption process are then analyzed using FT-IR to determine the functional group changes occurring before and after the adsorption process.

2.4. Data Analysis
The adsorption process is studied through kinetic and thermodynamic parameters. The adsorption kinetic is studied by variation of adsorption time and adsorption rate calculated based on Langmuir-Hinnshelwood adsorption equation as follows:

$$\frac{\ln\left(\frac{C_0}{C}\right)}{C} = k_1 \frac{t}{C} + \mathrm{K} \qquad \text{...} \qquad (1)$$

where:
C_0 = the initial concentrations
K_1 = the rate of adsorption
C = Concentrate Congo red after time
t = Adsorption time
K = Adsorption equilibrium constant

The thermodynamic parameters studied through variations in the concentration of Congo red dye, adsorption capacity and adsorption energy were calculated on the basis of the Langmuir equation as follows:

$$\frac{C}{m} = \frac{1}{bK} + \frac{C}{b} \qquad \text{...} \qquad (2)$$

$$E = - RT \ln K \qquad (3)$$

where:

C = The concentration of Congo red after adsorption reaches equilibrium
m = Mol Congo red adsorbed
K = Equilibrium constant
b = Adsorption Capacity
E = Adsorption Energy
R = Constanta
T = Temperature

While to find the value of coefficient of adsorbate distribution used equation as follows:

$$\ln Kd = \frac{\Delta S}{R} - \frac{\Delta H}{RT} \qquad (4)$$

.where:

Kd = Coefficient of adsorbate distribution (qe/Ce)
ΔH = Entalphi
ΔS = Entrophi
R = Constanta
T = Temperature

3. Results and Discussion

3.1. Effect of Adsorption Time

The adsorption time calculation data on the amount of adsorbed Congo red can be seen in Figure 1.

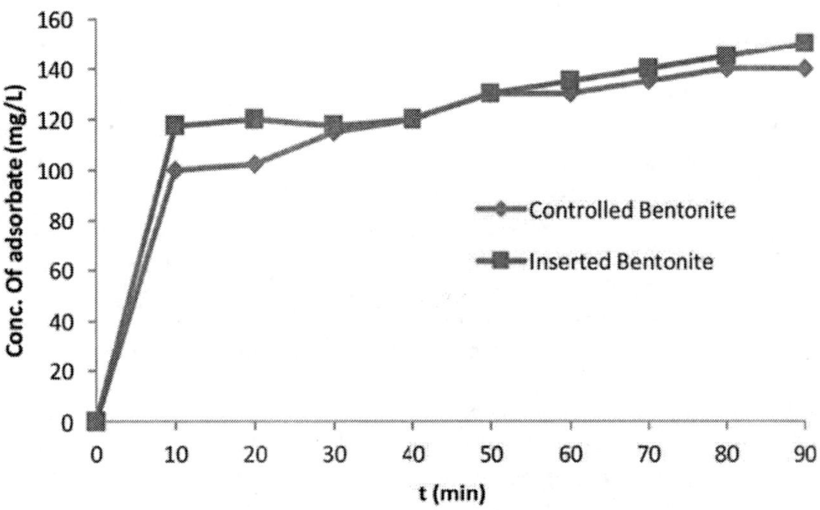

Figure 1. Influence of contact time of activated bentonite (control) and inserted bentonite to the amount of adsorbed Congo red

Figure 1 shows as the adsorption time increase, the greater the concentration of the adsorbed Congo red dye. In bentonite inserted to see the amount of the adsorbed red dye Congo increase within 20 minutes as it reaches the equilibrium time. This shows that bentonite inserted organometallic compound has reached the adsorption equilibrium. Adsorption time data can be used to determine the kinetic parameters calculation adsorption. Data of kinetic parameters presented in Figure 2 and b.

Based on the Langmuir-Hinshelwood equation, the adsorption constant rate can be calculated as the slope of the ln (Co/C)/C plot to t/C as the line equation. The value of each slope as the rate of adsorption constant (k1) and the correlation coefficient (R^2) of the plot (Co/C)/C to t/C is presented in Table1. The correlation coefficient obtained shows linear line equation. The equation of the line with the correlation coefficient (> 0.9) shows that the Congo adsorption kinetics credits the Langmuir-Hinshelwoodoold kinetics equation.

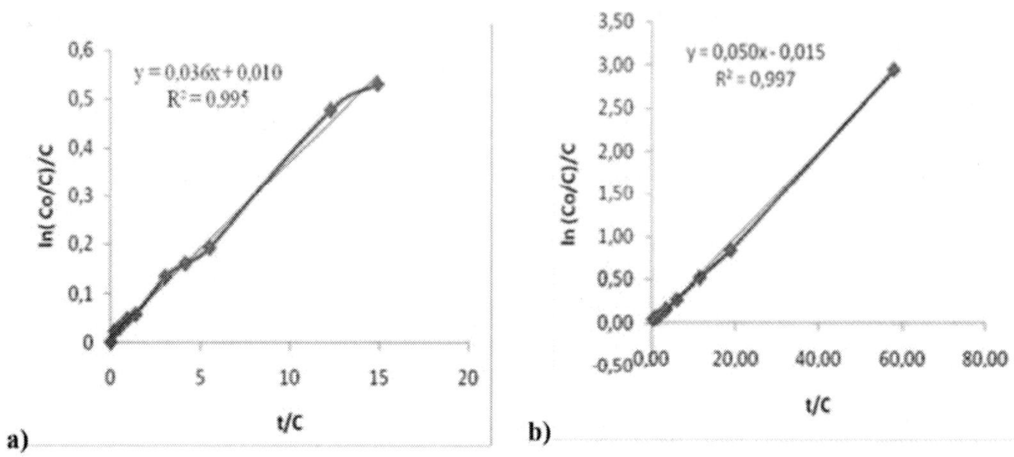

Figure 2. The bentonite adsorption kinetic parameters: a) controlled, and b) inserted bentonite

Table 1. Values of adsorption rates for Activated Bentonite (control) and Inserted Bentonite

Adsorbent	Rate of adsorption (min^{-1})	R^2
Activateded bentonite	0.036	0.995
Inserted bentonite	0.050	0.997

It is shown that the activated bentonite have the good ability as adsorbents. This is based on the basal spacing between the activation and inserted bentonite. From the previous study [6], It is found that the activated bentonite at 2 per angle shift has basal spacing 24 Å while inserted bentonite has basal spacing of 16,8 Å. The difference of basal spacing has only a slight gap. The more differences between them happened because of the adsorbate enter to inserted bentonite layer by layer. Solute attaches to the surface of montmorillonite and diffuses into the pores before finally being trapped inside of the adsorbent. The solute will be continuously deposited on the surface of the adsorbent and the layer will become thicker. The interaction force, between the solute and the adsorbent, means that the solutes cannot leave the adsorbent [7]. It is also known that the montmorillonite crystal structure consists of three layers: two units of the tetrahedral layer and one octahedral layer [8]. It can be said that the inserted bentonite rate of adsorption is higher than the activated one.

3.2. Effect of Concentration and Temperature

The curve of adsorption as the effect of concentration and temperature on the amount of adsorbed Congo red was presented in Figure 3a and b. The figures show the influence of temperature and concentration on the adsoprtion process of Congo red. It can be seen from the figure that the activated bentonite show the better performance than the inserted bentonite with it marked by the higher value of adsorbate concentration. This phenomenon was explained by another researcher[9,10]. The Congo red color has a S = O vibration with a very low intensity at 1072 cm^{-1} wave numbers. In addition, the aromatic C = C vibration appears at 1481 cm^{-1} wave numbers. The symmetric stretching vibration of the organometallic compound will reduced before the adsorption[9]. In addition, the adsoprtion capacity decreased with the increasing concentration. It is found that the higher temperature is to the advantage of adsorption and that the adsorption is an endothermic reaction. It is well known that increasing temperature may produce a swelling effect within the internal structure of absorbent, penetrating the large dye molecule also will block the pore of adsorbent. Finally, it will reduce the adsorbent performance[10].

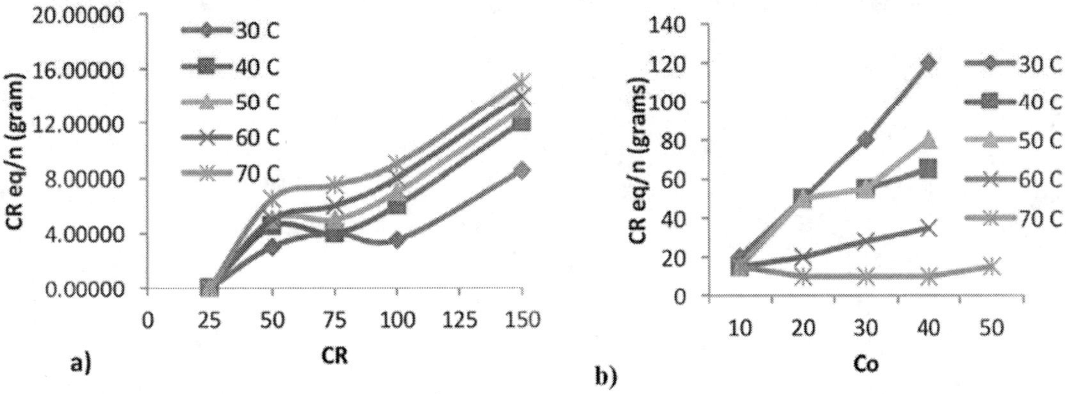

Figure 3. Effect of adsorption temperature and concentration: a) activated, and b) inserted bentonite

In addition, Figure 3 shows the number of Congo red adsorbed by activated bentonite greater than the inserted bentonite. This is due to the activated bentonite has a spacing between layers very small so when the adsorbate, Congo red directly entry to the inside of adsorbent, it is more easily filled compared to the inserted bentonite that has a greater spacing between layers. It can be concluded the activated bentonite is more reactive than inserted bentonite [11]. Furthermore, we calculate the kinetic and thermodinamic parameters using the equation 2,3 and 4. The result of kinetic parameter such as adsoprtion capacity and energy of adsorption are list in Table 2 while the thermodynamic parameters such as changes in Gibbs (ΔG°), enthalpy (ΔH°) Entropy (ΔS°) are list in Table 3.

Table 2. The kinetic parameter of adsorption process of activated and inserted bentonite

Adsorbent	Activated bentonite					Inserted bentonite				
Temp. (c)	30	40	50	60	70	30	40	50	60	70
b (mg/g)	15.6	43.4	71.4	62.5	58.8	0.35	0.59	0.72	1.33	4.48
E (kJ/mol)	-7.64	-5.32	2.21	7.02	10.9	-6.4	-3.83	-4.22	-1.93	-0.08

Table 3. The thermodynamic parameter of adsorption process of activated and inserted bentonite

Adsorbent	Activated bentonite					Inserted bentonite				
Conc. (ppm)	25	50	75	100	150	10	20	30	40	50
ΔH (kJ/mol)	69.5	81.8	59.1	54.6	42.2	59.4	153.2	190.2	198.3	124.2
ΔS (kJ/mol)	0.256	0.292	0.215	0.2	0.155	0.109	0.263	0.325	0.336	0.212

The adsorption capacity (b) of the activated bentonite (control) increased from 30 to 50°C but at 60 and 70°C, the adsorption capacity decreased. The inverted results was found to the inserted bentonite, the capacity of adsorption (b) has improved very well as temperature increases. This shows that in bentonite activated its optimum adsorption capacity at 50°C. For the adsorption energy, it is seen that the temperature of 70°C has the highest energy of adsorption. Further, the highest enthalphy and entropy occurs in inserted bentonite with Congo red concentration of 40 mg/L were 198.3 kJ/mol and

0.336 kJ/mol, respectively. This happens because adsorbent adsorbate irregularly occurs on the surface only and into the surface, it appears that even though the energy is large but the adsorption capacity is not so great and vice versa if the energy is small but it has a large amount of capacity.

3.3. Effect of pH

Effect of pH on the adsorption of Congo red dye is showed in Figure 4.

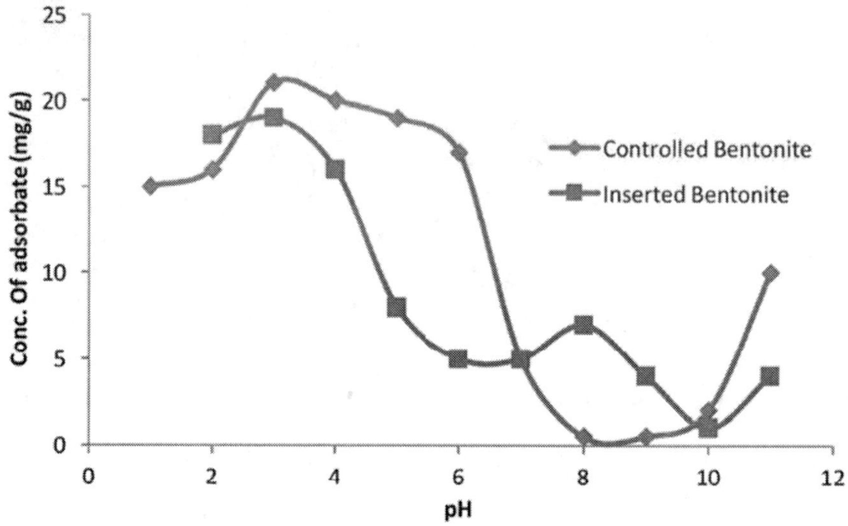

Figure 4. Effect of activated bentonite adsorption pH (control) and bentonite insertion of organometallic compound to the amount of Congo red adsorbed

The Congo red dye adsorption process by activated bentonite (control) has an initial pH of 4. After added the hydrochloric acid (HCl), the pH reached 3 and consequently the number of Congo red adsorbed was increased. In vice versa, after pH reahed 2 and 1, there was a decrease in the amount of adsorbed (Congo red). In the inverted process by adding process the sodium hydroxide (NaOH), the reducing number of Congo red was existed. However, in the of addition of sodium hydroxide (NaOH) so that the pH reaches 6 there is a decrease in the amount of adsorbed red Congo characterized by a concentration of Congo red dyestuff adsorbed by bentonite inserted into dark red. Two possible mechanisms of adsorption of Congo red may be considered: (a) electrostatic interaction between the protonated groups of bentonite and the dye, and (b) the chemical reaction between the adsorbate and the adsorbent [12].

As the pH was 8, the concentration of adsorbate adsorbed rises again. This is because of the more the addition of NaOH the more the amount of adsorbate is adsorbed and the solution is marked to be pink. Similarly, activated bentonite, Congo red adsorption by the adsorbent inserted bentonite organometallic compounds also showed a decrease in the amount of Congo red adsorbed at pH 8, 9, 10 followed by changes in the blend color to red tea. The existence of Congo red discoloration on adsorbed by the activated and inserted bentonite either in acidic or alkaline conditions based on pH of Congo red dye which naturally has effective pH values in the range 3-6. When the atmosphere is a mixture acid solution then changes color to purple, whereas under alkaline conditions the mixture becomes colorless Congo red.

3.4. Interaction Studies of Activated and Inserted Bentonite with Organometallic Compounds using FT-IR

The gradual changes occur in the vibrational frequency associated with a specific chemical bond. It concerning the peak position shift of IR spectra are based on the notion that the extent of frequency shift can be directly correlated with the level of specific molecular interactions, such as hydrogen bonding and dipole–dipole interactions. This also applies to the Congo red adsorption interaction study by activated and inserted adsorbent during before and after adsorption. Factors that affect the vibrational IR are acid activation, inserted organometallic and adsorption of congo red dye stuff process. During the acid activation process, the acid attack protons penetrate into the clay mineral layers and attack the structural OH groups[13]. By modification of bentonite through insertion of organo metallic will cause the basal spacing larger and consequently the adsorption capacity to increase. The existance of organometallic can be proven from shifting in IR spectra [14]. The vibration of the functional groups present in the Congo red dye will appear on the FT-IR spectrum with a typical wave number. The FTIR spectra before and after the activated bentonite adsorb Congo red dye contained was presented in Figure 5, while the spectra of inserted bentonite before and after Congo red dye adsorption was presented in Figure 6.

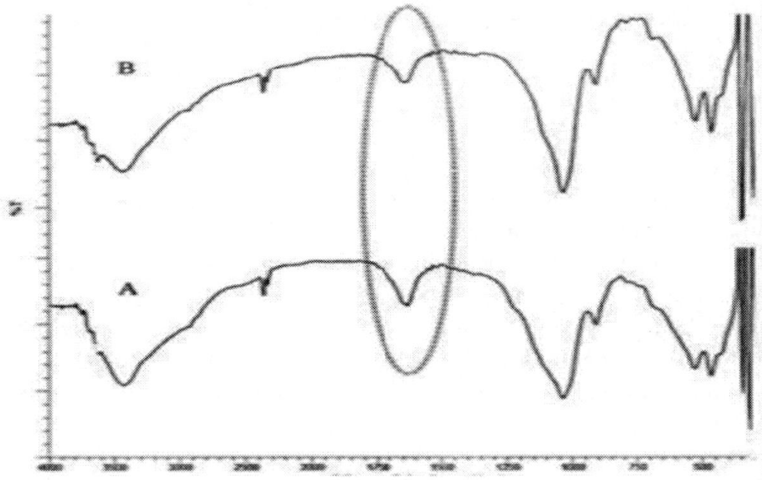

Figure 5. FTIR spectra of activated bentonite: a) before, and b) after adsorb the Congo red dye

Figure 5 shows the activated FT-IR bentonite spectrum changes before and after adsorb the non-contrasting Congo red substance. However, when viewed from the number of waves there is a difference of each bonding vibration. In the wave number 1041.5 cm^{-1} shows the vibration of Si-O-Si strain on bentonite and S = O of the Congo red dye with high intensity and sharpness. In addition, the bonding vibrations of bentonite after adsorbing Congo red appearing at the wave number 1635 cm^{-1} denotes the N=N vibration for the Congo red as a sign that the Congo red dye is an azo group.

Vibration NH contained in Congo red regional group appears at wave number 3618.4 cm^{-1}[15]. Figure 6 shows the change in the spectrum FT-IR bentonite inserts an organometallic compound before and after adsorb the visible Congo red dye. On the spectrum of inserted bentonite that have to adsorb dye Congo red visible vibration new bond indicates the group functional Congo red, which is the wave number 1635.6 cm^{-1}, which shows the absorption band for N=N as a marker of group azo.

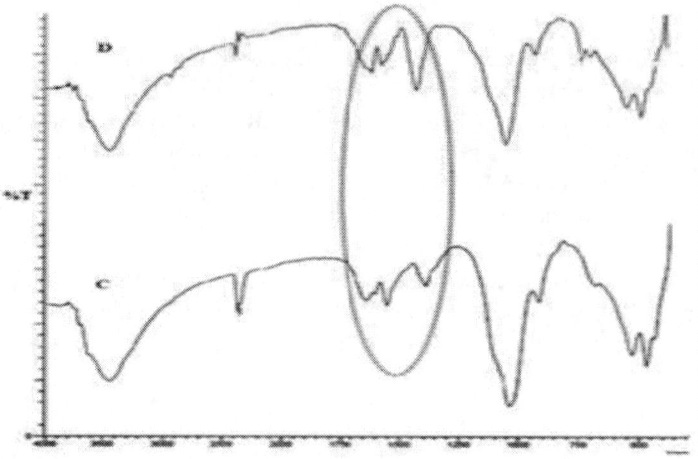

Figure 6. FTIR spectra of inserted bentonite: c) before, and d) after adsorb Congo red dye

4. Conclusion

The activated bentonite has smaller adsorption rate than the inserted bentonite i.e. 0.036 min^{-1} and 0.050 min^{-1}, respectively. The adsorption capacity of activated bentonite is greater than the inserted bentonite adsorbent as temperature increases. The greatest adsorption energy of activated bentonite was 11 kJ/mol at 60°C, while inserted bentonite was 6.4 kJ/mol at 30°C. The enthalpy of activated and inserted bentonite decreased with increasing the Congo red concentration. The lowest entropy was occured at 50 mg/L by activated bentonite of 0.155 kJ/mol, whereas in the lowest entropy of inserted bentonite was occured 0.109 kJ/mol at Congo red concentration of 10 mg/L. At pH 3 there was an increase in the amount of Congo red adsorbed by activated bentonite and inserted bentonite i.e. 20.45 mg/L and 19.52 mg/L, respectively.

Acknowledgment

The author would like to thank to Universitas Sriwijaya for the financial support through Hibah Unggulan Profesi 2017.

References

1. Ozcan S A and Ozcan A 2004 *J. Colloid & Interface Sci.* **276** 39
2. Abderrazek K, Najoua F S and Srasra E 2016 *App. Clay Sci.* **119** 229
3. Santos S C R, Oliveira Á F M and Boaventura R A R 2016 *J. Clean. Prod.* **126** 667
4. Sinaga L and Lesbani A 2017 *Sci.& Tech. Indonesia* **2(1)** 25
5. Said M and Palapa N R 2017 *Sci.& Tech. Indonesia* **2(1)** 17
6. Utami, H P and Siburian, D M 2016 *Sci.& Tech. Indonesia* **1(1)** 20
7. Said, M, Hasan, H A, Nor, M T M, Mohammad, A W 2016 Desal. & Water Treat. **57** 10490
8. Zhu H M and Vansant E E 2008 *Sep. Tech.***466** 153
9. Kaur R and Rajvir H K 2016 *Portugalle Acta* **3(3)** 185
10. Bertella F and Pergher S B C 2015 *Microporous and Mesoporous Mat.* **201** 116
11. Vitanesa, L, Purba, R N, Lesbani, A, Said, M 2017 Sci.& Tech. Indonesia **2(1)** 9
12. Namasivayam C and Kavitha D 2002 *Dyes Pigments* **54(1)** 47
13. Madejova, J 2003 *Vibrational Spectroscopy* **31** 1
14. Ikhtiyarova, G A, Ozcan, A S, Gok, O, and Ozcan, A 2012 *Clay Minerals* **47** 31
15. Wang L and Wang A 2007 *J. of Hazard. Mat.* **147** 979

Polypropylene Oil as a Fuel for Ni-YSZ|YSZ|LSCF Solid Oxide Fuel Cell

Andini W Pratiwi[1], Fitria Rahmawati[1]*, Refada A Rochman[2], Rahmat J E Syahputra[1], Arum P Prameswari[1]

[1] Research Group of Solid State Chemistry & Catalysis, Chemistry Dpartment, Sebelas Maret University, Jl. Ir. Sutami 36 A Kentingan, Surakarta 57126
[2] Chemistry Department, Faculty of Mathematics and Naatural Science, Sebelas Maret University, Jl.Ir. Sutami 36 A Kentingan Surakarta 57126

E-mail: fitria@mipa.uns.ac.id

Abstract. This research aims to convert polypropylene plastic to polypropylene oil through pyrolysis method and use the polypropylene oil as fuel for Solid Oxide Fuel Cell, SOFC, to produce electricity. The material for SOFC single cell are Ni-YSZ, YSZ, and LSCF as anode, electrolyte and cathode, respectively. YSZ is yttria-stabilized-zirconia. Meanwhile, LSCF is a commercial $La_{0.6}Sr_{0.4}Co_{0.2}Fe_{0.8}O_3$. The Ni-YSZ is a composite of YSZ with nickel powder. LSCF and Ni-YSZ slurry coated both side of YSZ electrolyte pellet through screen printing method. The result shows that, the produced polypropylene oil consist of C_8 to C_{27} hydrocarbon chain. Meanwhile, a single cell performance test at 673 K, 773 K and 873 K with polypropylene oil as fuel, found that the maximum power density is 1.729 μW. cm^{-2} at 673 K with open circuit voltage value of 9.378 mV.

1.Introduction

In the last few decades, plastic production in the world growth rapidly. It allows a serious environmental problems due to plastic waste needs almost 100 years to degrade [1]. Therefore, the pyrolysis waste plastic to produce a liquid fuel will be a good alternative to solve the environmental problems and also to produce an alternative fuel for energy supply [2]. The previous research found that the pyrolysis of polypropylene plastics in a vacuum reactor at 673 K – 773 K produces some long hydrocarbon chains that can be isolated as oil [3]. However, the produced oil quality is still insufficient to be directly used to fuel motor vehicles. Therefore, to increase the utilization of polypropylene oil, SOFC as an electrochemical based-energy conversion might become a promising technique to convert the polypropylene oil into electricity. It is known that SOFC can convert many type of fuels into electricity without any concern of fuel purity.

SOFC has emerged as a new alternative energy resources with a high efficient power generation, clean energy conversion and low noise [4].The SOFC operating temperatures depends on the functional material in the fuel cell, i.e., anode, cathode and electrolyte. The operating temperature can be defined into the low temperature SOFC (LT-SOFC) at 773 K – 923 K and the high temperature SOFC (HT-SOFC) at 1073 K - 1273 K [5]. In this research the polypropylene oil from polypropylene waste is directly used as fuel for Ni-YSZ | YSZ | LSCF single cell at 673 K, 773 K and 873 K to investigate the performance of the single cell with polypropylene oil fuel by measure the power density ($\mu W.cm^{-2}$) and the open circuit voltage (OCV).

2.Experiment

2.1.Pyrolisis of polypropylene oil

The drinking water bottle waste as polypropylene oil sources was cutted into little pieces and then was placed into the pyrolysis reactor and was heated at 673 K - 773 K [3]. The pyrolysis vapor was condensed by cooling water system and then collected in a container, as the polypropylene oil. The oil was analyzed by viscosity testing and Gas Chromatography Mass Spectroscopy (GCMS) (GC17A MSQP 5000 Shimadzu) to identify the molecular content in the oil.

2.2.Preparation of single cell

The YSZ (Yttria Stabilized-Zirconia) pellet were prepared by pressing with a hydraulic press. This research used a YSZ (Yttria Stabilized-Zirconia) (TOSOH), Nickel (Merck), (S)-(-)-α-Terpineol (Merck), Lantanum Stronsium Cobalt Ferrite (Aldrich). The green pellets then were sintered at 1523 K for 2 hours. The Ni-YSZ electrode was produced by mix Ni powder and YSZ powder at weight ratio of 0.2:0.8 and then mixed with α - terpineol as solvent, to form a homogeneous slurry. The sintered YSZ pellet was coated at the one sides with the Ni-YSZ slurry by screen printing method and then sinter at 1273 K for 2 hours. The LSCF ($La_{0.6}Sr_{0.4}Co_{0.2}Fe_{0.8}O_3$) electrode was produced by mix LSCF powder with α - terpineol as solvent, to form a homogeneous slurry. The YSZ pellet has diameter 10 mm and thickness 1.246 mm. The pellet was coated at the other sides with the LSCF slurry by screen printing method and then sinter at 1073 K for 2 hours, producing a single cell of Ni-YSZ | YSZ | LSCF. Figure 1 described the scheme of single cell Ni-YSZ | YSZ | LSCF.

Figure 1. Single cell of Ni-YSZ | YSZ | LSCF

2.3.Testing of single fuel cell with polypropylene oil as fuel

Ag mesh equipped with Ag wire, was attached on both side of single cell with Ag paste as sealer. The Ag wire connected the cell with I-V meter (Microampere-volt Multimeter, Constant 88) and a potentiometer. The fuel cell test was conducted at 673 K, 773 K and 873 K with polypropylene liquid as fuel and oxygen in the air as oxidant. The polypropylene liquid was converted into gas by heating at 418 K. The gas would then flowed into fuel cell compartment to undergo gas reforming under high operational temperature.

3.Result and discussion

The polypropylene oil that was produced from pyrolysis was analyzed by GC-MS. The spectrum is described in Figure 2.

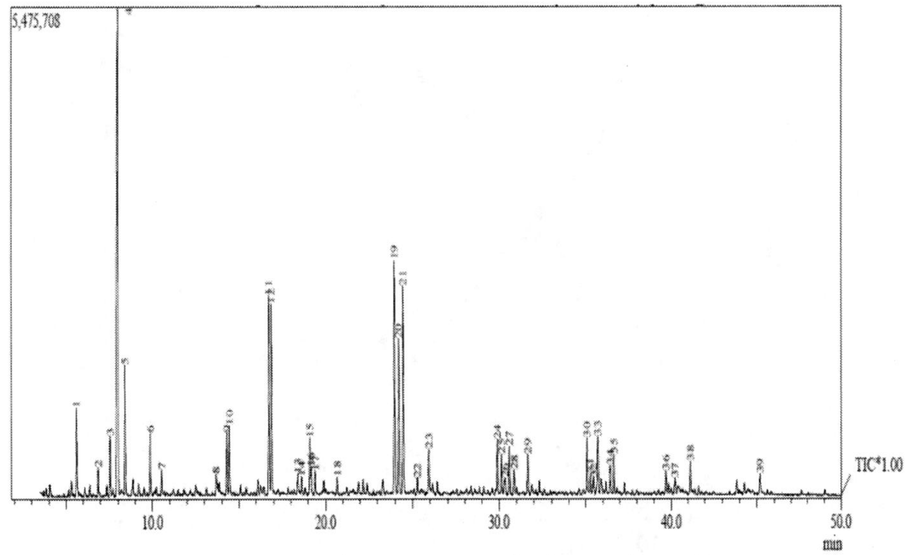

Figure 2. Chromatogram of polypropylene oil

The GC-MS spectrum in Figure 2 shows that the polypropylene oil consist of aliphatic and aromatic hydrocarbon chain compounds with carbon chains of C_8-C_9 dominates at 33.18%. Table 1 shows that the carbon chain ranging from C_8-C_{27}. Large range of chain number might be a problem during gas reforming due to different molecules size especially large molecules would be hard to degrade into smaller hydrocarbon molecules or even into hydrogen gas. Therefore, development on gas reforming process by supplying water steam should be considered. Increasing of gas reforming might increase redox reaction inside fuel cell, in which the oxygen in air reduce into oxygen ions and the hydrocarbon or the reformed product of hydrocarbon undergo oxidation at the anode side. As the reaction proceeds, the electrons flows would be detected as the electricity. The predicted reaction is described in equation (1) and (2) [6].

Cathode : $O_{2\,(g)} + 4e \longrightarrow 2O^{2-}$ .... (1)

Anode : $C_nH_{2n}{}^{+2} + (3n+1)\ O^{2-} \longrightarrow n\ CO_{2\,(g)} + (n+1)H_2O + (6n+2)^{e-}$ ....(2)

TABLE 1. The % area of carbon chain peaks as recorded in the polypropylene oil GC-MS

Compound	%Area
C_8-C_9	33.18
C_{10}-C_{11}	11.70
C_{12}-C_{13}	26.09
C_{14}-C_{15}	13.24
$\geq C_{16}$	16.42

The single fuel cell test produces I-V-P curves as shown in Figure 3, 4, and 5 at 673 K, 773 K, and 873 K. The I-V and I-P curves inform the open circuit voltage, OCV, value (mV) and the power density ($\mu W.cm^{-2}$) of fuel cell.

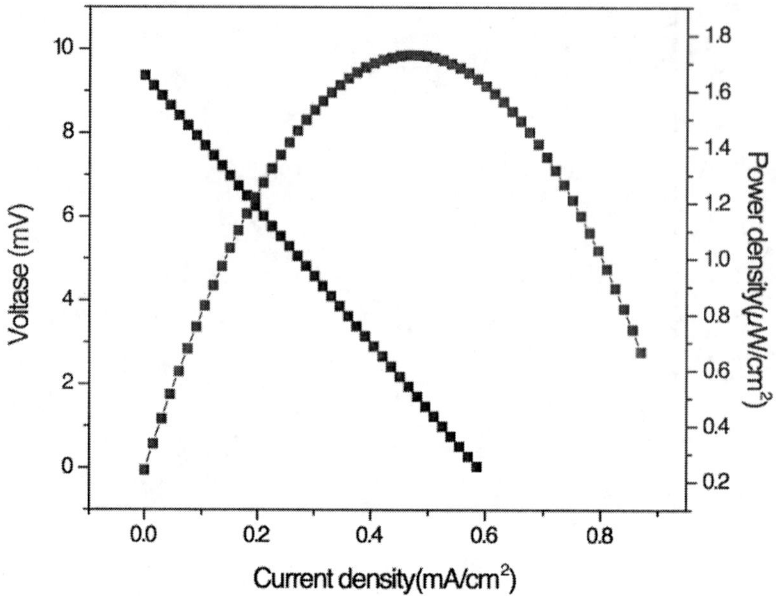

Figure 3. I-V-P curve as resulted from measurement of fuel cell test with polypropylene as fuel at 673 K

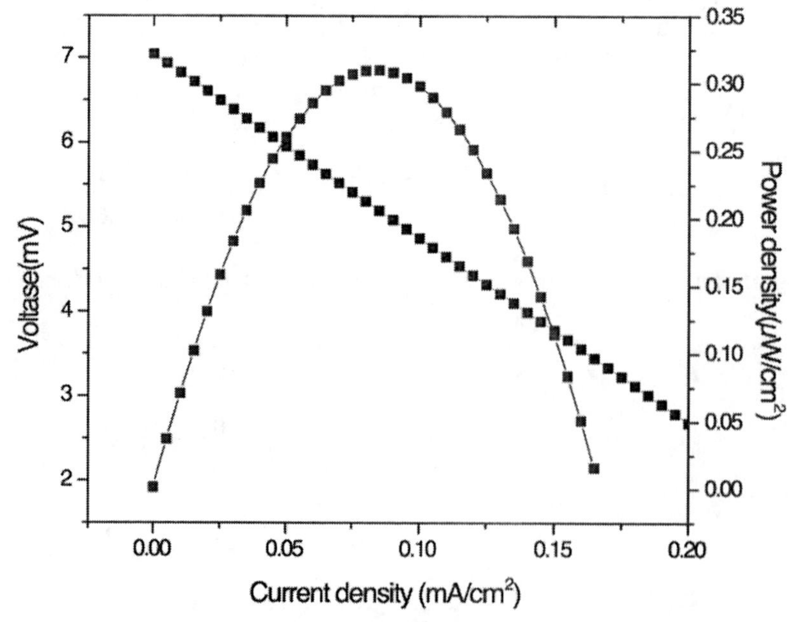

Figure 4. I-V-P curve as resulted from measurement of fuel cell test with polypropylene as fuel at 773 K

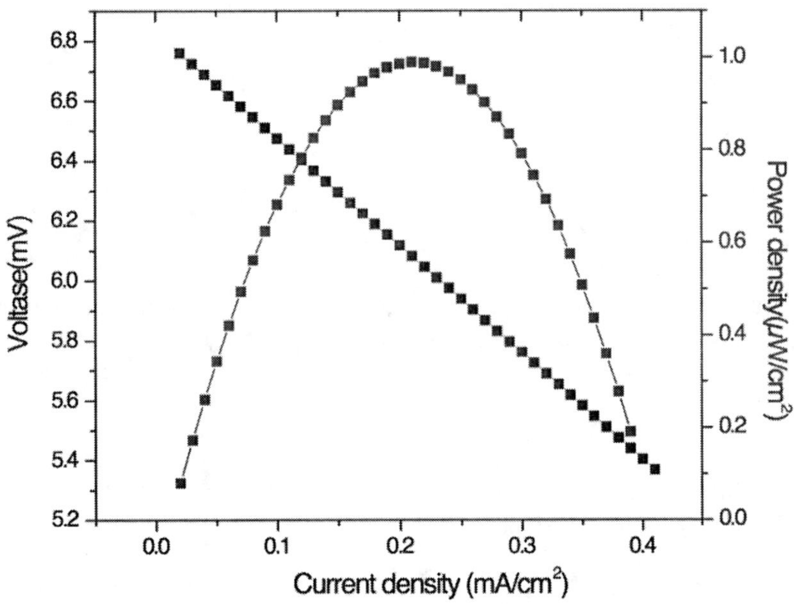

Figure 5. I-V-P curve as resulted from measurement of fuel cell test with polypropylene as fuel at 873 K

The result shows that the optimum power density is1.729 μW.cm^{-2}at 673 K with OCV of 9.378 mV (Figure 3). When the temperature is increased to 773 K, the power density decreased up to 0.309 μW.cm^{-2}. It also happen at the temperature 873 K, the power density decreased up to 0.989 μW.cm^{-2} as shown in Table 2. It indicates that the functional material started to drop their performance when the temperature reach over 673 K. It is because, at a temperature more than 673 K, cracking occurs in the electrolyte material so that instead of ion migration, it occurs ion migration, thus decreasing the conductivity value.

The power density is too small compare to the Samarium Doped-Ceria (SDC)-Carbonate as electrolyte, as it was studied in the previous research using the same polypropylene oil, producing the power density of 0.3736 mW.cm^{-2} and OCV of 0.770 V at 773 K [3]. It is because YSZ has lower ionic conductivity than SDC at 673 K. YSZ has ionic conductivity 5 x 10^{-6} S.cm^{-1} [7], SDC has 3.24 x 10^{-5} S.cm^{-1} [8]. The high molecular weight of hydrocarbon consisted in the polypropylene oil i.e., C_8-C_{27}, that was produced by pyrolysis might also became a problem for gas reforming. High molecular fuel also has less diffusion into the functional material producing slow reaction during fuel cell operation. In addition, the number of steam released by 30% part of the polypropylene-water mixture seems does not sufficient for gas reforming. Furthermore, redox reaction inside fuel cell can be increased by supplying water steam.

TABLE 2. The OCV and power density value at various operational temperature

Temperature	Open circuit voltage (mV)	Power density maximum (μW.cm^2)
673 K	9.378	1.729
773 K	7.048	0.309
873 K	6.831	0.989

4.Conclusion

Pyrolysis of polypropylene plastic waste produces liquid with fuel characteristics and has a hydrocarbon chemical compound ranging from C_8-C_{27}. The result of the test of the resulting power conversion, obtained the maximum value at the operating temperature of 673 K at 1.729 mW/cm^2.

References

[1] Mappiratu, Mujahidah, and Sikanna R 2013 *Online Jurnal Of Natural Science* **2** 25-34
[2] Khan M Z H, Sultana M, Al-Mamun M R, andHasan M R 2016 *Journal of Environmental and Public Health* **2016** 1
[3] Syahputra R J E, Rahmawati F, Prameswari A P, and Saktian R 2017 AIP Conference Proceedings **1823** 020061-1
[4] Bakar M S A, Ahmad S, A Muchtar, and H A Rahman 2014 *Journal of Advanced Review on Scientific Research* **1** 2289
[5] Sulistyo2015 Thesis Faculty of Mechanical and Manufacturing Engineering Universiti Tun Hussein Onn Malaysia
[6] Cimenti M and Hill J M 2009 *Energies* **2** 380
[7] Mills E M, Boyman M K, Janek J, Yang H, Browning N D , Takamura Y and Kim S 2016 *Phys Chem Chem Phys* **18** 10489
[8] Rahmawati F, Fajriati W, Heraldy E and Syarif D G 2015 *Procedia Chemistry* **14** 167

Performance of 7-cells Dye Sensitized Solar Module in Z-type Series Interconnection

Putri Nur Anggraini*, Lia Muliani, Natalita Maulani Nursam, Jojo Hidayat

Research Center for Electronics and Telecommunications, Indonesian Institute of Sciences (PPET – LIPI), Kampus LIPI, Jl. Sangkuriang, Bandung 40135, Indonesia

Email: putr014@lipi.go.id

Abstract. Dye sensitized solar cells (DSSC) is becoming attractive research topic as third generation solar cells technology since it provides clean energy and low cost fabrication. In this study, DSSC was fabricated into module scale, which is important for practical applications. The module was prepared in sandwich structure consisting of TiO_2 working electrode and Pt counter electrode on conductive substrate with an area of 100 mm x 100 mm, which was distributed into seven active cells. TiO_2 paste was deposited on FTO glass as working electrode with a size of 10 mm x 98 mm per unit cell by screen printing method. Each cell was connected in Z-type series that able to produce high voltage. $I - V$ measurement was applied in two methods consisting of laboratory testing using sun simulator under 500 W/m^2 of illumination and outdoor testing using a digital multimeter under direct sunlight. The result shows that DSSC module has photoconversion efficiency of 1.08% and 1.17% for laboratory and outdoor testing, respectively. The module was also tested in three different times to monitor its stability performance.

1. Introduction

Dye sensitized solar cells (DSSC) is becoming trending topic in renewable energy research area since Gratzel et al. invented it in 1991 [1]. DSSC offers alternative energy with low cost production, simple process fabrication and flexibility in design, which makes it part of third generation solar cells [2]. DSSC was also reported to have photoconversion efficiency over 11% [3] and it can potentially even reach higher efficiency with further research. Therefore, more research has been conducted to develop the practical implementation of DSSC by scaling up the area in module design.

Some aspects can affect the quality of DSSC such as the presence the additive in the electrolyte or improper sealing [4]. In fabrication of DSSC module particularly, sealing is one of the most challenging factors that determine the photovoltaic performance and lifetime stability. When DSSC module was applied for outdoor use, stability is considered as a major obstacle [5]. Moreover, it was confirmed that photovoltaic performance in module scale was lower than cell scale since the resistance become higher in larger area [6]. Gratzel had reported the fabrication of DSSC module with an efficiency 7% [7]. Wei et al. [8] fabricated grid type module with 7% efficiency. Han et al. [9] developed W-contact DSSC module and received photoconversion efficiency of 8.2% Nevertheless, the research in module designs have also been improved continuously to achieve optimum performance and could be considered to be widely implemented.

This work reports the fabrication of DSSC module using FTO glass substrate that consists of 7 cells interconnected in Z-type series. Series interconnection is able to produce higher voltage and lower current, which requires less metal fingers for module design [6]. The performance of DSSC module was evaluated by comparing the results from sun simulator testing as a standard laboratory measurement and outdoor testing to simulate real operating condition. Moreover, the stability of the module was also tested by measuring the performance of module over three different times.

2. Experiments

There were three parts in fabrication of DSSC module, which consisted of preparation of photoelectrode and counter electrode, assembly, and $I - V$ measurement.

Preparation of photoelectrode was preceded by making the module pattern consisting of 7 unit cells with an area of 10 mm x 98 mm for each cell, circuit contact, and sealant as they can be seen in Figure 1. In addition, screen for module was prepared using screen maker according to the patterns.

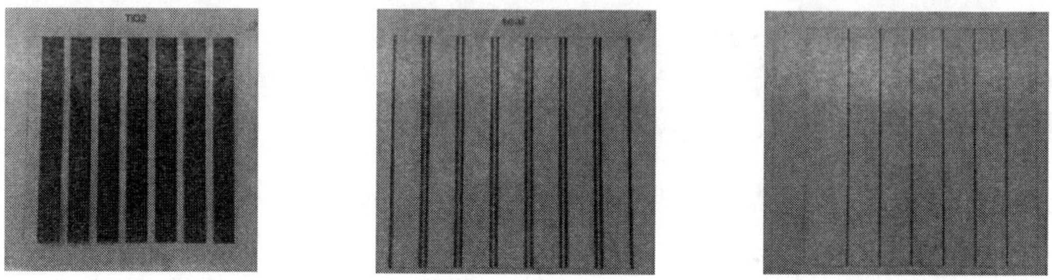

(a) (b) (c)

Figure 1. Pattern design for DSSC module screen for (a) TiO_2 active cells, (b) sealing area, and (c) circuit contact area.

FTO glass as conductive substrate was used with a size of 100 mm x 100 mm. First, FTO glass was scribed using laser to separate each area of cells. The substrate was then cleaned in DI water using ultrasonic cleaner and rinsed using ethanol for 10 minutes, and then dried naturally. Deposition process of TiO_2 semiconductor by screen printing method was applied on clean substrate using 7 cells screen (Figure 1.a) as made before. TiO_2 film was then dried in an oven at 100°C for 10 min. TiO_2 film was also sintered at 500 °C for 45 min for calcination process. The same process was also applied for working electrode on which platinum transparent paste was coated on to substrate by screen printing method using the same pattern as the photo electrode, dried and sintered with the same treatment as the photo electrode.

TiO_2 film was subsequently immersed in dye solution (N907 0.5 mM in ethanol) for 24 hours. After the dyes adsorption process, the dyed film was then rinsed with ethanol and dried naturally.

The next step was assembly process. Amosil 4R and amosil 4H (Solaronix) was mixed to obtain the sealant and then deposited on the sealant area (Figure 1.b) of the photo electrode using screen printing method, while Ag paste was deposited on the circuit contact area (Figure 1.c) of the counter electrode using the same method. Both photo electrode and counter electrode were constructed in sandwich structure. Electrolyte liquid was injected into the module through a predrilled hole, which was located on the counter electrode surface.

$I - V$ measurement was conducted on the module prototype to investigate its performance based on the generated curve and photovoltaic parameters. In this work, the module was tested in two conditions consisting of laboratory and outdoor testing. Laboratory testing was applied to the module using sun simulator with an intensity of 500 W/m^2 and the result was measured by National Instrumentation source meter. Meanwhile, outdoor testing was applied to the modul by using variable resistor under direct sunlight and the result was measured by digital multimeter.

3. Results and discussion

Herein, the fabricated DSSC module has Z-type series interconnection that has been widely used since it can reduce the area consumption for metal fingers, which consequently can be an effective way to increase the open area ratio [6]. Figure 2 shows the internal structure of DSSC module with series interconnection in cross section view, describing detailed construction of each material. In a module, sealant has an important role to protect each cell and avoid electrolyte leakage between the adjacent cells so that the module will be able to produce high stability. Assemblage technique also takes crucial role since it was done manually in order to construct the photoelectrode and counter electrode in sandwich structure. Top view of prototype of 7-cells DSSC module can be seen in Figure 3.

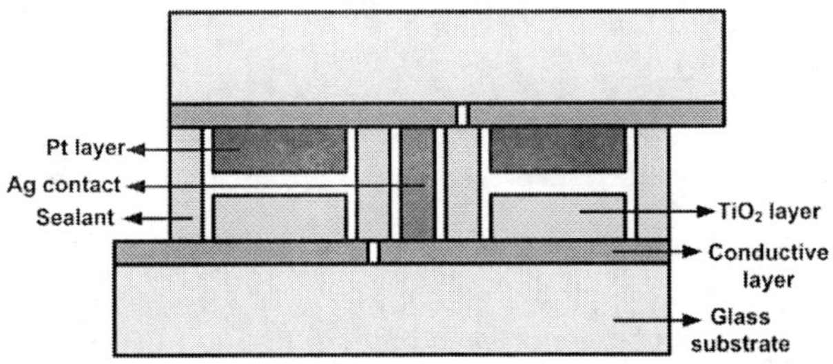

Figure 2. Internal structure of DSSC module with series interconnection.

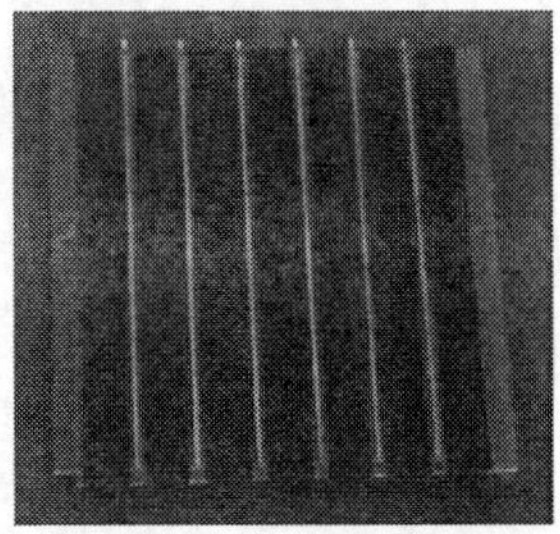

Figure 3. Prototype of 7-cell DSSC module designed in sandwich structure.

As mention before, DSSC module was tested in two conditions consisting of laboratory and outdoor testing. Laboratory testing provides controllable indoor test as it has stable light intensity using sun simulator. Outdoor testing was done to obtain the module performance in real condition, by using variable resistors and was measured using digital multimeter. The weakness of outdoor testing lies on the external factors, such as temperature and irradiance, which are uncontrollable and possibly affect the performance of solar cells [4].

Figure 4 shows the graph that compares the result of laboratory and outdoor testing. Both curves demonstrated photovoltaic performance of the module and it can be clearly seen that the outdoor testing exhibits better $I - V$ curve. However, the outdoor testing has inconstant $I - V$ data at the initial measurement that might be caused by unstable sunlight intensity. Laboratory testing result, which has lower performance than outdoor testing result, exhibits constant $I - V$ data that gives more accurate measurement of module performance.

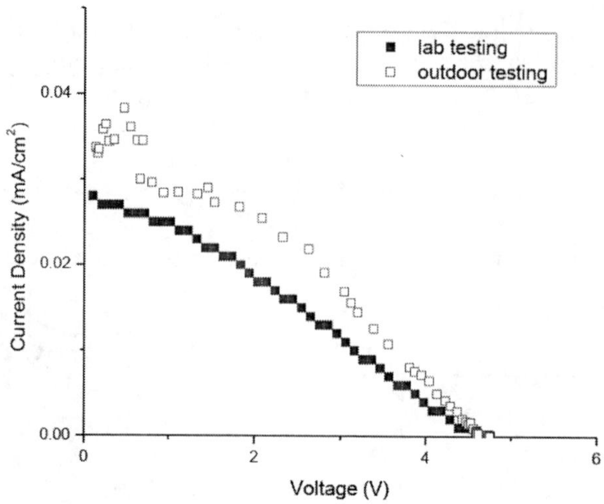

Figure 4. I-V curve of DSSC module from laboratory and outdoor testing.

Photovoltaic parameters obtained from laboratory and outdoor testing results are listed in Table 1, giving the performance value of each of 7 unit cells and as a module. The photovoltaic parameters show lower performance with nearly identical value for each cell. However, when the cells were accumulate as a module the performance has increased significantly. It indicates that the interconnection between adjacent cells had managed its function as path of electric current. Overall, the outdoor testing result has higher value of PV parameters with slight difference from the laboratory testing result. The highest value of short circuit current density (J_{sc}), open circuit voltage (V_{oc}) and photoconversion efficiency (η) was 0.028, 4.720 V, and 1.176%, respectively. The value of V_{oc} can be compared between cells and module form that shows significant difference. The result might be affected by the serial connection between cells since this type of connection can produce high voltage. In this case, serial connection in the module prototype could work properly and was able to increase the voltage in module form.

Table 1. Photovoltaic parameters of DSSC module from laboratory and outdoor testing.

Testing Method		Parameter		
		J_{sc} (mA/cm^2)	V_{oc} (V)	η (%)
Laboratory testing	Module	0.028	4.487	1.086
	Cell 1	0.012	0.672	0.416
	Cell 2	0.012	0.631	0.422
	Cell 3	0.011	0.632	0.399
	Cell 4	0.014	0.631	0.484
	Cell 5	0.013	0.632	0.456
	Cell 6	0.012	0.631	0.432
	Cell 7	0.012	0.631	0.434
Outdoor testing	Module	0.028	4.720	1.176
	Cell 1	0.005	0.500	0.367
	Cell 2	0.014	0.680	0.337
	Cell 3	0.015	0.660	0.428
	Cell 4	0.015	0.670	0.372
	Cell 5	0.017	0.650	0.348
	Cell 6	0.014	0.660	0.392
	Cell 7	0.015	0.650	0.402

Laboratory testing were performed three times to obtain the stability of the module. Figure 5 shows the comparison of IV curve from the 1st, 2nd, and 3rd laboratory testing. First testing was immediately applied to the module before it was stored under dark condition. Meanwhile, the second and third testing was performed with 5 days interval after the module was kept in storage with uncontrolled condition. Overall the module still could manage photovoltaic effect although the tests yielded less ideal curves.

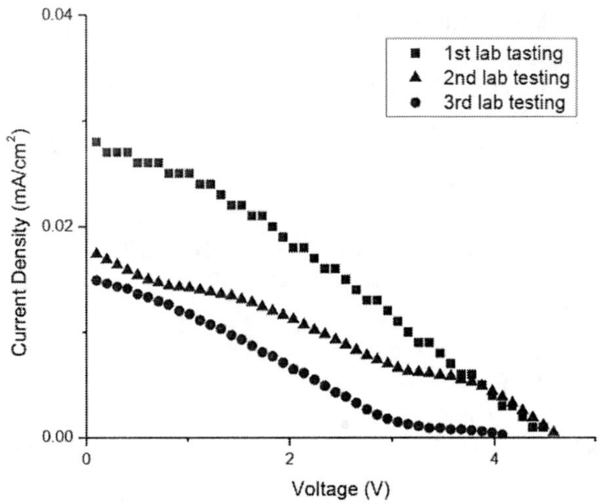

Figure 5. I-V curve of DSSC module from 1st, 2nd, and 3rd laboratory testing.

The photovoltaic parameters obtained after repeated tests are compared in Table 2. It can be seen that module performance was degraded over the storage time. The average of degradation was approximately 40% from the initial module efficiency after each testing. The degradation of the module was more likely caused by ineffective module sealing, which led to electrolyte leakage between the adjacent cells and possibly causing unwanted mass transport. Moreover, the uncontrolled condition during the storage affected the humidity and temperature of the module that can lower the module performance, in particular affecting the electrolyte reaction as it might be dried or reacted with sealant material. The degradation in the DSSC module performance indicates that more investigation on materials and more appropiate fabrication techniques are still required to improve the applicability of the DSSC modules.

Table 2. Photovoltaic parameters of DSSC module from 1st, 2nd, and 3rd laboratory testing.

Parameter	testing		
	1st	2nd	3rd
J_{SC} (mA/cm^2)	0.028	0.017	0.015
V_{OC} (V)	4.487	4.589	4.180
η (%)	1.086	0.666	0.398

4. Conclusions

DSSC module consisting of 7 active cells with Z-type series interconnection has been fabricated. The electrical measurement was conducted both in laboratory and outdoor testing and the results were compared. Photoconversion efficiency of the module for laboratory and outdoor testing was 1.086% and 1.176%, respectively. However, the laboratory testing gave more accurate result since it provided stable light intensity during the $I - V$ measurement. This research also compared the module performance over three repeated measurements. The results suggest that stability is still problematic since the degradation of the module performance occured after 2nd and 3rd testing, giving an approximately of 40% degradation in each testing.

References

[1] O'Regan B and Grätzel M 1991 *Nature* **353** pp 737–40
[2] Hagfeldt A, Boschloo G, Sun L, Kloo L and Pettersson H 2010 *Chem. Rev.* **110** pp 6595–663
[3] Mathew S, Yella A, Gao P, Humphry-Baker R, Curchod B F E, Ashari-Astani N, Tavernelli I, *Nat. Chem* **6** pp 242–7
[4] Asghar A, Emziane M, Pak H K and Oh S Y 2014 *Sol. Energy Mater. Sol. Cells* **128** pp 335–42
[5] Jun Y, Son J H, Sohn D and Kang M G 2008 *J. Photochem. Photobiol. A Chem.* **200** pp 314–7
[6] Zhang Y-D, Huang X-M, Li D-M, Luo Y-H and Meng Q-B 2012 *Sol. Energy Mater. Sol. Cells* **98** pp 417–23
[7] Grätzel M 2011 Photoelectrochemical cells *Nature* **414** pp 338–44
[8] Tzu-Chien Wei, Jo-Lin Lan, Chi-Chao Wan W-C H and Y-H C 2013 *Prog. Photovolt Res. Appl.* **21** pp 1625–33
[9] Han L, Fukui A, Chiba Y, Islam A, Komiya R, Fuke N, Koide N, Yamanaka R and Shimizu M 2009 *Appl. Phys. Lett.* **94** pp 1–4

Acknowledgement

This research was fully supported by DIPA LIPI funding program year of 2016. Author is fully supported by INSINAS Research Grant Program 2017 from the Ministry of Research, Technology and Higher Education of the Republic of Indonesia (contract no. 04/INS-2/PPK/E/E4/2017) to attend the conference and present the paper.

IOP Conf. Series: Materials Science and Engineering **299** (2018) 012089 doi:10.1088/1757-899X/299/1/012089

Theoretical study on the effect of solvents in chlorophyll solution for Dye-Sensitized Solar Cell

Mohamad Rodhi Faiz[1,2,*], **Denny Widhiyanuriyawan**[1], **Eko Siswanto**[1] **and I.N.G. Wardana**[1]

[1] Department of Mechanical Engineering, Brawijaya University, Malang, Indonesia
[2] Department of Electrical Engineering, Universitas Negeri Malang, Malang, Indonesia.
[*] mohamad.rodhi.ft@um.ac.id

Abstract. Using the density functional theory (DFT) and time-dependent DFT (TD-DFT), we have theoretically studied the electron excitation and absorption spectra of chlorophyll solution with several solvents for Dye-Sensitized Solar Cell (DSSC) application. The solvents were aceton, acetonitrile, DMSO, ethanol, hexane, methanol, THF and water. Also, energy level, oscilator strength, light harvesting efficiency and electron injection have been investigated. The studied dye in the presence of the solvents showed a smaller gap of the highest occupied molecular orbital (HOMO) - the lowest unoccupied molecular orbital (LUMO) and a higher light harvesting, oscillator strength and red shift in the absorption spectra. These changes facilitate the charge transfer phenomena in the nano structure of the chlorophyll as dye and improve the solar cell efficiency.

1. Introduction

Solar energy has the potential to be developed as a renewable energy source in the near future to replace fossil-based energy [1]. Devices that directly convert sunlight into electrical energy are solar cells. The dye-sensitized solar cell (DSSC) is new type of the solar cells that have many advantages compared to the silicon based ones which have been introduced by Gratzel [2] . The advantages of DSSC such as low-cost, flexibility, environmentally-friendly properties, and facile fabrication process [3]. DSSC consists of a porous layer of wide band gap semiconductor, such as TiO_2 coated with a dye that serves as light sensitizer on transparent conducting glass electrode, an electrolyte layer, and a counter electrode, typically coated with graphite or platinum [4].

Research related to DSSC efficiency is usually the discussion of three main components: dye, redox electrolyte reaction and semiconductor electrode [5]. The efficiency of a solar cell depends largely on how much light absorption can be made. In DSSC light absorption is performed by the sensitizer mixed on TiO2. Sensitizee role is very dominant in the energy harvesting process [6]. Besides the electron excitation process on dye due to solar radiation, electron injection ability from dye to semiconductor and electron collection capability by semiconductor also affect the efficiency of DSSC. The oxidized dye should be able to regenerate electrons where it is highly dependent on the type of dye and the type of electrolyte.

In DSSC, natural pigment can be used as dye sensitizer, which is produced from fruits, leaves, flowers, tree barks, roots etc. Natural pigments have several advantages: more available, simple to prepare, lower cost, non-toxic, environmentally friendly and biodegradable [7]. The plant pigments are electronic structure where if receive sunlight, it will convert the wavelengths that are transmitted or reflected by the plant tissue. This is called plant pigmentation and each pigment is represented by the absorbed wavelength and color that can be seen by humans [8]. Natural pigments that can be used as dye sensitizers include carotenoids, flavonoids, chlorophyll, and anthocyanins that can cheaply be obtained from plants [9].

Chlorophyll is widely studied as a sensitizer in the DSSC. They exhibit visible absorption properties because they have a porphyrin that consist of a conjugated pyroles system and composed of substituents on the sides. Thus chlorophyll contains a lot of non-bonding electrons and pairs of phi bonding electrons so that if radiated visible light, electrons in chlorophyll will be easily excited. Moreover, other properties such as HOMO-LUMO level energy with small bandgap, low cost, simple synthesis process, high molar absorption coefficient, good chemical and thermal stability make it perfect sensitizer [10]. Besides depending on the type of pigment, the absorption can capability be affected by the process during the extraction of the pigment. In this study, we will see the effect of the solvent used as the dye solution embedded in the semiconductor. We will see the electron transfer on the solvent media by using quantum mechanism chemistry simulation and the dye solution will be characterized. Quantum mechanism simulations have become an easy and effective to predict the DSSC device preformance [11].

2. Computational Method

The chlorophyll molecule model used in this work consist of 75 atoms. This model is a modification of the chlorophyll structure to reduce the number of atoms for simpler calculations, by eliminate the methyl groups on the phytyl ester side chain [12], also adding various solvents to see their effect on electron excitation ability.

Quantum mechanism calculations are performed using by the ORCA package program [13] and the geometry optimization was computed using DFT with def2-TZVP basis sets [14][15] were applied for chlorophylls with all solvents, in combination with full decontraction def2-TZVP/J auxiliary basis sets [16]. The DFT calculation uses the hybrid B3LYP Density which merges the three Becke functional parameters with the Lee-Yang-Parr parameter [17][18]. Furthermore, to show the excited states and optical properties, for producing the UV-Vis absorption spectrum,Time Dependent - Density Functional Theory (TD-DFT) method was engaged at B3LYP with basis set using def2-TZVP for the electronic transitions between HOMO and LUMO states. In the calculation, we used the RIJCosX algorithm [19][20][21]. The criterion of SCF convergence and to improve grid integration is used Grid5, Grid6 for all calculations. In order to have an accurate representation of the excitation energies in the realistic environment, the solvent effects have been evaluated by the non-equilibrium version of the conductor-like polarizable continuum model [22]. The solvent is represented as a dielectric polarizable continuum and the solute is placed in a cavity of approximately molecular shape. The solvent reaction field is described by polarization charges on the surface of the cavity.

3. Results And Analysis

3.1 Optimized Molecular Geometries

Chlorophyll molecule with various solvents were already analyzed with computational method (DFT/B3LYP). The parameter of the optimized geometrical in ground state as bond angles, bond lengths, also dihedral angles for all chlorophyll with various solvents are similar shown. As representation, molecul structure in the vacuum and water shown in Table 1. Solvents did not affect much of the optimized geometry structure of the chlorophyll molecule but the total optimized energy different.

Based on table 2, the total energy optimized by calculation on the DFT method has different value in the vacuum phase and solvents showed the solvent media can stabilize the chlorophyll structures through the intermolecular interactions of the chlorophyll and the solvent. The results similar to the research in the following reference [23]. And the lowest energy in the water and DMSO of -58393.99075 eV and -58393.96135 eV respectively.

Table 1. Optimized geometri and Frontier molecular orbital in the vacuum and water

Solvents	Optimized Geometry	Homo	Lumo
Vacuum	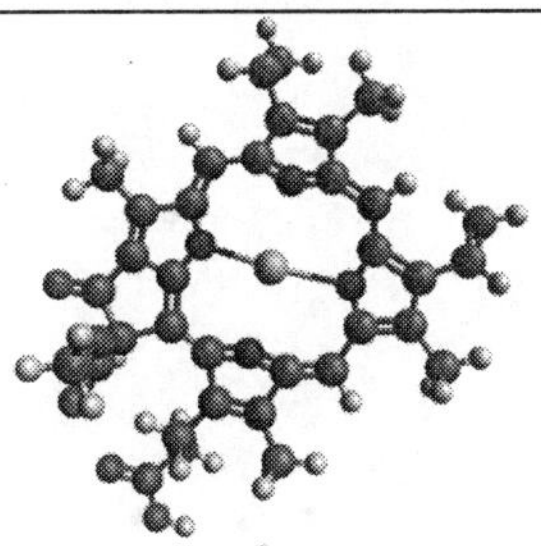	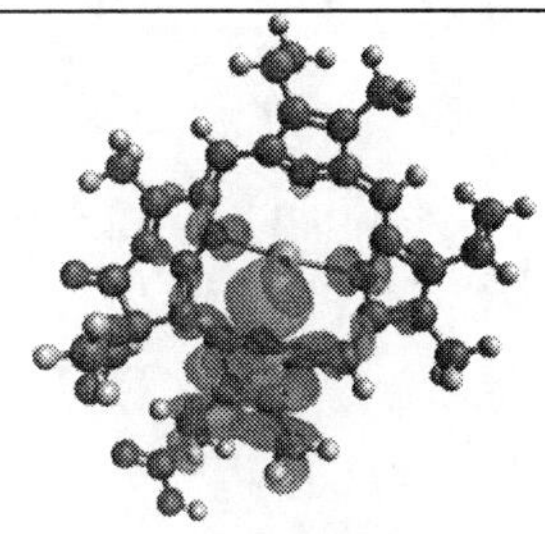	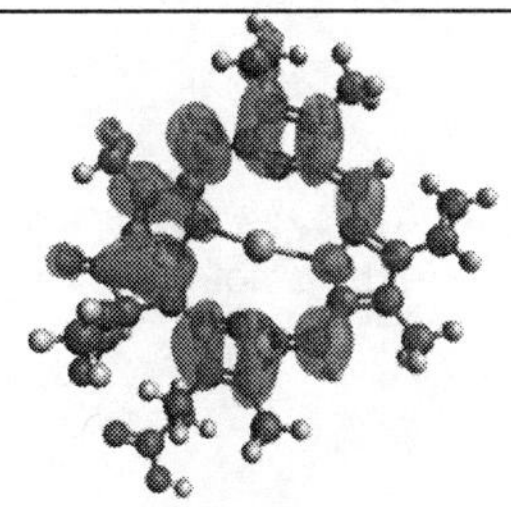
Water	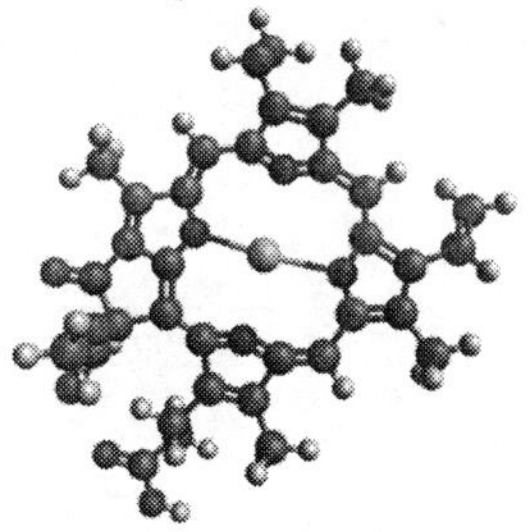	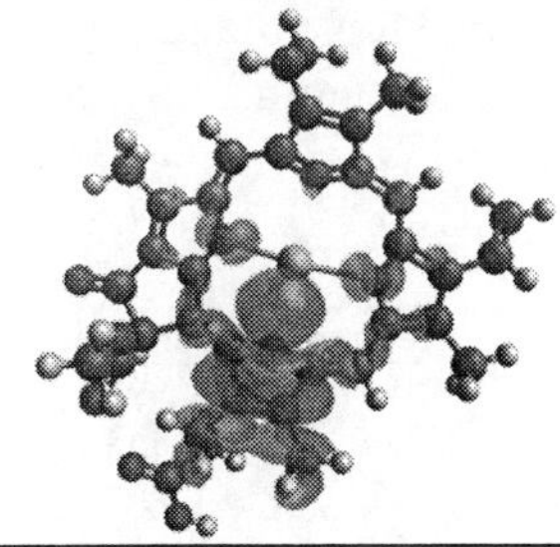	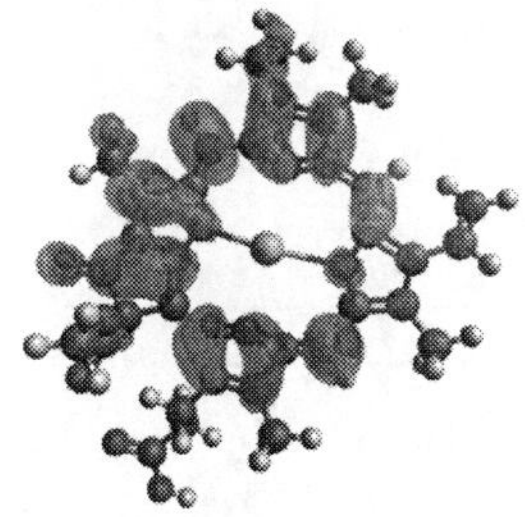

Table 2. Total energy optimized of the Chlorophyll with various solvents

Solvents	Total Energy	Solvents	Total Energy	Solvents	Total Energy
Vacuum	-58392.26201 eV	Ethanol	-58393.89611 eV	Water	-58393.99075 eV
Aceton	-58393.87333 eV	Hexane	-58392.82099 eV	THF	-58393.61213 eV
Acetonitrile	-58393.94103 eV	Methanol	-58393.93012 eV	DMSO	-58393.96135 eV

3.2 Electronic Properties

Electronic properties include HOMO level, LUMO level, bandgap energy, dipole moment, oscilatory strength (fosc), and electron injection capability. Based on table 3, the addition of solvents can improve HOMO - LUMO level energy which can increase the ability of electron injection and also reduce bandgap energy. All this results from an increase in dipole moment and oscillatory strength of chlorophyll molecules with solvents and it can improve the DSSC efficiency as equation (3). Based on the simulation results have shown that among the many solvents, water and DMSO are the solvents that have the most ability to improve chlorophyll performance. Water can improve chlorophyll performance although chlorophyll is not soluble in water. For the experiment it can be solved by mixing with solvents which may dissolve chlorophyll as examples mixed with ethanol as carried out in the following reference [24], where mixing can produce maximum absorption of light at ethanol-water mixing with a ratio of 2:1.

DSSC conversion efficiency (η), in Eq. (1), of the solar cell is described by the open-circuit photovoltage (V_{OC}), shortcircuit current density (J_{SC}), and fill factor (FF), as compared to the incident solar power (P_{inc}) [25].

$$\eta = \frac{J_{SC}\,V_{OC}\,FF}{P_{in}} \tag{1}$$

where, J_{SC} is determined by the Equation (2) [26]:

$$J_{sc} = \int_{\lambda} LHE(\lambda)\Phi_{inject}\eta_{collect}\,d\lambda \tag{2}$$

where, $LHE(\lambda)$ is the light harvesting efficiency at a given wavelength which could be measured by equation (3) [27], Φ_{inject} shows the electron injection efficiency, and $\eta_{collect}$ defines the charge collection efficiency. In the systems which are only different in sensitizer parameters, $\eta_{collect}$ can be wisely considered to be constant.

$$LHE(\lambda) = 1 - 10^{-f} \quad (3)$$

Φ_{inject} is related to the driving force, the free energy change of electron injection (ΔG_{inject}) from the excited states of dyes to the semiconductor substrate, which can be calculated by Equations (4) [28].

$$\Delta G_{inject} = E_{ox(dye*)} - E_{CB,TiO_2} \quad (4)$$

where, $E_{CB(TiO2)}$ is the conduction band energy of the TiO_2 (-4 eV) and $E_{OX(dye*)}$ is the excited state oxidation potential of the dye, which Rehm and Weller equation can showed its value by equation (5) [29].

$$E_{ox(dye*)} = E_{ox(dye)} - E_{0-0} + \omega_r \quad (5)$$

where, $E_{ox(dye)}$ is the dye oxidation potential in the ground state while $E_{0\text{-}0}$ is the vertical excitation energy and ω_r is a columbic stabilization term which is negligible [30]. Therefore $E_{ox(dye*)}$ is approximated by $E_{ox(dye)}$ - $E_{0\text{-}0}$.

Table 3.

	Vacuum	Acetone	Acetonitrile	Ethanol	Hexane	Methanol	Water	THF	DMSO
LUMO (eV)	-3,294	-3,007	-2,993	-3,002	-3,2	-2,995	-2,983	-3,057	-2,989
HOMO (eV)	-5,009	-4,49	-4,465	-4,482	-4,845	-4,469	-4,446	-4,585	-4,457
Bandgap (eV)	1,715	1,483	1,472	1,48	1,645	1,474	1,463	1,528	1,468
Momen Dipole (au)	3,579	5,125	5,204	5,152	4,056	5,191	5,263	4,834	5,228
Fosc	0,223	0,246	0,253	0,249	0,228	0,251	0,254	0,233	0,263
ΔG_{inject} (eV)	0,706	0,993	1,007	0,998	0,8	1,005	1,017	0,943	1,011

3.3 *Optical Characteristics*

According to figure 1 with the addition of solvents may increase the shift of the spectrum of several nm in the direction of infrared as compared with no solvents. This happens as a result of the decrease in energy gap and from the stabilization of the virtual orbitals in the solvent environment. The solvent stabilizes the contaminated substances in the electric field through the electrostatic interaction between the chlorophyll molecule. So it can be concluded that the transition energy is very sensitive to the effect of solvent.

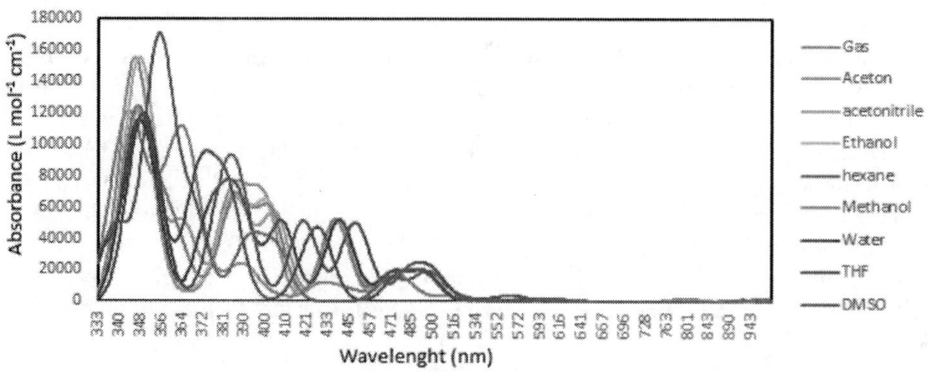

Figure 1. Spectrum absorbance of chloropbhyll solutions with different solvents

4. Conclusion

In this research, the electron excitation of chlorophyll sensitizer with various solvents have been investigated using the TD-DFT. The addition of solvent reduces the HOMO-LUMO bandgap and increases the dipole momen, oscillator strength and the intensity of the absorbtion spectrum shifting in the near infrared. Moreover, with increasing dipole momen and oscillator strength, it will have an impact on increasing light harvesting energy (LHE). Also the electron injection capability of the dye to the semiconductor have seen better with the solvents phase. All the parameter will have an effect on increasing efficiency of solar cell. The hope, this theoretical results can be used as a reference in experimental activities. In addition, these findings provide a more detailed understanding and the right tools to predict the efficiency of solar cells.

References

[1] N. Lewis and G. Crabtree, "Basic Research Needs for Solar Energy Utilization: report of the Basic Energy Sciences Workshop on Solar Energy Utilization," *US Dep. Energy, Off. Basic Energy Sci. Washington, DC*, 2005.
[2] B. O'Regan and M. Gratzel, "A low-cost, high-efficiency solar-cell based on dye sensitized colloidal TiO2 films," *Nature*, vol. 353, no. 40, p. 737, 1991.
[3] J. Maçaira, L. Andrade, and A. Mendes, "Review on nanostructured photoelectrodes for next generation dye-sensitized solar cells," *Renew. Sustain. Energy Rev.*, vol. 27, no. 2013, pp. 334–349, 2015.
[4] S. Shalini, R. Balasundara, S. Prasanna, T. K. Mallick, and S. Senthilarasu, "Review on natural dye sensitized solar cells : Operation , materials and methods," *Renew. Sustain. Energy Rev.*, vol. 51, pp. 1306–1325, 2015.
[5] J. Gong, J. Liang, and K. Sumathy, "Review on dye-sensitized solar cells (DSSCs): Fundamental concepts and novel materials," *Renew. Sustain. Energy Rev.*, vol. 16, no. 8, pp. 5848–5860, 2012.
[6] F. Odobel, Y. Pellegrin, E. A. Gibson, A. Hagfeldt, A. L. Smeigh, and L. Hammarström, "Recent advances and future directions to optimize the performances of p-type dye-sensitized solar cells," *Coord. Chem. Rev.*, vol. 256, no. 21–22, pp. 2414–2423, 2012.
[7] C. Giuseppe and G. Di Marco, "Red Sicilian orange and purple eggplant fruits as natural sensitizers for dye-sensitized solar cells," *Sol. Energy Mater. Sol. Cells*, vol. 92, pp. 1341–1346, 2008.
[8] K. Davies, "Plant Pigments and their manipulation.," *Annu. Plant Rev. 14th USA Blackwell Publ. Ltd*, 2004.
[9] N. T. R. N. Kumara *et al.*, "Layered co-sensitization for enhancement of conversion efficiency of natural dye sensitized solar cells," *J. Alloys Compd.*, vol. 581, pp. 186–191, 2013.
[10] M. C. Tsai, C. L. Wang, C. Y. Lin, C. iang Tsai, H. J. Yen, and H. C. You, "Novel Porphyrin-containing Polyimide for Memory Devices," *Polim. Chem.*, vol. 7, pp. 2780–4, 2016.
[11] N. Martsinovich and A. Troisi, "Theoretical studies of dye-sensitised solar cells : from electronic structure to elementary processes," *Environ. Sci.*, pp. 4473–4495, 2011.
[12] R. Bechaieb, A. Ben, and H. Gérard, "Quantum chemistry insight into Mg-substitution in chlorophyll by toxic heavy metals : Cd , Hg and Pb," *Chem. Phys. Lett.*, vol. 663, pp. 27–32, 2016.
[13] F. Neese, "The ORCA program system," *WIREs Comput Mol Sci*, vol. 2, pp. 73–78, 2012.
[14] F. Weigend and R. Ahlrichs, "Balanced basis sets of split valence, triple zeta valence and quadruple zeta valence quality for H to Rn: Design and assessment of accuracy," *Phys. Chem. Chem. Phys.*, vol. 7, pp. 3297–3305, 2005.
[15] A. Schäfer, H. Horn, and R. Ahlrichs, "Fully optimized contracted Gaussian basis sets for atoms Li to Kr," *Chem. Phys.*, vol. 97, p. 2571, 1992.
[16] F. Weigend, "Accurate Coulomb-fitting basis sets for H to Rn," *Phys. Chem. Chem. Phys.*, vol. 8, pp. 1057–1065, 2006.
[17] A. D. Becke, "Densityfunctional thermochemistry . III . The role of exact exchange," *Chem. Phys.*, vol. 98, pp. 5648–5652, 1993.
[18] C. Lee, W. Yang, and R. G. Parr, "Development of the Colle-Salvetti correlation-energy formula into a functional of the electron density," *Phys. Rev. B*, vol. 37, pp. 785–789, 1988.
[19] D. Andrae, U. Haeussermann, M. Dolg, H. Stoll, and H. Preuss, "Energy-adjusted ab initio pseudopotentials for the second and third row transition elements," *Theor. Chim. Acta*, vol. 77 (2), pp. 123–141, 1990.
[20] F. Neese, "An Improvement of the Resolution of the Identity," *Comput. Chem.*, vol. 24, pp. 1740–1747, 2003.

[21] F. Neese, F. Wennmohs, A. Hansen, and U. Becker, "Efficient , approximate and parallel Hartree – Fock and hybrid DFT calculations . A ' chain-of-spheres ' algorithm for the Hartree – Fock exchange," *Chem. Phys.*, vol. 356, pp. 98–109, 2009.
[22] M. Cossi and V. Barone, "Time-dependent density functional theory for molecules in liquid solutions," *Chem. Phys.*, vol. 115, pp. 4708–17, 2001.
[23] F. Arkan, M. Izadyar, and A. Nakhaeipour, "The role of the electronic structure and solvent in the dye-sensitized solar cells based on Zn-porphyrins : Theoretical study," *Energy*, vol. 114, pp. 559–567, 2016.
[24] M. A. M. Al-alwani, A. Bakar, A. Amir, H. Kadhum, and N. A. Ludin, "Effect of solvents on the extraction of natural pigments and adsorption onto TiO 2 for dye-sensitized solar cell applications," *Spectrochim. ACTA*, vol. 138, pp. 130–137, 2015.
[25] X. Ren, S. Jiang, M. Cha, G. Zhou, and Z. Wang, "Thiophene-Bridged Double D - π - A Dye for E ffi cient Dye-Sensitized Solar Cell," *Chem. Mater.*, vol. 24, pp. 3493–3499, 2012.
[26] J. Preat, D. Jacquemin, and E. A. Perp, "Towards new efficient dye-sensitised solar cells," *Energy Environ. Sci.*, vol. 22, no. i, pp. 891–904, 2010.
[27] J. Preat, D. Jacquemin, C. Michaux, and E. A. Perpète, "Improvement of the efficiency of thiophene-bridged compounds for dye-sensitized solar cells," *Chem. Phys.*, vol. 376, no. 1–3, pp. 56–68, 2010.
[28] R. Katoh *et al.*, "Efficiencies of Electron Injection from Excited N3 Dye into Nanocrystalline Semiconductor (ZrO2, TiO2, ZnO, Nb2O5, SnO2,In2O3) Films," *J Phys Chem*, vol. 108, pp. 4818–4822, 2004.
[29] D. Rehm and A. Weller, "Kinetics of Fluorescence Quenching by Electron and H-Atom Transfer," *J Chem*, vol. 8, pp. 259–71, 1970.
[30] J. L. Goodman and K. S. Peters, "Effect of Solvent and Salts on Ion Pair Energies in the Photoreduction of Benzophenone by DABCO," *J Am Chem Sos*, vol. 108, pp. 1700–1701, 1986.

Acknowledgments

We are grateful for the computing resources support from directorship of Max Planck Institute for Chemical Energy Conversion, Frank Neese, and for technical support from Conversion Energy Laboratory at Department of Electrical Engineering, Universitas Negeri Malang. We also gratefully acknowledge the support of DRPM RISTEK DIKTI.

IOP Conf. Series: Materials Science and Engineering **299** (2018) 012090 doi:10.1088/1757-899X/299/1/012090

Combustion characteristics of crude jatropha oil droplets using rhodium liquid as a homogeneous combustion catalyst

Hendry Y Nanlohy[1, 2], I N G Wardana[2], N Hamidi[2], L Yuliati[2]

[1]Department of Mechanical Engineering, University of Science and Technology Jayapura, 99351, Indonesia

[2]Department of Mechanical Engineering, Brawijaya University, Malang 65145, East Java, Indonesia

Email: hynanlohy@gmail.com

Abstract. Combustion characteristics of crude jatropha oil droplet at room temperature with and without catalyst have been studied experimentally. Its combustion characteristics have been observed by igniting the oil droplet on a junction of a thermocouple, and the combustion characteristics of oil droplets are observed using a high-speed camera. The results show that the uniqueness of crude jatropha oil as alternative fuel is evidenced by the different stages of combustion caused by thermal cracking in burning droplets. The results also show that the role of the catalyst is not only an accelerator agent, but there are other unique functions and roles as a stabilizer. Moreover, the results also found that the catalyst was able to shorten the ignition timing and burnout time. This phenomenon proves that the presence of catalysts alters and weakens the structure of the triglyceride geometry so that the viscosity and flash point is reduced, the fuel absorbs heat well and flammable.

1. Introduction

Crude vegetable oil such as jatropha is one of the alternative fuel sources that can be converted into biodiesel. Unfortunately, the use of biodiesel in the combustion process has not been maximized because it produces carbon dioxide causes global warming [1]. Moreover, the conversion process of crude vegetable oil into biodiesel requires additional energy, cost, and materials [2, 3]. Therefore, a reassessment of the conversion process such as the trans-esterification process and the esterification of crude vegetable oils become essential to finding energy-efficient conversion processes and costs resulting in energy-efficient fuels. In addition to the process of trans-esterification and esterification, there is also a hydrogenation process used to remove the double bond on the triglyceride chain because it can reduce the ignition properties such as oxidation stability [4]. Therefore, some researchers perform a partial process of hydrogenation using metal-based catalysts such as Rhodium [5-9], Rhodium and Ruthenium [10, 11], Rhodium and Iridium [12]. Unfortunately, the hydrogenation process has a very small influenced to fuel properties and improved fuel performance [13]. Moreover, the mechanism of homogeneous catalysts in the combustion process of diesel engine based crude jatropha oils has not been much revealed and very complex, including, the ignition process and the complex combustion chemical reaction between the fuel and air vapor. The unique characteristics of crude jatropha oil and the complexity of chemical processes make the effect of metal-based catalysts in combustion difficult to know through the performance of diesel engines. Therefore, more detailed research and observation of the catalyst mechanism and its effect on combustion characteristics using single droplet combustion method of crude jatropha oil using rhodium sulfate as homogenous combustion catalyst was performed.

2. Fuel preparations

The vegetable oil used is crude jatropha oil which is mixed with rhodium sulfate as a metal based combustion catalyst with dosing ratio of catalyst to oil volume of 0.001 ml: 100 ml. The fuel mixture was obtained by mixing the oil and catalyst in the test tube and shuffled manually. Meanwhile, the effect of the catalyst on the geometry of triglycerides chain is simulated using free chemistry software, which is Avogadro 1.1.1, and the result is displayed in figure 3. Fuel properties test results are presented in Table 1 and changes in the molecular mass of fatty acids are shown in Table 2

Table 1. Typical properties of crude jatropha oil.

Properties	With Catalyst	Without Catalyst
Flash Point(°C)	204	243
Caloric value (cal/gr)	8939	9400
Density at 15 °C (gr/ml)	0.916	0.917
Viscosity at 40 °C (cSt)	32.38	35.52

Table 2. Effect of catalyst on the weight of fatty acid molecules.

Chemical Composition	Cn:db	Formula	Composition %	Molecular mass, g/mole	
				Without Catalyst	With Catalyst
Caprylic	8:0	$C_8H_{16}O_2$	1.70	144.2114	132.1162
Capric	10:0	$C_{10}H_{20}O_2$	1.66	172.2646	160.1693
Lauric	12:0	$C_{12}H_{24}O_2$	7.71	200.3178	188.2225
Myristic	14:0	$C_{14}H_{28}O_2$	3.29	228.3709	216.2756
Palmitic	16:0	$C_{16}H_{32}O_2$	14.62	256.4241	244.3288
Palmitoleic	16:1	$C_{16}H_{30}O_2$	1.47	254.4082	242.3129
Stearic	18:0	$C_{18}H_{36}O_2$	7.36	284.4772	272.3820
Oleic	18:1	$C_{18}H_{34}O_2$	30.38	282.4614	270.3661
Linoleic	18:2	$C_{18}H_{32}O_2$	35.42	280.4455	268.3502

3. Experimental setup and procedures

The experimental apparatus is shown in figure 1. The oil droplet (7) was suspended at the junction of the thermocouple (6) made of a 13% Pt / Rh with a diameter of 0.1mm. The droplet diameter is about 0.6 - 1.1mm. The droplet is powered by an electric coil heater (5) 0.7mm diameter and made of Ni-Cr wire with a length of 30mm, a resistance of 1.02 Ω, has a voltage of 6 V and a current of 5 A (4). High-speed CCD camera (3) is used for taking pictures when the droplet is ignited. The shooting process takes place at a frame rate of 120 fps to allow in determining ignition time and burnout time. During the heating and ignition process, the temperature at the center of the droplet oil is recorded by the thermocouple sensor and acquired by a personal computer (1) connected to the data logger (2) with a frequency of 0.01 Hz. The process of taking data is repeated five times.

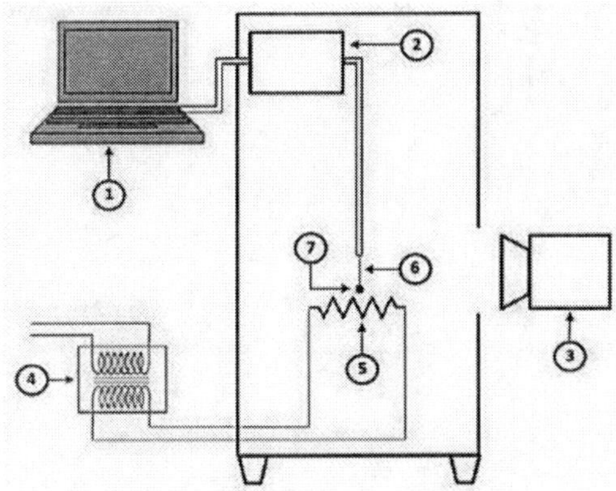

Figure 1. The experimental scheme

4. Result and discussion

Figures 2 and 3 show the triglyceride chains pulled by the catalyst and the adsorption models. The Rh^{3+} catalyst is an acceptor of 12 hydrogen bonds or a deficiency of 12 hydrogen atoms. With a larger amount of proton energy, the Rh^{3+} catalyst will attract 12 hydrogen atoms from the triglycerides hydrocarbon compound. The potential difference between the catalyst and the hydrocarbon chain allows for the transfer of electrons so that the mass of the carbon chain molecules decreases (see Table 2) causing weak van der Waals bonds marked by changing geometric structures and the distance between the triglyceride chains away from each other. This phenomenon is line with [14, 15] and proved by decreased viscosity and flash point (see table 1) and highly possible because it corresponds to Newton's law that force is directly proportional to mass and inversely proportional to distance.

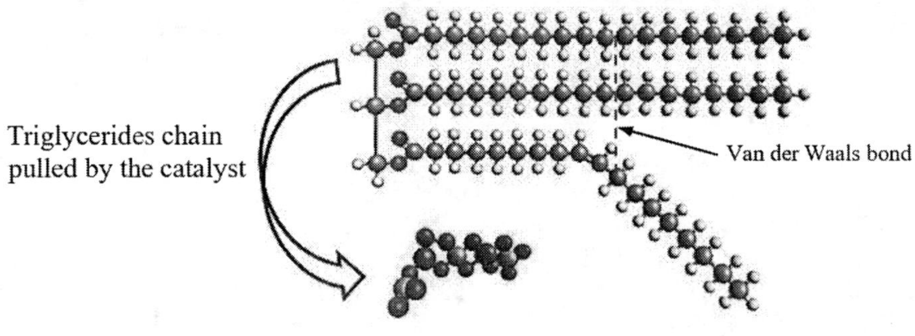

Figure 2. Interaction of the triglyceride chain with the catalyst.

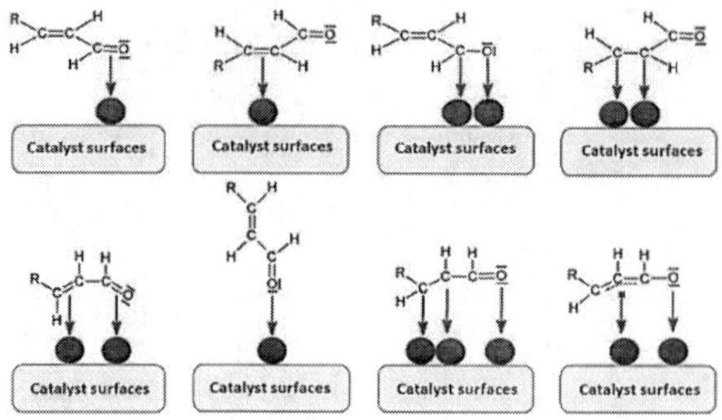

Figure 3. Hydrocarbon chain models when pulled by the catalyst. Adapted from [16]

Figure 4 shows the response of triglyceride chain jatropha crude oil to the catalyst work. It seems that when the triglyceride chain geometry catalyst is widened and the distance between the carbon chains away from each other. Changes in the geometry of triglycerides chain reduce and weaken the interaction of dipole and van der Waals bond between carbon chains and this phenomenon has an impact on combustion characteristics of crude Jatropha oil such as temperature evolution, micro-explosion, and burnout time.

Figure 5 shows the evolution of temperature in the center of oil droplets with and without the catalyst. The evolution of temperature consists of three main processes, namely the process of heating, evaporation, and combustion. The heating process begins with the heat transient that occurs in the droplet oil with and without catalyst which is about 0s to 1s and followed by steady heat transfer from the coil heater to the droplet that occurs at about the same time that is about seconds to 0.4 s. For oil without the catalyst, droplets ignite at temperatures of about 160 ^{o}C and accompanied by evaporation processes occurring at a constant temperature of about 1.22 s. As for the fuel with the catalyst, the droplet is ignited and accompanied by evaporation process occurs at around 148 ^{o}C temperature at 1.15 s second. For droplet oil without the catalyst, droplet ignite along with an increase in temperature at the beginning of the burning droplets reach around 355 ^{o}C, and at the same time, there is a second evaporation process around 2.1 s. As for the fuel with the catalyst, droplet burns along with an increase in temperature of about 320 ^{o}C and the second evaporation takes place around 1.96 s. The peak of the chart shows the highest temperature reached and simultaneously indicates the end of the combustion process, without the catalyst around 556 ^{o}C and with a catalyst 564 ^{o}C. From figure 5, it is also shown that the fuel with the catalyst has a shorter evolution time of 2.61 s, while the fuel with catalyst is about 2.81 s. These results are also confirmed and evident through figure 6 and figure 7, it shows that the amount of flame produced by oil with the catalyst is less than oil without catalyst and this phenomenon indicates that the fuel with the catalyst has greater power than fuel without the catalyst.

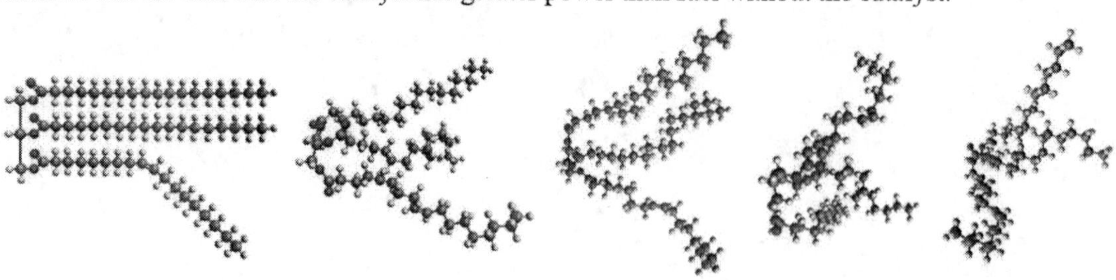

Figure 4. Effect of catalyst on geometry of triglyceride chains

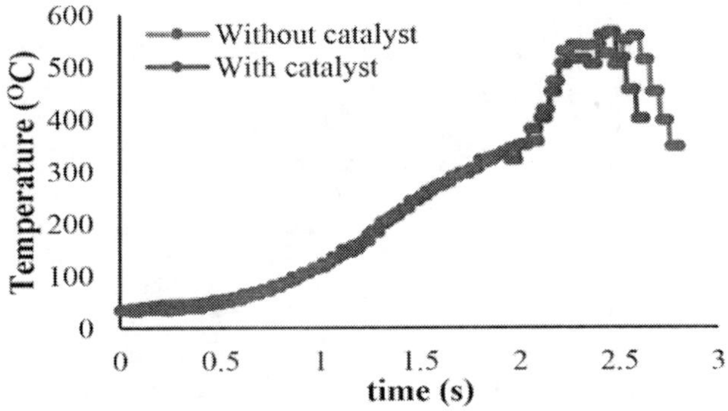

Figure 5. Temperature evolution in the center of the droplet with and without catalyst

Figures 6 and 7 show the evolution of flame oil droplets with and without the catalyst, such as the sequence of occurrence of micro-explosion of different intensities characterized by a spike-like flame form, this phenomenon consistent with the results of the study [17]. Figure 6 also shows that, the flame with catalyst looks calm and produces no smoke without catalysts. Whereas, the flame without catalyst look more volatile and produce more smoke (see figure 7). This phenomenon proves that the catalyst reduces the turmoil that occurs in the combustion process and produce emissions of clean and environmentally friendly.

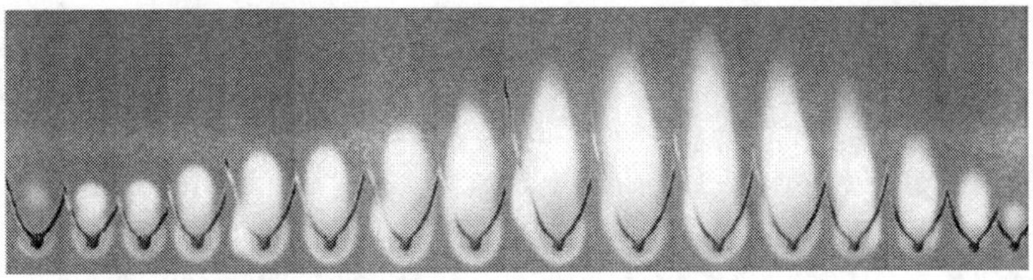

Figure 6. Flame evolution of oil droplet with catalyst

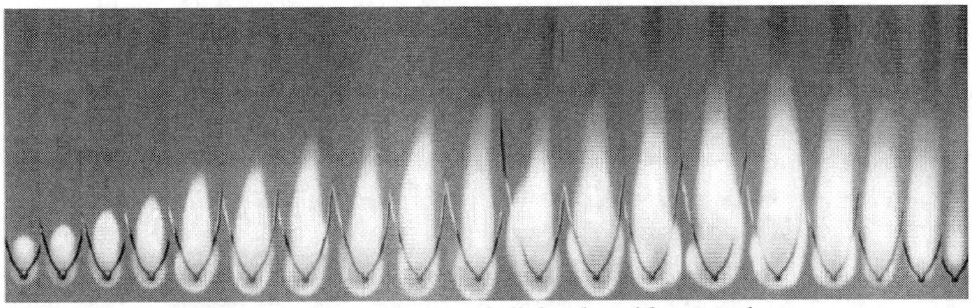

Figure 7. Flame evolution of oil droplet without catalyst

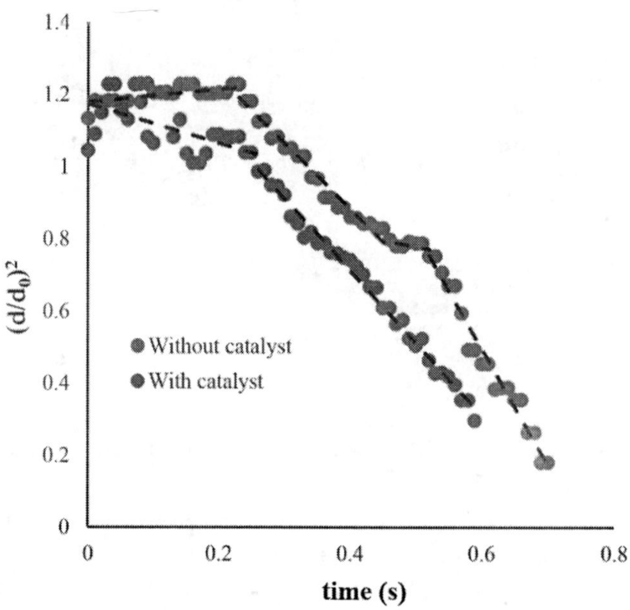

Figure 8. The evolution of burning droplet

Figure 8 shows the evolution of decreasing the volume of droplets during the burning points. Without a catalyst, it is shown that the droplet oil combustion process is unstable, this is indicated by the presence of several stages of combustion caused by thermal cracking. Crude jatropha oil is composed of multicomponent carbon chains such as saturated fatty acids, monounsaturated fatty acids, polyunsaturated fatty acids and glycerol, and each carbon chain burns at different times. This phenomenon is in line with [17]. Moreover, the result showed that fuel droplets without catalysts more volatile and oil droplets require more heat to evaporate and ignite that it impacts longer burnout time. In contrast, the fuel combustion process with the catalyst looks more stable and occurs only in one stage. This phenomenon shows that the catalyst is not only a reaction accelerator agent, but there is a new and unique role in the combustion process of crude jatropha oil that is as stabilizer and reducer. These results prove that the catalyst is able of altering the geometry of the triglyceride chain (see figure 3) and weakening the van der Waal bonds. This phenomenon is in line with [14, 15] which is discussed with different points of view. Therefore, oil droplets can absorb heat rapidly and accelerate the occurrence of molecular diffusion, and this makes the combustion process of crude jatropha oil more stable.

5. Conclusion and positive impacts for implementation

A comparative study of the effect of rhodium sulfate as a homogeneous combustion catalyst on jatropha crude oil combustion characteristics has been performed. The results showed that the addition of the catalyst was able to improve the combustion performance. Moreover, the result also found that jatropha crude oil has unique combustion characteristics and is shown by distinct stages of combustion. It was also found that the catalyst not only acted as an accelerator agent of the combustion reaction but found another unique role, namely as a stabilizer to the combustion process of crude jatropha oil. The results also show that the catalyst is capable of shortening ignition timing and burnout time, as well as being environmentally friendly. The results prove that the catalyst alters the molecular structure of the triglyceride chain and weakening the van der Waals bonds. Hence, the fuel rapid absorbs heat, molecule balance occurs and the combustion process of crude jatropha oil becomes stable.

These results prove that to produce an alternative fuel that is efficient and environmentally friendly, the process of trans-esterification, esterification, and hydrogenation can be ignored. Moreover, these results have a very positive impact to be applied as it saves production time, saves the use of oil and additives, and saves on production costs.

References

[1] COMMISSION E 2012 Amending Directive 98/70/EC relating to the quality of petrol and diesel fuels and amending Directive 2009/28/EC on the promotion of the use of energy from renewable sources.

[2] Ogden JM, Steinbugler MM, Kreutz TG 1999 *J Power Sources* **79** pp 143–68.

[3] Bautista Luis Fernando, Vicente Gemma, Rodriguez Rosalia, Pacheco Maria 2009 *Biomass Bioenergy* **33** pp 862–72.

[4] C.K. Westbrook, W.J. Pitz, S.M. Sarathy, M. Mehl 2013 *Proceedings of the Combustion Institute* **34** pp 3049–3056

[5] A. Bouriazos, K. Mouratidis, N. Psaroudakis, G. Papadogianakis 2008 *Catal. Lett.* **121** pp 158–164

[6] Achilleas Bouriazos, Sotiris Sotiriou, Constantinos Vangelis, Georgios Papadogianakis 2010 *Journal of Organometallic Chemistry* **695** pp 327–337.

[7] Constantinos Vangelis, Achilleas Bouriazos, Sotiris Sotiriou, Markus Samorski, Bernhard Gutsche, Georgios Papadogianakis 2010 *Journal of Catalysis* **274** pp 21–28

[8] Nikolaos Nikolaou, Christos E. Papadopoulos, Anastasia Lazaridou, Asimina Koutsoumba, Achilleas Bouriazos, Georgios Papadogianakis 2009 *Catalysis Communications* **10** pp 451–455.

[9] Hugo F. Ramalho, Karlla M.C. di Ferreira, Paula M.A. Machado, Renato S. Oliveira, Luciano P. Silva, Marcos J. Prauchner, Paulo A.Z. Suarez 2014 *Industrial Crops and Products* **52** pp 211–218.

[10] J. M. Grosselin," C. Mercier, G. Allmang, and F. Grass 1991 *Organometallics* **10** pp 7

[11] Christiana Vasiliou, Achilleas Bouriazos, Angeliki Tsichla, Georgios Papadogianakis 2014 *Applied Catalysis B: Environmental* **158–159** pp 373–381

[12] Kokil Saikia, Biswajit Deb, Dipak Kumar Dutta 2014 *Journal of Molecular Catalysis A: Chemical* **381** pp 188– 193

[13] Gerhard Knothe 2014 *Fuel* **119** pp 6–13

[14] Asep Kadarohman, Hernani, Ijang Rohman, Ririn Kusrini, Rizki Maryam Astuti 2012 *Fuel* **98** pp 73–79

[15] Satish Kumar, Wonsub Lim, Vinod Kumar Sharma, Sunil Kumar Garg, Il Moon 2013 *Journal of Molecular Liquids* **181** pp 55–61

[16] Claus, P 1998 *Catalysis* **5** pp 51–62.

[17] I.N.G. Wardana 2010 *Fuel* **89** pp 659–664

Acknowledgement

This research was supported by the Ministry of Research, Technology and Higher Education, The Republic of Indonesia, under the program of Graduate Education Scholarship with grant number SP DIPA-042.06.1.401516/2017

IOP Conf. Series: Materials Science and Engineering **299** (2018) 012091 doi:10.1088/1757-899X/299/1/012091

The flame characteristics of the biogas has produced through the digester method with various starters

Caturwati Ni Ketut[1)], Sudrajat Agung[1)], Permana Mekro[1)], Haryanto Heri[2)], Bachtiar[1)].

[1)]Mechanical Engineering Department,[2)]Electrical Engineering Department University of Sultan Ageng Tirtayasa
Jl. Jend. Sudirman km.3 Cilegon - Indonesia 42435

e-mail : n4wati@yahoo.co.id , n4wati@untirta.ac.id

Abstract. Increasing the volume of waste, especially in urban areas is a source of problems in realizing the comfort and health of the environment. It needs to do a good handling of garbage so as to provide benefits for the whole community. Organic waste processing through bio-digester method to produce a biogas as an energy source is an effort. This research was conducted to test the characteristics of biogas flame generated from organic waste processing through digester with various of the starter such as: cow dung, goat manure, and leachate that obtained from the landfill at Bagendung–Cilegon. The flame height and maximum temperature of the flame are measured for the same pressure of biogas. The measurements showed the flame produced by bio-digester with leachate starter has the lowest flame height compared to the other types of biogas, and the highest flame height is given by biogas from digester with cow dung as a starter. The maximum flame temperature of biogas produced by leachate as a starter reaches 1027 ^{0}C. This value is 7% lower than the maximum flame temperature of biogas produced by cow dung as a starter. Cow dung was observed to be the best starter compared to goat manure and leachate, but the use of leachate as a starter in producing biogas with biodigester method is not the best but it worked.

Key-words: biogas, bio-digester, starter, flame

1. Introduction

Biogas is a form of energy that is renewable, clean, environmentally friendly and very suitable for use in rural areas. The process of biogas formation is very good occurs in hot environmental conditions. Indonesia is a country located around the equator which has a higher environmental temperature than any other country located far from the equator. Thus Indonesia is very suitable to develop and utilize biogas as an energy source.

The handling of garbage in big cities today is becoming more complicated considering the increasing number of garbage and the narrowness of land used as shelters. Waste shelters that are getting closer to human settlements cause health problems for the surrounding environment. For that proper handling of municipal waste should get attention.

One of the breakthroughs that need to be developed and applied is the utilization of waste as an energy source. Various research and technology development shows that waste can be processed and used as energy source. [1,2,3,4,5].

Utilization of waste as an energy source can also be done by utilizing the biogas generated from the waste collection system by landfill method [3,5]. The utilization of leachate, wastewater from waste management by landfill method, for improve the quality and stabilize biogas production by recirculated of leachate with splashed leachate on the top of garbage pile [5].

In addition to landfill method of biogas, the formation can be done by anaerobic digester method. But that can be used in this case is only organic waste and animal or human waste. The process of forming biogas in an anaerobic digester is strongly influenced by the composition ratio of C / N in the raw material of biogas. To be able to utilize waste with a very high C / N value can be mixed with other materials with lower C / N values to optimize biogas product [6].

This paper shows the comparison of flame from biogas derived from organic waste with a mixture of starter: cow dung, goat manure, and leachate. This is done to determine the quality of each biogas as a source of energy, especially as fuel.

2. Experiment Details

In this research, we tested the flame characteristic of biogas combustion produced from organic waste biodigester with starter-type derived from:

1. Cow dung
2. Goat manure
3. Leachate

Biogas is produced from organic waste biodigester derived from a mixture of the following materials :

Type	Organic wastes	Water (liters)	Starter
Biogas I	50 liters	25	25 l cow dung
Biogas II	50 liters	25	25 l goat manure
Biogas III	50 liters	-	50 l leachate

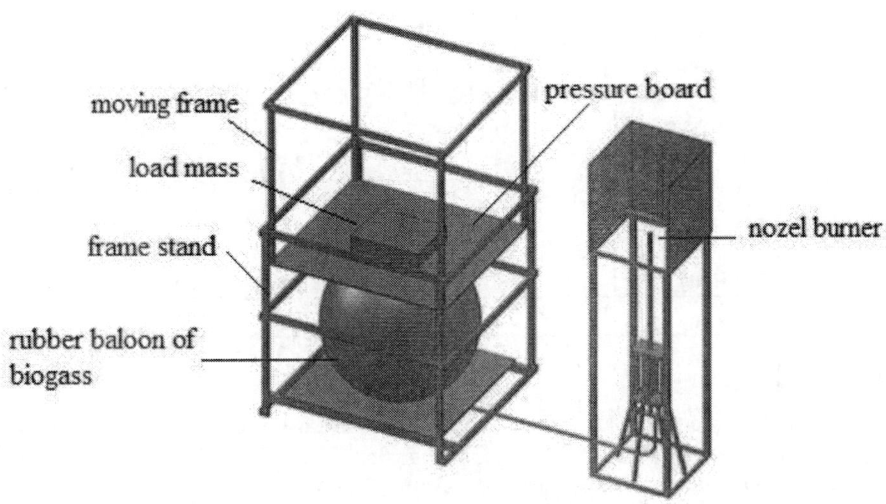

Figure 1.Skecth of flame measuring instrument.

Pressure board : 60 cm x 60 cm x 1,2 cm
Load Mass : 2690 grams
Moving Frame : 65 cm x 65 cm x 65 cm

Biogas that has been produced from biodigester is stored in a rubber balloon that has a diameter of 45 cm. In order for biogas to flow into the nozzle of the combustion chamber, it is necessary to increase the pressure inside the biogas balloon, with ballast in the form of a board placed over a gas balloon. By providing a fixed load of the board, it is expected that the gas pressure coming out to the burner is constant. The sketch of the biogas flame test apparatus is shown in Figure 1 .

The measurement of flame dimensions was done using photograph. Figure 2. Shows the dimensional determination of flame by using adobe photo shop from digital photos by comparing the number of pixels on the scale of the ruler with different pixel values on the height of the flame.

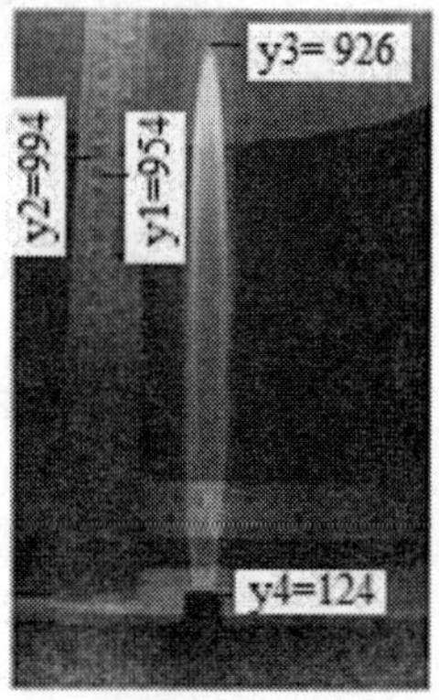

Figure 2. Determining the dimensions of flame through the photo

3. Results and Discussions

3.1 Flames dimensions

The length and width of the flame generated from burning biogas at the same pressure are shown in Table 1.

Table 1. Dimensions of flames: height (Hf) and width (Wf)

Type of starter	Dimension	1	2	3	average
Cow dung	Hf (cm)	21,6	25,4	20,5	22,50
	Wf (cm)	1,8	1,7	1,5	1,67
Goat manure	Hf (cm)	24,1	20,2	21,38	21,89
	Wf (cm)	1,8	2,3	2,25	2,12
Leachate	Hf (cm)	19,5	19,9	18,6	19,33
	Wf (cm)	2,8	2	2,3	2,37

The average flame dimensions are generated from the burning of different biogas from different starter types are shown in Figure 3. The flame of biogas produced from the type of cow dung starter has the highest flame length but has the smallest width of the flame base.

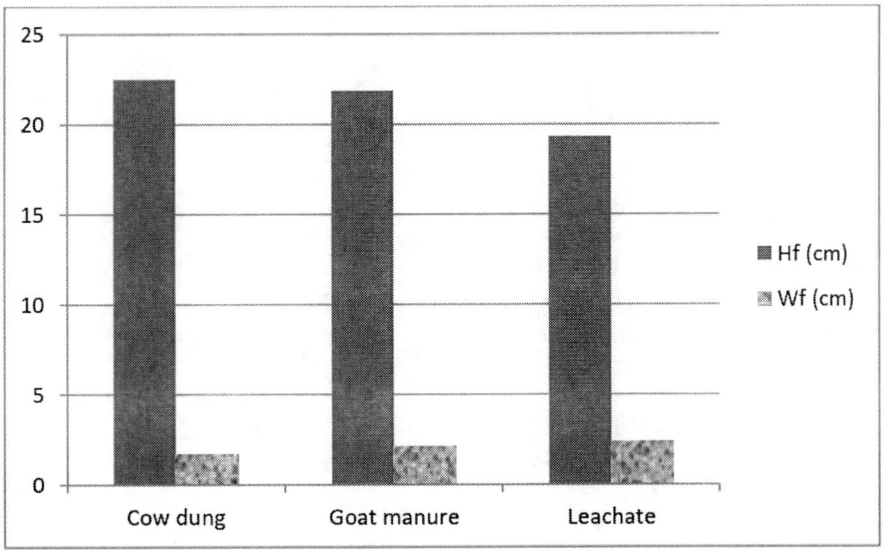

Figure 3. Height and width of flames from biogas with various starters

3.2 Maximum temperature of flames

The maximum temperature of flame from each biogas is measured using a thermocouple. The measurement results are shown in table 2. The table shows that maximum temperature is obtained at the flame of biogas resulting from goat manure starter and the lowest of maximum temperature flame is produced from biogas with leachate as a starter.

Table 2. Maximum temperature of flame

Type of starter	Tmax (^oC)
Cow dung	1089
Goat manure	1105
Leachate	1027

3.3 Flame color

In general, the color of the flame on the combustion of biogas from organic waste has a blue color with a reddish yellow mixture. The flame from burning biogas with a starter of cow dung and goat manure has more reddish color compared to biogas from organic waste with leachate as a starter.

4. Conclusions

The flame dimensions of biogas derived from organic waste with some starter variations show an inverse relationship between the height of the flame and the width of the flame. The highest flame will have the smallest flame width. The high flame difference resulting from organic waste biogas with a leachate starter 15% lower than the flame of biogas produced by the starter of cow dung at the same biogas pressure entering the combustion chamber. While the height of the flame from burning biogas with goat litter starter 2.7% lower than the high flame biogas organic waste with a starter of cow dung.

The color of flame from biogas with a starter of cow dung and goat manure has more red color than biogas with leachate starter. This shows that the biogas temperature of the leachate starter is lower than that of the two types of biogas produced from other starter types.

The maximum flame temperature of biogas with leachate as a starter has 7% below the maximum flame temperature of biogas with cow dung as a starter. So that leachate is feasible enough to be used as a starter on the production of biogas from organic waste by biodigester method.

References

[1] Arthur R, M. F. Baidoo, and E. Antwi 2011 *Renew. Energy* **36** pp. 1510–1516
[2] Nzihou Jean Fidele, Koulidiati Jean 2014 *Br. J. Appl. Sci. Technol* **4** pp. 450–464
[3] Caturwati NK, A. Sudrajat, H. Haryanto, P. Mekro, and Aminullah 2016 **2016** pp 364–371
[4] T. Zhang 2013 *PLoS One* **8** pp 1–8
[5] A. M. Sunil Kumar, Chart Chiemchaisri 2011 pp 77–79
[6] Sherwin Amini 2013 *International Journal of Agriculture and Crop Sciences* pp. 1453–1458

Acknowledgements

The authors would like to thank the Ministry of Research Technology and Higher Education Indonesia for financially supported through the Research Innovation Grant 2017.

Effect of Hydrochloric Acid Concentration on the Conversion of Sugarcane Bagasse to Levulinic Acid

Heni Anggorowati[1], Siti Jamilatun[1,2], Rochim B. Cahyono[1], and Arief Budiman[1,3,*)]
[1] Chemical Engineering Department, Gadjah Mada University, Jl. Grafika 2, Yogyakarta, Indonesia
[2] Chemical Engineering Department, Universitas Ahmad Dahlan, Jl. Kapas 9, Yogyakarta, Indonesia
[3] Center for Energy Studies, Gadjah Mada University, Sekip K1A, Yogyakarta, Indonesia
[*)] Corresponding author: abudiman@ugm.ac.id

Abstract. Levulinic acid is a new green platform chemical used to the synthesis of a variety of materials for numerous applications such as fuel additives, polymers and resins. It can be produced using renewable resources such as biomass like sugarcane bagasse which are cheap and widely available as waste in Indonesia. In this study, sugarcane bagasse was hydrolyzed using hydrochloric acid with a solid liquid ratio 1:10. The effects of hydrochloric acid concentration at temperature of 180 °C and reaction time of 30 min were studied. The presence of levulinic acid in product of hydrolysis was measured with gas chromatography (GC). It was found that the highest concentration of levulinic acid was obtained at 1 M hydrochloric acid in 25.56 yield%.

1. Introduction

In Indonesia the presence of biomass waste is very abundant. So far, the biomass is used for energy, such us oil palm empty fruit bunch (EFB) which converted to bio-oil [1], ulin wood which converted to bioenergy through pyrolysis process [2], sugarcane bagasse which converted to bioenergy by pyrolysis-gasification process [3] and utilize municipal solid waste (MSW) as the energy source [4]. Actually, biomass waste can be converted into platform chemicals like levulinic acid which has higher value.

Levulinic acid or 4-oxopentanoic acid is an organic compound which has ketone carbonyl group and acidic carboxyl groups. These functional groups turn levulinic acid into the building block for the synthesis of several organic compounds, such as: methyltetrahydrofuran (MTHF) for fuel additive, δ-amino levulinic acid (DALA) for herbicide, and diphenolic acid as a substitute for bisphenol in a polymer manufacture [5].

Traditionally, hexose sugar was the most convenient industrial route to produced levulinic acid [6]. Recently, the utilization of biomass and agricultural wastes for production of levulinic acid has been increased as research topic, since it would give low-cost material which widely available. Chang *et al* investigated wheat straw as a raw material for levulinic acid production and 19.86% yield of levulinic acid was obtained [7].

Levulinic acid can be produced from biomass by acid hydrolysis along with heating at high temperature. In this condition, biomass will be decomposed into glucose, then glucose is converted into hydroxy methyl furfural (HMF) which then produces levulinic acid as the main product and formic acid as a by-product [8].

One of the potential biomass sources to be processed into levulinic acid which is widely available in Indonesia is bagasse, which is a residue from the sugarcane industry. According to Association of Indonesian Sugar Experts (IKAGI), every year 11.38 million tons of bagasse is produced in Indonesia. Today, the utilization of bagasse is limited to animal feed, fertilizer (compost) material, pulp, and boiler fuel in sugar factory [9]. Like biomass in general, bagasse consists of polysaccharide which can be converted into a product or chemical compound which can be used to support production process of other industrial sectors. The polysaccharide in bagasse consists of cellulose (53.75%), hemicellulose (20.98%) and lignin (17.56%) [10]. High cellulose and hemicellulose contents are potential for processing into levulinic acid through hydrolysis.

The reaction of levulinic acid formation from biomass is a very complex series of reactions because it goes through several intermediate reaction stages and side reactions, such as humin formation. In certain condition, the more common reaction is side reactions such as lactone and humin formations, product repolymerization and intermediate repolymerization. Therefore, there should be an ideal reaction condition to obtain levulinic acid as a dominant product. According to Yan *et al*, factors which may influence reaction outcome includes acid concentration, temperature and reaction time [11]. Therefore, this paper will study the effect of hydrochloric acid concentration on levulinic acid yield.

2. Materials and Methods

2.1 Materials

The raw material used in this work is sugarcane bagasse collected from a local industry (PG Madukismo, Yogyakarta, Indonesia). It was air dried, milled, screened to select the fraction of particle size of less then 1 mm, homogenized in a single lot.

Hydrochloric acid 37% was purchased from Merc and levulinic acid 98% was purchased from Sigma Aldrich.

2.2 Methods

The hydrolysis reaction were carried out in a tube glass screw cap with an internal diameter of 13 mm and a length of 70 mm. Sugarcane bagasse and hydrochloric acid with ratio 1:10 placed in the tube. A series of tube was placed in oil bath and heated for (0-50) min at temperature of 180 °C. The oil bath temperature was monitored using thermocouple, with deviations maintened within ±1 °C. Insoluble humins, formed during the decomposition reaction, were separated from the solution by paper filtration. The liquid product was sampled and analyzed using a gas chromatography (GC).

2.3 Analysis

The composition of liquid phase was determined using a GC system by using a column HP-5/SH-Rxi-5Sil with He as carrier gas. The yield of levulinic acid was expressed in term of weight percent yield which calculated from a calibration curve plotting concentration of levulinic acid standart against areas obtained under the GC curve. The following formula was used for calculating the yield of levulinic acid

$$\text{Yield levulinic acid (wt\%)} = \frac{\text{weight of levulinic acid (g)}}{\text{weight of substrat feed (g)}} \times 100\%$$

3. Results and Discussion

3.1 Sugarcane bagasse composition

Sugarcane bagasse is lignocellulose biomass whose main composition is cellulose, hemicellulose and lignin. Table 1 shows the test result of cellulose, hemicellulose and lignin contents in the bagasse which used in this study. The method to determine bagasse composition was Chesson's method. The result shows that bagasse contains high cellulose and hemicellulose. Both can be converted into glucose which is an intermediet in the production of levulinic acid. Thus, glucose to be conferted to 5-HMF and then to levulinic acid as the final product. The conversion of a typical lignocellulosic biomass to levulinic acid is shown in Figure 1.

Table 1. Component of sugarcane bagasse.

Compound	Value (%)
Cellulose	41.9
Hemicellulose	23.9
Lignin	21.6
Others	12.6

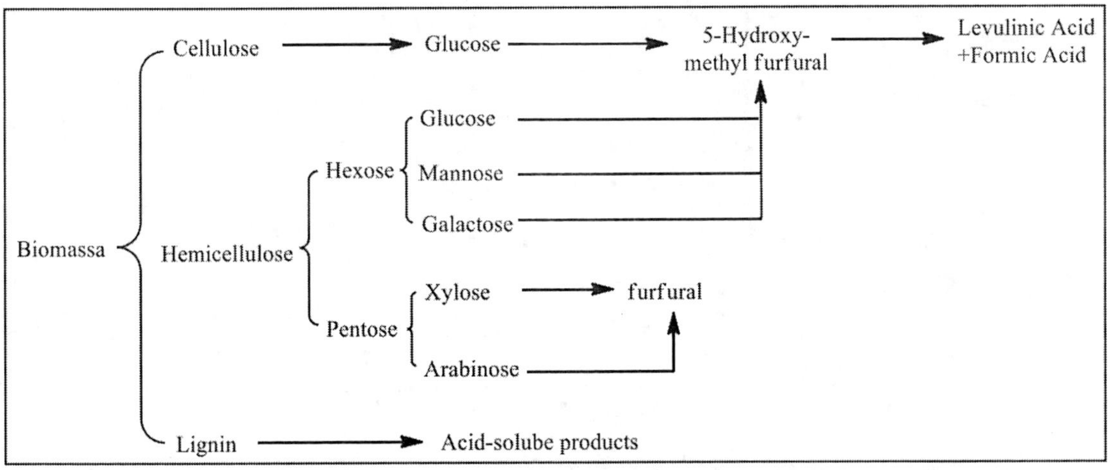

Figure 1 Simplified reaction scheme for the conversion of lignocellulosic biomass to levulinic acid [12]

3.2 Composition of the liquid product

Sugarcane bagasse was hydrolyzed with hydrochloric acid at 180 °C for 30 min, and after the reaction the products were analyzed by GC to measure the chemical compositions. The retention time of the standard sample of levulinic acid was at 10.380 min according to Figure 2.

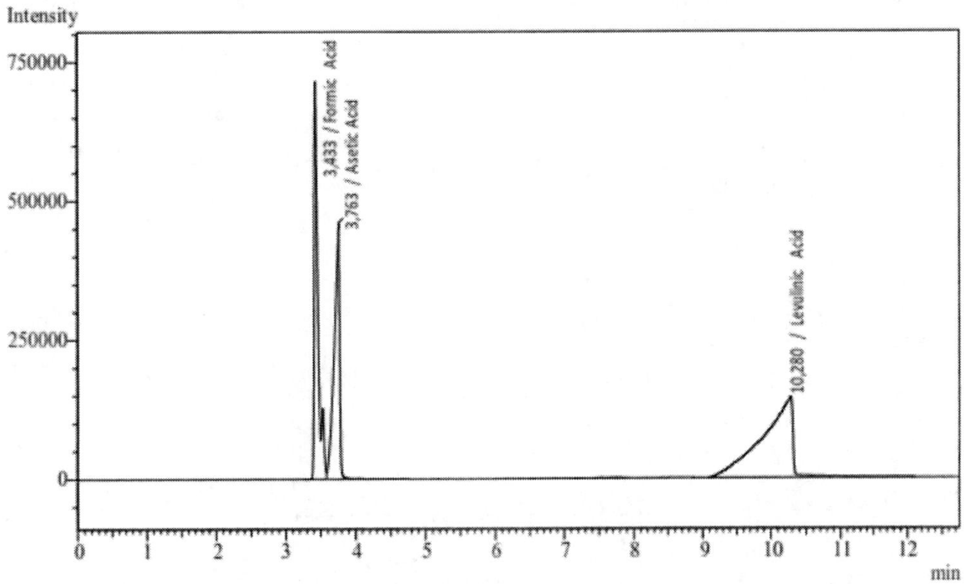

Figure 2 Representative GC profile of the standard samples

Figure 3 show the GC profiles of the products obtained from hydrolysis sugarcane bagasse. This shows that the peak for levulinic acid was determined at the retention time of 10.280, on the basis of the GC analysis of the standard sample. Besides levulinic acid, there were formic acid and acetic acid as by-product.

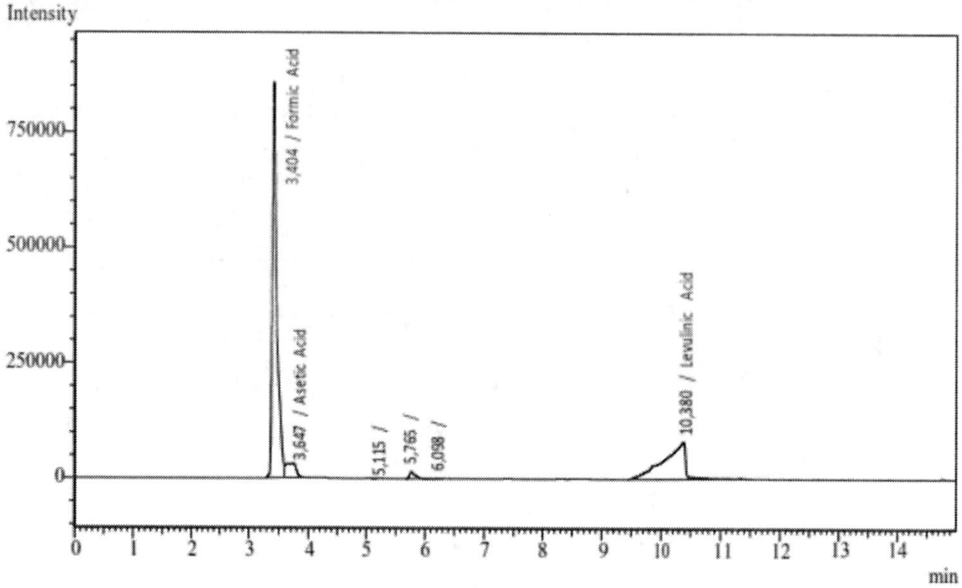

Figure 3 Representative GC profile of the liquefaction fraction from hydrolysis sugarcane bagasse

3.3 The effect of reaction time on levulinic acid yield

Figure 4 shows the effect of reaction time on the conversion sugarcane bagasse to levulinc acid with the concentration of hydrochloric acid of 1 M and temperature of 180 °C. This result show that the amount of levulinic acid increasing very fast during the 30 minute. Afterward it was observed that only a slight rise occurred after 30 min of reaction time. This happened because side reactions started to occur when the degradation products were exposed to high temperatures for a prolonged period of time.

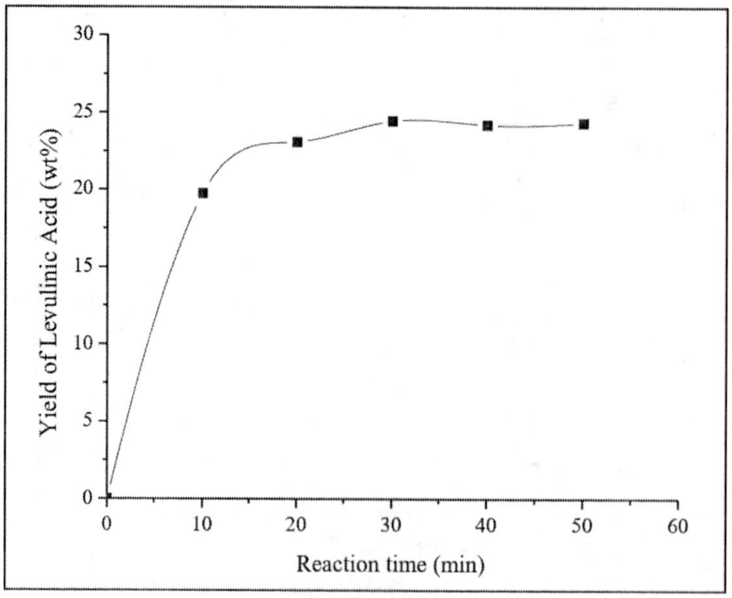

Figure 4 Effect of reaction time on yield levulinic acid

3.4 The effect of hydrochloric acid concentration on levulinic acid yield

Hydrochloric acid was selected to be acid catalyst because it is affordable and easily ionized, so it is more effective for the hydrolysis process of bagasse into levulinic acid. The effect of of hydrochloric acid concentration on the sugarcane conversion to levulinic acid was investigated in concentration range of 0.25-1M. Reaction was performed for 30 minutes at 180 °C and ratio of solid to liquid 1:10. Figure 5 shows effect of HCl concentration on yield levulinic acid.

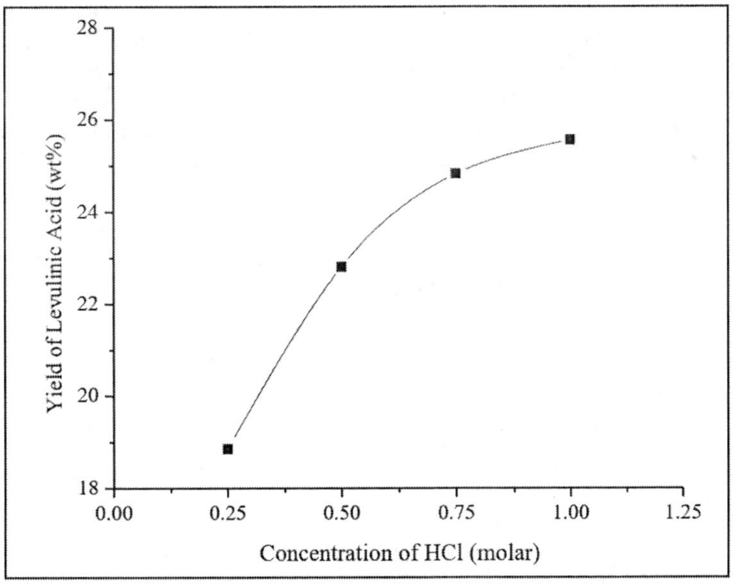

Figure 5 Effect of HCl concentration on yield levulinic acid

From Figure 5, we notice that the highest conversion was achieved at 1 M concentration of hydrochloric acid with yield of levulinic acid of 25.56 wt%. The result of this work showed that levulinic acid yield will increase as the concentration of hychloric acid catalyst increases until optimum condition was reached. Afterward, increasing catalyst concentration only has small effect on levulinic acid yield. Higher HCl concentration causing side reactions that may negatively affect the rate of hydrolysis cellulose to levulinic acid. Similar result with Fang and Hanna observed that higher sulfuric acid concentration also significantly increased the levulinic acid yield. Yield of levulinic acid was positively correlated with mineral acid concentration, which greatly boosted the yield of levulinic acid by the higher rate of hydrolysis of carbohydrates to organic acids [12].

4 Conclusion

Levulinic acid can be produced by hydrolysis reaction of sugarcane bagasse using hydrochloric acid catalyst. The conversion of bagasse into levulinic acid was affected by the concentration of hydrochloric acid as the catalyst. The yield of levulinic acid increased at higher consentration of hydrochloric acid. Maximum yield of levulinic acid (25.56wt%) was obtained at the concentration of hydrochloric acid of 1 M at 180 °C and reaction time of 30 minutes.

Acknowledgments

The authors are grateful to the financial support from Lembaga Pengelola Dana Pendidikan (LPDP), Kementerian Keuangan, Republik Indonesia.

References

[1] Pradana YS and Budiman A 2015 *Journal of Engineering Science and Technology* **10** pp 1-8
[2] Widiyannita AM, Cahyono RB, Budiman A, Sutijan and Akiyama T 2016 *AIP Conference Proceedings* 1755
[3] Daniyanto, Sutijan, Deendarlianto, and Budiman A 2016 *Journal of Engineering and Applied Sciences* **11** pp 9974-9980
[4] Sudibyo H, Majid AI, Pradana, YS, Budhijanto W, Deendarlianto, and Budiman A 2017 *Energy Procedia* **105** pp 263-269
[5] Morone A, Apte M and Pandey RA 2015 *Renewable and Sustainable Energy Reviews* 546-565
[6] Fang Q and Hanna MA 2002 *Industrial & Engineering Chemistry Research* **81** 187-192
[7] Chang C, Cen P and Ma X 2007 *Bioresource Technology* **98** 1448-1453
[8] Baugh KD and McCarty PL 1998 *Biotechnology and Bioengineering* **31** 50-61
[9] Setiati R., Wahyuningrum D, Siregar S and Marhaendrajana T 2016 *Jurnal Penelitian dan Pengabdian Masyarakat* pp 257-264
[10] Widyawati, Niken L and Argo B 2014 *Jurnal Teknologi Pertanian* **15** pp 1-6
[11] Yan L, Yang N, Pang H and Liao B 2008 *Journal Clean* **36** 158-163
[12] Girisuta B 2007 Levulinic acid from lignocellulosic Biomass *PhD Thesis* University of Groningen Netherlands

IOP Conf. Series: Materials Science and Engineering **299** (2018) 012093 doi:10.1088/1757-899X/299/1/012093

Utilization of Cacao Pod Husk (*Theobroma cacao l.*) as Activated Carbon and Catalyst in Biodiesel Production Process from Waste Cooking Oil

Devita Rachmat*, Lizda Johar Mawarani and Doty Dewi Risanti

Department of Engineering Physics, Faculty of Industrial Technology, Institut Teknologi Sepuluh Nopember, Jl. Arief Rahman Hakim, Surabaya 60111 Indonesia

*)devitarachmat@gmail.com

Abstract. Cocoa pod husk (Theobroma cacao l.) is a waste from cocoa beans processing. In this research we employ cocoa pod husk as activated carbon to decrease the value of FFA (Free Fatty Acid) in waste cooking oil and as K_2CO_3 catalyst in biodiesel production process from waste cooking oil. Cocoa pod husk was crusched and grounded into powder that passed thorugh 60 mesh-screen. As activated carbon, cocoa pod husk was firstly carbonized at three variant temperatures i.e 250°C, 300°C and 350°C. The activation process was done using HCl 2M as activator. Based on the results of XRD and FTIR, the carbonization at all variant temperatures does not cause a significant changes in terms of crystallite structure and water content. The pore of activated carbon started to form in sample that was carbonized at 350°C resulting in pore diameter of 5.14644 nm. This result was supported by the fact that the ability of this activated carbon in reducing the FFA of waste cooking oil was the most pronounced one, i.e. up to 86.7% of FFA. It was found that the performance of cocoa pod husk's activated carbon in reducing FFA is more effective than esterification using H_2SO_4 which can only decrease 80.8%. On the other hand, the utilization as K_2CO_3 catalyst was carried out by carbonization at temperature 650°C and extraction using aquadest solvent. The extraction of cocoa pod husk produced 7.067% K_2CO_3 catalyst. According to RD results the fraction of K_2CO_3 compound from the green catalysts is the same as the commercial (SAP, 99%) that is ≥ 60%. From the obtained results, the best yield percentage was obtained using K_2CO_3 catalyst from cacao pod husk extract, i.e. 73-85%. To cope with biodiesel conversion efficiency, a two-step process consisting pretreatment with activated carbon carbonized at 350°C and esterification with K_2CO_3 from cocoa pod husk catalyst was developed. This two-step process could reach a high conversion of 85%. From the results it was clear that the produced biodiesel fuel was within the recommended SNI 7182: 2015 standard.

1. Introduction

Cocoa pod husk are known to contain high levels of lignin, cellulose and hemicellulose that can be decomposed into carbon after calcination process [1]. There is also a potassium content that can be extracted as an element of K_2CO_3 [2]. The content makes cocoa waste to be further utilized as activated carbon and K_2CO_3 catalyst in biodiesel production process. Besides being able to cope with pollution due to cocoa waste, this can also overcome the operational costs that reach 75% in biodiesel production as a renewable energy [2].

In the biodiesel process production in addition of catalyst prices, the price of raw oil also leads the increasing cost in production that reach up to 2 times more expensive than petroleum diesel [3]. Therefore in this research, waste cooking oil is used as raw material to produce biodiesel. Because the using of waste cooking oil is not only can reduce the use of used cooking oil repeatedly, it also can be a solution to tackle the issue price of biodiesel. But the level of FFA in waste cooking oil can accommodate the resulting yield [4]. Therefore, the purpose of this study is utilizing cocoa pod husk as activated carbon to reduce FFA level in waste cooking oil and as a catalyst in biodiesel production process.

2. Experimental

2.1. Synthesis of Activated Carbon from Cacao Pod Husk
Cocoa pod husk is dried at 80°C for 24 hours to reduce its water content and refine into a 60 mesh powder. Then cacao pod husk charred in three temperature variations of 250°C, 300°C and 350 °C for 2 hours. Furthermore, 37.5 grams of cacao pod husk powder was mixed on 75 ml HCl 2M activator while stirring for 4 hours at 80°C. The mixture was allowed to stand for 24 hours and was activated on temperature of 200°C for 3 hours. Characterization of activated carbon is done through three testing methods, namely Brunauer Emmett Teller (BET) in Chemical Engineering Department in ITS to know the pores size of activated carbon, Fourier Transform Infrared (FTIR) and X-Ray Diffraction (XRD) to know the structure of the compound and the functional group of activated carbon.

2.2. Extraction of K_2CO_3 Element as Catalyst from Cocoa Pod Husk
Firstly, Cocoa pod husk is calcined at 650°C for 4 hours until it becomes ash. Furthermore, extraction was done by mixing 250 grams of ash with 1 liter of aquadest while stirring for 1 hour. Then proceed with filtering process using filter paper no.93 size 10 μm. The solution result of the filtrate was reheated at 140°C for 3 hours to gain K_2CO_3 brown crystal (crude potash). Finally, the K_2CO_3 brown crystals are washed in aquadest until the color turns into white. Characterization of K_2CO_3 catalysts was carried out through two methods of Fourier Transform Infrared (FTIR) and X-Ray Diffraction (XRD) that each used to identify functional groups formed on the catalyst and crystal fraction of K_2CO_3 compounds.

2.3. Pre-treatment of Wasted Cooking Oil
Sample preparation of waste cooking oil is done by heating the sample of cooking oil at a temperature of 120°C for 10 minutes to reduce its water content and filter out the residual solid waste of the frying process. Then the FFA content of waste cooking oil is calculated using the following equation [5],

$$\% \, FFA = \frac{volume\ of\ NaOH\ x\ N\ x\ molecular\ weight}{mass\ of\ cooking\ oil}\ x\ 100\% \qquad (1)$$

Pre-treatment is begun with soaked of 37.5 grams of activated carbon in 300 ml of waste cooking oil for 10 hours. Furthermore, filtration process using a rough filter paper to separate the cooking oil from activated carbon. As a comparison, pre-treatment is also done by esterification method using H_2SO_4. The esterification was carried out by steps, the waste cooking oil mixed with 0.5% (of oil volume) H_2SO_4 and 6: 1 molar ratio of methanol to oil, then heated to 60°C for 2 hours. After that, the oil is allowed to stand for 24 hours to gain 2 layers form, the top layer is crude biodiesel and the bottom layer is a mixture of glycerol and residual of H_2SO_4. The crude layer of biodiesel is separated from glycerol by a funnel separator. The last step wass recalculating the value of FFA using equation (1).

2.4. Production of Biodiesel
Firstly, lye solution was prepared by mixing 1 wt% catalyst (with methanol oil which ratio of molars to oil 6: 1 using magnetic stirrer for 5 minutes). Furthermore, transesterification was done by adding lye solution at 300 ml of waste cooking oil and dilution of stirring at 60°C for 1 hour, then the mixture is allowed to stand for 8 hours to form two layers (biodiesel and glycerol), then washed by aquadest.

2.5. Characterization of Biodiesel

The biodiesel characteristic values that observed here are density based on ASTM D 1298, kinematic viscosity according to ASTM D 445, flash point based on ASTM D 93-80 test procedure, and fog point and pour point appropriate with ASTM D 2500 and ASTM D 97 testing procedures.

3. Results and Discussion

3.1. Characteristics of Activated Carbon

The results XRD of carbon and activated carbon in Figure 1(a) shows that the process of banding and activation does not cause a change on the structure of the cocoa pod husk. It is characterized by two width peak between 15° and 35° as an indication of elements with amorphous structures [7]. This amorphous structure is derived from lignin and hemicellulose content in the cocoa pod husk which

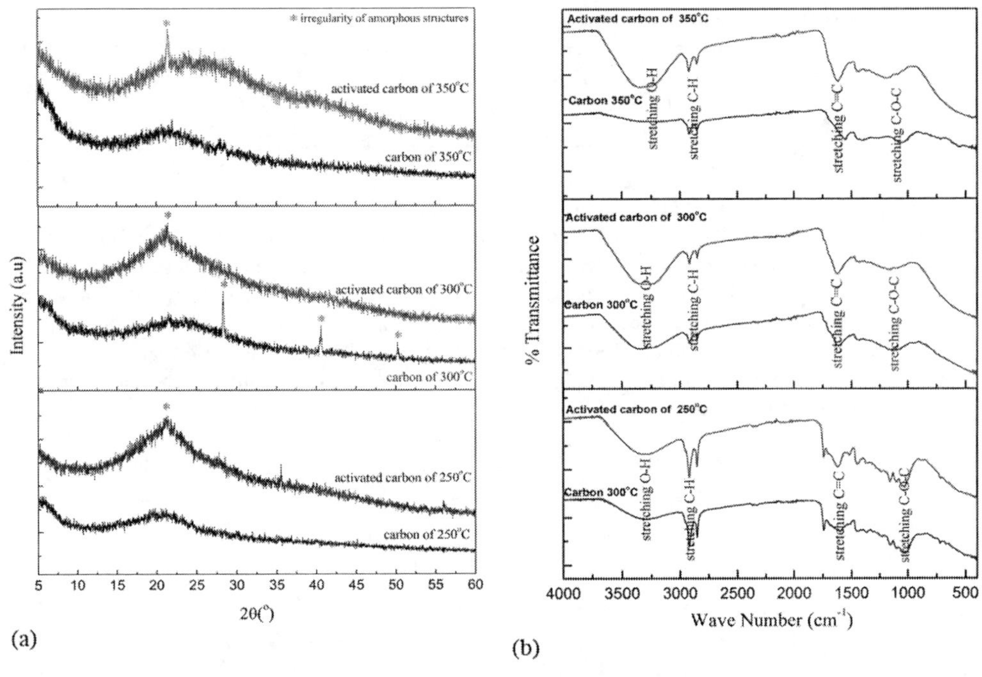

Fig. 1 Charactherization results (a) X-ray diffraction pattern from activated carnon and carbon of cocoa pod husk (b) Infrared spectra of unactivated and activated carbon of cocoa pod husk

also has amorphous structure [8]. So it also can be seen that samples at temperature 250°C, 300°C and 350°C does not have changes in the structure of lignin and hemicellulose cocoa pod husk. The increasing peak from carbon to activated carbon indicates that the activation process can lead to a little change to more crystal structure. But this it is not fully crystal just lead to the changing structure.

Based on FTIR results in Fig. 1(b) overall it is known that there is stretching of O-H bonds occurring in all charcoal and activated charcoal samples, this stretch indicates the presence of water molecules absorbed by the test sample [7,9]. The largest stretching of O-H bonding occurs between unactivated carbon and activated carbon at 350°C. This occurs because the function of HCl as an activator in cocoa pod husk can release the strong water bonds that exist in the pores of the carbon so as to increase its absorption [10]. This significant shift occurs only on the activated carbon sample temperature of 350°C so that the absorption will be greater than the other two samples.

From the results of BET testing on activated char was observed that the pore began to form at the time of refining at 350°C with pore size values as presented in Table 1. Whereas in the other two samples did not get pore size. Judging from the rise in the temperature of the furnace, this is in accordance with previous research [11] which states that the increase in carbonation temperature affects the increasing surface area of the activated carbon. While HCl as an activator in the manufacture of activated carbon can enlarge the surface area because it can evaporate substances flying and getting maximal with high temperature treatment between 700 to 900°C [12]. The invention is also supported by another invention [13], whereas in the activated charcoal making with cocoa cutaneous skin precursors the $ZnCl_2$ activator effectively increases its surface area to 780 m^2 / g with a carbonation temperature of 650°C.

Table 1. The pores size of activated carbon of cocoa pod husk at carbonization temperature at 350°C

Sample	*Activated Carbon 350°C*
Surface area (m^2/g)	9,662
Average pore diameter (nm)	5,14644
Micro pore surface area (m^2/g)	2,724
The external surface area of the micro pore (m^2/g)	6,938
Micro pore volume (cm^3/g)	0,001
Total pore volume (cm^3/g)	1,243 x 10^{-2}

3.2. FFA Decreasing of Waste Cooking Oil in Pre-treatment Process

From the FFA value data value in Table 2. It is known that the decrease of FFA using H_2SO_4 catalyst and activated carbon can decrease FFA to less than 2% in one process. The best result of the four treatments on the cooking oil was shown by the pre-treatment process using activated carbon with 350°C leaching temperature i.e. FFA content decreased to 0.539%. While esterification results have the same effectiveness to activated carbon at 300 °C 300°C in reducing FFA of 80.8%. So that waste cooking oil purification using both H_2SO_4 catalyst and activated carbon can fulfill the minimum requirement of FFA content as raw material of biodiesel.

Table 2. The FFA Value of Waste Cooking Oil before and after Pre-Treatment Process

Condition		*FFA Value*	*Percentage decrease*
before pre-*treatment*		4,045 %	-
Pre-*treatment*	250°C	1,348 %	66,7 %
	300°C	0,775 %	80,8 %
	350°C	0,539 %	86,7 %
Esterification		0,775 %	80,8 %

The effect of the carbonization temparature in cocoa pod husk on decreasing FFA value of cooking oil can indicate that the use of activated carbon can decrease 66.7% to 86.7% FFA on cooking oil. The temperature of cocoa pod husk carbonization affects the adsorption capacity of activated charbon to free fatty acids. In this research, it is found that the optimum carbonization temperature of cacao pod husk has the ability to decrease FFA by 86,7%. So it indicates that the higher carbonization temperature, the higher the absorption capacity of activated carbon to free fatty acids in cooking oil. This is because at high temperature heating, more and more carbon pores are formed [14].

3.3. Characteristics of K_2CO_3 Catalyst

Diffractogram of K_2CO_3 catalyst in Fig. 2 shows the presence of the crystallinity peaks formed between the K_2CO_3 catalysts from cocoa pod husk and commercial K_2CO_3 catalysts, indicating the presence of K_2CO_3, K_2O, SiO_2, Al_2O_3 and other impurities. K_2O comes from the decomposition that occurs at the time of K_2CO_3 extraction [12]. Whereas the presence of Al_2O_3 and SiO_2 compounds is the residue of ashes [9]. The presence of K_2O elements together with the Al-O-K functional groups which are strong bases are known to be activated by methanol to make effective catalysts used in the transesterification process [9], [15]. This occurs because at the time of transesterification, strong base elements can decrease the activation energy, accelerate the formation of methyl esters and increase the percentage of biodiesel conversion [15].

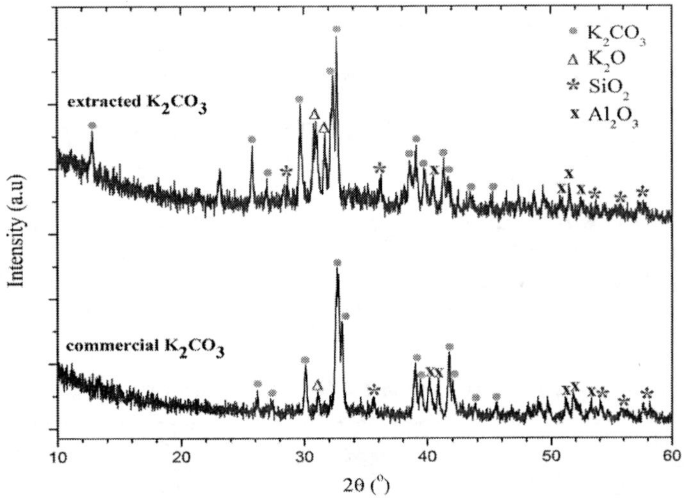

Fig. 2 Diffractogram of K_2CO_3 catalyst from cacao pod husk (blue) and commercial K_2CO_3 (black) catalyst

In view of the transmittance peak present in FTIR Fig.2, the catalyst has a spectrum of ~ 1440 cm^{-1} and ~ 1350 cm^{-1} wavelengths indicating the presence of carbonate compounds of K_2CO_3. In addition, the ~ 3120 cm^{-1} wavelength range indicates the presence of carbonate compounds in stretching of Al-O-K bonds. This carbonate compound refers to the content of K_2O in the catalyst due to the decomposition of K_2CO_3 [9]. Therefore, the result of catalyst characterization through XRD and FTIR method shows the same indication between K_2CO_3, K2O and Al_2O_3.

3.4. Yield of Biodiesel from Transesterification Process

From the initial volume of 150 ml waste cooking oil can be produced biodiesel as shown in Fig.5 which shows that the volume of biodiesel is most generated in the transesterification process using K_2CO_3 of cocoa pod husk as much as 110-128 ml. While the second largest biodiesel obtained by transesterification using commercial K_2CO_3 catalyst that is as much as 100-119 ml and the last yield of biodiesel volume by using KOH catalyst as much as 95-126 ml. Based on the volume of biodiesel produced, it can be concluded that the performance of K_2CO_3 from cacao pod husk is more effective cause it can produce higher amount of methyl ester of fatty acid (biodiesel) compared to two other commercial catalysts, ie K_2CO_3 (SAP, 99%) and KOH. This is related to the presence of active base elements of K_2O and Al-O-K [9]. Both of these are active components that can expand the active surface to obtain more methyl esters [16].

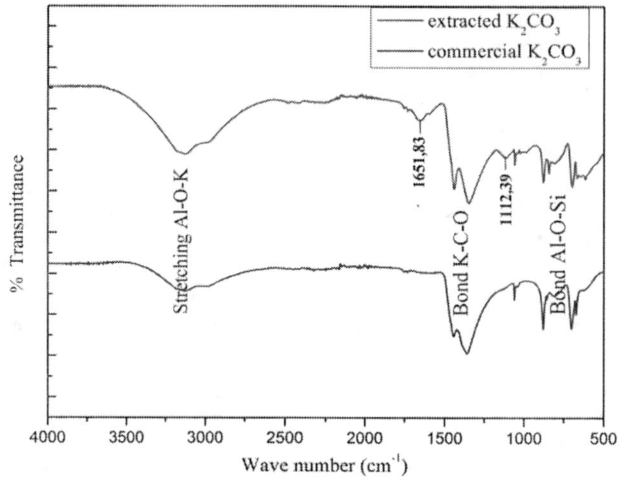

Fig.3 Infrared spectra of commercial K_2CO_3 and extracted K_2CO_3

3.5. Biodiesel Characteristics

In this research the value of biodiesel quality compared to the standard value of SNI quality requirements (Indonesian National Standard) and ASTM (American Society for Testing and Material) as in Table 3. The characteristic parameters tested are density, kinematic viscosity, flash point, fog and biodiesel pour point. Taken together, all the characteristic values of the biodiesel sample obtained meet the standards specified by ASTM.

Table 3. Comparison of characteristics of the biodiesel to characteristics value of B20, SNI and ASTM

Characteristics		*density (kg/m³)*	*Kinematic viscosity (mm²/s)*	*Point (°C)*		
				flash	**pour**	**fog**
Type of catalyst	KOH	850-880	3,91-4,89	171-179	13,0-15,6	8,3-12,4
	Commercial K_2CO_3	860-890	4,66-5,17	168-174	12,3-13,0	8,6-9,0
	Extracted K_2CO_3	820-890	5,36-5,51	172-175	14,2-14,5	9,7-10,7
ASTM		800-880	1,90-6,00	Min. 100	Maks. 18	Maks. 18
SNI		850-890	2,30-6,00	Min. 120	Maks. 18	Maks. 18

Table 3. shows that the fog points and pour points of all samples meet the SNI standard values. The fog point characteristics of biodiesel samples ranged from 8.3 to 12.4 °C, while the pour point ranged from 12.3-15.6 °C. The mass value of the biodiesel type is related to the calorific value and power to be generated by the diesel fuel unity volume [17]. This is because the fuel injection pump measures the fuel by volume. So the power generated in the combustion chamber is affected by the fuel density [18]. In addition, the kinematic viscosity value will affect the quality of the ratio of air and fuel mixture to the combustion engine [18]. Because of the high viscosity will cause the evaporation of the fuel so that the mixing process of air and fuel cannot proceed well [19].

4. Conclusion

Cocoa pod husk as activated carbon is potential to decrease the levels of FFA (Free Fatty Acid) in waste cooking oil. The pores on the surface of activated carbon from cocoa pod husk with 2M HCl activator begin to form at 350 °C carbonization temperature and succeeded in decreasing FFA in waste cooking oil up to 86.7%. Cocoa pod husk is potential to be utilized as a potassium carbonate (K_2CO_3) catalyst in the process of producing biodiesel from waste cooking oil. The use of K_2CO_3 catalyst from cocoa pod husk in transesterification process is capable to resulting the larger biodiesel yields compared to commercial K_2CO_3 catalysts and KOH, ie by 85%. From XRD result, fraction of K_2CO_3 compound which is produced from the cocoa pod husk is close to fraction value of commercial K_2CO_3 (SAP, 99%), that is 60.76%. Characteristics of density, kinematic viscosity, flash point, fog point and pour point of cooking oil biodiesel that has undergone pre-treatment with activated carbon as well as K_2CO_3 catalyst from cocoa pod husk can meet SNI and ASTM values.

References

[1] I. Rasjiddin. "Pembuatan Arang Aktif dari Tempurung Jambu Mede (Anacaradium occidentale) Sebagai Adsorben pada Pemurnian Minyak Goreng Bekas". Bogor: Fakultas Teknologi Pertanian Institut Pertanian Bogor (2006).

[2] C. O. Boateng, K. T. Lee. "The potential of using cocoa pod husks as green solid base catalys for the transesterification of soybean oil into biodiesel: Effects of biodiesel on engine performance". *Chemical Engineering Journal* (2013) 395-401.

[3] M. S. Kocak, E. Ileri, Z. Utlu, (2007). "Experimental study of emission parameters of biodiesel fuels obtained from canola, hazelnut, and waste cooking oils". *Energy Fuels*, 21: 3622-3626.

[4] E. I. Rhofita. "Penurunan Kadar Free Fatty Acid (FFA) pada Reaksi Esterifikasi dalam Proses Produksi Biodiesel : Kajian Waktu Reaksi dan Temperatur Reaksi". *Jurnal Ilmu-Ilmu Teknik – Sistem* Vol.11 No.1 (2015) 39-44.

[5] Z. Ainy. "Pengaruh Konsentrasi Aktivator Arang Aktif Serbuk Kayu terhadap Karakteristik Biodiesel Minyak Jelantah". Surabaya: Jurusan Teknik Fisika, Fakultas Teknologi Industri Institut teknologi Sepuluh Nopember (2015).

[6] T. M. Labib, S. I. Hawash, K. M. El-Khatib, A. M. Sharaky, G. I. E. Diwani, E. A. Kader. "Kinetic Study and Techno-economic Indicators for Base Catalyzed Transesterification of Jatropha Oil". Egyptian *Journal of Petroleum* 22 (2013) 9-16.

[7] M. G. A. Vieira, A. F. D. A. Neto, M. G. C. D. Silva, C. C. Nobrega, A. A. M. Filho. "Characterization and Use of *In Natura* and Calcined Rice Husk for Biosorption of Heavy Metals Ions from Aqueous Effluents". *Brazilian Journal of Chem. Eng*. Vol.29 No.03 (2012) 619-633.

[8] N. H. Rambat, Aprilita, B. Rusdiarso. "Aplikasi Limbah Kulit Buah Kakao Sebagai Media Fermentasi Asam Laktat untuk Bahan Baku Bioplastik". *Jurnal Kimia Kemasan*, Vol.37 No.2 (2015) 111-122.

[9] H. Liu, L. Su, F. Liu, C. Li, U. U. Solomon. "Cinder supported K_2CO_3 as catalyst for biodiesel production". *Applied Catalysis B: Environmental 106* (2011) 550-558.

[10] M. H. Dahlan, H. P. Siregar, M. Yusra. "Penggunaan Karbon Aktif dari Biji Kelor Dapat Memurnikan Minyak Jelantah". *Jurnal Teknik Kimia* No.3 Vol.19 (2013) 44-53.

[11] S. E. Abechi, C. F. Gimba, A. Uzairu, Y. A. Dallatu. "Preparation and Characterization of Activated Carbon from Palm Kernel Shell by Chemical Activation". *Research Journal of Chemical Sciences* Vol.3 (2013) 54-61

[12] S. Somasundaram, K. Sekar, V. K. Gupta, S. Ganesan. "Synthesis and characterization of mesoporous activated carbon from rice husk for adsorption of glycine from alcohol-aqueous mixture". *Journal of Molecular Liquids 177* (2013) 416–425.

[13] G. Cruz, M. Pirilia, M., Huuhtanen, L. Carrion, E. Alvarenga, R. L. Keiski. "Production of Activated Carbon from Cocoa (*Theobroma cacao*) Pod Husk". *Journal Civil Environment Eng* (2012) Vol.2 Issue 2.

[14] Z. A. Nasution, S. M. Rambe. "Pengaruh Temperatur Terhadap Pembentukan Struktur Pori Arang Cangkang Sawit". *Jurnal Dinamika Penelitian Industri* Vol.22 No.2 (2011) 100-105.

[15] M. Imaduddin, Yoeswono, K. Wijaya, I. Tahir, I. "Ekstraksi Kalium Abu Tandan Kosong Sawit sebagai Katalis pada Reaksi Transesterifikasi Minyak Sawit". *Bulletin of Chem. Reaction Eng. and Catalysis 3* (2008) 14-20.

[16] H. Husin, M. A. Taleb. "Studi Penggunaan Katalis Abu Sabut Kelapa, Abu Tandan Sawit dan K_2CO_3 untuk Konversi Minyak Jarak Menjadi Biodiesel". *Reaktor* Vol.13 No.4 (2011) 254-261.

[17] F. W. Arfika. "Pengaruh Waktu Perendaman Ampas Tebu Sebagai *Biomaterial Adsorbent* pada

Proses *Pretreatment* Terhadap Karakteristik Biodiesel Minyak Jelantah". Surabaya: Jurusan Teknik Fisika, Fakultas Teknologi Industri Institut Teknologi Sepuluh Nopember (2013).

[18] I. Barabas, I. A. Todorut. Biodiesel Quality, Standards and Properties. Biodiesel-Quality, Emissions and By-Products Dr. Gisela Montero Ed. Intech (2011) 3-28.

[19] J. Turnip. "Pengujian dan Analisa Performansi Motor Bakar Diesel Menggunakan Biodiesel Dimethil Ester B-01 dan B-02". Medan: Departemen Teknik Mesin, Program Pendidikan Ekstensi Fakultas Teknik Universitas Sumatera Utara (2009).

IOP Conf. Series: Materials Science and Engineering **299** (2018) 012094 doi:10.1088/1757-899X/299/1/012094

Phenols removal using ozonation-adsorption with granular activated carbon (GAC) in rotating packed bed reactor

E F Karamah[1], S Leonita[1] and S Bismo[1]

[1]Department of Chemical Engineering, Universitas Indonesia, Kampus UI Depok 16424, Indonesia

corresponding author: eva@che.ui.ac.id

Abstract. Synthetic wastewater containing phenols was treated using combination method of ozonation-adsorption with GAC (Granular Activated Carbon) in a packed bed rotating reactor. Ozone reacts quickly with phenol and activated carbon increases the oxidation process by producing hydroxyl radicals. Performance parameters evaluated are phenol removal percentage, the quantity of hydroxyl radical formed, changes in pH and ozone utilization, dissolved ozone concentration and ozone concentration in off gas. The performance of the combination method was compared with single ozonation and single adsorption. The influence of GAC dose and initial pH of phenols were evaluated in ozonation-adsorption method. The results show that ozonation-adsorption method generates more OH radicals than a single ozonation. Quantity of OH radical formation increases with increasing pH and quantity of the GAC. The combination method prove better performance in removing phenols. At the same operation condition, ozonation-adsorption method is capable of removing of 78.62% phenols as compared with single ozonation (53.15%) and single adsorption (36.67%). The increasing percentage of phenol removal in ozonation-adsorption method is proportional to the addition of GAC dose, solution pH, and packed bed rotator speed. Maximum percentage of phenol removal is obtained under alkaline conditions (pH 10) and 125 g of GAC

1. Introduction

Industrial development brings benefits and negative impacts of waste generated. Phenol is an important and main waste material in a many industries. Phenol compounds have high toxicity and are carcinogenic. United States Environmental Protection Agency (EPA) put phenol to the list of priority substances in the wastewater that must be eliminated [1]. Indonesian Environmental Minister by Decree No. KEP-51/MENLH/10/1995 stated that phenol in water safe for the environment within concentration range 0.5-1.0 mg/L and the threshold phenol in drinking water is 0.002 mg/L . Phenol wastes generated from the pharmaceutical industry and hospitals can reach 100-150 mg/L because of its extensive use to synthesize drugs and antiseptics.

Biological methods can only deal with the phenol concentration below 100 mg/L [2]. Other methods include such as, the activated carbon adsorption [3], wet air oxidation method [4], photocatalytic oxidation [5], supercritical water oxidation [6], as well as ozonation method [7]. Phenol degradation studies using ozonation-adsorption technique conclude that GAC could increase phenol removal percentage, and ozonation can oxidize heavy metals dissolved in water, organic compounds, as well as remove color, odor, and taste [8,9]. The addition of activated carbon not only as an adsorbent, but also as a catalyst in supporting ozone oxidation [9]. Ozone reacts quickly with phenol and activated carbon increases the oxidation process by producing hydroxyl radicals which are capable to mineralize substituents of wastewater [10].

In this study, phenol was removed by ozonation-adsorption techniques with GAC in rotating packed bed reactor. The reactor used is a modification from Lin and Wang study [11]. Rotator in packed bed serves to enlarge the mass transfer of ozone gas into the waste solution and GAC. Performance parameters evaluated are phenol removal percentage, the quantity of hydroxyl radical formed, changes in pH and ozone utilization, dissolved ozone concentration and ozone concentration in off gas. To evaluate the significance of combination method, the method configuration was varied (single adsorption, single ozonation, and combination of ozonation-adsorption). The variation of initial pH, stirrer speed in the reactor bed, and the quantity of GAC on ozonation-adsorption technique was conducted to evaluate the influence of these parameters to the phenol removal.

2. Experimental

Rotating packed bed reactor is used to minimize the damage and immobilizing the GAC particles, easy to control, and produce extensive contact between the gas phase and the solid or liquid so that the mass transfer and diffusion between adsorbate and adsorbent is faster. Bed reactor is connected to the rotor arms on stationary vessel with a diameter of 2.5 cm, and 28 cm axial height. Rotator connected to the rotor arms are also connected to the motor speed regulator. The reactor was made of acrylic material with a thickness of 0.5 cm, ID of 8 cm, 8 cm bed height, total reactor height of 35 cm, and a total reactor volume of 1.51 L. Figure 1 shows the scheme of the reactors.

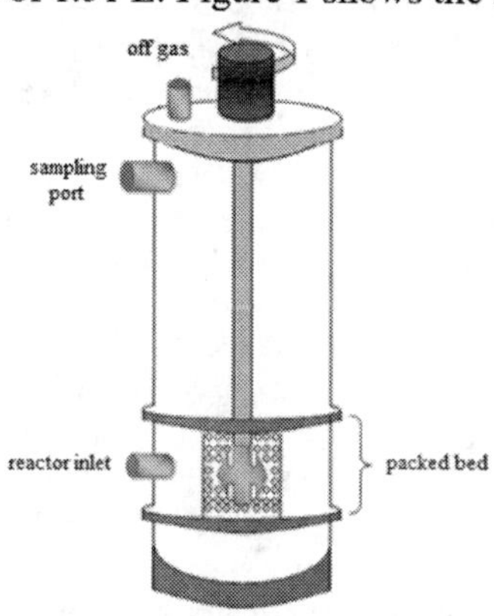

Figure 1. Scheme of Rotating Packed Bed Reactor

Before conducting phenol removal, quantity of OH radicals was measured for each configuration process, using iodometry. In addition, the productivity of ozonator was also tested by bubbling outlet gas from ozonator in a KI solution for a certain time. All of the experiment was conducted in a constant temperature of 20°C to prevent the decomposition of ozone into oxygen.

Phenol removal process was conducted in a series of research equipment as can be seen in Figure 2. The initial phenol concentration was 100 mg/L. Granular activated carbon (GAC) was dried at 100° C for 1-2 hours. The parameter variables in this experiment includes: initial solution pH (pH ≈4 acidic, alkaline pH ≈ 10, and phenol without pH adjustment), GAC quantity used as packed bed (75 g, 100 g and 125 g) and the rotator speed (150 rpm, 225 rpm, and 300 rpm).

The phenol removal process was conducted in the equipment as can be seen in the experimental set-up in Figure 2. Five liter of phenol solution in reservoir was circulated in a flow rate of 1.5 L/min and ozone gas as much as 3.31 mg/min supplied to the reactor. The process was carried out for 60 minutes and sampling was conducted every 15 minutes.

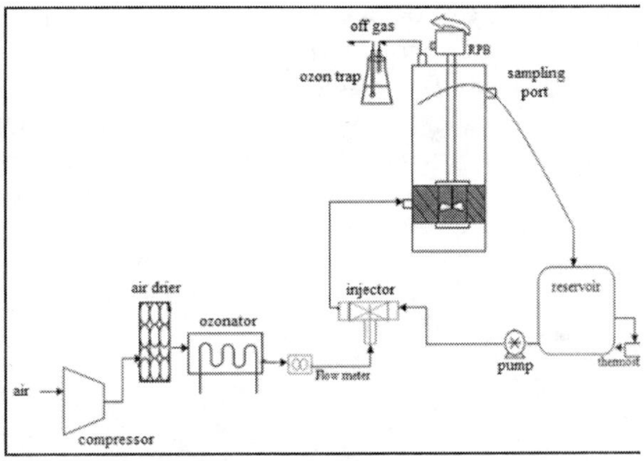

Figure 2. Experimental set-up for Ozonation-Adsorption process.

3. Results and Discussion

3.1. Quantification of OH radical

This quantification test aims to determine the effect of variations in configuration and operating conditions on OH radical (OH•) generated, using permanganometry. The following reactions occur during the test:

$$3H_2O_2 + 2KMnO_4 + 3H_2SO_4 \rightarrow 4O_2 + 2MnSO_4 + K_2SO_4 + 6H_2O \quad (1)$$

$$H_2O_2 \rightarrow 2OH\bullet \quad (2)$$

Figure 3 shows the results of quantification of OH radical for configurations of ozonation and its combination with adsorption. This figure shows that the amount of OH radical formed from ozonation-adsorption method is higher. This is because the existence of GAC with active site that allows the adsorption and surface reaction with ozone, thus accelerating the formation of the radical component when it reacts with ozone to form OH radical. However, starting from 15 minute to 45 minute, number of OH radical on single ozonation increased significantly. This is due to adjusting the pH of distilled water in accordance with phenol pH (pH ≈ 6) so that ozone can react directly and indirectly to form OH radical which needs longer time for its formation.

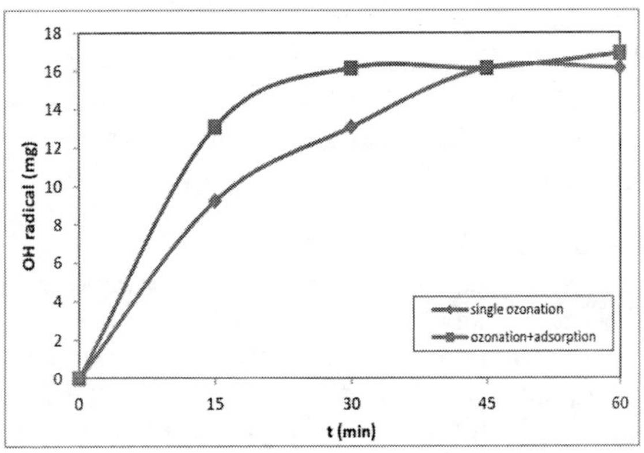

Figure 3. Quantity of OH radical for different configurations of treatment (GAC 75 gr, rotator speed 225 rpm, pH ≈ 6).

Figure 4 shows the effect of GAC dose on the number of OH radical. This figure shows that higher dosage of GAC produces higer numer of OH radical. This can occur because the ozone molecule is adsorbted on activated carbon hereinafter increasing residence time of ozone in the solution which will further increasing ozone decomposition into radicals. The steady of amount of OH radical starting from minute of 45 to 60 may be caused by saturation of activated carbon used. The addition of more GAC causes higher pressure drop in the bed so that the GAC can not be fluidized optimally. This reduces the mass transfer between ozone, solution, and activated carbon. Hence OH radical generated also decreased. Figure 5 explains that OH radical produced is higher in higher pH solution. The ozon decomposition reaction into OH radical is triggered by the existance of OH^- ions. The more number of OH^- ions in the solution, the more number of OH radical produced. In the acidic pH, direct ozone reaction dominates. While at pH ≈ 6, the direct (with ozon) and indirect (with OH radical) reactions occur.

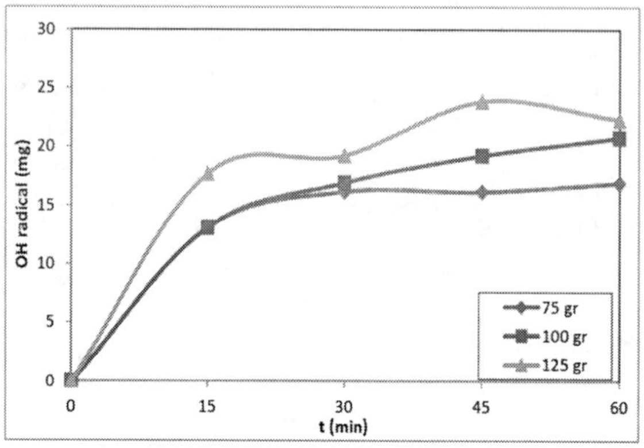

Figure 4. Effect of GAC quantity on production of OH radical (rotator speed 225 rpm, pH ≈ 6).

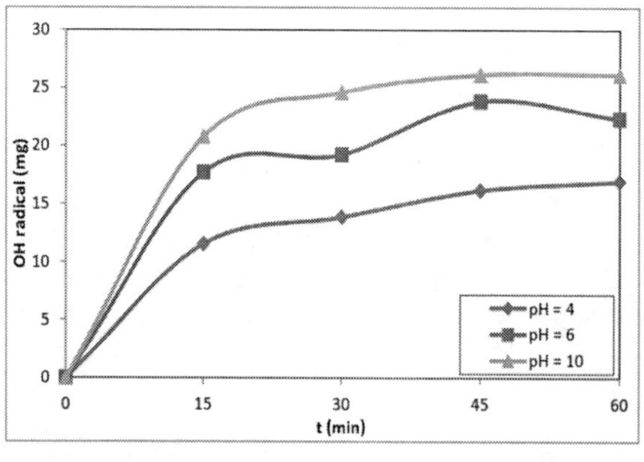

Figure 5. Effect of initial pH on production of OH radical (rotator speed 225 rpm, pH ≈ 6) (GAC 75 gr, rotator speed 225 rpm).

3.2. The percentage of Phenol Removal

Percentage of phenol removal for variation of process configurations and operating conditions are presented in Figures 6-9. Figure 6 and Figure 7 show that the removal of phenol achieves highest result in ozonation-adsorption method (78.62%), followed by single ozonation (53.15%) and single adsorption (36.87%). The difference of phenol percentage in these variety of configurations is influenced by the number of OH radical formed for each configuration as has been discussed in the previous sub-section.

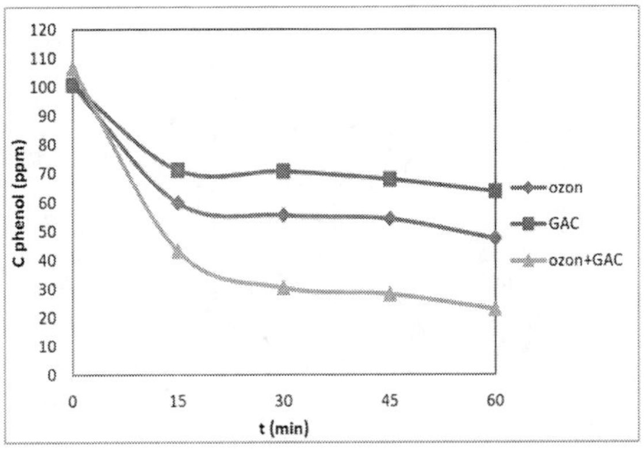

Figure 6. Effect of process configuration on Phenol Concentration (GAC 75 gr, rotator speed 225 rpm, pH ≈ 6).

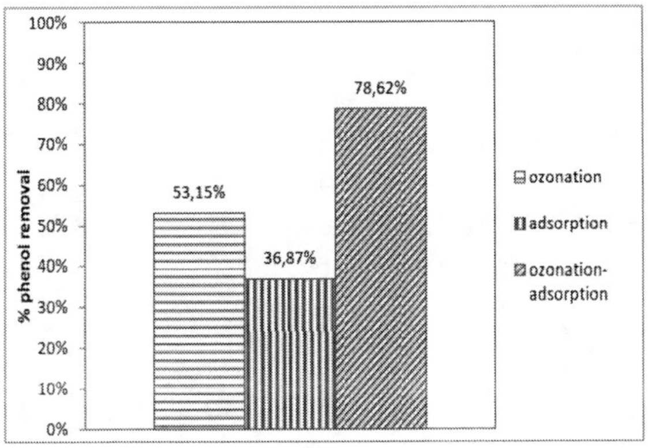

Figure 7. Phenol removal percentage for each configuration.

In the absence of GAC, ozonation of phenol tends to proceed directly through electrophilic reaction at the ortho- and para- position. In the single adsorption, degradation occurs physically involving intermolecular force between GAC and phenol. Through physical adsorption, phenol is not strongly bound to the surface of the GAC because of the absence of surface reactions between the active site GAC with phenol. In ozonation-adsorption method, GAC not only act as an adsorbent, but also as a catalyst to increase the decomposition of ozone to radical component via the following reactions [12].

$$O_3 + OH^- \rightarrow O_2{\bullet}^- + HO_2{\bullet} \quad (3)$$

$$O_3 + O_2{\bullet}^- \rightarrow O_3{\bullet}^- + O_2 \quad (4)$$

$$HO_3{\bullet} \leftrightarrow O_3{\bullet}^- + H^+ \quad (5)$$

$$HO_3{\bullet} \rightarrow OH{\bullet} + O_2 \quad (6)$$

Ozonation-adsorption mechanism consists of adsorption, surface reaction, desorption and other chemical reactions in solution of phenol [13]. Microporous GAC is able to minimize the diffusion limitations and accelerate the rate of adsorption and desorption that ozone molecules adsorbed to the GAC and transformed into active molecular radicals components and forming OH radical. In addition, phenol compounds intermediates with molecular size of 0.5-0.6 nm are also adsorbed on microporous GAC so that assist lowering the concentration of phenol in the solution [11].

Figure 8 shows the influence of GAC quantity (dose) on the percentage of phenol removal. It shows that the GAC quantity is directly proportional to the phenol removal percentage. The existence of GAC increases contact between the ozone and the active site to form the active radical.

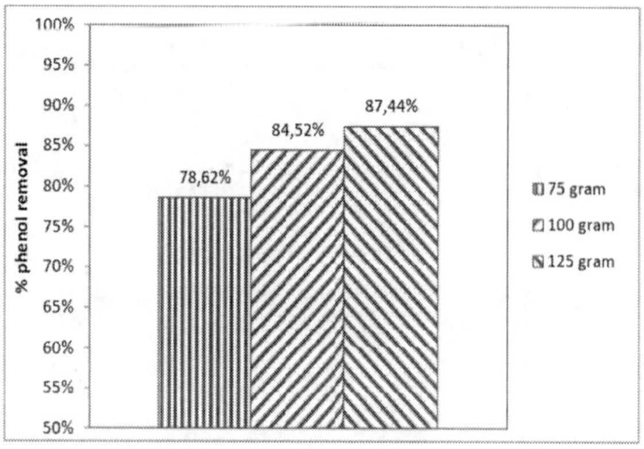

Figure 8. The influence of GAC quantity on the percentage of phenol removal (pH 6, 225 rpm).

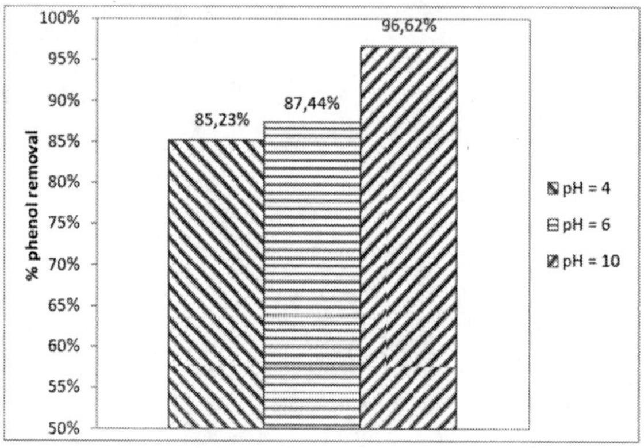

Figure 9. The influence of pH on the percentage of phenol removal (GAC 100 g, 225 rpm).

Figure 9 shows that the pH condition influences the performance of phenol removal in combination of ozonation and adsorption process. The highest removal achievement is obtained at the alkaline pH while the lowest at acidic pH. High levels of degradation under alkaline conditions due to the highest quantity of OH radical which enhances the indirect ozone reaction. OH radical reacts with phenols in non-selective oxidation fashion with higher reaction rate ($k = 10^8 - 10^{10}$ $M^{-1}s^{-1}$) and then the ozone ($k \cong 10^3$ $M^{-1}s^{-1}$). The decrease in phenol concentration on normal and acidic pH shows only a slight difference. This is due to a stable GAC adsorption process in a solution of pH < pKa (phenol pKa ≈ 10) [13]. At acidic pH, the degradation process begins with the adsorption of phenol by GAC lead up to an increase pH system. Increasing pH allows the indirect ozonation.

4. Conclusion

Ozonation-adsorption method generates more OH radicals than a single ozonation. Quantity of OH radical formation increases with increasing pH and quantity of the GAC. The presence of stirrer in bed reactor has an influence to increasing diffusion and mass transfer between the activated carbon, ozone molecules, and phenol solution. The increase of phenol removal percentage by ozonation-adsorption method is proportional to the increase in pH of the solution, GAC quantity, and rotator speed. Maximum percentage of phenol removal is obtained under alkaline conditions (pH 10), 125 grams of GAC quantity and the stirrer speed of 300 rpm.

References

[1] Figueiredo, J L 2011 Adsorption of Phenol on Supercritically Activated Carbon Fibers: Effect of Texture and Surface Chemistry *Journal of Colloid and Interface Science* **357** 210

[2] Bielefeldt, A R 999 Modeling Competitive Inhibition Effects During Biodegradation of BTEX Mixtures. *Water Res.* **33** 707

[3] Cooney D C, Xi Z 1994 Activated Carbon Catalyzes Reactions of Phenolics During Liquid-phase Adsorption. *AIChE J* **40** 361

[4] Lin, S H, Wu Y 1996 Catalytic Wet Air Oxidation of Phenolic Wastewater. *Environ. Technol.* **17** 175

[5] Chang, Chia-Chi 2009 Combined Photolysis and Catalytic Ozonation of Dimethyl phthalate in a High-gravity Rotating Packed Bed. *Journal of Hazardous Materials* **161** 287

[6] Ghaly, M Y, Hartel, G, Mayer, R, Hasender R 2001 Photochemical Oxidation of p-Chlorophenol by UV/H_2O_2 and Photo-Phenol Process A Comarative Study. *J. Wast Manag.* **21** 41

[7] Jung, O J 2001 Destruction of 2-Chlorophenol from Waste Water and Investigation of Product by Ozonation *J. Chem. Soc.* **22** 850

[8] Giraldez, I, Juan, F, Fernando, G, Beltran J F 2007 Activated Carbon Promoted Ozonation of Polyphenol Mixtures in Water: Comparison with Single Ozonation. *Ind. Eng. Chem. Res.* **46** 8241

[9] Beltran, J F, Giraldez, I, Juan F Garcia-Araya 2008 Kinetics of Activated Carbon Promoted Ozonation of Polyphenol Mixtures in Water. *Ind. Eng. Chem. Res.* **47** 1058

[10] Beltran, J F 2004 Ozone Reaction Kinetics for Water and Wastwater Systems. Florida: Lewis Publishers.

[11] Lin, S H, Ching H, Wang 2003 Ozonation of Phenolic Wastewater in a Gas-Induced Reaktor with a Fixed Granular Activated Carbon Bed. *Ind. Eng. Chem. Res.* **42** 1648

[12] Gottschalk, C, Libra, J A, Saupe A 2000 Ozonation of Water and Wastewater. Winhem: Wiley-VCH.

[13] Qu, X, Jingtang Z, Yanzong Z 2007 Catalytic Ozonation of Phenolic Wastewater with Activated Carbon Fiber in a Fluid Bed Reactor. *Journal of Colloid and Interface Science* **309** 429

Acknowledgments

This work was supported by Hibah PITTA 2017 UI. The authors would like to thank Directorate of Research and Community Service UI for facilitating this grant.

IOP Conf. Series: Materials Science and Engineering **299** (2018) 012095 doi:10.1088/1757-899X/299/1/012095

Microwave-assisted cationic polymerization of palm olein and their urea inclusion products

Bambang Soegijono[1], Muhamad Farid[1] and Zainal Alim Mas'ud[2]

[1] Department of Physics, University of Indonesia, Kampus UI Depok 16424

[2] Department of Chemistry, Bogor Agricultural University, Kampus IPB Dramaga Bogor 16680

Email: bambangsg11@yahoo.com

Abstract. Cationic polymerization is affected by the relative amount of unsaturated bond (C=C) in the compound. The enrichment of an unsaturated triglyceride fraction from oils may be performed using urea inclusion techniques. In this study, palm olein was enriched-unsaturated fraction using urea-methanol system. The palm olein and its urea-inclusion products were cationic polymerized with ethereal boron trifluoride catalyst and followed by irradiation using a commercial microwave (microwave-assisted). The microwave irradiated products were cured at 110 °C for 24 hours. Fatty acid composition of the palm olein and its urea-inclusion products were analyzed by gas chromatography. Iodine numbers, functional groups, and ultraviolet absorption spectra of all palm olein origin, urea inclusion products and polymerization products were analyzed using titrimetric, ultraviolet spectrophotometric, and Fourier Transform infrared spectrophotometric methods. Differential scanning calorimetric (DSC) was used to observe the thermal characteristics of the polymer. Urea-inclusion process increased the unsaturated fatty acid components as indicated by the increased iodine number, intensity of alkene band absorptions in the infrared spectra, and the absorbance of the ultraviolet spectra. The polymer formation is converting the C=C group to C-C, which is indicated by the opposite of the urea inclusion process. The curing process results in reformation of new C=C bonds that were similar to that of the urea inclusion process. The DSC thermogram curve shows that the enrichment process improves the thermal stability of the polymer formed.

1. Introduction

Cationic polymerization and cationic copolymerization process of various oil (except palm oil) have been investigated [1-6]. Microwaves have been widely used in facilitating the synthesis process, including in polymer synthesis (except palm oil) [7-15]. Previous studies (2017) showed that palm oil in several forms, i.e., crude palm oil (CPO), refined bleached deodorized palm oil (RBDPO), and refined bleached deodorized palm oil-olein (RBDPOO) can also be polymerized through microwaved-assisted cationic polymerization with boron trifluoride ethereal catalyst.

Cationic polymerization is affected by the relative amount of the unsaturated bond (C=C) in the materials. Therefore, the unsaturated components are required. The enrichment of an unsaturated triglyceride fraction from oils may be performed using urea inclusion techniques [16-22]. In this study, palm olein enriched the unsaturated fraction using urea-methanol system.

As physical process (urea inclusion) and chemistry process (cationic polymerization) were focused on the C=C bond, we observed the fatty acid composition (GC), the conversion of double bonds (iodine number, infrared and ultraviolet spectra) as well as the thermal properties (DSC thermogram) of the products.

2. Experiment

RBDPOO raw material was purchased from a market in Bogor, directly used without purification. Chemicals for analysis with analytical grade were obtained from Merck: $K_2Cr_2O_7$, HCl 32%, KI, $Na_2S_2O_3.5\ H_2O$, soluble starch, cyclohexane, CH_3CO_2H, Wijs solution, NaOH, CH_3OH, boron trifluoride-methanol complex 20%, NaCl, hexane, and anhydrous Na_2SO_4. Fatty acid methyl ester standard for fatty acid analysis was acquired from Supelco. Boron trifluoride ethyl ether complex ($BF_3.C_4H_{10}O$) for polymerization was acquired from TCI.

The urea inclusion was a modified procedure [22]. There were five treatments on urea (weight) - methanol (volume), i.e. (70:50), (60:50), (50:50), (40:50) and (30:50).

The iodine numbers followed an AOAC method [23]. The fatty acid composition was analyzed according to AOAC methods [24 and 25] using Shimadzu GC-2010 Plus gas chromatography, with stationary phase of Cyanopropyl methyl sil capillary column l = 60 m, $Ø_{in}$ = 0.25 mm, film thickness 0.25 µm. The ultraviolet spectra was taken using Shimadzu Pharmaspec UV-1700 spectrophotometer and the infrared spectra was taken using Shimadzu IRPrestige-21 FTIR spectrophotometer. Both spectra was taken following the manual instruction of the respective instrument. Thermal properties were analyzed using DSC Instruments Shimadzu DSC 60 under nitrogen atmosphere.

Polymerization reaction was carried out in 100 mL erlenmeyer, The irradiation processes were performed in glass vials ($Ø_{in}$ = 20 mm, h = 50 mm) inserted in threaded PTFE reactors (workshop made, $Ø_{in}$ = 60 mm, $Ø_{out}$ = 100 mm, h = 85 mm) as shown in Figure 1. The microwave ovens used was Samsung MG23H3185PK (230 V, 50 Hz, 2450 MHz; input 1400 W output 800 W).

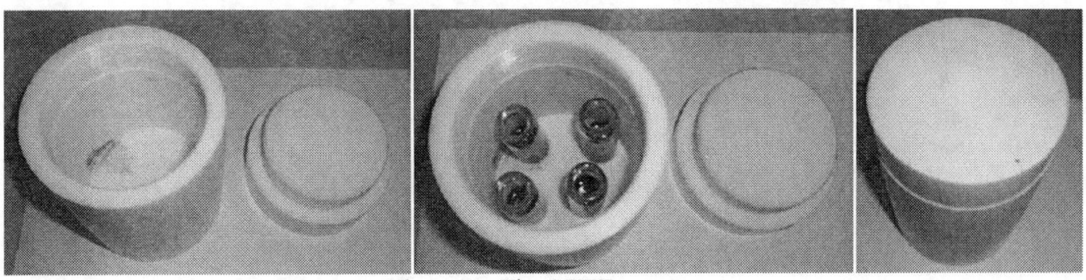

Figure 1. The PTFE reactor

3. Results and Discussion

3.1. Characteristics of the RBDPOO and its urea inclusion products

In accordance with previous result [17, 20, 22], the urea inclusion process increases the content of the unsaturated fatty acid fraction (Table 1, C18:1 and C18:2). Increasing the content of unsaturated fatty acids results in the increasing iodine number (Table 2), increased infrared absorption (Fig. 2) at υ 1580 cm^{-1}–1650 cm^{-1} [26-29], and ultraviolet absorption at λ 220 nm–280 nm (Fig. 3) as well. It is presumably due to the increasing number of the unsaturated carbon-carbon bond. The more the urea used in the mixture, the higher the content of the unsaturated fatty acids.

Table 1. Fatty acid compositions

Oil Type	Fatty Acid				
	C14:0	C16:0	C18:0	C18:1	C18:2
RBDPOO$_{(60:50)\text{-S}}$	0.70	28.53	3.02	39.06	11.80
RBDPOO$_{(60:50)\text{-U}}$	0.83	33.16	3.47	44.70	13.41
RBDPOO$_{(50:50)\text{-S}}$	0.72	29.35	2.97	38.62	11.49
RBDPOO$_{(50:50)\text{-U}}$	0.71	29.76	3.05	40.17	12.13
RBDPOO$_{(40:50)\text{-S}}$	0.65	26.67	2.68	35.06	10.47
RBDPOO$_{(40:50)\text{-U}}$	0.64	26.74	2.70	35.80	10.93

Table 2. Iodine numbers of palm oil olein, the urea inclusion products, and the derived polymers

Oil Type	Oils form	Polymer without curing	Polymer with curing
RBDPOO	54.47	29.07	41.50
RBDPOO$_{(60:50)\text{-U}}$	60.80	18.39	27.37
RBDPOO$_{(50:50)\text{-U}}$	57.48	23.70	30.96
RBDPOO$_{(40:50)\text{-U}}$	56.20	26.92	31.70

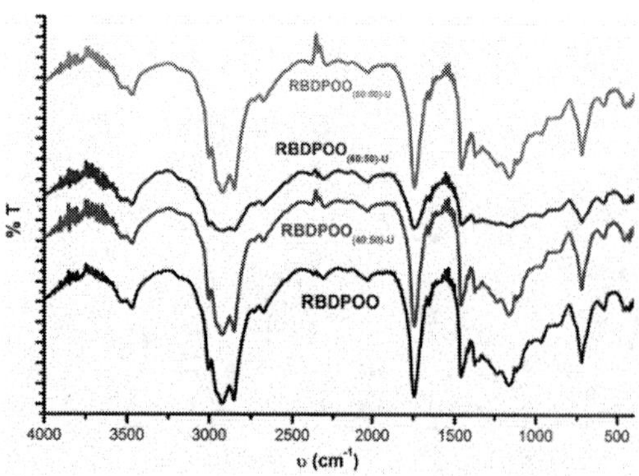

Figure 2. FTIR spectra of RBDPOO and its urea inclusion products

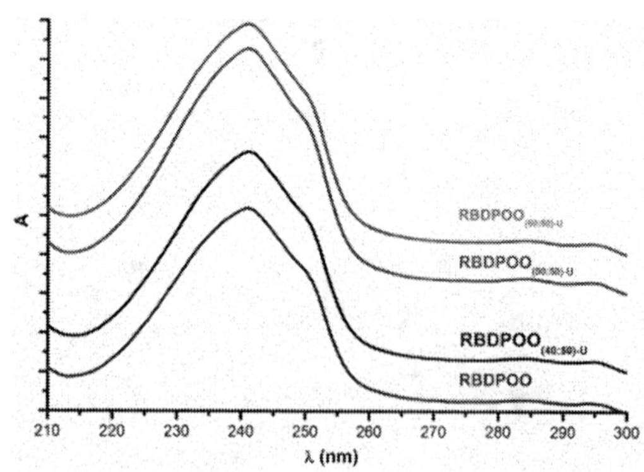

Figure 3. UV spectra of RBDPOO and its urea inclusions product

3.2. Characteristics of the RBDPOO polymers

The cationic polymerization process converts the C=C bonds to C-C bonds. The change is shown by decreasing the iodine number, the different infrared spectra (Fig. 4), and the shift in ultraviolet spectra (Fig. 5). There is a decreasing infrared absorption on wavenumbers C=C (1654 cm^{-1}, 1648 cm^{-1}) and =C-H (3025 cm^{-1}, 3006 cm^{-1}, 1418 cm^{-1}, 968 cm^{-1}, 914 cm^{-1}, and 723 cm^{-1}). In addition to the cationic polymerization, it is also suspected that the reaction of the formation of the double bond is conjugated resulting in an increase in the ultraviolet absorption.

The curing process resulted in the reaction of the reformation of double bonds again was marked by the increase of iodine number (Table 2: 29.07 to 41.50; 18.39 to 27.37; 23.70 to 30.96 and 26.92 to 31.70); and deformation the conjugated double bond so that the decrease of ultraviolet absorption because the conjugate bond increase the ultraviolet absorption. The curing also changes the thermal transition pattern (Fig. 6 and Fig. 7). Heating up to 150 °C deforms the resulted polymers.

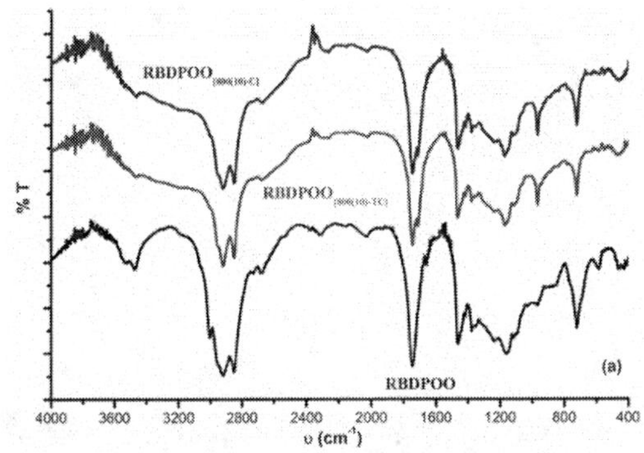

Figure 4. FTIR spectra of RBDPOO and its polymerization products

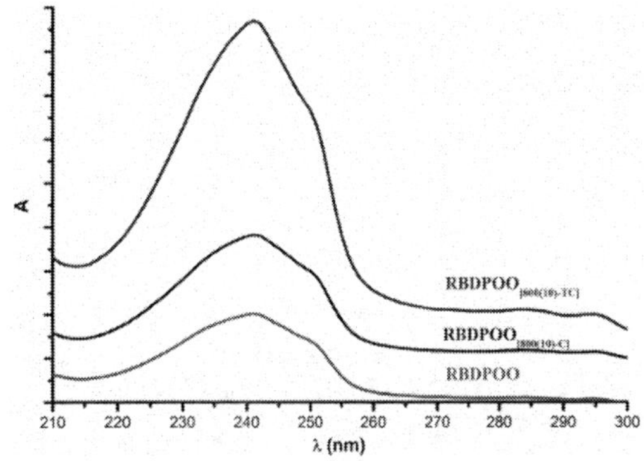

Figure 5. UV spectra of RBDPOO and its polymerization products

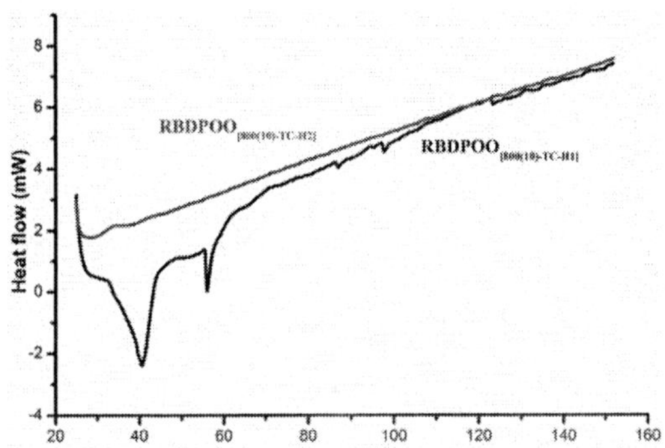

Figure 6. DSC thermograms of RBDPOO polymerization products without curing

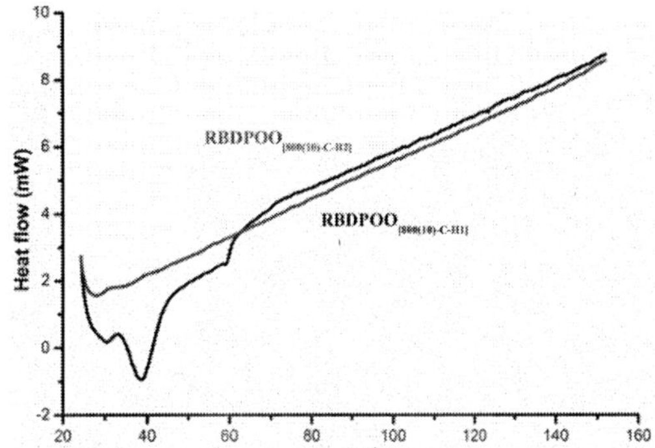

Figure 7. DSC thermograms of RBDPOO polymerization products with curing

3.3. Characteristics of polymers of the RBDPOO urea inclusion product

The polymerization of urea inclusion (60:50) of the RBDPOO shows some results similar to the original RBDPOO. The change is shown by the decreasing iodine number (60.80 to 18.39; Table 2); different infrared spectra (Fig. 8), and different ultraviolet spectra (Fig. 9). The curing also changes the thermal transition pattern (Fig.10 and Fig. 11). Without curing, heating up to 150 °C supposedly forms a new polymer because showed the different DSC thermogram pattern.

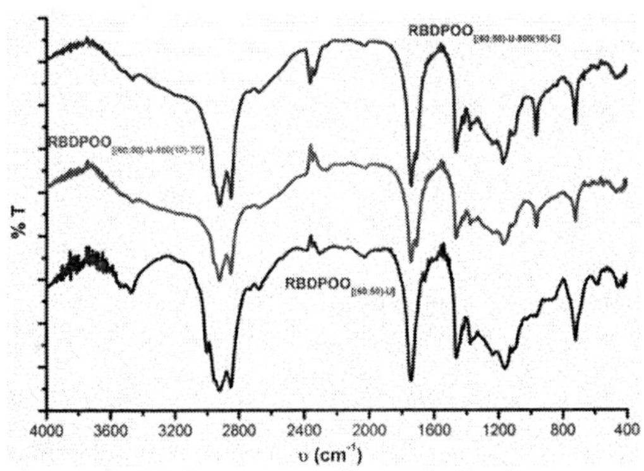

Figure 8. FTIR spectra of urea inclusion products (60:50) RBDPOO and its polymer

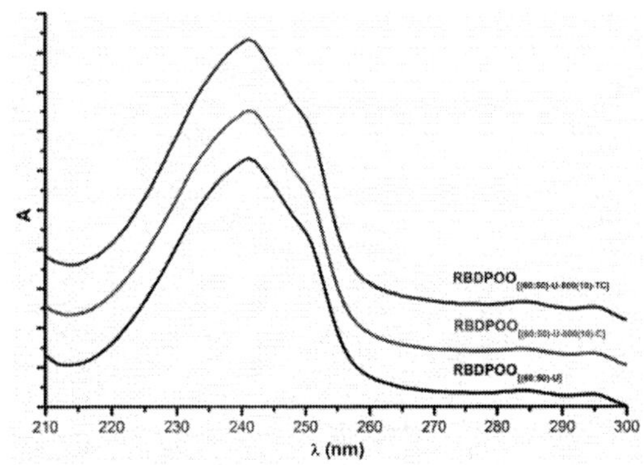

Figure 9. UV spectra of urea inclusion products (60:50) RBDPOO and its polymers

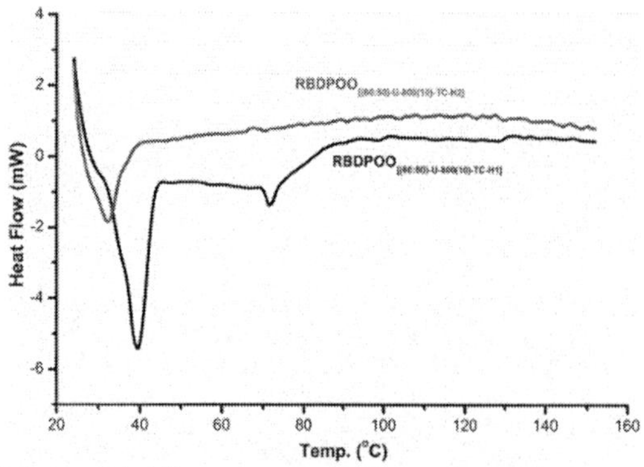

Figure 10. Thermograms of urea inclusion (60:50) RBDPOO polymerization products without curing

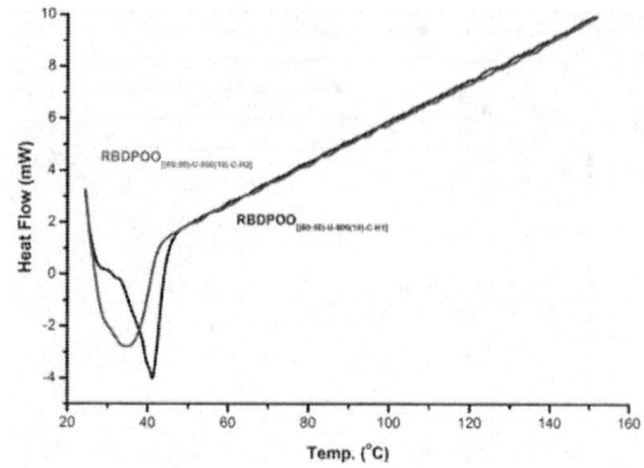

Figure 11. Thermograms of urea inclusion (60:50) RBDPOO polymerization products with curing

The polymerization of the urea inclusion (50:50) RBDPOO again shows some results similar to the original RBDPOO. The variation is shown by the decreasing iodine number, the different infrared spectra (Fig. 12), and the different ultraviolet spectra (Fig. 13). The curing also produces a different thermogram (Fig. 14). These patterns are similar to that of the polymerized urea inclusion (60:50) of the RBDPOO.

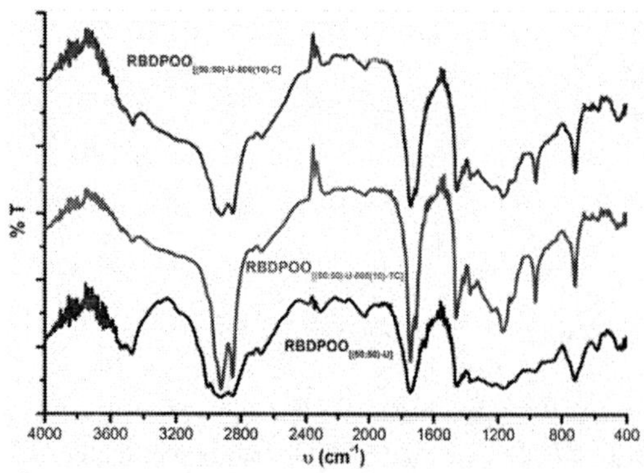

Figure 12. FTIR spectra of urea inclusion products (50:50) RBDPOO and its polymer

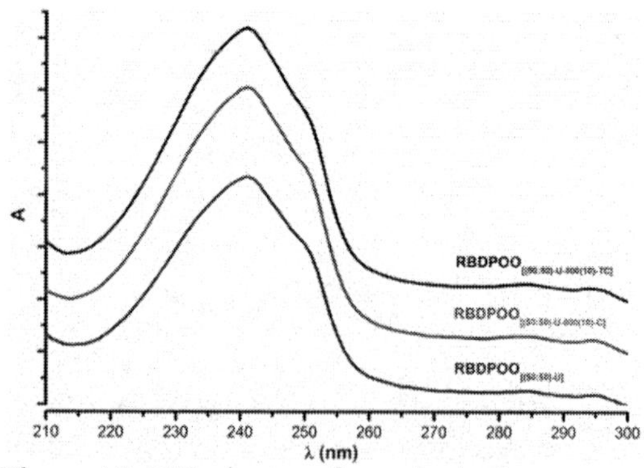

Figure 13. UV spectra of urea inclusion products (50:50) RBDPOO and its polymer

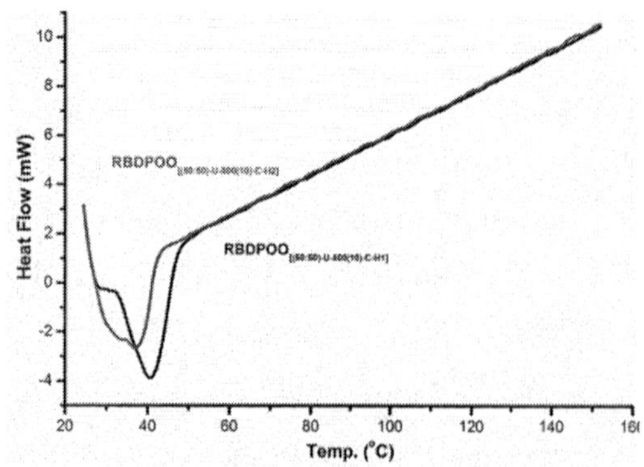

Figure 14. DSC Thermograms of urea inclusion (50:50) RBDPOO polymerization products with curing

The polymerization of urea inclusion (40:50) RBDPOO shows similar results with the starting RBDPOO. The change is shown by the decrease of iodine number, the change of the infrared spectra (Fig. 15), and the ultraviolet spectra (Fig. 16). The curing process also produces a thermogram (Fig. 17) with a completely different pattern with the previous thermograms.

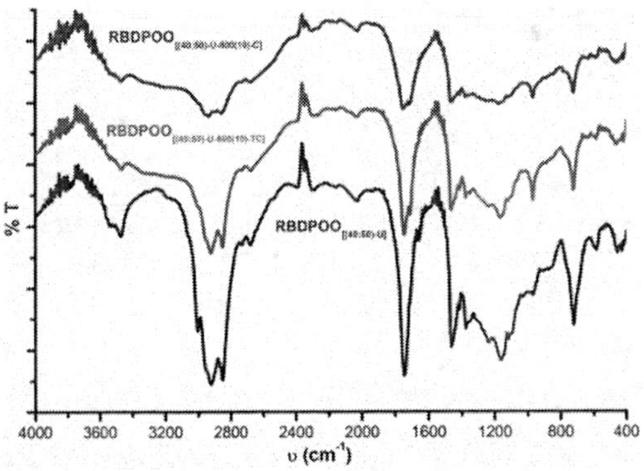

Figure 15. FTIR spectra of urea inclusion products (40:50) RBDPOO and its polymer

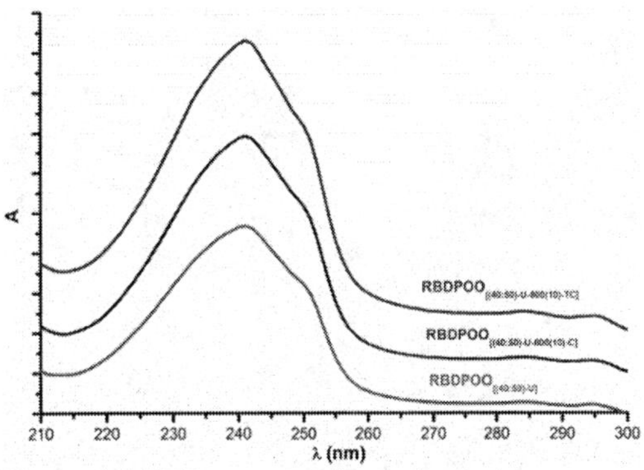

Figure 16. UV spectra of urea inclusion products (40:50) RBDPOO and its polymer

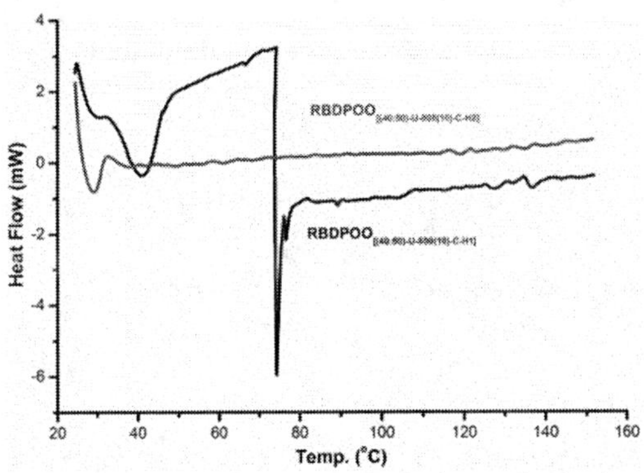

Figure 17. DSC thermograms of urea inclusion (40:50) RBDPOO polymerization products with curing

4. Conclusions

The urea inclusion process increases the unsaturated fatty acid content of the RBDPOO as evidenced by chromatography, spectrometry, and titrimetry techniques. This process is limited by the ratio of urea-methanol (60:50) to (40:50). The post-polymerization thermal treatment is required to complete the reaction. Further studies are needed to determine this thermal treatment in term of temperature and time of reaction. The proper enrichment process might improve the thermal stability of the polymer.

References

[1] Larock RC, Hanson M, and Li F 2002 US Patent 2002/0095007 A1
[2] Ionescu M and Petrović ZS 2008 US Patent 2008/0281071 A1
[3] Ionescu M and Petrović ZS 2009 US Patent 2009/0309064 A1
[4] Ionescu M and Petrović ZS 2009 US Patent 7501479 B2
[5] Ionescu M and Petrović ZS 2011 Polymerization of soybean oil with superacids *Soybean-Applications and Technology* ed TB Ng (Rijeka Croatia: InTech) pp 365–386
[6] Ionescu M and Petrović ZS 2011 US Patent 8013088 B2
[7] Zong L, Zhou S, Sgriccia N, Hawley MC, and Kempel LC 2003 *J Microwave Power EE* **38** (1)

49–74
[8] Wiesbrock F, Hoogenboom R, and Schubert US 2004 *Macromol Rapid Commun* **25** (20) 1739–1764
[9] Evalueserve 2005 *Developments in Microwave Chemistry* (London: Royal Society of Chemistry)
[10] Koopmans C, Iannelli M, Kerep P, Klink M, Schmitz S, Sinnwell S, and Ritter H 2006 *Tetrahedron* **62** (19) 4709–4714
[11] Hoogenboom R and Schubert US 2007 *Macromol Rapid Commun* **28** (4) 368–386
[12] Sinnwell S and Ritter H 2007 *Aust J Chem* **60** (10) 729–743
[13] Mallakpour S and Rafiee Z 2008 *Iran Polym J* **17** (12) 907–935
[14] Sosnik A, Gotelli G, and Abraham GA 2011 *Prog Polym Sci* **36** (8) 1050–1078
[15] Mishra A, Vats T, and Clark JH 2016 *Microwave-Assisted Polymerization* (Cambridge: The Royal Society of Chemistry)
[16] Redlich O, Gable CM, Dunlop AK, and Millar RW 1950 *J Am Chem Soc* **72** (09) 4153–4160
[17] Schlenk W and Holman RT 1950 *J Am Chem Soc* **72** (11) 5001–5004
[18] Zimmerschied WJ, Dimerstein RA, Wetkamp AW, and Marschner RF 1950 *Ind Eng Chem* **42** (07) 1300–1306
[19] Bishop R and Dance G 1988 New types of helical canal inclusion networks *Molecular Inclusion and Molecular Recognition-Clathrathes II* " ed E Weber (Berlin: Springer-Verlag) p 137–188
[20] Schlenk H 1954 Urea inclusion compounds of fatty acids *Progress in the Chemistry of Fats and Other Lipids* (Vol 2) eds R T Holman W O Lundberg and T Malkin (London: Pergamon Press) p 243–267
[21] Wosch D and Vogtle F 1987 Separation of enantiomers by clathrate formation *Molecular Inclusion and Molecular Recognition-Clathrates I* ed E Weber (Berlin: Springer-Verlag) p 21–42
[22] Ikan R *Natural Products A Laboratory Guide* 2nd Ed (San Diego: Academic Press)
[23] AOAC 2012 OMA AOAC Int 19th ed. Official Method 993.20 (Arlington Virginia: AOAC)
[24] AOAC 2012 OMA AOAC Int 19th ed. Official Method 969.33 (Arlington Virginia: AOAC)
[25] AOAC 2012 OMA AOAC Int 19th ed. Official Method 963.22 (Arlington Virginia: AOAC)
[26] Chapman D 1965 *J Am Oil Chem Soc* **42** (5) 353–371
[27] Freeman NK 1968 *J Am Oil Chem Soc* **45** (11) 798–809
[28] Guillén MD and Cabo N 1997 *J Sci Food Agric* **75** (1) 1–11
[29] Guillén MD and Cabo N 1997 *J Am Oil Chem Soc* **74** (10) 1281–1286

Development of Superhydrophobic Material SS 17-4 PH for Bracket Orthodontic Application by Metal Injection Molding

S Supriadi[1*], B Suharno[2], T Widjaya and S T Ayuningtyas[2] and E R Baek[3]

[1] Mechanical Engineering Department, University of Indonesia, Depok, Indonesia
[2] Metallurgy and Materials Engineering Department, University of Indonesia, Depok, Indonesia
[3] School of Materials Science and Engineering, Yeungnam University, South Korea

E-mail : sugeng@eng.ui.ac.id

Abstract. Dental's plaque is a common problem that encountered during orthodontic treatment using bracket. It is caused by demineralization of enamel due to the activity of bacteria. The bacteria increase with remaining excess food which trapped in teeth and bracket. A hydrophobic surface could reduce food attachment on the bracket because of extremely low wettability properties that make it easy to clean with water. There are several methods to obtain hydrophobic surfaces, which are sol-gel, template replica and also etching. The propose of this work is to compare etching treatment and surface modification on sintered SS 17-4 PH as bracket material using $CuCl_2$ and HCl as an etchant while stearic acid was used for surface modification. Hydrophobic surfaces were produced under various etching time i.e 15,30,45 and 60 seconds for $CuCl_2$ and 40, 50, 60 and 70 minutes for HCl and also HCl concentration i.e 1,2 and 3 mol/L at room temperature. The hydrophobicity is observed using contact angle measurement while the microstructures observed by Scanning Electron Microscope. The result shows the contact angle could be achieved up to 60% higher than the as-sintered material. Hydrophobic structure has successfully fabricated using etching technique that might be applied to the orthodontic bracket.

1. Introduction

Dental's plaque is one of the common problem that we could encountered during orthodontic treatment using the bracket. This problem is caused by demineralization of enamel due to accumulated bacteria in a high number on bracket [1]. Demineralization is a process of loss mineral on enamel in ion state [2]. Plaque's microorganism could form organic acid which lowering pH value on dental's plaque and caused ion calcium and phosphate diffusion from enamel through pellicle. When the pH value is already low enough, it would increase acid condition between plaque and bracket until the acid could not come out and finally attack the enamel [3]. The bacteria could increase because of remaining excess food which trapped between teeth and bracket. A hydrophobic surface is proposed to reduce food attachment on the bracket because of its potential use in self-cleaning properties which inspired from lotus leaves [3,4]. Lotus leaves have ability to always remain clean even though it is in a dirty pond. This self-cleaning performance is known as 'Lotus Effect'[5]. These properties could occur on hydrophobic surfaces because of hierachial structure that has an extremely low wettability which depend on surface roughness and also free energy surface [6]. Self-cleaning properties is identified with high static contact angle (>150°), it is also called superhydrophobic surface [7]. There are several methods to fabricate superhydrophobic surfaces, such as plasma spray [8], sol-gel [9],

electrostatic spray deposition [10], template replica [11]. However most of these methods has a very slow process, poor bond strength and also difficult to apply for complex shape.
Previous work by Supriadi et al. shows that fabrication of orthodontic bracket can be done by investment casting process using SS 316L with acceptable geometry, but it still has porosity in its surface which could affected the surface roughness [12]. The propose in this work is using a simple and low-cost method to achieve superhydrophobic surfaces for orthodontic bracket 17-4 PH stainless steel (AISI 630) produced by metal injection molding which has less porosity than investment casting process. The methods are chemical etching and surface modification [13,14]. This method was never used for 17-4 PH stainless steel before. The samples were chemically etched using a mixture of $CuCl_2$ and various concentration of HCl and also HCl without $CuCl_2$ to obtain microscale roughness and followed by surface modification to obtain low-free-energy surfaces so the contact angle would increase. This study's aims to see the results of chemical etching and surface modification on orthodontic bracket surfaces and determine the optimum condition for superhydrophobic surfaces using wet chemical etching and we also compare the microstructure of samples that was already etched with and without $CuCl_2$.

2. Experimental

2.1. Materials

In this work, 17-4 PH stainless steel was used as substrate because it has excellent corrosion resistant. 17-4 PH stainless steel feedstock was manufactured by Ryer, Inc., HCl (37%), Ethanol (99.9%), $CuCl_2$ and stearic acid were bought from chemical industry in Indonesia. 17-4 PH SS chemical composition can be seen from Table 1.

Table 1. Chemical composition of 17-4 PH stainless steel

Fe (%)	C (%)	Cu (%)	Co (%)	Cr (%)	Mn (%)	Mo (%)
74.26	0.04	3.94	0.03	16.44	0.16	0.02
Nb (%)	**Ni (%)**	**O (%)**	**P (%)**	**S (%)**	**Si (%)**	
0.29	4.04	0.35	0.02	-	0.39	

2.2. Experimental Methods

In our research team, first the 17-4 PH stainless steel feedstock was injected into dies to form cubic shape (5 mm x 5 mm x 5 mm). The injection processed is done under certain conditions, which are injection pressure is about ±2700 psi, barrel's temperature for about ±100°C, nozzle's temperature for about ±200°C and mold's temperature is about ±50°C. The product was called green part since it still has binder and metal powder in it. After that, we begin the debinding process for green part. Debinding process was performed in two stages, which are solvent debinding and also thermal debinding. In solvent debinding process, the samples were immersed in hexane at temperature of 50°C for 90 minutes with agitation method. The samples were then drying using vacuum atmosphere at temperatures of 50°C for 1 hour. Thermal debinding process was carried out after solvent debinding process, the samples were put into MTI tube furnace with temperature of 510°C, heat rate of 1°C/min for 60 minutes and then cooled by furnace cooling [15]. After thermal debinding process done, argon gas was flowed into MTI tube furnace to remove oxygen gas inside the tube furnace and then sintering process began. It was performed in vacuum condition with temperature of 1360°C, heat rate 5°C/min for 90 minutes [16].
Before chemical etching begin, the samples were cleaned with distilled water and grind with abrasive paper #400 to roughened the surface. The samples were immersed into the solution of HCl (1, 2, 3 mol/L) with and without addition of $CuCl_2$ (1 mol/L). After etching, the samples were washed in distilled water and alcohol to remove grease and residual solution and then, it was dried with hair dryer for about 5 min. Then, the dried samples were immersed into the solution of stearic acid (1 M) and ethanol for about 30 min. The samples were dried with hair dryer for 30 min.

2.3. Characterization

The contact angle (CA) were captured by digital camera and it was measured by averaging the angles that was formed on the left and right borders of the samples using imageJ. The average contact angle (CA) were measured at three times with the same sample. The droplet volume of water was 5 μL. Scanning electron microscope (SEM) and energy dispersive spectroscopy (EDS) were carried out to observed surface morphology and also elemental composition that was formed on the surface.

3. Results and Discussion

3.1. Wettability Properties of the stainless steel 17-4PH

3.1.1. Effect of HCl concentration on contact angle (with and without surface modification)

Based on **Figure 1a**, etching at 1 mol/L, 2 mol/L and 3 mol/L HCl with 40, 50, 60 and 70 minutes will cause decreasing contact angle at first, but it will increase with the increasing of etching time and decrease again after it reached the optimum time. Optimum time for 1 mol/L HCl is about 60 minutes, but for 2 mol/L and 3 mol/L HCl is about 50 minutes. From this experiment, 1 mol/L has better hydrophobic surface at 40 minutes than 2 mol/L solution concentration, but when the etching time is increase, samples will become more hydrophobic at 2 mol/L solution concentration in 50 minutes than 1 mol/L solution concentration. However, after 50 minutes the contact angle will decreased, it is because with the increasing of etching time, surface roughness will be destroyed. For 3 mol/L solution concentration, it has the lowest contact angle at beginning stage. it is in accordance with literature, when the HCl concentration is increased, it will affect the surface roughness which is it would be destroyed and reduce the contact angle [17]. Therefore, 2 mol/L was the optimal solution concentration to obtain hydrophobic surfaces with etching time is about 50 minutes with contact angle (θ) is about 96.4^0.

Figure 1b shows the relationship between etching time and contact angles on the samples surface. It shows that the contact angle will increase after the samples were immersed in a 1 mol/L of stearic acid and ethanol. The experiment was carried out with different HCl concentration i.e 1 mol/L, 2 mol/L and 3 mol/L and etching time i.e 40, 50, 60 and 70 minutes. The result showed, contact angle (θ) will increase with the increasing of etching time up to the optimum time. From this experiment, 2 mol/L solution concentration at 70 minutes has super-hydrophobic surfaces with optimum contact angle is 142° while for 1 mol/L and 3 mol/L solution concentration is about 123° and 136° in 60 minutes. After the contact angle reached optimum time, it will be decreased. This is because the etching time will affect micro- and nano structures on the stainless-steel surface. So, when it is too long the structures will be destroyed and reduced the contact angle, because the sample will be corroded [17]. The comparison between **Figure 1a** and **Figure 1b** shows that after surface modification, the contact angle will always higher than 100°, it is because stearic acid has low free energy surface so it could increase the contact angle. **Figure 2** shows a pits photograph from optical microscope on 17-4 PH stainless steel surface. We could see from the figure that with the increasing etching time the pits' diameter would become larger. It is because of the etching solution would attack the grain and created pits. So, the longer etching time will make the samples has a larger pit's diameter and make the sample corroded enough that it could lead to decreasing contact angle.**Figure 3** shows an optical photograph of water droplet on samples surface after etching and without surface modification.

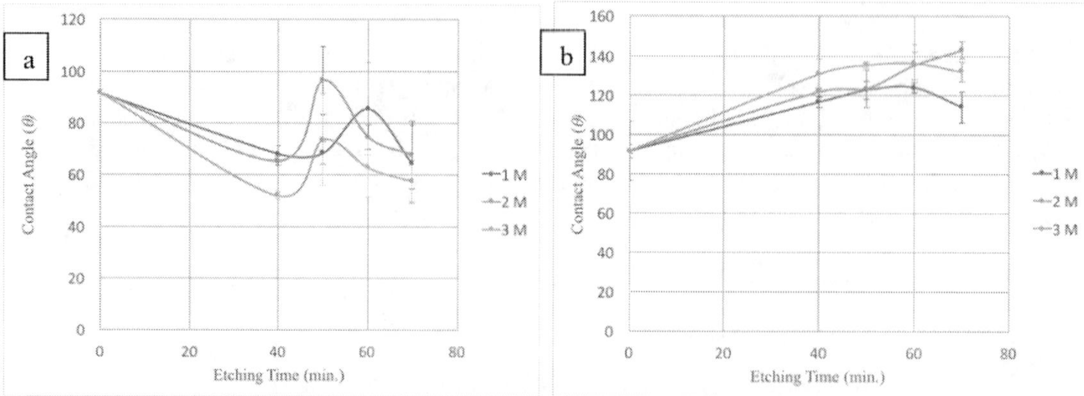

Figure 1. Water contact angles on 17-4 PH SS surface etched with different HCl concentration and time (a) without surface modification (b) with surface modification

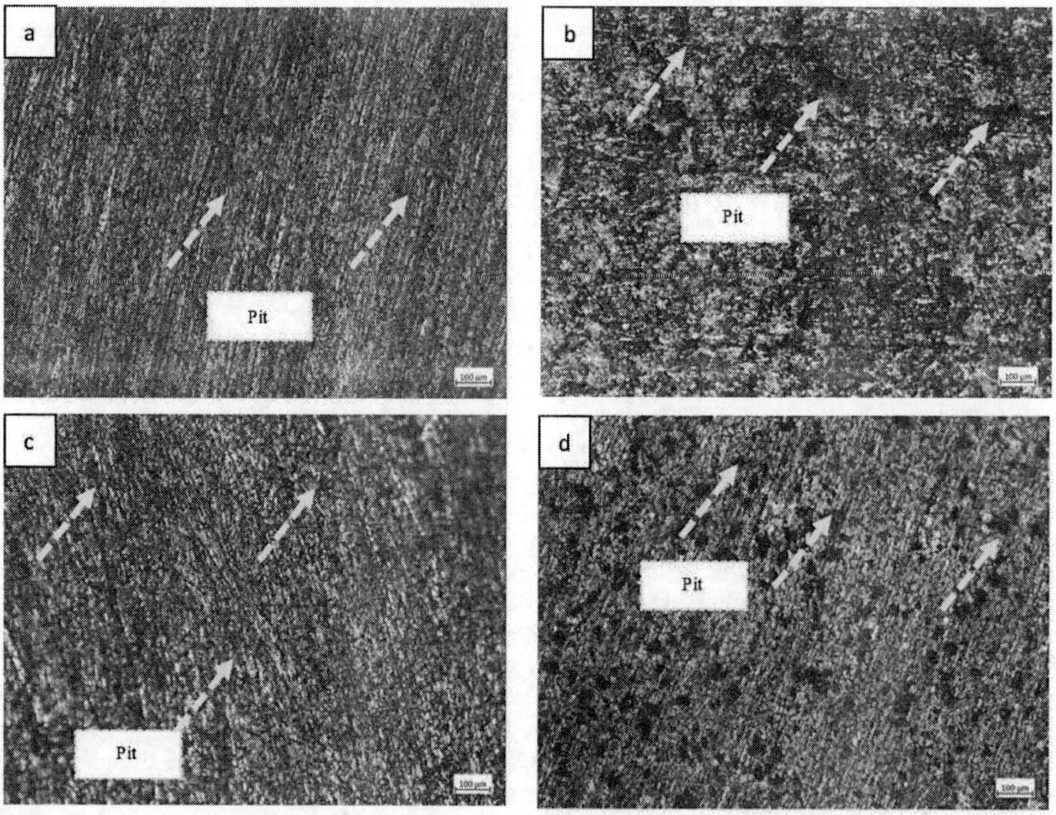

Figure 2. Pits photograph from optical microscope of 2 mol/L HCl on 17-4 PH stainless steel surface (a) 40 minutes, (b) 50 minutes, (c) 60 minutes and 70 minutes

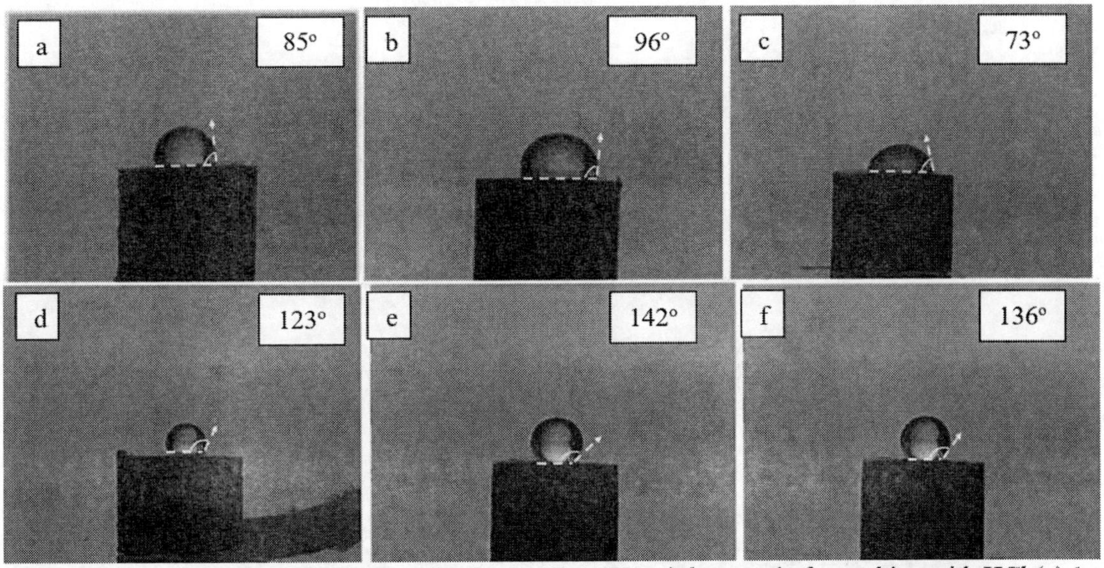

Figure 3. Optical photograph of water droplet on 17-4 PH stainless steel after etching with HCl (a) 1 mol/L for 60 minutes, (b) 2 mol/L for 50 minutes and (c) 3 mol/L for 50 minutes without surface modification and (d) 1 mol/L for 60 minutes, (e) 2 mol/.L for 70 minutes and (f) 3 mol/L for 60 minutes after surface modification

3.1.2. Effect of $CuCl_2$ and HCl in contact angle (without and with surface modification)

Figure 4a shows relationship between processing time and contact angle on samples surface. The contact angle was increased after it was immersed in mixture of $CuCl_2$ and various HCl concentration without surface modification. The contact angle was increased up to the optimum time and after that it would decrease. It is because the etching solution and time will affect micro and nanoscale surface roughness. By immersing the samples in to those solution, Cu particles will be deposited on to sample surface. Time and solution concentration play an important role in crystal Cu growth. The optimum contact angle that could be obtained is about 125° in 1 mol/L $CuCl_2$ and 1 mol/L HCl at 15 seconds. From this experiment by variating HCl solution concentration, the contact angle would decrease with increasing concentration.

Figure 4b shows an increasing contact angle with an increasing time after surface modification. This is due to the surface roughness that was created after etching and stearic acid which has low free energy surface, so that the contact angle value will higher than the one without surface modification. The contact angle reached 123° in 1 mol/L HCl after 15 seconds, with a longer time up to 45 seconds, the contact angle reached 142°. However, after 45 seconds the contact angle will decreased again and it is because the Cu crystal is destroyed with a longer time it could lead to decreasing of surface roughness [18]. Solution concentration is also an important factor that could affect contact angle. From this experiment, 3 mol/L and 2 mol/L HCl concentration has higher contact angle than 1 mol/L HCl at beginning stage but, decreased with a longer time. It shows that with higher concentration, the contact angle would be increased, but soon after that it would decreased.

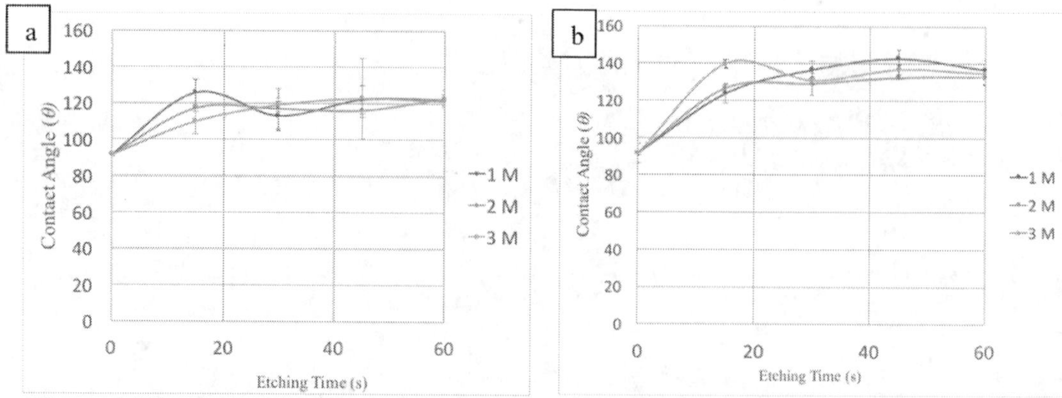

Figure 4. Water contact angles on 17-4 PH SS surface etched with 1 M $CuCl_2$ and different HCl concentration (a) without surface modification (b) with surface modification

3.2. Surface Morphology

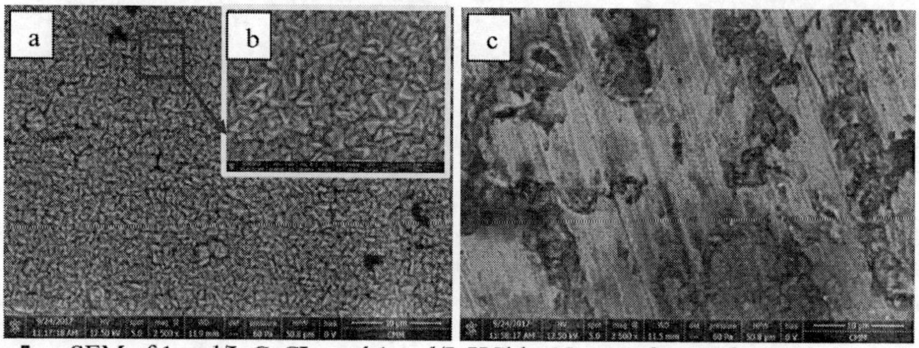

Figure 5. a. SEM of 1 mol/L $CuCL_2$ and 1 mol/L HCl in 15 seconds sample, (scale bar: 10 μm, b. 2μm), c. SEM of 2 mol/L HCl in 70 minutes sample

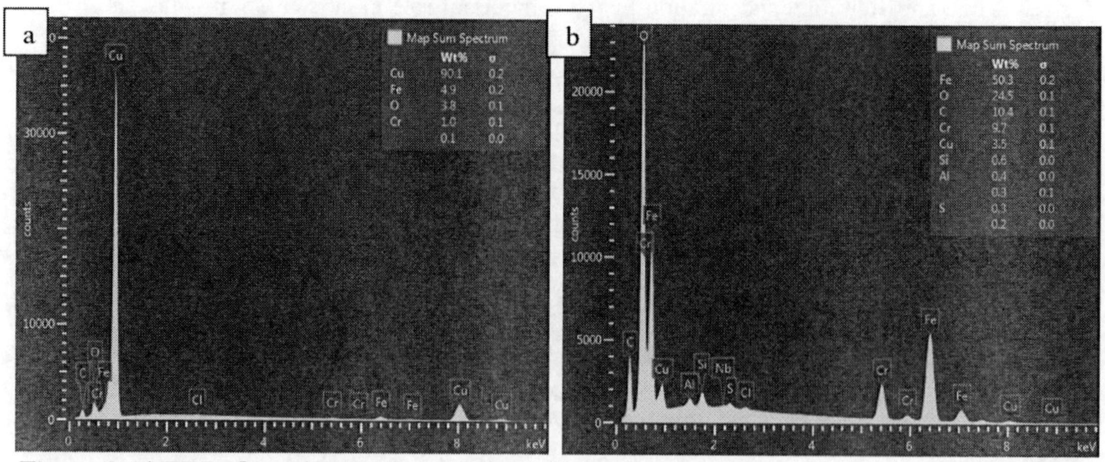

Figure 6. (a) EDS of 1 mol/L $CuCl_2$ and 1 mol/L HCl in 15 seconds sample, (b) EDS of 2 mol/L HCl in 70 minutes sample

SEM characterization was conducted by comparing the samples with and without $CuCl_2$, **Figure 5** Test result showed high content of Cu up to 90 wt% in sample surface which was immersed in 1 mol/L $CuCl_2$ and 1 mol/L HCl for 15 seconds, **Figure 6a**. This indicated that Cu was deposited on sample surface and formed crystal Cu layer. In small quantities, Cu^{2+} ions can be helpful for metabolic processes in human body, but Cu^{2+} ions in higher concentration can be harmful to cells in human body, which will damage cells and eventually the cell will die [19][20]. While, **Figure 6b** showed high content of Fe, Cr, O and C in the sample. This indicated that during etching process with 2 mol/L HCl there is oxidation reaction with possibility formed chromium oxide layer (Cr_2O_3) and the rest of O and C signal was from stearic acid modification [14][18].

4. Conclusion

Wet chemical etching followed by surface modification process for superhydrophobic bracket orthodontic was successfully conducted in this study. By only using etching, the superhydrophobic properties could not obtained, but after it was modified using stearic acid material the contact angle could increase up to >140°. The most higher contact angle that could be obtained in this study is about 142° by using 1 mol/L $CuCl_2$ and 1 mol/L HCl for 15 seconds and also 2 mol/L HCl for 70 minutes. This study still has not obtained an appropriate standard for self-cleaning properties, which has contact angle >150°. It is because the surface roughness was not optimal enough to achieve contact angle >150°. As etching solution concentration and processing time increase, the contact angle will also increase up to its optimum contact angle but after that it would decrease, because of destroyed structures. Etching using $CuCl_2$ solution is not acceptable for orthodontic application because it would have formed Cu layer on stainless steel surface, however by using HCl solution it was not formed Cu layer and might be applied in bracket orthodontic application.

5. Reference

[1] Türkkahraman H, Sayin MÖ, Bozkurt FY, Yetkin Z, Kaya S and Önal S 2005 Archwire ligation techniques, microbial colonization, and periodontal status in orthodontically treated patients *Angle Orthod.* **75**

[2] Sabel N, Robertson A, Nietzsche S and Nor JG 2012 Demineralization of Enamel in Primary Second Molars Related to Properties of the Enamel *Sci World J.*

[3] Shin S, Seo J, Han H, Kang S, Kim H and Lee T 2016 Bio-Inspired Extreme Wetting Surfaces for Biomedical Application *Materials.* **9**

[4] Gurav AB, Xu Q, Latthe SS, Vhatkar RS, Liu S, Yoon H, et al 2014 Superhydrophobic Coatings Prepared from Methyl- modified Silica Particles Using Simple Dip-coating Method *Ceram Int.*

[5] Shirtcliffe NJ, Mchale G and Newton MI 2009 Learning from Superhydrophobic Plants : The Use of Hydrophilic Areas on Superhydrophobic Surfaces for Droplet Control *Langmuir* **25**

[6] Yeo I, Kim H, Lim KS, and Han J 2012 Implant surface factors and bacterial adhesion : a review of the literature *Int. J. Artif. Organs.* **35**

[7] Zhang M, Feng S, Wang L and Zheng Y 2016 Lotus effect in wetting and self-cleaning *Biotribology* **5**

[8] Sharifi N, Pugh M, Moreau C and Dolatabadi A 2016 Developing hydrophobic and superhydrophobic TiO2 coatings by plasma spraying *Surf. Coat. Technol.*

[9] Mahadik SA, Sarika FP, Brahmanand SM, and Thorat SS 2016 Biocompatible superhydrophobic coating material for biomedical applications J. *Sol-Gel Sci. Technol.*

[10] Hu C, Liu S, Li B, Yang H, Fan C and Cui W 2013 Micro- / Nanometer Rough Structure of a Superhydrophobic Biodegradable Coating by Electrospraying for Initial. *Adv. Healthc. Mater.*

[11] Liu Y, Song Y, Niu S, Zhang Y, Han Z and Ren L 2016 Integrated super-hydrophobic and antireflective PDMS bio- templated from nano-conical structures of cicada wings *RSC Adv.*

[12] Supriadi S, Sitanggang TW, Irawan B, Suharno B, Kiswanto G and Prasetyadi T 2015 *Orthodontic Bracket Fabrication using the Investment Casting Process. Int. J. Technol.*

[13] Barshilia HC and Gupta N 2014 Superhydrophobic polytetrafluoroethylene surfaces with leaf-like micro-protrusions through Ar + O2 plasma etching process *Vacuum* **99**

[14] Alonso Frank M, Boccaccini AR and Virtanen S 2014 A facile and scalable method to produce superhydrophic stainless steel surface. *Appl. Surf. Sci.* **311**

[15] Supriadi S, Suharno B, Hidayatullah R, Maulana G and Baek E. 2017 Thermal Debinding Process of SS 17-4 PH in Metal Injection Molding Process with Variation of Heating Rate,

Temperature, and Holding Time. *Solid State Phenom.* **266**

[16] Suharno B, Ferdian D, Saputro HR, Suharno LP, Baek E and Supriadi S 2017 Vacuum Sintering Process in Metal Injection Molding for 17-4 PH Stainless Steel as Material for Orthodontic Bracket *Solid State Phenom.* **266**

[17] Zhang HF, Liu XW, Zhao M, Wang DB, and Shi CZ 2013 Fabrication of Super-Hydrophobic Surface on Stainless Steel Using Chemical Etching Method. *Key Eng Mate*r

[18] Liu Y, Zhang K, Yao W, Liu J, Han Z and Re L 2016 Bioinspired structured superhydrophobic and superoleophilic stainless steel mesh for efficient oil-water separation. *Colloids Surfaces A Physicochem Eng Asp.* **500**

[19] Hadidi M, Bigham A, Saebnoori E, Hassanzadeh-tabrizi SA and Rahmati S 2017 Electrophoretic-deposited hydroxyapatite-copper nanocomposite as an antibacterial coating for biomedical applications *Surf. Coat. Technol.* **321**

[20] Gonzalez-mateos A, Bulcke F, Santofimia-casta P and Dringen R 2015 Modulation of copper accumulation and copper-induced toxicity by antioxidants and copper chelators in cultured primary brain astrocytes *J Trace Elem. Med. Biol.* **32**

Acknowledgments

Authors would like to thanks to Universitas Indonesia for financial support through HIBAH PITTA (747/UN2R3.1/HKP.50.00/2017) and Directorate Research and Community Services Universitas Indonesia (DRPM UI)